T0384873

FINITE-DIMENSIONAL LINEAR ALGEBRA

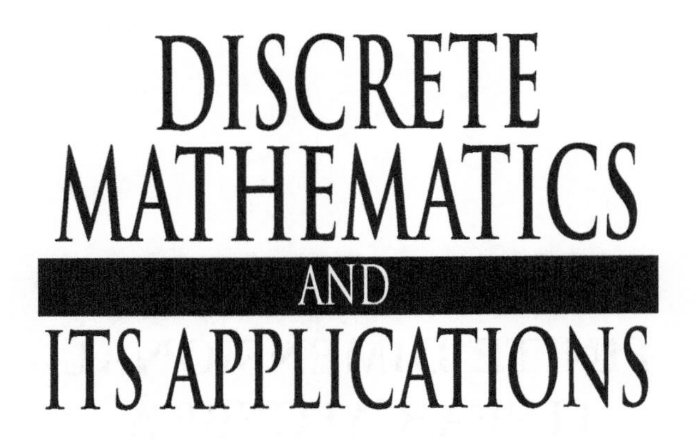

Series Editor

Kenneth H. Rosen, Ph.D.

Juergen Bierbrauer, Introduction to Coding Theory

Francine Blanchet-Sadri, Algorithmic Combinatorics on Partial Words

Richard A. Brualdi and Dragoš Cvetković, A Combinatorial Approach to Matrix Theory and Its Applications

Kun-Mao Chao and Bang Ye Wu, Spanning Trees and Optimization Problems

Charalambos A. Charalambides, Enumerative Combinatorics

Gary Chartrand and Ping Zhang, Chromatic Graph Theory

Henri Cohen, Gerhard Frey, et al., Handbook of Elliptic and Hyperelliptic Curve Cryptography

Charles J. Colbourn and Jeffrey H. Dinitz, Handbook of Combinatorial Designs, Second Edition

Martin Erickson, Pearls of Discrete Mathematics

Martin Erickson and Anthony Vazzana, Introduction to Number Theory

Steven Furino, Ying Miao, and Jianxing Yin, Frames and Resolvable Designs: Uses, Constructions, and Existence

Mark S. Gockenbach, Finite-Dimensional Linear Algebra

Randy Goldberg and Lance Riek, A Practical Handbook of Speech Coders

Jacob E. Goodman and Joseph O'Rourke, Handbook of Discrete and Computational Geometry, Second Edition

Jonathan L. Gross, Combinatorial Methods with Computer Applications

Jonathan L. Gross and Jay Yellen, Graph Theory and Its Applications, Second Edition

Jonathan L. Gross and Jay Yellen, Handbook of Graph Theory

Darrel R. Hankerson, Greg A. Harris, and Peter D. Johnson, Introduction to Information Theory and Data Compression, Second Edition

Darel W. Hardy, Fred Richman, and Carol L. Walker, Applied Algebra: Codes, Ciphers, and Discrete Algorithms, Second Edition

Daryl D. Harms, Miroslav Kraetzl, Charles J. Colbourn, and John S. Devitt, Network Reliability: Experiments with a Symbolic Algebra Environment

Silvia Heubach and Toufik Mansour, Combinatorics of Compositions and Words

Leslie Hogben, Handbook of Linear Algebra

Titles (continued)

Derek F. Holt with Bettina Eick and Eamonn A. O'Brien, Handbook of Computational Group Theory

David M. Jackson and Terry I. Visentin, An Atlas of Smaller Maps in Orientable and Nonorientable Surfaces

Richard E. Klima, Neil P. Sigmon, and Ernest L. Stitzinger, Applications of Abstract Algebra with Maple™ and MATLAB®, Second Edition

Patrick Knupp and Kambiz Salari, Verification of Computer Codes in Computational Science and Engineering

William Kocay and Donald L. Kreher, Graphs, Algorithms, and Optimization

Donald L. Kreher and Douglas R. Stinson, Combinatorial Algorithms: Generation Enumeration and Search

C. C. Lindner and C. A. Rodger, Design Theory, Second Edition

Hang T. Lau, A Java Library of Graph Algorithms and Optimization

Elliott Mendelson, Introduction to Mathematical Logic, Fifth Edition

Alfred J. Menezes, Paul C. van Oorschot, and Scott A. Vanstone, Handbook of Applied Cryptography

Richard A. Mollin, Advanced Number Theory *with Applications*

Richard A. Mollin, Algebraic Number Theory

Richard A. Mollin, Codes: The Guide to Secrecy from Ancient to Modern Times

Richard A. Mollin, Fundamental Number Theory with Applications, Second Edition

Richard A. Mollin, An Introduction to Cryptography, Second Edition

Richard A. Mollin, Quadratics

Richard A. Mollin, RSA and Public-Key Cryptography

Carlos J. Moreno and Samuel S. Wagstaff, Jr., Sums of Squares of Integers

Dingyi Pei, Authentication Codes and Combinatorial Designs

Kenneth H. Rosen, Handbook of Discrete and Combinatorial Mathematics

Douglas R. Shier and K.T. Wallenius, Applied Mathematical Modeling: A Multidisciplinary Approach

Alexander Stanoyevitch, Introduction to Cryptography with Mathematical Foundations and Computer Implementations

Jörn Steuding, Diophantine Analysis

Douglas R. Stinson, Cryptography: Theory and Practice, Third Edition

Roberto Togneri and Christopher J. deSilva, Fundamentals of Information Theory and Coding Design

W. D. Wallis, Introduction to Combinatorial Designs, Second Edition

Lawrence C. Washington, Elliptic Curves: Number Theory and Cryptography, Second Edition

DISCRETE MATHEMATICS AND ITS APPLICATIONS

Series Editor KENNETH H. ROSEN

FINITE-DIMENSIONAL LINEAR ALGEBRA

Mark S. Gockenbach

Michigan Technological University

Houghton, U.S.A.

CRC Press is an imprint of the
Taylor & Francis Group, an informa business
A CHAPMAN & HALL BOOK

CRC Press
Taylor & Francis Group
6000 Broken Sound Parkway NW, Suite 300
Boca Raton, FL 33487-2742

International Standard Book Number: 978-1-4398-1563-2 (Hardback)

Library of Congress Cataloging-in-Publication Data

Gockenbach, Mark S.
Finite-dimensional linear algebra / Mark S. Gockenbach.
p. cm. -- (Discrete mathematics and its applications)
Includes bibliographical references and index.
ISBN 978-1-4398-1563-2 (hardcover : alk. paper)
1. Algebras, Linear. 2. Dimensional analysis. 3. Finite fields (Algebra) 4. Vector spaces. I. Title. II. Series.

QA184.2.G63 2010
512'.5--dc22 2010008253

Visit the Taylor & Francis Web site at
http://www.taylorandfrancis.com

and the CRC Press Web site at
http://www.crcpress.com

Mark S. Gockenbach

Finite-Dimensional Linear Algebra

Dedicated to the memory of my son David,
who lived a full life in eleven short years.

Contents

Preface

Linear algebra forms the basis for much of modern mathematics—theoretical, applied, and computational. The purpose of this book is to provide a broad and solid foundation for the study of advanced mathematics. A secondary aim is to introduce the reader to many of the interesting applications of linear algebra.

Detailed outline of the book

Chapter 1 is optional reading; it provides a concise exposition of three main emphases of linear algebra: linear equations, best approximation, and diagonalization (that is, decoupling variables). No attempt is made to give precise definitions or results; rather, the intent is to give the reader a preview of some of the questions addressed by linear algebra before the abstract development begins.

Most students studying a book like this will already know how to solve systems of linear algebraic equations, and this knowledge is a prerequisite for the first three chapters. Gaussian elimination with back substitution is not presented until Section 3.7, where it is used to illustrate the theory of linear operator equations developed in the first six sections of Chapter 3. The discussion of Gaussian elimination was delayed advisedly; this arrangement of the material emphasizes the nature of the book, which presents the theory of linear algebra and does not emphasize mechanical calculations. However, if this arrangement is not suitable for a given class of students, there is no reason that Section 3.7 cannot be presented early in the course.

Many of the examples in the text involve spaces of functions and elementary calculus, and therefore a course in calculus is needed to appreciate much of the material.

The core of the book is formed by Chapters 2, 3, 4, and 6. They present an axiomatic development of the most important elements of finite-dimensional linear algebra: vector spaces, linear operators, norms and inner products, and determinants and eigenvalues. Chapter 2 begins with the concept of a field, of which the primary examples are **R** (the field of real numbers) and **C** (the field of complex numbers). Other examples are finite fields, particularly $\mathbf{Z}_p$, the field of integers modulo p (where p is a prime number). As much as possible, the results in the core part of the book (particularly Chapters 2–4) are phrased in terms of an arbitrary field, and examples are given that involve finite fields as well as the more standard fields of real and complex numbers.

Once fields are introduced, the concept of a vector space is introduced, along with the primary examples that will be studied in the text: Euclidean n-space and various spaces of functions. This is followed by the basic ideas necessary to describe vector spaces, particularly finite-dimensional vector spaces: subspace, spanning sets, linear independence, and basis. Chapter 2 ends with two optional application sections, Lagrange polynomials (which form a special basis for the space of polynomials) and piecewise polynomials (which are useful in many computational problems, particularly in solving differential equations). These topics are intended to illustrate why we study the common properties of vector spaces and bases: In a variety of applications, common issues arise, so it is convenient to study them abstractly. In addition, Section 2.8.1 presents an application to discrete mathematics: Shamir's scheme for secret sharing, which requires interpolation in a finite field.

Chapter 3 discusses linear operators, linear operator equations, and inverses of linear operators. Central is the fact that every linear operator on finite-dimensional spaces can be represented by a matrix, which means that there is a close connection between linear operator equations and systems of linear algebraic equations. As mentioned above, it is assumed in Chapter 2 that the reader is familiar with Gaussian elimination for solving linear systems, but the algorithm is carefully presented in Section 3.7, where it is used to illustrate the abstract results on linear operator equations. Applications for Chapter 3 include linear ordinary differential equations (viewed as linear operator equations), Newton's method for solving systems of nonlinear equations (which illustrates the idea of *linearization*), the use of matrices to represent graphs, binary linear block codes, and linear programming.

Eigenvalues and eigenvectors are introduced in Chapter 4, where the emphasis is on diagonalization, a technique for decoupling the variables in a system so that it can be more easily understood or solved. As a tool for studying eigenvalues, the determinant function is first developed. Elementary facts about permutations are needed; these are developed in Appendix B for the reader who has not seen them before. Results about polynomials form further background for Chapter 4, and these are derived in Appendix C. Chapter 4 closes with two interesting applications in which linear algebra is key: systems of constant coefficient linear ordinary differential equations and integer programming.

Chapter 4 shows that some matrices can be diagonalized, but others cannot. After this, there are two natural directions to pursue, given in Chapters 5 and 8. One is to try to make a nondiagonalizable matrix as close to diagonal form as possible; this is the subject of Chapter 5, and the result is the Jordan canonical form. As an application, the matrix exponential is presented, which completes the discussion of systems of ordinary differential equations that was begun in Chapter 4. A brief introduction to the spectral theory of graphs is also presented in Chapter 5. The remainder of the text does not depend on Chapter 5.

Chapter 6 is about orthogonality and its most important application, best approximation. These concepts are based on the notion of an inner product and the norm it defines. The central result is the projection theorem, which shows how to find the best approximation to a given vector from a finite-dimensional subspace (an infinite-dimensional version appears in Chapter 10). This is applied to problems such as solving overdetermined systems of linear equations and approximating functions by polynomials. Orthogonality is also useful for representing vector spaces in terms of orthogonal subspaces; in particular, this gives a detailed understanding of the four fundamental subspaces defined by a linear operator. Application sections address weighted polynomial approximation, the Galerkin method for approximating solutions to differential equations, Gaussian quadrature (that is, numerical integration), and the Helmholtz decomposition for vector fields.

Symmetric (and Hermitian) matrices have many special properties, including the facts that all their eigenvalues are real, their eigenvectors can be chosen to be orthogonal to one another, and every such matrix can be diagonalized. Chapter 7 develops these facts and includes applications to optimization and spectral methods for differential equations.

Diagonalization is an operation applied to square matrices, in which one tries to choose a special basis (a basis of eigenvectors) that results in diagonal form. In fact, it is always possible to obtain a diagonal form, provided two bases are used (one for the domain and another for the co-domain). This leads to the singular value decomposition (SVD) of a matrix, which is the subject of Chapter 8. The SVD has many advantages over the Jordan canonical form. It exists even for non-square matrices; it can be computed in finite-precision arithmetic (whereas the Jordan canonical form is unstable and typically completely obscured by the round-off inherent to computers); the bases involved are orthonormal (which means that operations with them are stable in finite-precision arithmetic). All of these advantages make the SVD a powerful tool in computational mathematics, whereas the Jordan canonical form is primarily a theoretical tool. As an application of the SVD, Chapter 8 includes a brief study of linear inverse problems. It also includes a discussion of the Smith normal form, which is used in discrete mathematics to study properties of integer matrices.

To use linear algebra in practical applications (whether they be to other areas of mathematics or to problems in science and engineering), it is typically necessary to do one or both of the following: Perform linear algebraic calculations on a computer (in finite-precision arithmetic), and introduce ideas from analysis about convergence. Chapter 9 includes a brief survey of the most important facts from numerical linear algebra, the study of computer algorithms for problems in linear algebra. Chapter 10 extends some results from single-variable analysis to Euclidean n-space, with an emphasis on the fact that all norms define the same notion of convergence on a finite-dimensional vector space. It then presents a very brief introduction to functional analysis, which is the study of linear algebra in infinite-dimensional vector spaces. In such settings, analysis is critical.

Exercises

Each section in the text contains exercises, which range from the routine to quite challenging. The results of some exercises are used later in the text; these are labeled "essential exercises," and the student should at least read these to be familiar with the results. Each section contains a collection of "miscellaneous exercises," which illustrate, verify, and extend the results of the section. Some sections contain "projects," which lead the student to develop topics that had to be omitted from the text for lack of space.

Figures

Figures appearing in the text were prepared using MATLAB®. For product information, please contact:

The MathWorks, Inc.
3 Apple Hill Drive
Natick, MA 01760-2098 USA
Tel: 508-647-7000
Fax: 508-647-7001
E-mail: info@mathworks.com
Web: www.mathworks.com

Applications

Twenty optional sections introduce the reader to various applications of linear algebra. In keeping with the goal of this book (to prepare the reader for further studies in mathematics), these applications show how linear algebra is essential in solving problems involving differential equations, optimization, approximation, and combinatorics. They also illustrate why linear algebra should be studied as a distinct subject: Many different problems can be addressed using vector spaces and linear operators.

Here is a list of the application sections in the text:

2.8 Polynomial interpolation and the Lagrange basis; includes a discussion of Shamir's method of secret sharing

2.9 Continuous piecewise polynomial functions

3.8 Newton's method

3.9 Linear ordinary differential equations

3.10 Graph theory

3.11 Coding theory

3.12 Linear programming

4.8 Systems of linear ODEs

4.9 Integer programming

5.5 The matrix exponential

5.6 Graphs and eigenvalues

6.8 More on polynomial approximation

6.9 The energy inner product and Galerkin's method

6.10 Gaussian quadrature

6.11 The Helmholtz decomposition

7.3 Optimization and the Hessian matrix

7.4 Lagrange multipliers

7.5 Spectral methods for differential equations

8.4 The SVD and linear inverse problems

8.5 The Smith normal form of a matrix

Possible course outlines

A basic course includes Sections 2.1—2.7, 3.1–3.7, 4.1–4.6, 6.1–6.7, 7.1–7.2, and either 5.1–5.4 or 8.1–8.3. I cover all the material in these sections, except that I only summarize Sections 4.1–4.4 (determinants) in two lectures to save more time for applications. I otherwise cover one section per day, so this material requires 30 or 31 lectures. Allowing up to five days for exams and review, this leaves about six or seven days to discuss applications (in a 14-week, three-credit course). An instructor could cover fewer applications to allow time for a complete discussion of the material on determinants and the background material in Appendices B and C. In this way, all the material can be developed in a rigorous fashion. An instructor with more class meetings has many more options, including dipping into Chapters 9 and 10. Students should at least be aware of this material.

The book's web site (`www.math.mtu.edu/~msgocken/fdlabook`) includes solutions to selected odd-numbered exercises and an up-to-date list of errors with corrections. Readers are invited to alert me of suspected errors by email.

Mark S. Gockenbach
`msgocken@mtu.edu`

About the Author

Mark S. Gockenbach is a professor in the Department of Mathematical Sciences at Michigan Technological University, where he currently serves as department chair. He received his Ph.D. from Rice University in 1994, and held faculty positions at Indiana University, the University of Michigan, and Rice University prior to joining Michigan Tech in 1998. His research interests include inverse problems, computational optimization, and mathematical software. His two previous books, *Partial Differential Equations: Analytical and Numerical Methods* and *Understanding and Implementing the Finite Element Method*, were published by the Society for Industrial and Applied Mathematics.

1

Some problems posed on vector spaces

This book is about finite-dimensional vector spaces and the associated concepts that are useful for solving interesting problems. Vector spaces form a mathematical subject because many different problems, when viewed abstractly, have a common foundation: the vector space operations of scalar multiplication and vector addition, and the related idea of linearity.

Each section in this initial chapter describes seemingly different problems that are actually closely related when viewed from an abstract point of view. The purpose of Chapter 1 is to hint at some of the important concepts—linearity, subspace, basis, and so forth—that will be carefully developed in the rest of the book. The informal nature of Chapter 1 means that it can be omitted without harm; all of the precise definitions and theorems appear in the following chapters, and the reader should bear in mind that all of the assertions appearing in this discussion will be justified in later chapters. Chapter 1 should be read lightly, with the goal of gaining a general sense of the theoretical issues arising in certain practical problems.

1.1 Linear equations

1.1.1 Systems of linear algebraic equations

Here is a familiar problem: Find real numbers x_1, x_2, x_3, x_4, x_5 satisfying the following equations:

$$\begin{aligned} -3x_1 + 2x_2 - 2x_3 - 3x_4 - 2x_5 &= 0, \\ -3x_1 + 2x_2 - x_3 - 2x_4 - x_5 &= 5, \\ 12x_1 - 7x_2 + 9x_3 + 14x_4 + 12x_5 &= 7. \end{aligned} \tag{1.1}$$

The usual algorithm for solving a system of this type is called *Gaussian elimination*; it consists of systematically adding multiples of one equation to another to reduce the system to an equivalent but simpler system. In this case,

Gaussian elimination yields

$$\begin{aligned} x_1 + \qquad\qquad x_4 + 2x_5 &= -2, \\ x_2 + \qquad x_4 + 3x_5 &= 2, \\ x_3 + x_4 + x_5 &= 5, \end{aligned}$$

or

$$\begin{aligned} x_1 &= -2 - x_4 - 2x_5, \\ x_2 &= 2 - x_4 - 3x_5, \\ x_3 &= 5 - x_4 - x_5. \end{aligned}$$

These equations determine values for x_1, x_2, and x_3, given any values for x_4 and x_5. If we choose $x_4 = s$ and $x_5 = t$ and use vector notation, then the solution is $(x_1, x_2, x_3, x_4, x_5) = (-2 - s - 2t, 2 - s - 3t, 5 - s - t, s, t)$. In other words, the solution set is

$$\{(-2 - s - 2t, 2 - s - 3t, 5 - s - t, s, t) \, : \, s, t \in \mathbf{R}\}.$$

We see that the original system has infinitely many solutions, and we can say that the solution set is determined by two degrees of freedom.

To understand the structure of the solution set further, it is helpful to write the original system in matrix-vector form. If we define

$$A = \begin{bmatrix} -3 & 2 & -2 & -3 & -2 \\ -3 & 2 & -1 & -2 & -1 \\ 12 & -7 & 9 & 14 & 12 \end{bmatrix}, \; x = \begin{bmatrix} x_1 \\ x_2 \\ x_3 \\ x_4 \\ x_5 \end{bmatrix}, \; b = \begin{bmatrix} 0 \\ 5 \\ 7 \end{bmatrix},$$

then (1.1) can be written as $Ax = b$, where Ax represents matrix-vector multiplication. From this point of view, it is natural to regard the 3×5 matrix A as defining an operator T which takes a 5-vector x as input and produces a 3-vector $y = Ax$ as output:$T : \mathbf{R}^5 \to \mathbf{R}^3$, $T(x) = Ax$. The problem posed by (1.1) is then equivalent to the question of whether there exists a vector x in the domain of T such that $T(x)$ equals the given b.

As shown above, there are many such vectors x; in fact, any x of the form[1]

$$x = \begin{bmatrix} -2 - s - 2t \\ 2 - s - 3t \\ 5 - s - t \\ s \\ t \end{bmatrix} = \begin{bmatrix} -2 \\ 2 \\ 5 \\ 0 \\ 0 \end{bmatrix} + s \begin{bmatrix} -1 \\ -1 \\ -1 \\ 1 \\ 0 \end{bmatrix} + t \begin{bmatrix} -2 \\ -3 \\ -1 \\ 0 \\ 1 \end{bmatrix}$$

[1]Throughout this book, a Euclidean vector will be written in either of two equivalent notations:

$$(x_1, x_2, x_3, x_4, x_5) = \begin{bmatrix} x_1 \\ x_2 \\ x_3 \\ x_4 \\ x_5 \end{bmatrix}.$$

is a solution of $T(x) = b$. We can write this formula for solutions of $T(x) = b$ as $x = \hat{x} + sy + tz$, where the vectors $\hat{x}$, y, and z are defined by $\hat{x} = (-2, 2, 5, 0, 0)$, $y = (-1, -1, -1, 1, 0)$, and $z = (-2, -3, -1, 0, 1)$.

We will now interpret the general solution $x = \hat{x} + sy + tz$ in light of the fact that T is *linear*:

$$T(\alpha x + \beta y) = \alpha T(x) + \beta T(y) \text{ for all } x, y \in \mathbf{R}^5, \ \alpha, \beta \in \mathbf{R}.$$

It follows from the linearity of T that

$$T(\hat{x} + sy + tz) = T(\hat{x}) + sT(y) + tT(z).$$

Moreover,

$$T(\hat{x}) = A\hat{x} = \begin{bmatrix} -3 & 2 & -2 & -3 & -2 \\ -3 & 2 & -1 & -2 & -1 \\ 12 & -7 & 9 & 14 & 12 \end{bmatrix} \begin{bmatrix} -2 \\ 2 \\ 5 \\ 0 \\ 0 \end{bmatrix} = \begin{bmatrix} 0 \\ 5 \\ 7 \end{bmatrix},$$

$$T(y) = Ay = \begin{bmatrix} -3 & 2 & -2 & -3 & -2 \\ -3 & 2 & -1 & -2 & -1 \\ 12 & -7 & 9 & 14 & 12 \end{bmatrix} \begin{bmatrix} -1 \\ -1 \\ -1 \\ 1 \\ 0 \end{bmatrix} = \begin{bmatrix} 0 \\ 0 \\ 0 \end{bmatrix},$$

$$T(z) = Az = \begin{bmatrix} -3 & 2 & -2 & -3 & -2 \\ -3 & 2 & -1 & -2 & -1 \\ 12 & -7 & 9 & 14 & 12 \end{bmatrix} \begin{bmatrix} -2 \\ -3 \\ -1 \\ 0 \\ 1 \end{bmatrix} = \begin{bmatrix} 0 \\ 0 \\ 0 \end{bmatrix}.$$

We see from these calculations that $\hat{x}$ is one solution to $T(x) = b$, while y and z both solve $T(x) = 0$. Then $\hat{x} + sy + tz$ is a solution to $T(x) = b$ for any values of s and t, as verified by the following calculation:

$$T(\hat{x} + sy + tz) = T(\hat{x}) + sT(y) + tT(z) = b + s \cdot 0 + t \cdot 0 = b.$$

Finally, suppose we wanted to solve the special linear algebraic system represented by $T(x) = 0$ (where 0 is the zero vector). The reader will find it easy to believe that the general solution is $x = sy + tz$, where y and z are the vectors given above. At least we know that such an x is a solution for any values of s and t:

$$T(sy + tz) = sT(y) + tT(z) = s \cdot 0 + t \cdot 0 = 0.$$

We have not verified that every solution of $T(x) = 0$ is of the form $x = sy + tz$ for some $s, t \in \mathbf{R}$, but this can be shown to be true by performing Gaussian elimination on the system represented by $T(x) = 0$.

A linear equation is called *homogeneous* if the right-hand side is zero, and *inhomogeneous* otherwise. Thus $T(x) = 0$ is a homogeneous equation and $T(x) = b$ is inhomogeneous if $b \neq 0$. The general solution of an inhomogeneous linear equation can be written as any one solution of the equation plus the general solution of the corresponding homogeneous equation.

1.1.2 Linear ordinary differential equations

We now turn to a very different problem. Consider the ordinary differential equation

$$x'' + 3x' + 2x = t. \tag{1.2}$$

The problem is to find a function $x = x(t)$, defined for all real numbers t, that satisfies (1.2). The general solution can be found using elementary methods:[2]

$$x(t) = \frac{1}{2}t - \frac{3}{4} + c_1 e^{-t} + c_2 e^{-2t}.$$

Here c_1 and c_2 can be any real constants.

The reader may have already noticed the similarity of the form of this general solution to that of the solution to $Ax = b$. To bring out this similarity, let us define the differential operator L by

$$\begin{aligned} &L : C^2(\mathbf{R}) \to C(\mathbf{R}), \\ &L(x) = x'' + 3x' + 2x. \end{aligned}$$

Inputs to L must be functions with two continuous derivatives, so the domain of L is $C^2(\mathbf{R})$, the collection of all functions $x : \mathbf{R} \to \mathbf{R}$ such that x' and x'' exist and are continuous. If $x \in C^2(\mathbf{R})$, then $L(x)$ is at least continuous, so the co-domain of L is $C(\mathbf{R})$, the collection of all functions $x : \mathbf{R} \to \mathbf{R}$ that are continuous.

If we define $f \in C(\mathbf{R})$ by $f(t) = t$, then (1.2) can be written concisely as $L(x) = f$. Moreover, like the operator T encountered above, L is linear:

$$L(\alpha x + \beta y) = \alpha L(x) + \beta L(y) \text{ for all } x, y \in C^2(\mathbf{R}), \alpha, \beta \in \mathbf{R}.$$

This is shown using the linearity of the derivative operator, as follows:

$$\begin{aligned} L(\alpha x + \beta y) &= (\alpha x + \beta y)'' + 3(\alpha x + \beta y)' + 2(\alpha x + \beta y) \\ &= \alpha x'' + \beta y'' + 3(\alpha x' + \beta y') + 2\alpha x + 2\beta y \\ &= \alpha x'' + \beta y'' + 3\alpha x' + 3\beta y' + 2\alpha x + 2\beta y \\ &= \alpha(x'' + 3x' + 2x) + \beta(y'' + 3y' + 2y) \\ &= \alpha L(x) + \beta L(y). \end{aligned}$$

[2] See any undergraduate text on differential equations, such as [46].

We now write

$$\hat{x}(t) = \frac{1}{2}t - \frac{3}{4}, \; y(t) = e^{-t}, \; z(t) = e^{-2t}.$$

Then it is straightforward to verify the following calculations:

$$L(\hat{x}) = f, \; L(y) = 0, \; L(z) = 0$$

(where $f(t) = t$). From this it follows that $x = \hat{x} + c_1 y + c_2 z$ is a solution of $L(x) = f$ for any values of c_1, c_2:

$$L(\hat{x} + c_1 y + c_2 z) = L(\hat{x}) + c_1 L(y) + c_2 L(z) = f + c_1 \cdot 0 + c_2 \cdot 0 = f.$$

Also, for any $c_1, c_2 \in \mathbf{R}$, $x = c_1 y + c_2 z$ is a solution of the corresponding homogeneous equation $L(x) = 0$:

$$L(c_1 y + c_2 z) = c_1 L(y) + c_2 L(z) = c_1 \cdot 0 + c_2 \cdot 0 = 0.$$

We will show in Section 3.9 that $x = \hat{x} + c_1 y + c_2 z$ and $x = c_1 y + c_2 z$ represent all possible solutions of $L(x) = f$ and $L(x) = 0$, respectively.

1.1.3 Some interpretation: The structure of the solution set to a linear equation

In the following discussion, T will denote a generic linear operator, not necessarily the specific linear algebraic operator defined in Section 1.1.1. The identity

$$T(\alpha x + \beta y) = \alpha T(x) + \beta T(y) \text{ for all } x, y, \alpha, \beta$$

must hold, which says something about the domain: If X is the domain of T and $x, y \in X$, then $\alpha x + \beta y$ must belong to X for any scalars α and β. The same property must hold true for the co-domain U of T: $\alpha u + \beta v$ must belong to U whenever $u, v \in U$ and α, β are scalars.

These considerations lead to the abstract notion of a *vector space*. A vector space is simply a nonempty set of objects that can be added together and multiplied by scalars to yield objects of the same type. In this sense, functions are vectors just as much as the more familiar Euclidean vectors. For instance, $f(t) = \cos(t)$ and $g(t) = e^t$ are functions, and so is $2f + 3g$, which is defined by

$$(2f + 3g)(t) = 2\cos(t) + 3e^t.$$

In order for a given space to qualify as a vector space, there are certain algebraic identities that must be satisfied by the operations of addition and scalar multiplication. These are all obvious in practice for the specific spaces mentioned above ($\mathbf{R}^5$, $\mathbf{R}^3$, $C^2(\mathbf{R})$, and $C(\mathbf{R})$), and will be carefully defined in the next chapter.

A concept closely associated with vector space is that of a *subspace*, a nonempty subset S that is a vector space in its own right by virtue of being closed under addition and scalar multiplication:

$$x, y \in S \Rightarrow x + y \in S,$$
$$x \in S, \alpha \in \mathbf{R} \Rightarrow \alpha x \in S.$$

The concept of subspace naturally arises when discussing linear equations because the solution set of the homogeneous equation $T(x) = 0$ is a subspace of the domain of T. To be precise, if X, U are vector spaces, $T : X \to U$ is a linear operator, and

$$S = \{x \in X \, : \, T(x) = 0\},$$

then S is a subspace:

$$\begin{aligned} x, y \in S &\Rightarrow T(x) = 0, T(y) = 0 \\ &\Rightarrow T(x + y) = T(x) + T(y) = 0 + 0 = 0 \\ &\Rightarrow x + y \in S, \\ x \in S, \alpha \in \mathbf{R} &\Rightarrow T(x) = 0 \\ &\Rightarrow T(\alpha x) = \alpha T(x) = \alpha \cdot 0 = 0 \\ &\Rightarrow \alpha x \in S. \end{aligned}$$

The reader should notice how the linearity of T was used in both parts of this proof. Linearity also implies that $T(0) = 0$, and hence that S is nonempty because $0 \in S$. If T is a linear operator, the solution set of $T(x) = 0$ is called the *kernel* of T.

In the example of Section 1.1.1, the general solution of the homogeneous equation was written as $x = \alpha y + \beta z$. The precise meaning of this is that the kernel of the linear operator is

$$\{\alpha y + \beta z \, : \, \alpha, \beta \in \mathbf{R}\}.$$

An expression of the form $\alpha y + \beta z$ is called a *linear combination* of the vectors y and z, so the kernel of T (in the example from Section 1.1.1) can be described as the set of all linear combinations of y and z, or the *span* of $\{y, z\}$, denoted $\mathrm{sp}\{y, z\}$. Alternatively, we say that $\{y, z\}$ is a spanning set for the kernel of T.

Most nontrivial vector spaces contain infinitely many vectors,[3] so it is important to find a finite representation for a vector space (or subspace) whenever possible. Most of this book deals with finite-dimensional vector spaces, which are simply those that can be written as the span of a finite number of vectors. When choosing a spanning set for a vector space, it is natural to include as few vectors as possible. For example, if y, z are two vectors in a given space, then $\mathrm{sp}\{y, z\} = \mathrm{sp}\{y, z, y+z\}$. It is more convenient to work with the spanning set $\{y, z\}$ rather than the larger set $\{y, z, y + z\}$. A spanning set containing the smallest possible number of vectors is called a *basis*.

[3]The only exception arises when the scalars are chosen from a finite field. This case is discussed in Section 1.1.4.

1.1.4 Finite fields and applications in discrete mathematics

In the foregoing discussion, it has been implicitly assumed that the scalars are real numbers. Complex numbers are also often used as scalars, partly because the easiest way to solve many problems originally posed in terms of real numbers is to take advantage of certain properties of complex numbers. Both $\mathbf{R}$ (the space of real numbers) and $\mathbf{C}$ (the space of complex numbers) are examples of *fields*. A field is a set on which are defined two operations, addition and multiplication, that satisfy certain conditions.

In discrete mathematics, certain finite fields often arise. The most common is the simplest field of all, $\mathbf{Z}_2$. The underlying set has just two elements, 0 and 1, and addition and multiplication are defined as follows:

$$\begin{gathered} 0+0=0,\ 0+1=1,\ 1+0=1,\ 1+1=0, \\ 0\cdot 0=0,\ 0\cdot 1=0,\ 1\cdot 0=0,\ 1\cdot 1=1. \end{gathered}$$

These operations are addition and multiplication *modulo* 2. In modular 2 arithmetic, one computes the ordinary sum or product and then takes the remainder after division by 2. For instance, $1+1=2$, but the remainder when 2 is divided by 2 is 0; hence $1+1=0$ in $\mathbf{Z}_2$.

Although $\mathbf{Z}_2$ is quite simple, it satisfies the same algebraic properties as do the infinite fields $\mathbf{R}$ and $\mathbf{C}$, and it is often used in applications in discrete mathematics, such as *coding theory* (see Section 3.11). In such an application, we might have to solve such a system as

$$\begin{aligned} x_1 \phantom{{}+x_2} + x_3 + x_4 \phantom{{}+x_5} &= 1, \\ x_1 + x_2 \phantom{{}+x_3} + x_4 + x_5 &= 0, \\ x_1 \phantom{{}+x_2+x_3+x_4} + x_5 &= 1. \end{aligned}$$

We can solve this system using Gaussian elimination, being careful to perform all operations in modular 2 arithmetic. In particular, if $x, y \in \mathbf{Z}_2$, then $x+x=0$, and $x+y=0$ implies $x=y$. Since such arithmetic is probably unfamiliar, we will solve the system given above explicitly. Adding the first equation to the second and third equations yields

$$\begin{aligned} x_1 \phantom{{}+x_2} + x_3 + x_4 \phantom{{}+x_5} &= 1, \\ x_2 + x_3 \phantom{{}+x_4} + x_5 &= 1, \\ x_3 + x_4 + x_5 &= 0. \end{aligned}$$

Next, adding the third equation to the first and second equations yields

$$\begin{aligned} x_1 \phantom{{}+x_2+x_3+x_4} + x_5 &= 1, \\ x_2 \phantom{{}+x_3} + x_4 \phantom{{}+x_5} &= 1, \\ x_3 + x_4 + x_5 &= 0, \end{aligned}$$

or

$$\begin{aligned} x_1 &= 1 + x_5, \\ x_2 &= 1 + x_4, \\ x_3 &= x_4 + x_5. \end{aligned}$$

The variables x_4 and x_5 can be assigned any values; taking $x_4 = \alpha$ and $x_5 = \beta$, we obtain the solution

$$x = (1+\beta, 1+\alpha, \alpha+\beta, \alpha, \beta) = (1,1,0,0,0) + \alpha(0,1,1,1,0) + \beta(1,0,1,0,1).$$

The general solution is $(1,1,0,0,0)$ plus any linear combination of $(0,1,1,1,0)$ and $(1,0,1,0,1)$. The reader should notice that this general solution has the same form as in the problems of Sections 1.1.1 and 1.1.2. In this case, though, since α and β can take only the values 0 and 1, there are exactly four solutions, corresponding to $(\alpha,\beta) = (0,0)$, $(\alpha,\beta) = (0,1)$, $(\alpha,\beta) = (1,0)$, and $(\alpha,\beta) = (1,1)$.

1.2 Best approximation

1.2.1 Overdetermined linear systems

In Section 1.1.1, we considered a system of linear equations with three equations and five unknowns. It was not surprising that the system had many solutions; the reason is simply that each of the three equations could be solved for one variable, but that still left two variables that could assume any value whatever. A system with more unknowns than equations is called *underdetermined.*

We will now consider the case of an *overdetermined* linear system, one with more equations than unknowns:

$$\begin{aligned} x_1 + x_2 - 2x_3 &= 2, \\ 3x_1 + 3x_2 - 5x_3 &= 7, \\ 3x_1 + 2x_2 - 10x_3 &= 2, \\ x_1 + x_2 - 7x_3 &= -3, \\ -4x_1 - 4x_2 + 11x_3 &= -4. \end{aligned} \tag{1.3}$$

In matrix-vector form, this system takes the form $Ax = b$, where

$$A = \begin{bmatrix} 1 & 1 & -2 \\ 3 & 3 & -5 \\ 3 & 2 & -10 \\ 1 & 1 & -7 \\ -4 & -4 & 11 \end{bmatrix}, \quad x = \begin{bmatrix} x_1 \\ x_2 \\ x_3 \end{bmatrix}, \quad b = \begin{bmatrix} 2 \\ 7 \\ 2 \\ -3 \\ -4 \end{bmatrix}.$$

We should not expect to find a solution to $Ax = b$ in this case; if this example follows the typical pattern for a system with five equations and only three unknowns, we could solve the first three equations for the three unknowns, and it would only be by coincidence that these values for the unknowns also satisfied the last two equations.

It might seem that a system that is not expected to have a solution is uninteresting, but, in fact, overdetermined systems arise in many applications. The typical example occurs when a certain linear relationship among variables is expected to hold, but the numbers in the equations are measured in an experiment. Because of measurement error, the equations do not hold exactly. In such a case, the standard practice is to measure more data points than there are degrees of freedom in the model, in which case an overdetermined system results. It is then desired to find values of the unknowns that come as close to satisfying the equations as possible.

If $Ax = b$ has no solution, then no x will make the *residual* $Ax - b$ equal to the zero vector. However, it would be sensible to choose x to make the residual as small as possible. This requires a *norm*, a measure of the size of vectors. The most common vector norm is the Euclidean norm:

$$\|x\|_2 = \sqrt{\sum_{i=1}^{n} x_i^2} \text{ for all } x \in \mathbf{R}^n.$$

In general, $\|\cdot\|$ denotes a generic norm, and the subscript "2" comes from the exponent in the formula for the Euclidean norm. Using the Euclidean norm, we "solve" an overdetermined system by choosing x to make $\|Ax - b\|_2$ as small as possible. Minimizing $\|Ax - b\|$ is equivalent to minimizing

$$\|Ax - b\|_2^2 = \sum_{i=1}^{m} \left((Ax)_i - b_i\right)^2,$$

and a minimizer is called a *least-squares* solution to $Ax = b$.

The Euclidean norm is related to the (Euclidean) *dot product*, defined by

$$x \cdot y = \sum_{i=1}^{n} x_i y_i \text{ for all } x, y \in \mathbf{R}^n.$$

The relationship is simply $\|x\|_2 = \sqrt{x \cdot x}$. The dot product is important because vectors x and y are *orthogonal* if $x \cdot y = 0$. Orthogonality is a generalization of perpendicularity. In two or three dimensions (that is, $\mathbf{R}^2$ or $\mathbf{R}^3$), Euclidean geometry can be used to prove that nonzero vectors x and y are perpendicular if and only if $x \cdot y = 0$.[4] In higher dimensions, orthogonality is defined by the dot product condition.

---

[4]Here we are thinking of a Euclidean vector as an arrow joining the origin to the point defined by the components of the vectors.

Orthogonality is important because of the following fact: x is a least-squares solution of $Ax = b$ if and only if $b - Ax$ is orthogonal to every vector in the *column space* of A. The column space of A is defined by

$$\text{col}(A) = \{Ax : x \in \mathbf{R}^n\};$$

it is the same as the *range* of the linear operator $T : \mathbf{R}^n \to \mathbf{R}^m$ defined by $T(x) = Ax$. The range of any linear operator is a subspace of the co-domain, so, in this case, col(A) is a subspace of $\mathbf{R}^m$.

The orthogonality condition mentioned above is illustrated in Figure 1.1. The value of Ax that we seek is the orthogonal projection of b onto col(A), which, as a subspace, is a "flat" set passing through the origin. This picture should agree with the reader's geometric intuition.

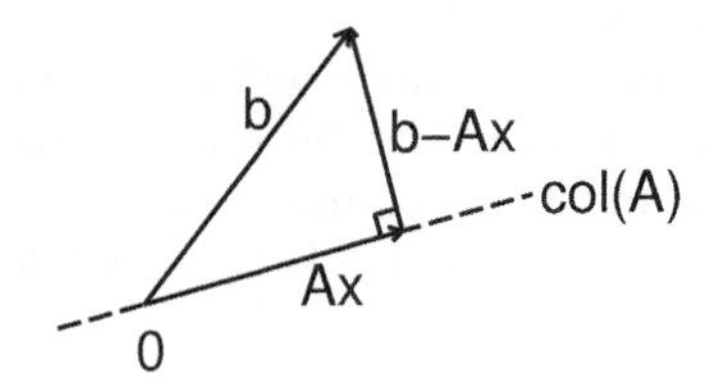

FIGURE 1.1
The orthogonal projection of b onto col(A).

Besides being geometrically plausible, the orthogonality condition leads to a computational approach for computing x. It is based on the *transpose* of a matrix. If A is an $m \times n$ matrix (that is, a matrix with m rows and n columns), then A^T is the $n \times m$ matrix whose rows are the columns of A. The transpose of a matrix is important because of the following relationship:

$$(Ax) \cdot y = x \cdot (A^T y) \text{ for all } x \in \mathbf{R}^n, y \in \mathbf{R}^m. \tag{1.4}$$

The orthogonality condition described above (x is a least-squares solution to $Ax = b$ if and only if $b - Ax$ is orthogonal to every vector in col(A)) is used as follows:

$$\begin{aligned} &(b - Ax) \cdot z = 0 \text{ for all } z \in \text{col}(A) \\ \Leftrightarrow &(b - Ax) \cdot Ay = 0 \text{ for all } y \in \mathbf{R}^n \\ \Leftrightarrow &A^T(b - Ax) \cdot y = 0 \text{ for all } y \in \mathbf{R}^n. \end{aligned}$$

Since the only vector in $\mathbf{R}^n$ that is orthogonal to every vector in $\mathbf{R}^n$ is the zero vector, we obtain

$$A^T(b - Ax) = 0$$

or

$$A^T Ax = A^T b.$$

This is an $n \times n$ system of equations (that is, a system with n equations and n unknowns) that can be solved for x. It can be shown that this system is guaranteed to have a solution. The equations represented by $A^T Ax = A^T b$ are referred to as the *normal equations* for $Ax = b$.

We can now solve the system (1.3) (in the least-squares sense) as follows. We first compute

$$A^T A = \begin{bmatrix} 1 & 3 & 3 & 1 & -4 \\ 1 & 3 & 2 & 1 & -4 \\ -2 & -5 & -10 & -7 & 11 \end{bmatrix} \begin{bmatrix} 1 & 1 & -2 \\ 3 & 3 & -5 \\ 3 & 2 & -10 \\ 1 & 1 & -7 \\ -4 & -4 & 11 \end{bmatrix}$$

$$= \begin{bmatrix} 36 & 33 & -98 \\ 33 & 31 & -88 \\ -98 & -88 & 299 \end{bmatrix},$$

$$A^T b = \begin{bmatrix} 1 & 3 & 3 & 1 & -4 \\ 1 & 3 & 2 & 1 & -4 \\ -2 & -5 & -10 & -7 & 11 \end{bmatrix} \begin{bmatrix} 2 \\ 7 \\ 2 \\ -3 \\ -4 \end{bmatrix} = \begin{bmatrix} 42 \\ 40 \\ -82 \end{bmatrix}.$$

Solving $A^T Ax = A^T b$ yields

$$x \doteq \begin{bmatrix} 4.4619 \\ -0.52603 \\ 1.0334 \end{bmatrix}$$

(where "$\doteq$" means "approximately equals"). The reader can also check that, with the above value of x, $\|Ax - b\| \doteq 0.61360$ (while the norm of b itself is about 9.0554). In this particular case $A^T Ax = A^T b$ has a unique solution, so the given x is the unique least-squares solution to $Ax = b$.

1.2.2 Best approximation by a polynomial

We will now consider a problem that is apparently quite different, but which turns out to have the same abstract form as the least-squares problem from the previous section. Suppose we have a continuous real-valued function f defined on an interval $[a, b]$, and we wish to approximate it by a polynomial. For definiteness, we will consider $f : [0, 1] \to \mathbf{R}$ defined by $f(x) = e^x$, and approximate f by a polynomial of degree at most three.

For this problem, the relevant vector spaces are $C[0, 1]$, the space of all continuous real-valued functions defined on $[0, 1]$, and the space $\mathcal{P}_3$ of all polynomials of degree three or less, regarded as a subspace of $C[0, 1]$. When approximating f by a polynomial from $\mathcal{P}_3$, we might as well ask for the best approximation. However, this requires a norm on the space $C[0, 1]$. To this end, we define the $L^2(0,1)$ inner product

$$\langle f,g\rangle_{L^2(0,1)} = \int_0^1 fg \text{ for all } f,g \in C[0,1],$$

and the associated norm $\|f\|_{L^2(0,1)} = \sqrt{\langle f,f\rangle_{L^2(0,1)}}$. Later in the book, it will be shown that the L^2 inner product is the natural generalization to functions of the Euclidean dot product.

Since $\mathcal{P}_3$ is a subspace of $C[0,1]$, we can apply the orthogonality condition used in the previous section (using the L^2 inner product in place of the Euclidean dot product): $p \in \mathcal{P}_3$ is the best approximation from $\mathcal{P}_3$ to f, in the L^2 norm, if and only if

$$\langle f-p,q\rangle_{L^2(0,1)} = 0 \text{ for all } q \in \mathcal{P}_3. \tag{1.5}$$

A polynomial $q \in \mathcal{P}_3$ has the form $c_1 + c_2x + c_3x^2 + c_4x^3$, and it is a fact from algebra that such a representation is unique (that is, for a given $p \in \mathcal{P}_3$, there is only one choice of c_1, c_2, c_3, c_4). It follows that $\{1, x, x^2, x^3\}$ is a basis for $\mathcal{P}_3$, a fact that we use in two ways. First of all, finding the best approximation $p \in \mathcal{P}_3$ means finding the values of c_1, c_2, c_3, c_4 so that $p(x) = c_1 + c_2x + c_3x^2 + c_4x^3$ satisfies (1.5). This means that the stated problem has four unknowns. Second, (1.5) must hold for all $q \in \mathcal{P}_3$ if it holds for q equal to the four basis functions. This gives four equations to determine the four unknowns. Let us write $q_1(x) = 1$, $q_2(x) = x$, $q_3(x) = x^2$, $q_4(x) = x^3$; then $p = \sum_{j=1}^4 c_jq_j$ and

$$\begin{aligned}
&\langle f-p,q_i\rangle_{L^2(0,1)} = 0,\ i=1,2,3,4 \\
&\Rightarrow \left\langle f - \sum_{j=1}^4 c_jq_j, q_i \right\rangle_{L^2(0,1)} = 0,\ i=1,2,3,4 \\
&\Rightarrow \langle f,q_i\rangle_{L^2(0,1)} - \sum_{j=1}^4 c_j \langle q_j,q_i\rangle_{L^2(0,1)} = 0,\ i=1,2,3,4 \\
&\Rightarrow \sum_{j=1}^4 c_j \langle q_j,q_i\rangle_{L^2(0,1)} = \langle f,q_i\rangle_{L^2(0,1)},\ i=1,2,3,4.
\end{aligned}$$

We now have four linear equations in four unknowns, and we would expect a unique solution for c_1, c_2, c_3, c_4. We see that

$$\langle q_j,q_i\rangle_{L^2(0,1)} = \int_0^1 x^{i-1}x^{j-1}\,dx = \int_0^1 x^{i+j-2}\,dx = \frac{1}{i+j-1}$$

and

$$\langle f, q_1 \rangle_{L^2(0,1)} = \int_0^1 e^x \, dx = e - 1,$$

$$\langle f, q_2 \rangle_{L^2(0,1)} = \int_0^1 x e^x \, dx = 1,$$

$$\langle f, q_3 \rangle_{L^2(0,1)} = \int_0^1 x^2 e^x \, dx = e - 2,$$

$$\langle f, q_4 \rangle_{L^2(0,1)} = \int_0^1 x^3 e^x \, dx = 6 - 2e.$$

The best approximation problem has now been reduced to solving $Ax = b$, where A is the 4×4 matrix

$$A = \begin{bmatrix} 1 & \frac{1}{2} & \frac{1}{3} & \frac{1}{4} \\ \frac{1}{2} & \frac{1}{3} & \frac{1}{4} & \frac{1}{5} \\ \frac{1}{3} & \frac{1}{4} & \frac{1}{5} & \frac{1}{6} \\ \frac{1}{4} & \frac{1}{5} & \frac{1}{6} & \frac{1}{7} \end{bmatrix}$$

and b is the 4-vector $b = (e - 1, 1, e - 2, 6 - 2e)$. The solution is

$$c \doteq \begin{bmatrix} 0.99906 \\ 1.0183 \\ 0.42125 \\ 0.27863 \end{bmatrix},$$

so the best cubic approximation to $f(x) = e^x$ on $[0, 1]$ is given by

$$p(x) \doteq 0.99906 + 1.0183x + 0.42125x^2 + 0.27863x^3.$$

The function f and approximation p are graphed together in Figure 1.2; for purposes of comparison, the third-degree Taylor polynomial of f around $x = 0$ is also shown. The errors in the two approximations are shown in Figure 1.3.

1.3 Diagonalization

Matrices are central in the study of finite-dimensional vector spaces, for the following reason: Any linear operator mapping one finite-dimensional vector space into another can be represented by a matrix. This fact will be explained in detail later in the book.

A special kind of matrix is a *diagonal* matrix, which has zero for every

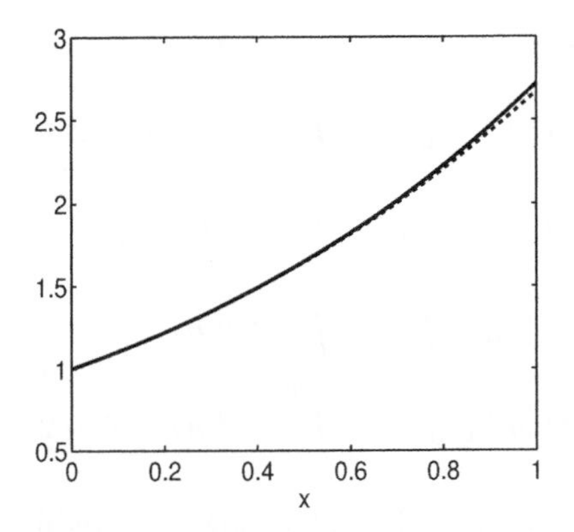

FIGURE 1.2
The function $f(x) = e^x$ on the interval $[0,1]$ (solid line), together with the best cubic polynomial approximation (dashed line) and the cubic Taylor polynomial of f around $x = 0$ (dotted line). At this scale, the best approximation cannot be distinguished from the function it approximates.

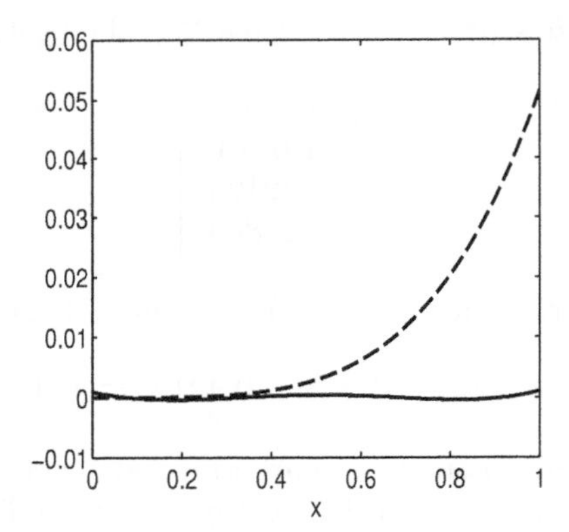

FIGURE 1.3
The errors in the best cubic approximation to $f(x) = e^x$ on $[0,1]$ (solid line) and the cubic Taylor polynomial of f around $x = 0$ (dashed line).

entry except those on the diagonal:

$$\begin{bmatrix} A_{11} & 0 & \cdots & 0 \\ 0 & A_{22} & \cdots & 0 \\ \vdots & \vdots & \ddots & \vdots \\ 0 & 0 & \cdots & A_{nn} \end{bmatrix}.$$

We shall see that almost every matrix-based calculation is simple if the matrix involved is diagonal. Here is one example: If A is an $n \times n$ diagonal matrix, then the linear system $Ax = b$ reduces to the following equations:

$$A_{11}x_1 = b_1, A_{22}x_2 = b_2, \ldots, A_{nn}x_n = b_n.$$

This system can be described as *decoupled* because each equation involves only one unknown. The system can be solved directly—no Gaussian elimination required—to yield

$$x_1 = b_1/A_{11}, x_2 = b_2/A_{22}, \ldots, x_n = b_n/A_{nn}.$$

For most square matrices, there is a change of variables (to be precise, a change of basis) that transforms the matrix into a diagonal matrix. We now present an example where this fact is useful. Consider the following system of linear ordinary differential equations (ODEs):

$$\begin{aligned} x_1' &= -\frac{7}{6}x_1 + \frac{1}{3}x_2 - \frac{1}{6}x_3, \\ x_2' &= \frac{1}{3}x_1 - \frac{5}{3}x_2 + \frac{1}{3}x_3, \\ x_3' &= -\frac{1}{6}x_1 + \frac{1}{3}x_2 - \frac{7}{6}x_3, \end{aligned}$$

The problem is to find functions $x_1(t)$, $x_2(t)$, and $x_3(t)$ that satisfy all three equations. We can write the system in matrix-vector form as $x' = Ax$, where

$$x = x(t) = \begin{bmatrix} x_1(t) \\ x_2(t) \\ x_3(t) \end{bmatrix}, \; x' = x'(t) = \begin{bmatrix} x_1'(t) \\ x_2'(t) \\ x_3'(t) \end{bmatrix}, \; A = \begin{bmatrix} -\frac{7}{6} & \frac{1}{3} & -\frac{1}{6} \\ \frac{1}{3} & -\frac{5}{3} & \frac{1}{3} \\ -\frac{1}{6} & \frac{1}{3} & -\frac{7}{6} \end{bmatrix}.$$

The system $x' = Ax$ can be solved by transforming it to an equivalent system $y' = Dy$ in which the matrix D is diagonal. The fact that the matrix is diagonal means that the system is decoupled and therefore that each equation can be solved for one of the unknowns.

The key to transforming A into a diagonal matrix is to identify the eigenvalues and eigenvectors of A. A scalar λ is called an *eigenvalue* of A if there exists a nonzero vector x such that $Ax = \lambda x$. The vector x is called an *eigenvector* of A corresponding to the eigenvalues λ. Eigenvectors of A are special vectors on which the action of A is particularly simple:

$$Ax = \begin{bmatrix} A_{11}x_1 + A_{12}x_2 + \ldots + A_{1n}x_n \\ A_{21}x_1 + A_{22}x_2 + \ldots + A_{2n}x_n \\ \vdots \\ A_{n1}x_1 + A_{n2}x_2 + \ldots + A_{nn}x_n \end{bmatrix}$$

reduces to simply

$$\lambda x = \begin{bmatrix} \lambda x_1 \\ \lambda x_2 \\ \vdots \\ \lambda x_n \end{bmatrix}.$$

In the case of the matrix A from this example, there are three *eigenpairs*: the eigenvalues are $\lambda_1 = -1$, $\lambda_2 = -1$, $\lambda_3 = -2$, and the corresponding eigenvectors are

$$x^{(1)} = \begin{bmatrix} \frac{1}{\sqrt{3}} \\ \frac{1}{\sqrt{3}} \\ \frac{1}{\sqrt{3}} \end{bmatrix}, \; x^{(2)} = \begin{bmatrix} \frac{1}{\sqrt{2}} \\ 0 \\ -\frac{1}{\sqrt{2}} \end{bmatrix}, \; x^{(3)} = \begin{bmatrix} \frac{1}{\sqrt{6}} \\ -\frac{2}{\sqrt{6}} \\ \frac{1}{\sqrt{6}} \end{bmatrix}.$$

We define U to be the 3×3 matrix whose columns are the eigenvectors of A and D to be the the 3×3 diagonal matrix whose diagonal entries are the corresponding eigenvalues:

$$U = \left[x^{(1)}|x^{(2)}|x^{(3)}\right] = \begin{bmatrix} \frac{1}{\sqrt{3}} & \frac{1}{\sqrt{2}} & \frac{1}{\sqrt{6}} \\ \frac{1}{\sqrt{3}} & 0 & -\frac{2}{\sqrt{6}} \\ \frac{1}{\sqrt{3}} & -\frac{1}{\sqrt{2}} & \frac{1}{\sqrt{6}} \end{bmatrix}, \; D = \begin{bmatrix} -1 & 0 & 0 \\ 0 & -1 & 0 \\ 0 & 0 & -2 \end{bmatrix}.$$

The reader can verify that the following manipulations are valid:

$$\begin{aligned} AU = A\left[x^{(1)}|x^{(2)}|x^{(3)}\right] &= \left[Ax^{(1)}|Ax^{(2)}|Ax^{(3)}\right] \\ &= \left[\lambda_1 x^{(1)}|\lambda_2 x^{(2)}|\lambda_3 x^{(3)}\right] = UD. \end{aligned}$$

In this particular case, the multiplicative inverse of the matrix U is simply its transpose, and $AU = UD$ becomes $A = UDU^T$.

We now substitute this relationship into $x' = Ax$:

$$\begin{aligned} x' = Ax &\Rightarrow x' = UDU^T x \\ &\Rightarrow U^T x' = DU^T x \quad \text{(multiplying both sides by the inverse of } U) \\ &\Rightarrow (U^T x)' = D(U^T x) \quad (U^T \text{ is constant with respect to } t) \\ &\Rightarrow y' = Dy \quad \text{where } y = U^T x. \end{aligned}$$

In the new variables, the system is

$$\begin{aligned} y_1' &= -y_1, \\ y_2' &= -y_2, \\ y_3' &= -2y_3. \end{aligned}$$

The simple ODE $y' = \alpha y$ has solution $y(t) = y(0)e^{\alpha t}$, so we obtain

$$y(t) = \begin{bmatrix} y_1(0)e^{-t} \\ y_2(0)e^{-t} \\ y_3(0)e^{-2t} \end{bmatrix},$$

and thus

$$
\begin{aligned}
x(t) &= Uy(t) \\
&= \begin{bmatrix} \frac{1}{\sqrt{3}} & \frac{1}{\sqrt{2}} & \frac{1}{\sqrt{6}} \\ \frac{1}{\sqrt{3}} & 0 & -\frac{2}{\sqrt{6}} \\ \frac{1}{\sqrt{3}} & -\frac{1}{\sqrt{2}} & \frac{1}{\sqrt{6}} \end{bmatrix} \begin{bmatrix} y_1(0)e^{-t} \\ y_2(0)e^{-t} \\ y_3(0)e^{-2t} \end{bmatrix} \\
&= y_1(0)e^{-t} \begin{bmatrix} \frac{1}{\sqrt{3}} \\ \frac{1}{\sqrt{3}} \\ \frac{1}{\sqrt{3}} \end{bmatrix} + y_2(0)e^{-t} \begin{bmatrix} \frac{1}{\sqrt{2}} \\ 0 \\ -\frac{1}{\sqrt{2}} \end{bmatrix} + y_3(0)e^{-2t} \begin{bmatrix} \frac{1}{\sqrt{6}} \\ -\frac{2}{\sqrt{6}} \\ \frac{1}{\sqrt{6}} \end{bmatrix} \\
&= \begin{bmatrix} \frac{y_1(0)}{\sqrt{3}}e^{-t} + \frac{y_2(0)}{\sqrt{2}}e^{-t} + \frac{y_3(0)}{\sqrt{6}}e^{-2t} \\ \frac{y_1(0)}{\sqrt{3}}e^{-t} - \frac{2y_3(0)}{\sqrt{6}}e^{-2t} \\ \frac{y_1(0)}{\sqrt{3}}e^{-t} - \frac{y_2(0)}{\sqrt{2}}e^{-t} + \frac{y_3(0)}{\sqrt{6}}e^{-2t} \end{bmatrix}.
\end{aligned}
$$

We now see that the solutions are

$$
\begin{aligned}
x_1(t) &= \frac{y_1(0)}{\sqrt{3}}e^{-t} + \frac{y_2(0)}{\sqrt{2}}e^{-t} + \frac{y_3(0)}{\sqrt{6}}e^{-2t}, \\
x_2(t) &= \frac{y_1(0)}{\sqrt{3}}e^{-t} - \frac{2y_3(0)}{\sqrt{6}}e^{-2t}, \\
x_3(t) &= \frac{y_1(0)}{\sqrt{3}}e^{-t} - \frac{y_2(0)}{\sqrt{2}}e^{-t} + \frac{y_3(0)}{\sqrt{6}}e^{-2t}.
\end{aligned}
$$

The values of $y_1(0)$, $y_2(0)$, and $y_3(0)$ can be obtained from $y(0) = U^T x(0)$ if $x(0)$ is known, and otherwise regarded as arbitrary constants in the formula for the general solution of $x' = Ax$.

The reader will recall the earlier contention that calculations are easy if the matrix involved is diagonal. In this case, we were able to transform the coupled system $x' = Ax$ to the decoupled system $y' = Dy$. The manipulations involved in the transformation were far from simple, but the resulting system $y' = Dy$ was indeed easy to solve. On the other hand, it is not at all clear how one would go about solving $x' = Ax$ directly.

1.4 Summary

The examples presented in this chapter illustrate the main applications and techniques that have driven the development of linear algebra as a subject in its own right. The primary applications are various kinds of linear equations and best approximation in its various forms. In most problems posed on vector spaces, a linear operator or matrix is involved in the statement or solution of the problem, and diagonalization frequently comes into play.

The purpose of the rest of the book is to carefully develop the theory hinted at by these examples and to show, by brief vignettes, how the theory can be used to address applications. Hopefully this last feature of the book will show the reader why the theory is necessary and worth studying.

To close this preview, it might be helpful to list some of the concepts that we have encountered and which will be explained in the succeeding chapters:

1. vector space (abstract space that encompasses both ordinary Euclidean vectors and functions; based on the two linear operations of addition and scalar multiplication)
2. field (the space from which the scalars are drawn; often $\mathbf{R}$ or $\mathbf{C}$, but can also be a finite collection of integers under modular arithmetic)
3. subspace (a subset of a vector space that is a vector space in its own right)
4. spanning set (finite representation of a (finite-dimensional) vector space or subspace)
5. basis (minimal spanning set)
6. linear operator (particularly simple kind of operator; preserves linear relationships among vectors)
7. kernel, range (special subspaces associated with a linear operator)
8. norm (measure of the size of a vector, for example, the size of an error vector)
9. inner product (related to the angle between two vectors; used to define orthogonality)
10. orthogonal (perpendicular) vectors
11. projection (related to best approximation; defined by an orthogonality condition)
12. coupled versus decoupled systems (a decoupled system has only one unknown or variable per equation and is easy to solve)
13. eigenvalues and eigenvectors (related to a special change of variables that can decouple a system)

2

Fields and vector spaces

The purpose of this book is to give a careful development of vector spaces and linear operators, and to show how these abstract concepts are useful in a variety of practical applications. All of the properties discussed and used will be proved, which means that we must choose a starting point—a set of axioms. For our purposes, the starting point will be the axioms of a *field*. Scalars are intrinsically linked to vectors, and the scalars used must belong to a field.

2.1 Fields

2.1.1 Definition and examples

The reader is naturally familiar with the set $\mathbf{R}$ of real numbers (these are almost certainly the objects that come to mind when the word "number" is used without a modifier). Complex numbers are also likely to be familiar objects; they have the form $\alpha + \beta i$, where α, β are real numbers and i is the square root of -1. The set $\mathbf{C}$ of complex numbers has come to be indispensable in higher mathematics because these numbers often arise naturally even when the original problem is posed entirely in terms of real numbers.[1] We will see an important example of this when studying the eigenvalues of a (real) matrix in Chapter 4. In addition, some application problems are most naturally posed in terms of complex numbers; for instance, the amplitude and phase of a periodic signal can be represented conveniently by a complex number, and therefore these numbers are fundamental to the study of signal processing.

A less familiar set of numbers is formed by the integers modulo a prime number p. These are the numbers $0, 1, \ldots, p-1$, which are added and multiplied modulo p. This means that, for any $\alpha, \beta \in \{0, 1, \ldots, p-1\}$, $\alpha + \beta$ is the remainder when the ordinary sum $\alpha + \beta$ is divided by p, and similarly for $\alpha\beta$. Thus, for instance, $6 + 5 = 4$ and $6 \cdot 5 = 2$ when these operations are interpreted modulo 7. The set $\{0, 1, \ldots, p-1\}$, with addition and multiplication

---

[1] "The shortest path between two truths in the real domain passes through the complex domain." This quote is usually attributed to Jacques Hadamard; for example, see [24].

defined modulo p, is denoted by $\mathbf{Z}_p$. There are many applications in discrete mathematics in which $\mathbf{Z}_p$ plays the central role (instead of $\mathbf{R}$ or $\mathbf{C}$).

In turns out that $\mathbf{R}$, $\mathbf{C}$, and $\mathbf{Z}_p$ have certain fundamental properties in common, and these properties qualify any of the sets to furnish the scalars operating on a vector space. Since all of these types of scalars, and the corresponding vector spaces, are important, it is natural to develop as much of the theory as possible without specifying the particular scalars to be used. This requires that we define precisely the properties that the abstract scalars are assumed to have.

Definition 1 *Let F be a nonempty set on which are defined two operations, called* addition *and* multiplication*:*

$$\alpha, \beta \in F \;\Rightarrow\; \alpha + \beta \in F, \alpha\beta \in F.$$

We say that F is a field *if and only if these operations satisfy the following properties:*

1. *$\alpha + \beta = \beta + \alpha$ for all $\alpha, \beta \in F$ (commutative property of addition);*
2. *$(\alpha + \beta) + \gamma = \alpha + (\beta + \gamma)$ for all $\alpha, \beta, \gamma \in F$ (associative property of addition);*
3. *there exists an element 0 of F such that $\alpha + 0 = \alpha$ for all $\alpha \in F$ (existence of an additive identity);*
4. *for each $\alpha \in F$, there exists an element $-\alpha \in F$ such that $\alpha + (-\alpha) = 0$ (existence of additive inverses);*
5. *$\alpha\beta = \beta\alpha$ for all $\alpha, \beta \in F$ (commutative property of multiplication);*
6. *$(\alpha\beta)\gamma = \alpha(\beta\gamma)$ for all $\alpha, \beta, \gamma \in F$ (associative property of multiplication);*
7. *there exists a nonzero element 1 of F such that $\alpha \cdot 1 = \alpha$ for all $\alpha \in F$ (existence of a multiplicative identity);*
8. *for each $\alpha \in F$, $\alpha \neq 0$, there exists an element $\alpha^{-1} \in F$ with the property that $\alpha\alpha^{-1} = 1$ (existence of multiplicative inverses);*
9. *$\alpha(\beta + \gamma) = \alpha\beta + \alpha\gamma$ for all $\alpha, \beta, \gamma \in F$ (distributive property of multiplication over addition).*

The reader should notice that we normally indicate multiplication by juxtaposition, as in $\alpha\beta$, but we sometimes use a "$\cdot$" when it seems clearer, as in $\alpha \cdot 1$. Also, we use the usual convention that multiplication has a higher precedence than addition; thus, in the distributive property, $\alpha\beta + \alpha\gamma$ mean $(\alpha\beta) + (\alpha\gamma)$ and not $((\alpha\beta) + \alpha)\gamma$ or $\alpha(\beta + \alpha)\gamma$.

It is important to notice the requirement that, in any field, $1 \neq 0$. This guarantees that a field has at least two distinct elements and will be important later.

The properties (1.1–1.9) are the axioms defining a field. We will take it for granted that these familiar properties are satisfied when $F = \mathbf{R}$ (the set of real numbers) with the usual operations of addition and multiplication. Exercise 5 asks that reader to show that $\mathbf{C}$ (the set of complex numbers) is also a field under the usual definitions of addition and multiplication.

Exercise 17 asks that reader to prove that $\mathbf{Z}_p$ is a field. In particular, the additive identity is 0 and the multiplicative identity is 1. It is easy to show the existence of additive inverses, while proving the existence of multiplicative inverses is not quite so straightforward (see the hint in Exercise 17). To actually compute the multiplicative inverse of a given element of $\mathbf{Z}_p$, the Euclidean algorithm can be used. This is explained in Appendix A.

In addition to $\mathbf{Z}_p$ (which has p elements), there exist finite fields with p^n elements, where p is prime and $n > 1$ is an integer. We will not develop these fields here, although Exercise 2.6.18 asks the reader to show that the number of elements in a finite field is necessarily p^n for some prime p and integer $n \geq 1$. For a construction of finite fields with p^n elements, $n > 1$, the reader can consult [30].

The field $\mathbf{Z}_2$ is the smallest possible field. The elements are 0 and 1, and Definition 1 states that every field must have at least the two distinct elements 0 and 1.

2.1.2 Basic properties of fields

For any field F, the operations of subtraction and division are defined by $\alpha - \beta = \alpha + (-\beta)$ and $\alpha/\beta = \alpha\beta^{-1}$ ($\beta \neq 0$), respectively. Similarly, we can define α^k for an arbitrary integer k by repeated multiplication (for example, $\alpha^3 = \alpha \cdot \alpha \cdot \alpha$ and $\alpha^{-3} = \alpha^{-1} \cdot \alpha^{-1} \cdot \alpha^{-1}$).

Some properties of fields are so immediate that we do not bother to write out the proofs, such as $0 + \alpha = \alpha$ and $1 \cdot \alpha = \alpha$, or the fact that the distributive law holds in the form $(\beta + \gamma)\alpha = \beta\alpha + \gamma\alpha$. The following theorem collects some properties that we will use constantly and whose proofs are not quite immediate.

Theorem 2 *Let F be a field. Then:*

1. *The additive and multiplicative identities of F are unique. That is, there is only one element 0 of F such that $\alpha + 0 = \alpha$ for all $\alpha \in F$, and similarly for 1.*

2. *The additive inverse $-\alpha$ is unique for each $\alpha \in F$.*

3. *The multiplicative inverse α^{-1} is unique for each $\alpha \neq 0$ in F.*

4. *(The cancellation property of addition.) If $\alpha, \beta, \gamma \in F$ and $\alpha + \gamma = \beta + \gamma$, then $\alpha = \beta$.*

5. *(The cancellation property of multiplication.) If* $\alpha, \beta, \gamma \in F$, $\alpha\gamma = \beta\gamma$, *and* $\gamma \neq 0$, *then* $\alpha = \beta$.

6. *For each* $\alpha \in F$, $0 \cdot \alpha = 0$ *and* $-1 \cdot \alpha = -\alpha$.

Proof

1. Suppose that F contains two additive identities, say 0 and z. Then, since z is an additive identity, we have $0 + z = 0$. On the other hand, since 0 is an additive identity, we also have $z + 0 = z$. But then, since addition is commutative, we have $z = z + 0 = 0 + z = 0$. Thus the additive identity is unique. The proof that the multiplicative identity is unique is exactly analogous (see Exercise 2).

2. Let $\alpha \in F$. Suppose that $\beta \in F$ is an additive inverse of α, that is, suppose β satisfies $\alpha + \beta = 0$. Adding $-\alpha$ to both sides of this equation yields

$$\begin{aligned} -\alpha + (\alpha + \beta) = -\alpha + 0 &\Rightarrow (-\alpha + \alpha) + \beta = -\alpha \\ &\Rightarrow 0 + \beta = -\alpha \\ &\Rightarrow \beta = -\alpha. \end{aligned}$$

(The reader should notice how the associative property of addition and the definition of 0 were used in the above steps.) Thus there is only one additive inverse for α.

3. The proof that the multiplicative inverse is unique is analogous to the previous proof (see Exercise 3).

4. Suppose $\alpha, \beta, \gamma \in F$ and $\alpha + \gamma = \beta + \gamma$. Then

$$\begin{aligned} (\alpha + \gamma) + (-\gamma) = (\beta + \gamma) + (-\gamma) &\Rightarrow \alpha + (\gamma + (-\gamma)) = \beta + (\gamma + (-\gamma)) \\ &\Rightarrow \alpha + 0 = \beta + 0 \\ &\Rightarrow \alpha = \beta. \end{aligned}$$

5. The proof is analogous to the proof of the cancellation property of addition (see Exercise 6).

6. Let $\alpha \in F$. We know that $0 + \alpha = \alpha$; multiplying both sides of this equation by α yields
$$(0 + \alpha)\alpha = \alpha \cdot \alpha.$$
The distributive property then yields
$$0 \cdot \alpha + \alpha \cdot \alpha = \alpha \cdot \alpha,$$
or
$$0 \cdot \alpha + \alpha \cdot \alpha = 0 + \alpha \cdot \alpha.$$

By the cancellation property, this implies that $0 \cdot \alpha = 0$.

To show that $-1 \cdot \alpha = -\alpha$, it suffices to show that $\alpha + (-1) \cdot \alpha = 0$. We have $\alpha = 1 \cdot \alpha$, so

$$\alpha + (-1) \cdot \alpha = 1 \cdot \alpha + (-1) \cdot \alpha = (1 + (-1))\alpha = 0 \cdot \alpha = 0.$$

Since α has only one additive inverse, this shows that $-1 \cdot \alpha = -\alpha$.

QED

Here are some further properties of a field:

Theorem 3 *Let F be a field. Then, for any $\alpha, \beta, \gamma \in F$:*

1. $-(-\alpha) = \alpha$;
2. $-(\alpha + \beta) = -\alpha + (-\beta) = -\alpha - \beta$;
3. $-(\alpha - \beta) = -\alpha + \beta$;
4. $\alpha(-\beta) = -(\alpha\beta)$;
5. $(-\alpha)\beta = -(\alpha\beta)$;
6. $(-\alpha)(-\beta) = \alpha\beta$;
7. $(-\alpha)^{-1} = -\left(\alpha^{-1}\right)$;
8. $(\alpha\beta)^{-1} = \alpha^{-1}\beta^{-1}$;
9. $\alpha(\beta - \gamma) = \alpha\beta - \alpha\gamma$;
10. $(\beta - \gamma)\alpha = \beta\alpha - \gamma\alpha$.

Proof Exercise 4.

Besides being essential for algebraic manipulations, various properties given above allow us to use simplified notation without ambiguity. For example, the expression $-\alpha\beta$ is formally ambiguous: Does it mean $(-\alpha)\beta$ or $-(\alpha\beta)$? In fact, by the previous theorem, these two expressions are equal, so the ambiguity disappears.

The distributive property extends to a sum with an arbitrary number of terms:

$$\beta(\alpha_1 + \alpha_2 + \ldots + \alpha_n) = \beta\alpha_1 + \beta\alpha_2 + \ldots + \beta\alpha_n \tag{2.1}$$

(Exercise 10 asks the reader to prove this fact). Using summation notation,[2]

[2]The following examples illustrate summation notation:

$$\begin{aligned}\sum_{i=1}^{n} \alpha_i &= \alpha_1 + \alpha_2 + \cdots + \alpha_n, \\ \sum_{i=1}^{m}\sum_{j=1}^{n} \alpha_{ij} &= \sum_{i=1}^{m} (\alpha_{i1} + \alpha_{i2} + \cdots + \alpha_{in}) \\ &= (\alpha_{11} + \alpha_{12} + \cdots + \alpha_{1n}) + (\alpha_{21} + \alpha_{22} + \cdots + \alpha_{2n}) + \cdots + \\ &\quad (\alpha_{m1} + \alpha_{m2} + \cdots + \alpha_{mn}).\end{aligned}$$

(2.1) is written as

$$\beta\sum_{i=1}^{n}\alpha_i=\sum_{i=1}^{n}\beta\alpha_i. \tag{2.2}$$

Repeated use of the commutative and associative properties of addition shows that

$$\sum_{i=1}^{n}(\alpha_i+\beta_i)=\sum_{i=1}^{n}\alpha_i+\sum_{i=1}^{n}\beta_i. \tag{2.3}$$

If $\alpha_{ij}\in F$ for each pair of integers i,j with $1\le i\le m$ and $1\le j\le n$, then the following also follows from repeated use of the commutative and associative properties:

$$\sum_{i=1}^{m}\sum_{j=1}^{n}\alpha_{ij}=\sum_{j=1}^{n}\sum_{i=1}^{m}\alpha_{ij}. \tag{2.4}$$

Application of this relationship is called *changing the order of summation.*

As an illustration, we will prove (2.3) by induction,[3] and leave the proof of (2.4) to the exercises (see Exercise 11). We wish to prove that, for any positive integer n and any elements $\alpha_1,\alpha_2,\ldots,\alpha_n$ and $\beta_1,\beta_2,\ldots,\beta_n$ of F,

$$\sum_{i=1}^{n}(\alpha_i+\beta_i)=\sum_{i=1}^{n}\alpha_i+\sum_{i=1}^{n}\beta_i.$$

For $n=1$, the result is obvious:

$$\sum_{i=1}^{1}(\alpha_i+\beta_i)=\alpha_1+\beta_1=\sum_{i=1}^{1}\alpha_1+\sum_{i=1}^{1}\beta_1.$$

Let us suppose, by way of induction, that the result holds for some positive integer n, and let $\alpha_1,\alpha_2,\ldots,\alpha_{n+1}$ and $\beta_1,\beta_2,\ldots,\beta_{n+1}$ belong to F. Then

$$\begin{aligned}\sum_{i=1}^{n+1}(\alpha_i+\beta_i) &= \left(\sum_{i=1}^{n}(\alpha_i+\beta_i)\right)+(\alpha_{n+1}+\beta_{n+1})\\ &= \left(\sum_{i=1}^{n}\alpha_i+\sum_{i=1}^{n}\beta_i\right)+(\alpha_{n+1}+\beta_{n+1}).\end{aligned}$$

The last step holds by the induction hypothesis. Applying the associative and

---

[3]The *principle of induction* is the following theorem:

Theorem *Suppose that, for each positive integer n, P_n is a statement, and assume:*

1. *P_1 is true;*
2. *For any positive integer n, if P_n is true, then P_{n+1} is also true.*

Then P_n is true for all positive integers n.

The principle of induction can be proved from an appropriate set of axioms for the natural numbers (that is, positive integers).

commutative properties of addition, we can rearrange this last expression into the form

$$\left(\sum_{i=1}^{n} \alpha_i + \alpha_{n+1}\right) + \left(\sum_{i=1}^{n} \beta_i + \beta_{n+1}\right),$$

which then equals

$$\sum_{i=1}^{n+1} \alpha_i + \sum_{i=1}^{n+1} \beta_i,$$

as desired.

Exercises

Essential exercises

1. Let F be a field.
   (a) Prove that $-1 \neq 0$. (Every field has an element -1; it is the additive inverse of the multiplicative identity 1.)
   (b) Must it be the case that $-1 \neq 1$? If the answer is yes, then prove it. If not, give an example of a field in which $-1 = 1$.

Miscellaneous exercises

2. Let F be a field. Prove that the multiplicative identity 1 is unique.
3. Let F be a field, and suppose $\alpha \in F$, $\alpha \neq 0$. Prove that α has a unique multiplicative inverse.
4. Prove Theorem 3.
5. Recall that each element in $\mathbf{C}$ (the space of complex numbers) is of the form $a + bi$, where $a, b \in \mathbf{R}$ and i is a new symbol, intended to represent the square root of -1. Addition and multiplication are defined by

$$\begin{aligned}(a+bi)+(c+di) &= (a+c)+(b+d)i,\\ (a+bi)(c+di) &= (ac-bd)+(ad+bc)i.\end{aligned}$$

   (a) Prove that this definition of multiplication implies that $i^2 = -1$. Note that i is the complex number $0 + 1 \cdot i$.
   (b) Prove that $\mathbf{C}$ is a field. The only part of the proof that might be challenging is verifying the existence of multiplicative inverses. Given $a + bi \in \mathbf{C}$, $a + bi \neq 0$ (meaning that $a \neq 0$ or $b \neq 0$), the equation $(a + bi)(x + yi) = 1$ can be solved for x and y to find the multiplicative inverse of $a + bi$.
6. Let F be a field and suppose $\alpha, \beta, \gamma \in F$, with $\gamma \neq 0$. Prove that if $\alpha\gamma = \beta\gamma$, then $\alpha = \beta$.

7. Let F be a field and let α, β be elements of F. Prove that the equation $\alpha + x = \beta$ has a unique solution $x \in F$. What is it?

8. Let F be a field and let α, β be elements of F. Does the equation $\alpha x = \beta$ always have a unique solution? If so, what is it? If not, explain why not and under what conditions the equation does have a unique solution.

9. Let F be a field. Recall that division in F is defined by multiplication by the multiplicative inverse: $\alpha/\beta = \alpha\beta^{-1}$. Prove:

$$\begin{aligned}
\frac{\alpha}{\beta} + \frac{\gamma}{\delta} &= \frac{\alpha\delta + \beta\gamma}{\beta\delta} \text{ for all } \alpha, \beta, \gamma, \delta \in F,\ \beta, \delta \neq 0, \\
\frac{\alpha}{\beta} \cdot \frac{\gamma}{\delta} &= \frac{\alpha\gamma}{\beta\delta} \text{ for all } \alpha, \beta, \gamma, \delta \in F,\ \beta, \delta \neq 0, \\
\frac{\alpha/\beta}{\gamma/\delta} &= \frac{\alpha\delta}{\beta\gamma} \text{ for all } \alpha, \beta, \gamma, \delta \in F,\ \beta, \gamma, \delta \neq 0.
\end{aligned}$$

10. Use the principle of mathematical induction to prove (2.1).

11. Prove (2.4) by using (2.3) and induction on n.

12. Determine if each of the following sets is a field under the usual definitions of multiplication and addition:

    (a) $\mathbf{Z}$, the set of all integers;

    (b) $\mathbf{Q}$, the set of all rational numbers;

    (c) $(0, \infty)$, the set of all positive real numbers.

13. Let F be the set of all ordered pairs (α, β) of real numbers:

$$F = \{(\alpha, \beta) : \alpha, \beta \in \mathbf{R}\}.$$

Define addition and multiplication on F as follows:

$$\begin{aligned}
(\alpha, \beta) + (\gamma, \delta) &= (\alpha + \gamma, \beta + \delta), \\
(\alpha, \beta) \cdot (\gamma, \delta) &= (\alpha\gamma, \beta\delta).
\end{aligned}$$

Does F, together with these operations, form a field? If it does, prove that the definition is satisfied; otherwise, state which parts of the definition fail.

14. In this exercise, F will represent the set of all positive real numbers and the operations defined on F will not be the standard addition and multiplication. For this reason, addition and multiplication on F will be denoted by $\oplus$ and $\odot$, respectively. We define $x \oplus y = xy$ for all $x, y \in F$ (that is, $x \oplus y$ is the ordinary product of x and y) and $x \odot y = x^{\ln y}$. Prove that F, with these operations, is a field.

15. In principle, the axioms for any mathematical object should be chosen to be a minimal set, in that no property is assumed as an axiom if it can be proved from the other axioms. However, this principle is sometimes violated when the result is a set of axioms that is easier to understand. Suppose we omit part 1 (commutativity of addition) of Definition 1. Use the remaining properties of a field to prove that addition must be commutative. (Hint: If addition is not assumed to be commutative, part 2 of Theorem 3 becomes $-(\alpha+\beta) = -\beta + (-\alpha)$. On the other hand, we can still prove $-\gamma = -1 \cdot \gamma$ for all $\gamma \in F$. Then $-(\alpha+\beta) = -1 \cdot (\alpha+\beta)$, and we can apply the distributive law and manipulate to get the desired result. However, care must be taken to use only results that are proved without the use of commutativity of addition.)

16. (a) Write out all possible sums and products in $\mathbf{Z}_2$. Identify the additive inverse of each element and the multiplicative inverse of each nonzero element.

    (b) Repeat for $\mathbf{Z}_3$ and $\mathbf{Z}_5$.

17. Prove that $\mathbf{Z}_p$ is a field for any prime number p. (Hint: The difficult part is proving the existence of multiplicative inverses. For any nonzero $\alpha \in \mathbf{Z}_p$, the elements $\alpha, \alpha^2, \alpha^3, \ldots$ cannot all be distinct because $\mathbf{Z}_p$ contains a finite number of elements. Therefore, $\alpha^k = \alpha^\ell \pmod{p}$ for some positive integers $k > \ell$. Prove that $\alpha^{k-\ell-1}$ is the multiplicative inverse of α. Notice that the statement $\alpha^k = \alpha^\ell \pmod{p}$ means that $\alpha^k = \alpha^\ell + np$ for some integer n.)

18. Explain why $\mathbf{Z}_p$ does not define a field if p is not prime.

19. Suppose F is a finite field.

    (a) Prove that there exists a smallest positive integer n such that $1 + 1 + \cdots + 1$ (n terms) is the additive identity 0. The integer n is called the *characteristic* of the field.

    (b) Prove that if n is the characteristic of F, then for each $\alpha \in F$, $\alpha + \alpha + \ldots + \alpha$ (n terms) is 0.

    (c) Prove that the characteristic of a finite field must be a prime number. (Hint: Suppose $n = k\ell$, where $1 < k, \ell < n$. Define $\alpha = 1 + 1 + \cdots + 1$ (k terms), $\beta = 1 + 1 + \cdots + 1$ (ℓ terms), and prove that $\alpha\beta = 0$. Why is this a contradiction?)

Project: Quaternions

20. When we construct the complex numbers from the real numbers, we introduce a new symbol i to obtain numbers of the form $a + bi$, $a, b \in \mathbf{R}$, and then define addition and multiplication for numbers of this form. In this exercise, we construct another number system from $\mathbf{R}$ by introducing three new symbols, i, j, k. Each number of the form

$x_1 + x_2 i + x_3 j + x_4 k$, $x_1, x_2, x_3, x_4 \in \mathbf{R}$, is called a *quaternion*, and the space of all quaternions is denoted by $\mathbf{H}$. Addition and multiplication in $\mathbf{H}$ are defined by

$$\begin{aligned} & (x_1 + x_2 i + x_3 j + x_4 k) + (y_1 + y_2 i + y_3 j + y_4 k) \\ = \; & (x_1 + y_1) + (x_2 + y_2)i + (x_3 + y_3)j + (x_4 + y_4)k, \\ & (x_1 + x_2 i + x_3 j + x_4 k)(y_1 + y_2 i + y_3 j + y_4 k) \\ = \; & (x_1 y_1 - x_2 y_2 - x_3 y_3 - x_4 y_4) + \\ & (x_1 y_2 + x_2 y_1 + x_3 y_4 - x_4 y_3)i + \\ & (x_1 y_3 - x_2 y_4 + x_3 y_1 + x_4 y_2)j + \\ & (x_1 y_4 + x_2 y_3 - x_3 y_2 + x_4 y_1)k. \end{aligned}$$

As with complex numbers, if $x = x_1 + x_2 i + x_3 j + x_4 k$ and any of x_1, x_2, x_3, x_4 are zero, we usually do not write the corresponding terms. In particular, $i = 0 + 1i + 0j + 0k$, and similarly for j and k.

(a) Compute each of the following:

$$i^2, j^2, k^2, ij, ik, jk, ji, ki, kj, ijk$$

(as usual, $i^2 = ii$, and similarly for j^2, k^2).

(b) For a given $x = x_1 + x_2 i + x_3 j + x_4 k$, define $\overline{x} = x_1 - x_2 i - x_3 j - x_4 k$. Compute $x\overline{x}$ and $\overline{x}x$.

(c) It is obvious that addition on $\mathbf{H}$ satisfies properties 1–4 of a field. What is the additive identity? What is the additive inverse of $x = x_1 + x_2 i + x_3 j + x_4 k$?

(d) Show that multiplication in $\mathbf{H}$ is not commutative.

(e) It is straightforward but tedious to verify that multiplication in $\mathbf{H}$ is associative and distributes over addition; you may take these properties for granted. Show that there is a multiplicative identity in $\mathbf{H}$. What is it?

(f) Prove that if $x \in \mathbf{H}$, $x \neq 0$, then there exists $x^{-1} \in \mathbf{H}$ such that $xx^{-1} = x^{-1}x = 1$.

We thus see that $\mathbf{H}$ satisfies all the properties of a field except commutativity of multiplication. Such a number system is called a *division ring*.

21. A *subfield* S of a field F is a nonempty subset of F that is a field in its own right (under the operations of addition and multiplication defined on F). Since the properties of commutativity and associativity of both addition and multiplication, as well as the distributive property, are

inherited by S, one can verify that S is a subfield by verifying that $0, 1 \in S$ and that the following properties are satisfied:

$$\begin{aligned} \alpha, \beta \in S &\Rightarrow \alpha + \beta \in S, \\ \alpha, \beta \in S &\Rightarrow \alpha\beta \in S, \\ \alpha \in S &\Rightarrow -\alpha \in S, \\ \alpha \in S, \alpha \neq 0 &\Rightarrow \alpha^{-1} \in S. \end{aligned}$$

(a) Prove that **R** is a subfield of **C**.

(b) Note that **R** can regarded as a subset of **H** (see the previous exercise) by regarding $x \in \mathbf{H}$, $x = x_1 + 0i + 0j + 0k$, as a real number. Show that the addition and multiplication on **H** reduce to the usual addition and multiplication of real numbers when restricted to **R**. In this sense, **R** is a subfield of the division ring **H**.

(c) Similarly, **C** can be regarded as a subset of **H** by regarding $x \in \mathbf{H}$, $x = x_1 + x_2 i + 0j + 0k$, as a complex number. Prove that **C** is a subfield of of the division ring **H**. (Again, you will have to prove that the addition and multiplication on **H** reduce to the usual addition and multiplication of complex numbers when restricted to **C**.)

(d) Define a subset of **H** as follows:

$$S = \{a + bi + cj \; : \; a, b, c \in \mathbf{R}\}.$$

Is S a subfield of **H**? Prove your answer.

2.2 Vector spaces

Mathematics derives its power largely from its ability to find the common features of various problems and study them abstractly. As we described in Chapter 1, there are many problems that involve the related concepts of addition, scalar multiplication, and linearity. To study these properties abstractly, we introduce the notion of a vector space.

Definition 4 *Let F be a field and let V be a nonempty set. Suppose two operations are defined with respect to these sets, addition and scalar multiplication:*

$$\begin{aligned} u, v \in V &\Rightarrow u + v \in V, \\ \alpha \in F, v \in V &\Rightarrow \alpha v \in V. \end{aligned}$$

We say that V is a vector space over F *if and only if the following properties are satisfied:*

1. $u + v = v + u$ *for all* $u, v \in V$ *(commutative property of addition);*

2. $u + (v + w) = (u + v) + w$ *for all* $u, v, w \in V$ *(associative property of addition);*

3. *There exists an element* 0 *of* V *such that* $u + 0 = u$ *for all* $u \in V$ *(existence of an additive identity);*

4. *For each* $u \in V$*, there exists an element* $-u$ *of* V *such that* $u + (-u) = 0$ *(existence of additive inverses);*

5. $\alpha(\beta u) = (\alpha\beta)u$ *for all* $\alpha, \beta \in F$*,* $u \in V$ *(associative property of scalar multiplication);*

6. $\alpha(u + v) = \alpha u + \alpha v$ *for all* $\alpha \in F$*,* $u, v \in V$ *(distributive property);*

7. $(\alpha + \beta)u = \alpha u + \beta u$ *for all* $\alpha, \beta \in F$*,* $u \in V$ *(distributive property);*

8. $1 \cdot u = u$ *for all* $u \in V$*, where* 1 *is the multiplicative identity of* F*.*

The elements of a vector space V are called *vectors*, and the elements of the corresponding field F are called *scalars*. We normally use Greek letters for scalars and Roman letters for vectors; however, the scalar 0 and the vector 0 are denoted by the same symbol. The context determines which is intended.

The following elementary properties of vector spaces are proved in much the same way as they were for fields.

Theorem 5 *Let* V *be a vector space over a field* F*. Then:*

1. *The additive identity* 0 *of* V *is unique.*

2. *For each* $u \in V$*, the additive inverse* $-u$ *is unique.*

3. *If* $u, v \in V$*, then* $-(u + v) = -u + (-v)$*.*

4. *If* $u, v, w \in V$ *and* $u + v = u + w$*, then* $v = w$ *(cancellation property).*

5. *If* $\alpha \in F$ *and* 0 *is the zero vector in* V*, then* $\alpha \cdot 0 = 0$*.*

6. *If* $\alpha \in F$*,* $u \in V$*, and* $\alpha u = 0$*, then* $\alpha = 0$ *or* $u = 0$*.*

7. *For each* $u \in V$*,* $0 \cdot u = 0$ *and* $(-1) \cdot u = -u$*.*

Proof Exercise 3.

As in the case of a field, we define subtraction as addition of the additive inverse:

$$u, v \in V \ \Rightarrow \ u - v = u + (-v).$$

Similarly, division of a vector by a nonzero scalar is defined by multiplication by the multiplicative inverse:

$$\alpha \in F, \alpha \neq 0, v \in V \ \Rightarrow \ \frac{v}{\alpha} = \alpha^{-1} v.$$

The axioms defining a vector space are all straightforward and merely guarantee that natural algebraic manipulations are valid. We will now consider the most important examples of vector spaces, of which there are two types. The first example is Euclidean n-space, which consists of vectors in the usual sense of the word. In fact, most students first encounter vectors in the concrete case of two- and three-dimensional Euclidean space. This example can be extended to allow Euclidean spaces of any dimension, and also to encompass vectors whose components belong to any given field (including a finite field). The second class of vector spaces consists of various spaces of functions, often distinguished by the amount of regularity (continuity and/or differentiability) required of the functions in a particular space.

2.2.1 Examples of vector spaces

1. Real Euclidean n-space, $\mathbf{R}^n$, is defined to be the collection of all n-tuples of real numbers:

$$\mathbf{R}^n = \left\{ \begin{bmatrix} x_1 \\ x_2 \\ \vdots \\ x_n \end{bmatrix} : x_1, x_2, \ldots, x_n \in \mathbf{R} \right\}.$$

Given $x \in \mathbf{R}^n$, the numbers $x_1, x_2, \ldots, x_n$ are called the *components* of x. Scalar multiplication and addition are defined componentwise:

$$\alpha \in \mathbf{R}, x \in \mathbf{R}^n \quad \Rightarrow \quad \alpha x = \alpha \begin{bmatrix} x_1 \\ x_2 \\ \vdots \\ x_n \end{bmatrix} = \begin{bmatrix} \alpha x_1 \\ \alpha x_2 \\ \vdots \\ \alpha x_n \end{bmatrix},$$

$$x, y \in \mathbf{R}^n \quad \Rightarrow \quad x + y = \begin{bmatrix} x_1 \\ x_2 \\ \vdots \\ x_n \end{bmatrix} + \begin{bmatrix} y_1 \\ y_2 \\ \vdots \\ y_n \end{bmatrix} = \begin{bmatrix} x_1 + y_1 \\ x_2 + y_2 \\ \vdots \\ x_n + y_n \end{bmatrix}.$$

Vectors in $\mathbf{R}^n$ will also be written in the alternate notation

$$x = (x_1, x_2, \ldots, x_n).$$

It can be shown without difficulty that $\mathbf{R}^n$, together with these operations, satisfies the definition of a vector space over the field $\mathbf{R}$ (see Exercise 4).

Euclidean 2-space, $\mathbf{R}^2$, can be visualized as the plane with rectangular (Cartesian) coordinates. A vector $x \in \mathbf{R}^2$ is usually visualized as an arrow with its tail at the origin and its head at the point with coordinates (x_1, x_2), as in Figure 2.1. Similarly, a vector $x \in \mathbf{R}^3$ is visualized as an

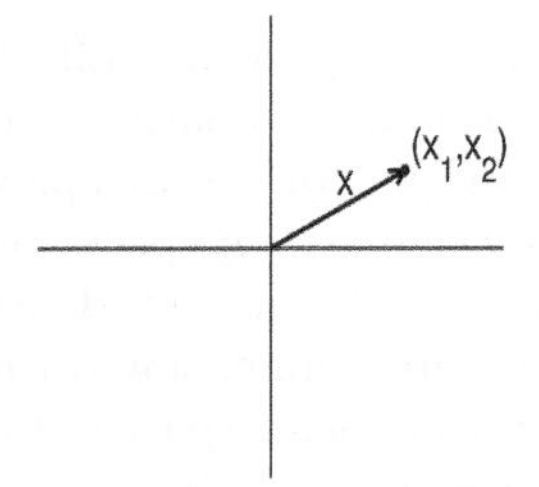

FIGURE 2.1
A vector in $\mathbf{R}^2$.

arrow in space, with its tail at the origin of a rectangular coordinate system and its head at the point with coordinates (x_1, x_2, x_3).

From this point of view, scalar multiplication αx stretches the vector x if $\alpha > 0$ and shrinks the vector x if $0 < \alpha < 1$. When $\alpha > 0$, the direction of the vector does not change. If $\alpha < 0$, then αx points in the opposite direction of x. The vector sum $x + y$ is visualized as the vector from the tail of x to the head of y, after y has been translated so that its tail coincides with the head of x (see Figure 2.2).

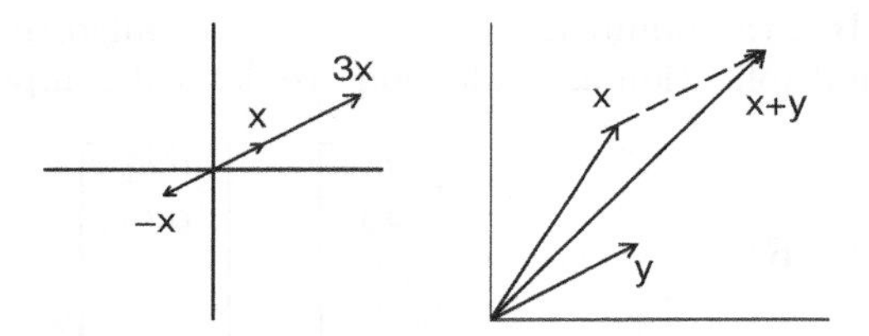

FIGURE 2.2
Visualizing the basic operations in $\mathbf{R}^2$.

It is also possible to visualize the vectors in $\mathbf{R}^2$ or $\mathbf{R}^3$ as the points themselves. There is not such a natural picture of the vector space operations as when using arrows. On the other hand, it is easier to visualize a given set of vectors as a set of points, particularly when the set forms a familiar geometric object such as a line or a plane.

From a mathematical point of view, the visualization is incidental to the definitions and is not essential.

2. Complex Euclidean n-space $\mathbf{C}^n$ is analogous to $\mathbf{R}^n$; the only difference is that the scalars and the components of $x \in \mathbf{C}^n$ are complex numbers. Scalar multiplication and addition are defined componentwise, as in the case of $\mathbf{R}^n$.

In general, given any field F, we can define the space F^n as a vector space over F:

$$F^n = \left\{ \begin{bmatrix} x_1 \\ x_2 \\ \vdots \\ x_n \end{bmatrix} : x_1, x_2, \ldots, x_n \in F \right\},$$

$$\alpha \in F, x \in F^n \Rightarrow \alpha x = \alpha \begin{bmatrix} x_1 \\ x_2 \\ \vdots \\ x_n \end{bmatrix} = \begin{bmatrix} \alpha x_1 \\ \alpha x_2 \\ \vdots \\ \alpha x_n \end{bmatrix},$$

$$x, y \in F^n \Rightarrow x + y = \begin{bmatrix} x_1 \\ x_2 \\ \vdots \\ x_n \end{bmatrix} + \begin{bmatrix} y_1 \\ y_2 \\ \vdots \\ y_n \end{bmatrix} = \begin{bmatrix} x_1 + y_1 \\ x_2 + y_2 \\ \vdots \\ x_n + y_n \end{bmatrix}.$$

The vector space $\mathbf{Z}_p^n$ arises in many application in discrete mathematics. The space $\mathbf{Z}_2^n$ is especially important, since a vector in $\mathbf{Z}_2^n$ can be regarded as a *bit string*.[4]

Any field F can be regarded as a vector space over itself. Except for notation, this is equivalent to taking $n = 1$ in the above definition.

3. Let a and b be real numbers, with $a < b$. There are a variety of spaces of real-valued functions defined on the interval $[a, b]$ that are useful in practical applications. The largest possible space (over the field $\mathbf{R}$) consists of all real-valued functions f defined on $[a, b]$:

$$F[a, b] = \{f \mid f : [a, b] \to \mathbf{R}\}.$$

Scalar multiplication and addition are defined as follows:

$$\alpha \in \mathbf{R}, f \in F[a, b] \Rightarrow (\alpha f)(x) = \alpha f(x),\ x \in [a, b],$$
$$f, g \in F[a, b] \Rightarrow (f + g)(x) = f(x) + g(x),\ x \in [a, b].$$

A word about notation will be helpful here. The correct way to define a function is to specify its domain and co-domain[5] and give the rule

---

[4] A *bit* is simply a zero or a one; the word "bit" is a contraction of *binary digit.* Bit strings are fundamental to modern digital computers, which are based on binary arithmetic. An important application is described in Section 3.11.

[5] For any function, the *domain* is the set of all allowable inputs, while the *co-domain* is the set in which the output values must lie. We write $f : D \to C$ to indicate that the function named f has domain D and co-domain C. Although $f(x)$ is defined for all $x \in D$, it is not guaranteed that, for each $y \in C$, there exists $x \in D$ such that $y = f(x)$. In other words, not every element of the co-domain need be an output of the function for some given input. See Section 3.5 for more details.

defining the function, that is, specify how an input from the domain is mapped unambiguously to an output in the co-domain. Often the rule is given by a formula, as in the following definition:

$$f : \mathbf{R} \to \mathbf{R}, \ f(x) = x^2.$$

Because we often define functions by formulas, it is common to avoid giving a name to a function, and just refer to the formula, as in "the function x^2." This specification omits the domain and co-domain, but frequently this is harmless because they are known by context.

However, specifying a function by a formula is formally ambiguous and sometimes causes confusion. For a given real number x, x^2 is another real number, not a function. It would be more precise to write "the function $x \mapsto x^2$" and reserve "x^2" to refer to a number, but this is rarely done. In this book, we will follow convention and sometimes use "x^2" to refer to a function, particularly when discussing spaces of polynomials, but the reader should be alert to this abuse of notation.

The above discussion should help the reader to understand notation such as

$$f, g \in F[a, b] \Rightarrow (f + g)(x) = f(x) + g(x), \ x \in [a, b].$$

If $f, g \in F[a, b]$, then $f+g$ must be another function, so it is necessary to specify the rule defining it. The symbol $(f + g)(x)$ indicates the output value of the function for the input $x \in [a, b]$; this value is the number $f(x) + g(x)$.

With the above definitions, $F[a, b]$ is a vector space over $\mathbf{R}$. In this vector space, the "vectors" are functions, and this is frequently a powerful idea. However, $F[a, b]$ contains so many functions, most of them not related to any interesting phenomenon, that it is not useful for practical applications. Interesting spaces of functions arise when we impose regularity conditions on the functions. The definitions of addition and scalar multiplication remain unchanged.

The space of all continuous, real-valued functions defined on $[a, b]$ is denoted $C[a, b]$:

$$C[a, b] = \{f : [a, b] \to \mathbf{R} \mid f \text{ is continuous}\}.$$

It is a theorem from analysis that the sum of two continuous functions is also continuous, and similarly that the scalar multiple of a continuous function is also continuous. These facts are beyond the scope of this book and will be taken for granted, but they are essential: The basic operations must produce elements of the same space.

We can also consider the space $C(a, b)$, in which the domain of the functions is the open interval (a, b) instead of the closed interval $[a, b]$.

4. Related to the previous example are the spaces $C^k[a,b]$, where k is a positive integer. Elements of $C^k[a,b]$ are real-valued functions defined on $[a,b]$ having continuous derivatives up to order k. These spaces, or the related spaces $C^k(a,b)$, are frequently useful in solving differential equations. Once again, it is a theorem from analysis that the sum of two differentiable functions is also differentiable, and similarly for scalar multiplication.

5. Our final example is $\mathcal{P}_n$, the space of all polynomial functions of degree n or less. By definition, the zero polynomial belongs to $\mathcal{P}_n$ for all n, although (again by definition) the zero polynomial has no degree. Assuming we restrict ourselves to polynomials having real coefficients, $\mathcal{P}_n$ is a vector space over the field $\mathbf{R}$.

By definition, the function spaces described above, including $C[a,b]$, $C^k[a,b]$, and $\mathcal{P}_n$, all refer to real-valued functions. Complex-valued functions defined on an interval of real numbers are also important, and we write $C([a,b];\mathbf{C})$, $C^k([a,b];\mathbf{C})$, and so forth to denote these spaces. Thus, for instance,

$$C([a,b];\mathbf{C}) = \{f : [a,b] \to \mathbf{C} \mid f \text{ is continuous}\}.$$

We also might wish to discuss polynomial functions defined on various fields. We will write $\mathcal{P}_n(F)$ for the space of polynomial functions defined on the field F. Thus $p \in \mathcal{P}_n(F)$ means that $p : F \to F$ has the form

$$p(x) = c_0 + c_1 x + c_2 x^2 + \cdots + c_n x^n \text{ for all } x \in F,$$

where $c_0, c_1, \ldots, c_n$ are fixed elements of F.

Exercises

Essential exercises

1. Let F be a field and let V be a set consisting of a single element 0: $V = \{0\}$. Let addition on V be defined by $0 + 0 = 0$ and let scalar multiplication be defined by $\alpha \cdot 0 = 0$ for all $\alpha \in F$. Prove that V is a vector space over F. Note: This vector space is called the *trivial* vector space. A vector space containing at least one nonzero vector is called a *nontrivial* vector space.

2. Let F be an infinite field (that is, a field with an infinite number of elements) and let V be a nontrivial vector space over F. Prove that V contains infinitely many vectors.

Miscellaneous exercises

3. Prove Theorem 5.

4. Let F be a field. Prove that F^n is a vector space under the operations of componentwise scalar multiplication and vector addition.

5. Show that $F[a, b]$ satisfies the definition of a vector space over $\mathbf{R}$.

6. (a) How many vectors belong to $\mathbf{Z}_p^n$?

   (b) List all the vectors in $\mathbf{Z}_2^2$ and compute the sum of each pair of vectors (including the sum of each vector with itself).

7. (a) List all the elements of $\mathcal{P}_1(\mathbf{Z}_2)$ and compute the sum of each pair of elements (including the sum of each polynomial with itself).

   (b) Show that $\mathcal{P}_2(\mathbf{Z}_2) = \mathcal{P}_1(\mathbf{Z}_2)$. (It is essential to remember that the elements of $\mathcal{P}_n(F)$ are functions; if two functions give the same output value for every possible input, then those two functions are the same.)

   (c) Let V be the vector space consisting of *all* functions from $\mathbf{Z}_2$ into $\mathbf{Z}_2$. Prove that $V = \mathcal{P}_1(\mathbf{Z}_2)$.

8. In this exercise, we will define operations on $V = (0, \infty)$ in a nonstandard way; to avoid confusion with the standard operations, we will denote vector addition by $\oplus$ and scalar multiplication by $\odot$. We define

$$\begin{aligned} u \oplus v &= uv \text{ for all } u, v \in V, \\ \alpha \odot u &= u^{\alpha} \text{ for all } \alpha \in \mathbf{R}, u \in V. \end{aligned}$$

   Is V a vector space over $\mathbf{R}$ under these operations? If so, prove that the properties of a vector space are all satisfied and identify the additive identity in V and the additive inverse of each element of V. If not, determine which properties of a vector space fail and which hold.

9. In this exercise, we will define vector addition on $V = \mathbf{R}^2$ in a nonstandard way; to avoid confusion with the standard addition, we will denote vector addition by $\oplus$. Scalar multiplication will be defined as usual:

$$\begin{aligned} u \oplus v &= (u_1 + v_1, u_2 + v_2 + 1) \text{ for all } u, v \in V, \\ \alpha u &= (\alpha u_1, \alpha u_2) \text{ for all } \alpha \in \mathbf{R}, u \in V. \end{aligned}$$

   Is V a vector space over $\mathbf{R}$ under these operations? If so, prove that the properties of a vector space are all satisfied and identify the additive identity in V and the additive inverse of each element of V. If not, determine which properties of a vector space fail and which hold.

10. Let $V = \mathbf{R}^2$ and suppose vector addition on V is defined as

$$u \oplus v = (\alpha_1 u_1 + \beta_1 v_1, \alpha_2 u_2 + \beta_2 v_2) \text{ for all } u, v \in V,$$

where α_1, β_1, α_2, β_2 are fixed real numbers. Assume scalar multiplication on V is defined by the usual componentwise formula. What values of α_1, β_1, α_2, β_2 will make V a vector space (over $\mathbf{R}$) under these operations?

11. Recall that $\mathcal{P}_n$ is defined as the space of all polynomials of degree *at most* n. Explain why we could not define $\mathcal{P}_n$ as the space of all polynomials of degree *exactly* n (together with the zero polynomial).

12. (a) Find a function that belongs to $C(0,1)$ but not to $C[0,1]$.
    (b) Find a function that belongs to $C[-1,1]$ but not to $C^1[-1,1]$.

13. Let V be the space of all infinite sequences of real numbers. We denote a sequence $x_1, x_2, x_3, \ldots$ as $\{x_n\}_{n=1}^{\infty}$ or simply $\{x_n\}$ for brevity. Prove that V is a vector space when the operations are defined as follows:

$$\begin{aligned} \{x_n\} + \{y_n\} &= \{x_n + y_n\}, \\ \alpha\{x_n\} &= \{\alpha x_n\}. \end{aligned}$$

14. A function $f : [a,b] \to \mathbf{R}$ is called *piecewise continuous* if there exist a finite number of points $a = x_0 < x_1 < x_2 < \cdots < x_{k-1} < x_k = b$ such that

    (a) f is continuous on (x_{i-1}, x_i) for $i = 1, 2, \ldots, k$;
    (b) the one-sided limits

$$\lim_{x \to x_i^+} f(x), \ i = 0, 1, \ldots, k-1,$$

$$\lim_{x \to x_i^-} f(x), \ i = 1, 2, \ldots, k,$$

    exist (as finite numbers).

    Let V be the set of all piecewise continuous functions on $[a,b]$. Prove that V is a vector space over $\mathbf{R}$, with addition and scalar multiplication defined as usual for functions.

15. If X and Y are any two sets, then the *(Cartesian) product* of X and Y is the set

$$X \times Y = \{(x,y) : x \in X, y \in Y\}.$$

    If U and V are two vector spaces over a field F, then we define operations on $U \times V$ by

$$\begin{aligned} (u,v) + (w,z) &= (u+w, v+z) \text{ for all } (u,v), (w,z) \in U \times V, \\ \alpha(u,v) &= (\alpha u, \alpha v) \text{ for all } \alpha \in F, (u,v) \in U \times V. \end{aligned}$$

    Prove that $U \times V$ is a vector space over F.

2.3 Subspaces

Many of the vector spaces that arise in practice are subsets of other vector spaces. For instance, considering the examples from the previous section, we have $C^k[a,b] \subset C[a,b]$ for any $k \geq 1$, and we often treat $\mathcal{P}_n$ as a subspace of $C[a,b]$ (in this context, each polynomial is regarded as defining a function on $[a,b]$). We will also encounter subsets of $\mathbf{R}^n$ that are vector spaces in their own right.

This situation is so common that there is special terminology describing it.

Definition 6 *Let V be a vector space over a field F, and let S be a subset of V. Then S is a* subspace *of V if and only if the following are true:*

1. *$0 \in S$.*
2. *If $\alpha \in F$ and $u \in S$, then $\alpha u \in S$ (that is, S is* closed *under scalar multiplication).*
3. *If $u, v \in S$, then $u + v \in S$ (that is, S is* closed *under addition).*

This definition implies the following theorem:

Theorem 7 *Let V be a vector space over a field F, and let S be a subspace of V. Then S is a vector space over F, where the operations on S are the same as the operations on V.*

Proof The algebraic properties of addition and scalar multiplication (properties 1, 2, 5, 6, 7, and 8 from Definition 4) automatically hold on S since the operations are inherited from the vector space V. Moreover, the zero vector belongs to S by assumption. Therefore, it remains only to verify that additive inverses exist in S, that is, if $u \in S$, then $-u$ is also in S. But this follows from Theorem 5 and the fact that S is closed under scalar multiplication: $-u = (-1) \cdot u \in S$.

QED

Given any vector space V, the set containing only the zero vector, $S = \{0\}$, is a subspace of V (see Exercise 1). This subspace is called the *trivial subspace* of V; a subspace containing nonzero vectors is called a *nontrivial* subspace.

The entire space V is a subspace of itself, as it can be shown to satisfy the three properties of a subspace. Any subspace of V that is neither the trivial subspace nor all of V is called a *proper* subspace of V.

We now present two examples of proper subspaces.

Example 8 *A plane in $\mathbf{R}^3$ is described by an equation of the form*

$$ax_1 + bx_2 + cx_3 = d,$$

where a, b, c, d *are real constants. The exact interpretation of the previous statement is that a plane in* $\mathbf{R}^3$ *is the set of all points (vectors) with coordinates* (x_1, x_2, x_3) *that satisfy the equation* $ax_1 + bx_2 + cx_3 = d$*, where* a, b, c, d *are fixed constants.*

Let us suppose that constants a, b, c, d *are given and that* P *is the plane in* $\mathbf{R}^3$ *they determine:*

$$P = \left\{(x_1, x_2, x_3) \in \mathbf{R}^3 \,:\, ax_1 + bx_2 + cx_3 = d\right\}.$$

Is P *a subspace of* $\mathbf{R}^3$*?*

We can answer this question by testing P *against the three defining properties of a plane. The first is that the zero vector* $(0, 0, 0)$ *must belong to every subspace. We see immediately that* P *cannot be a subspace unless* $d = 0$*, since otherwise* $(0, 0, 0)$ *does not satisfy the equation defining* P*.*

Let us therefore assume $d = 0$*:*

$$P = \left\{(x_1, x_2, x_3) \in \mathbf{R}^3 \,:\, ax_1 + bx_2 + cx_3 = 0\right\}.$$

Then $(0, 0, 0) \in P$*, and the first property of a subspace is satisfied. If* $x \in P$ *and* α *is any real number, then*

$$ax_1 + bx_2 + cx_3 = 0$$

and

$$a(\alpha x_1) + b(\alpha x_2) + c(\alpha x_3) = \alpha(ax_1 + bx_2 + cx_3) = \alpha \cdot 0 = 0.$$

It follows that $\alpha x \in P$*, and* P *satisfies the second defining property of a subspace.*

Finally, if x *and* y *both belong to* P*, then*

$$ax_1 + bx_2 + cx_3 = 0 \text{ and } ay_1 + by_2 + cy_3 = 0.$$

It follows that

$$\begin{aligned} & a(x_1 + y_1) + b(x_2 + y_2) + c(x_3 + y_3) \\ = \;& ax_1 + ay_1 + bx_2 + by_2 + cx_3 + cy_3 \\ = \;& (ax_1 + bx_2 + cx_3) + (ay_1 + by_2 + cy_3) \\ = \;& 0 + 0 = 0. \end{aligned}$$

Since $x + y = (x_1 + y_1, x_2 + y_2, x_3 + y_3)$*, this shows that* $x + y \in P$*, and therefore* P *satisfies the final property of a subspace.*

The conclusion of this example is that not every plane is a subspace of $\mathbf{R}^3$*, but only those that pass through the origin.*

Example 9 *In this example, we will consider the vector space* $C[a, b]$*, the space of continuous real-valued functions defined on the interval* $[a, b]$*. Let us consider two subsets of* $C[a, b]$*:*

$$\mathcal{S}_3 = \{p \in C[a, b] \,:\, p \text{ is a polynomial function of degree exactly 3}\},$$
$$\mathcal{P}_3 = \{p \in C[a, b] \,:\, p \text{ is a polynomial function of degree at most 3}\}.$$

The first subset, $\mathcal{S}_3$, is not a subspace, since it does not contain the zero function (the zero polynomial does not have degree three). We can also see that $\mathcal{S}_3$ is not a subspace as follows: $1+x-2x^2+x^3$ *and* $2-3x+x^2-x^3$ *both belong to $\mathcal{S}_3$, but their sum,*

$$(1+x-x^2+x^3)+(2-3x+x^2-x^3)=3-2x-x^2,$$

has degree only two and hence is not in $\mathcal{S}_3$. This shows that $\mathcal{S}_3$ is not closed under addition and hence is not a subspace.

On the other hand, $\mathcal{P}_3$ is a subspace of $C[a,b]$. By definition, the zero polynomial belongs to $\mathcal{P}_3$ (see Example 5 on page 35). Multiplying a polynomial by a scalar cannot increase the degree, nor can adding two polynomials. Therefore, $\mathcal{P}_3$ is closed under addition and scalar multiplication, and $\mathcal{P}_3$ is a subspace.

It is important to recognize the difference between a *subset* of a vector space and a *subspace*. A subspace is a special kind of subset; every subspace is a subset, but not every subset is a subspace. The previous example presented a subset of $C[a,b]$ that is not a subspace. Here is another example.

Example 10 *Consider the subset $S=\{(0,0),(1,0),(1,1)\}$ of $\mathbf{Z}_2^2$. Two of the properties of a subspace are satisfied by S: $0\in S$ and S is closed under scalar multiplication. However,*

$$(1,0)+(1,1)=(1+1,0+1)=(0,1)\not\in S,$$

and therefore S is not closed under addition. This shows that S is not a subspace of $\mathbf{Z}_2^2$.

Exercises

Essential exercises

1. Let V be a vector space over a field F.
   (a) Let $S=\{0\}$ be the subset of V containing only the zero vector. Prove that S is a subspace of V.
   (b) Prove that V is a subspace of itself.

Miscellaneous exercises

2. An alternate definition of subspace is: A subspace S of a vector space V is a nonempty subset of V that is closed under addition and scalar multiplication. (In other words, "$0\in S$" is replaced by "S is nonempty.") Prove that this is equivalent to Definition 6.

3. Let V be a vector space over $\mathbf{R}$ and let $v\in V$ be a nonzero vector. Prove that the subset $\{0,v\}$ is not a subspace of V.

4. Consider the vector space $\mathbf{Z}_2^3$. Which of the following subsets of $\mathbf{Z}_2^3$ are subspaces? Prove your answers.

   (a) $\{(0,0,0),(1,0,0)\}$

   (b) $\{(0,0,0),(0,1,0),(1,0,1),(1,1,1)\}$

   (c) $\{(0,0),(1,0,0),(0,1,0),(1,1,1)\}$

5. Suppose S is a subset of $\mathbf{Z}_2^n$. Show that S is a subspace of $\mathbf{Z}_2^n$ if and only if $0 \in S$ and S is closed under addition.

6. Define $S = \{x \in \mathbf{R}^2 : x_1 \geq 0, x_2 \geq 0\}$ (S is the first quadrant of the Euclidean plane). Is S a subspace of $\mathbf{R}^2$? Why or why not?

7. Define $S = \{x \in \mathbf{R}^2 : ax_1 + bx_2 = 0\}$, where $a, b \in \mathbf{R}$ are constants (S is a line through the origin). Is S a subspace of $\mathbf{R}^2$? Why or why not?

8. Define $S = \{x \in \mathbf{R}^2 : x_1^2 + x_2^2 \leq 1\}$ (S is the disk of radius one, centered at the origin). Is S a subspace of $\mathbf{R}^2$? Why or why not?

9. (a) Find a subset S of $\mathbf{R}^n$ that is closed under scalar multiplication but not under addition.

   (b) Find a subset S of $\mathbf{R}^n$ that is closed under addition but not under scalar multiplication.

10. Let V be a vector space over a field F, let $u \in V$, and define $S \subset V$ by $S = \{\alpha u : \alpha \in F\}$. Is S a subspace of V? Why or why not?

11. Regard $\mathbf{R}$ as a vector space over $\mathbf{R}$. Prove that $\mathbf{R}$ has no proper subspaces.

12. Describe all proper subspaces of $\mathbf{R}^2$.

13. Find a proper subspace of $\mathbf{R}^3$ that is not a plane.

14. Notice that $\mathbf{R}^n$ is a subset of $\mathbf{C}^n$. Is $\mathbf{R}^n$ a subspace of $\mathbf{C}^n$? If not, explain which part of the definition fails to hold.

15. Define S to be the following subset of $C[a,b]$:

$$S = \{u \in C[a,b] \, : \, u(a) = u(b) = 0\}.$$

    Prove that S is a subspace of $C[a,b]$.

16. Define S to be the following subset of $C[a,b]$:

$$S = \{u \in C[a,b] \, : \, u(a) = 1\}.$$

    Prove that S is not a subspace of $C[a,b]$.

17. Define S to be the following subset of $C[a,b]$:

$$S = \left\{ u \in C[a,b] \, : \, \int_a^b u(x)\, dx = 0 \right\}.$$

Prove that S is a subspace of $C[a,b]$.

18. (This exercise requires an understanding of convergence of sequences and series.) Let V be the vector space of all infinite sequences of real numbers (cf. Exercise 2.2.13). Prove that each of the following subsets of V is a subspace of V:

    (a) $Z = \{\{x_n\} \in V \, : \, \lim_{n\to\infty} x_n = 0\}$.
    (b) $S = \{\{x_n\} \in V \, : \, \sum_{n=1}^{\infty} x_n < \infty\}$.
    (c) $L = \{\{x_n\} \in V \, : \, \sum_{n=1}^{\infty} x_n^2 < \infty\}$. (In proving that L is closed under addition, you might find the following inequality helpful:

    $$|xy| \le \frac{1}{2}(x^2 + y^2).$$

    This inequality follows from the fact that $(x \pm y)^2 \ge 0$. It is elementary but very useful.)

    Are any of the subspaces Z, S, L a subspace of another? If so, which one(s)? In other words, is Z a subspace of S, or S a subspace of Z, or S a subspace of L, or ...?

19. Let V be a vector space over a field F, and let X, Y be two subspaces of V. Prove or give a counterexample:

    (a) $X \cap Y$ is a subspace of V.
    (b) $X \cup Y$ is a subspace of V.

    Note: "Prove" means prove that the statement is always true. If the statement is not always true, it may not always be false, but there must exist one or more particular situations in which the statement is false. A counterexample is one specific instance in which the statement is false. For instance, if "$X \cap Y$ is a subspace of V" fails to be a true statement, then one can demonstrate this by producing a specific vector space V and subspaces X, Y of V such that $X \cap Y$ is *not* a subspace of V.

20. Let V be a vector space over a field and let S be a nonempty subset of V. Let T be the intersection of all subspaces of V that contain S. Prove:

    (a) T is a subspace of V;
    (b) T is the smallest subspace of V containing S, in the following sense: If U is a subspace of V and $S \subset U$, then $T \subset U$.

21. Let V be vector space over a field F, and let S and T be subspaces of V. Define
$$S+T=\{s+t\,:\,s\in S,t\in T\}.$$
Prove that $S+T$ is a subspace of V.

2.4 Linear combinations and spanning sets

When we combine vectors using the operations of addition and scalar multiplication, the result is called a linear combination:

Definition 11 *Let V be a vector space over a field F, let $u_1,u_2,\ldots,u_k$ be vectors in V, and let $\alpha_1,\alpha_2,\ldots,\alpha_k$ be scalars in F. Then*
$$\alpha_1u_1+\alpha_2u_2+\cdots+\alpha_ku_k$$
is called a linear combination *of the vectors $u_1,u_2,\ldots,u_k$. The scalars $\alpha_1,\ldots,\alpha_k$ are called the* weights *in the linear combination.*

The following theorem can be proved by induction.

Theorem 12 *Let V be a vector space over a field F and let S be a subspace of V. Then, for any $u_1,u_2,\ldots,u_k\in S$ and $\alpha_1,\alpha_2,\ldots,\alpha_k\in F$, the linear combination $\alpha_1u_1+\alpha_2u_2+\cdots+\alpha_ku_k$ belongs to S.*

Proof Exercise 14

Linear combination is an important concept not just because subspaces are closed under the operation of taking linear combinations, but also because subspaces are often defined or represented by linear combinations.

Theorem 13 *Let V be a vector space over a field F, let $u_1,u_2,\ldots,u_n$ be vectors in V, where $n\geq 1$, and let S be the set of all linear combinations of $u_1,u_2,\ldots,u_n$:*
$$S=\{\alpha_1u_1+\alpha_2u_2+\cdots+\alpha_nu_n\,:\,\alpha_1,\alpha_2,\ldots,\alpha_n\in F\}.$$
Then S is a subspace of V.

Proof We must show that $0\in S$ and that S is closed under scalar multiplication and addition. If we choose $\alpha_1=\alpha_2=\ldots=\alpha_n=0$, then the linear combination
$$0\cdot u_1+0\cdot u_2+\cdots+0\cdot u_n=0$$
belongs to S, which shows that S satisfies the first property of a subspace.

Next, suppose $v\in S$ and $\beta\in F$. Since v belongs to S, it must be a linear combination of $u_1,u_2,\ldots,u_n$; that is,
$$v=\alpha_1u_1+\alpha_2u_2+\cdots+\alpha_nu_n$$

for some choice of $\alpha_1, \alpha_2, \ldots, \alpha_n \in F$. Then

$$\begin{aligned}\beta v &= \beta(\alpha_1 u_1 + \alpha_2 u_2 + \cdots + \alpha_n u_n) \\ &= \beta(\alpha_1 u_1) + \beta(\alpha_2 u_2) + \cdots + \beta(\alpha_n u_n) \\ &= (\beta\alpha_1)u_1 + (\beta\alpha_2)u_2 + \cdots + (\beta\alpha_n)u_n.\end{aligned}$$

This shows that βv is also a linear combination of $u_1, u_2, \ldots, u_n$, and hence that βv belongs to S. Thus S is closed under scalar multiplication.

Finally, suppose $v, w \in S$. Then

$$\begin{aligned}v &= \alpha_1 u_1 + \alpha_2 u_2 + \cdots + \alpha_n u_n, \\ w &= \beta_1 u_1 + \beta_2 u_2 + \cdots + \beta_n u_n\end{aligned}$$

for some $\alpha_1, \alpha_2, \ldots, \alpha_n \in F$, $\beta_1, \beta_2, \ldots, \beta_n \in F$, and so

$$\begin{aligned}v + w &= \alpha_1 u_1 + \alpha_2 u_2 + \cdots + \alpha_n u_n + \beta_1 u_1 + \beta_2 u_2 + \cdots + \beta_n u_n \\ &= (\alpha_1 u_1 + \beta_1 u_1) + (\alpha_2 u_2 + \beta_2 u_2) + \cdots + (\alpha_n u_n + \beta_n u_n) \\ &= (\alpha_1 + \beta_1)u_1 + (\alpha_2 + \beta_2)u_2 + \cdots + (\alpha_n + \beta_n)u_n.\end{aligned}$$

Therefore $v + w \in S$, and S is closed under addition. This completes the proof.

QED

The above construction is so important that there is terminology to describe it.

Definition 14 *Let V be a vector space over a field F, let $u_1, u_2, \ldots, u_n$ be vectors in V, where $n \geq 1$. The set of all linear combinations of $u_1, u_2, \ldots, u_n$ is called the* span *of $u_1, u_2, \ldots, u_n$, and is denoted by $\text{sp}\{u_1, u_2, \ldots, u_n\}$. We say that $\{u_1, u_2, \ldots, u_n\}$ is a* spanning set *for $S = \text{sp}\{u_1, u_2, \ldots, u_n\}$ and that $u_1, u_2, \ldots, u_n$* span *S.*

A common question is the following: Given a subspace S of a vector space V and a vector $v \in V$, does v belong to S? When S is defined by a spanning set, this question reduces to an equation to be solved, and often, when the equation is manipulated, to a system of linear algebraic equations. We now present several examples.

Example 15 *Let S be the subspace of $\mathbf{R}^3$ defined by $S = \text{sp}\{u_1, u_2\}$, where*

$$u_1 = \begin{bmatrix} 1 \\ 1 \\ -3 \end{bmatrix}, \quad u_2 = \begin{bmatrix} 2 \\ 3 \\ 2 \end{bmatrix},$$

and let $v \in \mathbf{R}^3$ be the vector

$$v = \begin{bmatrix} 1 \\ 2 \\ 3 \end{bmatrix}.$$

Does v belong to S?

Since S is the set of all linear combinations of u_1, u_2, the question is whether we can write v as a linear combination of u_1, u_2. That is, do there exist scalars α_1, α_2 satisfying $\alpha_1 u_1 + \alpha_2 u_2 = v$? This equation takes the form

$$\alpha_1 \begin{bmatrix} 1 \\ 1 \\ -3 \end{bmatrix} + \alpha_2 \begin{bmatrix} 2 \\ 3 \\ 2 \end{bmatrix} = \begin{bmatrix} 1 \\ 2 \\ 3 \end{bmatrix},$$

or

$$\begin{aligned} \alpha_1 + 2\alpha_2 &= 1, \\ \alpha_1 + 3\alpha_2 &= 2, \\ -3\alpha_1 + 2\alpha_2 &= 3. \end{aligned}$$

Gaussian elimination yields the equivalent system

$$\begin{aligned} \alpha_1 + 2\alpha_2 &= 1, \\ \alpha_2 &= 1, \\ 0 &= -2. \end{aligned}$$

The last equation could be written as $0 \cdot \alpha_1 + 0 \cdot \alpha_2 = -2$, which is not satisfied by any α_1, α_2. It follows that the original system has no solution, and therefore v does not belong to S.

In the following example, we need some fundamental facts about polynomials, which are reviewed in Appendix C. A polynomial of degree n has at most n roots (see Theorem 505). This has an important implication that we now develop. Let us suppose that

$$p(x) = a_0 + a_1 x + a_2 x^2 + \cdots + a_n x^n$$

is a polynomial function in $\mathcal{P}_n(F)$, where F is a given field, and that

$$a_0 + a_1 x + a_2 x^2 + \cdots + a_n x^n = 0 \text{ for all } x \in F. \tag{2.5}$$

(Equation (2.5) is equivalent to stating that $p = 0$ in $\mathcal{P}_n(F)$, that is, p is the zero function.) If F contains at least $n + 1$ elements, then (2.5) states that p has at least $n + 1$ roots, which is impossible if p has degree n. This does not mean that (2.5) cannot hold, only that it implies that $a_0 = a_1 = \ldots = a_n = 0$ (in which case p is the zero polynomial, which does not have degree n). In particular, if $F = \mathbf{R}$ or $F = \mathbf{C}$, then F contains infinitely many elements, and (2.5) implies that $a_0 = a_1 = \ldots = a_n = 0$.

A corollary of the above reasoning is that if

$$a_0 + a_1 x + a_2 x^2 + \cdots + a_n x^n = b_0 + b_1 x + b_2 x^2 + \cdots + b_n x^n \text{ for all } x \in F$$

and F contains at least $n + 1$ elements, then $a_0 = b_0, a_1 = b_1, \ldots, a_n = b_n$ (since subtracting the polynomial on the right from both sides yields (2.5), with each a_i replaced by $a_i - b_i$). In particular, this is true for $F = \mathbf{R}$ or $F = \mathbf{C}$. We will use these facts repeatedly in future examples.

Example 16 *In this example, the vector space is $\mathcal{P}_2$, the space of polynomials of degree two or less, and the subspace is $S = \mathrm{sp}\{1-3x+4x^2, -2+7x-7x^2\}$. If $p(x) = x + x^2$, does p belong to S?*

We must determine if there are scalars α_1, α_2 such that

$$\alpha_1(1-3x+4x^2) + \alpha_2(-2+7x-7x^2) = x + x^2 \text{ for all } x \in \mathbf{R}.$$

The equation can be rewritten as

$$(\alpha_1 - 2\alpha_2) + (-3\alpha_1 + 7\alpha_2)x + (4\alpha_1 - 7\alpha_2)x^2 = x + x^2 \text{ for all } x \in \mathbf{R}.$$

As discussed above, this equation implies that the coefficients of the two polynomials must be equal, which yields the following system of equations:

$$\begin{aligned} \alpha_1 - 2\alpha_2 &= 0, \\ -3\alpha_1 + 7\alpha_2 &= 1, \\ 4\alpha_1 - 7\alpha_2 &= 1. \end{aligned}$$

(The numbers on the right-hand sides of these equations are the coefficients of p: $p(x) = 0 + 1\cdot x + 1\cdot x^2$.) Using Gaussian elimination, one can show that this system has a unique solution, $\alpha_1 = 2$, $\alpha_2 = 1$. Therefore,

$$x + x^2 = 2(1-3x+4x^2) + 1(-2+7x-7x^2),$$

and p does belong to S.

Example 17 *Let $v_1 = (1,0,1,1)$ and $v_2 = (1,1,0,0)$ be vectors in $\mathbf{Z}_2^4$, and let $S = \mathrm{sp}\{v_1, v_2\}$. Does the vector $u = (1,1,1,1)$ belong to S?*

To answer this question, we must determine if there are any solutions to $\alpha_1 v_1 + \alpha_2 v_2 = u$, with $\alpha_1, \alpha_2 \in \mathbf{Z}_2$. In trying to solve this equation, we must remember to perform all operations in the arithmetic of $\mathbf{Z}_2$, that is, modulo 2. The equations simplify as follows:

$$\begin{aligned} \alpha_1 v_1 + \alpha_2 v_2 = u \quad &\Leftrightarrow \quad \alpha_1(1,0,1,1) + \alpha_2(1,1,0,0) = (1,1,1,1) \\ &\Leftrightarrow \quad (\alpha_1+\alpha_2, \alpha_2, \alpha_1, \alpha_1) = (1,1,1,1) \\ &\Leftrightarrow \quad \left\{ \begin{aligned} \alpha_1 + \alpha_2 &= 1, \\ \alpha_2 &= 1, \\ \alpha_1 &= 1, \\ \alpha_1 &= 1. \end{aligned} \right. \end{aligned}$$

We see that these equations are inconsistent, since the last three indicate that $\alpha_1 = \alpha_2 = 1$, but $1+1=0$ in $\mathbf{Z}_2$, which violates the first equation. Therefore, there is no solution and thus $u \notin S$.

It is interesting to note that S contains only four distinct vectors:

$$S = \{0, v_1, v_2, v_1+v_2\} = \{(0,0,0,0), (1,0,1,1), (1,1,0,0), (0,1,1,1)\}.$$

This follows from the fact that α_1, α_2 can have only the values 0 or 1. From this we can see immediately that $u \notin S$.

Example 18 *Let* $v_1 = (1, -2, 1, 2)$, $v_2 = (-1, 1, 2, 1)$, *and* $v_3 = (-7, 9, 8, 1)$ *be vectors in* $\mathbf{R}^4$, *and let* $S = \mathrm{sp}\{v_1, v_2, v_3\}$. *Does the vector* $u = (1, -4, 7, 8)$ *belong to* S*?*

We answer this question by trying to solve $\alpha_1 v_1 + \alpha_2 v_2 + \alpha_3 v_3 = u$*:*

$$\alpha_1 \begin{bmatrix} 1 \\ -2 \\ 1 \\ 2 \end{bmatrix} + \alpha_2 \begin{bmatrix} -1 \\ 1 \\ 2 \\ 1 \end{bmatrix} + \alpha_3 \begin{bmatrix} -7 \\ 9 \\ 8 \\ 1 \end{bmatrix} = \begin{bmatrix} 1 \\ -4 \\ 7 \\ 8 \end{bmatrix}$$

$$\Leftrightarrow \quad \left\{ \begin{array}{r} \alpha_1 - \alpha_2 - 7\alpha_3 = 1, \\ -2\alpha_1 + \alpha_2 + 9\alpha_3 = -4, \\ \alpha_1 + 2\alpha_2 + 8\alpha_3 = 7, \\ 2\alpha_1 + \alpha_2 + \alpha_3 = 8. \end{array} \right.$$

Applying Gaussian elimination, this last system reduces to

$$\begin{aligned} \alpha_1 &= 3 + 2\alpha_3, \\ \alpha_2 &= 2 - 5\alpha_3, \end{aligned}$$

which shows that there are many values of $\alpha_1, \alpha_2, \alpha_3$ *that satisfy the given equation. In fact, we can specify any value for* α_3 *and find corresponding values of* α_1, α_2 *such that* $\alpha_1, \alpha_2, \alpha_3$ *satisfy the equation. For instance, if we take* $\alpha_3 = 0$, *then we obtain* $\alpha_1 = 3$, $\alpha_2 = 2$, *and therefore* $u = 3v_1 + 2v_2$. *Therefore,* $u \in S$.

In the last example, given any $x \in S = \mathrm{sp}\{v_1, v_2, v_3\}$, it is possible to write $x = \alpha_1 v_1 + \alpha_2 v_2$ for certain choices of α_1, α_2 (Exercise 9 asks the reader to prove this). In other words, v_3 is not needed to represent any vector in S, and we can just as well write $S = \mathrm{sp}\{v_1, v_2\}$. If the goal is to represent S as efficiently as possible, we would prefer the spanning set $\{v_1, v_2\}$ over $\{v_1, v_2, v_3\}$. Using the smallest possible spanning set to represent a subspace is the topic of the next section.

Exercises

Miscellaneous exercises

1. Let S be the following subspace of $\mathbf{R}^4$:

$$S = \mathrm{sp}\{(-1, -2, 4, -2), (0, 1, -5, 4)\}.$$

Determine if each vector v belongs to S:

(a) $v = (-1, 0, -6, 6)$;

(b) $v = (1, 1, 1, 1)$.

2. Let S be the following subspace of $C[0,1]$:

$$S = \text{sp}\{e^x, e^{-x}\}.$$

Determine if each function f belongs to S:

(a) $f(x) = \cosh(x)$;

(b) $f(x) = 1$.

3. Let S be the following subspace of $\mathcal{P}_2$:

$$S = \text{sp}\left\{1 + 2x + 3x^2, x - x^2\right\}.$$

Determine if each polynomial p belongs to S:

(a) $p(x) = 2 + 5x + 5x^2$;

(b) $p(x) = 1 - x + x^2$.

4. Let $S = \text{sp}\{u_1, u_2\} \subset \mathbf{C}^3$, where $u_1 = (1+i, i, 2)$, $u_2 = (1, 2i, 2-i)$. Does $v = (2+3i, -2+2i, 5+2i)$ belong to S?

5. Let S be the following subspace of $\mathbf{Z}_3^4$:

$$S = \text{sp}\{(1,2,0,1), (2,0,1,2)\}.$$

Determine if each of the following vectors belongs to S:

(a) $(1,1,1,1)$;

(b) $(1,0,1,1)$.

6. Let S be the following subspace of $\mathcal{P}_3(\mathbf{Z}_3)$:

$$S = \text{sp}\{1 + x, x + x^2, 2 + x + x^2\}.$$

Determine if each of the following polynomial functions belongs to S:

(a) $p(x) = 1 + x + x^2$;

(b) $q(x) = x^3$.

(Hint: The reader should understand the discussion about polynomials on page 45, and take into account that $\mathbf{Z}_3$ has only three elements.)

7. Let $u = (1,1,-1)$, $v = (1,0,2)$. Show that $S = \text{sp}\{u, v\}$ is a plane in $\mathbf{R}^3$ by showing there exist constants $a, b, c \in \mathbf{R}$ such that

$$S = \left\{x \in \mathbf{R}^3 \,:\, ax_1 + bx_2 + cx_3 = 0\right\}.$$

8. Does the previous exercise remain true for any $u, v \in \mathbf{R}^3$? Prove or give a counterexample.

9. Let v_1, v_2, v_3 be the vectors defined in Example 18, and let x belong to $S = \mathrm{sp}\{v_1, v_2, v_3\}$. Prove that x can be written as a linear combination of v_1 and v_2 alone.

10. Show that

$$S_1 = \mathrm{sp}\{(1,1,1),(1,-1,1)\} \text{ and } S_2 = \mathrm{sp}\{(1,1,1),(1,-1,1),(1,0,1)\}$$

are the same subspace of $\mathbf{R}^3$. (Note: To show that two sets S_1 and S_2 are equal, you must show that $S_1 \subset S_2$ and $S_2 \subset S_1$. Equivalently, you must show that $x \in S_1$ if and only if $x \in S_2$.)

11. Let S be the subspace of $\mathbf{R}^3$ defined by

$$S = \mathrm{sp}\{(-1,-3,3),(-1,-4,3),(-1,-1,4)\}.$$

Is S a proper subspace of $\mathbf{R}^3$ or not? In other words, do there exist vectors in $\mathbf{R}^3$ that do not belong to S, or is S all of $\mathbf{R}^3$?

12. Repeat the previous exercise for

$$S = \mathrm{sp}\{(-1,-5,1),(3,14,-4),(1,4,-2)\}.$$

13. Let S be the subspace of $\mathcal{P}_2$ defined by

$$S = \mathrm{sp}\{1-x, 2-2x+x^2, 1-3x^2\}.$$

Is S a proper subspace of $\mathcal{P}_2$ or not?

14. Prove Theorem 12.

15. Let V be a vector space over a field F, and let u be a nonzero vector in V. Prove that, for any scalar $\alpha \in F$, $\mathrm{sp}\{u\} = \mathrm{sp}\{u, \alpha u\}$.

16. Let V be a vector space over a field F, and suppose

$$x, u_1, \ldots, u_k, v_1, \ldots, v_\ell$$

are vectors in V. Assume $x \in \mathrm{sp}\{u_1, \ldots, u_k\}$ and $u_j \in \mathrm{sp}\{v_1, \ldots, v_\ell\}$ for $j = 1, 2, \ldots, k$. Prove that $x \in \mathrm{sp}\{v_1, \ldots, v_\ell\}$.

17. (a) Let V be a vector space over $\mathbf{R}$, and let u, v be any two vectors in V. Prove that $\mathrm{sp}\{u, v\} = \mathrm{sp}\{u+v, u-v\}$.

    (b) Does the same result hold if V is a vector space over an arbitrary field F? Prove it does or give a counter-example.

2.5 Linear independence

One way to represent certain subspaces is by a spanning set. If the underlying field F contains infinitely many elements (as do **R** and **C**), then any nontrivial subspace contains infinitely many vectors (see Exercise 2.2.2). In this case, a finite representation, such as a spanning set, is essential for many manipulations. Even if the field is finite and the subspace contains a finite number of vectors, a spanning set is often the best way to represent the subspace.

The main point of this section and the next is to describe the most favorable kind of spanning set. We beginning by showing that a subspace can have many different spanning sets.

Lemma 19 *Suppose V is a vector space, $u_1, u_2, \ldots, u_n$ are vectors in V, and $v \in \mathrm{sp}\{u_1, u_2, \ldots, u_n\}$. Then*

$$\mathrm{sp}\{u_1, u_2, \ldots, u_n, v\} = \mathrm{sp}\{u_1, u_2, \ldots, u_n\}.$$

Proof Suppose first that x is any vector in $\mathrm{sp}\{u_1, u_2, \ldots, u_n\}$. By definition, this means that x is a linear combination of $u_1, u_2, \ldots, u_n$; that is, there exist scalars $\alpha_1, \alpha_2, \ldots, \alpha_n$ such that

$$x = \alpha_1 u_1 + \alpha_2 u_2 + \ldots + \alpha_n u_n.$$

But we can then write

$$x = \alpha_1 u_1 + \alpha_2 u_2 + \ldots + \alpha_n u_n + 0v,$$

which shows that x also belongs to $\mathrm{sp}\{u_1, u_2, \ldots, u_n, v\}$.

On the other hand, suppose

$$x \in \mathrm{sp}\{u_1, u_2, \ldots, u_n, v\},$$

that is, that

$$x = \alpha_1 u_1 + \alpha_2 u_2 + \ldots + \alpha_n u_n + \alpha_{n+1} v \tag{2.6}$$

for some scalars $\alpha_1, \alpha_2, \ldots, \alpha_{n+1}$. Since $v \in \mathrm{sp}\{u_1, u_2, \ldots, u_n\}$, there exist scalars $\beta_1, \beta_2, \ldots, \beta_n$ such that

$$v = \beta_1 u_1 + \beta_2 u_2 + \ldots + \beta_n u_n. \tag{2.7}$$

Substituting (2.7) into (2.6) yields

$$\begin{aligned} x &= \alpha_1 u_1 + \alpha_2 u_2 + \ldots + \alpha_n u_n + \alpha_{n+1} v \\ &= \alpha_1 u_1 + \alpha_2 u_2 + \ldots + \alpha_n u_n + \alpha_{n+1}(\beta_1 u_1 + \beta_2 u_2 + \ldots + \beta_n u_n) \\ &= (\alpha_1 + \alpha_{n+1}\beta_1)u_1 + (\alpha_2 + \alpha_{n+1}\beta_2)u_2 + \ldots + (\alpha_n + \alpha_{n+1}\beta_n)u_n. \end{aligned}$$

This shows that x is a linear combination of $u_1, u_2, \ldots, u_n$, that is, that x belongs to $\mathrm{sp}\{u_1, u_2, \ldots, u_n\}$. The completes the second part of the proof.

QED

It seems only reasonable that we would want to use as few representatives as possible. This leads to the concept of a *basis*, a minimal spanning set (that is, a spanning set containing the fewest possible elements).

It turns out that, to be a basis, a spanning set must possess a property called *linear independence.* There are several equivalent ways to define linear independence; the standard definition, which is given next, is the easiest to use, though the least intuitive.

Definition 20 *Let V be a vector space over a field F, and let $u_1, u_2, \ldots, u_n$ be vectors in V. We say that $\{u_1, u_2, \ldots, u_n\}$ is* linearly independent *if and only if the only scalars $\alpha_1, \alpha_2, \ldots, \alpha_n$ such that*

$$\alpha_1 u_1 + \alpha_2 u_2 + \ldots + \alpha_n u_n = 0 \tag{2.8}$$

are $\alpha_1 = \alpha_2 = \ldots = \alpha_n = 0$.

The set $\{u_1, u_2, \ldots, u_n\}$ is called linearly dependent *if and only if it is not linearly independent. Linear dependence can be defined directly as follows: $\{u_1, u_2, \ldots, u_n\}$ is linearly dependent if and only if there exists scalars $\alpha_1, \alpha_2, \ldots, \alpha_n$, not all zero, such that (2.8) holds. We say that $\alpha_1, \alpha_2, \ldots, \alpha_n$ is a nontrivial solution to (2.8) if not all of the scalars α_i are zero.*

Example 21 *We will determine whether the following set of vectors in $\mathbf{R}^4$ is linearly independent: $\{(3,2,2,3),(3,2,1,2),(3,2,0,1)\}$. To do this, we must solve*

$$\alpha_1(3,2,2,3) + \alpha_2(3,2,1,2) + \alpha_3(3,2,0,1) = (0,0,0,0) \tag{2.9}$$

and determine whether there is a nontrivial solution. Equation (2.9) is equivalent to the following system:

$$\begin{aligned} 3\alpha_1 + 3\alpha_2 + 3\alpha_3 &= 0, \\ 2\alpha_1 + 2\alpha_2 + 2\alpha_3 &= 0, \\ 2\alpha_1 + \alpha_2 &= 0, \\ 3\alpha_1 + 2\alpha_2 + \alpha_3 &= 0. \end{aligned}$$

When Gaussian elimination is performed, the last two equations cancel and the result is

$$\begin{aligned} \alpha_1 - \alpha_3 &= 0, \\ \alpha_2 + 2\alpha_3 &= 0 \end{aligned}$$

or

$$\begin{aligned} \alpha_1 &= \alpha_3, \\ \alpha_2 &= -2\alpha_3. \end{aligned}$$

There are nontrivial solutions, such as $\alpha_1 = 1, \alpha_2 = -2, \alpha_3 = 1$. This shows that the given vectors are linearly dependent.

Example 22 *Consider the polynomials* $1-2x-x^2$, $1+x$, $1+x+2x^2$ *in* $\mathcal{P}_2$. *To determine whether these polynomials form a linearly independent set, we must solve*

$$\alpha_1\left(1-2x-x^2\right)+\alpha_2\left(1+x\right)+\alpha_3\left(1+x+2x^2\right)=0$$

to see if there is a nontrivial solution. The previous equation is equivalent to

$$(\alpha_1+\alpha_2+\alpha_3)+(-2\alpha_1+\alpha_2+\alpha_3)x+(-\alpha_1+2\alpha_2)x^2=0.$$

Since a polynomial function (over **R***) equals zero if and only all of its coefficients are zero, this is equivalent to the system*

$$\begin{aligned} \alpha_1+\ \alpha_2+\alpha_3&=0,\\ -2\alpha_1+\ \alpha_2+\alpha_3&=0,\\ -\alpha_1+2\alpha_2\qquad&=0. \end{aligned}$$

Gaussian elimination can be used to show that the only solution to this system is the trivial one, $\alpha_1=\alpha_2=\alpha_3=0$. *Therefore, the three polynomials are linearly independent.*

Example 23 *Consider the vectors* $u_1=(1,0,1)$, $u_2=(0,1,1)$, $u_3=(1,1,0)$ *in* $\mathbf{Z}_2^3$. *We determine whether* $\{u_1,u_2,u_3\}$ *is linearly dependent or independent by solving* $\alpha_1u_1+\alpha_2u_2+\alpha_3u_3=0$:

$$\begin{aligned} &\alpha_1(1,0,1)+\alpha_2(0,1,1)+\alpha_3(1,1,0)=(0,0,0)\\ \Leftrightarrow\ &(\alpha_1+\alpha_3,\alpha_2+\alpha_3,\alpha_1+\alpha_2)=(0,0,0)\\ \Leftrightarrow\ &\left\{\begin{array}{l} \alpha_1\qquad\ +\alpha_3=0,\\ \qquad\alpha_2+\alpha_3=0,\\ \alpha_1+\alpha_2\qquad\ =0 \end{array}\right.\\ \Leftrightarrow\ &\left\{\begin{array}{l} \alpha_1\qquad\ +\alpha_3=0,\\ \qquad\alpha_2+\alpha_3=0,\\ \qquad\alpha_2+\alpha_3=0 \end{array}\right.\\ \Leftrightarrow\ &\left\{\begin{array}{l} \alpha_1\qquad\ +\alpha_3=0,\\ \qquad\alpha_2+\alpha_3=0 \end{array}\right.\\ \Leftrightarrow\ &\left\{\begin{array}{l} \alpha_1=\alpha_3,\\ \alpha_2=\alpha_3. \end{array}\right. \end{aligned}$$

(We show all the steps in solving the above system because arithmetic in $\mathbf{Z}_2$ *may be unfamiliar. For instance,* $\alpha_3\in\mathbf{Z}_2$ *implies that* $-\alpha_3=\alpha_3$, *so* $\alpha_1=\alpha_3$ *is obtained from* $\alpha_1+\alpha_3=0$ *by adding* α_3 *to both sides.) This shows that there is a nontrivial solution,* $\alpha_1=\alpha_2=\alpha_3=1$ *(and this is the only nontrivial solution, since* 1 *is the only nonzero element of* $\mathbf{Z}_2$*). Therefore,* $u_1+u_2+u_3=0$, *and* $\{u_1,u_2,u_3\}$ *is linearly dependent.*

The following theorem shows why the terms "linear independence" and "linear dependence" are used.

Theorem 24 *Let V be a vector space over a field F, and let $u_1, u_2, \ldots, u_n$ be vectors in V, where $n \geq 2$. Then $\{u_1, u_2, \ldots, u_n\}$ is linearly dependent if and only if at least one of $u_1, u_2, \ldots, u_n$ can be written as a linear combination of the remaining $n-1$ vectors.*

Proof If $\{u_1, u_2, \ldots, u_n\}$ is linearly dependent, then there exist scalars

$$\alpha_1, \alpha_2, \ldots, \alpha_n,$$

not all zero, such that

$$\alpha_1 u_1 + \alpha_2 u_2 + \ldots + \alpha_n u_n = 0.$$

Suppose $\alpha_i \neq 0$. Then

$$\begin{aligned} \alpha_1 u_1 + \alpha_2 u_2 + \ldots + \alpha_n u_n = 0 \Rightarrow \alpha_i u_i &= -\alpha_1 u_1 - \ldots - \alpha_{i-1} u_{i-1} - \\ &\quad \alpha_{i+1} u_{i+1} - \ldots - \alpha_n u_n \\ \Rightarrow u_i &= -\alpha_i^{-1} \alpha_1 u_1 - \ldots - \alpha_i^{-1} \alpha_{i-1} u_{i-1} - \\ &\quad \alpha_i^{-1} \alpha_{i+1} u_{i+1} - \ldots - \alpha_i^{-1} \alpha_n u_n, \end{aligned}$$

which shows that u_i is a linear combination of $u_1, \ldots, u_{i-1}, u_{i+1}, \ldots, u_n$. The reader should notice that it is possible to solve for u_i because the fact that α_i is nonzero implies that α_i^{-1} exists.

On the other hand, suppose one of the vectors $u_1, u_2, \ldots, u_n$, say u_i, is a linear combination of the others:

$$u_i = \alpha_1 u_1 + \ldots + \alpha_{i-1} u_{i-1} + \alpha_{i+1} u_{i+1} + \ldots + \alpha_n u_n.$$

We can rearrange this equation as

$$\alpha_1 u_1 + \ldots + \alpha_{i-1} u_{i-1} + (-u_i) + \alpha_{i+1} u_{i+1} + \ldots + \alpha_n u_n = 0,$$

or

$$\alpha_1 u_1 + \ldots + \alpha_{i-1} u_{i-1} + (-1) u_i + \alpha_{i+1} u_{i+1} + \ldots + \alpha_n u_n = 0.$$

This shows that $\{u_1, u_2, \ldots, u_n\}$ is linearly dependent, since at least one of the weights, namely the -1 that multiplies u_i, is nonzero. We use here two of the properties of a field: $-u_i = (-1)u_i$ and $-1 \neq 0$.

QED

The above theorem can be equivalently stated as follows: $\{u_1, u_2, \ldots, u_n\}$ is linearly independent if and only if none of the vectors $u_1, u_2, \ldots, u_n$ can be written as a linear combination of the others, that is, each vector is independent of the others.

The reader should appreciate that it is difficult to use Theorem 24 directly to decide that a given set is linearly dependent or independent, since it requires checking, one at a time, whether each vector in the set can be written as a linear combination of the rest. It is easiest to use Definition 20 to check whether $\{u_1, u_2, \ldots, u_n\}$ is linear dependent. If there is a nontrivial solution $\alpha_1, \alpha_2, \ldots, \alpha_n$ of

$$\alpha_1 u_1 + \alpha_2 u_2 + \cdots + \alpha_n u_n = 0,$$

then each vector u_i whose corresponding weight α_i is nonzero can be written as a linear combination of the other vectors.

Example 25 *We will show that* $u_1 = (1, 1, 3)$, $u_2 = (3, 2, 11)$, $u_3 = (0, 1, -2)$ *form a linearly dependent subset of* $\mathbf{R}^3$, *and show how to write one vector as a linear combination of the other two.*

We begin by solving

$$\alpha_1 \begin{bmatrix} 1 \\ 1 \\ 3 \end{bmatrix} + \alpha_2 \begin{bmatrix} 3 \\ 2 \\ 11 \end{bmatrix} + \alpha_3 \begin{bmatrix} 0 \\ 1 \\ -2 \end{bmatrix} = 0 \Rightarrow \left\{ \begin{array}{rcl} \alpha_1 + 3\alpha_2 & = & 0, \\ \alpha_1 + 2\alpha_2 + \alpha_3 & = & 0, \\ 3\alpha_1 + 11\alpha_2 - 2\alpha_3 & = & 0. \end{array} \right.$$

Gaussian elimination reduces this last system to

$$\begin{aligned} \alpha_1 &= -3\alpha_3, \\ \alpha_2 &= \alpha_3, \end{aligned}$$

from which we see that there are nontrivial solutions, one for each nonzero value of α_3. *For instance, if* $\alpha_3 = 1$, *then* $\alpha_1 = -3$ *and* $\alpha_2 = 1$, *and therefore* $-3u_1 + u_2 + u_3 = 0$. *Since all three weights are nonzero, we can write any of the three vectors as a linear combination of the other two:*

$$\begin{aligned} u_1 &= \frac{1}{3} u_2 + \frac{1}{3} u_3, \\ u_2 &= 3u_1 - u_3, \\ u_3 &= 3u_1 - u_2. \end{aligned}$$

Here is another property of a linearly independent set:

Theorem 26 *Let* V *be a vector space over a field* F, *and let* $u_1, u_2, \ldots, u_n$ *be vectors in* V. *Then* $\{u_1, u_2, \ldots, u_n\}$ *is linearly independent if and only if each vector in* $\mathrm{sp}\{u_1, u_2, \ldots, u_n\}$ *can be written uniquely as a linear combination of* $u_1, u_2, \ldots, u_n$.

Proof Let us suppose first that $\{u_1, u_2, \ldots, u_n\}$ is linearly independent. Suppose $v \in \mathrm{sp}\{u_1, u_2, \ldots, u_n\}$ can be written as

$$v = \alpha_1 u_1 + \alpha_2 u_2 + \ldots + \alpha_n u_n \tag{2.10}$$

and also as

$$v = \beta_1 u_1 + \beta_2 u_2 + \ldots + \beta_n u_n. \tag{2.11}$$

We must show that $\alpha_1 = \beta_1, \alpha_2 = \beta_2, \ldots, \alpha_n = \beta_n$.

Equations (2.10) and (2.11) imply that

$$\alpha_1 u_1 + \alpha_2 u_2 + \ldots + \alpha_n u_n = \beta_1 u_1 + \beta_2 u_2 + \ldots + \beta_n u_n.$$

Rearranging yields

$$(\alpha_1 - \beta_1)u_1 + (\alpha_2 - \beta_2)u_2 + \ldots + (\alpha_n - \beta_n)u_n = 0.$$

Since $\{u_1, u_2, \ldots, u_n\}$ is linearly independent, it follows from the definition that $\alpha_1 - \beta_1 = 0, \alpha_2 - \beta_2 = 0, \ldots, \alpha_n - \beta_n = 0$, and hence that $\alpha_1 = \beta_1, \alpha_2 = \beta_2, \ldots, \alpha_n = \beta_n$. This proves that the representation of v as a linear combination of $u_1, u_2, \ldots, u_n$ is unique.

On the other hand, suppose every vector in $\mathrm{sp}\{u_1, u_2, \ldots, u_n\}$ is uniquely represented as a linear combination of $u_1, u_2, \ldots, u_n$. Then, since 0 belongs to $\mathrm{sp}\{u_1, u_2, \ldots, u_n\}$, the only way to write 0 in the form

$$0 = \alpha_1 u_1 + \alpha_2 u_2 + \ldots + \alpha_n u_n$$

is

$$0 = 0 \cdot u_1 + 0 \cdot u_2 + \ldots + 0 \cdot u_n.$$

(This is one way to do it, and, by the assumed uniqueness, it must be the only way.) This proves that $\{u_1, u_2, \ldots, u_n\}$ is linearly independent.

QED

Exercises

Essential exercises

1. Let V be a vector space over a field F, and let u_1, u_2 be two vectors in V. Prove $\{u_1, u_2\}$ is linearly dependent if and only if one of u_1, u_2 is a multiple of the other.

2. Let V be a vector space over a field F, and suppose $v \in V$. Prove that $\{v\}$ is linearly dependent if and only if $v = 0$. (Hint: Use Theorem 2.2.5, part 6.)

3. Let V be a vector space over a field F, and let $u_1, u_2, \ldots, u_n$ be vectors in V. Prove that, if one of the vectors is the zero vector, then $\{u_1, u_2, \ldots, u_n\}$ is linearly dependent.

4. Let V be a vector space over a field F and let $\{u_1, \ldots, u_k\}$ be a linearly independent subset of V. Prove that if $v \notin \mathrm{sp}\{u_1, \ldots, u_k\}$, then $\{u_1, \ldots, u_k, v\}$ is also linearly independent.

Miscellaneous exercises

5. Determine whether each of the following sets of vectors is linearly dependent or independent:

(a) $\{(1,2),(1,-1)\} \subset \mathbf{R}^2$.

(b) $\{(-1,-1,4),(-4,-4,17),(1,1,-3)\} \subset \mathbf{R}^3$.

(c) $\{(-1,3,-2),(3,-10,7),(-1,3,-1)\} \subset \mathbf{R}^3$.

6. Determine whether each of the following sets of polynomials is linearly dependent or independent:

   (a) $\{1-x^2, x+x^2, 3+3x-4x^2\} \subset \mathcal{P}_2$.

   (b) $\{1+x^2, 4+3x^2+3x^3, 3-x+10x^3, 1+7x^2-18x^3\} \subset \mathcal{P}_3$.

7. Determine whether the subset $\{e^x, e^{-x}, \cosh x\}$ of $C[0,1]$ is linearly dependent or independent.

8. Determine whether the subset $\{(0,1,2),(1,2,0),(2,0,1)\}$ of $\mathbf{Z}_3^3$ is linearly dependent or independent.

9. Show that $\{1, x, x^2\}$ is linearly dependent in $\mathcal{P}_2(\mathbf{Z}_2)$.

10. Let $u_1 = (i, 1, 2i)$, $u_2 = (1, 1+i, i)$, $u_3 = (1, 3+5i, -4+3i)$. Is the set $\{u_1, u_2, u_3\} \subset \mathbf{C}^3$ linearly independent?

11. In Example 21, we showed that $\{(3,2,2,3),(3,2,1,2),(3,2,0,1)\}$ is linearly dependent.

    (a) Write one of the vectors as a linear combination of the others.

    (b) Show that $v = (-3,-2,2,1)$ belongs to

    $$\mathrm{sp}\{(3,2,2,3),(3,2,1,2),(3,2,0,1)\}$$

    and write v in two different ways as a linear combination of the three vectors.

12. Show that $\{(-1,1,3),(1,-1,-2),(-3,3,13)\} \subset \mathbf{R}^3$ is linearly dependent by writing one of the vectors as a linear combination of the others.

13. Show that $\{p_1, p_2, p_3\}$, where

    $$p_1(x) = 1 - x^2,\ p_2(x) = 1 + x - 6x^2,\ p_3(x) = 3 - 2x^2,$$

    is linearly independent and spans $\mathcal{P}_2$ by showing that each polynomial $p \in \mathcal{P}_2$ can be written uniquely as a linear combination of p_1, p_2, p_3.

14. Let V be a vector space over a field F and let $\{u_1, \ldots, u_k\}$ be a linearly independent subset of V. Prove or give a counterexample: If v and w are vectors in V such that $\{v, w\}$ is linearly independent and $v, w \notin \mathrm{sp}\{u_1, \ldots, u_k\}$, then $\{u_1, \ldots, u_k, v, w\}$ is also linearly independent.

15. Let V be a vector space over a field F, and suppose S and T are subspaces of V satisfying $S \cap T = \{0\}$. Suppose $\{s_1, s_2, \dots, s_k\} \subset S$ and $\{t_1, t_2, \dots, t_\ell\} \subset T$ are both linearly independent sets. Prove that
$$\{s_1, s_2, \dots, s_k, t_1, t_2, \dots, t_\ell\}$$
is a linearly independent subset of V.

16. Let V be a vector space over a field F, and let $\{u_1, \dots, u_k\}$ and $\{v_1, \dots, v_\ell\}$ be two linearly independent subsets of V. Find a condition that implies that
$$\{u_1, \dots, u_k, v_1, \dots, v_\ell\}$$
is linearly independent. Is your condition necessary as well as sufficient? Prove your answers.

17. (a) Prove: Let V be a vector space over $\mathbf{R}$, and suppose $\{x, y, z\}$ is a linearly independent subset of V. Then $\{x + y, y + z, x + z\}$ is also linearly independent.
    (b) Show that the previous result is not necessarily true if V is a vector space over an arbitrary field F.

18. Let U and V be vector spaces over a field F, and define $W = U \times V$ (see Exercise 2.2.15). Suppose $\{u_1, \dots, u_k\} \subset U$ and $\{v_1, \dots, v_\ell\} \subset V$ are linearly independent. Prove that
$$\{(u_1, 0), \dots, (u_k, 0), (0, v_1), \dots, (0, v_\ell)\}$$
is a linearly independent subset of W.

19. Let V be a vector space over a field F, and let $u_1, u_2, \dots, u_n$ be vectors in V. Suppose a nonempty subset of $\{u_1, u_2, \dots, u_n\}$ is linearly dependent. Prove that $\{u_1, u_2, \dots, u_n\}$ itself is linearly dependent.

20. Let V be a vector space over a field F, and suppose $\{u_1, u_2, \dots, u_n\}$ is a linearly independent subset of V. Prove that every nonempty subset of $\{u_1, u_2, \dots, u_n\}$ is also linearly independent.

21. Let V be a vector space over a field F, and suppose $\{u_1, u_2, \dots, u_n\}$ is linearly dependent. Prove that, given any i, $1 \le i \le n$, either u_i is a linear combination of $\{u_1, \dots, u_{i-1}, u_{i+1}, \dots, u_n\}$ or $\{u_1, \dots, u_{i-1}, u_{i+1}, \dots, u_n\}$ is linearly dependent.

2.6 Basis and dimension

We mentioned the concept of basis in the preceding section. Here is the precise definition.

Definition 27 *Let V be a vector space over a field F, and let $u_1, u_2, \ldots, u_n$ be vectors in V. We say that $\{u_1, u_2, \ldots, u_n\}$ is a* basis *for V if and only if $\{u_1, u_2, \ldots, u_n\}$ spans V and is linearly independent.*

A subspace of a vector space is a vector space in its own right, so we often speak of a basis for a subspace. We should also note that the plural of "basis" is "bases."

Here is an equivalent definition of basis:

Theorem 28 *Let V be a vector space over a field F, and suppose $u_1, u_2, \ldots, u_n$ are vectors in V. Then $\{u_1, u_2, \ldots, u_n\}$ is a basis for V if and only if each $v \in V$ can be written uniquely as a linear combination of $u_1, u_2, \ldots, u_n$.*

Proof By the definition of "span," $\{u_1, u_2, \ldots, u_n\}$ spans V if and only if each vector in V can be written as a linear combination of $\{u_1, u_2, \ldots, u_n\}$. By Theorem 26, $\{u_1, u_2, \ldots, u_n\}$ is linearly independent if and only if each vector in $\mathrm{sp}\{u_1, u_2, \ldots, u_n\}$ can be written uniquely as a linear combination of $u_1, u_2, \ldots, u_n$. Putting these two facts together yields the theorem.

QED

Checking Theorem 28 is sometimes the easiest way to check whether a given set of vectors is a basis for a vector space.

Example 29 *Euclidean n-space $\mathbf{R}^n$ has a basis $\{e_1, e_2, \ldots, e_n\}$ defined by*

$$e_1 = \begin{bmatrix} 1 \\ 0 \\ 0 \\ \vdots \\ 0 \end{bmatrix}, \; e_2 = \begin{bmatrix} 0 \\ 1 \\ 0 \\ \vdots \\ 0 \end{bmatrix}, \; \cdots, e_n = \begin{bmatrix} 0 \\ 0 \\ 0 \\ \vdots \\ 1 \end{bmatrix}.$$

The fact that $\{e_1, e_2, \ldots, e_n\}$ is a basis for $\mathbf{R}^n$ is easily proved by Theorem 28: $x \in \mathbf{R}^n$ can be written as

$$x = x_1 e_1 + x_2 e_2 + \ldots + x_n e_n,$$

and this is clearly the only way to write x as a linear combination of $e_1, \ldots, e_n$. This basis is called the standard *or* canonical *basis for $\mathbf{R}^n$.*

The standard basis for $\mathbf{R}^3$ is $\{(1,0,0), (0,1,0), (0,0,1)\}$. Here is an alternate basis for $\mathbf{R}^3$: $\{u_1, u_2, u_3\}$, where $u_1 = (1,0,2)$, $u_2 = (3,0,7)$, $u_3 = (4,1,10)$. To show that this really is a basis, we can apply Theorem 28 and show that, for each $x \in \mathbf{R}^3$, there is a unique choice of $\alpha_1, \alpha_2, \alpha_3$ such that $\alpha_1 u_1 + \alpha_2 u_2 + \alpha_3 u_3 = x$. This equation takes the form

$$\alpha_1 \begin{bmatrix} 1 \\ 0 \\ 2 \end{bmatrix} + \alpha_2 \begin{bmatrix} 3 \\ 0 \\ 7 \end{bmatrix} + \alpha_3 \begin{bmatrix} 4 \\ 1 \\ 10 \end{bmatrix} = x$$

or

$$\begin{aligned} \alpha_1 + 3\alpha_2 + \ 4\alpha_3 &= x_1, \\ \alpha_3 &= x_2, \\ 2\alpha_1 + 7\alpha_2 + 10\alpha_3 &= x_3. \end{aligned}$$

The reader can verify that Gaussian elimination yields the unique solution

$$\begin{aligned} \alpha_1 &= 7x_1 + 2x_2 - 3x_3, \\ \alpha_2 &= -2x_1 - 2x_2 + x_3, \\ \alpha_3 &= x_2, \end{aligned}$$

which shows that $\{u_1, u_2, u_3\}$ *really is a basis for* $\mathbf{R}^3$.

We will see that if a vector space has a basis, then it has many different bases, so there is nothing unusual about the above example.

Given any field F, we can define vectors $e_1, e_2, \ldots, e_n$ in F^n just as we did in Example 29 (e_i is the vector whose components are all 0, except the ith component, which is 1). The same proof shows that $\{e_1, e_2, \ldots, e_n\}$ is a basis for F^n.

Example 30 *The standard basis for the space* $\mathcal{P}_n$ *of polynomials of degree at most* n *is* $\{1, x, x^2, \ldots, x^n\}$. *It is clear that this set spans* $\mathcal{P}_n$, *since* $p \in \mathcal{P}_n$ *can be written*

$$p(x) = c_0 + c_1 x + c_2 x^2 + \ldots + c_n x^n = c_0 \cdot 1 + c_1 x + c_2 x^2 + \ldots + c_n x^n$$

(if the degree of p *is* $k < n$, *then* $c_{k+1} = \ldots = c_n = 0$*).*

Now suppose

$$c_0 \cdot 1 + c_1 x + c_2 x^2 + \ldots + c_n x^n = 0.$$

This equation means that the left side equals the zero function, that is, the left side is zero for every value of x. *As we discussed earlier (see page 45), this is possible only if* $c_0 = c_1 = \ldots = c_n = 0$, *and hence* $\{1, x, x^2, \ldots, x^n\}$ *is linearly independent.*

The monomials $1, x, x^2, \ldots, x^n$ may or may not be linearly independent functions in $\mathcal{P}_n(F)$ if F is a finite field. Exercise 11 explores this question.

We have seen that every vector space contains the trivial subspace $\{0\}$. There is no basis for the trivial subspace because it contains no linearly independent subset.

We now study the number of vectors in a basis. We will use the following terminology.

Definition 31 *Let* V *be a vector space over a field* F. *We say that* V *is* finite-dimensional *if* V *has a basis or* V *is the trivial vector space* $\{0\}$. *If* V *is not finite-dimensional, then it is called* infinite-dimensional.

By definition, a basis contains a finite number of vectors, which explains the above terminology.

Example 32 *1. The space* $\mathbf{R}^n$ *is finite-dimensional, since* $\{e_1, e_2, \ldots, e_n\}$ *is a basis.*

2. The space $\mathcal{P}_n$ *is also finite-dimensional, since* $\{1, x, x^2, \ldots, x^n\}$ *is a basis.*

An example of an infinite-dimensional vector space is $\mathcal{P}$, the space of all polynomials (over $\mathbf{R}$). It appears intuitively obvious that $\mathcal{P}$ has no basis, since a basis must have only finitely many elements, and it would seem that all (infinitely many) monomials $1, x, x^2, \ldots$ are needed to be able to write all of the possible polynomials. We will show that this reasoning is correct. We first need the following lemma:

Lemma 33 *Let* V *be a vector space over a field* F*, and let* $v_1, v_2, \ldots, v_n$ *be nonzero vectors in* V*. If* $\{v_1, v_2, \ldots, v_n\}$ *is linearly dependent, then there exists an integer* k*, with* $2 \le k \le n$*, such that* v_k *is a linear combination of* $v_1, v_2, \ldots, v_{k-1}$*.*

Proof Let k be the smallest positive integer such that $\{v_1, v_2, \ldots, v_k\}$ is linearly dependent. By assumption $k \le n$, and $k \ge 2$ because the singleton set $\{v_1\}$ is linearly dependent only if v_1 is the zero vector, which is not the case. By Theorem 24, one of the vectors $v_1, v_2, \ldots, v_k$ is a linear combination of the others. If it is v_k, then the proof is complete, so suppose v_t, $1 \le t < k$, is a linear combination of $v_1, \ldots, v_{t-1}, v_{t+1}, \ldots, v_k$:

$$v_t = \alpha_1 v_1 + \ldots + \alpha_{t-1} v_{t-1} + \alpha_{t+1} v_{t+1} + \ldots + \alpha_k v_k. \qquad (2.12)$$

We must have $\alpha_k \ne 0$, since otherwise $\{v_1, v_2, \ldots, v_{k-1}\}$ would be linearly dependent by Theorem 26, contradicting that $\{v_1, v_2, \ldots, v_\ell\}$ is linearly independent for $\ell < k$. But, with $\alpha_k \ne 0$, we can solve (2.12) for v_k:

$$v_k = -\alpha_k^{-1}\alpha_1 v_1 - \ldots - \alpha_k^{-1}\alpha_{t-1} v_{t-1} + \alpha_k^{-1} v_t - \alpha_k^{-1}\alpha_{t+1} v_{t+1} - \ldots - \alpha_k^{-1}\alpha_{k-1} v_{k-1}.$$

Therefore v_k is a linear combination of $v_1, v_2, \ldots, v_{k-1}$.

QED

Theorem 34 *Let* V *be a finite-dimensional vector space over a field* F*, and let* $\{u_1, u_2, \ldots, u_m\}$ *be a basis for* V*. If* $v_1, v_2, \ldots, v_n$ *are any* n *vectors in* V*, with* $n > m$*, then* $\{v_1, v_2, \ldots, v_n\}$ *is linearly dependent.*

Proof We can assume that none of $v_1, v_2, \ldots, v_n$ is the zero vector, since otherwise the conclusion of the theorem follows from Exercise 2.5.3. Since v_1 is a linear combination of $u_1, u_2, \ldots, u_m$, it follows from Theorem 24 that the set

$$\{v_1, u_1, u_2, \ldots, u_m\}$$

is linearly dependent. It follows from Lemma 33 that there exists an integer k such that $1 \leq k \leq m$ and u_k is a linear combination of $v_1, u_1, \ldots, u_{k-1}$. By Lemma 19, we can discard u_k and still have a spanning set for V.

We rename $u_1, \ldots, u_{k-1}, u_{k+1}, \ldots, u_m$ as $u_1^{(1)}, u_2^{(1)}, \ldots, u_{m-1}^{(1)}$; then

$$\mathrm{sp}\left\{v_1, u_1^{(1)}, u_2^{(1)}, \ldots, u_{m-1}^{(1)}\right\} = \mathrm{sp}\left\{u_1, u_2, \ldots, u_m\right\} = V.$$

Now consider the set

$$\left\{v_2, v_1, u_1^{(1)}, \ldots, u_{m-1}^{(1)}\right\}.$$

Since $v_2 \in V$, it follows that v_2 is a linear combination of $v_1, u_1^{(1)}, \ldots, u_{m-1}^{(1)}$, and hence that

$$\left\{v_2, v_1, u_1^{(1)}, \ldots, u_{m-1}^{(1)}\right\}$$

is linearly dependent. It follows that one of the vectors $v_1, u_1^{(1)}, \ldots, u_{m-1}^{(1)}$ is a linear combination of the preceding vectors. If it is v_1, then $\{v_2, v_1\}$ is linearly dependent and hence so is $\{v_1, v_2, \ldots, v_n\}$. In that case, the proof is complete. If it is $u_j^{(1)}$, $1 \leq j \leq m-1$, then we can remove it from the set $\left\{v_2, v_1, u_1^{(1)}, \ldots, u_{m-1}^{(1)}\right\}$ and the result will still span V. We rename the vectors that remain from the original basis as $u_1^{(2)}, u_2^{(2)}, \ldots, u_{m-2}^{(2)}$, and have

$$\mathrm{sp}\left\{v_2, v_1, u_1^{(2)}, u_2^{(2)}, \ldots, u_{m-2}^{(2)}\right\} = V.$$

Reasoning as above, the set

$$\left\{v_3, v_2, v_1, u_1^{(2)}, u_2^{(2)}, \ldots, u_{m-2}^{(2)}\right\}$$

must be linearly dependent. Continuing in this fashion, we can remove one of the vectors u_i from the original basis and add a v_j to produce a new linearly dependent set with $m+1$ elements. By doing this, we either learn that $\{v_\ell, \ldots, v_2, v_1\}$ is linearly dependent for some $\ell \leq m$, or we remove the final u_i and add v_{m+1}, and find that $\{v_{m+1}, \ldots, v_2, v_1\}$ is linearly dependent. In either case, the conclusion of the theorem follows.

QED

We can now verify that $\mathcal{P}$ is infinite-dimensional. Suppose, by way of contradiction, that $\{p_1, p_2, \ldots, p_m\}$ is a basis for $\mathcal{P}$. Then, by Theorem 34, $\{1, x, x^2, \ldots, x^{n-1}\}$ is linearly dependent for any $n > m$. But we already know from Example 30 that $\{1, x, x^2, \ldots, x^{n-1}\}$ is linearly independent for any positive integer n. This contradiction shows that $\mathcal{P}$ cannot have a basis, so it is an infinite-dimensional space.

Since $\mathcal{P}$ can be regarded as a subspace of any of the usual spaces of continuous or differentiable functions, such as $C[a,b]$, $C(a,b)$, $C^k[a,b]$, and $C^k(a,b)$, it follows that these spaces are also infinite-dimensional.

Theorem 34 has the following fundamental consequence:

Corollary 35 *Let V be a vector space over a field F, and let $\{u_1, u_2, \ldots, u_n\}$ and $\{v_1, v_2, \ldots, v_m\}$ be two bases for V. Then $m = n$.*

Proof Since $\{u_1, u_2, \ldots, u_n\}$ is a basis for V and $\{v_1, v_2, \ldots, v_m\}$ is linearly independent, Theorem 34 implies that $m \leq n$. But the same reasoning implies that $n \leq m$, and so $m = n$ must hold.

QED

We can now make the following definition:

Definition 36 *Let V be a finite-dimensional vector space. If V is the trivial vector space, then we say that the* dimension *of V is zero; otherwise, the dimension of V is the number of vectors in a basis for V.*

Since Corollary 35 shows that every basis for a finite-dimensional vector space has the same number of elements, the dimension of a vector space is well-defined.

Example 37 *The dimension of $\mathbf{R}^n$ is n, since $\{e_1, e_2, \ldots, e_n\}$ is a basis. Indeed, the dimension of F^n is n for any field F (including finite fields).*

Example 38 *The dimension of $\mathcal{P}_n$ is $n+1$, since $\{1, x, x^2, \ldots, x^n\}$ is a basis. In Section 2.8, we will study an important alternate basis for $\mathcal{P}_n$.*

Exercises 11 and 2.7.13 ask the reader to show that, if F is a finite field with q distinct elements, then

$$\dim\left(\mathcal{P}_n(F)\right) = \begin{cases} n+1, & q \geq n+1, \\ q, & q \leq n. \end{cases}$$

Thus, for example, $\dim\left(\mathcal{P}_n(\mathbf{Z}_2)\right) = 2$ for all $n \geq 1$. This might seem surprising, but the reader must bear in mind that $\mathcal{P}_n(\mathbf{Z}_2)$ is a space of functions. As functions mapping $\mathbf{Z}_2$ into $\mathbf{Z}_2$, $x^k = x$ for all $k \geq 1$. (The field $\mathbf{Z}_2$ has only two elements, 0 and 1, and $p(x) = x^k$ satisfies $p(0) = 0$ and $p(1) = 1$ for all $k \geq 1$.)

Example 39 *Let S be the subspace of $\mathbf{R}^4$ spanned by*

$$v_1 = \begin{bmatrix} -1 \\ 2 \\ 3 \\ 4 \end{bmatrix},\ v_2 = \begin{bmatrix} 0 \\ 1 \\ -3 \\ 5 \end{bmatrix},\ v_3 = \begin{bmatrix} -3 \\ 8 \\ 3 \\ 22 \end{bmatrix},\ v_4 = \begin{bmatrix} 3 \\ -3 \\ -19 \\ 1 \end{bmatrix}.$$

In this example, we will determine the dimension of S. We begin by determining whether $\{v_1, v_2, v_3, v_4\}$ is linearly independent. If it is, then $\{v_1, v_2, v_3, v_4\}$ is a basis for S and the dimension of S is 4.

The equation $\alpha_1 v_1 + \alpha_2 v_2 + \alpha_3 v_3 + \alpha_4 v_4 = 0$ can be written as

$$\alpha_1 \begin{bmatrix} -1 \\ 2 \\ 3 \\ 4 \end{bmatrix} + \alpha_2 \begin{bmatrix} 0 \\ 1 \\ -3 \\ 5 \end{bmatrix} + \alpha_3 \begin{bmatrix} -3 \\ 8 \\ 3 \\ 22 \end{bmatrix} + \alpha_4 \begin{bmatrix} 3 \\ -3 \\ -19 \\ 1 \end{bmatrix} = \begin{bmatrix} 0 \\ 0 \\ 0 \\ 0 \end{bmatrix}$$

or

$$\begin{aligned}
-\alpha_1 + 2\alpha_2 + 3\alpha_3 + 4\alpha_4 &= 0,\\
\alpha_2 - 3\alpha_3 + 5\alpha_4 &= 0,\\
-3\alpha_1 + 8\alpha_2 + 3\alpha_3 + 22\alpha_4 &= 0,\\
3\alpha_1 - 3\alpha_2 - 19\alpha_3 + \alpha_4 &= 0.
\end{aligned}$$

Applying Gaussian elimination, we find that this system has infinitely many nontrivial solutions; one such solution is $\alpha_1 = -3$, $\alpha_2 = -2$, $\alpha_3 = 1$, $\alpha_4 = 0$. *Therefore,* $\{v_1, v_2, v_3, v_4\}$ *is linearly dependent and is not a basis. However,*

$$-3v_1 - 2v_2 + v_3 = 0 \Rightarrow v_3 = 3v_1 + 2v_2.$$

It follows from Lemma 19 that

$$\text{sp}\{v_1, v_2, v_4\} = \text{sp}\{v_1, v_2, v_3, v_4\} = S.$$

If we apply Gaussian elimination to the equation $\beta_1 v_1 + \beta_2 v_2 + \beta_3 v_4 = 0$, *we see that the only solution is the trivial solution, and hence* $\{v_1, v_2, v_4\}$ *is linearly independent. It follows that* $\{v_1, v_2, v_4\}$ *is a basis for* S, *and thus* S *has dimension* 3.

Example 40 *Consider the subspace* $S = \text{sp}\{u_1, u_2, u_3\}$ *of* $\mathbf{Z}_5^4$, *where* $u_1 = (3, 2, 4, 1)$, $u_2 = (1, 0, 3, 2)$, *and* $u_3 = (2, 2, 0, 4)$. *We wish to determine the dimension of* S. *We begin by determining whether the given spanning set is linearly independent:*

$$\begin{aligned}
& \alpha_1(3,2,4,1) + \alpha_2(1,0,3,2) + \alpha_3(2,2,0,4) = (0,0,0,0)\\
\Leftrightarrow\ & (3\alpha_1 + \alpha_2 + 2\alpha_3, 2\alpha_1 + 2\alpha_3, 4\alpha_1 + 3\alpha_2, \alpha_1 + 2\alpha_2 + 4\alpha_3) = (0,0,0,0)\\
\Leftrightarrow\ & \left\{\begin{array}{l}
3\alpha_1 + \alpha_2 + 2\alpha_3 = 0,\\
2\alpha_1 + 2\alpha_3 = 0,\\
4\alpha_1 + 3\alpha_2 = 0,\\
\alpha_1 + 2\alpha_2 + 4\alpha_3 = 0.
\end{array}\right.
\end{aligned}$$

Gaussian elimination (applied modulo 5*) shows that this system of equations has only the trivial solution, and hence* $\{u_1, u_2, u_3\}$ *is linearly independent. Therefore,* $\{u_1, u_2, u_3\}$ *is a basis for* S *and* $\dim(S) = 3$.

Exercises

Essential exercises

1. Suppose $\{v_1, v_2, \ldots, v_n\}$ is a basis for a vector space V.

   (a) Show that if any v_j is removed from the basis, the resulting set of $n - 1$ vectors does not span V and hence is not a basis.

(b) Show that if any vector $u \in V$, $u \notin \{v_1, v_2, \ldots, v_n\}$, is added to the basis, the resulting set of $n+1$ vectors is linearly dependent.

Miscellaneous exercises

2. Consider the following vectors in $\mathbf{R}^3$: $v_1 = (-1, 4, -2)$, $v_2 = (5, -20, 9)$, $v_3 = (2, -7, 6)$. Is $\{v_1, v_2, v_3\}$ a basis for $\mathbf{R}^3$?

3. Repeat the previous exercise for $v_1 = (-1, 3, -1)$, $v_2 = (1, -2, -2)$, $v_3 = (-1, 7, -13)$.

4. Let $S = \text{sp}\{e^x, e^{-x}\}$ be regarded as a subspace of $C(\mathbf{R})$. Show that $\{e^x, e^{-x}\}$ and $\{\cosh(x), \sinh(x)\}$ are two different bases for S.

5. Let $p_1(x) = 1 - 4x + 4x^2$, $p_2(x) = x + x^2$, $p_3(x) = -2 + 11x - 6x^2$. Is $\{p_1, p_2, p_3\}$ a basis for $\mathcal{P}_2$?

6. Let $p_1(x) = 1 - x^2$, $p_2(x) = 2 + x$, $p_3(x) = x + 2x^2$. Is $\{p_1, p_2, p_3\}$ a basis for $\mathcal{P}_2$?

7. Consider the subspace $S = \text{sp}\{p_1, p_2, p_3, p_4, p_5\}$ of $\mathcal{P}_3$, where

$$
\begin{aligned}
p_1(x) &= -1 + 4x - x^2 + 3x^3, \ p_2(x) = 2 - 8x + 2x^2 - 5x^3, \\
p_3(x) &= 3 - 11x + 3x^2 - 8x^3, \ p_4(x) = -2 + 8x - 2x^2 - 3x^3, \\
p_5(x) &= 2 - 8x + 2x^2 + 3x^3.
\end{aligned}
$$

(a) Without doing any calculations, explain why $\{p_1, p_2, p_3, p_4, p_5\}$ must be linearly dependent.

(b) Find the dimension of S.

8. Find a basis for $\text{sp}\{(1, 2, 1), (0, 1, 1), (1, 1, 0)\} \subset \mathbf{R}^3$.

9. Find a basis for $\text{sp}\{(1, 2, 1, 2, 1), (1, 1, 2, 2, 1), (0, 1, 2, 0, 2)\} \subset \mathbf{Z}_3^5$.

10. Show that $\{1 + x + x^2, 1 - x + x^2, 1 + x + 2x^2\}$ is a basis for $\mathcal{P}_2(\mathbf{Z}_3)$.

11. The purpose of this exercise is to begin to determine the dimension of $\mathcal{P}_n(F)$, where F is a finite field. Suppose F has q distinct elements.

(a) Prove that if $n \le q - 1$, then $\{1, x, x^2, \ldots, x^n\}$ is linearly independent. It follows that $\dim(\mathcal{P}_n(F)) = n + 1$ in this case.

(b) Prove that if $n \ge q$, then $\{1, x, x^2, \ldots, x^{q-1}\}$ is linearly independent. This implies that $\dim(\mathcal{P}_n(F)) \ge q$ in this case.

In fact, in the case $n \ge q$, $\dim(\mathcal{P}_n(F))$ is exactly q. To prove this, we need a result from the next section (see Exercise 2.7.13).

12. Suppose V is a vector space over a field F, and S, T are two n-dimensional subspaces of V. Prove that if $S \subset T$, then in fact $S = T$.

13. Suppose V is a vector space over a field F, and S, T are two finite-dimensional subspaces of V. Prove that

$$S \subset T \Rightarrow \dim(S) \leq \dim(T).$$

14. Let V be a vector space over a field F, and let S and T be finite-dimensional subspaces of V. Prove that

$$\dim(S+T) = \dim(S) + \dim(T) - \dim(S \cap T).$$

15. Let V be a vector space over a field F, and let S and T be finite-dimensional subspaces of V. Consider the four subspaces

$$X_1 = S,\ X_2 = T,\ X_3 = S+T,\ X_4 = S \cap T.$$

For every choice of i, j with

$$1 \leq i < j \leq 4,$$

determine if $\dim(X_i) \leq \dim(X_j)$ or $\dim(X_i) \geq \dim(X_j)$ (or neither) must hold. Prove your answers.

16. Let V be a vector space over a field F, and suppose S and T are subspaces of V satisfying $S \cap T = \{0\}$. Suppose $\{s_1, s_2, \ldots, s_k\} \subset S$ and $\{t_1, t_2, \ldots, t_\ell\} \subset T$ are bases for S and T, respectively. Prove that

$$\{s_1, s_2, \ldots, s_k, t_1, t_2, \ldots, t_\ell\}$$

is a basis for $S+T$ (cf. Exercise 2.3.21).

17. Let U and V be vector spaces over a field F, and let $\{u_1, \ldots, u_n\}$ and $\{v_1, \ldots, v_m\}$ be bases for U and V, respectively. Prove that

$$\{(u_1, 0), \ldots, (u_n, 0), (0, v_1), \ldots, (0, v_m)\}$$

is a basis for $U \times V$ (cf. Exercise 2.2.15).

18. The purpose of this exercise is to show that the number of elements in a finite field must be p^n, where p is a prime number and n is a positive integer.

Let F be a finite field.

(a) Let p be the characteristic of F (see Exercise 2.1.19a). Then

$$0, 1, 1+1, 1+1+1, \ldots, 1+1+\cdots+1$$

($p-1$ terms in the last sum) are distinct elements of F. Write $2 = 1+1$, $3 = 1+1+1$, and so forth. Prove that $\{0, 1, 2, \ldots, p-1\}$ is a subfield of F isomorphic to $\mathbf{Z}_p$. (That is, prove the obvious mapping between $\{0, 1, 2, \ldots, p-1\} \subset F$ and $\mathbf{Z}_p$ preserves addition and multiplication.)

(b) Identifying the subfield $\{0, 1, 2, \ldots, p-1\} \subset F$ with $\mathbf{Z}_p$, prove that F is a vector space over $\mathbf{Z}_p$.

(c) Since F has only a finite number of elements, it must be a finite-dimensional vector space over $\mathbf{Z}_p$. Let the dimension be n. By choosing a basis for F, prove that the number of elements in F is p^n.

2.7 Properties of bases

We begin with two fundamental facts about bases.

Theorem 41 *Let V be a nontrivial vector space over a field F, and suppose $\{u_1, u_2, \ldots, u_m\}$ spans V. Then a subset of $\{u_1, u_2, \ldots, u_m\}$ is a basis for V.*

Proof If $\{u_1, u_2, \ldots, u_m\}$ is linearly independent, then it is a basis by definition. Otherwise, by Theorem 24, at least one of the vectors $u_1, u_2, \ldots, u_m$, say u_i, is a linear combination of the rest. By Lemma 19, it follows that

$$\text{sp}\{u_1, \ldots, u_{i-1}, u_{i+1}, \ldots, u_m\} = \text{sp}\{u_1, u_2, \ldots, u_m\} = V.$$

We can continue to remove vectors from the spanning set in this fashion as long as it remains linearly dependent. The process ends with a spanning set containing at least one vector. If there is exactly one vector in the final spanning set, then it must be nonzero because the zero vector cannot span a nontrivial vector space. The set consisting of a single nonzero vector is linearly independent by Exercise 2.5.2, and therefore the final spanning set is a basis in this case. On the other hand, if the final spanning set contains more than one vector, then it must be linearly independent. Otherwise, by Theorem 24, we could remove another vector and produce a proper subset that still spans S, contradicting that the given spanning set is the final one. Therefore, in this case also the process described above produces a basis that is a subset of the original spanning set.

QED

The preceding theorem states that every spanning set contains a basis. If we have a spanning set for a vector space (or subspace) and wish to find a basis, we can follow the reasoning of the proof.

Example 42 *Let $S = \text{sp}\{u_1, u_2, u_3, u_4\} \subset \mathbf{R}^4$, where*

$$u_1 = (1, -1, 0, -1),\ u_2 = (-2, 2, -1, 2),$$
$$u_3 = (5, -5, 1, -5),\ u_4 = (-4, 4, -1, 4).$$

We first check whether the given spanning set is a basis, that is, whether it is linearly independent. We therefore solve

$$\alpha_1 u_1 + \alpha_2 u_2 + \alpha_3 u_3 + \alpha_4 u_4 = 0,$$

which is equivalent to the system

$$\begin{aligned} \alpha_1 - 2\alpha_2 + 5\alpha_3 - 4\alpha_4 &= 0, \\ -\alpha_1 + 2\alpha_2 - 5\alpha_3 + 4\alpha_4 &= 0, \\ -\alpha_2 + \alpha_3 - \alpha_4 &= 0, \\ -\alpha_1 + 2\alpha_2 - 5\alpha_3 + 4\alpha_4 &= 0. \end{aligned}$$

Elimination leads to the equivalent system

$$\begin{aligned} \alpha_1 \quad + 3\alpha_3 - 2\alpha_4 &= 0, \\ \alpha_2 - \alpha_3 + \alpha_4 &= 0 \end{aligned}$$

(two equations cancel completely) or

$$\begin{aligned} \alpha_1 &= -3\alpha_3 + 2\alpha_4, \\ \alpha_2 &= \alpha_3 - \alpha_4. \end{aligned}$$

Two solutions for $(\alpha_1, \alpha_2, \alpha_3, \alpha_4)$ *are* $(-3, 1, 1, 0)$ *and* $(2, -1, 0, 1)$. *The first solution implies*

$$-3u_1 + u_2 + u_3 = 0 \text{ or } u_3 = 3u_1 - u_2.$$

The second yields

$$2u_1 - u_2 + u_4 = 0 \text{ or } u_4 = -2u_1 + u_2.$$

Therefore, $u_4 \in \mathrm{sp}\{u_1, u_2, u_3\}$ *(in fact,* $u_4 \in \mathrm{sp}\{u_1, u_2\}$*), and so* u_4 *can be discarded:* $\mathrm{sp}\{u_1, u_2, u_3\} = \mathrm{sp}\{u_1, u_2, u_3, u_4\} = S$. *Similarly,* $u_3 \in \mathrm{sp}\{u_1, u_2\}$ *and so* u_3 *can be discarded:* $\mathrm{sp}\{u_1, u_2\} = \mathrm{sp}\{u_1, u_2, u_3\} = S$. *We now see that* $\{u_1, u_2\}$ *is a spanning set for* S. *Moreover,* $\{u_1, u_2\}$ *is linearly independent (since neither* u_1 *nor* u_2 *is a multiple of the other). Thus* $\{u_1, u_2\}$ *is a basis for* S *(which shows that* S *has dimension 2).*

Here is the companion to Theorem 41; it states that every linearly independent subset can be extended to form a basis.

Theorem 43 *Let* V *be a finite-dimensional vector space over a field* F, *and suppose* $\{u_1, u_2, \ldots, u_k\} \subset V$ *is linearly independent. If* $\{u_1, u_2, \ldots, u_k\}$ *does not span* V, *then there exist vectors* $u_{k+1}, \ldots, u_n$ *such that*

$$\{u_1, u_2, \ldots, u_k, u_{k+1}, \ldots, u_n\}$$

is a basis for V.

Proof Since $\{u_1, u_2, \ldots, u_k\}$ fails to span V, there must exist a vector, call it u_{k+1}, in V such that

$$v_{k+1} \notin \text{sp}\{u_1, u_2, \ldots, u_k\}. \tag{2.13}$$

It then follows from Exercise 2.5.4 that $\{u_1, u_2, \ldots, u_k, u_{k+1}\}$ is linearly independent.

By hypothesis, V is finite-dimensional; let us suppose that the dimension of V is n. We can continue to add vectors $u_{k+2}, \ldots, u_m$ until there no longer exists a vector v that does not belong to $\text{sp}\{u_1, u_2, \ldots, u_m\}$. (Notice that only finitely many vectors can be added to $\{u_1, u_2, \ldots, u_k\}$ in this manner, since we know from Theorem 34 that any set containing more than n vectors is linearly dependent. Thus, not only is m finite, but $m \leq n$ holds.) At this point $\{u_1, u_2, \ldots, u_m\}$ is linearly independent by construction and spans V since there is no vector in V that does not belong to $\text{sp}\{u_1, u_2, \ldots, u_m\}$. Therefore, $\{u_1, u_2, \ldots, u_m\}$ is a basis for V (and $m = n$ must hold).

QED

Corollary 44 *Every nontrivial subspace of a finite-dimensional vector space has a basis.*

Proof Exercise 12.

To extend a linearly independent subset to a basis, we must repeatedly find vectors outside the span of a given set of vectors. This is easier than it might sound, for the following reason: A proper subspace of a vector space represents a very small part of that space. For example, a two-dimensional subspace of $\mathbf{R}^3$ forms a plane in three-dimensional space. If a vector is chosen at random from the larger space, there is almost no chance that it will lie in the subspace (mathematically, it can be shown that the probability is zero). Therefore, one way to extend a linearly independent subset to a basis is to add random vectors one by one, verifying that linear independence is maintained at each step; alternatively, to extend a linearly independent set of k vectors to a basis for an n-dimensional space, one can add $n - k$ vectors chosen randomly and then verify linear independence. (However, the reader should note that this procedure may fail for textbook examples in which all the vectors, including the ones supposedly chosen at random, have components that are small integers! It might also fail if the underlying field is finite, in which case the above reasoning is not valid.)

When applying either Theorem 41 or Theorem 43, it is convenient to know the following results.

Theorem 45 *Let V be an n-dimensional vector space over a field F, and let $u_1, u_2, \ldots, u_n$ be vectors in V.*

1. *If $\{u_1, u_2, \ldots, u_n\}$ spans V, then $\{u_1, u_2, \ldots, u_n\}$ is linearly independent and hence is a basis for V.*

2. *If $\{u_1, u_2, \ldots, u_n\}$ is linearly independent, then $\{u_1, u_2, \ldots, u_n\}$ spans V and hence is a basis for V.*

Proof

1. Suppose $\{u_1, u_2, \ldots, u_n\}$ spans V. Then, by Theorem 41, there is a subset of $\{u_1, u_2, \ldots, u_n\}$ that is a basis for V. But every basis of V contains n vectors, so the only subset that could be a basis is the entire set $\{u_1, u_2, \ldots, u_n\}$ itself.

2. Suppose $\{u_1, u_2, \ldots, u_n\}$ is linearly independent. As in the proof of Theorem 43, if there exists

$$v \notin \mathrm{sp}\{u_1, u_2, \ldots, u_n\},$$

then we could add that vector v to the set to obtain a set of $n+1$ linearly independent vectors. However, since V has dimension n, any set of $n+1$ vectors is linearly dependent. This shows that every $v \in V$ belongs to $\mathrm{sp}\{u_1, u_2, \ldots, u_n\}$. Hence $\{u_1, u_2, \ldots, u_n\}$ spans V and is therefore a basis for V.

QED

The interpretation of Theorem 45 is quite simple: To verify that a given set is a basis for a vector space, the definition suggests that we must check two properties, that the set spans the vector space and that it is linearly independent. Theorem 45 implies that we need only check one of these two properties, provided we know that the proposed basis contains the correct number of vectors.

Example 46 *The vectors $u_1 = (-1, 2, 5, 3)$, $u_2 = (0, 1, -3, 4)$ form a linearly independent set in $\mathbf{R}^4$. To extend $\{u_1, u_2\}$ to a basis for $\mathbf{R}^4$, we need two more vectors, u_3 and u_4. We will try $u_3 = e_3 = (0, 0, 1, 0)$ and $u_4 = e_4 = (0, 0, 0, 1)$. Since we already know that $\dim(\mathbf{R}^4) = 4$, Theorem 45 implies that we need only check that $\{u_1, u_2, u_3, u_4\}$ is linearly independent, that is, that*

$$\alpha_1 u_1 + \alpha_2 u_2 + \alpha_3 u_3 + \alpha_4 u_4 = 0$$

has only the trivial solution. This equation is equivalent to the system

$$\begin{aligned} -\alpha_1 \qquad\qquad\qquad &= 0, \\ 2\alpha_1 + \alpha_2 \qquad\qquad &= 0, \\ 5\alpha_1 - 3\alpha_2 + \alpha_3 \qquad &= 0, \\ 3\alpha_1 + 4\alpha_2 \qquad + \alpha_4 &= 0. \end{aligned}$$

It is easy to see that the only solution is the trivial one (the first equation shows that $\alpha_1 = 0$, whereupon the second equation implies $\alpha_2 = 0$, and so on). Thus $\{u_1, u_2, u_3, u_4\}$ is a basis for $\mathbf{R}^4$.

Example 47 *The vectors* $u_1 = (1,1,0)$ *and* $u_2 = (1,1,1)$ *form a linearly independent set in* $\mathbf{Z}_2^3$. *Since* $\dim(\mathbf{Z}_2^3) = 3$, *we need one more vector to extend* $\{u_1, u_2\}$ *to a basis. We will try* $u_3 = (0,1,1)$. *We know from Theorem 45 that* $\{u_1, u_2, u_3\}$ *is a basis for* $\mathbf{Z}_2^3$ *if it is linearly independent, and therefore we solve* $\alpha_1 u_1 + \alpha_2 u_2 + \alpha_3 u_3 = 0$ *for* $\alpha_1, \alpha_2, \alpha_3 \in \mathbf{Z}_2$:

$$\begin{aligned} & \alpha_1(1,1,0) + \alpha_2(1,1,1) + \alpha_3(0,1,1) = (0,0,0) \\ \Leftrightarrow \quad & (\alpha_1 + \alpha_2, \alpha_1 + \alpha_2 + \alpha_3, \alpha_2 + \alpha_3) = (0,0,0) \\ \Leftrightarrow \quad & \left\{ \begin{array}{rl} \alpha_1 + \alpha_2 \phantom{{}+\alpha_3} & = 0, \\ \alpha_1 + \alpha_2 + \alpha_3 & = 0, \\ \alpha_2 + \alpha_3 & = 0. \end{array} \right. \end{aligned}$$

Gaussian elimination shows that this last system has only the trivial solution. Therefore, $\{u_1, u_2, u_3\}$ *is linearly independent and hence is a basis for* $\mathbf{Z}_2^3$.

Exercises

Miscellaneous exercises

1. Consider the following vectors in $\mathbf{R}^3$: $v_1 = (1,5,4)$, $v_2 = (1,5,3)$, $v_3 = (17,85,56)$, $v_4 = (1,5,2)$, $v_5 = (3,16,13)$.

   (a) Show that $\{v_1, v_2, v_3, v_4, v_5\}$ spans $\mathbf{R}^3$.

   (b) Find a subset of $\{v_1, v_2, v_3, v_4, v_5\}$ that is a basis for $\mathbf{R}^3$.

2. Consider the following vectors in $\mathbf{R}^4$:

   $$u_1 = (1,3,5,-1), \ u_2 = (1,4,9,0), \ u_3 = (4,9,7,-5).$$

   (a) Show that $\{u_1, u_2, u_3\}$ is linearly independent.

   (b) Find a vector $u_4 \in \mathbf{R}^4$ such that $\{u_1, u_2, u_3, u_4\}$ is a basis for $\mathbf{R}^4$.

3. Let $p_1(x) = 2 - 5x$, $p_2(x) = 2 - 5x + 4x^2$.

   (a) Explain why it is obvious that $\{p_1, p_2\}$ is linearly independent (cf. Exercise 2.5.1).

   (b) Find a polynomial $p_3 \in \mathcal{P}_2$ such that $\{p_1, p_2, p_3\}$ is a basis for $\mathcal{P}_2$.

4. Define $p_1, p_2, p_3, p_4, p_5 \in \mathcal{P}_2$ by

   $$\begin{aligned} & p_1(x) = x, \ p_2(x) = 1 + x, \ p_3(x) = 3 + 5x, \\ & p_4(x) = 5 + 8x, \ p_5(x) = 3 + x - x^2. \end{aligned}$$

   (a) Show that $\{p_1, p_2, p_3, p_4, p_5\}$ spans $\mathcal{P}_2$.

   (b) Find a subset of $\{p_1, p_2, p_3, p_4, p_5\}$ that forms a basis for $\mathcal{P}_2$.

5. Let $u_1 = (1, 4, 0, -5, 1)$, $u_2 = (1, 3, 0, -4, 0)$, $u_3 = (0, 4, 1, 1, 4)$ be vectors in $\mathbf{R}^5$.

   (a) Show that $\{u_1, u_2, u_3\}$ is linearly independent.

   (b) Extend $\{u_1, u_2, u_3\}$ to a basis for $\mathbf{R}^5$.

6. Consider the following vectors in $\mathbf{R}^5$:

$$u_1 = \begin{bmatrix} 1 \\ 2 \\ 0 \\ 1 \\ 1 \end{bmatrix}, \quad u_2 = \begin{bmatrix} -1 \\ 3 \\ 2 \\ 1 \\ -1 \end{bmatrix}, \quad u_3 = \begin{bmatrix} 1 \\ 7 \\ 2 \\ 3 \\ 1 \end{bmatrix},$$

$$u_4 = \begin{bmatrix} 1 \\ -2 \\ -1 \\ 1 \\ 1 \end{bmatrix}, \quad u_5 = \begin{bmatrix} 2 \\ 10 \\ 3 \\ 6 \\ 2 \end{bmatrix}.$$

   Let $S = \text{sp}\{u_1, u_2, u_3, u_4, u_5\}$. Find a subset of $\{u_1, u_2, u_3, u_4, u_5\}$ that is a basis for S.

7. Consider the following polynomials in $\mathcal{P}_3$:

$$\begin{aligned} &p_1(x) = 1 - 4x + x^2 + x^3, \; p_2(x) = 3 - 11x + x^2 + 4x^3, \\ &p_3(x) = -x + 2x^2 - x^3, \; p_4(x) = -x^2 + 2x^3, \\ &p_5(x) = 5 - 18x + 2x^2 + 5x^3. \end{aligned}$$

   What is the dimension of $\text{sp}\{p_1, p_2, p_3, p_4, p_5\}$?

8. Let $S = \text{sp}\{v_1, v_2, v_3, v_4\} \subset \mathbf{C}^3$, where

$$\begin{aligned} &v_1 = (1 - i, 3 + i, 1 + i), \; v_2 = (1, 1 - i, 3), \\ &v_3 = (i, -2 - 2i, 2 - i), \; v_4 = (2 - i, 7 + 3i, 2 + 5i). \end{aligned}$$

   Find a basis for S.

9. Consider the vectors $u_1 = (3, 1, 0, 4)$ and $u_2 = (1, 1, 1, 4)$ in $\mathbf{Z}_5^4$.

   (a) Prove that $\{u_1, u_2\}$ is linearly independent.

   (b) Extend $\{u_1, u_2\}$ to a basis for $\mathbf{Z}_5^4$.

10. Let $S = \text{sp}\{v_1, v_2, v_3\} \subset \mathbf{Z}_3^3$, where

$$v_1 = (1, 2, 1), \; v_2 = (2, 1, 2), \; v_3 = (1, 0, 1).$$

   Find a subset of $\{v_1, v_2, v_3\}$ that is a basis for S.

11. Let F be a field. The following exercises show how to produce different bases for a nontrivial, finite-dimensional vector space over F.

    (a) Let V be a 1-dimensional vector space over F, and let $\{u_1\}$ be a basis for V. Prove that $\{\alpha u_1\}$ is a basis for V for any $\alpha \neq 0$.

    (b) Let V be a 2-dimensional vector space over F, and let $\{u_1, u_2\}$ be a basis for V. Prove that $\{\alpha u_1, \beta u_1 + \gamma u_2\}$ is a basis for V for any $\alpha \neq 0$, $\gamma \neq 0$.

    (c) Let V be a vector space over F with basis $\{u_1, \ldots, u_n\}$. By generalizing the construction of the previous two parts, show how to produce a collection of different bases for V.

    The reader should notice that, if F has infinitely many elements, then this exercise shows that a nontrivial, finite-dimensional vector space over F has infinitely many different bases.

12. Prove Corollary 44.

13. Let F be a finite field with q distinct elements, and let n be positive integer, $n \geq q$. Exercise 2.6.11 states that $\dim(\mathcal{P}_n(F)) \geq q$. Prove that $\dim(\mathcal{P}_n(F)) = q$. (Hint: Suppose $F = \{\alpha_1, \alpha_2, \ldots, \alpha_q\}$. Any function $p \in \mathcal{P}_n(F)$ is completely determined by the corresponding vector $(p(\alpha_1), p(\alpha_2), \ldots, p(\alpha_q)) \in F^q$. Use the facts that $\{1, x, \ldots, x^{q-1}\}$ is linearly independent in $\mathcal{P}_n(F)$ (see Exercise 2.6.11) and that $\dim(F^q) = q$ to prove that $\{1, x, \ldots, x^{q-1}\}$ is a basis for $\mathcal{P}_n(F)$.)

14. Let V be an n-dimensional vector space over a field F, and suppose S and T are subspaces of V satisfying $S \cap T = \{0\}$. Suppose that $\{s_1, s_2, \ldots, s_k\}$ is a basis for S, $\{t_1, t_2, \ldots, t_\ell\}$ is a basis for T, and $k+\ell = n$. Prove that $\{s_1, s_2, \ldots, s_k, t_1, t_2, \ldots, t_\ell\}$ is a basis for V.

15. Let V be a vector space over a field F, and let $\{u_1, \ldots, u_n\}$ be a basis for V. Let $v_1, \ldots, v_k$ be vectors in V, and suppose

    $$v_j = \alpha_{1,j} u_1 + \ldots + \alpha_{n,j} u_n, \ j = 1, 2, \ldots, k.$$

    Define the vectors $x_1, \ldots, x_k$ in F^n by

    $$x_j = (\alpha_{1,j}, \ldots, \alpha_{n,j}), \ j = 1, 2, \ldots, k.$$

    Prove:

    (a) $\{v_1, \ldots, v_k\}$ is linearly independent if and only if $\{x_1, \ldots, x_k\}$ is linearly independent;

    (b) $\{v_1, \ldots, v_k\}$ spans V if and only if $\{x_1, \ldots, x_k\}$ spans F^n.

16. Consider the polynomials $p_1(x) = -1 + 3x + 2x^2$, $p_2(x) = 3 - 8x - 4x^2$, and $p_3(x) = -1 + 4x + 5x^2$ in $\mathcal{P}_2$. Using the previous exercise explicitly (with the standard basis for $\mathcal{P}_2$), determine if $\{p_1, p_2, p_3\}$ is linearly independent.

2.8 Polynomial interpolation and the Lagrange basis

We have now spent considerable effort in carefully defining the concept of basis and exploring some of the common properties of bases. Exercise 2.7.11 shows that a finite-dimensional vector space has many bases, and in applications we often take advantage of this fact by choosing a basis that is particularly well-suited to the problem at hand. In this section, we will explore one such example.

A *polynomial interpolation* problem takes the following form: We are given $n+1$ points $(x_0, y_0), (x_1, y_1), \ldots, (x_n, y_n)$, and we wish to find a polynomial $p(x) = c_0 + c_1x + c_2x^2 + \ldots + c_nx^n$ that interpolates the given data. In other words, p is to satisfy the equations

$$p(x_i) = y_i, \ i = 0, 1, \ldots, n.$$

These equations form a system of $n+1$ equations in $n+1$ unknowns, namely, the coefficients $c_0, c_1, \ldots, c_n$:

$$\begin{aligned} c_0 + c_1x_0 + c_2x_0^2 + \ldots + c_nx_0^n &= y_0, \\ c_0 + c_1x_1 + c_2x_1^2 + \ldots + c_nx_1^n &= y_1, \\ \vdots \qquad\qquad &\quad \vdots \\ c_0 + c_1x_n + c_2x_n^2 + \ldots + c_nx_n^n &= y_n. \end{aligned}$$

The numbers $x_0, x_1, \ldots, x_n$ are called the *interpolation nodes*, and they are assumed to be distinct. The reader should notice that the degree of p is chosen to be n so that the the number of equations and unknowns would match; in such a case, we hope to find a unique solution. We will discuss the existence and uniqueness question for this problem below, but first we work out an example.

Example 48 *Consider the following points in* $\mathbf{R}^2$*:*

$$(-2, 10), \ (-1, -3), \ (0, 2), \ (1, 7), \ (2, 18).$$

In this example, $n = 4$ *and we are seeking a polynomial*

$$p(x) = c_0 + c_1x + c_2x^2 + c_3x^3 + c_4x^4.$$

The linear system determining the coefficients is

$$\begin{aligned} c_0 - 2c_1 + 4c_2 - 8c_3 + 16c_4 &= 10, \\ c_0 - \ c_1 + \ c_2 - \ c_3 + \ c_4 &= -3, \\ c_0 \qquad\qquad\qquad\qquad &= 2, \\ c_0 + \ c_1 + \ c_2 + \ c_3 + \ c_4 &= 7, \\ c_0 + 2c_1 + 4c_2 + 8c_3 + 16c_4 &= 18. \end{aligned}$$

Gaussian elimination leads to the unique solution

$$c_0 = 2, c_1 = 6, c_2 = -1, c_3 = -1, c_4 = 1,$$

or $p(x) = 2 + 6x - x^2 - x^3 + x^4$.

The original data points and the interpolating polynomial are shown in Figure 2.3.

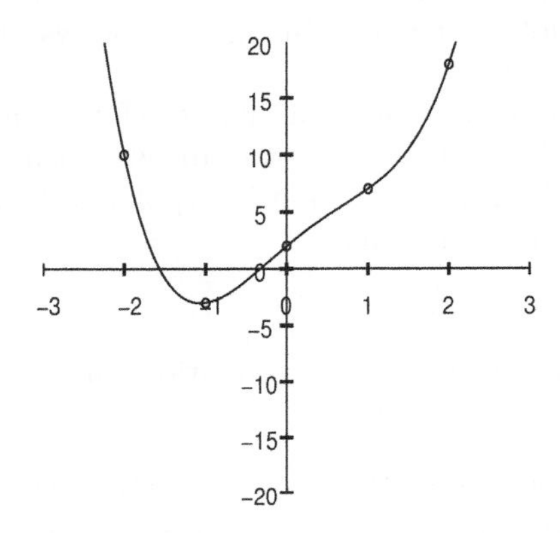

FIGURE 2.3
The data points from Example 48 and the fourth degree interpolating polynomial.

The polynomial interpolation problem commonly arises in the context of real numbers (so that each interpolation node x_i is a real number, as is each value y_i). However, there are many applications in which polynomial interpolation over a finite field is required; we will discuss one such application, secret sharing, below. Here is an example of polynomial interpolation in a finite field.

Example 49 *We wish to find a quadratic polynomial* p *having coefficients in* $\mathbf{Z}_5$ *and satisfying*

$$p(1) = 2,\ p(2) = 1,\ p(3) = 4.$$

In this example, all numbers are to be interpreted as elements of $\mathbf{Z}_5$. *We write* $p(x) = c_0 + c_1x + c_2x^2$ *and solve for* $c_0, c_1, c_2 \in \mathbf{Z}_5$. *We have the following equations:*

$$\begin{aligned}
p(1) = 2 \quad &\Rightarrow \quad c_0 + c_1 + c_2 = 2,\\
p(2) = 1 \quad &\Rightarrow \quad c_0 + 2c_1 + 4c_2 = 1,\\
p(3) = 4 \quad &\Rightarrow \quad c_0 + 3c_1 + 4c_2 = 4.
\end{aligned}$$

The reader should notice that, in computing $p(3) = c_0 + 3c_1 + 4c_2$, *we used the fact that* $3^2 = 4$ *in* $\mathbf{Z}_5$. *We now solve this system of equations using elimination:*

$$\begin{array}{rcl} c_0 + c_1 + c_2 &=& 2 \\ c_0 + 2c_1 + 4c_2 &=& 1 \\ c_0 + 3c_1 + 4c_2 &=& 4 \end{array} \rightarrow \begin{array}{rcl} c_0 + c_1 + c_2 &=& 2 \\ c_1 + 3c_2 &=& 4 \\ 2c_1 + 3c_2 &=& 2 \end{array}$$

$$\rightarrow \begin{array}{rcl} c_0 + c_1 + c_2 &=& 2 \\ c_1 + 3c_2 &=& 4 \\ 2c_2 &=& 4 \end{array} \rightarrow \begin{array}{rcl} c_0 + c_1 + c_2 &=& 2 \\ c_1 + 3c_2 &=& 4 \\ c_2 &=& 2 \end{array}$$

$$\rightarrow \begin{array}{rcl} c_0 + c_1 &=& 0 \\ c_1 &=& 3 \\ c_2 &=& 2 \end{array} \rightarrow \begin{array}{rcl} c_0 &=& 2 \\ c_1 &=& 3 \\ c_2 &=& 2 \end{array}.$$

The desired interpolating polynomial is $p(x) = 2 + 3x + 2x^2$. *The reader can check that this polynomial satisfies the desired conditions; for example,*

$$p(3) = 2 + 3 \cdot 3 + 2 \cdot 3^2 = 2 + 4 + 2 \cdot 4 = 2 + 4 + 3 = 4.$$

We will show that the polynomial interpolation problem on $n+1$ nodes can be posed and uniquely solved in the context of an arbitrary field F (provided F has at least $n+1$ elements). The above examples illustrate one way to solve the problem: form and solve the linear system that determines the coefficients $c_0, c_1, \ldots, c_n$ of $p \in \mathcal{P}_n(F)$. In this approach, we are representing a polynomial in terms of the standard basis $\{1, x, x^2, \ldots, x^n\}$ for $\mathcal{P}_n(F)$. The shortcomings of this method are that it is not obvious that the linear system will always have a unique solution and that, if it does, we must go to the trouble of solving the system.[6]

With an alternate basis for $\mathcal{P}_n(F)$, the problem becomes much easier. Given $n+1$ distinct elements $x_0, x_1, \ldots, x_n \in F$, we define the following polynomials:

$$\begin{aligned} L_0(x) &= \frac{(x - x_1)(x - x_2)(x - x_3) \cdots (x - x_n)}{(x_0 - x_1)(x_0 - x_2)(x_0 - x_3) \cdots (x_0 - x_n)}, \\ L_1(x) &= \frac{(x - x_0)(x - x_2)(x - x_3) \cdots (x - x_n)}{(x_1 - x_0)(x_1 - x_2)(x_1 - x_3) \cdots (x_1 - x_n)}, \\ L_2(x) &= \frac{(x - x_0)(x - x_1)(x - x_3) \cdots (x - x_n)}{(x_2 - x_0)(x_2 - x_1)(x_2 - x_3) \cdots (x_2 - x_n)}, \\ &\vdots \\ L_n(x) &= \frac{(x - x_0)(x - x_1)(x - x_2) \cdots (x - x_{n-1})}{(x_n - x_0)(x_n - x_1)(x_n - x_2) \cdots (x_n - x_{n-1})}. \end{aligned}$$

[6]The system of linear equations derived on page 73 is called a *Vandermonde system*. It is possible to prove directly, using the theory of determinants, that a Vandermonde system always has a unique solution. See Exercise 4.3.11.

It is important to note that if F is a finite field, then F must contain at least $n+1$ elements to perform this construction.

Each L_i has degree n and satisfies

$$L_i(x_j) = \begin{cases} 1, & j = i, \\ 0, & j \neq i. \end{cases} \tag{2.14}$$

If $j \neq i$, then the numerator of $L_i(x_j)$ contains a factor of $(x_j - x_j)$, and hence $L_i(x_j) = 0$. On the other hand, the numerator of $L_i(x_i)$ is identical to its denominator, and so $L_i(x_i) = 1$.

We can now show that $\{L_0, L_1, \ldots, L_n\}$ is an alternate basis for $\mathcal{P}_n(F)$. Since the dimension of $\mathcal{P}_n(F)$ is $n+1$ (cf. Exercise 2.7.13 in the case of a finite field), the proposed basis contains the correct number of elements, and it suffices by Theorem 45 to show that $\{L_0, L_1, \ldots, L_n\}$ is linearly independent. So suppose

$$\alpha_0 L_0 + \alpha_1 L_1 + \ldots + \alpha_n L_n = 0.$$

This means that the polynomial $\alpha_0 L_0 + \alpha_1 L_1 + \ldots + \alpha_n L_n$ is the zero polynomial, and so

$$\alpha_0 L_0(x) + \alpha_1 L_1(x) + \ldots + \alpha_n L_n(x) = 0 \text{ for all } x \in F.$$

But then, in particular,

$$\begin{aligned} & \alpha_0 L_0(x_i) + \ldots + \alpha_i L_i(x_i) + \ldots + \alpha_n L_n(x_i) = 0, \ i = 0, 1, \ldots, n \\ \Rightarrow \quad & \alpha_0 \cdot 0 + \ldots + \alpha_i \cdot 1 + \ldots + \alpha_n \cdot 0 = 0, \ i = 0, 1, \ldots, n \\ \Rightarrow \quad & \alpha_i = 0, \ i = 0, 1, \ldots, n. \end{aligned}$$

Since $\alpha_i = 0$ for $i = 0, 1, \ldots, n$, this shows that $\{L_0, L_1, \ldots, L_n\}$ is linearly independent and hence is a basis for $\mathcal{P}_n(F)$. This basis is called a *Lagrange basis.*

The special property of the Lagrange basis is that we can find the interpolating polynomial for the points $(x_0, y_0), (x_1, y_1), \ldots, (x_n, y_n)$ without any additional work; it is simply

$$p(x) = y_0 L_0(x) + y_1 L_1(x) + \ldots + y_n L_n(x).$$

This follows from (2.14); for each $i = 0, 1, \ldots, n$, we have

$$\begin{aligned} p(x_i) &= y_0 L_0(x_i) + y_1 L_1(x_i) + \ldots + y_i L_i(x_i) + \ldots + y_n L_n(x_i) \\ &= y_0 \cdot 0 + y_1 \cdot 0 + \ldots + y_i \cdot 1 + \ldots + y_n \cdot 0 \\ &= y_i. \end{aligned}$$

By this method, we represent p as a linear combination of $L_0, L_1, \ldots, L_n$, and the Lagrange basis is easy to use for this application because no work is required to find the weights in the linear combination.

The reader should notice that the use of the Lagrange basis shows immediately that the polynomial interpolation problem has a solution, provided the interpolation nodes are distinct. The uniqueness of the interpolating polynomial is the subject of Exercises 6 and 7.

We also remark that, if F is a finite field with q elements $x_0, x_1, \ldots, x_{q-1}$, then there are q^q functions $f : F \to F$. In fact, every one of these functions can be represented as a polynomial; we simply find $p \in \mathcal{P}_{q-1}(F)$ satisfying

$$p(x_0) = f(x_0), p(x_1) = f(x_1), \ldots, p(x_{q-1}) = f(x_{q-1}).$$

Such a p exists, as we have just shown, and $p(x) = f(x)$ for every $x \in F$. Thus p and f are equal.

Example 50 *We will compute the interpolating polynomial of Example 49 using the Lagrange basis. The interpolation nodes are* $x_0 = 1$, $x_1 = 2$, *and* $x_2 = 3$, *and therefore*

$$\begin{aligned} L_0(x) &= \frac{(x-x_1)(x-x_2)}{(x_0-x_1)(x_0-x_2)} = \frac{(x-2)(x-3)}{(1-2)(1-3)} = \frac{(x-2)(x-3)}{2}, \\ L_1(x) &= \frac{(x-x_0)(x-x_2)}{(x_1-x_0)(x_1-x_2)} = \frac{(x-1)(x-3)}{(2-1)(2-3)} = \frac{(x-1)(x-3)}{4}, \\ L_2(x) &= \frac{(x-x_0)(x-x_1)}{(x_2-x_0)(x_2-x_1)} = \frac{(x-1)(x-2)}{(3-1)(3-2)} = \frac{(x-1)(x-2)}{2}. \end{aligned}$$

We want $p(1) = 2$, $p(2) = 1$, *and* $p(3) = 4$ *(in* $\mathbf{Z}_5$*), and thus*

$$\begin{aligned} p(x) &= 2L_0(x) + L_1(x) + 4L_2(x) \\ &= (x-2)(x-3) + \frac{(x-1)(x-3)}{4} + \frac{4(x-1)(x-2)}{2} \\ &= (x^2+1) + 4(x^2+x+3) + 2(x^2+2x+2) \\ &= x^2 + 1 + 4x^2 + 4x + 2 + 2x^2 + 4x + 4 \\ &= (1+2+4) + (4+4)x + (1+4+2)x^2 \\ &= 2 + 3x + 2x^2. \end{aligned}$$

Using the Lagrange basis yields the same result computed earlier.

2.8.1 Secret sharing

An interesting application of polynomial interpolation in finite fields arises in an algorithm for *secret sharing* due to Shamir [41]. There is some sensitive information that should be available to a group (such as the board of directors of a company), but for reasons of security, it should not be available to any of the individuals in the group acting alone. Here is a precise specification of the problem. We assume that there are n individuals in the group and that the secret should be available to any k individuals who agree to access it together.

In addition, the information given to each individual must be worthless by itself; it must not give partial information about the secret.

Here is Shamir's method of secret sharing.

1. Assume that the secret has been encoded by some means as a positive integer N (the method of encoding is not secret).

2. Choose a prime number p greater than $\max\{N, n\}$ and choose $k-1$ random numbers $c_1, c_2, \ldots, c_k$ in $\mathbf{Z}_p$. Construct the polynomial p in $\mathcal{P}_{k-1}(\mathbf{Z}_p)$ defined by

$$p(x) = N + c_1 x + c_2 x^2 + \cdots + c_{k-1} x^{k-1}.$$

   Anyone who can compute the polynomial p has access to the secret; it is just the constant term in the polynomial.

3. Choose n distinct elements $x_1, x_2, \ldots, x_n$ in $\mathbf{Z}_p$, and, for $i = 1, \ldots, n$, compute $y_i = p(x_i) \in \mathbf{Z}_p$. Each person in the group is given one pair (x_i, y_i). Any k individuals collectively have k interpolation points, enough information to determine $p \in \mathcal{P}_k(\mathbf{Z}_p)$ and hence to compute N.

A critical aspect of the secret-sharing method just described is that if fewer than k individuals pool their information, they cannot deduce anything about the secret. Given $\ell < k$ interpolation points, we can add the pair $(0, \tilde{N})$ for *any* $\tilde{N} \in \mathbf{Z}_p$, and there is a polynomial q in $\mathcal{P}_k(\mathbf{Z}_p)$ satisfying the $\ell + 1$ interpolation conditions. The polynomial q satisfies $q(0) = \tilde{N}$; in other words, $\tilde{N}$ is the constant term in q. The point of this is *every* "secret" $\tilde{N} \in \mathbf{Z}_p$ is consistent with the $\ell < k$ interpolation conditions, and therefore those ℓ conditions provide no information about the secret.

Example 51 *Suppose that $n = 4$, $k = 3$, and the "secret" is a two-digit integer N. This means that any three people from a group of four, working together, are to be allowed access to the secret. We will take $p = 101$ (the smallest prime number larger than all two-digit integers). With $k = 3$, we must choose a polynomial p in $\mathcal{P}_2(\mathbf{Z}_p)$; we take*

$$p(x) = N + 23x + 61x^2.$$

(Recall that the coefficients of x and x^2 are to be chosen at random from $\mathbf{Z}_p$; we have chosen 23 and 61, respectively.) To give a concrete example of the rest of the calculations, let us assume that $N = 57$.

We choose four interpolation nodes,

$$x_1 = 14,\ x_2 = 37,\ x_3 = 75,\ x_4 = 90$$

(again, these numbers are to be chosen at random from $\mathbf{Z}_p$, avoiding using 0

as an interpolation node). We now must compute $y_i = p(x_i)$*,* $i = 1, 2, 3, 4$*, remembering to perform all calculations in* $\mathbf{Z}_{101}$*. We obtain*

$$\begin{aligned} y_1 &= p(x_1) = 13, \\ y_2 &= p(x_2) = 82, \\ y_3 &= p(x_3) = 93, \\ y_4 &= p(x_4) = 14. \end{aligned}$$

Thus the four individuals who are to share the secret are each given one of the data points $(14, 13)$*,* $(37, 82)$*,* $(75, 93)$*, and* $(90, 14)$*.*

Now let us assume that the first three individuals pool their data points ($(14, 13)$*,* $(37, 82)$*,* $(75, 93)$*) and wish to determine the secret. They can do this by computing the unique polynomial* $p \in \mathcal{P}_n(\mathbf{Z}_{101})$ *that interpolates the three points, and we now proceed to compute* p*. We will use the Lagrange basis, which we compute as follows:*

$$\begin{aligned} L_0(x) &= \frac{(x-37)(x-75)}{(14-37)(14-75)} = \frac{(x-37)(x-75)}{90} = 55(x-37)(x-75), \\ L_1(x) &= \frac{(x-14)(x-75)}{(37-14)(37-75)} = \frac{(x-14)(x-75)}{35} = 26(x-14)(x-75), \\ L_2(x) &= \frac{(x-14)(x-37)}{(75-14)(75-37)} = \frac{(x-14)(x-37)}{96} = 20(x-14)(x-37). \end{aligned}$$

The reader should note that, in computing $L_0(x)$*, we needed to divide by* 90*, which is equivalent to multiplying by the multiplicative inverse of* 90 *in* $\mathbf{Z}_{101}$*, which is* 55*. Similar calculations were performed to compute* $L_1(x)$ *and* $L_2(x)$*. Appendix A explains how to find multiplicative inverses in* $\mathbf{Z}_p$*.*

We now have

$$\begin{aligned} p(x) &= 13L_0(x) + 82L_1(x) + 93L_2(x) \\ &= 13 \cdot 55(x-37)(x-75) + 82 \cdot 26(x-14)(x-75) + \\ &\quad\; 93 \cdot 20(x-14)(x-37) \\ &= 8(x-37)(x-75) + 11(x-14)(x-75) + 42(x-14)(x-37). \end{aligned}$$

Finally, we wish to find the constant term in $p(x)$*, which is simply* $p(0)$*. Computing* $p(0)$ *using* $\mathbf{Z}_{101}$ *arithmetic yields* 57*, as expected.*

Exercises

Unless otherwise stated, the underlying field should be taken to be $\mathbf{R}$. For problems involving an arbitrary field F, we implicitly assume F has at least as many distinct elements as there are interpolation nodes.

1. (a) Find the Lagrange polynomials for the interpolation nodes $x_0 = 1$, $x_1 = 2$, $x_2 = 3$.

(b) Construct the quadratic polynomial interpolating $(1,0)$, $(2,2)$, $(3,1)$.

2. (a) Write down the formulas for the Lagrange polynomials for the interpolation nodes of Example 48.

   (b) Use the Lagrange basis to solve Example 48. Show that the result is the interpolating polynomial calculated in the text.

3. Consider the data $(1,5)$, $(2,-4)$, $(3,-4)$, $(4,2)$. Find an interpolating polynomial for these data two ways:

   (a) Using the standard basis (that is, by forming and solving a system of equations);

   (b) Using the Lagrange polynomials.

4. Let $\{L_0, L_1, \ldots, L_n\}$ be the Lagrange basis constructed on the interpolation nodes $x_0, x_1, \ldots, x_n \in F$. Prove that

$$p(x) = p(x_0)L_0(x) + p(x_1)L_1(x) + \ldots + p(x_n)L_n(x)$$

   for any polynomial $p \in \mathcal{P}_n(F)$.

5. Write $p_2(x) = 2 + x - x^2$ as a linear combination of the Lagrange polynomials constructed on the nodes $x_0 = -1$, $x_1 = 1$, $x_2 = 3$.

6. Prove that, given the points $(x_0, y_0), (x_1, y_1), \ldots, (x_n, y_n) \in F^2$, the polynomial interpolation problem has at most one solution, provided the interpolation nodes $x_0, x_1, \ldots, x_n$ are distinct. (Hint: Suppose there are two different interpolating polynomials $p, q \in \mathcal{P}_n(F)$. Then $p - q$ is a nonzero polynomial of degree at most n. Consider its roots.)

7. Use the Lagrange basis, and (2.14) in particular, to prove the uniqueness result of the previous exercise. (Hint: Any interpolating polynomial in $\mathcal{P}_n(F)$ can be written in terms of the Lagrange basis $\{L_0, L_1, \ldots, L_n\}$).

8. Suppose $x_0, x_1, \ldots, x_n$ are distinct real numbers. Prove that, for any real numbers $y_0, y_1, \ldots, y_n$, the system

$$\begin{aligned}
c_0 + c_1 x_0 + c_2 x_0^2 + \ldots + c_n x_0^n &= y_0, \\
c_0 + c_1 x_1 + c_2 x_1^2 + \ldots + c_n x_1^n &= y_1, \\
\vdots \qquad\qquad & \quad \vdots \\
c_0 + c_1 x_n + c_2 x_n^2 + \ldots + c_n x_n^n &= y_n
\end{aligned}$$

   has a unique solution $c_0, c_1, \ldots, c_n$.

9. List all possible functions $f : \mathbf{Z}_2 \to \mathbf{Z}_2$ by completing the following table, and then find a polynomial in $\mathcal{P}_1(\mathbf{Z}_2)$ representing each function.

x	$f_1(x)$	$f_2(x)$	$\cdots$
0	0	1	
1	0	0	

10. The following table defines three functions mapping $\mathbf{Z}_3 \to \mathbf{Z}_3$. Find a polynomial in $\mathcal{P}_2(\mathbf{Z}_3)$ representing each one.

x	$f_1(x)$	$f_2(x)$	$f_3(x)$
0	1	0	2
1	2	0	2
2	0	1	1

11. Consider a secret sharing scheme, as described in Section 2.8.1, in which five individuals will receive information about the secret, and any two of them, working together, will have access to the secret. Assume that the secret is a two-digit integer, and that p is chosen to be 101. What will be the degree of the polynomial? Choose an appropriate polynomial and generate the five data points.

12. An integer N satisfying $1 \leq N \leq 256$ represents a secret to be shared among five individuals. Any three of the individuals are allowed access to the information. The secret is encoded in a polynomial p, according to the secret sharing scheme described in Section 2.8.1, lying in $\mathcal{P}_2(\mathbf{Z}_{257})$. Suppose three of the individuals get together, and their data points are $(15, 13)$, $(114, 94)$, and $(199, 146)$. What is the secret?

13. Suppose we wish to solve the following interpolation problem: Given $v_1, v_2, d_1, d_2 \in \mathbf{R}$, find $p \in \mathcal{P}_3$ such that

$$p(0) = v_1,\ p(1) = v_2,\ p'(0) = d_1,\ p'(1) = d_2.$$

  (a) Explain how to solve this problem using the standard basis. (That is, represent p as $p(x) = c_0 + c_1 x + c_2 x^2 + c_3 x^3$ and explain how to find c_0, c_1, c_2, c_3.)

  (b) Find a special basis $\{q_1, q_2, q_3, q_4\}$ of $\mathcal{P}_3$ satisfying the following conditions:

$$\begin{aligned}
q_1(0) = 1,\ q_1'(0) = 0,\ q_1(1) = 0,\ q_1'(1) = 0,\\
q_2(0) = 0,\ q_2'(0) = 1,\ q_2(1) = 0,\ q_2'(1) = 0,\\
q_3(0) = 0,\ q_3'(0) = 0,\ q_3(1) = 1,\ q_3'(1) = 0,\\
q_4(0) = 0,\ q_4'(0) = 0,\ q_4(1) = 0,\ q_4'(1) = 1.
\end{aligned}$$

  What is the solution to the interpolation problem in terms of the basis $\{q_1, q_2, q_3, q_4\}$?

14. This exercise explores a general interpolation problem, of which the preceding exercise is a special case. We are given $n+1$ interpolation nodes, $x_0, x_1, \ldots, x_n$. Given $v_0, v_1, \ldots, v_n \in \mathbf{R}$, $d_0, d_1, \ldots, d_n \in \mathbf{R}$, we wish to find $p \in \mathcal{P}_{2n+1}$ such that

$$p(x_i) = v_i, \ p'(x_i) = d_i, \ i = 0, 1, \ldots, n.$$

This is called a *Hermite* interpolation problem.

We can construct a special basis for $\mathcal{P}_{2n+1}$, analogous to the Lagrange basis, which makes the Hermite interpolation problem easy to solve. This basis is defined in terms of the Lagrange polynomials $L_0, L_1, \ldots, L_n$ (relative to the nodes $x_0, x_1, \ldots, x_n$). Each L_i has degree n. We now define, for $i = 0, 1, \ldots, n$,

$$\begin{aligned} A_i(x) &= (1 - 2(x - x_i)L_i'(x_i))\, L_i^2(x), \\ B_i(x) &= (x - x_i)L_i^2(x). \end{aligned}$$

The reader is asked to prove the following properties of these polynomials:

(a) $A_i, B_i \in \mathcal{P}_{2n+1}$ for all $i = 0, 1, \ldots, n$.

(b) Each A_i satisfies

$$\begin{aligned} A_i(x_j) &= \begin{cases} 1, & i = j, \\ 0, & i \neq j, \end{cases} \\ A_i'(x_j) &= 0, \ j = 0, 1, \ldots, n. \end{aligned}$$

(c) Each B_i satisfies

$$\begin{aligned} B_i(x_j) &= 0, \ j = 0, 1, \ldots, n, \\ B_i'(x_j) &= \begin{cases} 1, & i = j, \\ 0, & i \neq j. \end{cases} \end{aligned}$$

(d) $\{A_0, \ldots, A_n, B_0, \ldots, B_n\}$ is a basis for $\mathcal{P}_{2n+1}$.

What is the solution to the Hermite interpolation problem in terms of $v_0, \ldots, v_n$, $d_0, \ldots, d_n$, and the basis defined above?

2.9 Continuous piecewise polynomial functions

In this section, we discuss the problem of approximating real-valued functions by polynomials. Polynomials are often used to approximate more general functions because polynomials are easy to evaluate, integrate, and differentiate. One way to construct such approximations is by interpolation. Often this works well, as the following example shows.

Example 52 *Suppose we wish to approximate $f(x) = e^x$ on the interval $[0,1]$. We can construct $n+1$ evenly spaced interpolation nodes as follows:*

$$x_i = i\Delta x, \ i = 0, 1, \ldots, n,$$

where $\Delta x = 1/n$. We then compute the polynomial p_n of degree n that interpolates $(x_0, f(x_0)), (x_1, f(x_1)), \ldots, (x_n, f(x_n))$ and view p_n as an approximation to f. The results, for $n = 2, 4, 6, 8$, are shown in Figure 2.4, which shows the errors $y = f(x) - p_n(x)$. The errors go to zero quickly as n increases.

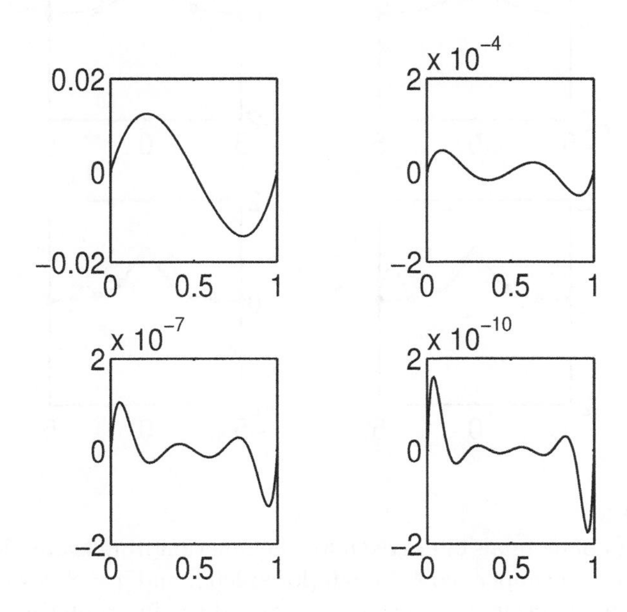

FIGURE 2.4
The errors in interpolating $f(x) = e^x$ at $n+1$ evenly spaced interpolation nodes: $n = 2$ (upper left), $n = 4$ (upper right), $n = 6$ (lower left), and $n = 8$ (lower right).

Unfortunately, interpolation at evenly spaced nodes does not always work as well as in the previous example.

Example 53 *Let $f : \mathbf{R} \to \mathbf{R}$ be defined by*

$$f(x) = \frac{1}{1+x^2},$$

and suppose we wish to approximate f on the interval $[-5,5]$. Once again, we construct evenly spaced interpolation nodes:

$$x_i = -5 + i\Delta x, \ i = 0, 1, \ldots, n, \ \Delta x = \frac{10}{n}.$$

Figure 2.5 shows f and the interpolating polynomial p_n constructed on these $n+1$ nodes. For no value of n is the interpolating polynomial close to f over the entire interval, and the interpolating polynomials exhibit increasingly large oscillations as the degree increases.

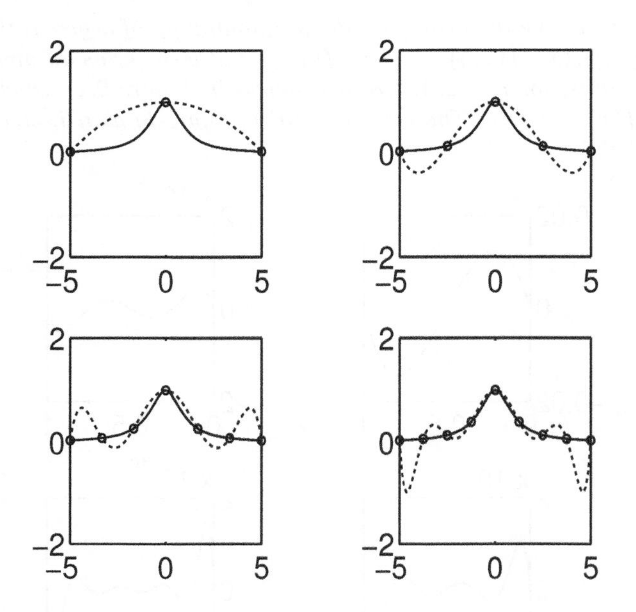

FIGURE 2.5
Interpolating polynomials of degree n for the function from Example 53: $n=2$ (upper left), $n=4$ (upper right), $n=6$ (lower left), and $n=8$ (lower right). In each graph, the function f is the solid curve and the interpolating polynomial is the dashed curve. The interpolation data is also indicated.

The preceding example, which is called *Runge's example*, is commonly used in books on numerical analysis to show one of the pitfalls of polynomial interpolation. When using many evenly-spaced interpolation nodes, the resulting interpolating polynomials can exhibit unwanted oscillations.

2.9.1 Continuous piecewise linear functions

It is possible to avoid the phenomenon illustrated by Runge's example by choosing the interpolation nodes more judiciously. Another approach to approximation of functions is to use *piecewise polynomials* in place of ordinary polynomials. A piecewise polynomial is defined relative to a *mesh*. For an interval $[a,b]$, a mesh is a collection of subintervals whose union is $[a,b]$:

$$\{[x_{i-1},x_i]\,:\,i=1,2,\ldots,n\},$$

where

$$a = x_0 < x_1 < \cdots < x_n = b.$$

For convenience, we always use a uniform mesh, in which $x_i = i\Delta x$, so that all the *elements* $[x_{i-1}, x_i]$ of the mesh have the same length $\Delta x = (b-a)/n$. Meshes and piecewise polynomials are most commonly used in the finite element method (see [14]), where the notation $h = \Delta x$ is standard. Accordingly, we will write h in place of Δx.

The simplest piecewise polynomials are piecewise linear functions. A piecewise linear function v has the property that v is linear on each element $[x_{i-1}, x_i]$. In other words, there exist scalars a_i, b_i, $i = 1, 2, \ldots, n$, such that

$$v(x) = a_i + b_i x, \ x_{i-1} < x < x_i, \ i = 1, 2, \ldots, n.$$

We will consider only continuous piecewise linear functions (a typical example is shown in Figure 2.6), although discontinuous piecewise polynomials also find application in numerical methods (see, for example, Johnson [22] or Rivière [36]). We will write $\mathcal{P}_h^{(1)}$ for the space of all continuous piecewise linear functions relative to the mesh defined by $h = (b-a)/n$. It should be easy to verify that $\mathcal{P}_h^{(1)}$ is a vector space over $\mathbf{R}$; a linear combination of two functions in $\mathcal{P}_h^{(1)}$ is still linear on each element and therefore belongs to $\mathcal{P}_h^{(1)}$. We now wish to determine a convenient basis for $\mathcal{P}_h^{(1)}$.

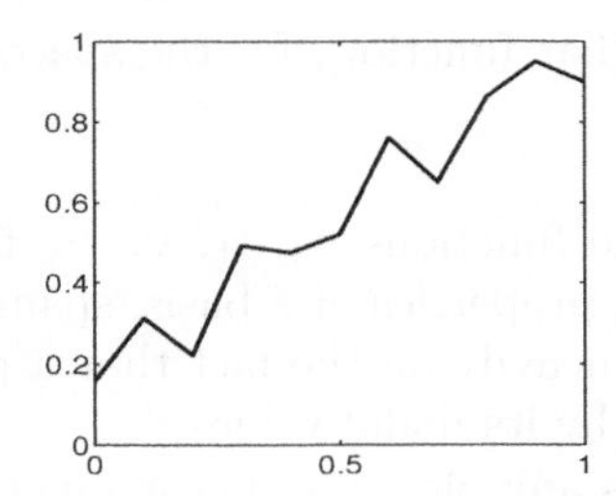

FIGURE 2.6
A piecewise linear function defined on a mesh with ten elements.

Since $v \in \mathcal{P}_h^{(1)}$ must be continuous, the $2n$ scalars a_i, b_i, $i = 1, 2, \ldots, n$, are not independent. Instead, continuity requires

$$a_{i-1} + b_{i-1}x_{i-1} = a_i + b_i x_{i-1}, \ i = 2, 3, \ldots, n.$$

Since these equations define $n-1$ constraints on the $2n$ parameters, this suggests that there are $2n - (n-1) = n+1$ degrees of freedom in defining $v \in \mathcal{P}_h^{(1)}$, and hence that the dimension of $\mathcal{P}_h^{(1)}$ is $n+1$. In fact, this is true, since $v \in \mathcal{P}_h^{(1)}$ is entirely determined by its *nodal values*, that is, by $v(x_0), v(x_1), \ldots, v(x_n)$. Given these values, all other values $v(x)$ are determined by linear interpolation.

We define a basis $\{\phi_0, \phi_1, \ldots, \phi_n\}$ for $\mathcal{P}_h^{(1)}$ by the conditions

$$\phi_j(x_i) = \begin{cases} 1, & i = j, \\ 0, & i \neq j. \end{cases} \tag{2.15}$$

As explained above, specifying the nodal values of ϕ_j uniquely determines it as an element of $\mathcal{P}_h^{(1)}$, so $\phi_0, \phi_1, \ldots, \phi_n \in \mathcal{P}_h^{(1)}$ are well defined. The graph of a typical ϕ_j is shown in Figure 2.7. Because of the characteristic shape of the graph, $\phi_0, \phi_1, \ldots, \phi_n$ are sometimes called *hat functions.*

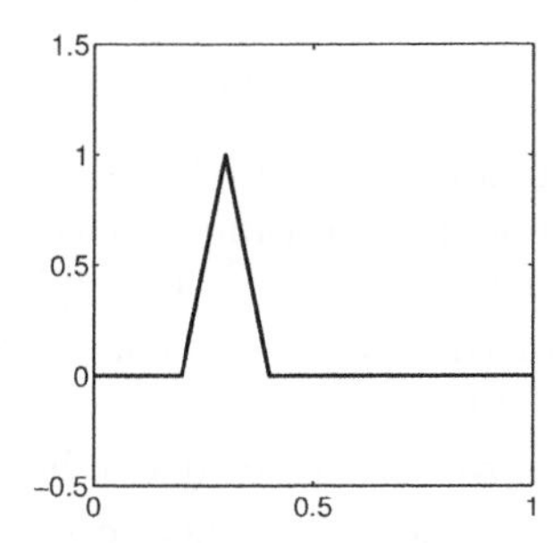

FIGURE 2.7
A typical basis function (hat function) for the space of continuous piecewise linear functions.

We can prove that the functions $\phi_0, \phi_1, \ldots, \phi_n$ form a basis for $\mathcal{P}_h^{(1)}$ by verifying the two defining properties of a basis, spanning and linear independence. These proofs rely heavily on the fact that a piecewise linear function is determined completely by its nodal values.

First suppose that v is any piecewise linear function defined on the given mesh. We claim that

$$v = \sum_{i=0}^{n} v(x_i)\phi_i.$$

To see this, notice that, for any $j = 0, 1, \ldots, n$,

$$\begin{aligned} \sum_{i=0}^{n} v(x_i)\phi_i(x_j) &= v(x_0)\phi_0(x_j) + \ldots + v(x_j)\phi_j(x_j) + \ldots + v(x_n)\phi_n(x_j) \\ &= v(x_0) \cdot 0 + \ldots + v(x_j) \cdot 1 + \ldots + v(x_n) \cdot 0 \\ &= v(x_j). \end{aligned}$$

The two piecewise linear functions, v and $\sum_{i=0}^{n} v(x_i)\phi_i$, have the same nodal values and are therefore equal. Thus each element of V_n can be written as a linear combination of $\phi_0, \phi_1, \ldots, \phi_n$.

Now suppose there exists scalars $\alpha_0, \alpha_1, \ldots, \alpha_n$ such that

$$\sum_{i=0}^{n} \alpha_i \phi_i = 0,$$

that is, that $\sum_{i=0}^{n} \alpha_i \phi_i$ is the zero function. Evaluating this function at x_j gives

$$\sum_{i=0}^{n} \alpha_i \phi_i(x_j) = 0.$$

But, using (2.15) as in the previous paragraph, the expression on the left simplifies to α_j, and so $\alpha_j = 0$. Since this holds for all $j = 0, 1, \ldots, n$, we see that $\{\phi_0, \phi_1, \ldots, \phi_n\}$ is linearly independent and hence a basis for $\mathcal{P}_h^{(1)}$. Because of the defining property (2.15), $\{\phi_0, \phi_1, \ldots, \phi_n\}$ is referred to as a *nodal basis* (or sometimes a *Lagrange basis*).

The reader should appreciate how simple it is to work with the nodal basis $\{\phi_0, \phi_1, \ldots, \phi_n\}$. Normally it is necessary to solve a system of equations to express a given vector in terms of a basis. In this case, however, there is essentially no work involved in determining the necessary weights representing $v \in \mathcal{P}_h^{(1)}$ in terms of the $\{\phi_0, \phi_1, \ldots, \phi_n\}$. Similarly, if the goal is to approximate $u \notin \mathcal{P}_h^{(1)}$ by $v \in \mathcal{P}_h^{(1)}$, a reasonable approximation can be defined immediately:

$$v = \sum_{i=0}^{n} u(x_i)\phi_i$$

is the piecewise linear interpolant of u. Figure 2.8 shows several piecewise linear interpolants of Runge's function.

2.9.2 Continuous piecewise quadratic functions

It is possible to construct continuous piecewise polynomials of any degree. The basic principle underlying this construction is the fact that $k+1$ nodal values uniquely determine a polynomial of degree k. Therefore, if a function v is to reduce to a polynomial of degree k on each element in a mesh, and if these polynomial pieces are to be determined by nodal values, then there must be $k+1$ nodes on each element. To show how this works, we will describe the construction of continuous piecewise quadratic functions relative to a given mesh, and leave the extension to higher degree piecewise polynomials to the reader.

Given the mesh described above, with elements $[x_{i-1}, x_i]$, $i = 1, 2, \ldots, n$, we place nodes $m_1, m_2, \ldots, m_n$ at the midpoints of the n elements. For simplicity of notation, we then rename the nodes

$$x_0, m_1, x_1, m_2, x_2, \ldots, x_{n-1}, m_n, x_n$$

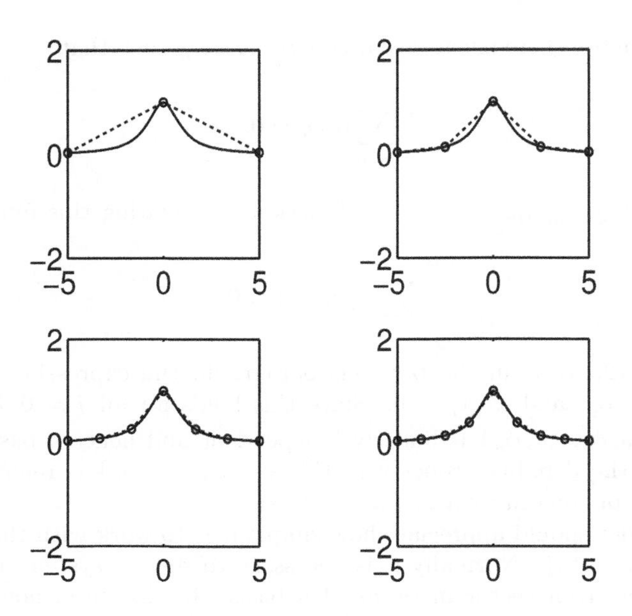

FIGURE 2.8
Piecewise linear interpolants (defined on a uniform mesh of n elements) for the function $f(x) = 1/(1+x^2)$: $n = 2$ (upper left), $n = 4$ (upper right), $n = 6$ (lower left), and $n = 8$ (lower right). In each graph, the function f is the solid curve and the interpolant is the dashed curve.

as

$$x_0, x_1, \ldots, x_{2n},$$

so that the ith element is now denoted $[x_{2i-2}, x_{2i}]$, $i = 1, 2, \ldots, n$. If the mesh is uniform, with element length h, then $x_i = ih/2$, $i = 0, 1, \ldots, 2n$.

We will write $\mathcal{P}_h^{(2)}$ for the space of continuous piecewise quadratic functions, relative to the given mesh. Then $v \in \mathcal{P}_h^{(2)}$ is determined by its $2n + 1$ nodal values $v(x_0), v(x_1), \ldots, v(x_{2n})$. A nodal basis $\{\phi_0, \phi_1, \ldots, \phi_{2n}\}$ can be defined exactly as in the case of piecewise linear functions:

$$\phi_j(x_i) = \begin{cases} 1, & i = j, \\ 0, & i \neq j. \end{cases} \tag{2.16}$$

As in the case of the hat functions, each of these nodal basis functions corresponds to one of the nodes in the mesh. However, since there are two kinds of nodes—element midpoints and element endpoints—there are two types of basis functions in the piecewise quadratic case. If ϕ_i corresponds to an element midpoint (which is the case if i is odd), then ϕ_i is nonzero on exactly one element and consists of a single nonzero quadratic piece. On the other hand, if ϕ_i corresponds to an element endpoint (which holds when i is even), then ϕ_i is nonzero on two elements, and consists of two nonzero quadratic pieces.

The nodal basis for $\mathcal{P}_h^{(2)}$ has many of the same properties as the nodal basis for $\mathcal{P}_h^{(1)}$. For example, if $v \in \mathcal{P}_h^{(2)}$, then

$$v = \sum_{i=0}^{2n} v(x_i)\phi_i.$$

If $u \notin \mathcal{P}_h^{(2)}$, then the piecewise quadratic interpolant of u is

$$v = \sum_{i=0}^{2n} u(x_i)\phi_i.$$

Piecewise quadratic interpolants of Runge's function are shown in Figure 2.9. The errors in the piecewise linear and piecewise quadratic interpolants of Runge's function, on the same mesh of eight elements, are shown in Figure 2.10.

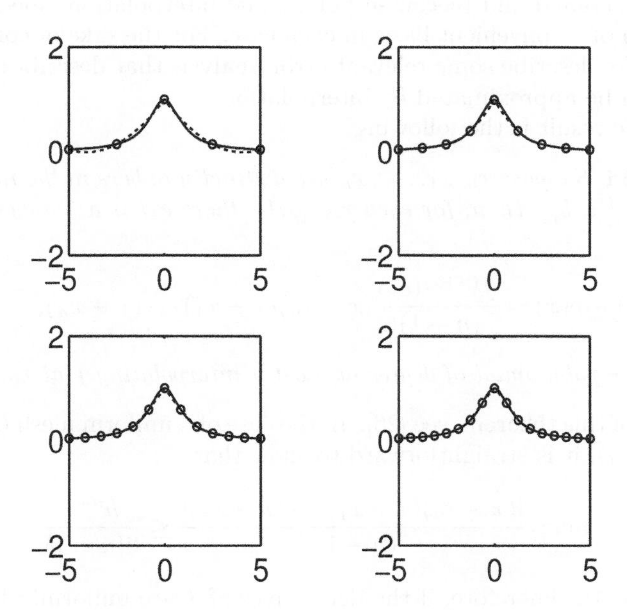

FIGURE 2.9
Piecewise quadratic interpolants (defined on a uniform mesh of n elements) for the function $f(x) = 1/(1 + x^2)$: $n = 2$ (upper left), $n = 4$ (upper right), $n = 6$ (lower left), and $n = 8$ (lower right). In each graph, the function f is the solid curve and the interpolant is the dashed curve.

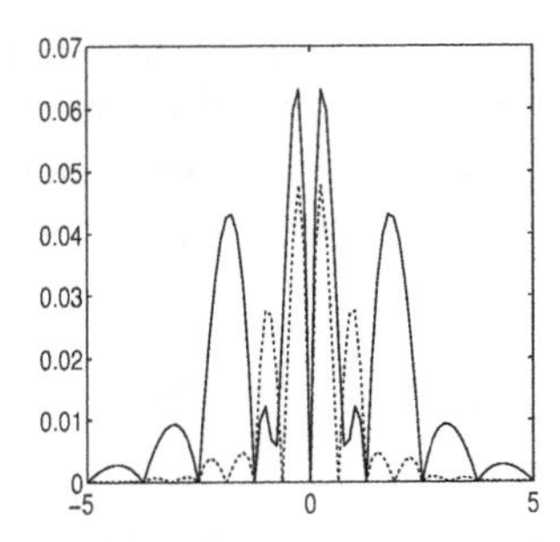

FIGURE 2.10
The error in the piecewise linear (solid curve) and piecewise quadratic (dashed curves) interpolants of $f(x) = 1/(1+x^2)$ on a uniform mesh of eight elements.

2.9.3 Error in polynomial interpolation

The focus of this section and the previous one is on the linear algebraic aspects of polynomial and piecewise polynomial interpolation, specifically, the construction of a convenient basis in each case. For the sake of completeness, we will briefly describe some relevant error analysis that describes how well a function can be approximated by interpolation.

The basic result is the following:

Theorem 54 *Suppose $x_0, x_1, \ldots, x_n$ are distinct numbers in the interval $[a,b]$ and $f \in C^{n+1}[a,b]$. Then, for each $x \in [a,b]$, there exists a number $c_x \in (a,b)$ such that*

$$f(x) - p(x) = \frac{f^{(n+1)}(c_x)}{(n+1)!}(x-x_0)(x-x_1)\cdots(x-x_n), \tag{2.17}$$

where p is the polynomial of degree at most n interpolating f at $x_0, x_1, \ldots, x_n$.

For a proof of this theorem, see [26]. In the case of a uniform mesh ($x_i = a+ih$, $h = (b-a)/n$), it is straightforward to show that

$$\max_{x\in[a,b]} \frac{|(x-x_0)(x-x_1)\cdots(x-x_n)|}{(n+1)!} \le \frac{h^{n+1}}{2(n+1)} \tag{2.18}$$

(see Exercise 4). Therefore, if the derivatives of f are uniformly bounded on the interval $[a,b]$, that is, if

$$\max_{x\in[a,b]} \left| f^{(n+1)}(x) \right| \le M \text{ for all } n = 1, 2, \ldots,$$

then the polynomial interpolants converge to f as the number of (uniformly-spaced) interpolation nodes goes to infinity. This certainly holds for many functions, such as $f(x) = e^x$ in Example 52. Convergence is also obtained if the derivatives of f do not grow too fast as n increases. However, there are many functions (such as Runge's function) for which the growth in the derivatives is fast enough to preclude convergence.

If one wishes to approximate smooth functions by interpolating polynomials, uniformly-spaced interpolation nodes are not the best choice. To obtain a choice of interpolation nodes that is optimal in one sense, one can choose $x_0, x_1, \ldots, x_n$ to minimize

$$\max_{x\in[a,b]} |(x-x_0)(x-x_1)\ldots(x-x_n)|.$$

The result leads to consideration of the Chebyshev polynomials, which form a fascinating study in themselves, albeit one that is largely beyond the scope of this book (but see Section 6.8). (For an introduction, the reader can consult [26]. A detailed study is given in [37].) The main result is that, with a proper choice of the interpolation nodes, one can obtain

$$\max_{x\in[a,b]} |(x-x_0)(x-x_1)\cdots(x-x_n)| \leq \frac{1}{2^n}$$

and hence

$$\max_{x\in[a,b]} |f(x)-p(x)| \leq \frac{1}{2^n(n+1)!} \max_{x\in[a,b]} \left| f^{(n+1)}(x) \right|,$$

where p is the interpolating polynomial defined on the given nodes. Nevertheless, there still exist smooth functions f whose derivatives grow so fast that the polynomial interpolants fail to converge to f as $n \to \infty$.

When approximating functions by interpolating polynomials, one hopes to obtain convergence by increasing the degree of the polynomial, that is, by increasing the number of interpolation nodes. As explained above, this does not necessarily work, at least when using a predetermined pattern of nodes.[7] However, the increasing number of nodes could be used to define piecewise polynomial interpolants, and it is intuitively obvious that such an approach will yield convergence to any smooth function. Indeed, one of the main results of differential calculus is that every smooth function looks like a low degree polynomial function when restricted to a small enough interval.

In terms of the above analysis, the point about using low-degree piecewise polynomial interpolation rather than polynomial interpolation of increasing degree is that one avoids increasing the order of the derivative that appears in the error bound. For instance, if f is approximated by its piecewise linear interpolant p, then, on each element $[x_{i-1}, x_i]$,

$$|f(x)-p(x)| \leq \frac{\max_{x\in[x_{i-1},x_i]} |f''(x)|}{2} \max_{x\in[x_{i-1},x_i]} |(x-x_{i-1})(x-x_i)|.$$

[7]Give a specific function f, it is possible to choose a sequence of sets of interpolation nodes *for that* f such that the corresponding interpolating polynomials converge. It is not possible, though, to choose the interpolation nodes so that convergence is obtained for every possible f. See Section 6.1 of [26] for a discussion.

If we consider a uniform mesh with $h = x_i - x_{i-1}$ for all i, then it is easy to show that

$$\max_{x\in[x_{i-1},x_i]} |(x - x_{i-1})(x - x_i)| \le \frac{h^2}{4}.$$

Writing

$$M = \max_{x\in[a,b]} |f''(x)|,$$

we obtain

$$|f(x) - p(x)| \le \frac{M}{8}h^2, \tag{2.19}$$

and we have convergence as $h \to 0$. Moreover, f need only belong to $C^2[a, b]$ for this result to hold.

Exercises

1. (This exercise requires the use of computer software.) Let $f(x) = e^x$. Using a uniform mesh on $[0, 1]$, approximate f by both polynomial interpolation and piecewise linear interpolation for a sequence of values of n. Make a table showing the values of n and the maximum errors in each approximation. Do you see the error in piecewise linear interpolation decrease as predicted by (2.19)? (Take n from 1 to 10 or higher, as time allows. It is recommended that you do not try to compute the maximum error exactly, but just estimate it by evaluating the error on a finer mesh.)

2. Repeat the previous exercise for Runge's function $f(x) = 1/(1 + x^2)$ on $[-5, 5]$.

3. Repeat the previous two exercises, using piecewise quadratic interpolation instead of piecewise linear interpolation.

4. Let $x_0, x_1, \ldots, x_n$ define a uniform mesh on $[a, b]$ (that is, $x_i = a + ih$, $i = 0, 1, \ldots, n$, where $h = (b - a)/n$). Prove (2.18). (Hint: We have

$$|(x - x_0)(x - x_1)\cdots(x - x_n)| = |x - x_0||x - x_1|\cdots|x - x_n|,$$

which is the product of the distances from x to x_i for $i = 0, 1, \ldots, n$. Argue that the distance from x to the nearest node is at most $h/2$, while the distance to the next nearest node is at most h, and then $2h, 3h, \ldots, nh$.)

5. Derive the bound analogous to (2.19) for the case of piecewise quadratic interpolation on a uniform mesh.

3

Linear operators

An operator maps one vector space into another: $L : X \to U$, where X and U are vector spaces. The simplest kind of operator is one that preserves the operations of addition and scalar multiplication; such operators are called linear and are the subject of this chapter. Linear operators are fundamental in many areas of mathematics, for two reasons: Many important operators are linear, and even when a given operator is nonlinear, it is often approximated by a linear operator.

In the context of finite-dimensional vector spaces, linear operators are intrinsically linked with matrices: Every matrix defines a linear operator, and every linear operator mapping one finite-dimensional vector space to another can be represented by a matrix. For this reason, much of this chapter, and the remainder of the book, concerns matrices.

We should point out that operator is one of several synonymous terms: function, mapping, operator, transformation. In the context of vector spaces, we prefer not to use the word "function" to describe $L : X \to U$, because often the elements of X and/or U are themselves functions. The word mapping is acceptable, but the most commonly used terms are operator and transformation. It is entirely a matter of taste that we adopt "operator" instead of "transformation."

3.1 Linear operators

The following definition states precisely what it means for an operator to preserve addition and scalar multiplication of vectors.

Definition 55 *Let X and U be vector spaces over a field F, and let $L : X \to U$. We say that L is* linear *if and only if it satisfies the following conditions:*

1. *$L(\alpha x) = \alpha L(x)$ for all $\alpha \in F$ and $x \in X$;*
2. *$L(x + y) = L(x) + L(y)$ for all $x, y \in X$.*

If L is not linear, then it is called nonlinear.

We note that the symbol $L : X \to U$ is used in two ways, as a noun and as a sentence (or clause). Using it as a noun, we might write "Let $L : X \to U$

be a linear operator." On the other hand, in the above definition, "...let $L : X \to U$" should be read as "...let L map X into U."

The following theorem can be established by induction; its proof is left as an exercise.

Theorem 56 *Let X and U be vector spaces over a field F, and let $L : X \to U$ be a linear operator. If $x_1, x_2, \ldots, x_k$ are vectors in X and $\alpha_1, \alpha_2, \ldots, \alpha_k$ are scalars, then*

$$L(\alpha_1 x_1 + \alpha_2 x_2 + \ldots + \alpha_k x_k) = \alpha_1 L(x_1) + \alpha_2 L(x_2) + \ldots + \alpha_k L(x_k).$$

Proof Exercise 15.

Here is an elementary property of linear operators that is nonetheless very useful.

Theorem 57 *Let X and U be vector spaces over a field F, and let $L : X \to U$ be a linear operator. Then $L(0) = 0$. (The first "0" is the zero vector in X, while the second "0" is the zero vector in U.)*

Proof Let x be any vector in X. By linearity, we have $L(0 \cdot x) = 0 \cdot L(x)$. But, for any vector in any vector space, the zero scalar times that vector yields the zero vector (see Theorem 5). Therefore, $0 \cdot x = 0$ and $0 \cdot L(x) = 0$, so we obtain

$$L(0) = L(0 \cdot x) = 0 \cdot L(x) = 0.$$

QED

The next theorem concerns the composition of linear operators. Given three vector spaces X, U, Z and operators $L : X \to U$ and $M : U \to Z$, the composition $M \circ L$ is the operator $M \circ L : X \to Z$, $(M \circ L)(x) = M(L(x))$. This definition applies to any operators M, L, provided the domain of M equals the co-domain of L. In the case of linear operators, we usually write simply ML instead of $M \circ L$.

Theorem 58 *Let X, U, and Z be vector spaces over a field F and let*

$$L : X \to U \text{ and } M : U \to Z$$

be linear operators. Then ML is also linear.

Proof The proof is a straightforward verification of the two conditions defining a linear operator:

$$\begin{aligned}
(ML)(\alpha x) = M(L(\alpha x)) = M(\alpha L(x)) &= \alpha M(L(x)) \\
&= \alpha(ML)(x) \text{ for all } \alpha \in F, x \in X, \\
(ML)(x+y) = M(L(x+y)) &= M(L(x) + L(y)) \\
&= M(L(x)) + M(L(y)) \\
&= (ML)(x) + (ML)(y) \text{ for all } x, y \in X.
\end{aligned}$$

QED

3.1.1 Matrix operators

The most important linear operators on finite-dimensional vector spaces are defined by matrices. Indeed, as we will see in the next sections, there is a sense in which these are the only linear operators on finite-dimensional spaces.

Definition 59 *An $m \times n$* matrix *A is a collection of mn scalars from a field F, denoted A_{ij}, $i = 1, 2, \ldots, m$, $j = 1, 2, \ldots, n$. A matrix is written as a two-dimensional array:*

$$A = \begin{bmatrix} A_{11} & A_{12} & \cdots & A_{1n} \\ A_{21} & A_{22} & \cdots & A_{2n} \\ \vdots & \vdots & \ddots & \vdots \\ A_{m1} & A_{m2} & \cdots & A_{mn} \end{bmatrix}.$$

We say that A has m rows, *the vectors*

$$(A_{11}, A_{12}, \ldots, A_{1n}), (A_{21}, A_{22}, \ldots, A_{2n}), \ldots, (A_{m1}, A_{m2}, \ldots, A_{mn}),$$

and n columns, *the vectors*

$$(A_{11}, A_{21}, \ldots, A_{m1}), (A_{12}, A_{22}, \ldots, A_{m2}), \ldots, (A_{1n}, A_{2n}, \ldots, A_{mn}).$$

The rows are vectors in F^n, while the columns are vectors in F^m (see Section 2.2.1).

The columns of A will frequently be denoted as $A_1, A_2, \ldots, A_n$, and we write $A = [A_1|A_2|\cdots|A_n]$ to indicate this. We do not have a standard notation for the rows of A, but if we need to name them we can write, for example,

$$A = \begin{bmatrix} r_1 \\ \hline r_2 \\ \hline \vdots \\ \hline r_m \end{bmatrix}.$$

The set of all $m \times n$ matrices of scalars from F is denoted $F^{m\times n}$. When $m = n$, we call matrices in $F^{m\times n}$ square.

The most important operation involving matrices is matrix-vector multiplication.

Definition 60 *Let $A \in F^{m\times n}$ and let $x \in F^n$. The* matrix-vector product *Ax is the vector in F^m defined by*

$$Ax = \sum_{j=1}^{n} x_j A_j,$$

where $x_1, x_2, \ldots, x_n$ are the components of x and $A_1, A_2, \ldots, A_n$ are the columns of A. Thus Ax is a linear combination of the columns of A, where the components of x are the weights in the linear combination.

Since addition and scalar multiplication of vectors in F^m are both defined componentwise, we have

$$(Ax)_i = \left(\sum_{j=1}^{n} x_j A_j\right)_i = \sum_{j=1}^{n} A_{ij} x_j, \; i = 1, 2, \ldots, m, \tag{3.1}$$

which gives an equivalent definition of matrix-vector multiplication. (The reader should notice that the components of A_j are $A_{1j}, A_{2j}, \ldots, A_{mj}$; this fact was used in the above calculation.) The column-wise definition is usually more useful, though.

Theorem 61 *Let F be a field and define $L : F^n \to F^m$ by $L(x) = Ax$, where $A \in F^{m \times n}$ is a given matrix. Then L is a linear operator.*

Proof For any $x \in F^n$ and $\alpha \in F$, we have $\alpha x = (\alpha x_1, \alpha x_2, \ldots, \alpha x_n)$. Therefore,

$$A(\alpha x) = \sum_{j=1}^{n} (\alpha x_j) A_j = \sum_{j=1}^{n} \alpha (x_j A_j) = \alpha \sum_{j=1}^{n} x_j A_j = \alpha (Ax),$$

which shows that $L(\alpha x) = \alpha L(x)$. Thus the first property of a linear operator is satisfied.

If $x, y \in F^n$, then

$$\begin{aligned} L(x+y) = A(x+y) = \sum_{j=1}^{n} (x_j + y_j) A_j &= \sum_{j=1}^{n} (x_j A_j + y_j A_j) \\ &= \sum_{j=1}^{n} x_j A_j + \sum_{j=1}^{n} y_j A_j \\ &= Ax + Ay \\ &= L(x) + L(y). \end{aligned}$$

This verifies that the second property of a linear operator is satisfied, and hence that L is linear.

QED

One reason that matrix operators are so important is that any system of linear algebraic equations,

$$\begin{aligned} \alpha_{11} x_1 + \alpha_{12} x_2 + \ldots + \alpha_{1n} x_n &= b_1, \\ \alpha_{21} x_1 + \alpha_{22} x_2 + \ldots + \alpha_{2n} x_n &= b_2, \\ \vdots \qquad\qquad & \quad \vdots \\ \alpha_{m1} x_1 + \alpha_{m2} x_2 + \ldots + \alpha_{mn} x_n &= b_m, \end{aligned}$$

can be written as $Ax = b$, where $A_{ij} = \alpha_{ij}$. Since A defines a linear operator, this puts the system into the form of a *linear operator equation*. In Section 3.4, we develop the general theory of linear operator equations.

Given two matrices $A \in F^{m\times n}$ and $B \in F^{p\times m}$, we have linear operators $L : F^n \to F^m$ and $M : F^m \to F^p$ defined by $L(x) = Ax$ and $M(u) = Bu$, respectively. There is also the composite operator, $ML : F^n \to F^p$. We might expect that there would be a matrix defining ML, just as A defines L and B defines M. We can find this matrix by computing $(ML)(x)$ for an arbitrary x and recognizing the result as the linear combination of certain vectors, where the weights in the linear combination are the components of x. Those vectors will then form the columns of the desired matrix. Here is the calculation:

$$\begin{aligned}(ML)(x) = M(L(x)) = B(Ax) &= B\left(\sum_{j=1}^{n} x_j A_j\right) \\ &= \sum_{j=1}^{n} x_j (BA_j).\end{aligned}$$

In the last step, we used the linearity of matrix-vector multiplication. We see that $(ML)(x)$ is the linear combination of the vectors $BA_1, BA_2, \ldots, BA_n$, where $A_1, A_2, \ldots, A_n$ are the columns of A. Therefore, $(ML)(x) = Cx$, where $C = [BA_1|BA_2|\cdots|BA_n]$. Since B is $p \times m$ and each A_j is an m-vector, it follows that each BA_j, $j = 1, 2, \ldots, n$, is a p-vector. Therefore, C is $p \times n$. This tells us how we ought to define the product of two matrices.

Definition 62 *Let F be a field and suppose $A \in F^{m\times n}$, $B \in F^{p\times q}$. If $q = m$, then the* matrix-matrix product *$BA \in F^{p\times n}$ is defined by*

$$BA = [BA_1|BA_2|\cdots|BA_n].$$

If $q \neq m$, then BA is undefined.

A little thought shows that matrix multiplication cannot be commutative; in most cases when BA is defined, AB is undefined. It is possible that both BA and AB are defined but of different sizes. Even when BA and AB are defined and of the same size, usually $BA \neq AB$. The following examples illustrate these remarks.

Example 63 *1. Let $A \in \mathbf{R}^{2\times 2}$ and $B \in \mathbf{R}^{3\times 2}$ be defined by*

$$A = \begin{bmatrix} 1 & 3 \\ -2 & 2 \end{bmatrix}, \quad B = \begin{bmatrix} 0 & 2 \\ -2 & -3 \\ 1 & 1 \end{bmatrix}.$$

Then BA is defined:

$$\begin{aligned} BA &= \begin{bmatrix} 0 & 2 \\ -2 & -3 \\ 1 & 1 \end{bmatrix} \begin{bmatrix} 1 & 3 \\ -2 & 2 \end{bmatrix} \\ &= \left[1 \cdot \begin{bmatrix} 0 \\ -2 \\ 1 \end{bmatrix} - 2 \cdot \begin{bmatrix} 2 \\ -3 \\ 1 \end{bmatrix} \middle| 3 \cdot \begin{bmatrix} 0 \\ -2 \\ 1 \end{bmatrix} + 2 \cdot \begin{bmatrix} 2 \\ -3 \\ 1 \end{bmatrix} \right] \\ &= \begin{bmatrix} -4 & 4 \\ 4 & -12 \\ -1 & 5 \end{bmatrix}. \end{aligned}$$

On the other hand, AB is not defined because the columns of B are 3-vectors, which cannot be multiplied by the 2×2 matrix A.

2. *If $A \in \mathbf{R}^{2\times 3}$ is defined by*

$$A = \begin{bmatrix} -1 & -2 & -2 \\ 2 & 1 & -1 \end{bmatrix}$$

and B is the 3×2 matrix defined above, then both AB and BA are defined:

$$\begin{aligned} AB &= \begin{bmatrix} 2 & 2 \\ -3 & 0 \end{bmatrix}, \\ BA &= \begin{bmatrix} 4 & 2 & -2 \\ -4 & 1 & 7 \\ 1 & -1 & -3 \end{bmatrix} \end{aligned}$$

(the reader can check these results). However, AB and BA are obviously unequal since they have different sizes.

3. *When A and B are square and of the same size, then AB and BA are both defined and of the same size. However, AB and BA need not be equal, as the following example shows:*

$$\begin{aligned} A &= \begin{bmatrix} 1 & 1 \\ -1 & -2 \end{bmatrix}, \\ B &= \begin{bmatrix} -1 & 3 \\ 3 & 1 \end{bmatrix}, \\ AB &= \begin{bmatrix} 2 & 4 \\ -5 & -5 \end{bmatrix}, \\ BA &= \begin{bmatrix} -4 & -7 \\ 2 & 1 \end{bmatrix}. \end{aligned}$$

In fact, when $A, B \in F^{n\times n}$, the typical case is that $AB \neq BA$, although there are special cases in which the two products are equal. This point will be discussed further later in the book.

Exercises

Miscellaneous exercises

1. In elementary algebra, we call a function $f : \mathbf{R} \to \mathbf{R}$ linear if it is of the form $f(x) = mx + b$. Is such a function linear according to Definition 55? Discuss completely.

2. Prove that $f : \mathbf{R} \to \mathbf{R}$ defined by $f(x) = x^2$ is not linear. (Note: Here $\mathbf{R}$ should be regarded as a one-dimensional vector space.)

3. Let $L : \mathbf{R}^3 \to \mathbf{R}^3$ be defined by the following conditions:

   (a) L is linear;

   (b) $L(e_1) = (1, 2, 1)$;

   (c) $L(e_2) = (2, 0, -1)$;

   (d) $L(e_3) = (0, -2, -3)$.

   Here $\{e_1, e_2, e_3\}$ is the standard basis for $\mathbf{R}^3$. Prove that there is a matrix $A \in \mathbf{R}^{3\times3}$ such that $L(x) = Ax$ for all $x \in \mathbf{R}^3$. What is A? (Hint: Any $x \in \mathbf{R}^3$ can be written as $x = x_1e_1 + x_2e_2 + x_3e_3$. Since L is linear, it follows that $L(x) = x_1L(e_1) + x_2L(e_2) + x_3L(e_3)$. On the other hand, if $A \in \mathbf{R}^{3\times3}$, what is Ax?)

4. Consider the operator $M : \mathcal{P}_n \to \mathcal{P}_{n+1}$ defined by $M(p) = q$, where $q(x) = xp(x)$. (That is, given any polynomial $p \in \mathcal{P}_n$, $M(p)$ is the polynomial obtained by multiplying $p(x)$ by x.) Prove that M is linear.

5. Which of the following real-valued functions defined on $\mathbf{R}^n$ is linear?

   (a) $f : \mathbf{R}^n \to \mathbf{R}$, $f(x) = \sum_{i=1}^n x_i$.

   (b) $g : \mathbf{R}^n \to \mathbf{R}$, $g(x) = \sum_{i=1}^n |x_i|$.

   (c) $h : \mathbf{R}^n \to \mathbf{R}$, $h(x) = \prod_{i=1}^n x_i$. ($\prod_{i=1}^n x_i$ denotes the product of $x_1, x_2, \ldots, x_n$.)

6. Let $\mathcal{A} : C[a, b] \to C^1[a, b]$ be defined by the condition $\mathcal{A}(f) = F$, where $F' = f$ (that is, $\mathcal{A}(f)$ is an antiderivative of f). Is $\mathcal{A}$ linear or nonlinear? Prove your answer.

7. Let $Q : \mathcal{P}_n \to \mathcal{P}_{n+1}$ be defined by

   $$Q\left(c_0 + c_1x + \ldots + c_nx^n\right) = c_0x + \frac{c_1}{2}x^2 + \ldots + \frac{c_n}{n+1}x^{n+1}.$$

   Is Q linear or nonlinear? Prove your answer.

8. Let $L : \mathcal{P}_n \to \mathcal{P}_{2n-1}$ be defined by $L(p) = pp'$. Is L linear or nonlinear? Prove your answer.

9. Compute the matrix-vector product Ax in each of the following cases:

   (a) $A \in \mathbf{C}^{2\times 3}$, $x \in \mathbf{C}^3$ are defined by

$$A = \begin{bmatrix} 1+i & 1-i & 2i \\ 2-i & 1+2i & 3 \end{bmatrix}, \quad x = \begin{bmatrix} 3 \\ 2+i \\ 1-3i \end{bmatrix}.$$

   (b) $A \in \mathbf{Z}_2^{3\times 3}$, $x \in \mathbf{Z}_2^3$ are defined by

$$A = \begin{bmatrix} 1 & 1 & 0 \\ 1 & 0 & 1 \\ 0 & 1 & 1 \end{bmatrix}, \quad x = \begin{bmatrix} 1 \\ 1 \\ 1 \end{bmatrix}.$$

10. For each matrix A and vector b given below, there exists a unique vector x such that $Ax = b$. Find it.

    (a) $A \in \mathbf{R}^{3\times 3}$, $b \in \mathbf{R}^3$,

$$A = \begin{bmatrix} 1 & -2 & -4 \\ 5 & -11 & -15 \\ -2 & 6 & -1 \end{bmatrix}, \quad b = \begin{bmatrix} 4 \\ 23 \\ -14 \end{bmatrix}.$$

    (b) $A \in \mathbf{Z}_3^{3\times 3}$, $b \in \mathbf{Z}_3^3$,

$$A = \begin{bmatrix} 2 & 1 & 2 \\ 1 & 0 & 1 \\ 1 & 2 & 0 \end{bmatrix}, \quad b = \begin{bmatrix} 0 \\ 1 \\ 2 \end{bmatrix}.$$

11. Let $k : [a,b] \times [c,d] \to \mathbf{R}$ be continuous (that is $k(s,t)$ is a continuous function of $s \in [a,b]$ and $t \in [c,d]$), and define $K : C[c,d] \to C[a,b]$ by

$$K(x) = y, \; y(s) = \int_c^d k(s,t)x(t)\,dt, \; s \in [a,b].$$

    Prove that K is linear. An operator of the form K is called an *integral operator*, and the function k is called the *kernel* of the operator.[1]

12. Using (3.1), give a formula for $(AB)_{ij}$, assuming $A \in F^{m\times n}$, $B \in F^{n\times p}$.

13. Using the previous exercise, give a formula for the ith row of AB, assuming $A \in F^{m\times n}$, $B \in F^{n\times p}$.

14. Is it possible for a linear operator to be constant? That is, do there exist any operators of the form $L : X \to U$, where X and U are vector spaces and L is linear, such that $L(x) = v$ for a constant vector v and all $x \in X$?

15. Prove Theorem 56 by induction on k.

[1]The word "kernel" is used in a very different sense in most of this book; see Section 3.4 for the standard meaning of the word kernel in linear algebra. However, the use of the word kernel in the context of integral operators is also well established.

3.2 More properties of linear operators

3.2.1 Vector spaces of operators

In the previous chapter, we described how a collection of functions with the same domain and co-domain and similar properties can form a vector space. Given two vectors spaces X and U over a common field F, linear operators mapping X into U are functions (although, as mentioned on page 93, we use the word operator instead of function). To show that the set of all linear operators from X into U is a vector space, we just need the fact that addition and scalar multiplication preserve linearity.

Theorem 64 *Let X and U be vector spaces over a field F.*

1. *If $L : X \to Y$ is linear and $\alpha \in F$, then αL is also linear.*
2. *If $L : X \to Y$ and $M : X \to U$ are linear, then so is $L + M$.*

Proof Exercise 7.

Corollary 65 *Let X and U be vector spaces over a field F. Then the set of all linear operators mapping X into U is a vector space over F.*

The space of linear operators mapping X into U will be denoted by $\mathcal{L}(X, U)$.

In this book, we do not use the vector space structure of $\mathcal{L}(X, U)$ directly (except in a few exercises). However, we will sometimes refer to the fact that a linear combination of linear operators is linear. The reader will recall that a composition of linear operators is also linear, and thus we now have two ways of combining linear operators to form more linear operators.

3.2.2 The matrix of a linear operator on Euclidean spaces

We have already seen that $A \in F^{m\times n}$ defines a linear operator $T : F^n \to F^m$ by $T(x) = Ax$. We recall that, for any $x \in F^n$,

$$Ax = x_1A_1 + x_2A_2 + \ldots + x_nA_n,$$

where $A_1, A_2, \ldots, A_n$ are the columns of A. In particular, if $\{e_1, e_2, \ldots, e_n\}$ is the standard basis for F^n, then

$$Ae_j = A_j, \; j = 1, 2, \ldots, n.$$

It follows that $T(e_j) = A_j$, that is, T maps the standard basis vectors of F^n to the columns of A. From this fact we obtain the following theorem.

Theorem 66 *Let F be a field and let $T : F^n \to F^m$ be linear. Then there exists a unique matrix $A \in F^{m\times n}$ such that $T(x) = Ax$ for all $x \in F^n$, namely,*

$$A = [T(e_1)|T(e_2)| \cdots |T(e_n)], \tag{3.2}$$

where $\{e_1, e_2, \ldots, e_n\}$ is the standard basis for F^n.

Proof Define the matrix A by (3.2). Then, for any $x \in F^n$, we have

$$Ax = x_1A_1 + x_2A_2 + \ldots + x_nA_n = x_1T(e_1) + x_2T(e_2) + \ldots + x_nT(e_n).$$

On the other hand, $x = x_1e_1 + x_2e_2 + \ldots + x_ne_n$, and hence, by linearity,

$$T(x) = T(x_1e_1 + x_2e_2 + \ldots + x_ne_n) = x_1T(e_1) + x_2T(e_2) + \ldots + x_nT(e_n).$$

This shows that $T(x) = Ax$ for all $x \in F^n$.

To prove uniqueness of the matrix A, suppose there is another $B \in F^{m\times n}$ such that $T(x) = Bx$ for all $x \in F^n$. Then, for any $j = 1, 2, \ldots, n$, $T(e_j) = B_j$. This shows that A and B have the same columns, and hence A and B are the same matrix.

QED

Example 67 *A rotation of the plane defines a linear operator $R : \mathbf{R}^2 \to \mathbf{R}^2$ (see Exercise 2). Let R be the rotation of $45°$ in the positive (counterclockwise) direction. According to the above theorem, there is a matrix $A \in \mathbf{R}^{2\times 2}$ such that $R(x) = Ax$ for all $x \in \mathbf{R}^2$. Moreover, to determine A, it suffices to determine the effect of R on the two standard basis vectors $e_1 = (1, 0)$ and $e_2 = (0, 1)$.*

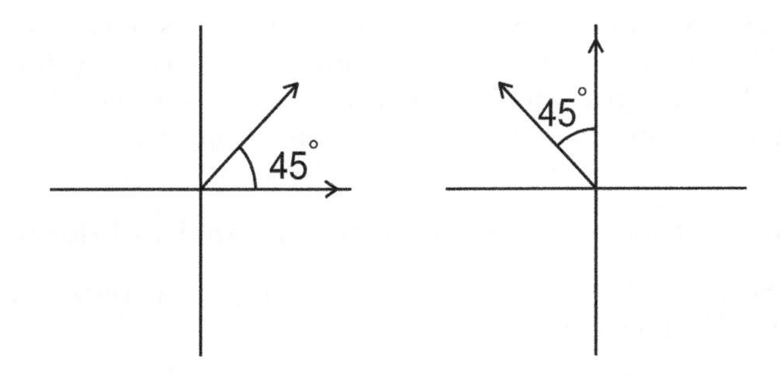

FIGURE 3.1
The standard basis vectors in $\mathbf{R}^2$ and their images under a $45°$ rotation.

This is easily done by elementary geometry; the results are

$$R(e_1) = (\sqrt{2}/2, \sqrt{2}/2),\ R(e_2) = (-\sqrt{2}/2, \sqrt{2}/2)$$

(see Figure 3.1). It follows that the matrix representing R is

$$A = [R(e_1)|R(e_2)] = \begin{bmatrix} \frac{\sqrt{2}}{2} & -\frac{\sqrt{2}}{2} \\ \frac{\sqrt{2}}{2} & \frac{\sqrt{2}}{2} \end{bmatrix}.$$

3.2.3 Derivative and differential operators

We have already discussed the vector space $C^1(a,b)$ of continuously differentiable functions defined on the interval (a,b) (see page 35), and similarly the space $C(a,b)$ of continuous functions. The derivative operator can be regarded as an operator mapping $C^1(a,b)$ into $C(a,b)$:

$$D: C^1(a,b) \to C(a,b),$$
$$D(f) = f'.$$

It is a theorem of calculus that D is linear (although this term is often not used in calculus textbooks). For example,

$$D\left(3\sin(t) + 2t^2\right) = 3D(\sin(t)) + 2D\left(t^2\right) = 3\cos(t) + 2 \cdot 2t = 3\cos(t) + 4t.$$

The second derivative operator is simply D composed with itself. Since we use product notation to denote composition of linear operators ($ML = M \circ L$), it is natural to denote the second derivative operator as D^2:

$$D^2(f) = f''.$$

By Theorem 58, D^2 is linear, as is any higher derivative D^k. Also, by Theorem 64, any linear combination of derivative operators is a linear operator.

Many important applications (in electrical circuits and mechanical systems, for example) are modeled by *linear ordinary differential equations with constant coefficients*. An example would be

$$2x'' + x' + 200x = 0.$$

Using operator notation, this is written as

$$2D^2(x) + D(x) + 200I(x) = 0,$$

or

$$(2D^2 + D + 200I)(x) = 0,$$

where I represents the identity operator: $I(x) = x$. The identity operator is linear, so $2D^2 + D + 200I$ is a linear combination of linear operators and hence linear. We call an operator such as $2D^2 + D + 200I$ a linear differential operator.

Linear differential equations form an important class of linear operator equations. Although superficially linear differential equations have little in common with the algebraic equations represented by $Ax = b$, in fact all linear operator equations share many important properties.

3.2.4 Representing spanning sets and bases using matrices

As we have already seen, given a matrix $A \in F^{m \times n}$ and a vector $x \in F^n$, the matrix-vector product Ax is a linear combination of the columns of A, where the components of x are the weights in the linear combination. We can approach this from the opposite direction: If we are given a set of vectors $u_1, u_2, \ldots, u_n \in F^m$, we can form the matrix $A \in F^{m\times n}$ whose columns are the given vectors:

$$A = [u_1|u_2|\cdots|u_n].$$

Then every linear combination of $u_1, u_2, \ldots, u_n$ can be represented as a matrix-vector product. In other words,

$$\text{sp}\{u_1, u_2, \ldots, u_n\} = \{Ax \,:\, x \in F^n\}.$$

As we will see in Section 3.6, the set $\{Ax \,:\, x \in F^n\}$ is called the *column space* of A; as the span of the columns of A, it is a subspace of F^m.

If $n = m$ and $u_1, u_2, \ldots, u_n$ forms a basis for F^n, then not only can every vector in F^n be written as a matrix-vector product Ax, but the representation is unique by Theorem 28 (each vector in the space can be written uniquely as a linear combination of the basis vectors). Thus, in this case, for each $y \in F^n$, there exists a unique $x \in \mathbf{R}^n$ such that $y = Ax$.

The use of matrix-vector products to represent linear combinations is a significant notational convenience, and it allows us to bring the facts about matrices that we develop in the rest of this chapter to bear on questions about spans, spanning sets, and bases.

There is a similar relationship between matrix-vector products and linear independence of vectors in F^m. The set $\{u_1, u_2, \ldots, u_n\} \subset F^m$ is linearly independent if and only if there is a nontrivial solution $x_1, x_2, \ldots, x_n$ to

$$x_1u_1 + x_2u_2 + \cdots + x_nu_n = 0. \tag{3.3}$$

If $A = [u_1|u_2|\cdots|u_n]$, then a nontrivial solution to (3.3) is equivalent to a nonzero vector $x \in F^n$ satisfying $Ax = 0$.

Example 68 *Let* $u_1 = (3, 2, -4, 1, -2)$, $u_2 = (1, 2, -4, 0, 1)$, $u_3 = (0, 0, 3, 1, -3)$ *be vectors in* $\mathbf{R}^5$ *and define* $A \in \mathbf{R}^{5\times 3}$ *by*

$$A = [u_1|u_2|u_3] = \begin{bmatrix} 3 & 1 & 0 \\ 2 & 2 & 0 \\ -4 & -4 & 3 \\ 1 & 0 & 1 \\ -2 & 1 & -3 \end{bmatrix}.$$

If $y = (1, 1, 1, 1, 1) \in \mathbf{R}^5$, *then* $y \in \text{sp}\{u_1, u_2, u_3\}$ *if and only if the equation* $Ax = y$ *has a solution* $x \in \mathbf{R}^3$. *Also,* $\{u_1, u_2, u_3\}$ *is linearly independent if and only if the only solution to* $Ax = 0$ *is* $x = 0$.

3.2.5 The transpose of a matrix

Given a matrix $A \in F^{m\times n}$, we can define a new matrix $A^T \in F^{n\times m}$, called the *transpose* of A, by

$$\left(A^T\right)_{ij} = A_{ji},\ i = 1, 2, \ldots, n,\ j = 1, 2, \ldots, m.$$

According to this definition, the rows of A^T are the columns of A, and vice versa. For example, if

$$A = \begin{bmatrix} 1 & -1 & 3 \\ 4 & 2 & -1 \end{bmatrix},$$

then

$$A^T = \begin{bmatrix} 1 & 4 \\ -1 & 2 \\ 3 & -1 \end{bmatrix}.$$

The matrix A defines an operator $L : F^n \to F^m$ by $L(x) = Ax$ for all $x \in F^n$, and the transpose matrix defines a related operator mapping F^m back into F^n. We explore the significance of this related operator in Section 6.2.

Exercises

Essential exercises

1. Let A be an $m \times n$ matrix with real entries, and suppose $n > m$. Prove that $Ax = 0$ has a nontrivial solution $x \in \mathbf{R}^n$. (Hint: Recall that Ax is a linear combination of the columns of A. Interpret the system $Ax = 0$ as a statement about linear independence/dependence of the columns of A.)

Miscellaneous exercises

In the following exercises, if asked to find the matrix representing a linear operator, use Theorem 66 (as in Example 67).

2. Let $R : \mathbf{R}^2 \to \mathbf{R}^2$ be the rotation of angle θ about the origin (a positive θ indicates a counterclockwise rotation).
   (a) Give a geometric argument that R is linear. (That is, give geometric arguments that the defining conditions $R(\alpha x) = \alpha R(x)$ and $R(x + y) = R(x) + R(y)$ are satisfied.)
   (b) Find the matrix A such that $R(x) = Ax$ for all $x \in \mathbf{R}^2$.
3. Consider the linear operator mapping $\mathbf{R}^2$ into itself that sends each vector (x, y) to its projection onto the x-axis, namely, $(x, 0)$. Find the matrix representing this linear operator.
4. A (horizontal) *shear* acting on the plane maps a point (x, y) to the point $(x + ry, y)$, where r is a real number. Find the matrix representing this operator.
5. A linear operator $L : \mathbf{R}^n \to \mathbf{R}^n$ defined by $L(x) = rx$ is called a *dilation* if $r > 1$ and a *contraction* if $0 < r < 1$. What is the matrix of L?
6. Let $w = \alpha + i\beta$ be a fixed complex number and define $f : \mathbf{C} \to \mathbf{C}$ by $f(z) = wz$.

(a) Regarding **C** as a vector space (over the field **C**), prove that f is linear.

(b) Now regard the set **C** as identical with $\mathbf{R}^2$, writing (x, y) for $x + iy$. Represent the function f by multiplication by a 2×2 matrix.

7. Prove Theorem 64.

8. The discrete Fourier transform (DFT) is the mapping $\mathcal{F} : \mathbf{C}^N \to \mathbf{C}^N$ defined by

$$(\mathcal{F}(x))_n = \frac{1}{N} \sum_{j=0}^{N-1} x_j e^{-2\pi inj/N}, \quad n = 0, 1, \ldots, N-1,$$

where i is the complex unit ($i = \sqrt{-1}$).[2] Find the matrix $A \in \mathbf{C}^{N \times N}$ such that $\mathcal{F}(x) = Ax$ for all $x \in \mathbf{C}^N$.[3] Notice that, in the above notation, the vectors $x \in \mathbf{C}^N$ are written as $x = (x_0, x_1, \ldots, x_{N-1})$ instead of the usual $x = (x_1, x_2, \ldots, x_N)$.

9. Let $x \in \mathbf{R}^N$ be denoted as $x = (x_0, x_1, \ldots, x_{N-1})$. Given $x, y \in \mathbf{R}^N$, the *convolution* of x and y is the vector $x * y \in \mathbf{R}^N$ defined by

$$(x * y)_n = \sum_{m=0}^{N-1} x_m y_{n-m}, \quad n = 0, 1, \ldots, N-1.$$

In this formula, y is regarded as defining a periodic vector of period N; therefore, if $n - m < 0$, we take $y_{n-m} = y_{N+n-m}$. For instance, $y_{-1} = y_{N-1}$, $y_{-2} = y_{N-2}$, and so forth. Prove that if $y \in \mathbf{R}^N$ is fixed, then the mapping $x \mapsto x * y$ is linear. Find the matrix representing this operator.

10. Let $L : C^2(\mathbf{R}) \to C(\mathbf{R})$ be the differential operator $L = D^2 + \omega^2 I$, where $\omega > 0$ is a real number.

(a) Which of the functions $x(t) = \sin(\omega t)$, $x(t) = \cos(\omega t)$, $x(t) = e^{\omega t}$, $x(t) = e^{-\omega t}$ satisfy $L(x) = 0$?

(b) Does either of the functions $x(t) = \sin(\omega t) + t^2$, $x(t) = t^2 \sin(\omega t)$ satisfy $L(x) = f$, where $f(t) = \omega^2 t^2 + 2$?

11. Consider the operator $F : C^2(a, b) \to C(a, b)$ defined by $F(x) = y$, where

$$y(t) = p(t)x''(t) + q(t)x'(t) + r(t)x(t),$$

[2] There are other, essentially equivalent, forms of the DFT. See [4] for a complete discussion.

[3] Because of the symmetries present in the matrix A, it is possible to compute Ax very efficiently by an algorithm called the *fast Fourier transform* (FFT). The FFT is one of the most important algorithms in computational science, and the original paper [6] announcing it is reputed to be the most widely cited mathematical paper of all time—see [25], page 295.

where p, q, and r are continuous functions ($p, q, r \in C(a, b)$). Is F linear or nonlinear? Prove your answer.

12. Let F be a field and $A \in F^{m\times n}$, $B \in F^{n\times p}$. Prove that $(AB)^T = B^T A^T$. (Hint: The result of Exercise 3.1.12 might be helpful.)

13. Exercise 3.1.11 introduced integral operators of the form $K(x) = y$,

$$y(s) = \int_c^d k(s,t)x(t)\,dt, \ a \le s \le b.$$

The purpose of this exercise is to show that an integral operator is the natural generalization to functions of a matrix operator on Euclidean vectors.

Suppose $k : [a,b] \times [c,d] \to \mathbf{R}$ is given and we establish a *rectangular grid* on the domain of k by choosing numbers

$$a < s_1 < s_2 < \cdots < s_m < b, \ c < t_1 < t_2 < \cdots < t_n < d.$$

The grid consists of the mn *nodes* (s_i, t_j), $i = 1, \ldots, m$, $j = 1, \ldots, n$. For reasons that will become clear below, we define $\Delta s = (b-a)/m$ and assume $a + (i-1)\Delta s \le s_i \le a + i\Delta s$. Similarly, $\Delta t = (d-c)/n$, $c + (j-1)\Delta t \le t_j \le c + j\Delta t$. We then approximate a function $x \in C[c,d]$ by a vector $X \in \mathbf{R}^n$, where $X_j = x(t_j)$ (or, depending on the context, $X_j \doteq x(t_j)$). Similarly, we approximate $y \in C[a,b]$ by $Y \in \mathbf{R}^m$, where $Y_i = y(s_i)$ or $Y_i \doteq y(s_i)$.

The integral defining $y = K(x)$ can be approximated by a Riemann sum in terms of X and the values of k at the nodes of the grid:

$$y(s_i) \doteq \sum_{j=1}^{m} k(s_i, t_j)x(t_j)\Delta t.$$

Find a matrix $A \in \mathbf{R}^{m\times n}$ such that this equation corresponds to the matrix-vector equation $Y = AX$.

3.3 Isomorphic vector spaces

We wish to show that all linear operators between finite-dimensional vector spaces can be represented as matrix operators, even if the vector spaces are not Euclidean. To do this, we must show that any finite-dimensional vector space is *isomorphic* (that is, essentially equivalent) to a Euclidean space.

3.3.1 Injective and surjective functions; inverses

We begin by reviewing some results applying to all functions, including linear operators.

Definition 69 *Let X and Y be any sets and let $f : X \to Y$ be a function.*

1. *We say that f is* injective *(or* one-to-one*) if and only if, for all $x_1, x_2 \in X$, $f(x_1) = f(x_2)$ implies $x_1 = x_2$.*

2. *We say that f in* surjective *(or* onto*) if and only if, for each $y \in Y$, there exists an $x \in X$ such that $f(x) = y$.*

3. *If f is both injective and surjective, then it is called* bijective *(or one-to-one and onto).*

The reader will notice that the above definition is not limited to the vector space setting; it applies to any sets and functions. The same is true of the following theorem.

Theorem 70 *Let X and Y be sets and let $f : X \to Y$ be a given function. Then f is bijective if and only if there exists a function $f^{-1} : Y \to X$, called the* inverse *of f, such that*

$$f^{-1}(f(x)) = x \text{ for all } x \in X, \tag{3.4}$$

and

$$f\left(f^{-1}(y)\right) = y \text{ for all } y \in Y. \tag{3.5}$$

Proof First we assume that f is bijective. Let us define $f^{-1} : Y \to X$ by the following condition: $f^{-1}(y) = x$, where x is the element of X such that $y = f(x)$. The first part of the proof is to show that f^{-1} is well-defined. This follows from the bijectivity of f: Since f is surjective, for any $y \in Y$, there exists $x \in X$ such that $f(x) = y$. Moreover, since f is injective, there is only one such x, and hence f^{-1} is well-defined.

Now consider any $x \in X$, and let $y = f(x)$. Then $f^{-1}(y) = x$, that is,

$$f^{-1}(f(x)) = x.$$

On the other hand, consider any $y \in Y$ and let $x = f^{-1}(y)$. Then $f(x) = y$, that is,

$$f\left(f^{-1}(y)\right) = y.$$

Thus f^{-1} satisfies (3.4) and (3.5).

Conversely, suppose the inverse function f^{-1} exists. If $x_1, x_2 \in X$ and $f(x_1) = f(x_2)$, then

$$f^{-1}(f(x_1)) = f^{-1}(f(x_2)).$$

But this implies, by (3.4), that $x_1 = x_2$, which shows that f is injective. Moreover, if y is any element of Y, then $x = f^{-1}(y)$ satisfies

$$f(x) = f\left(f^{-1}(y)\right) = y$$

by (3.5), and hence f is surjective. Therefore, the existence of f^{-1} implies that f is bijective.

QED

The following theorem shows that f can have only one inverse f^{-1}.

Theorem 71 *Let X and Y be sets and let $f : X \to Y$ be a bijection. Then the inverse function f^{-1} is unique. In other words, if there exists $g : Y \to X$ such that*

$$g(f(x)) = x \text{ for all } x \in X \tag{3.6}$$

and

$$f(g(y)) = y \text{ for all } y \in Y, \tag{3.7}$$

then $g = f^{-1}$.

Proof Exercise 2

Definition 72 *Let X and Y be sets and $f : X \to Y$ be a function. We say that f is* invertible *if the inverse function f^{-1} exists.*

If it is known that a function f is invertible, then only one of (3.6) or (3.7) is sufficient to show that g is the inverse of f.

Theorem 73 *Let X and Y be sets and let $f : X \to Y$ be invertible.*

1. *If $g : Y \to X$ satisfies*

$$g(f(x)) = x \text{ for all } x \in X,$$

then $g = f^{-1}$.

2. *If $g : Y \to X$ satisfies*

$$f(g(y)) = y \text{ for all } y \in Y,$$

then $g = f^{-1}$.

Proof Exercise 3

If there exists a bijection between two sets, then the two sets are the same except for the names of the elements. When the sets in question have additional structure, then the sets are considered to be the "same" if there is bijection that preserves this additional structure. To be precise, we say that the two sets are *isomorphic* if this condition holds.

A linear operator preserves the vector space operations, which explains the following definition.

Definition 74 *Let X and Y be two finite-dimensional vector spaces over a field F. We say that X and Y are* isomorphic *if and only if there exists a bijective linear operator E mapping X onto Y. The operator $E : X \to Y$ is called an* isomorphism.

We will use the following straightforward theorem.

Theorem 75 *Let X, Y, and Z be vector spaces over a field F, and suppose X and Y are isomorphic and Y and Z are also isomorphic. Then X and Z are isomorphic.*

Proof Exercise 15.

Here is the fundamental theorem about isomorphisms between finite-dimensional subspaces.

Theorem 76 *Let X and Y be n-dimensional vector spaces over a field F. Then X and Y are isomorphic.*

Proof Let $\{x_1, x_2, \ldots, x_n\}$, $\{y_1, y_2, \ldots, y_n\}$ be bases for X and Y, respectively. Define $E : X \to Y$ by the condition that

$$E(\alpha_1 x_1 + \alpha_2 x_2 + \ldots + \alpha_n x_n) = \alpha_1 y_1 + \alpha_2 y_2 + \ldots + \alpha_n y_n.$$

Since each $x \in X$ can be represented uniquely in the form

$$x = \alpha_1 x_1 + \alpha_2 x_2 + \ldots + \alpha_n x_n,$$

the operator E is well-defined. The proof that E is linear is left as an exercise (see Exercise 16). We must prove that E is a bijection.

First we show that E is surjective. Let y be an arbitrary element of Y. Then $y = \alpha_1 y_1 + \alpha_2 y_2 + \ldots + \alpha_n y_n$ for some scalars $\alpha_1, \alpha_2, \ldots, \alpha_n$. Define $x = \alpha_1 x_1 + \alpha_2 x_2 + \ldots + \alpha_n x_n$; then, by definition of E, $E(x) = y$, and we have shown that E is surjective.

Next we must show that E is injective. Suppose vectors $u, v \in X$ satisfy $E(u) = E(v)$. There exist scalars $\alpha_1, \alpha_2, \ldots, \alpha_n$ and $\beta_1, \beta_2, \ldots, \beta_n$ such that $u = \alpha_1 x_1 + \alpha_2 x_2 + \ldots + \alpha_n x_n$ and $v = \beta_1 x_1 + \beta_2 x_2 + \ldots + \beta_n x_n$. But then

$$E(u) = E(v) \Rightarrow \alpha_1 y_1 + \alpha_2 y_2 + \ldots + \alpha_n y_n = \beta_1 y_1 + \beta_2 y_2 + \ldots + \beta_n y_n,$$

which implies that $\alpha_1 = \beta_1, \alpha_2 = \beta_2, \ldots, \alpha_n = \beta_n$ by Theorem 26. This in turn implies that $u = v$, and we have shown that E is injective.

QED

Corollary 77 *Let X be an n-dimensional vector space over a field n. Then X is isomorphic to F^n.*

When the standard basis $\{e_1, e_2, \ldots, e_n\}$ is used to represent F^n and a basis $\mathcal{X} = \{x_1, x_2, \ldots, x_n\}$ is used to represent a given n-dimensional vector space X over F, we obtain the following isomorphism $E : X \to F^n$:

$$E(a_1x_1 + a_2x_2 + \ldots + a_nx_n) = a_1e_1 + a_2e_2 + \ldots + a_ne_n = (a_1, a_2, \ldots, a_n).$$

The standard notation for this isomorphism is

$$[x]_{\mathcal{X}} = a,$$

where $x = \sum_{i=1}^{n} a_ix_i$, $a = (a_1, a_2, \ldots, a_n)$. Thus $[x]_{\mathcal{X}}$ represents the vector in F^n whose components are the weights required to represent the vector $x \in X$ in terms of the basis X. We call $[x]_{\mathcal{X}}$ the *coordinate vector* of x relative to the basis $\mathcal{X}$, or simply the $\mathcal{X}$-coordinate vector of x.

Example 78 *Consider the vector space* $\mathbf{R}^2$ *and the basis*

$$\mathcal{X} = \left\{ \begin{bmatrix} 1 \\ 1 \end{bmatrix}, \begin{bmatrix} 1 \\ -1 \end{bmatrix} \right\}.$$

Given any $x \in \mathbf{R}^2$*, we can find* $[x]_{\mathcal{X}}$ *by solving*

$$y_1 \begin{bmatrix} 1 \\ 1 \end{bmatrix} + y_2 \begin{bmatrix} 1 \\ -1 \end{bmatrix} = \begin{bmatrix} x_1 \\ x_2 \end{bmatrix} \iff \begin{cases} y_1 + y_2 = x_1, \\ y_1 - y_2 = x_2. \end{cases}$$

We can easily solve the system to find

$$[x]_{\mathcal{X}} = y = \begin{bmatrix} \frac{1}{2}x_1 + \frac{1}{2}x_2 \\ \frac{1}{2}x_1 - \frac{1}{2}x_2 \end{bmatrix}.$$

An interesting question then arises as to the relationship between $[x]_{\mathcal{X}}$ and $[x]_{\mathcal{Y}}$, where $\mathcal{X}$ and $\mathcal{Y}$ are two different bases for X. This is explored in Exercise 3.6.23.

3.3.2 The matrix of a linear operator on general vector spaces

Now we will consider a linear operator $L : X \to U$, where X is an n-dimensional vector space over a field F and U is an m-dimensional vector space over the same field F. We have seen that X is isomorphic to F^n and U is isomorphic to F^m. Let $\mathcal{X} = \{x_1, x_2, \ldots, x_n\}$ and $\mathcal{U} = \{u_1, u_2, \ldots, u_n\}$ be bases for X and U, respectively. We then have isomorphisms $E_X : X \to F^n$ and $E_U : U \to F^m$ defined as follows:

$$E_X(\alpha_1x_1 + \alpha_2x_2 + \ldots + \alpha_nx_n) = (\alpha_1, \alpha_2, \ldots, \alpha_n),$$
$$E_U(\beta_1u_1 + \beta_2u_2 + \ldots + \beta_nu_n) = (\beta_1, \beta_2, \ldots, \beta_n).$$

The linear operator L can then be represented by a matrix A in the following sense: Given any $a = \alpha_1x_1 + \alpha_2x_2 + \ldots + \alpha_nx_n \in X$, let b be the corresponding vector in F^n: $b = (\alpha_1, \alpha_2, \ldots, \alpha_n)$. Then we can find a matrix $A \in F^{m\times n}$ such that Ab is the vector in F^m corresponding to $L(a) \in U$. The following diagram illustrates this construction:

$$\begin{array}{ccc} F^n & \xrightarrow{A} & F^m \\ E_X \uparrow & & \uparrow E_U \\ X & \xrightarrow[L]{} & U \end{array}$$

Formally, A is the matrix defining the linear operator $E_U L E_X^{-1}$; the existence of A is guaranteed by Theorem 66. It follows that the columns of A are the following vectors in F^m:

$$E_U L E_X^{-1}(e_j) = E_U L(x_j), \ j = 1, 2, \ldots, n.$$

We can express the relationship between L and A using coordinate vectors relative to the bases $\mathcal{X}$ and $\mathcal{U}$ for X and U, respectively:

$$A[x]_{\mathcal{X}} = [L(x)]_{\mathcal{U}} \text{ for all } x \in X. \tag{3.8}$$

If we substitute $x = x_j$ in this equation, we obtain

$$A[x_j]_{\mathcal{X}} = [L(x_j)]_{\mathcal{U}} \Rightarrow A_j = Ae_j = [L(x_j)]_{\mathcal{U}}.$$

Thus the entries in the jth column of A are the weights required to represent $L(x_j)$ in terms of $u_1, u_2, \ldots, u_m$.

Example 79 *Differentiation defines a linear operator D mapping $\mathcal{P}_n$ into $\mathcal{P}_{n-1}$:*

$$D(p) = p' \text{ for all } p \in \mathcal{P}_n.$$

Bases for $\mathcal{P}_n$ and $\mathcal{P}_{n-1}$ are $\{1, x, x^2, \ldots, x^n\}$ and $\{1, x, x^2, \ldots, x^{n-1}\}$, respectively. It follows that $\mathcal{P}_n$ is isomorphic to $\mathbf{R}^{n+1}$ and $\mathcal{P}_{n-1}$ to $\mathbf{R}^n$. We will write E_n for the isomorphism mapping $\mathcal{P}_n$ into $\mathbf{R}^{n+1}$ and E_{n-1} for the isomorphism mapping $\mathcal{P}_{n-1}$ into $\mathbf{R}^n$. To find the matrix $A \in \mathbf{R}^{n\times n+1}$ that represents the operator D, we must compute $E_{n-1}D$ applied to each of the standard basis functions $1, x, x^2, \ldots, x^n$:

$$\begin{aligned} D(1) = 0 &\Rightarrow E_{n-1}(D(1)) = (0,0,0,\ldots,0), \\ D(x) = 1 &\Rightarrow E_{n-1}(D(x)) = (1,0,0,\ldots,0), \\ D\left(x^2\right) = 2x &\Rightarrow E_{n-1}(D\left(x^2\right)) = (0,2,0,\ldots,0), \\ &\vdots \qquad\qquad \vdots \\ D\left(x^n\right) = nx^{n-1} &\Rightarrow E_{n-1}(D\left(x^n\right)) = (0,0,0,\ldots,n). \end{aligned}$$

It follows that A is the matrix

$$A = \begin{bmatrix} 0 & 1 & 0 & \cdots & 0 \\ 0 & 0 & 2 & \cdots & 0 \\ \vdots & \vdots & \vdots & \ddots & \vdots \\ 0 & 0 & 0 & \cdots & n \end{bmatrix}.$$

It is important to recognize that the matrix representing a given linear operator depends on the bases chosen for the domain and co-domain. If L is a linear operator mapping X into U, $\mathcal{X} = \{x_1, x_2, \ldots, x_n\}$ is a basis for X, and $\mathcal{U} = \{u_1, u_2, \ldots, u_m\}$ is a basis for U, then we write $[L]_{\mathcal{X},\mathcal{U}}$ for the matrix representing L with respect to the bases $\mathcal{X}$, $\mathcal{U}$. The fundamental relationship between the coordinate vectors and the matrix representing L is (3.8), which can be written as

$$[L]_{\mathcal{X},\mathcal{U}}[x]_{\mathcal{X}} = [L(x)]_{\mathcal{U}} \text{ for all } x \in X. \tag{3.9}$$

Example 80 *Let $A \in \mathbf{R}^{2\times 2}$ be defined by*

$$A = \begin{bmatrix} 2 & 1 \\ 1 & 2 \end{bmatrix}$$

and define $L : \mathbf{R}^2 \to \mathbf{R}^2$ by $L(x) = Ax$. We will find $[L]_{\mathcal{X},\mathcal{X}}$, where $\mathcal{X}$ is the basis for $\mathbf{R}^2$ that was used in Example 78.

The two columns of $[L]_{\mathcal{X},\mathcal{X}}$ are

$$[L(y_1)]_{\mathcal{X}},\ [L(y_2)]_{\mathcal{X}},$$

where $y_1 = (1,1)$ and $y_2 = (1,-1)$ are the basis vectors comprising $\mathcal{X}$. We know from Example 78 that

$$[x]_{\mathcal{X}} = \begin{bmatrix} \frac{1}{2}x_1 + \frac{1}{2}x_2 \\ \frac{1}{2}x_1 - \frac{1}{2}x_2 \end{bmatrix}.$$

We have

$$L(y_1) = \begin{bmatrix} 3 \\ 3 \end{bmatrix} \Rightarrow [L(y_1)]_{\mathcal{X}} = \begin{bmatrix} \frac{1}{2}\cdot 3 + \frac{1}{2}\cdot 3 \\ \frac{1}{2}\cdot 3 - \frac{1}{2}\cdot 3 \end{bmatrix} = \begin{bmatrix} 3 \\ 0 \end{bmatrix}$$

and

$$L(y_2) = \begin{bmatrix} 1 \\ -1 \end{bmatrix} \Rightarrow [L(y_2)]_{\mathcal{X}} = \begin{bmatrix} \frac{1}{2}\cdot 1 + \frac{1}{2}\cdot(-1) \\ \frac{1}{2}\cdot 1 - \frac{1}{2}\cdot(-1) \end{bmatrix} = \begin{bmatrix} 0 \\ 1 \end{bmatrix}.$$

Therefore,

$$[L]_{\mathcal{X},\mathcal{X}} = \begin{bmatrix} 3 & 0 \\ 0 & 1 \end{bmatrix}.$$

An interesting question is how $[L]_{\mathcal{X},\mathcal{U}}$ changes if the bases are changed. In other words, if $\mathcal{Y} = \{y_1, y_2, \dots, y_n\}$ and $\mathcal{V} = \{v_1, v_2, \dots, v_m\}$ are alternate bases for X and U, respectively, then what is the relationship between $[L]_{\mathcal{X},\mathcal{U}}$ and $[L]_{\mathcal{Y},\mathcal{V}}$? This is explored in Exercise 3.6.23.

In the previous section, we introduced the matrix of a linear operator mapping one Euclidean space into another. In that context, it was implicitly assumed that the standard bases were used for both the domain and co-domain. To make this precise, let us write $\mathcal{S}_n$ for the standard basis of F^n, that is, $\mathcal{S}_n = \{e_1, e_2, \dots, e_n\}$. Given a linear operator $L : F^n \to F^m$, the matrix representing it, as introduced in the previous section, is $[L]_{\mathcal{S}_n,\mathcal{S}_m}$. However, there is no reason that different bases could not be chosen for F^n and F^m, in which case the matrix of L would be different.

Since every linear operator on finite-dimensional vector spaces can be represented by a matrix operator, it follows that, on finite-dimensional spaces, every linear operator equation $L(x) = b$ is equivalent to a matrix-vector equation, that is, to a system of linear algebraic equations. This explains the importance of matrices in linear algebra. Much of the remainder of the book will involve the analysis of matrices.

Exercises

Miscellaneous exercises

1. For each of the following functions, decide if it is invertible or not. If it is, find the inverse function; if it is not, determine which property of an invertible function fails to hold, injectivity or surjectivity (or both). Prove your conclusions.

   (a) $f : \mathbf{R} \to \mathbf{R}$, $f(x) = 2x + 1$;

   (b) $f : \mathbf{R} \to (0, \infty)$, $f(x) = e^x$;

   (c) $f : \mathbf{R}^2 \to \mathbf{R}^2$, $f(x) = (x_1 + x_2, x_1 - x_2)$;

   (d) $f : \mathbf{R}^2 \to \mathbf{R}^2$, $f(x) = (x_1 - 2x_2, -2x_1 + 4x_2)$.

2. Prove Theorem 71.

3. Prove Theorem 73.

4. Let X, Y, and Z be sets, and suppose $f : X \to Y$, $g : Y \to Z$ are bijections. Show that $g \circ f$ is a bijection mapping X onto Z. What is $(g \circ f)^{-1}$?

5. Let X, Y, and Z be sets, and suppose $f : X \to Y$, $g : Y \to Z$ are given functions. For each statement below, prove it or give a counterexample:

   (a) If f and $g \circ f$ are invertible, then g is invertible.

   (b) If g and $g \circ f$ are invertible, then f is invertible.

(c) If $g \circ f$ is invertible, then f and g are invertible.

6. Consider the operator $M : \mathcal{P}_n \to \mathcal{P}_{n+1}$ defined by $M(p) = q$, where $q(x) = xp(x)$. In Exercise 3.1.4, you were asked to show that M is linear. Find the matrix representing M, using the standard bases for both $\mathcal{P}_n$ and $\mathcal{P}_{n+1}$.

7. Let $L : \mathbf{R}^2 \to \mathbf{R}^2$ be defined by $L(x) = Ax$, where

$$A = \begin{bmatrix} 1 & 1 \\ 1 & 1 \end{bmatrix}.$$

Let $\mathcal{S}$ be the standard basis, $\mathcal{S} = \{(1,0),(0,1)\}$, and let $\mathcal{X}$ be the alternate basis $\{(1,1),(1,2)\}$. As discussed above, we have $[L]_{\mathcal{S},\mathcal{S}} = A$. Find $[L]_{\mathcal{X},\mathcal{X}}$.

8. Consider two bases for $\mathbf{R}^3$:

$$\mathcal{S} = \{(1,0,0),(0,1,0),(0,0,1)\},\ \mathcal{X} = \{(1,1,1),(0,1,1),(0,0,1)\}.$$

Let $I : \mathbf{R}^3 \to \mathbf{R}^3$ be the identity operator: $Ix = x$ for all $x \in \mathbf{R}^3$. Obviously I is a linear operator. Find each of the following: $[I]_{\mathcal{S},\mathcal{S}}$, $[I]_{\mathcal{S},\mathcal{X}}$, $[I]_{\mathcal{X},\mathcal{S}}$, $[I]_{\mathcal{X},\mathcal{X}}$.

9. Find the matrix $[L]_{\mathcal{X},\mathcal{X}}$ of Example 80 by an alternate method: Let $x \in \mathbf{R}^2$ be arbitrary, compute

$$[L(x)]_{\mathcal{X}}$$

and solve (3.9 for $[L]_{\mathcal{X},\mathcal{X}}$ (notice that $[x]_{\mathcal{X}}$ is already known for this example).

10. Let $T : \mathcal{P}_2 \to \mathcal{P}_2$ be defined by $T(p) = q$, where $q(x) = p(x-1)$. Prove that T is linear and find $[T]_{\mathcal{M},\mathcal{M}}$, where $\mathcal{M}$ is the standard basis for $\mathcal{P}_2$: $\mathcal{M} = \{1, x, x^2\}$.

11. Is the operator T from the previous exercise an isomorphism? Prove your answer.

12. Let $\mathcal{X} = \{(1,0,0),(1,1,0),(1,1,1)\} \subset \mathbf{Z}_2^3$.

    (a) Prove that $\mathcal{X}$ is a basis for $\mathbf{Z}_2^3$.

    (b) Find $[x]_{\mathcal{X}}$ for an arbitrary vector x in $\mathbf{Z}_2^3$.

13. Let $\mathcal{X}$ be the basis for $\mathbf{Z}_2^3$ from the previous exercise, let $A \in \mathbf{Z}_2^{3\times 3}$ be defined by

$$A = \begin{bmatrix} 1 & 1 & 0 \\ 1 & 0 & 1 \\ 0 & 1 & 1 \end{bmatrix},$$

and define $L : \mathbf{Z}_2^3 \to \mathbf{Z}_2^3$ by $L(x) = Ax$. Find $[L]_{\mathcal{X},\mathcal{X}}$.

14. Consider the basis $\mathcal{S} = \{1, x, x^2\}$ for $\mathcal{P}_3(\mathbf{Z}_3)$. Let c_0, c_1, c_2, c_3 be arbitrary elements of $\mathbf{Z}_3$ and find

$$\left[c_0 + c_1 x + c_2 x^2 + c_3 x^3\right]_{\mathcal{S}}.$$

15. Prove Theorem 75.

16. Prove that the operator E from the proof of Theorem 76 is linear.

17. (a) Let F be a field. Show that $F^{m\times n}$ is a vector space over F and determine the dimension of $F^{m\times n}$. Addition of two elements of $F^{m\times n}$ is defined entrywise: $(A+B)_{ij} = A_{ij} + B_{ij}$. Similarly, $(\alpha A)_{ij} = \alpha A_{ij}$.
    (b) Prove that $F^{m\times n}$ is isomorphic to F^{mn} by finding an isomorphism.
    (c) Let X and U be vector spaces over a field F, and suppose the dimensions of X and U are n and m, respectively. Prove that $\mathcal{L}(X, U)$ is isomorphic to $F^{m\times n}$. (Recall that $\mathcal{L}(X, U)$ is the space of all linear operators mapping X into U; see page 101.)

18. Let $U = \mathbf{R}$, regarded as a vector space over $\mathbf{R}$, and let V be the vector space of Exercise 2.2.8. Prove that U and V are isomorphic, and find an isomorphism from U to V.

19. Is the operator D defined in Example 79 an isomorphism? Prove your answer.

20. Let F be a field and let X and U be finite-dimensional vector spaces over F. Let $\mathcal{X} = \{x_1, x_2, \ldots, x_n\}$ and $\mathcal{U} = \{u_1, u_2, \ldots, u_m\}$ be bases for X and U, respectively. Let $A \in F^{m\times n}$ be given. Prove that there exists a unique linear operator $L : X \to U$ such that $[L]_{\mathcal{X},\mathcal{U}} = A$.

3.4 Linear operator equations

In this section we consider an abstract linear operator equation $L(x) = u$, where X and U are vector spaces over a common field F, $L : X \to U$ is linear, and u is an element of U. If $u \neq 0$, the equation $L(x) = u$ is called *inhomogeneous*, while $L(x) = 0$ is called *homogeneous*. We point out that the terms homogeneous and inhomogeneous are not used for nonlinear equations.

There is a special name for the solution set of a homogeneous equation.

Definition 81 *Let X, U be vector spaces over a field F, and let $L : X \to U$ be linear. The* kernel $\ker(L)$ *of L is the set of all solutions to $L(x) = 0$:*

$$\ker(L) = \{x \in X \, : \, L(x) = 0\}.$$

We also remind the reader of the definition of the *range* of L:

$$\mathcal{R}(L) = \{L(x) \in U : x \in X\} = \{u \in U : u = L(x) \text{ for some } x \in X\}.$$

An operator is a function, and this is the usual definition of the range of a function.

Theorem 82 *Let X and U be vector spaces over a field F, and let $L : X \to U$ be linear. Then $\ker(L)$ is a subspace of X and $\mathcal{R}(L)$ is a subspace of U.*

Proof By Theorem 57, $L(0) = 0$, which shows both that $0 \in \ker(L)$ and that $0 \in \mathcal{R}(L)$. If $x \in \ker(L)$, then $L(x) = 0$. For any $\alpha \in F$,

$$L(\alpha x) = \alpha L(x) = \alpha \cdot 0 = 0,$$

which shows that $\alpha x \in \ker(L)$. Similarly, if $x, y \in \ker(L)$, then $L(x) = 0$, $L(y) = 0$, and thus

$$L(x + y) = L(x) + L(y) = 0 + 0 = 0.$$

This shows that $x + y \in \ker(L)$, and hence $\ker(L)$ is a subspace of X.

Suppose $u \in \mathcal{R}(L)$ and $\alpha \in F$. By definition of range, there must exist $x \in X$ such that $L(x) = u$. But then $L(\alpha x) = \alpha L(x) = \alpha u$, which shows that $\alpha u \in \mathcal{R}(L)$. Similarly, if $u, v \in \mathcal{R}(L)$, then there exist $x, y \in X$ such that $L(x) = u$ and $L(y) = v$. It follows that

$$L(x + y) = L(x) + L(y) = u + v,$$

and hence that $u + v \in \mathcal{R}(L)$. Therefore, $\mathcal{R}(L)$ is a subspace of U.

QED

3.4.1 Homogeneous linear equations

A homogeneous linear equation always has at least one solution, namely, the zero vector. Other solutions are called nontrivial solutions. In the case that the field F is infinite, such as $F = \mathbf{R}$ or $F = \mathbf{C}$, if there is at least one nontrivial solution, there are in fact infinitely many nontrivial solutions.

Theorem 83 *Let X and U be vector spaces over a field F, and assume F contains infinitely many elements. If $L : X \to U$ is linear, $L(x) = 0$ has either exactly one solution or infinitely many solutions. (In other words, $\ker(L)$ contains either exactly one or infinitely many elements.)*

Proof We know that $L(x) = 0$ always has the solution $x = 0$. If this is the only solution, then the theorem holds. It remains to show that if $L(x) = 0$ has a nontrivial solution, then it has infinitely many solutions. But if $L(x) = 0$ has a nontrivial solution y, then $\ker(L)$ is a nontrivial subspace of X. By an earlier exercise, every nontrivial subspace contains infinitely many vectors, provided F has infinitely many elements (see Exercise 2.2.2).

QED

3.4.2 Inhomogeneous linear equations

An inhomogeneous equation need not have a solution, as the following example shows.

Example 84 *Let A be the 3×2 matrix*

$$A = \begin{bmatrix} 1 & 2 \\ 1 & 2 \\ 1 & 2 \end{bmatrix},$$

and let $L : \mathbf{R}^2 \to \mathbf{R}^3$ be the linear operator defined by $L(x) = Ax$. For any $x \in \mathbf{R}^2$, we have

$$Ax = \begin{bmatrix} 1 & 2 \\ 1 & 2 \\ 1 & 2 \end{bmatrix} \begin{bmatrix} x_1 \\ x_2 \end{bmatrix} = x_1 \begin{bmatrix} 1 \\ 1 \\ 1 \end{bmatrix} + x_2 \begin{bmatrix} 2 \\ 2 \\ 2 \end{bmatrix} = (x_1 + 2x_2) \begin{bmatrix} 1 \\ 1 \\ 1 \end{bmatrix}.$$

This shows that every vector $L(x)$ is a scalar multiple of $(1, 1, 1)$; that is,

$$\mathcal{R}(L) = \mathrm{sp}\{(1, 1, 1)\}.$$

It follows that $L(x) = b$ fails to have a solution if b is not a multiple of $(1, 1, 1)$. For example, with $b = (1, 2, 1)$, $L(x) = b$ has no solution.

The following results characterize the solution set of $L(x) = u$ when there is at least one solution.

Lemma 85 *Let X and U be vector spaces over a field F, let $L : X \to U$ be linear, and suppose $u \in U$. If $\hat{x} \in X$ is a solution to $L(x) = u$ and $y \in \ker(L)$, then $\hat{x} + y$ is another solution of $L(x) = u$.*

Proof By hypothesis, $L(\hat{x}) = u$ and $L(y) = 0$. Therefore, by linearity,

$$L(\hat{x} + y) = L(\hat{x}) + L(y) = u + 0 = u.$$

QED

Lemma 86 *Let X and U be vector spaces over a field F, let $L : X \to U$ be linear, and suppose $u \in U$. If x_1 and x_2 are two solutions of $L(x) = u$, then $x_1 - x_2 \in \ker(L)$.*

Proof By hypothesis, $L(x_1) = u$ and $L(x_2) = u$. Therefore, by linearity,

$$L(x_1 - x_2) = L(x_1) - L(x_2) = u - u = 0.$$

QED

Example 87 *Let* $L : C^2(\mathbf{R}) \to C(\mathbf{R})$ *be defined by*

$$L(x) = x'' - 2x' + 2x,$$

and define $f(t) = e^t$, $x_1(t) = e^t(1 + \sin(t))$, $x_2(t) = e^t(1 + \cos(t))$. *The functions* x_1 *and* x_2 *are both solutions of* $L(x) = f$. *For instance,*

$$x_1'(t) = e^t(1 + \cos(t) + \sin(t)),\ x_1''(t) = e^t(1 + 2\cos(t)),$$

and therefore

$$\begin{aligned} & x_1''(t) - 2x_1'(t) + 2x_1(t) \\ = \; & e^t(1 + 2\cos(t)) - 2e^t(1 + \cos(t) + \sin(t)) + 2e^t(1 + \sin(t)) \\ = \; & e^t\,(1 + 2\cos(t) - 2 - 2\cos(t) - 2\sin(t) + 2 + 2\sin(t)) \\ = \; & e^t. \end{aligned}$$

The verification that $L(x_2) = f$ *is similar. By Lemma 86,* $z = x_1 - x_2$ *must belong to* $\ker(L)$. *We have*

$$z(t) = x_1(t) - x_2(t) = e^t(1 + \sin(t)) - e^t(1 + \cos(t)) = e^t(\sin(t) - \cos(t)).$$

The reader can verify that z *satisfies* $L(z) = 0$.

For the main result, we will use the following notation.

Definition 88 *Let* U *be a vector space over a field* F, *and let* S, T *be two subsets of* U. *The (algebraic)* sum *of* S *and* T *is the set*

$$S + T = \{s + t \,:\, s \in S, t \in T\}.$$

If S *is a singleton set,* $S = \{x\}$, *we write* $x + T$ *instead of* $\{x\} + T$. *Thus*

$$x + T = \{x + t \,:\, t \in T\}.$$

The following theorem appeared in an earlier exercise.

Theorem 89 *Let* U *be a vector space over a field* F, *and let* V, W *be subspaces of* U. *Then* $V + W$ *is also a subspace of* U.

Proof Exercise 2.3.21.

Here is the main result about inhomogeneous linear equations.

Theorem 90 *Let* X *and* U *be vector spaces over a field* F, *let* $L : X \to U$ *be linear, and let* $u \in U$. *If* $\hat{x} \in X$ *is a solution to* $L(x) = u$, *then the set of all solutions to* $L(x) = u$ *is*

$$\hat{x} + \ker(L).$$

Proof Let S denote the solution set of $L(x) = u$. By hypothesis, $\hat{x} \in S$. We must show that $S = \hat{x} + \ker(L)$.

Suppose first that $y \in S$. Then, since y and $\hat{x}$ are both solutions of $L(x) = u$, it follows from Lemma 86 that $z = y - \hat{x}$ belongs to $\ker(L)$. But then $y = \hat{x} + z$ with $z \in \ker(L)$, that is, $y \in \hat{x} + \ker(L)$. Thus $S \subset \hat{x} + \ker(L)$.

On the other hand, if $y \in \hat{x} + \ker(L)$, then $y = \hat{x} + z$ for some $z \in \ker(L)$. By Lemma 85, it follows that y is a solution of $L(x) = u$, that is, $y \in S$. This shows that $\hat{x} + \ker(L) \subset S$.

QED

3.4.3 General solutions

Suppose $L : X \to U$ is a linear operator, $L(x) = u$ has a solution $\hat{x}$, and $\ker(L)$ is finite-dimensional. If $\{x_1, x_2, \ldots, x_k\}$ is a basis for $\ker(L)$, then the solution set of $L(x) = u$ is

$$\begin{aligned}\hat{x} + \ker(L) &= \hat{x} + \mathrm{sp}\{x_1, x_2, \ldots, x_k\} \\ &= \hat{x} + \{\alpha_1 x_1 + \alpha_2 x_2 + \ldots + \alpha_k x_k \,:\, \alpha_1, \alpha_2, \ldots, \alpha_k \in F\} \\ &= \{\hat{x} + \alpha_1 x_1 + \alpha_2 x_2 + \ldots + \alpha_k x_k \,:\, \alpha_1, \alpha_2, \ldots, \alpha_k \in F\}.\end{aligned}$$

It is common to drop the set notation and simply say that

$$x = \hat{x} + \alpha_1 x_1 + \alpha_2 x_2 + \ldots + \alpha_k x_k$$

is the *general solution* of $L(x) = u$. This is understood to mean both that $\hat{x} + \alpha_1 x_1 + \alpha_2 x_2 + \ldots + \alpha_k x_k$ is a solution of $L(x) = u$ for any choice of the scalars $\alpha_1, \alpha_2, \ldots, \alpha_k$, and also that every solution can be represented in this form by choosing appropriate values of $\alpha_1, \alpha_2, \ldots, \alpha_k$.

Example 91 *Let $A \in \mathbf{Z}_3^{3\times 5}$, $b \in \mathbf{Z}_3^3$ be defined by*

$$A = \begin{bmatrix} 1 & 2 & 1 & 2 & 0 \\ 2 & 0 & 1 & 2 & 0 \\ 2 & 1 & 0 & 2 & 2 \end{bmatrix}, \quad b = \begin{bmatrix} 1 \\ 0 \\ 0 \end{bmatrix}.$$

The equation $Ax = b$ is equivalent to the system

$$\begin{aligned} x_1 + 2x_2 + x_3 + 2x_4 \phantom{{}+2x_5} &= 1, \\ 2x_1 + x_3 + 2x_4 \phantom{{}+2x_5} &= 0, \\ 2x_1 + x_2 + 2x_4 + 2x_5 &= 0, \end{aligned}$$

and Gaussian elimination yields

$$\begin{array}{rcl} x_1 \phantom{{}+x_2+x_3} + 2x_4 + 2x_5 &=& 1, \\ x_2 \phantom{{}+x_3} + x_4 + x_5 &=& 1, \\ x_3 + x_4 + 2x_5 &=& 1 \end{array} \quad \Leftrightarrow \quad \begin{array}{rcl} x_1 &=& 1 + x_4 + x_5, \\ x_2 &=& 1 + 2x_4 + 2x_5, \\ x_3 &=& 1 + 2x_4 + x_5. \end{array}$$

Setting x_4 and x_5 equal to α_1 and α_2, respectively, we obtain

$$x = (1 + \alpha_1 + \alpha_2, 1 + 2\alpha_1 + 2\alpha_2, 1 + 2\alpha_1 + \alpha_2, \alpha_1, \alpha_2),$$

or

$$x = (1, 1, 1, 0, 0) + \alpha_1(1, 2, 2, 1, 0) + \alpha_2(1, 2, 1, 0, 1),$$

which is the general solution of the equation $Ax = b$.

Exercises

Miscellaneous exercises

1. Suppose $L : \mathbf{R}^3 \to \mathbf{R}^3$ is linear, $b \in \mathbf{R}^3$ is given, and $y = (1, 0, 1)$, $z = (1, 1, -1)$ are two solutions to $L(x) = b$. Find two more solutions to $L(x) = b$.

2. Let $L : C^2(\mathbf{R}) \to C(\mathbf{R})$ be a linear differential operator, and let f in $C(\mathbf{R})$ be defined by $f(t) = 2(1 - t)e^t$. Suppose $x_1(t) = t^2e^t$ and $x_2(t) = (t^2 + 1)e^t$ are solutions of $L(x) = f$. Find two more solutions of $L(x) = f$.

3. Suppose $T : \mathbf{R}^4 \to \mathbf{R}^4$ has kernel

$$\ker(T) = \operatorname{sp}\{(1, 2, 1, 2), (-2, 0, 0, 1)\}.$$

   Suppose further that $T(y) = b$, where

$$y = (1, 2, -1, 1), \; b = (3, 1, -2, -1).$$

   Find three distinct solutions, each different from y, to $T(x) = b$.

4. Let T and b be defined as in the previous exercise. Is $z = (0, 4, 0, 1)$ a solution of $T(x) = b$?

5. Let $L : \mathbf{R}^3 \to \mathbf{R}^3$ satisfy $\ker(L) = \operatorname{sp}\{(1, 1, 1)\}$ and $L(u) = v$, where $u = (1, 1, 0)$ and $v = (2, -1, 2)$. Which of the following vectors is a solution of $L(x) = v$?

   (a) $x = (1, 2, 1)$

   (b) $x = (3, 3, 2)$

   (c) $x = (-3, -3, -2)$

6. (a) Let X and U be vector spaces over a field F, and let $L : X \to U$ be linear. Suppose that $b, c \in U$, $y \in X$ is a solution to $L(x) = b$, and $z \in X$ is a solution to $L(x) = c$. Find a solution to $L(x) = \beta b + \gamma c$, where $\beta, \gamma \in F$.

(b) Suppose $F : \mathbf{R}^3 \to \mathbf{R}^3$ is a linear operator, and define $b = (1, 2, 1)$, $c = (1, 0, 1)$. Assume $y = (1, 1, 0)$ solves $F(x) = b$, while $z = (2, -1, 1)$ solves $F(x) = c$. Find a solution to $F(x) = d$, where $d = (1, 4, 1)$. (Hint: Write d as a linear combination of b and c and apply the preceding result.)

7. Let $A \in \mathbf{Z}_2^{3\times 3}$, $b \in \mathbf{Z}_2^3$ be defined by

$$A = \begin{bmatrix} 1 & 1 & 0 \\ 1 & 0 & 1 \\ 0 & 1 & 1 \end{bmatrix}, \ b = \begin{bmatrix} 1 \\ 0 \\ 1 \end{bmatrix}.$$

List all solutions to $Ax = 0$ and $Ax = b$. Explain how your results illustrate Theorem 90.

8. Let $A \in \mathbf{Z}_3^{3\times 3}$, $b \in \mathbf{Z}_3^3$ be defined by

$$A = \begin{bmatrix} 1 & 2 & 0 \\ 2 & 0 & 1 \\ 0 & 1 & 2 \end{bmatrix}, \ b = \begin{bmatrix} 1 \\ 1 \\ 1 \end{bmatrix}.$$

List all solutions to $Ax = 0$ and $Ax = b$. Explain how your results illustrate Theorem 90.

9. Let V be a vector space over a field F, let $\hat{x} \in V$, and let S be a subspace of V. Prove that if $\tilde{x} \in \hat{x} + S$, then $\tilde{x} + S = \hat{x} + S$. Interpret this result in terms of the solution set $\hat{x} + \ker(T)$ for a linear operator equation $T(x) = b$.

10. Let $D : C^1(\mathbf{R}) \to C(\mathbf{R})$ be the derivative operator: $D(F) = F'$.

(a) What is the kernel of D?

(b) We usually write indefinite integrals in the form

$$\int f(x)\, dx = F(x) + C,$$

where C is the *constant of integration.* Indefinite integration is equivalent to solving the linear operator equation $D(F) = f$. Interpret the constant of integration in terms of the results of this section on linear operator equations.

11. Let $F : \mathbf{R}^2 \to \mathbf{R}$ be defined by $F(x) = x_1^2 + x_2^2 - 1$. (Here the co-domain is regarded as a one-dimensional vector space.) The following statements would be true if F were linear:

(a) The range of F is a subspace of $\mathbf{R}$.

(b) The solution set of $F(x) = 0$ (that is, $\{x \in \mathbf{R}^2 : F(x) = 0\}$) is a subspace of $\mathbf{R}^2$.

(c) The solution set of $F(x) = c$ (where $c \in \mathbf{R}$ is a given constant) is $\hat{x} + \{x \in \mathbf{R}^2 : F(x) = 0\}$, where $\hat{x}$ is any one solution to $F(x) = c$.

Which of these is true and which is false for the given nonlinear operator F? Prove your answers.

12. Let X and U be vector spaces over a field F, let $T : X \to U$ be linear, and suppose $u \in U$ is a nonzero vector. Is the solution set to $T(x) = u$ a subspace of X? Prove your answer.

Project: Quotient spaces and linear operators

This project describes one way to define a new vector space from a given vector space and a proper subspace of it. It then shows that every linear transformation defines a related isomorphism.

The definition of quotient space, given below, is based on the concept of an equivalence relation.[4]

13. Let V be a vector space over a field F, and let S be a proper subspace of V. Prove that the relation $\sim$ defined by $u \sim v$ if and only if $u - v \in S$ is an equivalence relation on V.

14. Let V be a vector space over a field F, and let S be a proper subspace of V. For any vector $u \in V$, let $[u]$ denote the equivalence class of u under the equivalence relation defined in the previous part of the exercise. We denote the set of all equivalence classes by V/S (the *quotient space* of V over S) and define addition and scalar multiplication on V/S by

$$\begin{aligned} [u] + [v] &= [u + v] \text{ for all } [u], [v] \in V/S, \\ \alpha[u] &= [\alpha u] \text{ for all } [u] \in V/S, \alpha \in F. \end{aligned}$$

(a) Notice that $[u] + [v]$ and $\alpha[u]$ are defined with reference to given representatives of the equivalent classes, so it is not obvious that these operations are well-defined. In other words, if $[u] = [w]$ and $[v] = [z]$, then $[u] + [v]$ must equal $[w] + [z]$ if addition is to be well-defined, and similarly for scalar multiplication. Prove that both addition and scalar multiplication are well-defined.

(b) Prove that V/S, defined in the previous part of the exercise, is a vector space under the given operations.

---

[4]Let X be a set. A *relation* on X is a subset R of $X \times X$. We usually choose a binary symbol, such as $\sim$, and write $x \sim y$ if and only if $x, y \in X$ and $(x, y) \in R$.

Let X be a set and let $\sim$ denote a relation defined on X. We say that $\sim$ is an *equivalence relation* on X if and only if the following three conditions are true:

1. For all $x \in X$, $x \sim x$.
2. If x and y belong to X and $x \sim y$, then $y \sim x$.
3. If x, y, and z belong to X and $x \sim y$, $y \sim z$, then $x \sim z$.

Thus an equivalence relation satisfies the same basic properties as the relation "is equal to."

15. Now let X and U be vector spaces over a field F, and let $L : X \to U$ be a linear operator. Define $T : X/\ker(L) \to \mathcal{R}(L)$ by

$$T([x]) = L(x) \text{ for all } [x] \in X/\ker(L).$$

   (a) Prove that T is a well-defined linear operator.

   (b) Prove that T is an isomorphism.

   (c) Let $u \in \mathcal{R}(L)$ be given, and let $\hat{x} \in X$ be a solution to $L(x) = u$. In terms of $X/\ker(L)$, what is the solution set to $L(x) = u$? How can you describe $X/\ker(L)$ in terms of linear operator equations of the form $L(x) = v$, $v \in \mathcal{R}(L)$?

3.5 Existence and uniqueness of solutions

The reader will recall from Theorem 70 that a function $f : X \to Y$ has an inverse f^{-1} if and only if f is bijective (that is, both injective and surjective). For linear operators, both surjectivity and injectivity are equivalent to conditions that are easy to understand. In this section, we will study these conditions and apply them to the questions of existence and uniqueness of solutions to linear operator equations.

3.5.1 The kernel of a linear operator and injectivity

Injectivity is easy to characterize for a linear operator.

Theorem 92 *Let X and U be vector spaces over a field F, and let $T : X \to U$ be linear. Then T is injective if and only if $\ker(T)$ is trivial (that is, equals the trivial subspace $\{0\}$ of X).*

Proof Since T is linear, $T(0) = 0$. Suppose T is injective. Then

$$x \in \ker(T) \Rightarrow T(x) = 0 \Rightarrow T(x) = T(0) \Rightarrow x = 0$$

(the last step follows from the injectivity of T). Thus $\ker(T) = \{0\}$.

On the other hand, assume $\ker(T) = \{0\}$, and suppose $x, y \in X$. Because T is linear, we can reason as follows:

$$\begin{aligned} T(x) = T(y) \Rightarrow T(x) - T(y) = 0 &\Rightarrow T(x - y) = 0 \\ &\Rightarrow x - y \in \ker(T) = \{0\} \\ &\Rightarrow x - y = 0 \\ &\Rightarrow x = y. \end{aligned}$$

This shows that T is injective.

QED

The following theorem describes an important restriction on linear operators.

Theorem 93 *Let X and U be finite-dimensional vector spaces over a field F, and let $T : X \to U$ be linear and injective. Then* $\dim(X) \leq \dim(U)$.

Proof Let $\{x_1, x_2, \ldots, x_n\}$ be a basis for X (so that $\dim(X) = n$). It suffices to prove that $\{T(x_1), T(x_2), \ldots, T(x_n)\}$ is a linearly independent subset of U, since then Theorem 43 implies that $\dim(U) \geq n = \dim(X)$. But

$$\begin{aligned} & \alpha_1 T(x_1) + \alpha_2 T(x_2) + \cdots + \alpha_n T(x_n) = 0 \\ \Rightarrow\ & T(\alpha_1 x_1 + \alpha_2 x_2 + \cdots + \alpha_n x_n) = 0 \\ \Rightarrow\ & \alpha_1 x_1 + \alpha_2 x_2 + \cdots + \alpha_n x_n \in \ker(T) = \{0\} \\ \Rightarrow\ & \alpha_1 x_1 + \alpha_2 x_2 + \cdots + \alpha_n x_n = 0 \\ \Rightarrow\ & \alpha_1 = \alpha_2 x_2 = \ldots = \alpha_n = 0, \end{aligned}$$

where the last step follows from the linear independence of $\{x_1, x_2, \ldots, x_n\}$. This shows that $\{T(x_1), T(x_2), \ldots, T(x_n)\}$ is linearly independent.

QED

The significance of the previous theorem is more obvious when it is stated in the contrapositive:[5] If $\dim(X) > \dim(U)$, then $T : X \to U$ cannot be injective.

Example 94 *Let $D : \mathcal{P}_n \to \mathcal{P}_{n-1}$ be defined by $D(p) = p'$. Theorem 93 implies immediately that D is not injective, since $\dim(\mathcal{P}_n) > \dim(\mathcal{P}_{n-1})$. Indeed, we know from calculus that the derivative of any constant function is zero; therefore, two polynomials that differ by a constant have the same derivative.*

It should be emphasized that Theorem 93 does not hold for nonlinear operators.

Since the dimension of $\ker(T)$ is fundamental to the above results and some that follow, we have a name for it.

Definition 95 *Let X, U be vector spaces over a field F, and let $T : X \to U$ be linear. The dimension of* $\ker(T)$ *is called the* nullity *of T and is denoted by* $\text{nullity}(T)$.

We will also use the following terms.

Definition 96 *Let X, U be vector spaces over a field F, and let $T : X \to U$ be linear. We say that T is* singular *if and only if* $\ker(T)$ *is nontrivial and* nonsingular *if and only if* $\ker(T)$ *is trivial.*

---

[5]If a logical statement is "If P, then Q," its contrapositive is "If not Q, then not P." A little thought shows that a statement is true if and only if its contrapositive is true.

3.5.2 The rank of a linear operator and surjectivity

The *range* of a function $f : X \to Y$ is the set of all actual outputs of the function:

$$\mathcal{R}(f) = \{f(x) : x \in X\} \subset Y.$$

A function is surjective if and only if its range is all of its co-domain. When the function in question is a linear operator on finite-dimensional spaces, this leads to a simple equivalent condition for surjectivity. We will express this condition in terms of the following definition.

Definition 97 *Let X, U be vector spaces over a field F, and let $T : X \to U$ be linear. The dimension of the range $\mathcal{R}(T)$ of T is called the* rank *of T and is denoted by* $\text{rank}(T)$.

Theorem 98 *Let X and U be finite-dimensional vector spaces over a field F, and let $T : X \to U$ be linear. Then T is surjective if and only if the rank of T equals the dimension of U.*

Proof If T is surjective, then $\mathcal{R}(T) = U$, and hence $\text{rank}(T) = \dim(U)$. On the other hand, if $\text{rank}(T) = \dim(U) = m$, then $\mathcal{R}(T)$ contains a set of m linearly independent vectors, say $\{T(x_1), T(x_2), \ldots, T(x_m)\}$. Theorem 45 guarantees that any linearly independent set of m vectors in U is a basis for U, so

$$U = \text{sp}\{T(x_1), T(x_2), \ldots, T(x_m)\} \subset \mathcal{R}(T).$$

Since we know that $\mathcal{R}(T) \subset U$, this shows that $\mathcal{R}(T) = U$.

QED

We now derive the companion to Theorem 93.

Theorem 99 *Let X and U be vector spaces over a field F, and let $T : X \to U$ be linear. If T is surjective, then* $\dim(U) \leq \dim(X)$.

Proof If T is surjective, then there exists a basis for U consisting of vectors in the range of T, say $\{T(x_1), T(x_2), \ldots, T(x_m)\}$. We must show that $\dim(X) \geq m$. To do this, it suffices to prove that $\{x_1, x_2, \ldots, x_m\}$ is a linearly independent subset of X. But this is straightforward:

$$\begin{aligned}
& \alpha_1 x_1 + \alpha_2 x_2 + \cdots + \alpha_m x_m = 0 \\
\Rightarrow \quad & T(\alpha_1 x_1 + \alpha_2 x_2 + \cdots + \alpha_m x_m) = 0 \\
\Rightarrow \quad & \alpha_1 T(x_1) + \alpha_2 T(x_2) + \cdots + \alpha_m T(x_m) = 0 \\
\Rightarrow \quad & \alpha_1 = \alpha_2 = \ldots = \alpha_m = 0,
\end{aligned}$$

where the last step follows from the fact that $\{T(x_1), T(x_2), \ldots, T(x_m)\}$ is linearly independent. We have shown that $\{x_1, x_2, \ldots, x_m\}$ is linearly independent, and the proof is complete.

QED

As in the case of Theorem 93, the previous theorem is more understandable in its contrapositive form: If $\dim(U) > \dim(X)$, then no linear operator $T : X \to U$ can be surjective.

Example 100 *Let $S : \mathcal{P}_{n-1} \to \mathcal{P}_n$ be defined by $S(p) = q$, where $q(x) = \int_0^x p(t)\,dt$ or, equivalently,*

$$S\left(a_0 + a_1 x + \ldots + a_{n-1}x^{n-1}\right) = a_0 x + \frac{a_1}{2}x^2 + \ldots + \frac{a_{n-1}}{n}x^n.$$

Since $\dim(\mathcal{P}_n) > \dim(\mathcal{P}_{n-1})$*, Theorem 99 implies that S is not surjective. Exercise 12 asks the reader to identify elements of $\mathcal{P}_n$ not belonging to $\mathcal{R}(S)$.*

Like Theorem 93, Theorem 99 is not true for nonlinear operators (cf. Exercise 16).

The proof of Theorem 99 shows that the largest possible value for $\text{rank}(T)$ is $\dim(X)$, which explains the following terminology.

Definition 101 *Let X and U be finite-dimensional vector spaces over a field F, and let $T : X \to U$ be linear. If* $\text{rank}(T) = \dim(X)$*, then we say that T has* full rank.

Theorems 93 and 99 lead immediately to the following fundamental result.

Theorem 102 *Let X and U be finite-dimensional vector spaces over a field F and let $T : X \to U$ be linear. If T is bijective, then* $\dim(X) = \dim(U)$.

Proof If T is injective, then Theorem 93 shows that $\dim(X) \leq \dim(U)$, while if T is surjective, then Theorem 99 shows that $\dim(X) \geq \dim(U)$. The result follows immediately.

QED

This implies that the converse of Theorem 76 holds, and we thus obtain the following result.

Corollary 103 *Let X and U be finite-dimensional vectors spaces over a field F. Then X and U are isomorphic if and only if* $\dim(X) = \dim(U)$.

Therefore, if $\dim(X) \neq \dim(U)$, then X and U are not isomorphic.

The reader should notice that the converse of Theorem 102 itself is not true, namely, if $T : X \to U$ is linear and $\dim(X) = \dim(U)$, this does not imply that T is bijective (see Exercise 1). In other words, $\dim(X) = \dim(U)$ is a necessary and sufficient condition that there exists *some* isomorphism $T : X \to U$; however, for a particular linear operator $T : X \to U$, $\dim(X) = \dim(U)$ tells us nothing except that it is possible that T is an isomorphism.

3.5.3 Existence and uniqueness

We can now discuss existence and uniqueness for a linear operator equation $T(x) = u$, where $T : X \to U$ and x is the unknown. There are two senses in which we can discuss existence and uniqueness questions:

1. Given a specific $u \in U$, does there exist an $x \in X$ such that $T(x) = u$ and, if so, is it unique?

2. Does $T(x) = u$ have at least one solution $x \in X$ for every $u \in U$? Does $T(x) = u$ have at most one solution $x \in X$ for every $u \in U$?

We wish to discuss existence and uniqueness in the second sense. Our discussion consists only of expressing results obtained above in different language, and in this sense it might be thought to offer little insight. However, the terminology of linear algebra offers several different ways to express many of the key concepts, and it is important to be fluent in this language.

Given a linear operator $T : X \to U$, we say that T satisfies the *existence property* if there exists at least one solution $x \in X$ to $T(x) = u$ for every $u \in U$. The existence property is equivalent to the surjectivity of T and therefore, by Theorem 98, to the condition $\mathrm{rank}(T) = \dim(U)$.

We say that T satisfies the *uniqueness property* if there exists at most one solution $x \in X$ for each $u \in U$. The uniqueness property is equivalent to the injectivity of T and therefore, by Theorem 92, to the condition $\ker(T) = \{0\}$.

What is the significance of these results? Taking the second result first, we see that we can settle the uniqueness of solutions to $T(x) = u$, for all possible u, by solving the single equation $T(x) = 0$. If $T(x) = 0$ has only the trivial solution, then, for any $u \in U$, $T(x) = u$ has at most one solution. For nonlinear operators, such a procedure is not possible; if T is nonlinear, it is perfectly possible that $T(x) = 0$ has exactly one solution and $T(x) = u$, for some $u \neq 0$, has two solutions (or any other number of solutions).

We also see that the existence question can be answered by computing the $\mathrm{rank}(T)$. This might not seem so easy, but we will see in the next section that the fundamental theorem of linear algebra allows us to compute $\mathrm{rank}(T)$ easily from $\mathrm{nullity}(T) = \dim(\ker(T))$ (provided $\dim(X)$ is known). Therefore, solving the equation $T(x) = 0$ answers, not only the uniqueness question, but also the question of existence of solutions to $T(x) = u$ for every $u \in U$.

Exercises

Miscellaneous exercises

1. Give an example of a matrix operator $T : \mathbf{R}^n \to \mathbf{R}^n$ that is not bijective.

2. Each part of this exercise describes an operator with certain properties. If there exists such an operator, give an example. If it is impossible for such an example to exist, explain why.

(a) A linear operator $T : \mathbf{R}^3 \to \mathbf{R}^2$ that is nonsingular.
(b) A linear operator $T : \mathbf{R}^2 \to \mathbf{R}^3$ that is nonsingular.
(c) A linear operator $T : \mathbf{R}^3 \to \mathbf{R}^2$ having full rank.
(d) A linear operator $T : \mathbf{R}^2 \to \mathbf{R}^3$ having full rank.
(e) A linear operator $T : \mathbf{R}^2 \to \mathbf{R}^3$ that is nonsingular but not invertible.
(f) An invertible linear operator $T : \mathbf{R}^2 \to \mathbf{R}^2$.

3. Each part of this exercise describes an operator with certain properties. If there exists such an operator, give an example. If it is impossible for such an example to exist, explain why.

   (a) A linear operator $T : \mathbf{R}^3 \to \mathbf{R}^2$ such that $T(x) = b$ has a solution for all $b \in \mathbf{R}^2$.
   (b) A linear operator $T : \mathbf{R}^2 \to \mathbf{R}^3$ such that $T(x) = b$ has a solution for all $b \in \mathbf{R}^3$.
   (c) A linear operator $T : \mathbf{R}^3 \to \mathbf{R}^2$ such that, for some $b \in \mathbf{R}^2$, the equation $T(x) = b$ has infinitely many solutions.
   (d) A linear operator $T : \mathbf{R}^2 \to \mathbf{R}^3$ such that, for some $b \in \mathbf{R}^3$, the equation $T(x) = b$ has infinitely many solutions.
   (e) A linear operator $T : \mathbf{R}^2 \to \mathbf{R}^3$ with the property that $T(x) = b$ does not have a solution for all $b \in \mathbf{R}^3$, but when there is a solution, it is unique.

4. Each part of this exercise describes an operator with certain properties. If there exists such an operator, give an example. If it is impossible for such an example to exist, explain why.

   (a) A linear operator $T : \mathbf{Z}_2^3 \to \mathbf{Z}_2^2$ that is surjective.
   (b) A linear operator $T : \mathbf{Z}_2^2 \to \mathbf{Z}_2^3$ that is surjective.
   (c) A linear operator $T : \mathbf{Z}_3^3 \to \mathbf{Z}_3^2$ that is injective.
   (d) A linear operator $T : \mathbf{Z}_3^2 \to \mathbf{Z}_3^3$ that is injective.
   (e) A linear operator $T : \mathbf{Z}_2^2 \to \mathbf{Z}_2^3$ that is injective but not bijective.

5. Define $M : \mathbf{R}^4 \to \mathbf{R}^3$ by

$$M(x) = \begin{bmatrix} x_1 + 3x_2 - x_3 - x_4 \\ 2x_1 + 7x_2 - 2x_3 - 3x_4 \\ 3x_1 + 8x_2 - 3x_3 - 16x_4 \end{bmatrix}.$$

   Find the rank and nullity of M.

6. Define $T : \mathcal{P}_n \to \mathcal{P}_{n+1}$ by $T(p)(x) = xp(x)$ (that is, for each $p \in \mathcal{P}_n$, $T(p)$ is the polynomial q defined by $q(x) = xp(x)$). Find the rank and nullity of T.

7. Define $S : \mathcal{P}_n \to \mathcal{P}_n$ by $S(p)(x) = p(2x+1)$ (that is, for each $p \in \mathcal{P}_n$, $S(p)$ is the polynomial q defined by $q(x) = p(2x+1)$). Find the rank and nullity of S.

8. Define $L : \mathbf{Z}_5^3 \to \mathbf{Z}_5^3$ by $L(x) = (x_1+x_2+x_3, x_1-2x_2+x_3, x_1+4x_2+x_3)$. Find the rank and nullity of L.

9. Let p be a prime number. Given a positive integer n, define

$$T : \mathbf{Z}_p^{n+1} \to \mathcal{P}_n(\mathbf{Z}_p),$$
$$T(c) = c_1 + c_2x + c_3x^2 + \cdots + c_{n+1}x^n.$$

   For which values of n is T an isomorphism? Prove your answer.

10. Is the following statement a theorem?

    Let X and U be vector spaces over a field F, and let $T : X \to U$ be linear. Then $\{x_1, x_2, \ldots, x_n\} \subset X$ is linearly independent if and only if $\{T(x_1), T(x_2), \ldots, T(x_n)\} \subset U$ is linearly independent.

    If it is, prove it. If it is not, give a counterexample.

11. Suppose X and U are vector spaces over a field F, with U finite-dimensional, and $L : X \to U$ is linear. Let $\{u_1, u_2, \ldots, u_m\}$ be a basis for U and assume that, for each j, $L(x) = u_j$ has a solution $x \in X$. Prove that L is surjective.

12. Let S be the operator of Example 100. Determine the polynomials in $\mathcal{P}_n$ not belonging to $\mathcal{R}(S)$.

13. (a) Suppose X and U are finite-dimensional vector spaces over a field F and $T : X \to U$ is an injective linear operator. Prove that T defines an isomorphism between X and a subspace of U. What is the subspace?

    (b) We have already seen that the operator $S : \mathcal{P}_{n-1} \to \mathcal{P}_n$ from Example 100 is injective. Therefore, by the previous part of this exercise, S defines an isomorphism between $\mathcal{P}_{n-1}$ and a subspace of $\mathcal{P}_n$. Give a basis for this subspace.

14. Let V be a vector space over a field F, and let S, T be subspaces of V. Since S and T are vector spaces in their own right, we can define the product $S \times T$ (see Exercise 2.2.15). We can also define the subspace $S + T$ of V (see Exercise 2.3.21). Define a linear operator $L : S \times T \to S + T$ by $L((s,t)) = s + t$.

    (a) Prove that $\ker(L)$ is isomorphic to $S \cap T$ and find an isomorphism.

    (b) Suppose $S \cap T = \{0\}$. Prove that $S \times T$ is isomorphic to $S + T$ and that L is an isomorphism.

15. Let $K : C[c,d] \to C[a,b]$ be an integral operator of the form

$$K(x) = y, \ y(s) = \int_c^d k(s,t)x(t)\,dt, \ a \le s \le b$$

(cf. Exercise 3.1.11). Suppose the kernel[6] k has the special form

$$k(s,t) = \sum_{i=1}^{N} f_i(s)g_i(t),$$

where $f_i \in C[a,b]$, $g_i \in C[c,d]$ for $i = 1, 2, \ldots, N$ (such a kernel is said to be *separable*). Prove that $\text{rank}(K) \le N$. Can you find conditions under which $\text{rank}(K) = N$?

16. Every real number x can be written uniquely in a decimal expansion of the form

$$x = \ldots x_k x_{k-1} \ldots x_0 . x_{-1} x_{-2} \ldots,$$

where each digit x_i belongs to $\{0, 1, 2, \ldots, 9\}$, $x_i \neq 0$ for only finitely many positive integers i, and $x_i \neq 0$ for infinitely many negative integers i. The last requirement is required for uniqueness; otherwise, for example, the real number $1/2$ would have two decimal expansions, $0.5000\ldots$ and $0.4999\ldots$. The above conditions ensure that the second form is chosen. Define $f : \mathbf{R}^2 \to \mathbf{R}$ by

$$f(x,y) = \ldots x_k y_k x_{k-1} y_{k-1} \ldots x_0 y_0 . x_{-1} y_{-1} x_{-2} y_{-2} \ldots.$$

Prove that f is a bijection. (This example shows that Theorems 93 and 99 do not hold for nonlinear operators.)

3.6 The fundamental theorem; inverse operators

We now have a necessary condition for the existence of T^{-1} when T is linear, namely, that $\dim(X) = \dim(U)$. The following result, which is sometimes called the *fundamental theorem of linear algebra*, will lead to useful necessary and sufficient conditions.

Theorem 104 *Let X and U be vector spaces over a field F, where X is finite-dimensional, and let $T : X \to U$ be linear. Then*

$$\text{rank}(T) + \text{nullity}(T) = \dim(X).$$

[6]The reader should recall from Exercise 3.1.11 that the word "kernel" has a special meaning in the context of integral operators; it does not refer to the kernel of a linear operator.

Proof Let $\{x_1, \ldots, x_k\}$ be a basis for $\ker(T)$. By Theorem 43, there exist vectors $x_{k+1}, \ldots, x_n$ such that $\{x_1, \ldots, x_k, x_{k+1}, \ldots, x_n\}$ is a basis for X. With this notation, $\text{nullity}(T) = k$ and $\dim(X) = n$, so we must show that $\text{rank}(T) = n - k$. For now, we assume that $0 < k < n$ (so that $\ker(T)$ is a proper subspace of X).

First of all, we show that $\{T(x_{k+1}), \ldots, T(x_n)\}$ is linearly independent. Suppose scalars $\alpha_{k+1}, \ldots, \alpha_n \in F$ satisfy

$$\alpha_{k+1}T(x_{k+1}) + \ldots + \alpha_n T(x_n) = 0.$$

By linearity, we have

$$T(\alpha_{k+1}x_{k+1} + \ldots + \alpha_n x_n) = 0,$$

and hence $\alpha_{k+1}x_{k+1} + \ldots + \alpha_n x_n \in \ker(T)$. But then, since $\{x_1, \ldots, x_k\}$ is a basis for $\ker(T)$, it follows that there exist scalars $\alpha_1, \ldots, \alpha_k$ such that

$$\alpha_{k+1}x_{k+1} + \ldots + \alpha_n x_n = \alpha_1 x_1 + \ldots + \alpha_k x_k,$$

whence

$$\alpha_1 x_1 + \ldots + \alpha_k x_k - \alpha_{k+1}x_{k+1} - \ldots - \alpha_n x_n = 0.$$

The linear independence of $\{x_1, \ldots, x_n\}$ then implies that all of the scalars, including $\alpha_{k+1}, \ldots, \alpha_n$, are zero, and thus $\{T(x_{k+1}), \ldots, T(x_n)\}$ is linearly independent.

We now show that $\{T(x_{k+1}), \ldots, T(x_n)\}$ spans $\mathcal{R}(T)$. Consider any u belonging to $\mathcal{R}(T)$. By definition, there exists $x \in X$ such that $T(x) = u$. Since $\{x_1, \ldots, x_n\}$ is a basis for X, there exist $\alpha_1, \ldots, \alpha_n \in F$ such that

$$x = \alpha_1 x_1 + \ldots + \alpha_n x_n.$$

But then

$$\begin{aligned} u &= T(\alpha_1 x_1 + \ldots + \alpha_n x_n) \\ &= \alpha_1 T(x_1) + \ldots + \alpha_k T(x_k) + \alpha_{k+1}T(x_{k+1}) + \ldots + \alpha_n T(x_n) \\ &= \alpha_{k+1}T(x_{k+1}) + \ldots + \alpha_n T(x_n), \end{aligned}$$

since $T(x_1) = \ldots = T(x_k) = 0$. Therefore u is a linear combination of $T(x_{k+1}), \ldots, T(x_n)$, and hence $\{T(x_{k+1}), \ldots, T(x_n)\}$ spans $\mathcal{R}(T)$. We have thus shown that $\{T(x_{k+1}), \ldots, T(x_n)\}$ is a basis for $\mathcal{R}(T)$.

The reader can verify that, if $k = 0$ and the kernel of T is trivial, then the above proof remains valid with only slight changes. If $k = n$, then $\mathcal{R}(T)$ is the trivial subspace of U, and we have $\text{nullity}(T) = n$ and $\text{rank}(T) = 0$. Thus the theorem holds in this case as well.

QED

The reader should notice that $\ker(T)$ is a subspace of X, while $\mathcal{R}(T)$ is a subspace of U, which makes it surprising on the surface that there is such a strong connection between their dimensions.

Corollary 105 *Let X and U be n-dimensional vector spaces over a field F, and let $T : X \to U$ be linear. Then T is surjective if and only if it is injective.*

Proof Exercise 1.

As discussed in Section 3.5.3, the fundamental theorem gives $\text{rank}(T)$ immediately from $\text{nullity}(T)$:

$$\text{rank}(T) = \dim(X) - \text{nullity}(T).$$

Therefore, solving $T(x) = 0$ is the key to understanding both the existence and uniqueness questions for $T(x) = u$.

Example 106 *Let $A \in \mathbf{R}^{3\times 5}$ be defined by*

$$A = \begin{bmatrix} 1 & -5 & 4 & -2 & 2 \\ 1 & -6 & 5 & -3 & 2 \\ -2 & 11 & -8 & 5 & -2 \end{bmatrix}$$

and let $T : \mathbf{R}^5 \to \mathbf{R}^3$ be the linear operator defined by $T(x) = Ax$ for all $x \in \mathbf{R}^5$. Applying Gaussian elimination to the system represented by $T(x) = 0$ (that is, $Ax = 0$), we obtain

$$\begin{aligned} x_1 + 3x_4 + 4x_5 &= 0, \\ x_2 + x_4 + 2x_5 &= 0, \\ x_3 + 2x_5 &= 0 \end{aligned} \quad \Longleftrightarrow \quad \begin{aligned} x_1 &= -3x_4 - 4x_5, \\ x_2 &= -x_4 - 2x_5, \\ x_3 &= -2x_5. \end{aligned}$$

The general solution is therefore

$$x = (-3\alpha - 4\beta, -\alpha - 2\beta, -2\beta, \alpha, \beta) = \alpha(-3, -1, 0, 1, 0) + \beta(-4, -2, -2, 0, 1),$$

which shows that $\ker(T)$ *has dimension 2. By the fundamental theorem,*

$$\text{rank}(T) + \text{nullity}(T) = \dim(\mathbf{R}^5) \ \Rightarrow \ \text{rank}(T) = 5 - 2 = 3.$$

Since $\text{rank}(T) = \dim(\mathbf{R}^3)$*, we see that T is surjective. Therefore, for each $u \in \mathbf{R}^3$, there exists a solution $x \in \mathbf{R}^5$ to $T(x) = u$. However, the solution is not unique since* $\ker(T)$ *is nontrivial.*

3.6.1 The inverse of a linear operator

We recall that $T : X \to U$ is invertible if and only if T is bijective. The previous corollary simplifies this result for linear operators.

Theorem 107 *Let X and U be n-dimensional vector spaces over a field F, and let $T : X \to U$ be linear.*

1. *T is invertible if and only if it is injective.*
2. *T is invertible if and only if it is surjective.*

Of course, if $\dim(X) \neq \dim(U)$, then T cannot be invertible.

The first result of Theorem 107 can also be expressed as "T is invertible if and only if it is nonsingular," while the second can be stated as "T is invertible if and only if it has full rank."

The inverse of a linear operator is always linear.

Theorem 108 *Let X and U be vector spaces over a field F, and let $T : X \to U$ be linear. If T is invertible, then T^{-1} is also linear.*

Proof Exercise 2.

Theorem 108 is often called the *principle of superposition*, especially in the context of (differential equations arising in) physics and engineering.

Example 109 *The differential equation*

$$mx'' + kx = f(t),\ t > 0,$$

together with the initial conditions $x(0) = 0$, $x'(0) = 0$, *describes a simple mass-spring system. An object of mass m hanging on a spring is initially at rest (this is the meaning of the initial conditions) and an external force f is applied to the mass, beginning at $t = 0$. (The constant k is the spring constant, describing how much the spring resists stretching.) The solution to the differential equation is the displacement $x = x(t)$ of the mass from equilibrium.*

To pose the problem of finding x given f in linear algebraic terms, we define

$$X = \left\{x \in C^2[0,\infty) : x(0) = x'(0) = 0\right\}$$

and $T : X \to C[0,\infty)$ by

$$T(x) = mx'' + kx.$$

Then, given $f \in C[0,\infty)$, we wish to solve $T(x) = f$. It is a fact from elementary differential equations (see, for example, [46], Section 4.1) that there is a unique solution for each f; that is, T is invertible. Since T is a linear operator, T^{-1} is also linear by Theorem 108. Therefore, given two forces f and g, we have $T^{-1}(f+g) = T^{-1}(f) + T^{-1}(g)$. This means that the response to the superposition of two forces can be obtained by computing the response to each force individually and adding them together. In particular, $T^{-1}(2f) = 2T^{-1}(f)$, meaning that if the applied force is doubled in magnitude, the resulting displacement is also doubled.

3.6.2 The inverse of a matrix

A matrix $A \in F^{m\times n}$, where F is a field, defines a linear operator $T : F^n \to F^m$ by matrix-vector multiplication. Since T^{-1} is also linear (when it exists), it must also be represented by a matrix.

Before we study the inverse of a matrix, we make some definitions that allow us to avoid defining the operator T when this is not desirable.

Definition 110 *Let A be an $m \times n$ matrix with entries in a field F.*

1. *The* null space $\mathcal{N}(A)$ *of A is the set of all solutions in F^n to $Ax = 0$:*

$$\mathcal{N}(A) = \{x \in F^n \, : \, Ax = 0\}.$$

We say that A is nonsingular *if and only if $\mathcal{N}(A)$ is trivial, and* singular *otherwise.*

2. *The* column space $\mathrm{col}(A)$ *of A is the set of all vectors in F^m of the form Ax:*

$$\mathrm{col}(A) = \{Ax \, : \, x \in F^n\}.$$

If $A \in \mathbf{R}^{m \times n}$ and $T : F^n \to F^m$ is the linear operator defined by $T(x) = Ax$, then $\mathcal{N}(A) = \ker(T)$ and $\mathrm{col}(A) = \mathcal{R}(T)$. Therefore, we know that both of these sets are subspaces. We will use the terms *nullity* and *rank* to refer to either the matrix or the operator it defines:

$$\begin{aligned} \mathrm{nullity}(A) &= \dim(\mathcal{N}(A)) = \dim(\ker(T)) = \mathrm{nullity}(T), \\ \mathrm{rank}(A) &= \dim(\mathrm{col}(A)) = \dim(\mathcal{R}(T)) = \mathrm{rank}(T). \end{aligned}$$

A matrix A has full rank if and only if its rank equals the number of columns (which is the dimension of the domain of the matrix operator). Equivalently, a matrix has full rank if and only if its columns are linearly independent.

As we saw in Section 3.1.1, matrix-matrix multiplication is not commutative. It is, however, associative, and we will need this fact below.

Theorem 111 *Let F be a field and let $A \in F^{m \times n}$, $B \in F^{n \times p}$, and $C \in F^{p \times q}$. Then*

$$(AB)C = A(BC).$$

Proof Matrix-matrix multiplication was defined so that $(AB)x = A(Bx)$ for all vectors x. We have

$$BC = [BC_1|BC_2|\cdots|BC_q],$$

where $C_1, C_2, \ldots, C_q$ are the columns of C. Therefore,

$$\begin{aligned} A(BC) &= [A(BC_1)|A(BC_2)|\cdots|A(BC_q)] \\ &= [(AB)C_1|(AB)C_2|\cdots|(AB)C_q] \\ &= (AB)C. \end{aligned}$$

QED

When we restrict ourselves to square matrices, there is a multiplicative identity for matrix-matrix multiplication.

Definition 112 *Let F be a field. The $n \times n$* identity matrix *over F is the matrix $I = [e_1|e_2|\cdots|e_n]$, where $\{e_1, e_2, \ldots, e_n\}$ is the standard basis for F^n. In other words,*

$$I = \begin{bmatrix} 1 & 0 & \cdots & 0 \\ 0 & 1 & \cdots & 0 \\ \vdots & \vdots & \ddots & \vdots \\ 0 & 0 & \cdots & 1 \end{bmatrix}.$$

Theorem 113 *Let F be a field and let I be the $n \times n$ identity matrix over F.*

1. *For all $x \in F^n$, $Ix = x$.*
2. *For all $A \in F^{n \times n}$, $AI = A$ and $IA = A$.*

Proof

1. For any $x \in F^n$, we have $x = x_1e_1 + x_2e_2 + \ldots + x_ne_n$ and so
$$Ix = x_1e_1 + x_2e_2 + \ldots + x_ne_n = x.$$
2. If $A_1, A_2, \ldots, A_n$ are the columns of A, then
$$AI = [Ae_1|Ae_2|\cdots|Ae_n] = [A_1|A_2|\cdots|A_n] = A.$$
On the other hand, using the first part of the theorem,
$$IA = [IA_1|IA_2|\cdots|IA_n] = [A_1|A_2|\cdots|A_n] = A.$$

QED

The previous result is also valid if A is not square: If I is the $n \times n$ identity matrix, then $AI = A$ for all $A \in \mathbf{R}^{m \times n}$ and $IA = A$ for all $A \in \mathbf{R}^{n \times m}$.

Definition 114 *Let F be a field and $A \in F^{n \times n}$. We say that A is* invertible *if and only if there exists a matrix $B \in F^{n \times n}$ such that*

$$AB = I \text{ and } BA = I, \tag{3.10}$$

where I is the $n \times n$ identity matrix. If A is invertible, the matrix B satisfying (3.10) is called the inverse *of A and is denoted A^{-1}.*

Theorem 115 *Let F be a field, let $A \in F^{n \times n}$, and let $T : F^n \to F^n$ be defined by $T(x) = Ax$. Then A is invertible if and only if T is invertible. If A is invertible, then A^{-1} is the matrix defining T^{-1}; that is, $T^{-1}(u) = A^{-1}u$ for all $u \in F^n$.*

Proof Exercise 3.

Based on the previous theorem, we can express our earlier results, particularly Theorem 107, in terms of matrices instead of operators. For instance, $A \in F^{n\times n}$ is invertible if and only if it is nonsingular, that is, if and only if its null space is trivial. Similarly, $A \in F^{n\times n}$ is invertible if and only if it has full rank.

Here are a couple of simple examples of matrices, one invertible and one not.

Example 116 *Let*

$$A = \begin{bmatrix} 3 & 1 \\ 5 & 2 \end{bmatrix}, \ B = \begin{bmatrix} 2 & -1 \\ -5 & 3 \end{bmatrix}.$$

Direct calculation shows that $AB = I$ *and* $BA = I$*, so* A *is invertible and* $B = A^{-1}$.

Example 117 *Let*

$$A = \begin{bmatrix} 1 & 1 \\ 1 & 1 \end{bmatrix}.$$

Then $\mathrm{col}(A) = \mathrm{sp}\{(1,1)\}$*, which shows that* $\mathrm{rank}(A) = 1$. *Therefore,* A *does not have full rank and is not invertible.*

The following theorem leads to the standard method for computing the inverse of an invertible matrix.

Theorem 118 *Let* F *be a field and suppose* $A \in F^{n\times n}$.

1. *If there exists a matrix* $B \in F^{n\times n}$ *such that* $AB = I$*, where* $I \in F^{n\times n}$ *is the identity matrix, then* A *is invertible and* $B = A^{-1}$.

2. *If there exists a matrix* $B \in F^{n\times n}$ *such that* $BA = I$*, where* $I \in F^{n\times n}$ *is the identity matrix, then* A *is invertible and* $B = A^{-1}$.

Proof

1. We assume that there exists $B \in F^{n\times n}$ such that $AB = I$. If u is an element in F^n and $x = Bu$, then

$$Ax = A(Bu) = (AB)u = Iu = u.$$

It follows that A defines a surjective linear operator and hence, by the fundamental theorem of linear algebra and Theorem 115, A is invertible. Moreover,

$$\begin{aligned} AB = I \Rightarrow A^{-1}(AB) = A^{-1}I &\Rightarrow (A^{-1}A)B = A^{-1} \\ &\Rightarrow IB = A^{-1} \Rightarrow B = A^{-1}. \end{aligned}$$

This proves the first part of the theorem.

2. Exercise 4.

QED

According to this theorem, to compute A^{-1} (when it exists), it suffices to solve the matrix-matrix equation $AB = I$. In the next section, we show how to use this fact to compute A^{-1}.

We end this section with the following fundamental property of matrix inverses.

Theorem 119 *Suppose $A, B \in \mathbf{R}^{n\times n}$ are both invertible. Then AB is invertible and $(AB)^{-1} = B^{-1}A^{-1}$.*

Proof Exercise 5.

Exercises

Miscellaneous exercises

1. Prove Corollary 105.
2. Prove Theorem 108.
3. Prove Theorem 115.
4. Prove the second part of Theorem 118.
5. Prove Theorem 119.
6. Let $M : \mathcal{P}_n \to \mathcal{P}_{n+1}$ be defined by $M(p) = q$, where $q(x) = xp(x)$.
   (a) Is M surjective? If not, find the range of M.
   (b) Is M injective? If not, find the kernel of M.
7. Repeat the previous exercise for the following operators:
   (a) $M : \mathbf{R}^2 \to \mathbf{R}^3$ defined by $M(x) = Ax$, where
   $$A = \begin{bmatrix} 1 & 1 \\ 1 & 0 \\ 0 & 1 \end{bmatrix}.$$
   (b) $M : \mathbf{R}^3 \to \mathbf{R}^2$ defined by $M(x) = Ax$, where
   $$A = \begin{bmatrix} 1 & 2 & 1 \\ 1 & 0 & -1 \end{bmatrix}.$$
8. Repeat Exercise 6 for the following operators:

(a) $M : \mathbf{Z}_3^2 \to \mathbf{Z}_3^3$ defined by $M(x) = Ax$, where

$$A = \begin{bmatrix} 1 & 2 \\ 2 & 1 \\ 1 & 2 \end{bmatrix}.$$

(b) $M : \mathbf{Z}_3^3 \to \mathbf{Z}_3^2$ defined by $M(x) = Ax$, where

$$A = \begin{bmatrix} 1 & 0 & 2 \\ 1 & 2 & 1 \end{bmatrix}.$$

9. Let X and U be finite-dimensional vector spaces over a field F, let $\mathcal{X} = \{x_1, x_2, \ldots, x_n\}$ and $\mathcal{U} = \{u_1, u_2, \ldots, u_m\}$ be bases for X and Y respectively, and let $T : X \to U$ be linear. Prove that T is invertible if and only if $[T]_{\mathcal{X},\mathcal{U}}$ is an invertible matrix.

10. State and prove the counterpart of Theorem 118 for linear operators.

11. Is the following statement true or false? If it is true, prove it; if not, give a counterexample.

    Let X and U be vector spaces over a field F, where X and U are finite-dimensional, and let $T : X \to U$ be linear. Then T is nonsingular if and only if T has full rank.

12. Construct a different proof to Theorem 104, as follows: Choose vectors $x_1, x_2, \ldots, x_k$ in X such that $\{T(x_1), T(x_2), \ldots, T(x_k)\}$ is a basis for $\mathcal{R}(T)$, and choose a basis $\{y_1, \ldots, y_\ell\}$ for $\ker(T)$. Prove that $\{x_1, \ldots, x_k, y_1, \ldots, y_\ell\}$ is a basis for X. (Hint: First prove that the intersection of $\ker(T)$ and $\mathrm{sp}\{x_1, x_2, \ldots, x_k\}$ is trivial.)

13. Let F be a field and suppose $A \in F^{m\times n}$, $B \in F^{n\times p}$. Prove that $\mathrm{rank}(AB) \le \mathrm{rank}(A)$.

14. Consider the following *initial value problem*: Find $x \in C^1(a, b)$ satisfying

$$x'(t) + p(t)x(t) = f(t), a < t < b, \tag{3.11}$$

$$x(t_0) = 0. \tag{3.12}$$

Here $p \in C(a, b)$, $f \in C(a, b)$, and $a < t_0 < b$. We will consider p and t_0 fixed throughout this problem. It is a theorem that given any $f \in C(a, b)$, there exists a unique $x \in C^1(a, b)$ satisfying (3.11–3.12).

(a) Express the IVP (3.11–3.12) as a linear operator equation. (Hint: The domain of the linear operator should incorporate the initial condition.)

(b) Explain why the mapping from f to the solution x is linear.

(c) Suppose the initial condition is changed to $x(t_0) = x_0$, where x_0 is a fixed nonzero real number. The same theorem still holds: For each $f \in C(a, b)$, there exists a unique $x \in C^1(a, b)$ satisfying (3.11–3.12). Prove that the mapping from f to x is not linear in this case.

15. A matrix $A \in \mathbf{C}^{n\times n}$ is called *diagonally dominant* if each diagonal entry is at least as large as the sum of the off-diagonal entries in that row:

$$|A_{ii}| \geq \sum_{\substack{j=1 \\ j \neq i}}^{n} |A_{ij}|, \quad i = 1, 2, \ldots, n.$$

The matrix A is *strictly* diagonally dominant if the previous inequality is strict for each i. Prove that a strictly diagonally dominant matrix is nonsingular.

Project: Left and right inverses

16. Let X and U be vector spaces over a field F, and let $T : X \to U$.

(a) If there exists an operator $S : U \to X$ such that

$$S(T(x)) = x \text{ for all } x \in X,$$

then S is called a *left inverse* of T.

(b) If there exists an operator $S : U \to X$ such that

$$T(S(u)) = u \text{ for all } u \in U,$$

then S is called a *right inverse* of T.

Prove the following theorem: Let X and U be vector spaces over a field F, and let $T : X \to U$ be linear.

(a) There exists a left inverse S of T if and only if T is injective (and hence nonsingular).

(b) There exists a right inverse S of T if and only if T is surjective.

17. Let $A \in F^{m\times n}$, where F is a field. If there exists a matrix $B \in F^{n\times m}$ such that $AB = I$, then B is called a *left inverse* for A, while if $B \in F^{m\times n}$ satisfies $BA = I$, then B is called a *right inverse* for A.

Prove the following theorem: Let $A \in F^{m\times n}$.

(a) There exists a left inverse B of A if and only if $\mathcal{N}(A)$ is trivial.

(b) There exists a right inverse B of A if and only if $\text{col}(A) = F^m$.

18. Use the previous exercise to give a different proof for Theorem 118.

19. Consider the operators D and S from Examples 94 and 100. Determine whether each statement is true or false:

    (a) D is a left inverse of S.

    (b) D is a right inverse of S.

    Prove your answers.

20. Let $M : \mathcal{P}_n \to \mathcal{P}_{n+1}$ be the operator defined in Exercise 6.

    (a) Does M have a left inverse? If so, what is it?

    (b) Does M have a right inverse? If so, what is it?

21. Repeat the preceding exercise for the operators defined in Exercise 7.

22. Let X, Y, and Z be vector spaces over a field F, and let $T : X \to Y$ and $S : Y \to Z$ be linear operators.

    (a) Prove that if S and T have left inverses R and P, respectively, then ST also has a left inverse. What is it?

    (b) Prove that if S and T have right inverses R and P, respectively, then ST also has a right inverse. What is it?

Project: Change of coordinates

Let X be a finite-dimensional vector space over a field F, and let $\mathcal{X} = \{x_1, x_2, \ldots, x_n\}$ and $\mathcal{Y} = \{y_1, y_2, \ldots, y_n\}$ be two bases for X. Any vector x in X can be expressed in terms of $\mathcal{X}$ or $\mathcal{Y}$, yielding coordinate vectors $[x]_{\mathcal{X}}$ and $[x]_{\mathcal{Y}}$. Also, any linear operator $L : X \to X$ can be represented by a matrix, using either basis: $[L]_{\mathcal{X},\mathcal{X}}$ or $[L]_{\mathcal{Y},\mathcal{Y}}$. In this project, we will find the relationships between $[x]_{\mathcal{X}}$ and $[x]_{\mathcal{Y}}$, and between $[L]_{\mathcal{X},\mathcal{X}}$ and $[L]_{\mathcal{Y},\mathcal{Y}}$.

23. Since $\mathcal{Y}$ is a basis for $\mathcal{X}$, each of the basis vectors $x_j \in \mathcal{X}$ can be written uniquely as

    $$x_j = \sum_{i=1}^{n} C_{ij} y_i$$

    for scalars $C_{1j}, C_{2j}, \ldots, C_{nj} \in F$. Let $C \in F^{n \times n}$ be defined by the scalars C_{ij}, $i, j = 1, 2, \ldots, n$. Prove that for all $x \in X$, $C[x]_{\mathcal{X}} = [x]_{\mathcal{Y}}$. We call C the *change of coordinates* matrix from basis $\mathcal{X}$ to basis $\mathcal{Y}$.

24. Prove that the change of coordinate matrix C is invertible. It then follows that $[x]_{\mathcal{X}} = C^{-1}[x]_{\mathcal{Y}}$.

25. Let $A = [L]_{\mathcal{X},\mathcal{X}}$ and $B = [L]_{\mathcal{Y},\mathcal{Y}}$. Find a formula for B in terms of A and the change of coordinate matrix C.

26. Generalize the preceding results to a linear operator $L : X \to U$, where X and U are two finite-dimensional vector spaces. To be specific, suppose X has bases $\{x_1, x_2, \ldots, x_n\}$ and $\mathcal{Y} = \{y_1, y_2, \ldots, y_n\}$, while U has bases $\mathcal{U} = \{u_1, u_2, \ldots, u_m\}$ and $\mathcal{V} = \{v_1, v_2, \ldots, v_m\}$. Let C and D be the change of coordinate matrices from $\mathcal{X}$ to $\mathcal{Y}$ and from $\mathcal{U}$ to $\mathcal{V}$, respectively. Find a formula for $[L]_{\mathcal{Y},\mathcal{V}}$ in terms of $[L]_{\mathcal{X},\mathcal{U}}$, C, and D.

3.7 Gaussian elimination

We have been assuming that the reader is familiar with the standard Gaussian elimination algorithm for solving a system of linear algebraic equations. In this section, we will review this algorithm and the terminology associated with it. Initially, all of our examples will involve real numbers.

A system of linear algebraic equations, written in matrix-vector form, is an example of a linear operator equation and the solutions of such systems illustrate the results of the preceding sections. Moreover, since every linear operator equation on finite-dimensional space is equivalent to a matrix-vector equation, it is essential to be able to solve such systems.

The standard algorithm for solving linear systems is properly called *Gaussian elimination* with *back substitution* and consists of two stages, as the name suggests. The Gaussian elimination stage of the algorithm systematically eliminates variables by adding multiples of one equation to other equations. We illustrate with the example

$$\begin{aligned} 2x_1 + 4x_2 - 2x_3 &= 4, \\ -4x_1 - 7x_2 + 8x_3 &= 1, \\ 6x_1 + 11x_2 - 7x_3 &= 9. \end{aligned}$$

For this 3×3 system, Gaussian elimination consists of two steps:

1. Eliminate x_1 from equations 2 and 3 by adding 2 times equation 1 to equation 2 and -3 times equation 1 to equation 3:

$$\begin{aligned} 2x_1 + 4x_2 - 2x_3 &= 4 \\ x_2 + 4x_3 &= 9 \\ -x_2 - x_3 &= -3. \end{aligned}$$

2. Eliminate x_2 from equation 3 by adding equation 2 to equation 3:

$$\begin{aligned} 2x_1 + 4x_2 - 2x_3 &= 4 \\ x_2 + 4x_3 &= 9 \\ 3x_3 &= 6. \end{aligned}$$

In the ideal case (such as the above example), Gaussian elimination transforms the system to an equivalent system in which the final equation contains a single unknown, the next-to-last equation contains only two unknowns, and so forth. Back substitution solves the last equation for its unknown, substitutes this value into the next-to-last equation so that that equation can be solved for its other unknown, and so forth. In this example, the last equation is solved for x_3, the second equation is solved for x_2 (using the known value of x_3), and the first equation is solved for x_1 (using the known values of x_2 and x_3):

$$\begin{aligned} 3x_3 = 6 &\quad\Rightarrow\quad x_3 = 2 \\ x_2 + 4x_3 = 9 &\quad\Rightarrow\quad x_2 + 4 \cdot 2 = 9 \Rightarrow x_2 = 1 \\ 2x_1 + 4x_2 - 2x_3 = 4 &\quad\Rightarrow\quad 2x_1 + 4 \cdot 1 - 2 \cdot 2 = 4 \\ &\quad\Rightarrow\quad 2x_1 + 4 - 4 = 4 \\ &\quad\Rightarrow\quad 2x_1 = 4 \\ &\quad\Rightarrow\quad x_1 = 2 \end{aligned}$$

The solution is $x = (2, 1, 2)$.

For hand computation, it is simplest to regard back substitution as elimination. Solving the third equation is accomplished by multiplying through by $1/3$:

$$\begin{aligned} 2x_1 + 4x_2 - 2x_3 &= 4, \\ x_2 + 4x_3 &= 9, \\ x_3 &= 2. \end{aligned}$$

Substituting the value of x_3 into the other two equations is essentially equivalent to eliminating x_3 by adding -4 times equation 3 to equation 2 and 2 times equation 3 to equation 1:

$$\begin{aligned} 2x_1 + 4x_2 \qquad &= 8, \\ x_2 \qquad &= 1, \\ x_3 &= 2. \end{aligned}$$

The second equation is already solved for x_2, so the next step is to eliminate x_2 from equation 1 by adding -4 times equation 2 to equation 1:

$$\begin{aligned} 2x_1 \qquad\qquad &= 4, \\ x_2 \qquad &= 1, \\ x_3 &= 2. \end{aligned}$$

The final step is to multiply the first equation by $1/2$, which yields the solution:

$$\begin{aligned} x_1 \qquad\qquad &= 2, \\ x_2 \qquad &= 1, \\ x_3 &= 2. \end{aligned}$$

After applying Gaussian elimination with back substitution a few times, one realizes all the calculations are determined by the coefficients, and we might as well write the numbers only. In matrix-vector form, the system

$$\begin{aligned} 2x_1 + 4x_2 - 2x_3 &= 4, \\ -4x_1 - 7x_2 + 8x_3 &= 1, \\ 6x_1 + 11x_2 - 7x_3 &= 9 \end{aligned}$$

is written as $Ax = b$, where

$$A = \begin{bmatrix} 2 & 4 & -2 \\ -4 & -7 & 8 \\ 6 & 11 & -7 \end{bmatrix}, \; b = \begin{bmatrix} 4 \\ 1 \\ 9 \end{bmatrix}.$$

We can solve the system by applying the elimination algorithm directly to the *augmented matrix* $[A|b]$. Here is the previous example, solved again using this shorthand.

$$\left[\begin{array}{ccc|c} 2 & 4 & -2 & 4 \\ -4 & -7 & 8 & 1 \\ 6 & 11 & -7 & 9 \end{array}\right] \rightarrow \left[\begin{array}{ccc|c} 2 & 4 & -2 & 4 \\ 0 & 1 & 4 & 9 \\ 0 & -1 & -1 & -3 \end{array}\right] \rightarrow \left[\begin{array}{ccc|c} 2 & 4 & -2 & 4 \\ 0 & 1 & 4 & 9 \\ 0 & 0 & 3 & 6 \end{array}\right]$$
$$\rightarrow \left[\begin{array}{ccc|c} 2 & 4 & -2 & 4 \\ 0 & 1 & 4 & 9 \\ 0 & 0 & 1 & 2 \end{array}\right] \rightarrow \left[\begin{array}{ccc|c} 2 & 4 & 0 & 8 \\ 0 & 1 & 0 & 1 \\ 0 & 0 & 1 & 2 \end{array}\right] \rightarrow \left[\begin{array}{ccc|c} 2 & 0 & 0 & 4 \\ 0 & 1 & 0 & 1 \\ 0 & 0 & 1 & 2 \end{array}\right]$$
$$\rightarrow \left[\begin{array}{ccc|c} 1 & 0 & 0 & 2 \\ 0 & 1 & 0 & 1 \\ 0 & 0 & 1 & 2 \end{array}\right]$$

In the ideal case of a unique solution, we succeed in reducing the augmented matrix $[A|b]$ to $[I|x]$, where I is the identity matrix and x is the solution to $Ax = b$.

In the above example, we used two operations, which are called *elementary row operations*:

- Add a multiple of one row to another.
- Multiply a row by a nonzero constant.

In some problems, a third operation is needed to preserve the systematic nature of the algorithm:

- Interchange two rows.

The essential point about these elementary row operations is that each transforms the system to an equivalent system, that is, another system with the same solution set.

We will usually refer to Gaussian elimination with back substitution, when implemented using the augmented matrix notation, as simply *row reduction*.

Example 120 *Solve*

$$\begin{aligned} x_2 - 2x_3 &= -1 \\ x_1 + x_2 - x_3 &= 1, \\ 2x_1 + 2x_2 + 3x_3 &= 7. \end{aligned}$$

Notice that we cannot use the first equation to eliminate x_1 from equations 2 and 3, because x_1 is missing in equation 1. So we begin by interchanging rows 1 and 2:

$$\left[\begin{array}{ccc|c} 0 & 1 & -2 & -1 \\ 1 & 1 & -1 & 1 \\ 2 & 2 & 3 & 7 \end{array}\right] \rightarrow \left[\begin{array}{ccc|c} 1 & 1 & -1 & 1 \\ 0 & 1 & -2 & -1 \\ 2 & 2 & 3 & 7 \end{array}\right] \rightarrow \left[\begin{array}{ccc|c} 1 & 1 & -1 & 1 \\ 0 & 1 & -2 & -1 \\ 0 & 0 & 5 & 5 \end{array}\right]$$

$$\rightarrow \left[\begin{array}{ccc|c} 1 & 1 & -1 & 1 \\ 0 & 1 & -2 & -1 \\ 0 & 0 & 1 & 1 \end{array}\right] \rightarrow \left[\begin{array}{ccc|c} 1 & 1 & 0 & 2 \\ 0 & 1 & 0 & 1 \\ 0 & 0 & 1 & 1 \end{array}\right] \rightarrow \left[\begin{array}{ccc|c} 1 & 0 & 0 & 1 \\ 0 & 1 & 0 & 1 \\ 0 & 0 & 1 & 1 \end{array}\right].$$

The solution is $x = (1, 1, 1)$.

We can use Gaussian elimination with back substitution—systematic application of elementary row operations—to solve any linear system. If the system has no solution or has multiple solutions, this will become apparent in the course of the algorithm.

Example 121 *Consider the system*

$$\begin{aligned} x_1 + x_2 + 2x_3 &= 4, \\ 2x_1 - x_2 + 3x_3 &= 4, \\ 5x_1 + 2x_2 + 9x_3 &= 15. \end{aligned}$$

We try to solve by Gaussian elimination:

$$\left[\begin{array}{ccc|c} 1 & 1 & 2 & 4 \\ 2 & -1 & 3 & 4 \\ 5 & 2 & 9 & 15 \end{array}\right] \longrightarrow \left[\begin{array}{ccc|c} 1 & 1 & 2 & 4 \\ 0 & -3 & -1 & -4 \\ 0 & -3 & -1 & -5 \end{array}\right] \longrightarrow \left[\begin{array}{ccc|c} 1 & 1 & 2 & 4 \\ 0 & -3 & -1 & -4 \\ 0 & 0 & 0 & -1 \end{array}\right].$$

At this point, the transformed system is

$$\begin{aligned} x_1 + x_2 + 2x_3 &= 4, \\ -3x_2 - x_3 &= -4, \\ 0 &= -1. \end{aligned}$$

The last equation is impossible, so the system has no solution.

A system that has no solution, as in the previous example, is called *inconsistent.*

Example 122 *This is a variation on the last example:*

$$\begin{aligned} x_1 + \ x_2 + 2x_3 &= 4, \\ 2x_1 - \ x_2 + 3x_3 &= 4, \\ 5x_1 + 2x_2 + 9x_3 &= 16. \end{aligned}$$

We try to solve by Gaussian elimination:

$$\left[\begin{array}{ccc|c} 1 & 1 & 2 & 4 \\ 2 & -1 & 3 & 4 \\ 5 & 2 & 9 & 16 \end{array}\right] \longrightarrow \left[\begin{array}{ccc|c} 1 & 1 & 2 & 4 \\ 0 & -3 & -1 & -4 \\ 0 & -3 & -1 & -4 \end{array}\right] \longrightarrow \left[\begin{array}{ccc|c} 1 & 1 & 2 & 4 \\ 0 & -3 & -1 & -4 \\ 0 & 0 & 0 & 0 \end{array}\right].$$

Now the transformed equations are

$$\begin{aligned} x_1 + \ x_2 + 2x_3 &= 4, \\ -3x_2 - \ x_3 &= -4, \\ 0 &= 0. \end{aligned}$$

There is no inconsistency, but one of the equations canceled. We can continue the elimination to solve for x_2 and then x_1:

$$\left[\begin{array}{ccc|c} 1 & 1 & 2 & 4 \\ 0 & -3 & -1 & -4 \\ 0 & 0 & 0 & 0 \end{array}\right] \longrightarrow \left[\begin{array}{ccc|c} 1 & 1 & 2 & 4 \\ 0 & 1 & \frac{1}{3} & \frac{4}{3} \\ 0 & 0 & 0 & 0 \end{array}\right] \longrightarrow \left[\begin{array}{ccc|c} 1 & 0 & \frac{5}{3} & \frac{8}{3} \\ 0 & 1 & \frac{1}{3} & \frac{4}{3} \\ 0 & 0 & 0 & 0 \end{array}\right].$$

Now the transformed equations are

$$\begin{aligned} x_1 \qquad + \frac{5}{3}x_3 &= \frac{8}{3}, \\ x_2 + \frac{1}{3}x_3 &= \frac{4}{3}. \end{aligned}$$

or

$$\begin{aligned} x_1 &= \frac{8}{3} - \frac{5}{3}x_3, \\ x_2 &= \frac{4}{3} - \frac{1}{3}x_3. \end{aligned}$$

These equations tell us the values of x_1, x_2 given *the value of x_3, but x_3 can have any value. We say that x_3 is a* free variable.

Since x_3 can have any value, the system has infinitely many solutions. Setting $x_3 = \alpha$, we obtain the general solution:

$$x = \left(\frac{8}{3} - \frac{5}{3}\alpha, \frac{4}{3} - \frac{1}{3}\alpha, \alpha\right) = \left(\frac{8}{3}, \frac{4}{3}, 0\right) + \alpha\left(-\frac{5}{3}, -\frac{1}{3}, 1\right).$$

This expression can be written as $\hat{x} + \alpha y$, where

$$\hat{x} = \left(\frac{8}{3}, \frac{4}{3}, 0\right), \ y = \left(-\frac{5}{3}, -\frac{1}{3}, 1\right).$$

The general solution to the last example has the form predicted by Theorem 90. The reader should notice that $\{y\}$ is a basis for the null space of the matrix A. In fact, a little thought reveals that the general solution to $Ax = 0$ is precisely $x = \alpha y$, since $\hat{x}$ is the result of transforming the right-hand-side vector b during the course of the algorithm. If b is the zero vector, then $\hat{x}$ will be as well.

Example 123 *Let $L : \mathbf{R}^5 \to \mathbf{R}^3$ be the matrix operator defined by*

$$A = \begin{bmatrix} -1 & 2 & 4 & -19 & 82 \\ 4 & -9 & -20 & 113 & -375 \\ -2 & 4 & 7 & -28 & 156 \end{bmatrix},$$

and let $b = (-28, 134, -52)$. Then the inhomogeneous equation $L(x) = b$ is equivalent to the system

$$\begin{aligned} -x_1 + 2x_2 + \ 4x_3 - \ 19x_4 + \ 82x_5 &= -28, \\ 4x_1 - 9x_2 - 20x_3 + 113x_4 - 375x_5 &= 134, \\ -2x_1 + 4x_2 + \ 7x_3 - \ 28x_4 + 156x_5 &= -52. \end{aligned}$$

The reader can verify that row reduction transforms the augmented matrix

$$\left[\begin{array}{ccccc|c} -1 & 2 & 4 & -19 & 82 & -28 \\ 4 & -9 & -20 & 113 & -375 & 134 \\ -2 & 4 & 7 & -28 & 156 & -52 \end{array}\right]$$

to

$$\left[\begin{array}{ccccc|c} 1 & 0 & 0 & -15 & -20 & 0 \\ 0 & 1 & 0 & 3 & 15 & -6 \\ 0 & 0 & 1 & -10 & 8 & -4 \end{array}\right].$$

The transformed equations are

$$\begin{aligned} x_1 - \qquad\qquad 15x_4 - 20x_5 &= 0, \\ x_2 + \qquad 15x_4 - \ 6x_5 &= -6, \\ x_3 - 10x_4 + \ 8x_5 &= -4 \end{aligned}$$

or

$$\begin{aligned} x_1 &= 15x_4 + 20x_5, \\ x_2 &= -6 - 15x_4 + 6x_5, \\ x_3 &= -4 + 10x_4 - 8x_5. \end{aligned}$$

Setting $x_4 = \alpha$ and $x_5 = \beta$, we obtain the general solution:

$$\begin{aligned} x &= (15\alpha + 20\beta, -6 - 15\alpha + 6\beta, -4 + 10\alpha - 8\beta, \alpha, \beta) \\ &= (0, -6, -4, 0, 0) + \alpha(15, -15, 10, 1, 0) + \beta(20, 6, -8, 0, 1). \end{aligned}$$

If we define $\hat{x} = (0, -6, -4, 0, 0)$, $y = (15, -15, 10, 1, 0)$, and $z = (20, 6, -8, 0, 1)$, then the general solution is $x = \hat{x} + \alpha y + \beta z$. Here $\hat{x}$ is one solution to $Ax = b$, and $\mathcal{N}(A) = \mathrm{sp}\{y, z\}$.

When $\mathcal{N}(A)$ is nontrivial, row reduction always produces a basis for $\mathcal{N}(A)$ (see Exercise 1). This is true whether the algorithm is applied to $Ax = b$ $(b \neq 0)$ or $Ax = 0$.

3.7.1 Computing A^{-1}

The reader may have noticed that the sequence of row operations that arise in solving $Ax = b$ is determined entirely by the matrix A. Therefore, if we were to use row reduction to solve $Ax = B_1, Ax = B_2, \ldots, Ax = B_k$ for k different vectors $B_1, B_2, \ldots, B_k$, the only difference in the calculations would take place in the final column of the augmented matrix, which is initially the right-hand-side vector and ends up as the solution. We can perform these calculations all at once by applying row reduction to the augmented matrix $[A|B]$, where $B = [B_1|B_2|\cdots|B_k]$.

Example 124 *Let $A \in \mathbf{R}^{3\times 3}$, $B_1 \in \mathbf{R}^3$, and $B_2 \in \mathbf{R}^3$ be defined by*

$$A = \begin{bmatrix} -1 & -2 & -1 \\ -3 & -7 & 0 \\ 2 & 9 & -12 \end{bmatrix}, \; B_1 = \begin{bmatrix} 5 \\ 20 \\ -34 \end{bmatrix}, \; B_2 = \begin{bmatrix} 4 \\ 17 \\ -32 \end{bmatrix}.$$

We wish to find vectors $X_1, X_2 \in \mathbf{R}^3$ such that $AX_1 = B_1$, $AX_2 = B_2$. We do this by solving the matrix-matrix equation $AX = B$, where $B = [B_1|B_2]$, $X = [X_1|X_2]$, by row reducing $[A|B]$ to $[I|X]$:

$$\left[\begin{array}{ccc|cc} -1 & -2 & -1 & 5 & 4 \\ -3 & -7 & 0 & 20 & 17 \\ 2 & 9 & -12 & -34 & -32 \end{array}\right] \rightarrow \left[\begin{array}{ccc|cc} -1 & -2 & -1 & 5 & 4 \\ 0 & -1 & 3 & 5 & 5 \\ 0 & 5 & -14 & -24 & -24 \end{array}\right]$$

$$\rightarrow \left[\begin{array}{ccc|cc} -1 & -2 & -1 & 5 & 4 \\ 0 & -1 & 3 & 5 & 5 \\ 0 & 0 & 1 & 1 & 1 \end{array}\right] \rightarrow \left[\begin{array}{ccc|cc} -1 & -2 & 0 & 6 & 5 \\ 0 & -1 & 0 & 2 & 2 \\ 0 & 0 & 1 & 1 & 1 \end{array}\right]$$

$$\rightarrow \left[\begin{array}{ccc|cc} -1 & -2 & 0 & 6 & 5 \\ 0 & 1 & 0 & -2 & -2 \\ 0 & 0 & 1 & 1 & 1 \end{array}\right] \rightarrow \left[\begin{array}{ccc|cc} -1 & 0 & 0 & 2 & 1 \\ 0 & 1 & 0 & -2 & -2 \\ 0 & 0 & 1 & 1 & 1 \end{array}\right]$$

$$\rightarrow \left[\begin{array}{ccc|cc} 1 & 0 & 0 & -2 & -1 \\ 0 & 1 & 0 & -2 & -2 \\ 0 & 0 & 1 & 1 & 1 \end{array}\right]$$

We then see that $X_1 = (-2, -2, 1)$, $X_2 = (-1, -2, 1)$.

As the previous example shows, we can use row reduction to solve a matrix-matrix equation of the form $AX = B$, where A and B are given matrices and X is an unknown matrix. Theorem 118 shows that, to compute the inverse A^{-1} of an $n \times n$ matrix A, it suffices to solve the matrix-matrix equation $AB = I$, where now A is given, I is the $n \times n$ identity matrix, and $B \in \mathbf{R}^{n\times n}$ is unknown. If a solution B exists, Corollary 118 shows that $B = A^{-1}$. If a solution does not exist, it means that A is not invertible; the inconsistency of the equation $AB = I$ will become apparent during Gaussian elimination.

Example 125 *In this example, we compute A^{-1}, where $A \in \mathbf{R}^{3\times 3}$ is the matrix from the previous example. As we have explained, we find A^{-1} by row reducing $[A|I]$ to $[I|A^{-1}]$:*

$$\left[\begin{array}{ccc|ccc} -1 & -2 & -1 & 1 & 0 & 0 \\ -3 & -7 & 0 & 0 & 1 & 0 \\ 2 & 9 & -12 & 0 & 0 & 1 \end{array}\right] \rightarrow \left[\begin{array}{ccc|ccc} -1 & -2 & -1 & 1 & 0 & 0 \\ 0 & -1 & 3 & -3 & 1 & 0 \\ 0 & 5 & -14 & 2 & 0 & 1 \end{array}\right]$$

$$\rightarrow \left[\begin{array}{ccc|ccc} -1 & -2 & -1 & 1 & 0 & 0 \\ 0 & -1 & 3 & -3 & 1 & 0 \\ 0 & 0 & 1 & -13 & 5 & 1 \end{array}\right] \rightarrow \left[\begin{array}{ccc|ccc} -1 & -2 & 0 & -12 & 5 & 1 \\ 0 & -1 & 0 & 36 & -14 & -3 \\ 0 & 0 & 1 & -13 & 5 & 1 \end{array}\right]$$

$$\rightarrow \left[\begin{array}{ccc|ccc} -1 & -2 & 0 & -12 & 5 & 1 \\ 0 & 1 & 0 & -36 & 14 & 3 \\ 0 & 0 & 1 & -13 & 5 & 1 \end{array}\right] \rightarrow \left[\begin{array}{ccc|ccc} -1 & 0 & 0 & -84 & 33 & 7 \\ 0 & 1 & 0 & -36 & 14 & 3 \\ 0 & 0 & 1 & -13 & 5 & 1 \end{array}\right]$$

$$\rightarrow \left[\begin{array}{ccc|ccc} 1 & 0 & 0 & 84 & -33 & -7 \\ 0 & 1 & 0 & -36 & 14 & 3 \\ 0 & 0 & 1 & -13 & 5 & 1 \end{array}\right].$$

We thus see that

$$A^{-1} = \left[\begin{array}{ccc} 84 & -33 & -7 \\ -36 & 14 & 3 \\ -13 & 5 & 1 \end{array}\right].$$

A final note is in order: The examples in this section (and throughout the text) have been constructed so that the arithmetic is as simple as possible. This should allow the reader to focus on the concepts rather than tedious arithmetic. In actual practice, most computations involving linear algebra are destined to be performed by computers in finite-precision arithmetic, and of course there are no restrictions that the arithmetic be simple. On the other hand, finite-precision arithmetic always involves round-off error, which introduces other considerations, primarily whether the round-off can accumulate to such a degree as to ruin the accuracy of the final answer. In fact, this is possible on certain problems. The study of practical algorithms to perform the computations of linear algebra in finite-precision arithmetic is called *numerical linear algebra*, which is the topic of Chapter 9.

3.7.2 Fields other than R

Gaussian elimination works for matrices and vectors with entries from any field, not just the field of real numbers. The calculations by hand are fairly easy if the field is $\mathbf{Z}_p$ (at least for a small prime p), but quite tedious if the field is $\mathbf{C}$.

Example 126 *Let $A \in \mathbf{Z}_3^{3\times 3}$ be defined by*

$$A = \begin{bmatrix} 1 & 2 & 0 \\ 1 & 1 & 0 \\ 2 & 0 & 1 \end{bmatrix}.$$

We will compute A^{-1}. We apply row reduction as follows:

$$\left[\begin{array}{ccc|ccc} 1 & 2 & 0 & 1 & 0 & 0 \\ 1 & 1 & 0 & 0 & 1 & 0 \\ 2 & 0 & 1 & 0 & 0 & 1 \end{array}\right] \rightarrow \left[\begin{array}{ccc|ccc} 1 & 2 & 0 & 1 & 0 & 0 \\ 0 & 2 & 0 & 2 & 1 & 0 \\ 0 & 2 & 1 & 1 & 0 & 1 \end{array}\right]$$

$$\rightarrow \left[\begin{array}{ccc|ccc} 1 & 2 & 0 & 1 & 0 & 0 \\ 0 & 2 & 0 & 2 & 1 & 0 \\ 0 & 0 & 1 & 2 & 2 & 1 \end{array}\right] \rightarrow \left[\begin{array}{ccc|ccc} 1 & 2 & 0 & 1 & 0 & 0 \\ 0 & 1 & 0 & 1 & 2 & 0 \\ 0 & 0 & 1 & 2 & 2 & 1 \end{array}\right]$$

$$\rightarrow \left[\begin{array}{ccc|ccc} 1 & 0 & 0 & 2 & 2 & 0 \\ 0 & 1 & 0 & 1 & 2 & 0 \\ 0 & 0 & 1 & 2 & 2 & 1 \end{array}\right].$$

This calculation shows that A is invertible and that

$$A^{-1} = \begin{bmatrix} 2 & 2 & 0 \\ 1 & 2 & 0 \\ 2 & 2 & 1 \end{bmatrix}.$$

Example 127 *Let $A \in \mathbf{C}^{2\times 2}$, $b \in \mathbf{C}^2$ be defined by*

$$A = \begin{bmatrix} 1+i & 1-i \\ 1-i & 2 \end{bmatrix}, \ b = \begin{bmatrix} 1 \\ 1 \end{bmatrix}.$$

We solve $Ax = b$ by row reduction, but now hand calculations are more involved.[7] *The augmented matrix is*

$$\left[\begin{array}{cc|c} 1+i & 1-i & 1 \\ 1-i & 2 & 1 \end{array}\right].$$

For the first step, we must add a multiple of the first row to the second row to eliminate the $2,1$ entry. The necessary multiplier is

$$-\frac{1-i}{1+i} = -\frac{1-i}{1+i} \cdot \frac{1-i}{1-i} = -\frac{-2i}{2} = i.$$

Therefore, we add i times row 1 to row 2 to get

$$\left[\begin{array}{cc|c} 1+i & 1-i & 1 \\ 0 & 3+i & 1+i \end{array}\right].$$

[7] There are a number of software packages that will carry out the necessary calculations, eliminating the need for hand calculations. The most popular, at least among the interactive packages, are *Mathematica*®, Maple(TM), and MATLAB®.

Since the system is 2×2, that was the only step involved in Gaussian elimination, and we now apply back substitution. Dividing the third row by $3+i$ yields the value of x_2:

$$\frac{1+i}{3+i} = \frac{1+i}{3+i} \cdot \frac{3-i}{3-i} = \frac{4+2i}{10} = \frac{2}{5} + \frac{1}{5}i.$$

The matrix becomes

$$\left[\begin{array}{cc|c} 1+i & 1-i & 1 \\ 0 & 1 & \frac{2}{5}+\frac{1}{5}i \end{array}\right].$$

We now add $-(1-i)$ times row 2 to row 1, and then divide row 1 by $1+i$:

$$\left[\begin{array}{cc|c} 1+i & 1-i & 1 \\ 0 & 1 & \frac{2}{5}+\frac{1}{5}i \end{array}\right] \rightarrow \left[\begin{array}{cc|c} 1+i & 0 & \frac{2}{5}+\frac{1}{5}i \\ 0 & 1 & \frac{2}{5}+\frac{1}{5}i \end{array}\right] \rightarrow \left[\begin{array}{cc|c} 1 & 0 & \frac{3}{10}-\frac{1}{10}i \\ 0 & 1 & \frac{2}{5}+\frac{1}{5}i \end{array}\right].$$

The solution is

$$x = \left[\begin{array}{c} \frac{3}{10}-\frac{1}{10}i \\ \frac{2}{5}+\frac{1}{5}i \end{array}\right].$$

Exercises

Miscellaneous exercises

1. Suppose $A \in F^{m \times n}$, and row reduction, applied to $Ax = 0$, produces a general solution
$$x = \alpha_1 v_1 + \alpha_2 v_2 + \ldots + \alpha_k v_k,$$
where $x_{i_1}, x_{i_2}, \ldots, x_{i_k}$ are the free variables, the parameter α_j corresponds to x_{i_j} (that is, $x_{i_j} = \alpha_j$), and $v_1, v_2, \ldots, v_k \in F^n$. Prove that $\{v_1, v_2, \ldots, v_k\}$ is linearly independent.

2. Apply the row reduction algorithm to solve each of the following systems of equations. In each case, state whether the system has no solution, exactly one solution, or infinitely many solutions. Also, state the rank and nullity of A, where A is the coefficient matrix of the system, and find a basis for $\mathcal{N}(A)$, where possible.

   (a)
$$\begin{aligned} x + y + z &= 3 \\ 2x - 3y - z &= -8 \\ -x + 2y + 2z &= 3 \end{aligned}$$

   (b)
$$\begin{aligned} -2x_1 - 9x_2 - 12x_3 + x_4 &= 8 \\ 2x_1 + 9x_2 + 9x_3 - 4x_4 &= -5 \\ 2x_1 + 6x_2 - 10x_4 &= 4 \\ 4x_1 + 18x_2 + 36x_3 + 10x_4 &= 12 \end{aligned}$$

(c)

$$\begin{aligned} 4x_1 + 2x_2 + 18x_3 &= 4 \\ 4x_1 + x_2 + 15x_3 &= 2 \\ 8x_1 - x_2 + 21x_3 &= -2 \end{aligned}$$

(d)

$$\begin{aligned} 3x_1 - 9x_2 + 3x_3 &= -9 \\ -15x_1 + 45x_2 - 15x_3 &= 45 \\ -9x_1 + 27x_2 - 9x_3 &= 27 \end{aligned}$$

(e)

$$\begin{aligned} -x_1 - 5x_2 + 10x_3 - x_4 &= 2 \\ 2x_1 + 11x_2 - 23x_3 + 2x_4 &= -4 \\ -4x_1 - 23x_2 + 49x_3 - 4x_4 &= 8 \\ x_1 + 2x_2 - x_3 + x_4 &= -2 \end{aligned}$$

(f)

$$\begin{aligned} -x_1 + 2x_2 + x_3 + x_4 &= -4 \\ x_1 + x_2 + x_3 - 3x_4 &= -6 \\ -3x_1 + 18x_2 + 12x_3 - 4x_4 &= -48 \\ -2x_1 + x_2 + x_3 + 9x_4 &= 18 \end{aligned}$$

(g)

$$\begin{aligned} x_1 - 25x_2 - 3x_3 &= -24 \\ 4x_1 - 95x_2 - 12x_3 &= -91 \end{aligned}$$

3. Let $A \in \mathbf{R}^{3\times 3}$ be defined by

$$A = \begin{bmatrix} -1 & -1 & 5 \\ 4 & 4 & -21 \\ 4 & 4 & -23 \end{bmatrix}.$$

Find the inverse, if it exists, of A.

4. Let $A \in \mathbf{R}^{3\times 3}$ be defined by

$$A = \begin{bmatrix} 0 & 1 & 2 \\ 1 & 1 & 5 \\ 0 & -6 & -11 \end{bmatrix}.$$

Find the inverse, if it exists, of A.

5. Apply Gaussian elimination to solve the system

$$\begin{aligned} x_1 + x_2 &= 1, \\ x_1 + x_3 &= 1, \\ x_1 + x_2 + x_3 &= 0 \end{aligned}$$

for $x_1, x_2, x_3 \in \mathbf{Z}_2$.

6. Let $A \in \mathbf{Z}_3^{3\times 3}$ be defined by

$$A = \begin{bmatrix} 1 & 2 & 0 \\ 1 & 1 & 2 \\ 1 & 1 & 1 \end{bmatrix}.$$

Find the inverse, if it exists, of A.

7. Let $A \in \mathbf{Z}_3^{3\times 3}$ be defined by

$$A = \begin{bmatrix} 0 & 2 & 2 \\ 1 & 2 & 1 \\ 2 & 0 & 1 \end{bmatrix}.$$

Find the inverse, if it exists, of A.

8. Let $A \in \mathbf{C}^{2\times 2}$ be defined by

$$A = \begin{bmatrix} 1+i & 1-i \\ 1-i & 2 \end{bmatrix}.$$

Find the inverse, if it exists, of A.

3.8 Newton's method

One of the reasons that linear algebra is important is that many problems of practical interest are linear. Another, equally important, reason is that nonlinear problems are frequently solved by *linearization*, that is, by replacing a nonlinear function by a linear approximation. In this section, we will explain one example: the use of successive linear approximation to solve a system of nonlinear equations. We will concentrate on problems involving real numbers.

Here is the example we will use as illustration in this section: Find real number x_1, x_2 satisfying

$$x_1^2 + x_2^2 = 1, \tag{3.13a}$$
$$x_2 = x_1^2. \tag{3.13b}$$

This is a system of two nonlinear equations in two unknowns. Since the graph of the first equation is a circle and the graph of the second is a parabola, we can easily locate the solutions (points of intersection) graphically in Figure 3.2.

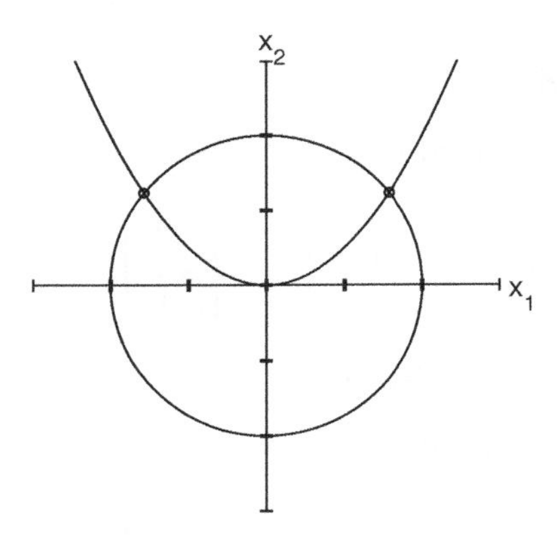

FIGURE 3.2
The graphs of $x_1^2 + x_2^2 = 1$ and $x_2 = x_1^2$, and the two points of intersection.

This particular nonlinear system is simple enough that we can find the two solutions shown in Figure 3.2 algebraically (see Exercise 1), but most nonlinear systems cannot be solved algebraically. For this reason, we will develop *Newton's method*, which computes an accurate solution by repeatedly solving linear systems in place of the original nonlinear system.

Before explaining Newton's method, we rewrite the system (3.13). It is equivalent to $F(x) = 0$,[8] where $F : \mathbf{R}^2 \to \mathbf{R}^2$ is defined by

$$F(x) = \begin{bmatrix} x_1^2 + x_2^2 - 1 \\ x_2 - x_1^2 \end{bmatrix}.$$

When dealing with nonlinear equations, we have to forget most of what we know about linear equations. For example, one of the most fundamental properties of a linear equation is that it must have no solution, exactly one, or infinitely many solutions. In the present example, we can see from Figure 3.2 that $F(x) = 0$ has exactly two solutions, a situation that cannot arise for a linear equation (involving real numbers).

We will focus our attention on the solution lying in the first quadrant, which we denote by x^*. We do not know the exact value of x^*, but can easily make a reasonable estimate of x^* by looking at Figure 3.2: x^* is not far from

[8]For a linear system, we distinguish between a homogeneous equation $T(x) = 0$ and an inhomogeneous equation $T(x) = u$. However, the standard form for any nonlinear system is $F(x) = 0$; there is no point in moving nonzero constants to the right-hand side, because, for a nonlinear operator F, $F(0)$ may be nonzero.

$x^{(0)} = (0.75, 0.5)$. We label this estimate with a superscript of 0 because we will shortly show how to produce a sequence of estimates $x^{(0)}, x^{(1)}, x^{(2)}, \ldots$ that converges to x^*.

Newton's method is based on the following observation. The derivative $F'\left(x^{(0)}\right)$ of F at $x^{(0)}$ defines a linear approximation to F that is accurate near $x = x^{(0)}$:

$$F(x) \doteq F\left(x^{(0)}\right) + F'\left(x^{(0)}\right)\left(x - x^{(0)}\right), \quad x \text{ near } x^{(0)}. \tag{3.14}$$

In this formula, $F'\left(x^{(0)}\right)$ denotes the *Jacobian matrix* of F at $x^{(0)}$:

$$F'(x) = \begin{bmatrix} \frac{\partial F_1}{\partial x_1}(x) & \frac{\partial F_1}{\partial x_2}(x) \\ \frac{\partial F_2}{\partial x_1}(x) & \frac{\partial F_2}{\partial x_2}(x) \end{bmatrix}.$$

The derivative operator $DF(x)$ is a linear operator mapping $\mathbf{R}^2$ into itself, and $F'(x)$ is the matrix representing $DF(x)$ (as in Section 3.2.2). In this example,

$$F'(x) = \begin{bmatrix} 2x_1 & 2x_2 \\ -2x_1 & 1 \end{bmatrix}.$$

Here is the idea of Newton's method: Since $F(x)$ is well approximated by $F\left(x^{(0)}\right) + F'\left(x^{(0)}\right)\left(x - x^{(0)}\right)$ for x near $x^{(0)}$, and since $x^{(0)}$ is close to x^*, the solution of

$$F\left(x^{(0)}\right) + F'\left(x^{(0)}\right)\left(x - x^{(0)}\right) = 0 \tag{3.15}$$

ought to be close to x^*. The equation (3.15) is linear in $x - x^{(0)}$:

$$F'\left(x^{(0)}\right)\left(x - x^{(0)}\right) = -F\left(x^{(0)}\right). \tag{3.16}$$

(The reader should notice that, in spite of the complicated notation, $F'\left(x^{(0)}\right)$ is just a known square matrix, and $-F\left(x^{(0)}\right)$ is just a known vector.) Assuming the Jacobian matrix is nonsingular, we can solve (3.15) for x and call the solution $x^{(1)}$, the new approximation to x^*:

$$x^{(1)} = x^{(0)} - F'\left(x^{(0)}\right)^{-1} F\left(x^{(0)}\right).$$

The formula for $x^{(1)}$ is expressed in terms of the inverse matrix, but in practice, one would apply Gaussian elimination to the square system represented by (3.16) rather than compute the inverse of the Jacobian matrix.

The first step of Newton's method is illustrated in Figure 3.3, which shows the graphs of the original (nonlinear) equation and the linearized equations (3.15).

Figure 3.4 shows a magnified view of the graphs near x^*. It can be seen that $x^{(1)}$ (the point of intersection of the lines, which are the graphs of the linearized equations) is much closer to x^* than is $x^{(0)}$.

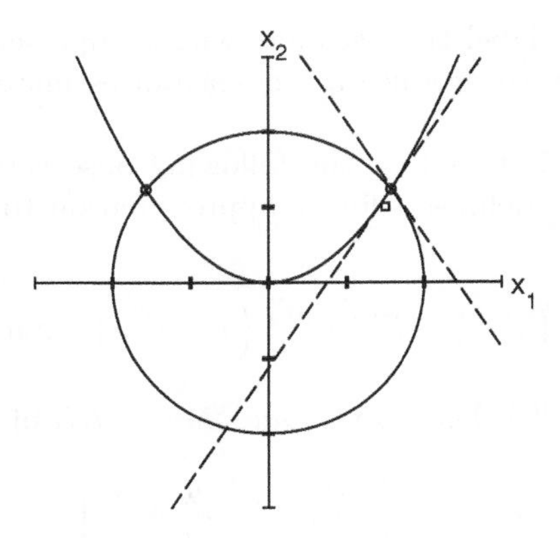

FIGURE 3.3
The graphs of $x_1^2+x_2^2=1$ and $x_2=x_1^2$ (the solid curves) and the linearizations of these equations near $x^{(0)}=(0.75,0.5)$ (the dashed lines). The point $x^{(0)}$ where the linearization is constructed is also indicated on the graph (by the small open square).

FIGURE 3.4
A close-up view of Figure 3.3 near x^*.

Newton's method repeats the above process of linearization to obtain a sequence $x^{(0)}, x^{(1)}, x^{(2)}, \ldots$ converging to x^*. The approximation $x^{(2)}$ is the solution of the following equation, obtained by linearizing F near $x=x^{(1)}$:

$$F\left(x^{(1)}\right)+F'\left(x^{(1)}\right)\left(x-x^{(1)}\right)=0.$$

In general, $x^{(k+1)}$ is the solution of

$$F\left(x^{(k)}\right)+F'\left(x^{(k)}\right)\left(x-x^{(k)}\right)=0, \tag{3.17}$$

k	$x_1^{(k)}$	$x_2^{(k)}$
0	0.75000000000000	0.50000000000000
1	0.79166666666667	0.62500000000000
2	0.78618421052632	0.61805555555556
3	0.78615137857530	0.61803398895790
4	0.78615137775742	0.61803398874989

TABLE 3.1
The results of four steps of Newton's method on the example of Section 3.8.

which is

$$x^{(k+1)} = x^{(k)} - F'\left(x^{(k)}\right)^{-1} F\left(x^{(k)}\right).$$

Table 3.1 shows the results of applying several steps of Newton's method to the example from this section. The results should be compared to the exact solution (see Exercise 1).

It can be shown that if $F'(x^*)$ is nonsingular and $x^{(0)}$ is sufficiently close to x^*, then the sequence generated by Newton's method converges quickly to x^*. To define "quickly" requires the language of norms, which is presented in the next chapter. However, we will not pursue a rigorous convergence analysis of Newton's method in this book.

It should be noted that Newton's method is not defined if it encounters a point $x^{(k)}$ such that $F'\left(x^{(k)}\right)$ is singular, and it converges slowly if $F'\left(x^*\right)$ is singular. A practical algorithm must deal with such cases. The book by Dennis and Schnabel [7] covers both the convergence theory and many practical details about implementation.

Exercises

Miscellaneous exercises
The use of computer software is recommended when applying Newton's method.

1. Solve (3.13) algebraically and find the exact values of the two solutions.
2. Consider the following system of nonlinear equations:

$$\begin{aligned} x_1^3 + x_1 - x_2 + 1 &= 0, \\ x_2^2 - 3x_2 - x_1 - 2 &= 0. \end{aligned}$$

   (a) By graphing the two equations in the x_1x_2-plane, determine how many solutions the system has and find a rough estimate of each solution.

   (b) Write the system in the standard form $F(x) = 0$ ($F : \mathbf{R}^2 \to \mathbf{R}^2$) and apply Newton's method to find each solution accurately.

3. Apply Newton's method to find a solution of the following system:

$$\begin{aligned} x_1^2 + x_2^2 + x_3^2 &= 1, \\ x_1^2 + x_2^2 &= x_3, \\ 3x_1 + x_2 + 3x_3 &= 0. \end{aligned}$$

4. Apply Newton's method to find a solution of the following system:

$$\begin{aligned} x_1^2 + x_2^2 + x_3^2 + x_4^2 &= 1, \\ x_2 &= x_1^2, \\ x_3 &= x_2^2, \\ x_4 &= x_3^2. \end{aligned}$$

5. Specialize Newton's method to the case of a single equation in a single unknown, $f(x) = 0$, $f : \mathbf{R} \to \mathbf{R}$. What simpler form does (3.17) take in this case? Apply the resulting iteration to solve $\cos(x) - x = 0$.

3.9 Linear ordinary differential equations

In this section we study the following linear ordinary differential equation (ODE) with constant coefficients:

$$\alpha_n x^{(n)} + \alpha_{n-1} x^{(n-1)} + \cdots + a_1 x' + \alpha_0 x = f(t), \tag{3.18}$$

where f is a given function and $\alpha_0, \alpha_1, \ldots, \alpha_n$ are real constants, with $\alpha_n \neq 0$. To put this in the form of a linear operator equation, we define the operator $L : C^n(\mathbf{R}) \to C(\mathbf{R})$ by

$$L = \alpha_n D^n + \alpha_{n-1} D^{n-1} + \cdots + \alpha_1 D + \alpha_0 I.$$

Here D is the derivative operator, $Df = f'$, and I is the identity operator, $If = f$. The reader will recall that D^k denotes D composed with itself k times. Since the composition of linear operators is linear (Theorem 58), each D^k is linear, and since a linear combination of linear operators is linear (Theorem 64), it follows that L is a linear operator.

The ODE (3.18) can then be posed as $L(x) = f$, where $f \in C(\mathbf{R})$. According to Theorem 90, the solution set of $L(x) = f$ is $\hat{x} + \ker(L)$, where $\hat{x}$ is any one solution and, as usual, the kernel $\ker(L)$ is the solution set of $L(x) = 0$.

3.9.1 The dimension of $\ker(L)$

A linear ODE never has a single solution; extra conditions are always needed to select one of the infinite number of solutions. The basic problem associated with the ODE (3.18) is the *initial value problem* (IVP), which is (3.18) together with the *initial conditions*

$$x(t_0) = v_1, x'(t_0) = v_2, \ldots, x^{(n-1)}(t_0) = v_n.$$

The following theorem, whose proof is beyond the scope of this book, shows that the IVP is a well-posed problem—it always has a unique solution.

Theorem 128 *Let $f, \alpha_0, \alpha_1, \ldots, \alpha_n \in C(a,b)$, let $v_1, v_2, \ldots, v_n \in \mathbf{R}$, and let $t_0 \in (a,b)$. Assume $\alpha_n(t) \neq 0$ for all $t \in (a,b)$. Then the IVP*

$$\begin{aligned} \alpha_n(t)x^{(n)} + \alpha_{n-1}(t)x^{(n-1)} + \cdots + a_1(t)x' + \alpha_0(t)x = f(t),\ a < t < b, \\ x(t_0) = v_1, \\ x'(t_0) = v_0, \\ \vdots \quad \vdots \\ x^{(n-1)}(t_0) = v_n \end{aligned}$$

has a unique solution $x \in C^n(a,b)$.

Although the theorem is true for nonconstant coefficients $\alpha_0(t), \ldots, \alpha_n(t)$, we will study only the case of constant coefficients. We also work on the interval $\mathbf{R} = (-\infty, \infty)$ rather than a more restricted interval.

According to the above theorem, there is a unique solution x_i to the IVP for each of the n sets of initial conditions $v = e_i$, $i = 1, 2, \ldots, n$, where $\{e_1, e_2, \ldots, e_n\}$ is the standard basis for $\mathbf{R}^n$. In other words, there is a unique solution to

$$\alpha_n x^{(n)} + \alpha_{n-1}x^{(n-1)} + \cdots + a_1 x' + \alpha_0 x = 0, \tag{3.19}$$

together with the initial conditions

$$x(0) = 1, x'(0) = 0, \ldots, x^{(n-1)}(0) = 0.$$

We call this solution x_1. There is also a unique solution to (3.19) subject to the initial conditions

$$x(0) = 0, x'(0) = 1, x''(0) = 0, \ldots, x^{(n-1)}(0) = 0,$$

which we call x_2. Continuing in this fashion, we find n different solutions $x_1, x_2, \ldots, x_n$ to (3.19). We will now show that $\{x_1, x_2, \ldots, x_n\}$ is a basis for $\ker(L)$.

First suppose that there exist real numbers $c_1, c_2, \ldots, c_n$ such that

$$c_1 x_1 + c_2 x_2 + \cdots + c_n x_n = 0.$$

This is a statement about the function $c_1 x_1 + c_2 x_2 + \cdots + c_n x_n$, namely, that it equals the zero function in $C^n(R)$. It follows that

$$c_1 x_1(t) + c_2 x_2(t) + \cdots + c_n x_n(t) = 0 \text{ for all } t \in \mathbf{R}$$

and, in particular,

$$c_1x_1(0) + c_2x_2(0) + \cdots + c_nx_n(0) = 0.$$

But, according to the initial conditions satisfied by $x_1, x_2, \ldots, x_n$, this reduces to $c_1 = 0$.

Next, since $c_1x_1 + c_2x_2 + \cdots + c_nx_n$ is the zero function and each x_i has n continuous derivatives, we obtain

$$\begin{aligned} c_1x_1'(0) + c_2x_2'(0) + \cdots + c_nx_n'(0) &= 0, \\ c_1x_1''(0) + c_2x_2''(0) + \cdots + c_nx_n''(0) &= 0, \\ &\vdots \\ c_1x_1^{(n-1)}(0) + c_2x_2^{(n-1)}(0) + \cdots + c_nx_n^{(n-1)}(0) &= 0. \end{aligned}$$

Again applying the initial conditions, the first equation yields $c_2 = 0$, the second $c_3 = 0$, and so forth to $c_n = 0$.

Since the only scalars $c_1, c_2, \ldots, c_n$ such that $c_1x_1+c_2x_2+\cdots+c_nx_n = 0$ are $c_1 = c_2 = \ldots = c_n = 0$, it follows that $\{x_1, x_2, \ldots, x_n\}$ is linearly independent.

We now show that $\{x_1, x_2, \ldots, x_n\}$ spans $\mathrm{ker}(L)$ by showing that every y in $\mathrm{ker}(L)$ can be written as $y = c_1x_1 + c_2x_2 + \cdots + c_nx_n$ for some constants $c_1, c_2, \ldots, c_n$. In fact, the necessary values of the constants are

$$c_1 = y(0), c_2 = y'(0), \ldots, c_n = y^{(n-1)}(0).$$

To see this, we note that the IVP

$$\begin{aligned} \alpha_nx^{(n)} + \alpha_{n-1}x^{(n-1)} + \cdots + a_1x' + \alpha_0x &= 0, \\ x(0) = c_1, x'(0) = c_2, \ldots, x^{(n-1)}(0) &= c_n, \end{aligned}$$

where $c_1 = y(0), c_2 = y'(0), \ldots, c_n = y^{(n-1)}(0)$, has a unique solution by Theorem 128. Obviously this solution is y; it satisfies the ODE because y belongs to $\mathrm{ker}(L)$, and the initial conditions hold because the initial values are precisely the initial values of y and its derivatives. On the other hand, it can be shown that $c_1x_1 + c_2x_2 + \cdots + c_nx_n$ also solves this IVP (see Exercise 6). By the uniqueness of the solution to the IVP, it must be the case that $y = c_1x_1 + c_2x_2 + \cdots + c_nx_n$. This completes the proof that $\{x_1, x_2, \ldots, x_n\}$ spans $\mathrm{ker}(L)$, and hence is a basis for $\mathrm{ker}(L)$.

The most important conclusion from this section is that the dimension of $\mathrm{ker}(L)$, where L is the nth-order differential operator

$$L = \alpha_nD^n + \alpha_{n-1}D^{n-1} + \cdots + \alpha_1D + \alpha_0I,$$

is n. This indicates a strategy for computing the general solution of $L(x) = f$:

1. Find n linearly independent solutions $x_1, x_2, \ldots, x_n$ of $L(x) = 0$ (that is, find a basis for $\mathrm{ker}(L)$).

2. Find any one solution $\hat{x}$ of $L(x) = f$.

3. The general solution is then

$$x = \hat{x} + c_1 x_1 + c_2 x_2 + \cdots + c_n x_n.$$

The usual solution process for the ODE $L(x) = f$, as presented in any introductory text on differential equations, applies this strategy directly: There is a direct method for computing a basis for $\ker(L)$ (although this method does not usually produce the special basis $\{x_1, x_2, \ldots, x_n\}$ described above, where $x_i^{(j)}(t_0) = 0$ for all $j \neq i-1$, $x_i^{(i-1)}(t_0) = 1$). There are also several methods for producing a solution $\hat{x}$ to the inhomogeneous equation. In most texts on ODEs, $\hat{x}$ is called a *particular* solution. Two popular methods for finding a particular solution are the methods of *variation of parameters* and *undetermined coefficients* (see, for example, [46]). The general solution of the homogeneous equation is called the *complementary solution*.

In the next section, we show how to compute a basis for $\ker(L)$ in the special case $n = 2$. This restriction will keep the discussion to a reasonable length; however, the techniques presented here extend directly to the case $n > 2$ (see [46]).

3.9.2 Finding a basis for $\ker(L)$

We wish to find two linearly independent solutions of

$$\alpha_2 x'' + \alpha_1 x' + \alpha_0 x = 0. \tag{3.20}$$

The computation is based on an inspired guess. We look for a solution of the form $x(t) = e^{mt}$, where m is constant. Noticing that

$$D^k e^{mt} = m^k e^{mt},$$

that is,

$$D^k x = m^k x,$$

we see that

$$\begin{aligned} L(x) = 0 \quad &\Leftrightarrow \quad \alpha_2 D^2 x + \alpha_1 D x + \alpha_0 x = 0 \\ &\Leftrightarrow \quad \alpha_2 m^2 x + \alpha_1 m x + \alpha_0 x = 0 \\ &\Leftrightarrow \quad \left(\alpha_2 m^2 + \alpha_1 m + \alpha_0\right) x = 0 \\ &\Leftrightarrow \quad \alpha_2 m^2 + \alpha_1 m + \alpha_0 = 0. \end{aligned}$$

The last step follows from the fact that $x(t) = e^{mt}$ is never zero. Therefore, if $m = r$ is any solution of the *auxiliary equation*

$$\alpha_2 m^2 + \alpha_1 m + \alpha_0 = 0, \tag{3.21}$$

then $x(t) = e^{rt}$ is a solution of $L(x) = 0$. The polynomial $a_2m^2 + \alpha_1 m + \alpha_0$ is called the *auxiliary polynomial* of the ODE.

We have seen that the dimension of $\ker(L)$ is 2, so any basis for $\ker(L)$ contains 2 solutions of $L(x) = 0$. Moreover, we know from Theorem 45 that any 2 linearly independent solutions form a basis. The auxiliary equation necessarily has 2 roots, provided we allow complex roots and count roots according to their multiplicity. The reader will recall that if r is a root of 3.21, then $(m - r)$ is a factor of $\alpha_2 m^2 + \alpha_1 m + \alpha_0$. If $(m - r)^2$ is a factor of $\alpha_2 m^2 + \alpha_1 m + \alpha_0$ (that is, if $\alpha_2 m^2 + \alpha_1 m + \alpha_0 = \alpha_2(m - r)^2$), then r is called a root of multiplicity 2.

In the case we consider here ($n = 2$), there are only three possibilities: the auxiliary equations has

1. two distinct real roots,
2. a single root of multiplicity 2, or
3. a pair of complex conjugate roots.

We now show how to find a basis for $\ker(L)$ in each case.

3.9.2.1 The easy case: Distinct real roots

If the auxiliary equation (3.21) has 2 distinct real roots r_1, r_2, then we have solutions

$$x_1(t) = e^{r_1 t}, x_2(t) = e^{r_2 t}.$$

These solutions can be shown to be linearly independent; we discuss this below. We notice that, in this case, the auxiliary polynomial factors as

$$\alpha_2 m^2 + \alpha_1 m + \alpha_0 = \alpha_2(m - r_1)(m - r_2).$$

3.9.2.2 The case of repeated real roots

Now let us suppose that the auxiliary polynomial has a single real root r. In this case, the above reasoning gives a single solution, $x_1(t) = e^{rt}$. There is a general method, called *reduction of order*, that uses a given solution to a linear ODE to find more solutions. This method assumes there are solutions of the form $x(t) = c(t)x_1(t)$, where x_1 is the known solution. In our case, $x_1(t) = e^{rt}$, so we assume there is a solution of the form $x_2(t) = c(t)e^{rt}$, and substitute into the differential equation:

$$\begin{aligned}
& \alpha_2 x_2'' + \alpha_1 x_2' + \alpha_0 x_2 = 0 \\
\Rightarrow\quad & \alpha_2 \left(r^2 c e^{rt} + 2rc' e^{rt} + c'' e^{rt}\right) + \alpha_1 \left(rce^{rt} + c' e^{rt}\right) + \alpha_0 c e^{rt} = 0 \\
\Rightarrow\quad & \left(\alpha_2 r^2 + \alpha_1 r + \alpha_0\right) c e^{rt} + \left(\alpha_2 c'' + (\alpha_1 + 2\alpha_2 r)c'\right) e^{rt} = 0.
\end{aligned}$$

Since r is a repeated root of the polynomial $\alpha_2 m^2 + \alpha_1 m + \alpha_0$, it follows that

$$\alpha_2 r^2 + \alpha_1 r + \alpha_0 = 0,\ \ \alpha_1 + 2\alpha_2 r = 0,$$

as the reader can easily verify. Therefore, the ODE reduces to

$$\alpha_2 c'' e^{rt} = 0,$$

or simply to $c'' = 0$ since $\alpha_2 \neq 0$, $e^{rt} \neq 0$. We just need one solution for c, so we can take $c(t) = t$, which implies that $x_2(t) = te^{rt}$ is a second solution. We will show below that $\{x_1, x_2\}$ is linearly independent, and therefore we have found a basis for $\ker(L)$ in this case.

3.9.2.3 The case of complex roots

If the auxiliary polynomial has a complex root $r + i\theta$, then, since the polynomial has real coefficients, the conjugate $r - i\theta$ must also be a root. This implies that two solutions are

$$\begin{aligned} y_1(t) &= e^{(r+i\theta)t} = e^{rt}e^{i\theta t} = e^{rt}(\cos(\theta t) + i\sin(\theta t)), \\ y_1(t) &= e^{(r-i\theta)t} = e^{rt}e^{-i\theta t} = e^{rt}(\cos(\theta t) - i\sin(\theta t)), \end{aligned}$$

where we have used Euler's formula:

$$e^{i\theta} = \cos(\theta) + i\sin(\theta).$$

The solutions y_1, y_2 are complex-valued, which might not be desirable, since the problem was originally posed in the space of real numbers. However, since the ODE is linear and homogeneous, any linear combination of y_1 and y_2 is also a solution. We have

$$\begin{aligned} \frac{1}{2}y_1(t) + \frac{1}{2}y_2(t) &= \frac{1}{2}e^{rt}(\cos(\theta t) + i\sin(\theta t)) + \frac{1}{2}e^{rt}(\cos(\theta t) - i\sin(\theta t)) \\ &= e^{rt}\cos(\theta t). \end{aligned}$$

Similarly,

$$-\frac{i}{2}y_1(t) + \frac{i}{2}y_2(t) = e^{rt}\sin(\theta t).$$

Therefore, two real-valued solutions are

$$x_1(t) = e^{rt}\cos(\theta t), \; x_2(t) = e^{rt}\sin(\theta t).$$

We will see below that $\{x_1, x_2\}$ is linearly independent and therefore is a basis for $\ker(L)$.

3.9.3 The Wronskian test for linear independence

In each of the three possible cases, we have found two solutions of the second-order linear ODE (3.20). It is actually simple to show that, in each case, the given solutions form a linearly independent set, since a set of two vectors is linearly dependent if and only if one is a multiple of the other. However, we prefer to use the following result, which provides a general method for testing linear independence of smooth functions, particularly solutions of linear ODEs.

Theorem 129 *Let a and b be real numbers with $a < b$, and let $x_1, x_2, \ldots, x_n$ be n elements of $C^{n-1}(a,b)$. If the matrix*

$$W = \begin{bmatrix} x_1(t_0) & x_2(t_0) & \cdots & x_n(t_0) \\ x_1'(t_0) & x_2'(t_0) & \cdots & x_n'(t_0) \\ \vdots & \vdots & \ddots & \vdots \\ x_1^{(n-1)}(t_0) & x_2^{(n-1)}(t_0) & \cdots & x_n^{(n-1)}(t_0) \end{bmatrix} \tag{3.22}$$

is nonsingular for any $t_0 \in (a,b)$, then $\{x_1, x_2, \ldots, x_n\}$ is linearly independent.

Proof We will prove the contrapositive: If $\{x_1, x_2, \ldots, x_n\}$ is linearly dependent, then the matrix W defined by (3.22) is singular for each $t_0 \in (a,b)$. If $\{x_1, x_2, \ldots, x_n\}$ is linearly dependent, then there exist scalars $c_1, c_2, \ldots, c_n$ in $\mathbf{R}$, not all zero, such that

$$c_1x_1 + c_2x_2 + \cdots + c_nx_n = 0.$$

The 0 in this equation is the zero function in $C^n(a,b)$; thus the equation can also be written as

$$c_1x_1(t) + c_2x_2(t) + \cdots + c_nx_n(t) = 0 \text{ for all } t \in (a,b).$$

Differentiating this equation $n-1$ times, and substituting $t = t_0$, we obtain the following system of equations:

$$\begin{aligned} c_1x_1(t_0) + c_2x_2(t_0) + \cdots + c_nx_n(t_0) &= 0, \\ c_1x_1'(t_0) + c_2x_2'(t_0) + \cdots + c_nx_n'(t_0) &= 0, \\ &\vdots \\ c_1x_1^{(n-1)}(t_0) + c_2x_2^{(n-1)}(t_0) + \cdots + c_nx_n^{(n-1)}(t_0) &= 0. \end{aligned}$$

In matrix-vector form, this system takes the form $Wc = 0$, where c is the vector $(c_1, c_2, \ldots, c_n)$. By assumption, c is not the zero vector, which shows that W is singular. This holds for each $t_0 \in (a,b)$, which completes the proof.

QED

The preceding theorem applies to any collection of n functions in $C^n(a,b)$. If the functions are all solutions to the same linear ODE, then the result can be strengthened.

Theorem 130 *Suppose $x_1, x_2, \ldots, x_n \in C^n(a,b)$ are solutions to the ODE*

$$\alpha_n(t)x^{(n)} + \alpha_{n-1}(t)x^{(n-1)} + \cdots + a_1(t)x' + \alpha_0(t)x = 0,$$

where $a_0, a_1, \ldots, a_n \in C(a,b)$ (this includes the constant coefficient equation as a special case). Then $\{x_1, x_2, \ldots, x_n\}$ is linearly independent if and only if the matrix W defined by (3.22) is nonsingular for all $t_0 \in (a,b)$.

Proof By the previous theorem, if W is nonsingular for all (or even for one) t_0, then $\{x_1, x_2, \ldots, x_n\}$ is linearly independent. Conversely, let t_0 be any element of (a, b) and suppose the matrix W defined by (3.22) is singular. By Theorem 128, there is a unique solution to the IVP

$$\begin{aligned} \alpha_n(t)x^{(n)} + \alpha_{n-1}(t)x^{(n-1)} + \cdots + a_1(t)x' + \alpha_0(t)x &= 0,\ a < t < b, \\ x(t_0) &= 0, \\ x'(t_0) &= 0, \\ \vdots\ &\ \ \vdots \\ x^{(n-1)}(t_0) &= 0. \end{aligned}$$

This solution is obviously the zero function. On the other hand, for any $c_1, c_2, \ldots, c_n \in \mathbf{R}$, $x = c_1x_1 + c_2x_2 + \cdots + c_nx_n$ is a solution of the ODE by linearity. (In linear algebraic terms, each $x_i \in \ker(L)$, where L is the differential operator $L = \alpha_nD^n + \alpha_{n-1}D^{n-1} + \cdots + \alpha_1D + \alpha_0I$, and so any linear combination of $x_1, x_2, \ldots, x_n$ also belongs to $\ker(L)$.) The derivatives of x are given by

$$\begin{aligned} x(t_0) &= c_1x_1(t_0) + c_2x_2(t_0) + \cdots + c_nx_n(t_0), \\ x'(t_0) &= c_1x_1'(t_0) + c_2x_2'(t_0) + \cdots + c_nx_n'(t_0), \\ \vdots\ &\qquad \vdots \\ x^{(n-1)}(t_0) &= c_1x_1^{(n-1)}(t_0) + c_2x_2^{(n-1)}(t_0) + \cdots + c_nx_n^{(n-1)}(t_0), \end{aligned}$$

so x satisfies the initial conditions if and only if

$$\begin{aligned} c_1x_1(t_0) + c_2x_2(t_0) + \cdots + c_nx_n(t_0) &= 0, \\ c_1x_1'(t_0) + c_2x_2'(t_0) + \cdots + c_nx_n'(t_0) &= 0, \\ \vdots\qquad &\quad \vdots \\ c_1x_1^{(n-1)}(t_0) + c_2x_2^{(n-1)}(t_0) + \cdots + c_nx_n^{(n-1)}(t_0) &= 0. \end{aligned}$$

This last system is equivalent to $Wc = 0$. Since W is singular, there is a nontrivial solution to $Wc = 0$, and the corresponding function

$$x = c_1x_1 + c_2x_2 + \cdots + c_nx_n$$

solves the IVP. Since x must be the zero function (the unique solution to the IVP), and not all of $c_1, c_2, \ldots, c_n$ are zero, this shows that $\{x_1, x_2, \ldots, x_n\}$ is linearly dependent. This completes the proof.

QED

The test for linear independence of solutions defined by Theorem 130 is called the *Wronskian* test.

We can apply the Wronskian test to each of the three cases for (3.20). In each case, it is simplest to take $t_0 = 0$. As an example, we examine the case of two distinct real roots. In this case, $x_1(t) = e^{r_1 t}$, $x_2(t) = e^{r_2 t}$, where $r_1 \neq r_2$. We have

$$\begin{bmatrix} x_1(t) & x_2(t) \\ x_1'(t) & x_2'(t) \end{bmatrix} = \begin{bmatrix} e^{r_1 t} & e^{r_2 t} \\ r_1 e^{r_1 t} & r_2 e^{r_2 t} \end{bmatrix}.$$

Therefore, with $t = t_0 = 0$,

$$W = \begin{bmatrix} 1 & 1 \\ r_1 & r_2 \end{bmatrix}.$$

Since $r_1 \neq r_2$, it is obvious that the columns of W form a linearly independent set, and therefore W is nonsingular. Thus $\{x_1, x_2\}$ is linearly independent.

The other cases (a repeated real root, a pair of complex conjugate roots) are left to Exercises 1 and 2.

3.9.4 The Vandermonde matrix

To conclude this discussion, we consider the original ODE (3.19) of order n. To find solutions, we can substitute $x(t) = e^{rt}$, just as in Section 3.9.2, and we find that x is a solution if r satisfies the auxiliary equation

$$\alpha_n r^n + \alpha_{n-1} r^{n-1} + \cdots + \alpha_1 r + \alpha_0 = 0.$$

We wish to discuss only the case in which this equation has n distinct real solutions $r_1, r_2, \ldots, r_n$. Then n solutions to (3.19) are

$$x_1(t) = e^{r_1 t}, x_2(t) = e^{r_2 t}, \ldots, x_n(t) = e^{r_n t}.$$

We can verify the linear independence of $\{x_1, x_2, \ldots, x_n\}$ by the Wronskian test; taking $t_0 = 0$, the matrix W is

$$W = \begin{bmatrix} x_1(0) & x_2(0) & \cdots & x_n(0) \\ x_1'(0) & x_2'(0) & \cdots & x_n'(0) \\ \vdots & \vdots & \ddots & \vdots \\ x_1^{(n-1)}(0) & x_2^{(n-1)}(0) & \cdots & x_n^{(n-1)}(0) \end{bmatrix}$$

$$= \begin{bmatrix} 1 & 1 & \cdots & 1 \\ r_1 & r_2 & \cdots & r_n \\ \vdots & \vdots & \ddots & \vdots \\ r_1^{n-1} & r_2^{n-1} & \cdots & r_n^{n-1} \end{bmatrix}.$$

This special matrix is called a *Vandermonde* matrix, and it is closely related to the matrix arising when doing polynomial interpolation (see Section 2.8).[9] In Chapter 4, we will use the theory of determinants to show that W is nonsingular when $r_1, r_2, \ldots, r_n$ are distinct real numbers (see Exercise 4.3.11).

[9]The rows of the matrix W are the columns of the Vandermonde matrix in Section 2.8, and vice versa. Depending on the author, one form or the other is designated as a Vandermonde matrix; we have elected to refer to both as Vandermonde matrices.

Exercises

1. Use the Wronskian test to show that $x_1(t) = e^{rt}$, $x_2(t) = te^{rt}$, where $r \in \mathbf{R}$, form a linearly independent set in $C(\mathbf{R})$.

2. Given $r, \theta \in \mathbf{R}$, define $x_1(t) = e^{rt}\cos(\theta t)$, $x_2(t) = e^{rt}\sin(\theta t)$. Use the Wronskian test to show that $\{x_1, x_2\}$ is a a linearly independent subset of $C(\mathbf{R})$.

3. Find the general solution of each of the following ODEs:

   (a) $x'' + 4x' + 4x = 0$

   (b) $x'' + 2x' + 2x = 0$

   (c) $x'' + 3x' + 2x = 0$

4. Consider the ODE

$$x'' + 3x' + 2x = \sin(t) + 3\cos(t).$$

   Given that $x(t) = \sin(t)$ is one solution, find the general solution.

5. Use the results of the previous exercise to solve the initial value problem

$$\begin{aligned} x'' + 3x' + 2x &= \sin(t) + 3\cos(t), \ t > 0, \\ x(0) &= 1, \\ x'(0) &= 0. \end{aligned}$$

6. Let $x_1, x_2, \ldots, x_n$ be the special solutions of

$$\alpha_n x^{(n)} + \alpha_{n-1} x^{(n-1)} + \cdots + a_1 x' + \alpha_0 x = 0$$

   defined in Section 3.9.1 (see page 159). Prove that, for any values of $c_1, c_2, \ldots, c_n$, the function $c_1 x_1 + c_2 x_2 + \cdots + c_n x_n$ solves the IVP

$$\begin{gathered} \alpha_n x^{(n)} + \alpha_{n-1} x^{(n-1)} + \cdots + a_1 x' + \alpha_0 x = 0, \\ x(0) = c_1, x'(0) = c_2, \ldots, x^{(n-1)}(0) = c_n. \end{gathered}$$

7. Consider the set $\{x_1, x_2, x_3\} \subset C(\mathbf{R})$, where $x_1(t) = t$, $x_2(t) = t^2$, $x_3(t) = t^3$.

   (a) Prove that $\{x_1, x_2, x_3\}$ is linearly independent.

   (b) Show that the Wronskian of $\{x_1, x_2, x_3\}$, at $t_0 = 0$, is singular.

   Why do these facts not violate Theorem 130?

3.10 Graph theory

Graph theory is central to *discrete mathematics*, the area of mathematics that studies problems with solutions characterized by integers or by quantities that are otherwise discrete. An important topic in discrete mathematics is the relationships among a given set of objects. For example, we might study a set of computers and the network connections that exist between particular computers in the set. Other examples include cities and the airline routes that connect them, and people and the relationships between them (for example, whether they are acquainted or not).

To study the types of situations described above abstractly, we define a graph as follows.

Definition 131 *A* graph *consists of two sets, a set V of* nodes *(also called* vertices*), and a set E of* edges*, each of which is a set of two distinct nodes. Thus an edge has the form $\{u, v\}$, where $u, v \in V$.*[10]

If a graph is called G, then we denote its node and edge sets by V_G and E_G, respectively.

We say that $u, v \in V$ are adjacent*if $\{u, v\} \in E$, in which case we say that there is an edge between u and v, or joining u and v.*

A graph is usually illustrated as in Figure 3.5, with nodes denoted by points in the plane and edges as line segments joining the nodes. Any given graph can be drawn in many ways, since the location of nodes is irrelevant—only the adjacency between nodes is important.

Graph theory is a vast area with many technical results; our purpose in this section and Section 5.6 is to provide a glimpse of how linear algebra is used to study graphs and their properties. A key way that this occurs is through the representation of graphs by matrices.

3.10.1 The adjacency matrix of a graph

The *order* of a graph is simply the number of nodes in the graph. A graph G of order n is uniquely represented by an $n \times n$ matrix, as described in the following definition.

Definition 132 *Let G be a graph with node set $V_G = \{v_1, v_2, \dots, v_n\}$. The* adjacency matrix *A_G of G is the $n \times n$ matrix defined by*

$$(A_G)_{ij} = \begin{cases} 1, & \{v_i, v_j\} \in E_G, \\ 0, & \{v_i, v_j\} \notin E_G. \end{cases}$$

[10]A graph as defined here is sometimes called a *simple graph*, while a general graph is allowed to contains *loops* (edges connecting a node to itself) and *multiple edges* (that is, more than one edge between a given pair of nodes). On the other hand, some authors adopt our definition of a graph, while the generalization that allows loops and multiple edges is called a *multigraph*. The terminology is not completely standardized, so the reader must be careful when consulting different references.

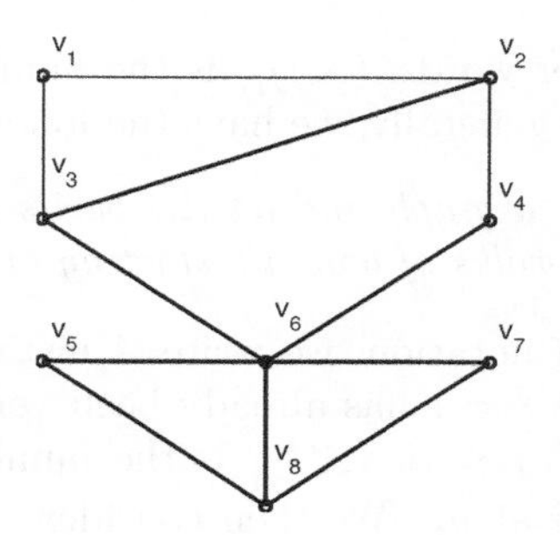

FIGURE 3.5
A graph with 8 nodes, $v_1, v_2, \ldots, v_8$, and 11 edges.

As an example, if G is the graph displayed in Figure 3.5, then the adjacency matrix is

$$A_G = \begin{bmatrix} 0 & 1 & 1 & 0 & 0 & 0 & 0 & 0 \\ 1 & 0 & 1 & 1 & 0 & 0 & 0 & 0 \\ 1 & 1 & 0 & 0 & 0 & 1 & 0 & 0 \\ 0 & 1 & 0 & 0 & 0 & 1 & 0 & 0 \\ 0 & 0 & 0 & 0 & 0 & 1 & 0 & 1 \\ 0 & 0 & 1 & 1 & 1 & 0 & 1 & 1 \\ 0 & 0 & 0 & 0 & 0 & 1 & 0 & 1 \\ 0 & 0 & 0 & 0 & 1 & 1 & 1 & 0 \end{bmatrix}.$$

3.10.2 Walks and matrix multiplication

Definition 133 *Let G be a graph. A* walk of length ℓ *in G is an alternating sequence of nodes and edges,*

$$v_{i_0}, e_{i_1}, v_{i_1}, e_{i_2}, v_{i_2}, \ldots, e_{i_\ell}, v_{i_\ell},$$

such that each edge e_{i_j} joins nodes $v_{i_{j-1}}$ and v_{i_j}. The nodes and nodes in the walk are not assumed to be distinct. (If the nodes are all distinct, then the walk is called a path.*) The reader should notice that the length of a walk is the number of edges in it.*

To simplify the notation, we will denote a walk by the sequence of nodes, $v_{i_0}, v_{i_1}, \ldots, v_{i_\ell}$, with the understanding that each $\{v_{i_{j-1}}, v_{i_j}\}$ must belong to E_G.

For example, if G is the graph of Figure 3.5, then v_1, v_3, v_6, v_8 is a walk of length 3 joining v_1 and v_8. As another example, $v_1, v_2, v_3, v_1, v_2, v_4$ is a walk of length 5 joining v_1 and v_4.

We can interpret the adjacency matrix A_G as follows: $(A_G)_{ij}$ is 1 if there is a walk of length 1 between v_i and v_j (a walk of length 1 is just an edge), and 0 otherwise. In other words, $(A_G)_{ij}$ is the number of walks of length 1 between v_i and v_j. More generally, we have the following result.

Theorem 134 *Let G be a graph and let A_G be its adjacency matrix. Then $\left(A_G^\ell\right)_{ij}$ is the number of walks of length ℓ starting at v_i and ending at v_j.*

Proof For simplicity of notation, we write A instead of A_G. We argue by induction. The base case $\ell = 1$ has already been verified above. We assume that, for each i, j, $1 \le i, j \le n$, $\left(A^\ell\right)_{ij}$ is the number of walks of length ℓ starting at v_i and ending at v_j. We then consider a specific pair of integers i, j, $1 \le i, j \le n$. We have

$$\left(A^{\ell+1}\right)_{ij} = \sum_{k=1}^{n} \left(A^\ell\right)_{ik} A_{kj}.$$

Now consider the expression

$$\left(A^\ell\right)_{ik} A_{kj}. \tag{3.23}$$

By the induction hypothesis, the factor $\left(A^\ell\right)_{ik}$ is the number of walks of length ℓ starting at v_i and ending at v_k, while the factor A_{kj} is 1 if there is an edge joining v_k and v_j, and 0 otherwise. It follows that (3.23) is the number of walks of length $\ell + 1$ from v_i to v_j that pass through v_k just before reaching v_j. Since any walk of length $\ell + 1$ from v_i to v_j must pass through some node v_k just before reaching v_j, the sum

$$\sum_{k=1}^{n} \left(A^\ell\right)_{ik} A_{kj}$$

is the total number of walks of length $\ell + 1$ joining v_i and v_j.

QED

For example, referring again to the graph displayed in Figure 3.5, we have

$$A_G^3 = \begin{bmatrix} 2 & 4 & 4 & 2 & 1 & 2 & 1 & 1 \\ 4 & 2 & 6 & 5 & 2 & 1 & 2 & 2 \\ 4 & 6 & 2 & 1 & 1 & 8 & 1 & 2 \\ 2 & 5 & 1 & 0 & 1 & 7 & 1 & 2 \\ 1 & 2 & 1 & 1 & 2 & 7 & 2 & 5 \\ 2 & 1 & 8 & 7 & 7 & 4 & 7 & 7 \\ 1 & 2 & 1 & 1 & 2 & 7 & 2 & 5 \\ 1 & 2 & 2 & 2 & 5 & 7 & 5 & 4 \end{bmatrix}.$$

By the previous theorem, we see that there are 4 walks of length 3 from v_1 to v_3 (since $\left(A_G^3\right)_{13} = 4$). These walks are

$$v_1, v_2, v_1, v_3; \;\; v_1, v_3, v_1, v_3; \;\; v_1, v_3, v_2, v_3; \;\; v_1, v_3, v_6, v_3.$$

3.10.3 Graph isomorphisms

The essential fact about a graph is the adjacency of nodes, not how the graph is drawn on paper or even how the nodes are labeled. Given two graphs G and H, if the two graphs become the same upon merely relabeling the nodes of H (in some way) to be the same as the nodes of G, then the two graphs are not essentially different. This is the idea behind the following definition.

Definition 135 *Suppose G and H are two graphs. If there exists a bijection $\phi : V_G \to V_H$ such that*

$$\{\phi(v), \phi(u)\} \in E_H \Leftrightarrow \{u, v\} \in E_G,$$

then G and H are said to be isomorphic.

According to this definition, G and H are isomorphic if and only if there exists a one-to-one correspondence between the nodes of G and the nodes of H such that two nodes in G are adjacent if and only if the corresponding pair of nodes is adjacent in H. It is clear that, if G and H are isomorphic, then the number of nodes in G is the same as the number of nodes in H, and likewise for the number of edges in the two graphs. Of course, these conditions are necessary but not sufficient for two graphs to be isomorphic.

It may not be easy to decide if two graphs are isomorphic, especially from drawings of the graphs. For instance, Figure 3.6 shows a graph that, like the graph in Figure 3.5, has 8 nodes and 11 edges. The graphs illustrated in

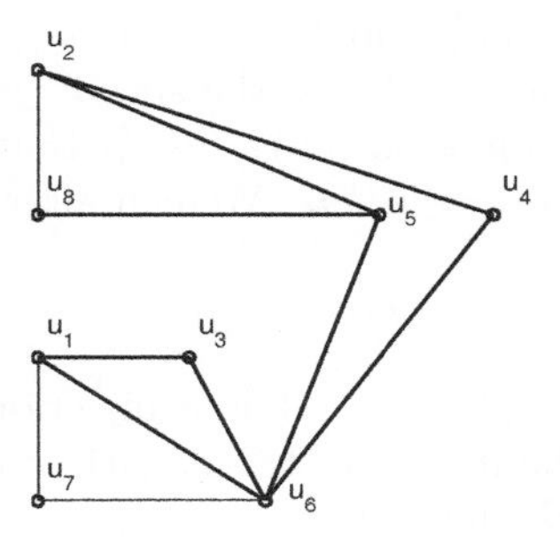

FIGURE 3.6
Another graph with 8 nodes and 11 edges (cf. Figure 3.5).

Figures 3.5 and 3.6 are isomorphic. Calling the graph of Figure 3.5 G and the graph of Figure 3.6 H, we have

$$\begin{aligned} V_G &= \{v_1, v_2, v_3, v_4, v_5, v_6, v_7, v_8\}, \\ E_G &= \{\{v_1, v_2\}, \{v_1, v_3\}, \{v_2, v_3\}, \{v_2, v_4\}, \{v_3, v_6\}, \{v_4, v_6\}, \\ &\quad \{v_5, v_6\}, \{v_5, v_8\}, \{v_6, v_7\}, \{v_6, v_8\}, \{v_7, v_8\}\}, \end{aligned}$$

while

$$\begin{aligned} V_H &= \{u_1, u_2, u_3, u_4, u_5, u_6, u_7, u_8\}, \\ E_H &= \{\{u_2, u_8\}, \{u_5, u_8\}, \{u_2, u_5\}, \{u_2, u_4\}, \{u_5, u_6\}, \{u_4, u_6\}, \\ &\quad \{u_6, u_7\}, \{u_1, u_7\}, \{u_3, u_6\}, \{u_1, u_6\}, \{u_1, u_3\}\}. \end{aligned}$$

The reader can verify that $\phi : V_G \to V_H$ defined by

$$\begin{aligned} &\phi(v_1) = u_8,\ \phi(v_2) = u_2,\ \phi(v_3) = u_5,\ \phi(v_4) = u_4,\ \phi(v_5) = u_7, \\ &\phi(v_6) = u_6,\ \phi(v_7) = u_3,\ \phi(v_8) = u_1 \end{aligned}$$

is an isomorphism. For example, $\{v_1, v_2\} \in E_G$ and

$$\{\phi(v_1), \phi(v_2)\} = \{u_8, u_2\} \in E_H.$$

As another example, $\{u_5, u_8\} \in E_H$, and

$$\{\phi^{-1}(u_5), \phi^{-1}(u_8)\}\} = \{v_3, v_1\} \in E_G.$$

We leave to the reader the rest of the verification that E_G and E_H are in one-to-one correspondence according to the rule $\{v_i, v_j\} \in E_G$ if and only if $\{\phi(v_i), \phi(v_j)\} \in E_H$.

Since a graph is completely defined by its adjacency matrix, there must be a way to describe a graph isomorphism in terms of the adjacency matrices of the isomorphic graphs. Let us assume that the two graphs are called G and H, with $V_G = \{v_1, v_2, \dots, v_n\}$ and $V_H = \{u_1, u_2, \dots, u_n\}$ (in the following discussion, G and H no longer refer to the specific graphs of Figures 3.5 and 3.6). An isomorphism is a function $\phi : V_G \to V_H$ that establishes a one-to-one correspondence $v_i \in V_G \leftrightarrow \phi(v_i) \in V_H$. We can alternately express this as

$$v_i \leftrightarrow u_{\psi(i)},$$

where $\psi : \{1, 2, \dots, n\} \to \{1, 2, \dots, n\}$ is a bijection. Such a bijection ψ is called a *permutation* of the integers $\{1, 2, \dots, n\}$ (see Appendix B). If A_G is the adjacency matrix of G, then

$$(A_G)_{ij} = \begin{cases} 1, & \text{if } \{v_i, v_j\} \in E_G, \\ 0, & \text{otherwise.} \end{cases}$$

Similarly,

$$(A_H)_{ij} = \begin{cases} 1, & \text{if } \{u_i, u_j\} \in E_H, \\ 0, & \text{otherwise,} \end{cases}$$

and since H is isomorphic to G, this can be expressed as

$$(A_H)_{ij} = \begin{cases} 1, & \text{if } \{v_{\psi^{-1}(i)}, v_{\psi^{-1}(j)}\} \in E_G, \\ 0, & \text{otherwise.} \end{cases}$$

Replacing i by $\psi(k)$ and j by $\psi(\ell)$, this becomes

$$(A_H)_{\psi(k),\psi(\ell)} = \begin{cases} 1, & \text{if } \{v_k, v_\ell\} \in E_G, \\ 0, & \text{otherwise.} \end{cases}$$

From this we see that

$$(A_H)_{\psi(k),\psi(\ell)} = (A_G)_{k\ell} \tag{3.24}$$

for all $k, \ell = 1, 2, \ldots, n$.

The relationship (3.24) shows that A_G is obtained by permuting (reordering) the rows and columns of A_H. We can express this using matrix multiplication by defining the notion of a *permutation matrix*.

Definition 136 *Let n be a positive integer. A* permutation matrix $P \in \mathbf{R}^{n\times n}$ *is a matrix obtained by permuting the rows of the $n \times n$ identity matrix I. More precisely, let $\psi : \{1, 2, \ldots, n\} \to \{1, 2, \ldots, n\}$ be a permutation. The corresponding permutation matrix P is defined by*

$$P_{ij} = \begin{cases} 1, & \text{if } j = \psi(i), \\ 0, & \text{otherwise.} \end{cases}$$

In other words,

$$P_{ij} = I_{\psi(i),j}, \quad i, j = 1, 2, \ldots, n.$$

A permutation matrix can be used to express compactly the result of permuting the rows or columns of a given matrix. If $A \in \mathbf{R}^{n\times n}$ and P is the permutation matrix of Definition 136, then

$$(PA)_{ij} = \sum_{k=1}^{n} P_{ik} A_{kj}.$$

As k runs from 1 to n, only one of the entries P_{ik} is not 0: $P_{ik} = 1$ if and only if $k = \psi(i)$. It follows that

$$(PA)_{ij} = A_{\psi(i),j},$$

and we see that row i of PA is row $\psi(i)$ of A. Similar reasoning shows that

$$\left(AP^T\right)_{ij} = \sum_{k=1}^{n} A_{ik} P_{jk} = A_{i,\psi(j)},$$

and thus column j of AP^T is column $\psi(j)$ of A.

Returning now to the relationship (3.24) between the adjacency matrices of isomorphic graphs G and H, we see that the k,ℓ-entry of A_G is the entry from row $\psi(k)$ and column $\psi(\ell)$ of A_H; in other words, A_G is obtained by permuting the rows and columns of A_H:

$$A_G = PA_H P^T. \tag{3.25}$$

We have proved half of the following theorem (the "only if" part):

Theorem 137 *Let G and H be graphs, each having n nodes, and let A_G, A_H be the adjacency matrices of G, H, respectively. Then G and H are isomorphic if and only if there exists a permutation matrix $P \in \mathbf{R}^{n\times n}$ such that*

$$A_G = PA_HP^T.$$

The proof of the "if" part of the theorem is left to the reader (see Exercise 5).

Theorem 137 is theoretically useful (cf. Section 5.6), but it does not give a computationally efficient way to decide if two graphs are isomorphic. Indeed, there is no efficient algorithm for deciding if two general graphs are isomorphic (although there are efficient algorithms for certain special classes of graphs). For details about the graph isomorphism and its computational demands, the reader can consult Garey and Johnson [10], especially Section 7.1.

Exercises

1. Let G be a graph, and let v_i, v_j be two nodes in V_G. The *distance* $d(v_i, v_j)$ between v_i and v_j is the smallest length of any walk joining v_i and v_j. Explain how to use the powers $A_G, A_G^2, A_G^3, \ldots$ to determine $d(v_i, v_j)$ for a given pair v_i, v_j of nodes.

2. A graph is *connected* if and only if there is a walk between every pair of nodes in the graph. The *diameter* of a connected graph G is

$$\text{diam}(G) = \max\{d(u,v) \,:\, u, v \in V_G\}.$$

Prove the following result: diam(G) is the smallest value of d such that, for each $i \neq j$, there exists ℓ such that $1 \leq \ell \leq d$ and $\left(A_G^\ell\right)_{ij} \neq 0$.

3. Let G be a graph, and let $v \in V_G$. By definition, the *degree* of v is the number of edges *incident* with v (that is, the number of edges having v as an endpoint). Let A_G be the adjacency matrix of G. Prove that

$$\left(A_G^2\right)_{ii}$$

is the degree of v_i for each $i = 1, 2, \ldots, n$.

4. The *complete graph* on n nodes $v_1, v_2, \ldots, v_n$ is a graph with edges joining every pair of distinct nodes.

   (a) What is the adjacency matrix A of a complete graph?

   (b) Can you find a formula for the number of walks of length ℓ between two vertices in a complete graph?

5. Prove the "if" part of Theorem 137.

6. Let G and H be the graphs of Figures 3.5 and 3.6, respectively. Find a permutation matrix P such that $A_G = PA_HP^T$.

3.11 Coding theory

Coding theory studies techniques for transmitting information (a message) correctly in the presence of noise. It should be distinguished from cryptography, which is the study of how to transmit information secretly. In coding theory, secrecy is not an issue; rather, one wishes to send some information over a communication channel that is intrinsically noisy, such as a telephone line. The message that is received is not identical to the one that was sent, and the goal is to be able to reconstruct the original. The schemes constructed by coding theory are therefore often called *error-correcting codes.*

There are many variations of codes, and for this brief introduction, we will focus on *binary linear block codes.* The word *binary* indicates that the information to be sent is a string of binary digits (bits), that is, a string of zeros and ones. It most cases, this is not actually the original message, but rather a preliminary step has been performed to transform the message into binary. This preliminary step is not part of coding theory. If the message is in English (or uses the familiar Roman alphabet), then one way to translate it into binary is to replace each character by its ASCII equivalent. ASCII assigns a number between 0 and 127 to characters including letters, digits, and punctuation marks. These numbers can be expressed in binary using 7 bits, although 8 bits is more common since computer memory is organized into 8-bit *bytes.* We use 8 bits in our examples below (the first bit in every 8-bit ASCII code is 0).

Example 138 *Let us suppose we wish to send the message "MEET ME AT 7." The ASCII codes for the 26 characters A through Z are the consecutive numbers 65–90, and the space is encoded as* 32. *The characters could be encoded in binary as follows:*

character	*ASCII*	*binary equivalent*
M	*77*	*01001101*
E	*69*	*01000101*
T	*84*	*01010100*
A	*65*	*01000001*
7	*55*	*00110111*
space	*32*	*00100000*

The binary message is then

$$\begin{gathered}010011010100010101000101010101000010000001001101\\010001010010000001000001010101000010000000110111\end{gathered}$$

(a total of $12 \cdot 8 = 96$ *bits).*

A *block code* assumes that the information is grouped in blocks of a given length, say k, and each block is represented by a *codeword.* Assuming we wish to construct a binary block code, in which the blocks have length k, we need 2^k codewords, one to represent each of the possible sequences of k bits. Each block is an element of $\mathbf{Z}_2^k$, where $\mathbf{Z}_2$ is the finite field whose only elements are 0 and 1. For instance, if we take $k = 4$, the message from Example 138 is written in the following blocks:

$$
\begin{array}{c}
0100\ 1101\ 0100\ 0101\ 0100\ 0101\ 0101\ 0100\ 0010\ 0000\ 0100\ 1101 \\
0100\ 0101\ 0010\ 0000\ 0100\ 0001\ 0101\ 0100\ 0010\ 0000\ 0011\ 0111
\end{array}
$$

A *linear code* is one in which the codewords form a proper subspace C of a vector space V. To remain in the binary realm, we take $V = \mathbf{Z}_2^n$ for some $n > k$. If C is to contain 2^k codewords, then the dimension of C must be k. The space C of codewords is also referred to simply as the *code.*

We can think of $\mathbf{Z}_2^n$ as the collection of all possible codewords, and the subspace C as the collection of all actual (or meaningful) codewords. The dimension n of the larger space $\mathbf{Z}_2^n$ must be strictly larger than k; otherwise, every possible codeword is an actual codeword, and it would be impossible to correct errors. To make this point clear, it is helpful to describe precisely the nature of errors. We send a codeword, which is an element c of $\mathbf{Z}_2^n$ (that is, a string of n zeros and ones), and some of the bits (components) of c might be "flipped" (switched from 0 to 1 or vice versa). In $\mathbf{Z}_2$, $0+1=1$ and $1+1=0$, which means that the flipping of a bit can be represented by addition of 1, and the totality of errors in the bits of c can be presented by addition of an error vector $e \in \mathbf{Z}_2^n$ to c ($e_i = 1$ if there is an error in the ith bit and $e_i = 0$ if there is not). In other words, the noisy vector is $c + e$ for some $e \in \mathbf{Z}_2^n$.

If we are to correct the errors in the received codeword (that is, recognize that $c+e$ should be c), a minimum requirement is that we be able to detect that $c + e$ contains errors. For this to be the case, $c + e$ should not be a codeword ($c + e \notin C$). Since $c + e \in \mathbf{Z}_2^n$, this means that C must be a proper subspace of $\mathbf{Z}_2^n$.

Of course, it is not possible to detect every possible error, since if c and d are two codewords ($c, d \in C$), then $d = c + e$ if e is defined as $d - c$. (Incidentally, since $1 + 1 = 0$ in $\mathbf{Z}_2$, $d - c = d + c$.) However, it is a reasonable assumption that the communication channel is fairly reliable and therefore that it is unlikely that an actual error vector e contains too many ones. Therefore, a key property of a successful code C is that one codeword in C cannot be obtained from another by flipping a small number of bits. We will be more precise about this below, when we specify what it means for a code to be able to correct a certain number of errors.

We summarize the above considerations in the following definition.

Definition 139 *A* binary linear block code *is a k-dimensional subspace C of $\mathbf{Z}_2^n$, where $n > k$ is the* length *of the code. We refer to such a code concisely as a (n, k) binary block linear code.*

3.11.1 Generator matrices; encoding and decoding

Since a (n, k) binary linear block code C is just a k-dimensional subspace of $\mathbf{Z}_2^n$, it can be represented by a basis consisting of k linearly independent vectors in $\mathbf{Z}_2^n$. These vectors are usually arranged as the rows of a matrix.

Definition 140 *Let C be a (n, k) binary linear block code. We say that the matrix $G \in \mathbf{Z}_2^{k\times n}$ is a* generator matrix *for C if the rows of G form a basis for C.*

If G is a generator matrix for C, then any linear combination of the rows of G is a codeword. Given $m \in \mathbf{Z}_2^k$ and $G \in \mathbf{Z}_2^n$, we define mG, the *vector-matrix product*, as the linear combination of the rows of G, where the weights are the components of m. In other words, if

$$G = \begin{bmatrix} c_1 \\ \hline c_2 \\ \hline \vdots \\ \hline c_k \end{bmatrix},$$

where $c_i \in \mathbf{Z}_2^n$ for each i, then

$$mG = \sum_{i=1}^{k} m_i c_i.$$

(In the context of this book, it would be more natural to take $G \in \mathbf{Z}_2^{n\times k}$, where the columns of G form the basis of C, but it is traditional in coding theory to represent codewords as row vectors.)

As the reader will recall, each block in the message is an element of $\mathbf{Z}_2^k$ (a string of k bits). We can therefore *encode* each block m as the codeword mG. Since the rows of G span C, mG is a codeword for each $m \in \mathbf{Z}_2^k$. Moreover, since the rows of G are linearly independent, there is a one-to-one correspondence between $m \in \mathbf{Z}_2^k$ and $c = mG \in C$.

We can now describe the entire process of sending, receiving, and interpreting a block m:

1. Encode m as $c = mG$.

2. Send c over the (noisy) channel, producing, at the receiver, $c + e$.

3. Replace $c + e$ by (hopefully) c; this step is called *error correction.*

4. Recover m by solving $mG = c$. (In terms of the column-oriented notation used in the rest of this book, this is equivalent to solving $G^T m = c$ for m given c.)

The last two steps together are referred to as *decoding.*

Apart from the unfamiliar row-oriented notation, the process of solving $mG = c$ is well understood. Since the rows of G span C, there is a unique solution m to $mG = c$ for each $c \in C$. Therefore, the key step is error-correction, that is, recognizing that $c + e$ represents c.

Here is an example of a binary block linear code.

Example 141 *Consider the* $(6, 4)$ *binary linear block code with generator matrix*

$$G = \begin{bmatrix} 1 & 0 & 0 & 0 & 1 & 1 \\ 0 & 1 & 0 & 0 & 0 & 1 \\ 0 & 0 & 1 & 0 & 1 & 0 \\ 0 & 0 & 0 & 1 & 1 & 1 \end{bmatrix}.$$

It is obvious that the rows of G *are linearly independent because the first four columns form the* 4×4 *identity matrix. We will call this code* C_1.

To encode the message from Example 138, we work with one block at a time. The first block (as an element of Z_2^4*) is* $m = (0, 1, 0, 0)$*, and* mG *is just the second row of* G*:*

$$mG = (0, 1, 0, 0, 0, 1).$$

Therefore, the block 0100 *corresponds to the codeword* 010001*. The second block is* $m = (1, 1, 0, 1)$*, and* mG *is the sum of the first, second, and fourth rows of* G*:*

$$mG = (1, 0, 0, 0, 1, 1) + (0, 1, 0, 0, 0, 1) + (0, 0, 0, 1, 1, 1) = (1, 1, 0, 1, 0, 1)$$

(recall that the arithmetic is done in $\mathbf{Z}_2$*; thus, in computing the fifth component of the sum,* $1+0+1 = 0$ *and, in computing the sixth,* $1+1+1 = 1$*). Continuing in this fashion, the entire message is*

010001 110101 010001 010110 010001 010110 010110 010001
001010 000000 010001 110101 010001 010110 001010 000000
010001 000111 010110 010001 001010 000000 001101 011100

Here is another binary block linear code; the length of the code is 7, in contrast to the code of the last example, which has a length of 6.

Example 142 *Consider the generator matrix*

$$G = \begin{bmatrix} 1 & 0 & 0 & 1 & 1 & 0 & 1 \\ 0 & 1 & 0 & 1 & 0 & 1 & 1 \\ 0 & 0 & 1 & 0 & 1 & 1 & 1 \\ 0 & 0 & 0 & 1 & 1 & 1 & 0 \end{bmatrix}.$$

Since the first four columns are obviously linearly independent, this matrix defines a $(7, 4)$ *binary block linear code, which we will call* C_2*. The message from Example 138 would be encoded as follows:*

0101011 1101000 0101011 0100101 0101011 0100101 0100101 0101011
0010111 0000000 0101011 1101000 0101011 0100101 0010111 0000000
0101011 0001110 0100101 0101011 0010111 0000000 0011001 0110010

The last two examples differ in the lengths of the codes, and this introduces an important concept in linear block codes.

Definition 143 *Let C be an (n, k) linear block code. The ratio k/n is called the* information rate *(or simply* rate*) of the code.*

The information rate is a measure of the redundancy in the code; the smaller the rate, the smaller the ratio of information to message length. All other things being equal, we prefer a larger information rate because a larger rate corresponds to more efficiency. However, the redundancy is introduced to enable the correction of errors, and a smaller information rate may be the price paid to make satisfactory error correction possible.

In Examples 141 and 142, the information rates are $4/6 = 2/3$ and $4/7$, respectively. By this measure, then, the first code is better. However, we will see that the second code is (much) better when error correction is taken into account, and this is the key feature of a code. (After all, we can achieve an information rate of 1—the highest possible—by simply sending the original message. But if we do that, every possible codeword is an actual codeword and it is not possible to detect errors, much less correct them.)

3.11.2 Error correction

We will assume that errors (the flipping of bits) are fairly unlikely and therefore that we will be satisfied if we can correct t errors per codeword, where t is a small number (such as 1 or 2). Below we will discuss a little probabilistic reasoning that allows us to make this assumption and to determine a value for t, but for now we will take t as given.

The following terminology will make it easier to discuss error correction.

Definition 144 *Let $e \in \mathbf{Z}_2^n$. The* weight[11] *$w(e)$ of e is simply the number of components of e equal to* 1.

We wish to be able to recognize that the intended codeword is c if $c + e$ is received and $w(e) \leq t$.

Definition 145 *Let $x, y \in \mathbf{Z}_2^n$. The* distance *$d(x, y)$ between x and y is the number of components in which x and y differ.*

Since $x_i + y_i = 0$ if x_i and y_i are the same, and $x_i + y_i = 1$ if they differ, we see that $d(x, y) = w(x + y)$.

Definition 146 *Let C be an (n, k) binary linear block code. The* minimum distance *of the code is the minimum distance between any two distinct elements of C. Thus, if the minimum distance of C is d, then*

[11]The weight $w(e)$ is often called the *Hamming weight* after Richard Hamming, a pioneer in coding theory. Similarly, the distance of Definition 146 can be called the *Hamming distance.*

1. $d(x,y) \geq d$ for all $x,y \in C$, $x \neq y$;

2. there exist $x,y \in C$ such that $d(x,y) = d$.

If an (n,k) code has minimum distance d, we call it an (n,k,d) binary linear block code.

The following fact follows from the comment preceding Definition 146, and makes computing the minimum distance of a code easier.

Lemma 147 *Let C be an (n,k,d) binary linear block code. Then*

$$d = \min\{w(x) \,:\, x \in C, x \neq 0\}.$$

As mentioned above, we are not concerned about the possibility of more than t errors (flipped bits) per codeword. Therefore, if we receive $\hat{c}$, we want it to be the case that there is a unique codeword $c \in C$ such that $d(\hat{c},c) \leq t$. This will be true if d, the minimum distance between any two codewords in C, is strictly greater than $2t$. (Here we are assuming that $\hat{c}$ contains at most t errors, so that there does exist some $c \in C$ with $d(\hat{c},c) \leq t$. The only question is about uniqueness.) To understand the significance of the requirement $d > 2t$, let us consider a contrary case in which $d = 2t$. Then there are codewords $c_1, c_2 \in C$ with $d(c_1,c_2) = 2t$. We can construct $\hat{c} \in \mathbf{Z}_2^n$ such that $\hat{c}$ agrees with c_1 and c_2 on the $n - 2t$ bits on which c_1 and c_2 agree, and $\hat{c}$ agrees with each of c_1 and c_2 on t of the remaining $2t$ bits. Then $d(\hat{c},c_1) = d(\hat{c},c_2) = t$, and there is no way to decide if $\hat{c}$ is supposed to be c_1 or c_2.

These considerations lead to the following terminology.

Definition 148 *Let C be an (n,k,d) binary linear block code. We say that C can* correct *up to t errors if $d > 2t$. For any value of d, we say that C can* detect *up to $d-1$ errors.*

If $\hat{c}$ contains at least one but not more than $d-1$ errors, then it is guaranteed that it does not equal any codeword $c \in C$. This is the point of error detection; in such a case we know that errors have occurred, although there may not be a unique codeword closest to $\hat{c}$.

The reader will recall that the minimum distance of a linear code C is simply the minimum weight of any nonzero codeword in C. This makes it easy (at least in principle) to compute the minimum distance of any given binary linear block code: just generate all the code words and compute the distance of each one. For example, the $(6,4)$ binary linear block code C_1 of Example 141 is seen to have a minimum distance of 2, while the $(7,4)$ code C_2 of Example 142 has a minimum distance of 3. (Thus the first is a $(6,4,2)$ code, while the second is a $(7,4,3)$ code.)

The reader can now appreciate the reason for the reduced information rate of C_2 over C_1. While C_1 can *detect* 1 error per codeword ($t = 1 = d - 1$ in this case), C_2 can *correct* 1 error per codeword (since, with $t = 1$, $2t < d = 3$). It makes sense to accept a lower information rate for the ability to correct errors.

3.11.3 The probability of errors

In order to assess the probability of flipped bits and thereby choose a value of t for the above analysis, we need a model of the communications channel. A common model is that of *binary symmetric channel*, which assumes that binary data (bits) are transmitted, and that the probability of a bit's being flipped is a constant p, where $0 < p < 1$. It is usually assumed that p is quite small ($p \ll 1/2$). The word "symmetric" refers to the fact that a flip from 0 to 1 is equally likely as a flip from 1 to 0. It is also assumed that the transmission of each bit is an independent event, so that an error in one bit does not affect the probability of an error in later bits.

Under these assumptions, the transmission of a codeword is described by a *binomial distribution*, a standard probability distribution described in any introductory text on probability and statistics (see, for example, [44]). The probability of *exactly* k errors in a codeword of length n is

$$\left(\begin{array}{c} n \\ k \end{array}\right) p^k(1-p)^{n-k}, \tag{3.26}$$

where

$$\left(\begin{array}{c} n \\ k \end{array}\right) = \frac{n!}{k!(n-k)!}$$

is the number of combinations of n objects taken k at a time. The symbol $\left(\begin{array}{c} n \\ k \end{array}\right)$ is usually read as "n choose k." From (3.26), we can compute the probability that at most t errors occur. For example, the probability of at most zero errors is

$$(1-p)^n \doteq 1 - np + O((np)^2),$$

which means that the probability of at least one error is

$$1 - (1-p)^n \doteq np + O((np)^2).$$

The probability of at most one error is

$$(1-p)^n + np(1-p)^{n-1} \doteq 1 - \frac{n^2-n}{2}p^2 + O((np)^3),$$

and hence the probability of more than one error is

$$1 - (1-p)^n - np(1-p)^{n-1} \doteq \frac{n^2-n}{2}p^2 + O((np)^3).$$

If p is sufficiently small that $np \ll 1$, then $(n^2-n)p^2/2$ is very small, so it might be perfectly reasonable to assume that no more than 1 error occurs. To give one more example, the probability of at most 2 errors is

$$(1-p)^n + np(1-p)^{n-1} + \frac{n(n-1)}{2}p^2(1-p)^{n-2} \doteq 1 - \frac{n^6-3n^2+2n}{6}p^3 + O((np)^4),$$

and therefore the probability of more than 2 errors is

$$1-(1-p)^n-np(1-p)^{n-1}-\frac{n(n-1)}{2}p^2(1-p)^{n-2} \doteq \frac{n^6-3n^2+2n}{6}p^3+O((np)^4).$$

This probability is $O((np)^3)$.

Assuming $pn \ll 1$, the probability of more than t errors decreases rapidly as t grows. Given specific values of p and n, we just choose t large enough that the probability of more than t errors is considered negligible.

Exercises

In these exercises, every message is originally expressed in 8-bit ASCII.

1. Consider the code C of Example 141.

   (a) Verify that $(0,1,0,1,0,0)$ is not a codeword.

   (b) Show that there is not a unique codeword closest to $(0,1,0,1,0,0)$.

2. The following message is received.

   010110 001101 010110 010001 010001
   111111 010110 000000 001010 000111

   It is known that the code of Example 141 is used. What is the original message (in English)?

3. The following message is received.

   010110 000101 010110 010110 010001 100100 010110 010001

   It is known that the code of Example 141 is used. Show that the message cannot be decoded unambiguously. (The original message is a single English word.)

4. The following message is received.

   0000101 0001110 0100101 0100101 0101011 1000011 0100101 0101011

   It is known that the code of Example 142 is used.

   (a) Show that the received message contains an error.

   (b) Show that, nevertheless, the message can be unambiguously decoded.

3.12 Linear programming

Linear programming problems form a particular class of optimization problems, in which it is desired to find the maximum or minimum value of an *objective function* subject to some constraints. In a linear programming problem, the objective function is linear and the constraints are defined by linear functions. In this section, we will briefly describe the theory of linear programming and how linear algebra is used in solving linear programming problems.

Incidentally, the word "programming" is not used in the sense of computer programming (although certainly computer programs are written to solve the problems). Rather, "programming" is used in the sense of "planning" or "scheduling," and the idea is to determine an optimal plan or schedule for accomplishing a certain task. Linear programming came into prominence due to George Dantzig, who invented the most popular algorithm (the simplex method) while addressing logistical problems for the US Air Force in the 1940s.

It is easiest to introduce linear programming with an example.

Example 149[12] *Consider a factory that produces four different products using three types of machines. For each product, the profit per unit and the machine time required per unit are known. This information is summarized in the following table:*

Machine type	*Prod. 1*	*Prod. 2*	*Prod. 3*	*Prod. 4*	*Hours available*
A	*1.5*	*1.0*	*2.4*	*1.0*	*2000*
B	*1.0*	*5.0*	*1.0*	*3.5*	*8000*
C	*1.5*	*3.0*	*3.5*	*1.0*	*5000*
Profit ($/unit)	*5.24*	*7.30*	*8.34*	*4.18*	

For instance, the profit per unit of product 1 is $5.24, and to produce one unit of product 1 requires 1.5 hours on machine A, 1.0 hours on machine B, and 1.5 hours on machine C. "Hours available" means the number of hours per week available on each type of machine.

If we write x_i, $i = 1, 2, 3, 4$, *for the number of units of product i produced each week, then the objective is to maximize the profit, which is given by*

$$5.24x_1 + 7.30x_2 + 8.34x_3 + 4.18x_4 = c \cdot x, \ c = (5.24, 7.30, 8.34, 4.18).$$

Here we have written the objective function using the Euclidean dot product:

$$c \cdot x = c_1x_1 + c_2x_2 + \cdots + c_nx_n.$$

In Chapter 6, we study the dot product as an example of an inner product; in

[12] Taken from Hadley [17], Section 1-3.

this section, we will use the dot product merely as a convenient notation. It is obvious that $x \mapsto c \cdot x$ defines a linear function of x.

We have constraints because there is limited machine time:

$$\begin{aligned} 1.5x_1 + 1.0x_2 + 2.4x_3 + 1.0x_4 &\leq 2000, \\ 1.0x_1 + 5.0x_2 + 1.0x_3 + 3.5x_4 &\leq 8000, \\ 1.5x_1 + 3.0x_2 + 3.5x_3 + 1.0x_4 &\leq 5000. \end{aligned}$$

We also have the constraint that each x_i must be nonnegative: $x_i \geq 0$, $i = 1, 2, 3, 4$. If we write

$$A = \begin{bmatrix} 1.5 & 1.0 & 2.4 & 1.0 \\ 1.0 & 5.0 & 1.0 & 3.5 \\ 1.5 & 3.0 & 3.5 & 1.0 \end{bmatrix}, \ b = \begin{bmatrix} 2000 \\ 8000 \\ 5000 \end{bmatrix},$$

then we can write the constraints as $Ax \leq b$ and $x \geq 0$. In writing these constraints in vector form, we use the convention that if $u, v \in \mathbf{R}^n$, then $u \leq v$ means $u_i \leq v_i$ for all $i = 1, 2, \ldots, n$, and similarly for other inequality signs such as $\geq$. In other words, inequalities between vectors are interpreted componentwise.

The problem is to find a value of $x \in \mathbf{R}^4$ satisfying $Ax \leq b$ and $x \geq 0$ that makes $c \cdot x$ as large as possible. We normally write the problem in the form

$$\begin{aligned} \max \quad & c \cdot x \\ s.t. \quad & Ax \leq b, \\ & x \geq 0. \end{aligned}$$

3.12.1 Specification of linear programming problems

A linear programming problem has a linear objective function, $\phi(x) = c \cdot x$, where $c \in \mathbf{R}^n$ is a constant vector. In addition to the constraints illustrated in the previous example, there can also be equality constraints $Ax = b$ or inequality constraints of the form $Ax \geq b$. However, a $\geq$ constraint can be changed to a $\leq$ constraint by multiplying both sides of the inequality by -1, so we will not discuss $\geq$ constraints any further.

The goal in a linear programming problem can be either to maximize or minimize the objective function. However, minimizing $c \cdot x$ is equivalent to maximizing $-c \cdot x$. For this reason, we will always assume that the objective function is to be maximized.

Inequality constraints can be converted to equality constraints by the addition of *slack variables*. For example, the first constraint from Example 149 is

$$1.5x_1 + 1.0x_2 + 2.4x_3 + 1.0x_4 \leq 2000. \tag{3.27}$$

If we define

$$x_5 = 2000 - 1.5x_1 - 1.0x_2 - 2.4x_3 - 1.0x_4,$$

then (3.27) becomes

$$1.5x_1 + 1.0x_2 + 2.4x_3 + 1.0x_4 + x_5 = 2000,$$

where x_5 must satisfy $x_5 \geq 0$. Adding slack variables to the other two equations yields

$$\begin{aligned} 1.5x_1 + 1.0x_2 + 2.4x_3 + 1.0x_4 + x_5 \qquad\qquad &= 2000, \\ 1.0x_1 + 5.0x_2 + 1.0x_3 + 3.5x_4 \qquad + x_6 \qquad &= 8000, \\ 1.5x_1 + 3.0x_2 + 3.5x_3 + 1.0x_4 \qquad\qquad + x_7 &= 5000, \end{aligned}$$

where now $x_i \geq 0$ for all $i = 1, 2, \ldots, 7$. Defining

$$\tilde{A} = [A|I], \ \tilde{c} = (c_1, c_2, c_3, c_4, 0, 0, 0),$$

and $\tilde{x} \in \mathbf{R}^7$ to be the original vector x, augmented by the three slack variables, the problem from Example 149 can now be written

$$\begin{aligned} \max \quad & \tilde{c} \cdot \tilde{x} \\ \text{s.t.} \quad & \tilde{A}\tilde{x} = b, \\ & \tilde{x} \geq 0. \end{aligned}$$

Based on the above discussion, we will recognize two standard forms for linear programming problems: the *standard inequality form*,

$$\begin{aligned} \max \quad & c \cdot x & \text{(3.28a)} \\ \text{s.t.} \quad & Ax \leq b, & \text{(3.28b)} \\ & x \geq 0, & \text{(3.28c)} \end{aligned}$$

and the *standard equality form*,

$$\begin{aligned} \max \quad & c \cdot x & \text{(3.29a)} \\ \text{s.t.} \quad & Ax = b, & \text{(3.29b)} \\ & x \geq 0. & \text{(3.29c)} \end{aligned}$$

Either problem is referred to as a *linear program* (LP).[13] We will assume $x \in \mathbf{R}^n$ and $A \in \mathbf{R}^{m \times n}$. In LP (3.28), there is no restriction on m and n, but in LP (3.29) it is always assumed that $m < n$ (otherwise, $Ax = b$ would normally—that is, if A has full rank—describe at most a single point).

We use the following terms in discussing LPs.

Definition 150 *Given either (3.28) or (3.29), $x \in \mathbf{R}^n$ is called a* feasible solution *if it satisfies the constraints. The set of all feasible solutions is called the* feasible set *of the LP. A feasible solution x is called an* optimal solution*if it gives the maximum value of the objective function $\phi(x) = c \cdot x$ over all feasible solutions.*

[13]There is no agreement in the literature as to what constitutes the standard form of an LP; moreover, some authors refer to a *canonical form*, and some label one form the standard form and another the canonical form.

3.12.2 Basic theory

We wish to sketch the theory of linear programs, and this is most easily done if we introduce the geometry of the feasible set.

Definition 151 *Given a constant vector $r \in \mathbf{R}^n$ and a constant $\beta \in \mathbf{R}$, the set*

$$\{x \in \mathbf{R}^n \, : \, r \cdot x \leq \beta\}$$

is called a closed half-space.

The intersection of a finite number of closed half-spaces is called a polyhedron. *A bounded polyhedron is called a* polytope.

Given an LP in the standard inequality form (3.28), the constraint $Ax \leq b$ represents m inequalities of the form $r_i \cdot x \leq b_i$ (where $r_1, r_2, \ldots, r_m$ are the rows of A), each of which defines a half-space. Moreover, each constraint $x_i \geq 0$ can be written as $-e_i \cdot x \leq 0$ and thus defines a half-space. A vector $x \in \mathbf{R}^n$ is feasible if and only if it lies in all of the $m + n$ half-spaces defined by these constraints; in other words, the feasible set of the LP (3.28) is a polyhedron.

Referring to the standard equality form LP (3.29), the constraint $Ax = b$ represents m equations of the form $r_i \cdot x = 0$. Each of these equations is equivalent to two inequalities,

$$r_i \cdot x \leq 0 \text{ and } -r_i \cdot x \leq 0,$$

and thus represents the intersection of two half-spaces. As a result, the feasible set for (3.29) is also a polyhedron. If the feasible set is bounded, then it is a polytope.

Below we will need the concept of convexity.

Definition 152 *Let x, y be two points in $\mathbf{R}^n$. A* convex combination *of x and y is a point of the form $\alpha x + \beta y$, where $\alpha, \beta \in [0, 1]$ and $\alpha + \beta = 1$.*

A convex combination of $x_1, x_2, \ldots, x_k \in \mathbf{R}^n$ is a point of the form

$$\alpha_1 x_1 + \alpha_2 x_2 + \cdots + \alpha_k x_k,$$

where $\alpha_1, \alpha_2, \ldots, \alpha_k \in [0, 1]$ and $\alpha_1 + \alpha_2 + \cdots + \alpha_k = 1$.

A set $C \subset \mathbf{R}^n$ is called convex *if, given any $x, y \in C$, all convex combinations of x and y belong to C.*

Geometrically (that is, in two or three dimensions, where we can visualize the points and sets), a convex combination of two points is a point lying on the line segment joining the two points. A convex set is one that contains the line segment joining any two points in the set. Figure 3.7 shows two sets, one of which is convex and one of which is not.

It is easy to show that a half-space is a convex set and that the intersection of convex sets is convex (see Exercises 7, 8). Therefore, a polyhedron (and hence the feasible set of any LP) is a convex set.

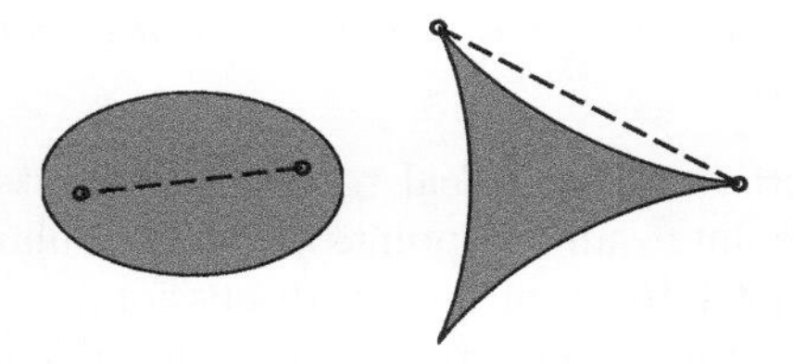

FIGURE 3.7
Two sets in $\mathbf{R}^2$. The one on the left is convex, while the one on the right is not. For the set C on the left, no matter how $c, y \in C$ are chosen, the line segment joining x and y lies entirely in C.

In two dimensions, a polytope is the region bounded by a polygon. For example, the feasible set of the LP

$$\begin{aligned} \max \quad & 3x_1 + 2x_2 & \text{(3.30a)} \\ \text{s.t.} \quad & -x_1 + 2x_2 \le 2, & \text{(3.30b)} \\ & x_1 + x_2 \le 4, & \text{(3.30c)} \\ & x_1 - x_2 \le 2, & \text{(3.30d)} \\ & x_1, x_2 \ge 0 & \text{(3.30e)} \end{aligned}$$

is the polytope displayed in Figure 3.8, the intersection of five half-spaces.

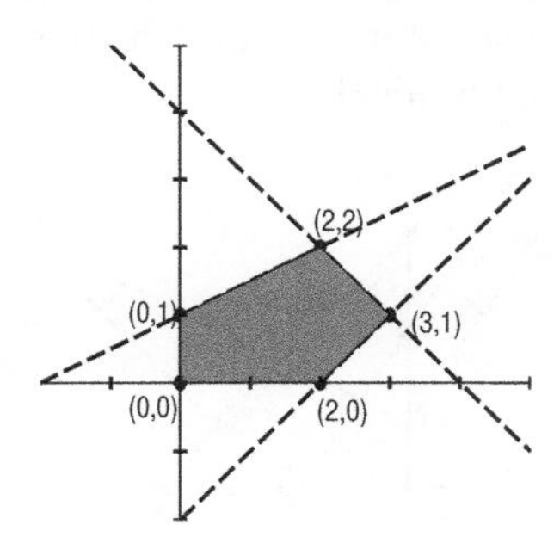

FIGURE 3.8
The feasible region for the LP (3.30).

In understanding polytopes in general and the theory of LPs in particular, the *vertices* of the polytope are of particular importance. The feasible set shown in Figure 3.8 has five vertices, $(0,0)$, $(0,1)$, $(2,2)$, $(3,1)$, and $(2,0)$. The vertices are the extreme points of the polytope.

Definition 153 *Let $C \in \mathbf{R}^n$ be a convex set. A point $x \in C$ is an* extreme point *of C if it not possible to write $x = \alpha u + \beta v$ for $u, v \in C$ and $\alpha, \beta \in (0,1)$,*

$\alpha+\beta=1$. *In other words, x is an extreme point if it does not lie in the interior of a line segment in C.*

(For a polytope, vertices are identical to extreme points. A convex set like a disk has extreme points—all the points on the circular boundary—but no vertices, but this is not important for our discussion.)

The significance of the extreme points of the feasible set of an LP is easily seen by considering the contours of the objective function. If the objective function is $\phi(x)=c\cdot x$, then the contours ϕ are sets of the form

$$\{x\in\mathbf{R}^n : \phi(x)=z\},$$

where $z\in\mathbf{R}$ is a constant. In other words, the contours of ϕ are the sets on which ϕ is constant. If we wish to maximize ϕ, then geometrically we must find the largest value of z such that the contour defined by z intersects the feasible set.

Since ϕ is linear, its contours are especially simple; they are lines in $\mathbf{R}^2$, planes in $\mathbf{R}^3$, and, in general, *hyperplanes* in $\mathbf{R}^n$. In Figure 3.9, we show the feasible set of the LP (3.30), along with several contours of the objective function. From this graph, we see that the maximum value of the objective function is $z=11$, and that it occurs at an extreme point. Intuitively, it should be clear that this will always be the case: Since the feasible set has "flat" faces joined at corners, and since the contours of the objective function are flat, a contour that just touches the feasible set (that is, such that increasing the constant defining the contour would cause the contour to move completely out of the feasible set) must intersect the feasible set at an extreme point. Below this fact is proven algebraically.

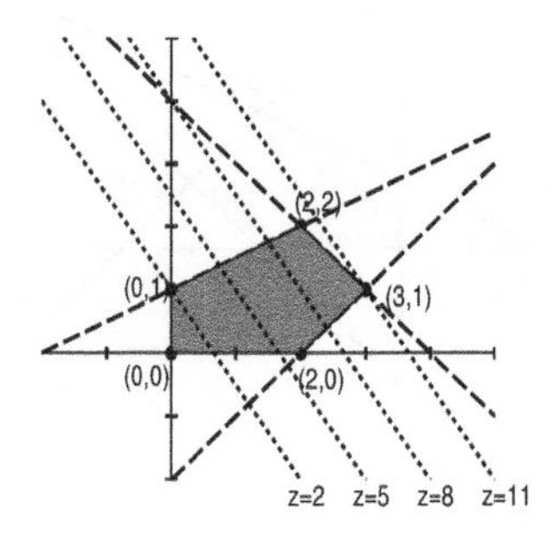

FIGURE 3.9
The feasible region for the LP (3.30), along with some of the contours of the objective function (the dotted lines).

The reader may wonder about the special case in which the contours of the objective function are parallel to one of the faces of the feasible region. For instance, in Figure 3.9, what would happen if the objective function were changed so that its contours were parallel to the line segment joining $(2, 2)$ and $(3, 1)$? The answer is that in that case, every point on the line segment would be an optimal solution; there would be optimal solutions that are not extreme points, but the extreme points $(2, 2)$ and $(3, 1)$ would also be optimal. Therefore, even in that special case, an extreme point is an optimal solution.

Let us consider the standard equality form (3.29) of the LP, and try to express these geometric considerations algebraically. In (3.29), we have n variables and a total of $m + n$ constraints (m equality constraints and n inequality constraints $x_1, \ldots, x_n \geq 0$), where $n > m$. The extreme points are points in $\mathbf{R}^n$, which are determined by the intersection of n (independent) hyperplanes. Each of the m equality constraints defines a hyperplane in $\mathbf{R}^n$, and there is a hyperplane associated with each inequality $x_i \geq 0$ (namely, the hyperplane corresponding to $x_i = 0$). Out of these $m+n$ hyperplanes, n must intersect to define an extreme point, which implies that (at least) $n - m$ of the inequalities $x_i \geq 0$ must be satisfied as equations. Thus, at an extreme point, $n-m$ of the variables are zero, and the other m variables will typically be nonzero. If more than $n - m$ variables are zero, this is a *degenerate* case.

The last few paragraphs have been rather imprecise and have appealed to geometric intuition. We will now justify the foregoing statements precisely. We begin by establishing some notation and and terminology. Let $x \in \mathbf{R}^n$ satisfy $Ax = b$ and have at least $n-m$ components equal to 0. Let the indices $\{1, 2, \ldots, n\}$ be partitioned into subsets B and N, where B contains m indices, N contains $n - m$ indices and $x_i = 0$ for $i \in N$. If the set $\{A_i : i \in B\}$ is a linearly independent subset of $\mathbf{R}^m$, then x is called a *basic solution* of the LP. If $x_i \geq 0$ for all $i \in B$, then x is a feasible solution and is therefore called a *basic feasible solution* (BFS). If there exists $i \in B$ such that $x_i = 0$, then x is called a *degenerate* BFS; otherwise, $x_i > 0$ for all $i \in B$ and x is a *nondegenerate* BFS.

Our discussion suggests that extreme points and BFSs are closely related. In fact, for an LP in standard equality form (3.29) with $\operatorname{rank}(A) = m$, the two concepts are the same: x is an extreme point of the feasible set if and only if x is a BFS. The reader is asked to prove this in Exercise 9. (Of course, if $\operatorname{rank}(A) < m$, then there are no basic solutions to $Ax = b$, $x \geq 0$.)

It appears, based on the geometric reasoning given above, that if an LP has an optimal solution, then it has an optimal BFS. In fact, this is true under a single assumption, namely, that $\operatorname{rank}(A) = m$. This is part of the following theorem, which is sometimes called the *fundamental theorem of linear programming*.

Theorem 154 *Consider the LP (3.29) in standard equality form, and assume that* $\operatorname{rank}(A) = m$.

1. *If the LP has a feasible solution, then it has a basic feasible solution.*

2. *If the LP has an optimal solution, then some basic feasible solution is optimal.*

Proof

1. Let $x \in \mathbf{R}^n$ be a feasible solution, and define

$$\begin{aligned} B &= \{i \in \mathbf{Z} : 1 \leq i \leq n, x_i > 0\} = \{i_1, i_2, \ldots, i_k\}, \\ N &= \{i \in \mathbf{Z} : 1 \leq i \leq n, x_i = 0\} = \{j_1, j_2, \ldots, j_\ell\}. \end{aligned}$$

The proof that there is a BFS is divided into two cases.

Case 1 $\{A_{i_1}, A_{i_2}, \ldots, A_{i_k}\}$ is linearly independent. In this case, x is a BFS. This is obvious if $k = m$ (since then x has exactly m nonzero components and the corresponding columns of A are linearly independent). If $k < m$, then, since $\text{rank}(A) = m$ and $\{A_{i_1}, A_{i_2}, \ldots, A_{i_k}\}$ is linearly independent, we can find $m - k$ columns among $A_{j_1}, A_{j_2}, \ldots, A_{j_\ell}$ that, together with $A_{i_1}, A_{i_2}, \ldots, A_{i_k}$, form a linearly independent set. This implies that x is a degenerate BFS.

Case 2 $\{A_{i_1}, A_{i_2}, \ldots, A_{i_k}\}$ is linearly dependent. We assume for now that this set contains at least one nonzero vector. We will show that we can produce another feasible solution x' with the property that $x'_i = 0$ for all $i \in N$ and $x'_i = 0$ for at least one $i \in B$, namely, an index i corresponding to a vector in the set $\{A_{i_1}, A_{i_2}, \ldots, A_{i_k}\}$ that is a linear combination of the remaining $k - 1$ vectors. By doing this, possibly several times, to remove dependent columns, we will produce a feasible solution that is described by Case 1 and is therefore a BFS.

Since $\{A_{i_1}, A_{i_2}, \ldots, A_{i_k}\}$ is linearly dependent, there exist scalars

$$u_{i_1}, u_{i_2}, \ldots, u_{i_k},$$

with at least one u_{i_s} positive, such that

$$u_{i_1} A_{i_1} + u_{i_2} A_{i_2} + \cdots + u_{i_k} A_{i_k} = 0. \tag{3.31}$$

Define $u \in \mathbf{R}^n$ by defining $u_{j_s} = 0$ for $s = 1, 2, \ldots, \ell$. Now, since $x_i > 0$ for all $i \in B$ and $u_i = 0$ for all $i \in N$, it follows that $x - \epsilon u \geq 0$ for all ϵ sufficiently small. Moreover, u has been chosen so that $Au = 0$, and therefore

$$A(x - \epsilon u) = Ax - \epsilon Au = b - \epsilon \cdot 0 = b.$$

Thus $x - \epsilon u$ is feasible for all ϵ sufficiently small. Define

$$\epsilon' = \min\left\{\frac{x_i}{u_i} : 1 \leq i \leq n, u_i > 0\right\},$$

and note that ϵ' is well-defined since at least one u_{i_s} is positive. It is straightforward to verify that $x' = x - \epsilon' u$ is feasible and that, if $i \in B$ satisfies

$$\frac{x_i}{u_i} = \epsilon',$$

then $x'_i = 0$. Thus we have shown how to produce another feasible solution whose nonzero variables are a proper subset of those of x; by the above argument, the proof is complete.

We still have to deal with the special case that $\{A_{i_1}, A_{i_2}, \ldots, A_{i_k}\}$ is linearly dependent because each of the columns of A in this set is the zero vector. But this would imply that b is the zero vector, and hence $x = 0$ is a feasible vector. Choosing any m linearly independent columns of A shows that this x is a BFS.

2. If x is an optimal solution, then we can prove that there exists an optimal BFS by an argument very similar to the one we just completed. We divide the proof into two cases, as above. In the first case, x itself is an optimal BFS (possibly degenerate). In the second case, the construction is the same as above. We just have to show that, if x is an optimal solution and u satisfies (3.31), $u_i = 0$ for $i \in N$, then $x - \epsilon u$ is not only feasible but also optimal for all ϵ sufficiently small. But this must be true: If $c \cdot u \neq 0$, then x is not optimal because there exists a feasible point $x - \epsilon u$ with

$$c \cdot (x - \epsilon u) = c \cdot x - \epsilon(c \cdot u) > c \cdot x$$

(the sign of ϵ is chosen according to the sign of $c \cdot u$ so that $\epsilon(c \cdot u) < 0$). Since x is assumed optimal, it must be the case that $c \cdot u = 0$, in which case $x - \epsilon u$ is optimal for all ϵ sufficiently small:

$$c \cdot (x - \epsilon u) = c \cdot x - \epsilon(c \cdot u) = c \cdot x - \epsilon \cdot 0 = c \cdot x.$$

Therefore, the construction outlined in Case 2 above will produce an optimal BFS from an optimal feasible solution.

QED

The fundamental theorem was phrased in terms of the standard equality form of an LP. However, the reader should recall that an LP in standard inequality form can be transformed into one in standard equality form by the addition of slack variables. Moreover, in that case, the constraint matrix becomes $\tilde{A} = [A|I]$, which is guaranteed to have rank m. Thus, the fundamental theorem applies to an LP in standard inequality form with no conditions whatever on the matrix A.

3.12.3 The simplex method

According to the fundamental theorem of linear programming, if an LP has an optimal solution, then it has an optimal BFS. Moreover, an LP in standard equality form has at most

$$\binom{n}{m} = \frac{n!}{m!(n-m)!} \tag{3.32}$$

BFSs,[14] and we can find an optimal solution by simply computing the objective function at each of the possible BFSs and choosing the largest value. Although this is a simple algorithm, it is prohibitively expensive for n, m large because (3.32) is so large. The simplex method improves on the exhaustive search by moving from one BFS to another BFS with a larger optimal value. In this way, it is not necessary to test all the possible BFSs, and the algorithm usually finds an optimal solution in a reasonable amount of time.

Before we describe the simplex algorithm, we explain explicitly how to compute a BFS from m linearly independent columns of A. We begin by partitioning $\{1, 2, \ldots, n\}$ into subsets

$$B = \{i_1, i_2, \ldots, i_m\},\ N = \{j_1, j_2, \ldots, j_{n-m}\}.$$

We assume that $\{A_{i_1}, A_{i_2}, \ldots, A_{i_m}\}$ is linearly independent, and we write

$$A_B = [A_{i_1}|A_{i_2}|\cdots|A_{i_m}],\ A_N = \left[A_{j_1}|A_{j_2}|\cdots|A_{j_{n-m}}\right].$$

For any $x \in \mathbf{R}^n$, we define

$$x_B = (x_{i_1}, x_{i_2}, \ldots, x_{i_m}),\ x_N = (x_{j_1}, x_{j_2}, \ldots, x_{j_{n-m}}).$$

We then have $Ax = A_B x_B + A_N x_B$ (both sides of this equation represent the same linear combination of the columns of A). Moreover, since the columns of A_B are linearly independent by assumption, A_B is invertible, and we have

$$\begin{aligned} Ax = b \Rightarrow A_B x_B + A_N x_N = b &\Rightarrow A_B x_B = b - A_N x_N \\ &\Rightarrow x_B = A_B^{-1} b - A_B^{-1} A_N x_N. \end{aligned}$$

This equation shows that we can choose any values for the components of x_N and determine uniquely values for the components of x_B such that x satisfies $Ax = b$. If we choose $x_N = 0$, then $x_B = A_B^{-1} b$ and x is a basic solution. If $x_B = A_B^{-1} b \geq 0$, then x is feasible and hence is a BFS.

Using the above notation, we can now introduce the simplex method. We assume that we have an initial BFS $x \in \mathbf{R}^n$ (below we will discuss how to find a BFS to get started). The value of the objective function at x is

$$\begin{aligned} c \cdot x = c_B \cdot x_B + c_N \cdot x_N &= c_B \cdot \left(A_B^{-1} b - A_B^{-1} A_N x_N\right) + c_N \cdot x_N \\ &= c_B \cdot \left(A_B^{-1} b\right) + c_N \cdot x_N - c_B \cdot \left(A_B^{-1} A_N x_N\right) \\ &= c_B \cdot \left(A_B^{-1} b\right) + c_N \cdot x_N - \left(A_N^T A_B^{-T} c_B\right) \cdot x_N \\ &= c_B \cdot \left(A_B^{-1} b\right) + \left(c_N - A_N^T A_B^{-T} c_B\right) \cdot x_N. \end{aligned}$$

Since $x_N = 0$ by assumption, we have $c \cdot x = c_B \cdot \left(A_B^{-1} b\right)$, which gives us the

[14]The expression $\left(\begin{array}{c} n \\ m \end{array}\right)$ is read as "n choose m"; it represents the number of ways to choose m objects from a collection of n objects. In our context, it is the number of ways to choose m columns out of the n columns of A. Each such choice leads to a BFS if these columns form a linearly independent set and the corresponding vector x satisfies $x \geq 0$.

value of the objective function at x. More importantly, though, the formula shows how the objective function changes if we change the nonbasic variables. We write

$$\tilde{c}_N = c_N - A_N^T A_B^{-T} c_B.$$

To maintain feasibility, we can only increase the components of x_N. If $\tilde{c}_N \leq 0$, then increasing components of x_N will not increase the value of the objective function, which means that x is already optimal.

Let us clarify this argument before continuing. The main point is that the conditions

$$Ax = b,\ x \geq 0,\ z = c \cdot x$$

are equivalent to

$$x_B = A_B^{-1}b - A_B^{-1}A_N x_N,\ x_B \geq 0,\ x_N \geq 0,\ z = c_B \cdot \left(A_B^{-1}b\right) + \tilde{c}_N \cdot x_N.$$

If $\tilde{c}_N \leq 0$, there is no value of $x_N \geq 0$ that leads to a larger value of z than the value corresponding to $x_N = 0$. Hence we cannot change the value of x so that it remains feasible and yields a larger value of z. Thus, as stated above, the current BFS x must be optimal.

Now let us suppose that $\tilde{c}_N$ does not satisfy $\tilde{c}_N \leq 0$, that is, that one or more components of $\tilde{c}_N$ is positive. The simplex method moves to another BFS with a larger value of z by increasing a single component of x_N (with the result that the corresponding nonbasic variable becomes basic) until one of the variables that is currently basic becomes zero (and hence that basic variable becomes nonbasic). In other words, the simplex method moves from one BFS to another by interchanging a nonbasic variable with a basic variable. The particular nonbasic variable is chosen by the criterion that increasing it must increase the value of the objective function. The particular basic variable is chosen as the basic variable that first becomes equal to zero when the nonbasic variable is increased.

We will call the nonbasic variable that becomes basic the *entering variable*, and the basic variable that becomes nonbasic, the *leaving* variable. The set of all basic variables is sometimes called the *basis* (a slight abuse of notation—it is the corresponding columns of A that form a basis for $\mathbf{R}^m$); thus the entering variable enters the basis, while the leaving variable leaves the basis. The process of exchanging the entering and leaving variables is referred to as *pivoting*.

We have described the basics of the simplex algorithm, which is based on using the equations

$$x_B = A_B^{-1}b - A_B^{-1}A_N x_N, \tag{3.33a}$$

$$z = c_B \cdot \left(A_B^{-1}b\right) + \tilde{c}_N \cdot x_N \tag{3.33b}$$

to decide how to move from one BFS to another. At this point, it will be helpful to illustrate with a concrete example.

Example 155 *Consider the LP*

$$\begin{aligned} \max \quad & c\cdot x \\ \text{s.t.} \quad & Ax = b, \\ & x \geq 0, \end{aligned}$$

where

$$A = \begin{bmatrix} -1 & 2 & 1 & 0 & 0 \\ 1 & 1 & 0 & 1 & 0 \\ 1 & -1 & 0 & 0 & 1 \end{bmatrix}, \ b = \begin{bmatrix} 2 \\ 4 \\ 2 \end{bmatrix}, \ c = \begin{bmatrix} 3 \\ 2 \\ 0 \\ 0 \\ 0 \end{bmatrix}$$

and $x \in \mathbf{R}^5$. This is the LP from Example 3.30, written in standard equality form (x_3, x_4, x_5 are slack variables). We have $m = 3$ and $n = 5$, so we must choose three basic variables to get started. In this case, there is an obvious choice: We choose $B = \{3, 4, 5\}$ and $N = \{1, 2\}$. Then $A_B = I$ and the columns of A corresponding to x_3, x_4, x_5 obviously form a linearly independent set. We now wish to form the equations $x_B = A_B^{-1}b - A_B^{-1}A_N x_N$, which simply means solving $Ax = b$ for the basic variables x_3, x_4, x_5:

$$x_3 = 2 + x_1 - 2x_2, \tag{3.34a}$$

$$x_4 = 4 - x_1 - x_2, \tag{3.34b}$$

$$x_5 = 2 - x_1 + x_2. \tag{3.34c}$$

The formula

$$z = c\cdot x = 3x_1 + 2x_2$$

expresses z in terms of the nonbasic variables x_1 and x_2, and thus is equivalent to $z = c_B \cdot \left(A_B^{-1}b\right) + \tilde{c}_N \cdot x_N$. (Note that the constant term is zero because, in this problem, $c_B = 0$ when $B = 3, 4, 5$.) Thus the needed equations are

$$x_3 = 2 + x_1 - 2x_2, \tag{3.35a}$$

$$x_4 = 4 - x_1 - x_2, \tag{3.35b}$$

$$x_5 = 2 - x_1 + x_2, \tag{3.35c}$$

$$z = c\cdot x = 3x_1 + 2x_2. \tag{3.35d}$$

The BFS is obtained by setting x_1 and x_2 to zero; we obtain $x = (0, 0, 2, 4, 2)$. We have $z = 0$ for this value of x. The vector $\tilde{c}_N$ is $(3, 2)$, and since $\tilde{c}_N \not\leq 0$, this initial BFS is not optimal.

We can increase z by increasing either x_1 or x_2 (or both, but in the simplex method, we only increase one nonbasic variable at the time). We choose x_1 as the entering variable, and now we must decide how much it can be increased. From the above equations, x_3 is positive for all positive values of x_1 (x_3 increases as x_1 is increased). However, both x_4 and x_5 decrease when x_1 is increased (bearing in mind that x_2 is held fixed at 0). The variable x_4 equals

0 when $x_1 = 4$, *while* $x_5 = 0$ *when* $x_1 = 2$. *To maintain feasibility, we must choose the smaller value of* x_1, *so we take* $x_1 = 2$, *which makes* x_5 *equal to zero; thus* x_5 *is the leaving variable.*

Now we have a second BFS, with $B = \{1, 3, 4\}$ *and* $N = \{2, 5\}$. *We must solve* $Ax = b$ *and* $z = c \cdot x$ *for* x_B *and* z *in terms of (the new)* x_N. *It is simplest to begin with (3.35), solve the third equation for* x_1 *in terms of* x_2 *and* x_5, *and then substitute into the other three equations:*

$$
\begin{aligned}
& x_5 = 2 - x_1 + x_2 \\
\Rightarrow \quad & x_1 = 2 + x_2 - x_5 \\
\Rightarrow \quad & \left\{ \begin{array}{lcl} x_3 & = & 2 + (2 + x_2 - x_5) - 2x_2 = 4 - x_2 - x_5, \\ x_4 & = & 4 - (2 + x_2 - x_5) - x_2 = 2 - 2x_2 + x_5, \\ z & = & 3(2 + x_2 - x_5) + 2x_2 = 6 + 5x_2 - 3x_5 \end{array} \right.
\end{aligned}
$$

Thus, for this new choice of B, N, *the equations (3.33) can be written as*

$$
\begin{aligned}
x_3 &= 4 - x_2 - x_5, && \text{(3.36a)} \\
x_4 &= 2 - 2x_2 + x_5, && \text{(3.36b)} \\
x_1 &= 2 + x_2 - x_5, && \text{(3.36c)} \\
z &= 6 + 5x_2 - 3x_5. && \text{(3.36d)}
\end{aligned}
$$

Recalling that $x_N = 0$, *that is,* $x_2 = x_5 = 0$, *we see that the BFS is* $x = (2, 0, 4, 2, 0)$ *and, for this value of* x, $z = 6$. *The new value of* $\tilde{c}_N$ *is* $(5, -3)$; *since there is a positive component,* x *is still not optimal.*

Since $\tilde{c}_N$ *has a single positive component, there is only one choice the entering variable, namely,* x_2. *Both* x_3 *and* x_4 *decrease when* x_2 *is increased;* x_3 *will be zero when* $x_2 = 4$, *while* $x_4 = 0$ *for* $x_2 = 1$. *To maintain feasibility, we must take* $x_2 = 1$, *which means that* x_4 *is the leaving variable. We solve the second equation in (3.36) for* x_2 *in terms of* x_4 *and then substitute to obtain the new form of (3.33), now with* $B = \{1, 2, 3\}$ *and* $N = \{4, 5\}$:

$$
\begin{aligned}
x_3 &= 3 + \frac{1}{2}x_4 - \frac{3}{2}x_5, && \text{(3.37a)} \\
x_2 &= 1 - \frac{1}{2}x_4 + \frac{1}{2}x_5, && \text{(3.37b)} \\
x_1 &= 3 - \frac{5}{2}x_4 - \frac{1}{2}x_5, && \text{(3.37c)} \\
z &= 11 - \frac{5}{2}x_4 - \frac{1}{2}x_5. && \text{(3.37d)}
\end{aligned}
$$

With $x_N = 0$, *that is,* $x_4 = x_5 = 0$, *we have* $x = (3, 1, 3, 0, 0)$ *and* $z = 11$. *Moreover, the value of* $\tilde{c}_N$ *for this BFS is* $(-5/2, -1/2)$. *Since* $\tilde{c}_N < 0$, *we know that this value of* z *is optimal, and that* x *is an optimal BFS.*

The preceding example shows all the features of the basic simplex method. We point out that the algorithm has not been completely specified, however.

When there is more than one positive component of $\tilde{c}_N$, there must be some way of choosing the entering variable. Similarly, the leaving variable is the basic variable that first becomes zero; however, there may be more than one such variable. In that case, again, there is a choice to be made.

There are three possible complications that must be addressed. In the last of these, the possibility of degeneracy leading to cycling, the problem can be resolved by making the right choices for the entering and leaving variables.

3.12.3.1 Finding an initial BFS

If the original LP is in inequality form and $b \geq 0$, then it is easy to find an initial BFS: The m slack variables will be the basic variables and the others (the original variables) are the nonbasic variables. With this choice, A_B is just the identity matrix, and $x_B = b$ (which is why $b \geq 0$ is necessary). This was the situation in Example 155.

If the LP is in inequality form (3.28) and $b \geq 0$ does not hold, then we can use a "phase one" procedure to find an initial BFS (if one exists). We introduce an *artificial variable* x_0, subtracting it from each inequality to get

$$Ax - x_0 e \leq b,$$

where e is the vector of all ones. The point of the artificial variable is that if x_0 is sufficiently large, then $x = 0$ will satisfy $Ax - x_0 e \leq b$.

We add slack variables, which for clarity we will denote $s_1, s_2, \ldots, s_m$ (rather than $x_{n+1}, x_{n+2}, \ldots, x_{n+m}$), and pose the LP

$$\min \quad x_0 \tag{3.38a}$$

$$\text{s.t.} \quad Ax - x_0 e + s = b, \tag{3.38b}$$

$$x \geq 0, x_0 \geq 0, s \geq 0. \tag{3.38c}$$

If the optimal solution of this LP has $x_0 = 0$, then the corresponding (x, s) is a BFS of the original LP and we can start the simplex method with that BFS. If the optimal solution of (3.38) has $x_0 > 0$, then there is no feasible solution (and hence no BFS) of the original LP.

The advantage of using (3.38) to find an initial BFS for (3.28) is that we can easily find an initial BFS for (3.38), and then apply the simplex method. We are assuming that there exist components of b that are negative; let b_i be the most negative component of b. We solve the corresponding equation (the ith equation in the system $Ax - x_0 e + s = b$) to get

$$x_0 = -b_i + s_i + \sum_{j=1}^{n} A_{ij} x_j.$$

We then substitute this into the other equations and solve for the slack variables:

$$\sum_{j=1}^{n} A_{kj} x_j - x_0 e + s_k = b_k \Rightarrow s_k = b_k - b_i + s_i - \sum_{j=1}^{n} (A_{kj} - A_{ij}) x_j.$$

Setting $x = 0$, $s_i = 0$, we find that $x_0 = -b_i > 0$ and $s_k = b_k - b_i \geq 0$, $k \neq i$ (since b_i is the most negative component of b). Thus we have an initial BFS for (3.38), where the basic variables are $x_0, s_1, \ldots, s_{i-1}, s_{i+1}, \ldots, s_m$. We can then proceed to solve (3.38), find a BFS for (3.28) (if one exists), and then it is possible to start the simplex method on (3.28). In this context, applying the simplex method to (3.38) is called phase one, while applying the simplex method to the original LP is called phase two. The entire procedure is called the *two-phase simplex method.*

We illustrate the phase one procedure with an example.

Example 156 *Consider the LP*

$$
\begin{aligned}
\max \quad & 2x_1 + x_2 && \text{(3.39a)}\\
s.t. \quad & -x_1 - x_2 \leq -1, && \text{(3.39b)}\\
& -x_1 + x_2 \leq 2, && \text{(3.39c)}\\
& 2x_1 - x_2 \leq 4, && \text{(3.39d)}\\
& x_1, x_2, x_3 \geq 0. && \text{(3.39e)}
\end{aligned}
$$

This LP is of the type described above; there is no obvious BFS because the value on the right-hand side of one of the inequality constraints is negative.

The phase one LP is

$$
\begin{aligned}
\min \quad & x_0 && \text{(3.40a)}\\
s.t. \quad & -x_1 - x_2 - x_0 \leq -1, && \text{(3.40b)}\\
& -x_1 + x_2 - x_0 \leq 2, && \text{(3.40c)}\\
& 2x_1 - x_2 - x_0 \leq 4, && \text{(3.40d)}\\
& x_0, x_1, x_2, x_3 \geq 0. && \text{(3.40e)}
\end{aligned}
$$

We add an artificial variable and slack variables to obtain

$$
\begin{aligned}
-x_1 - x_2 - x_0 + x_3 &= -1,\\
-x_1 + x_2 - x_0 + x_4 &= 2,\\
2x_1 - x_2 - x_0 + x_5 &= 4,
\end{aligned}
$$

with $z = x_0$. (Unlike in the general description of phase one, given above, we have returned to writing the slack variables as $x_{n+1}, \ldots, x_{n+m}$. This allows us to use our previous notation for the simplex method.) Since there is a single negative component of the vector b, there is only one choice for the initial basis: x_0, x_4, x_5. Solving the constraints for these variables yields

$$
\begin{aligned}
x_0 &= 1 - x_1 - x_2 + x_3,\\
x_4 &= 3 - 2x_2 + x_3,\\
x_5 &= 5 - 3x_1 + x_3,\\
z &= 1 - x_1 - x_2 + x_3.
\end{aligned}
$$

We choose x_1 as the entering variable; x_0 and x_5 both decrease as x_1 increases, but x_0 reaches zero first (at $x_1 = 1$), so it is the leaving variable. Therefore, we solve the first equation for x_1 in terms of x_2 and substitute into the other equations to obtain

$$\begin{aligned} x_1 &= 1 - x_0 - x_2 + x_3, \\ x_4 &= 3 - 2x_2 + x_3, \\ x_5 &= 2 + 3x_0 + 3x_2 - 2x_3, \\ z &= x_0. \end{aligned}$$

The BFS corresponding to these equation is

$$(x_0, x_1, x_2, x_3, x_4, x_5) = (0, 1, 0, 0, 3, 2),$$

and the equation for z shows that this BFS is an optimal solution for (3.40). Moreover, the optimal solution corresponds to $x_0 = 0$, and therefore the original LP (3.39) has feasible solutions. An initial BFS is found by simply dropping the artificial variable:

$$(x_1, x_2, x_3, x_4, x_5) = (1, 0, 0, 2, 2).$$

To begin the simplex method on (3.39), we merely express $z = 2x_1 + x_2$ in terms of the nonbasic variables:

$$z = 2x_1 + x_2 = 2(1 - x_2 + x_3) + x_2 = 2 - x_2 + 2x_3,$$

and then (with $B = \{1, 4, 5\}$ and $N = \{2, 3\}$), the crucial equations (3.33) are

$$\begin{aligned} x_1 &= 1 - x_2 + x_3, \\ x_4 &= 3 - 2x_2 + x_3, \\ x_5 &= 2 + 3x_2 - 2x_3, \\ z &= 2 - x_2 + 2x_3. \end{aligned}$$

We can now apply the simplex method to these equations for the original LP (this is phase two). For the first pivot, x_3 is the entering variable and x_5 is the leaving variable. Solving the third equation for x_3 in terms of x_3 and x_5 and substituting into the other equation yields

$$\begin{aligned} x_1 &= 2 + \frac{1}{2}x_2 - \frac{1}{2}x_5, \\ x_4 &= 4 - \frac{1}{2}x_2 - \frac{1}{2}x_5, \\ x_3 &= 1 + \frac{3}{2}x_2 - \frac{1}{2}x_5, \\ z &= 4 + 2x_2 - x_5. \end{aligned}$$

For the next pivot, x_2 enters and x_4 leaves the basis to yield

$$\begin{aligned}
x_1 &= 6 - x_4 - x_5, \\
x_2 &= 8 - 2x_4 - x_5, \\
x_3 &= 13 - 3x_4 - 2x_5, \\
z &= 20 - 4x_4 - 3x_5.
\end{aligned}$$

The equation for z shows that this BFS is optimal; we have $x = (6, 8, 13, 0, 0)$ with $B = \{1, 2, 3\}$ and $N = \{4, 5\}$. In terms of the original variables, the optimal solution is $x_1 = 6$ and $x_2 = 8$, with $z = 20$.

If the original LP is in equality form (3.29), it is possible to design a phase one procedure that involves adding artificial variables to each of the equations in the system $Ax = b$. We refer the reader to Chapter 8 of [5] for details.

3.12.3.2 Unbounded LPs

It is possible that an LP has no solution because the objective function can be made arbitrarily large by choosing feasible vectors. Fortunately, this situation is easily detected when the simplex method is applied. It is easiest to explain this in the course of an example.

Example 157

$$\max \quad x_1 + x_2 \tag{3.41a}$$

$$s.t. \quad -x_1 + x_2 \leq 2, \tag{3.41b}$$

$$x_1 - 2x_2 \leq 1, \tag{3.41c}$$

$$x_1, x_2 \geq 0. \tag{3.41d}$$

We add slack variables x_3, x_4 and choose $B = \{3, 4\}$, $N = \{1, 2\}$ (so that the slack variables are the initial basic variables). The equations (3.33) are

$$\begin{aligned}
x_3 &= 2 + x_1 - x_2, \\
x_4 &= 1 - x_1 + 2x_2, \\
z &= x_1 + x_2.
\end{aligned}$$

We take x_1 as the entering variable, in which case x_4 must leave the basis. Solving the first equation for x_1 and substituting into the other equations yields

$$\begin{aligned}
x_3 &= 3 + x_2 - x_4, \\
x_1 &= 1 + 2x_2 - x_4, \\
z &= 1 + 3x_2 - x_4.
\end{aligned}$$

For the next pivot, x_2 is the only choice for the entering variable. However, both x_1 and x_3 increase as x_2 is increased, which means that there is no restriction on how much x_2, and therefore z, can be increased. Thus z can be made arbitrarily large, and the LP has no solution. Exercise 1 asks the reader to sketch the feasible set for (3.41).

The previous example shows how an unbounded LP is detected during the simplex method: If the objective function can be increased by bringing a certain variable into the basis, and feasibility of the current basic variables imposes no limit to how much the entering variable can be increased, then the LP is unbounded and has no solution.

3.12.3.3 Degeneracy and cycling

If the simplex method encounters a degenerate BFS, there can be an unfortunate side effect, namely, that the algorithm gets "stuck" and cannot make progress. We illustrate with an example.

Example 158 *Consider the LP*

$$\begin{aligned} \max \quad & 10x_1 - 57x_2 - 9x_3 - 24x_4 && (3.42a)\\ s.t. \quad & 0.5x_1 - 5.5x_2 - 2.5x_3 + 9x_4 \leq 0, && (3.42b)\\ & 0.5x_1 - 1.5x_2 - 0.5x_3 + x_4 \leq 0, && (3.42c)\\ & x_1 \leq 1, && (3.42d)\\ & x_1, x_2, x_3, x_4 \geq 0 && (3.42e) \end{aligned}$$

(taken from Chvátal [5], Chapter 3). We add slack variables x_5, x_6, x_7, choose the slack variables to be the basic variables (so that $B = \{5,6,7\}$, $N = \{1,2,3,4\}$), and form the equations (3.33):

$$\begin{aligned} x_5 &= -0.5x_1 + 5.5x_2 + 2.5x_3 - 9x_4,\\ x_6 &= -0.5x_1 + 1.5x_2 + 0.5x_3 - x_4,\\ x_7 &= 1 - x_1,\\ z &= 10x_1 - 57x_2 - 9x_3 - 24x_4. \end{aligned}$$

We notice that the initial BFS, $x = (0,0,0,0,0,0,1)$, is degenerate. This will have an interesting consequence if we perform the pivots in a certain way. The reader will recall that the simplex method is not completely specified if there is more than one choice for the entering and/or leaving variable. In this example, we will make the following (perfectly reasonable) choices:

1. *If there are multiple variables eligible to enter the basis, we will choose the one corresponding to the largest coefficient in $\tilde{c}_N$.*

2. *If there are multiple variables eligible to leave the basis, we will choose the one with the smallest subscript.*

We now show how the simplex method progresses when pivoting according to these rules.

In the first iteration, x_1 is the entering variable. All three basic variables decrease when x_1 is increased; however, x_5 and x_6 are already 0 and hence *cannot decrease at all. One of them must leave the basis—it will be x_5 according to the above rule—and x_1 enters the basis with value 0:*

$$\begin{aligned} x_1 &= -2x_5 + 11x_2 + 5x_3 - 18x_4, \\ x_6 &= x_5 - 4x_2 - 2x_3 + 8x_4, \\ x_7 &= 1 + 2x_5 - 11x_2 - 5x_3 + 18x_4, \\ z &= -20x_5 + 53x_2 + 41x_3 - 204x_4. \end{aligned}$$

Now we have a different basis ($B = \{1, 6, 7\}$, $N = \{2, 3, 4, 5\}$), but the BFS is the same: $x = (0, 0, 0, 0, 0, 0, 1)$.

The subsequent pivots are similar in effect. Next, x_2 enters and x_6 leaves:

$$\begin{aligned} x_1 &= 0.75x_5 - 2.75x_6 - 0.5x_3 + 4x_4, \\ x_2 &= 0.25x_5 - 0.25x_6 - 0.5x_3 + 2x_4, \\ x_7 &= 1 - 0.75x_5 + 2.75x_6 + 0.5x_3 - 4x_4, \\ z &= -6.75x_5 - 13.25x_6 + 14.5x_3 - 98x_4. \end{aligned}$$

Then x_3 enters and x_1 leaves:

$$\begin{aligned} x_3 &= 1.5x_5 - 5.5x_6 - 2x_1 + 8x_4, \\ x_2 &= -0.5x_5 + 2.5x_6 + x_1 - 2x_4, \\ x_7 &= 1 - x_1, \\ z &= 15x_5 - 93x_6 - 29x_1 + 18x_4. \end{aligned}$$

Then x_4 enters and x_2 leaves:

$$\begin{aligned} x_3 &= -0.5x_5 + 4.5x_6 + 2x_1 - 4x_2, \\ x_4 &= -0.25x_5 + 1.25x_6 + 0.5x_1 - 0.5x_2, \\ x_7 &= 1 - x_1, \\ z &= 10.5x_5 - 70.5x_6 - 20x_1 - 9x_2. \end{aligned}$$

Then x_5 enters and x_3 leaves:

$$\begin{aligned} x_5 &= -2x_3 + 9x_6 + 4x_1 - 8x_2, \\ x_4 &= +0.5x_3 - x_6 - 0.5x_1 + 1.5x_2, \\ x_7 &= 1 - x_1, \\ z &= -21x_3 + 24x_6 + 22x_1 - 93x_2. \end{aligned}$$

Finally, x_6 enters and x_4 leaves:

$$\begin{aligned} x_5 &= +2.5x_3 - 9x_4 - 0.5x_1 + 5.5x_2, \\ x_6 &= +0.5x_3 - x_4 - 0.5x_1 + 1.5x_2, \\ x_7 &= 1 - x_1, \\ z &= -9x_3 - 24x_4 + 10x_1 - 57x_2. \end{aligned}$$

We now have $B = \{5,6,7\}$, $N = \{1,2,3,4\}$, *and* $x = (0,0,0,0,0,0,1)$, *which is where we started. The effect of the simplex method, in the face of degeneracy and with the particular pivoting rules we imposed, is to move variables in and out of the basis without ever changing the BFS. This will continue indefinitely and is referred to as* cycling.

It is easy to show that if the simplex method does not terminate, then cycling must occur. (For the method to terminate means that it either finds an optimal BFS or determines that the LP is unbounded.) Therefore, a pivoting rule that eliminates the possibility of cycling would ensure that the simplex method always works.

Fortunately, there are pivoting rules that make cycling impossible. The simplest is to use the smallest subscript rule for both the entering and leaving variable (see [5], page 37). Moreover, while cycling is a theoretical concern, it almost never happens in LPs that arise in real applications. Therefore, simply disregarding the possibility of cycling is a realistic strategy.

3.12.4 Variations on the standard LPs

For simplicity in this brief introduction, we have focused on the two standard LPs (3.28) and (3.29). The main restriction that this imposes on the LP is that all variables are assumed to be nonnegative. Theoretically, this is no restriction, since there are various tricks for transforming other bounds on the variables into this form. For instance, if one of the variables x_i is in fact unrestricted (any value of x_i is allowed), then we can replace x_i everywhere in the statement of the problem by $x_i^{(1)} - x_i^{(2)}$, where $x_i^{(1)}, x_i^{(2)} \geq 0$. Similarly, if x_i is constrained by $x_i \leq u_i$, then we can substitute $x_i = u_i - \overline{x}_i$, where $\overline{x}_i \geq 0$.

However, for practical (efficient) algorithms, it is better to deal with the various bounds on the variables directly. In a general linear program, we assume bounds of the form $\ell_i \leq x_i \leq u_i$, $i = 1, 2, \ldots, n$, where any ℓ_i can equal $-\infty$ and any u_i can equal ∞ (however, if, for instance, $\ell_i = -\infty$, the bound is interpreted as $-\infty < x_i$, not $-\infty \leq x_i$). We can define a notion of BFS for these bounds: If u_i is finite and $x_i = u_i$, this is equivalent to $x_i = 0$ when the bound is $x_i \geq 0$ and x_i would be considered a nonbasic variable (unless the BFS is degenerate). The simplex method can be modified to take into account this notion of BFS.

For readers who would like more information about linear programming, including general bounds on the variables, the text by Chvátal [5] is an excellent introduction.

Exercises

1. Sketch the feasible set for the LP (3.41).

2. Consider the LP

$$\begin{aligned} \max \quad & c \cdot x \\ \text{s.t.} \quad & Ax \leq b, \\ & x \geq 0, \end{aligned}$$

where

$$A = \begin{bmatrix} 1 & 1 \\ 1 & 5 \\ 2 & 1 \end{bmatrix}, \; b = \begin{bmatrix} 8 \\ 30 \\ 14 \end{bmatrix}, \; c = \begin{bmatrix} 1 \\ 2 \end{bmatrix}.$$

   (a) Sketch the feasible set.
   (b) Apply the simplex method to find the optimal solution.
   (c) Indicate all of the basic solutions on your sketch of the feasible set.
   (d) How many of the basic solutions are feasible? Which BFSs are visited by the simplex method?

3. Consider the LP from the previous exercise, with c changed to $c = (2, 1)$. How is the new LP significantly different from the original?

4. Consider the LP

$$\begin{aligned} \max \quad & c \cdot x \\ \text{s.t.} \quad & Ax \leq b, \\ & x \geq 0, \end{aligned}$$

where

$$A = \begin{bmatrix} -1 & -1 \\ -1 & 1 \\ 1 & 1 \end{bmatrix}, \; b = \begin{bmatrix} -2 \\ 2 \\ 4 \end{bmatrix}, \; c = \begin{bmatrix} 1 \\ 2 \end{bmatrix}.$$

   (a) Sketch the feasible set, and notice that the origin is infeasible. Explain why this means that the two-phase simplex method is needed.
   (b) Apply the two-phase simplex method to find the optimal solution.
   (c) On the sketch of the feasible set, indicate the BFSs visited by the simplex method.

5. Apply the simplex method to solve the LP

$$\begin{aligned} \max \quad & 2x_1 + 3x_2 \\ \text{s.t.} \quad & -x_1 + x_2 \leq 2, \\ & -x_1 + 2x_2 \leq 6, \\ & x_1 - 3x_2 \leq 3, \\ & x_1, x_2 \geq 0. \end{aligned}$$

6. Apply the simplex method to solve the LP

$$\begin{aligned} \max \quad & x_1 + 2x_2 + 3x_3 \\ \text{s.t.} \quad & x_1 + x_2 + x_3 \le 8, \\ & x_1 + 2x_2 + x_3 \le 6, \\ & 2x_1 + x_2 + 4x_3 \le 10, \\ & x_1, x_2, x_3 \ge 0. \end{aligned}$$

7. Prove that a half-space in $\mathbf{R}^n$ is a convex set.

8. Prove that the intersection of any number of convex sets is convex.

9. Consider the LP (3.29) in standard equality form, and let S be the feasible set of the LP (which we know is a polyhedron). Assume that $\text{rank}(A) = m$. Prove that x is an extreme point of S if and only if x is a BFS of the LP.

10. Let C be a convex subset of $\mathbf{R}^n$. Suppose

$$x_1, \ldots, x_k \in C, \ \alpha_1, \ldots, \alpha_k \ge 0, \ \alpha_1 + \cdots + \alpha_k = 1.$$

Prove that $\alpha_1 x_1 + \cdots + \alpha_k x_k$ belongs to C.

11. Apply the simplex method to the LP of Example 158, using the smallest subscript rule to choose both the entering and leaving variables, and show that cycling does not occur.

4

Determinants and eigenvalues

Problems in finite-dimensional linear algebra often reduce to the analysis of matrices. This is because of the fundamental fact that every linear operator from one finite-dimensional space to another can be represented by a matrix. The simplest kind of matrix is a *diagonal* matrix—a matrix whose only nonzero entries lie on the diagonal. To be precise, $A \in F^{m\times n}$ is diagonal if $A_{ij} = 0$ for all $i \neq j$. In this chapter, we will restrict ourselves to the square matrices. We pursue the case of nonsquare matrices, and the special case of nonsquare diagonal matrices, in Chapter 8.

It is not much of an exaggeration to say that everything in linear algebra is simple if the matrix involved is diagonal. For example, solving $Ax = b$ is easy if $A \in F^{n\times n}$ is diagonal because the equations are *decoupled* (each equation involves only one variable):

$$\begin{aligned} A_{11}x_1 &= b_1, \\ A_{22}x_2 &= b_2, \\ \vdots \quad & \quad \vdots \\ A_{nn}x_n &= b_n. \end{aligned}$$

Similarly it is easy to decide if a square, diagonal matrix is invertible: an $n\times n$ matrix A is invertible if and only if $A_{ii} \neq 0$ for all $i = 1, 2, \ldots, n$. When a diagonal matrix is invertible, then its inverse can be found immediately:

$$\left(A^{-1}\right)_{ij} = \begin{cases} A_{ii}^{-1}, & i = j, \\ 0, & i \neq j. \end{cases}$$

The nice properties of a diagonal matrix $A \in F^{n\times n}$ all follow from the fact that A, viewed as an operator, has a simple effect on the standard basis $\{e_1, e_2, \ldots, e_n\}$ for F^n:

$$Ae_j = A_{jj}e_j.$$

If $T : F^n \to F^n$ is the linear operator defined by A, then the special property of T is that there is a basis for F^n consisting of vectors x, each satisfying

$$T(x) = \lambda x$$

for a scalar λ. This leads to the most important definition of this chapter.

Definition 159 *Let X be a vector space over a field F, and let $T : X \to X$. If there exists a nonzero vector $x \in X$ and a scalar $\lambda \in F$ such that*

$$T(x) = \lambda x,$$

then we call λ an eigenvalue *of T and x is called a* eigenvector *of T corresponding to λ. Together, λ, x are called an* eigenpair *of T.*

We will be interested in finding the eigenpairs of a given linear operator T. With respect to the variables λ, x, the equation $T(x) = \lambda x$ is nonlinear, and the techniques we have studied so far are not applicable.[1] However, if the eigenvalue λ is known, the problem becomes linear in the unknown x.

We will begin by studying the eigenvalue problem in matrix form: Find $\lambda \in F$, $x \in F^n$, $x \neq 0$, such that

$$Ax = \lambda x,$$

where $A \in F^{n\times n}$ is a given matrix. We see that λ is an eigenvalue of A if and only if

$$\lambda x - Ax = 0 \text{ for some } x \neq 0,$$

which is equivalent to

$$(\lambda I - A)x = 0, \ x \neq 0.$$

We can therefore say that λ is an eigenvalue of A if and only if the matrix $\lambda I - A$ is singular.

We can find the eigenvalues of A by solving the equation $\det(\lambda I - A) = 0$, where the *determinant* function $\det : F^{n\times n} \to F$ is a special function with the property that $\det(B) = 0$ if and only if B is singular. We therefore prepare for our study of eigenvalues by developing the determinant function.

4.1 The determinant function

It is easiest to develop the determinant function if we regard it as a function of the columns of a matrix. We will therefore write $\det(A)$ and $\det(A_1, A_2, \ldots, A_n)$ interchangeably. The most interesting property of the determinant function is that $\det(A)$ is zero if and only if A is singular. This will follow from the fact that $\det(A_1, A_2, \ldots, A_n)$ is the the (signed) volume of the parallelopiped[2] in F^n determined by the vectors $A_1, A_2, \ldots, A_n$.

Considering the special case of $A \in \mathbf{R}^2$ for inspiration, a volume function should have the following properties:

---

[1]Newton's method for nonlinear systems is not directly applicable because we have $n+1$ unknowns but only n equations. Moreover, even if we were to add another equation, such as $x \cdot x = 1$, to obtain a square system, Newton's method would give no idea of the number of solutions to expect. We will do better with other methods.

[2]A parallelopiped is the multidimensional generalization of a parallelogram.

1. $\det(e_1, e_2) = 1$ since the vectors e_1, e_2 determine a square with side length one.

2. For any $\lambda > 0$, $\det(\lambda A_1, A_2) = \det(A_1, \lambda A_2) = \lambda\det(A_1, A_2)$, since stretching one side of a parallelogram multiplies the area of the parallelogram by the same amount. We require that these equations hold for all real numbers λ so that the formula for the determinant is simpler; this means that $\det(A_1, A_2)$ will be plus or minus the area of the parallelogram determined by A_1 and A_2.

3. For any $\lambda \in \mathbf{R}$, $\det(A_1, A_2 + \lambda A_1) = \det(A_1, A_2)$. This equation expresses the geometric fact that the area of a parallelogram is determined by the length of a base and the corresponding perpendicular height of the figure (see Figure 4.1).

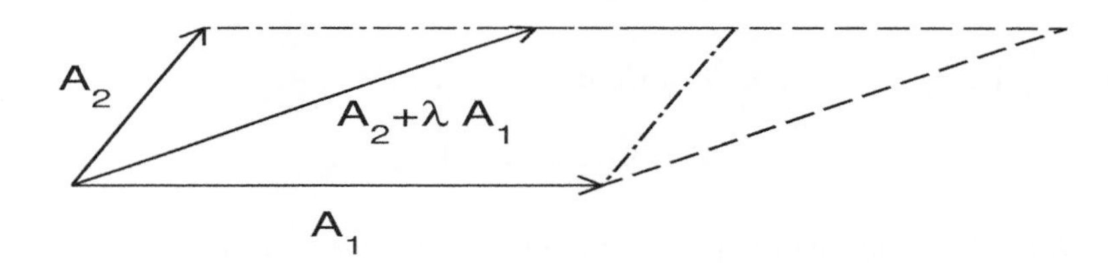

FIGURE 4.1
The parallelograms determined by A_1, A_2, and by A_1, $A_2 + \lambda A_1$. These parallelograms have the same area.

The above considerations motivate the following definition.

Definition 160 *Let F be a field and n a positive integer. The* determinant *function* $\det : (F^n)^n \to F$ *is defined by the following properties:*

1. *If $\{e_1, e_2, \ldots, e_n\}$ is the standard basis for F^n, then*
$$\det(e_1, e_2, \ldots, e_n) = 1. \tag{4.1}$$

2. *For any $A_1, A_2, \ldots, A_n \in F^n$, any $\lambda \in F$, and any j, $1 \le j \le n$,*
$$\det(A_1, \ldots, A_{j-1}, \lambda A_j, A_{j+1}, \ldots, A_n) = \lambda\det(A_1, \ldots, A_n). \tag{4.2}$$

3. *For any $A_1, A_2, \ldots, A_n \in F^n$, any $\lambda \in F$, and any $i \neq j$, $1 \le i, j \le n$,*
$$\det(A_1, \ldots, A_{j-1}, A_j + \lambda A_i, A_{j+1}, \ldots, A_n) = \det(A_1, \ldots, A_n). \tag{4.3}$$

We must show that the determinant function is well-defined, that is, that the above conditions define a unique function on $(F^n)^n$. This will take the remainder of this section. We begin by deriving some properties that det must have if it satisfies the definition.

Theorem 161 *Let F be a field, let n be a positive integer, and suppose*

$$\det : (F^n)^n \to F$$

satisfies Definition 160. Suppose $A_1, A_2, \ldots, A_n \in F^n$. Then:

1. *If j is between 1 and n and $\lambda_i \in F$ for $i \neq j$, then*

$$\det\left(A_1, \ldots, A_j + \sum_{i \neq j} \lambda_i A_i, \ldots, A_n\right) = \det(A_1, \ldots, A_j, \ldots, A_n).$$

2. *If one of the vectors $A_1, A_2, \ldots, A_n$ is the zero vector, then*

$$\det(A_1, A_2, \ldots, A_n) = 0.$$

3. *If $\{A_1, A_2, \ldots, A_n\}$ is linearly dependent, then*

$$\det(A_1, A_2, \ldots, A_n) = 0.$$

4. *Interchanging two arguments to* det *changes the sign of the output:*

$$\det(A_1, \ldots, A_j, \ldots, A_i, \ldots, A_n) = -\det(A_1, \ldots, A_i, \ldots, A_j, \ldots, A_n).$$

5. *If B_j is any vector in F^n, then*

$$\begin{aligned} \det(A_1, \ldots, A_j + B_j, \ldots, A_n) &= \det(A_1, \ldots, A_j, \ldots, A_n) + \\ &\quad \det(A_1, \ldots, B_j, \ldots, A_n). \end{aligned}$$

Putting together conclusion 5 with the second defining property of det, *we see that* det *is* multilinear, *that is, linear in any one argument when the other arguments are held fixed.*

Proof

1. The first conclusion results from the repeated application of (4.3):

$$\begin{aligned} & \det(A_1, \ldots, A_j, \ldots, A_n) \\ = \; & \det(A_1, \ldots, A_j + \lambda_1 A_1, \ldots, A_n) \\ = \; & \det(A_1, \ldots, A_j + \lambda_1 A_1 + \lambda_2 A_2, \ldots, A_n) \\ & \vdots \\ = \; & \det\left(A_1, \ldots, A_j + \sum_{i \neq j} \lambda_i A_i, \ldots, A_n\right). \end{aligned}$$

2. If $A_j = 0$, then, applying (4.2),

$$\begin{aligned}\det(A_1, \dots, A_j, \dots, A_n) &= \det(A_1, \dots, 0 \cdot A_j, \dots, A_n)\\ &= 0 \cdot \det(A_1, \dots, A_j, \dots, A_n)\\ &= 0.\end{aligned}$$

3. If $\{A_1, A_2, \dots, A_n\}$ is linearly dependent, then there exists j between 1 and n and scalars $\lambda_i \in F$, $i \neq j$, such that

$$A_j + \sum_{i \neq j} \lambda_i A_i = 0.$$

Then

$$\begin{aligned}\det(A_1, \dots, A_j, \dots, A_n) &= \det\left(A_1, \dots, A_j + \sum_{i \neq j} \lambda_i A_i, \dots, A_n\right)\\ &= \det(A_1, \dots, 0, \dots, A_n)\\ &= 0.\end{aligned}$$

4. Repeatedly applying (4.3), we see

$$\begin{aligned}&\det(A_1, \dots, A_j, \dots, A_i, \dots, A_n)\\ =\ &\det(A_1, \dots, A_j, \dots, A_i - A_j, \dots, A_n)\\ =\ &\det(A_1, \dots, A_j + A_i - A_j, \dots, A_i - A_j, \dots, A_n)\\ =\ &\det(A_1, \dots, A_i, \dots, A_i - A_j, \dots, A_n)\\ =\ &\det(A_1, \dots, A_i, \dots, A_i - A_j - A_i, \dots, A_n)\\ =\ &\det(A_1, \dots, A_i, \dots, -A_j, \dots, A_n)\\ =\ &-\det(A_1, \dots, A_i, \dots, A_j, \dots, A_n).\end{aligned}$$

The last step follows from (4.2).

5. We leave it as an exercise to show that, if $\{A_1, A_2, \dots, A_n\}$ is linearly dependent, then

$$\det(A_1, \dots, A_j + B_j, \dots, A_n) = \det(A_1, \dots, B_j, \dots, A_n).$$

Then, since $\det(A_1, A_2, \dots, A_n) = 0$, the result follows in this case.

We therefore assume that $\{A_1, A_2, \dots, A_n\}$ is linearly independent. Then there exist scalars $\lambda_1, \lambda_2, \dots, \lambda_n \in F$ such that

$$B_j = \sum_{i=1}^{n} \lambda_i A_i,$$

and so

$$\begin{aligned}
\det(A_1, \ldots, A_j + B_j, \ldots, A_n) &= \det\left(A_1, \ldots, A_j + \sum_{i=1}^{n} \lambda_i A_i, \ldots, A_n\right) \\
&= \det(A_1, \ldots, A_j + \lambda_j A_j, \ldots, A_n) \\
&= (1 + \lambda_j)\det(A_1, \ldots, A_j, \ldots, A_n).
\end{aligned}$$

But

$$\begin{aligned}
\det(A_1, \ldots, B_j, \ldots, A_n) &= \det\left(A_1, \ldots, \sum_{i=1}^{n} \lambda_i A_i, \ldots, A_n\right) \\
&= \det(A_1, \ldots, \lambda_j A_j, \ldots, A_n) \\
&= \lambda_j \det(A_1, \ldots, A_j, \ldots, A_n)
\end{aligned}$$

and thus

$$\begin{aligned}
& (1 + \lambda_j)\det(A_1, \ldots, A_j, \ldots, A_n) \\
= \; & \det(A_1, \ldots, A_j, \ldots, A_n) + \lambda_j \det(A_1, \ldots, A_j, \ldots, A_n) \\
= \; & \det(A_1, \ldots, A_j, \ldots, A_n) + \det(A_1, \ldots, B_j, \ldots, A_n).
\end{aligned}$$

This completes the proof.

QED

We can now use the above properties, particularly the fact that det is multilinear, to show that det is well-defined by deriving a formula for it. Denoting the components of A_j as $A_{1j}, A_{2j}, \ldots, A_{nj}$, we have

$$\begin{aligned}
& \det(A_1, A_2, \ldots, A_n) \\
= \; & \det\left(\sum_{i_1=1}^{n} A_{i_1,1} e_{i_1}, \sum_{i_2=1}^{n} A_{i_2,2} e_{i_2}, \ldots, \sum_{i_n=1}^{n} A_{i_n,n} e_{i_n}\right) \\
= \; & \sum_{i_1=1}^{n} \sum_{i_2=1}^{n} \cdots \sum_{i_n=1}^{n} \det(e_{i_1}, e_{i_2}, \ldots, e_{i_n}) A_{i_1,1} A_{i_2,2} \cdots A_{i_n,n}.
\end{aligned}$$

The key to understanding this formula is to recognize that $\det(e_{i_1}, e_{i_2}, \ldots, e_{i_n})$ must be zero if $e_{i_k} = e_{i_\ell}$ for some $k \neq \ell$, that is, if $i_k = i_\ell$ for $k \neq \ell$, because in that case, $\{e_{i_1}, e_{i_2}, \ldots, e_{i_n}\}$ is linearly dependent. Since each i_k takes on all values between 1 and n, most of the terms in the multiple sum are zero; the only ones that are not zero are those for which $(i_1, i_2, \ldots, i_n)$ is a permutation of $\{1, 2, \ldots, n\}$.

4.1.1 Permutations

We now summarize some basic facts about permutations; for completeness, the following results are derived in Appendix B. By definition, a *permutation* of $\mathbf{n} = \{1, 2, \ldots, n\}$ is simply a bijection $\tau : \mathbf{n} \to \mathbf{n}$. We usually write a permutation τ by writing its values as a finite sequence: $\tau = (i_1, i_2, \ldots, i_n)$, where $i_j = \tau(j)$. We denote the set of all permutations of $\mathbf{n}$ as S_n. For example,

$$S_3 = \{(1,2,3), (1,3,2), (2,1,3), (2,3,1), (3,1,2), (3,2,1)\}.$$

It is easy to see that S_n has $n!$ elements.

A *transposition* is a special permutation that fixes all but two integers,which are interchanged. For example, the following are some of the transpositions in S_5:

$$(2,1,3,4,5), (3,2,1,4,5), (4,2,3,1,5).$$

We use the special notation $[i, j]$ to denote the transposition in S_n that interchanges i and j; with this notation, the integer n must be understood from context. Thus, with $n = 5$,

$$(2,1,3,4,5) = [1,2], (3,2,1,4,5) = [1,3], (4,2,3,1,5) = [1,4].$$

It is important to keep in mind that both $(2,1,3,4,5)$ and $[1,2]$ represent functions. If $\tau = (2,1,3,4,5)$, then $\tau(1) = 2$, $\tau(2) = 1$, $\tau(3) = 3$, $\tau(4) = 4$, and $\tau(5) = 5$. To avoid introducing function names when they are not needed, we will write $[i, j](k)$ to indicate the image of k under the transposition $[i, j]$. Thus

$$[i,j](k) = \begin{cases} j, & k = i, \\ i, & k = j, \\ k, & \text{otherwise.} \end{cases}$$

We will use two basic theorems about permutations (see Theorems 486 and 489 in Appendix B). First of all, every permutation can be written as the composition of transpositions. For example, with $n = 5$,

$$(3,5,2,1,4) = [1,4][1,2][2,5][1,3]. \tag{4.4}$$

(We use product notation to indicate composition.) We can check (4.4) as follows:

$$\begin{aligned}
&[1,4][1,2][2,5][1,3](1) = [1,4][1,2][2,5](3) = [1,4][1,2](3) = [1,4](3) = 3, \\
&[1,4][1,2][2,5][1,3](2) = [1,4][1,2][2,5](2) = [1,4][1,2](5) = [1,4](5) = 5, \\
&[1,4][1,2][2,5][1,3](3) = [1,4][1,2][2,5](1) = [1,4][1,2](1) = [1,4](2) = 2, \\
&[1,4][1,2][2,5][1,3](4) = [1,4][1,2][2,5](4) = [1,4][1,2](4) = [1,4](4) = 1, \\
&[1,4][1,2][2,5][1,3](5) = [1,4][1,2][2,5](5) = [1,4][1,2](2) = [1,4](1) = 4.
\end{aligned}$$

This shows that (4.4) holds.

A representation of a permutation as a composition of transpositions is not unique. For example, the reader can verify that, with $n = 4$,

$$(2,4,1,3) = [3,4][1,3][1,2].$$

But the representation

$$(2,4,1,3) = [1,3][2,3][2,4][3,4][1,4]$$

is also valid. What is unique is the parity (oddness or evenness) of the number of transpositions required to represent a given permutation. This is the second theorem we need: Given a permutation, if one representation of the permutation as a composition of transpositions involves an even number of transpositions, then so do all such representations, and similarly for odd. We therefore describe a permutation τ as *even* or *odd* according to the number of transpositions required to represent it. If τ is a permutation, we define the *signature* $\sigma(\tau)$ of τ by

$$\sigma(\tau) = \begin{cases} 1, & \tau \text{ is even}, \\ -1, & \tau \text{ is odd}. \end{cases}$$

4.1.2 The complete expansion of the determinant

We can now complete the derivation of the determinant function. As mentioned above, if $(i_1, i_2, \ldots, i_n)$ is not a permutation of $(1, 2, \ldots, n)$, then

$$\det(e_{i_1}, e_{i_2}, \ldots, e_{i_n}) = 0.$$

On the other hand, if $(i_1, i_2, \ldots, i_n)$ is a permutation, then, since

$$\det(e_1, e_2, \ldots, e_n) = 1$$

and each interchange of two arguments changes the sign of the determinant, the value of $\det(e_{i_1}, e_{i_2}, \ldots, e_{i_n})$ is simply ± 1, depending on whether the standard basis vectors have been interchanged an even or odd number of times:

$$\det(e_{i_1}, e_{i_2}, \ldots, e_{i_n}) = \sigma(i_1, i_2, \ldots, i_n).$$

Therefore,

$$\begin{aligned} &\det(A_1, A_2, \ldots, A_n) \\ &= \sum_{i_1=1}^{n} \sum_{i_2=1}^{n} \cdots \sum_{i_n=1}^{n} \det(e_{i_1}, e_{i_2}, \ldots, e_{i_n}) A_{i_1,1} A_{i_2,2} \cdots A_{i_n,n} \\ &= \sum_{(i_1,\ldots,i_n)\in S_n} \sigma(i_1, i_2, \ldots, i_n) A_{i_1,1} A_{i_2,2} \cdots A_{i_n,n}. \end{aligned}$$

We can simplify the notation a little by writing τ for an arbitrary element of S_n; then $\tau = (\tau(1), \tau(2), \ldots, \tau(n))$.

At this point, we have proved that if $\det : (F^n)^n \to F$ satisfies Definition 160, then

$$\det(A_1, A_2, \ldots, A_n) = \sum_{\tau \in S_n} \sigma(\tau) A_{\tau(1),1} A_{\tau(2),2} \cdots A_{\tau(n),n}. \tag{4.5}$$

We have not yet proved that det, as defined by (4.5), does satisfy Definition 160. To be precise, we have proved uniqueness—there is at most one function det satisfying Definition 160—but we have not proved existence.

We refer to (4.5) as the *complete expansion* of the determinant. The following theorem shows that the complete expansion does satisfy the original definition.

Theorem 162 *Let F be a field, and let* $\det : (F^n)^n \to F$ *be defined by (4.5). Then* det *satisfies Definition 160.*

Proof We must show that det satisfies the three defining properties of a determinant function.

1. If $A_j = e_j$, $j = 1, 2, \ldots, n$, then

$$A_{ij} = (e_j)_i = \begin{cases} 1, & i = j, \\ 0, & i \neq j. \end{cases}$$

   Therefore, $A_{\tau(1),1}A_{\tau(2),2} \cdots A_{\tau(n),n} = 0$ unless $\tau = (1, 2, \ldots, n)$, in which case $A_{\tau(1),1}A_{\tau(2),2} \cdots A_{\tau(n),n} = 1$ and $\sigma(\tau) = 1$. Therefore, the sum defining $\det(e_1, e_2, \ldots, e_n)$ contains a single term, which has value 1, and therefore $\det(e_1, e_2, \ldots, e_n) = 1$.

2. The second defining property of the determinant function follows immediately from the fact that the complete expansion defines a multilinear form:

$$\begin{aligned} & \det(A_1, \ldots, \lambda A_j, \ldots, A_n) \\ = & \sum_{\tau \in S_n} \sigma(\tau) A_{\tau(1),1} A_{\tau(2),2} \cdots \lambda A_{\tau(j),j} \cdots A_{\tau(n),n} \\ = & \sum_{\tau \in S_n} \lambda \sigma(\tau) A_{\tau(1),1} A_{\tau(2),2} \cdots A_{\tau(j),j} \cdots A_{\tau(n),n} \\ = & \lambda \sum_{\tau \in S_n} \sigma(\tau) A_{\tau(1),1} A_{\tau(2),2} \cdots A_{\tau(n),n} \\ = & \lambda \det(A_1, \ldots, A_n). \end{aligned}$$

3. To prove that (4.5) satisfies the third defining property of the determinant, we first show that interchanging two vectors A_i and A_j changes the sign. For given integers $i \neq j$ and for any $\tau \in S_n$, we will write $\tau' = \tau[i, j]$. We leave it as an exercise to show that $\tau \mapsto \tau'$ defines a bijection from S_n onto itself (see Exercise 11). We also note that $\sigma(\tau') = -\sigma(\tau)$. It follows that, for any $A_1, A_2, \ldots, A_n \in F^n$,

$$\sum_{\tau \in S_n} \sigma(\tau) A_{\tau(1),1} \cdots A_{\tau(n),n} = \sum_{\tau \in S_n} \sigma(\tau') A_{\tau'(1),1} \cdots A_{\tau'(n),n}$$

   (the second sum contains exactly the same terms as the first sum, just given in a different order). By definition, $\tau'(k) = \tau(k)$ for all k other than i, j, while $\tau'(i) = \tau(j)$ and $\tau'(j) = \tau(i)$. We now perform the following calculation, where it is understood that in the initial expression $\det(A_1, \ldots, A_j, \ldots, A_i, \ldots, A_n)$, A_j appears as the ith argument and A_i as the jth argument:

$$
\begin{aligned}
& \det(A_1, \ldots, A_j, \ldots, A_i, \ldots, A_n) \\
= & \sum_{\tau \in S_n} \sigma(\tau) A_{\tau(1),1} \cdots A_{\tau(i),j} \cdots A_{\tau(j),i} \cdots A_{\tau(n),n} \\
= & \sum_{\tau \in S_n} \sigma(\tau) A_{\tau'(1),1} \cdots A_{\tau'(j),j} \cdots A_{\tau'(i),i} \cdots A_{\tau'(n),n} \\
= & \sum_{\tau \in S_n} (-\sigma(\tau')) A_{\tau'(1),1} \cdots A_{\tau'(j),j} \cdots A_{\tau'(i),i} \cdots A_{\tau'(n),n} \\
= & -\sum_{\tau \in S_n} \sigma(\tau') A_{\tau'(1),1} \cdots A_{\tau'(i),i} \cdots A_{\tau'(j),j} \cdots A_{\tau'(n),n} \\
= & -\det(A_1, \ldots, A_i, \ldots, A_j, \ldots, A_n).
\end{aligned}
$$

Finally, let $A_1, A_2, \ldots, A_n \in F^n$ and $\lambda \in F$ be given. We have

$$
\begin{aligned}
& \det(A_1, \ldots, A_i, \ldots, A_j + \lambda A_i, \ldots, A_n) \\
= & \sum_{\tau \in S_n} \sigma(\tau) A_{\tau(1),1} \cdots A_{\tau(i),i} \cdots (A_{\tau(j),j} + \lambda A_{\tau(j),i}) \cdots A_{\tau(n),n} \\
= & \sum_{\tau \in S_n} \sigma(\tau) A_{\tau(1),1} \cdots A_{\tau(i),i} \cdots A_{\tau(j),j} \cdots A_{\tau(n),n} + \\
& \lambda \sum_{\tau \in S_n} \sigma(\tau) A_{\tau(1),1} \cdots A_{\tau(i),i} \cdots A_{\tau(j),i} \cdots A_{\tau(n),n} \\
= & \det(A_1, A_2, \ldots, A_n) + \lambda \det(A_1, \ldots, A_i, \ldots, A_i, \ldots, A_n).
\end{aligned}
$$

It suffices therefore to prove that

$$\det(A_1, \ldots, A_i, \ldots, A_i, \ldots, A_n) = 0.$$

But, by the second property of det, proved above, interchanging the ith and jth arguments to det changes the sign of the output. In this case, though, the ith and jth arguments are identical, so we obtain

$$\det(A_1, \ldots, A_i, \ldots, A_i, \ldots, A_n) = -\det(A_1, \ldots, A_i, \ldots, A_i, \ldots, A_n),$$

that is,

$$\det(A_1, \ldots, A_i, \ldots, A_i, \ldots, A_n) = 0,$$

as desired.

QED

We now see that there is a unique function det : $(F^n)^n \to F$ satisfying Definition 160, namely, the function defined by (4.5). We also know that det has the properties listed in Theorem 161. In the next section, we derive additional properties of the determinant function.

Exercises

Miscellaneous exercises

1. Use the definition of the determinant and Theorem 161 to compute the determinants of the following matrices:

$$\begin{bmatrix} 0 & 1 & 0 \\ 0 & 0 & 1 \\ 1 & 0 & 0 \end{bmatrix}, \begin{bmatrix} 0 & 0 & 1 \\ 0 & 1 & 0 \\ 1 & 0 & 0 \end{bmatrix}.$$

   State the properties of the determinant used in the computations.

2. Repeat Exercise 1 for the following matrices:

$$\begin{bmatrix} 0 & 0 & 0 & 1 \\ 0 & 1 & 0 & 0 \\ 1 & 0 & 0 & 0 \\ 0 & 0 & 1 & 0 \end{bmatrix}, \begin{bmatrix} 0 & 0 & 1 & 0 \\ 0 & 0 & 0 & 1 \\ 0 & 1 & 0 & 0 \\ 1 & 0 & 0 & 0 \end{bmatrix}.$$

3. Repeat Exercise 1 for the following matrices:

$$\begin{bmatrix} a & 0 & 0 & 0 \\ 0 & b & 0 & 0 \\ 0 & 0 & c & 0 \\ 0 & 0 & 0 & d \end{bmatrix}, \begin{bmatrix} a & b & c & d \\ 0 & e & f & g \\ 0 & 0 & h & i \\ 0 & 0 & 0 & j \end{bmatrix}.$$

4. Repeat Exercise 1 for the following matrices:

$$\begin{bmatrix} 1 & 0 & \lambda & 0 \\ 0 & 1 & \lambda & 0 \\ 0 & 0 & \lambda & 0 \\ 0 & 0 & 0 & 1 \end{bmatrix}, \begin{bmatrix} 1 & 0 & a & 0 \\ 0 & 1 & b & 0 \\ 0 & 0 & c & 0 \\ 0 & 0 & d & 1 \end{bmatrix}.$$

5. Consider the permutation $\tau = (4, 3, 2, 1) \in S_4$. Write τ in two different ways as a product of permutations. What is $\sigma(\tau)$?

6. Repeat the previous exercise for $\tau = (2, 5, 3, 1, 4) \in S_5$.

7. Let $A \in F^{2\times 2}$, where F is any field. Write out the formula for

$$\det(A) = \det(A_1, A_2)$$

   explicitly in terms of the entries of A.

8. Repeat the previous exercise for $A \in F^{3\times 3}$.

9. Let $A \in F^{2\times 2}$, $b \in F^2$. When Gaussian elimination with back substitution is applied to solve $Ax = b$, a single step of Gaussian elimination is required (following which back substitution can be done); moreover, the $2,2$-entry of the matrix resulting from the single step of Gaussian elimination reveals whether A is singular or nonsingular. Show that, assuming $A_{11} \neq 0$, this $2,2$-entry is

$$\frac{\det(A)}{A_{11}},$$

and hence A is nonsingular if and only if $\det(A) \neq 0$. Does this conclusion change if $A_{11} = 0$? Explain.

10. Let F be a field and let $A_1, A_2, \ldots, A_n, B$ be vectors in F^n. Prove that, if $\{A_1, A_2, \ldots, A_n\}$ is linearly dependent and det satisfies Definition 160, then

$$\det(A_1, \ldots, A_j + B, \ldots, A_n) = \det(A_1, \ldots, B, \ldots, A_n).$$

11. Let n be a positive integer, and let i and j be integers satisfying

$$1 \leq i, j \leq n, \ i \neq j.$$

For any $\tau \in S_n$, define τ' by $\tau' = \tau[i, j]$ (that is, τ' is the composition of τ and the transposition $[i, j]$. Finally, define $f : S_n \rightarrow S_n$ by $f(\tau) = \tau'$. Prove that f is a bijection.

12. Let $(j_1, j_2, \ldots, j_n) \in S_n$. What is

$$\det(A_{j_1}, A_{j_2}, \ldots, A_{j_n})$$

in terms of $\det(A_1, A_2, \ldots, A_n)$?

13. Suppose that the following, alternate definition of the determinant is adopted: Let F be a field and n a positive integer. The *determinant* function $\det : (F^n)^n \rightarrow F$ is defined by the following properties:

    (a) If $\{e_1, e_2, \ldots, e_n\}$ is the standard basis for F^n, then

    $$\det(e_1, e_2, \ldots, e_n) = 1.$$

    (b) For any $A_1, A_2, \ldots, A_n \in F^n$ and any $i \neq j$, $1 \leq i, j \leq n$,

    $$\begin{aligned} & \det(A_1, \ldots, A_j, \ldots, A_i, \ldots, A_n) \\ = \ & -\det(A_1, \ldots, A_i, \ldots, A_j, \ldots, A_n). \end{aligned}$$

(c) For any $A_1, A_2, \ldots, A_n \in F^n$, any $\lambda \in F$, and any j, $1 \leq j \leq n$,

$$\det(A_1, \ldots, A_{j-1}, \lambda A_j, A_{j+1}, \ldots, A_n) = \lambda \det(A_1, \ldots, A_n).$$

(d) For any $A_1, A_2, \ldots, A_n, B_j \in F^n$, and any j, $1 \leq j \leq n$,

$$\begin{aligned}\det(A_1, \ldots, A_j + B_j, \ldots, A_n) &= \det(A_1, \ldots, A_j, \ldots, A_n) + \\ &\quad \det(A_1, \ldots, B_j, \ldots, A_n).\end{aligned}$$

Prove that this alternate definition is equivalent to Definition 160.

4.2 Further properties of the determinant function

We will now use the notations $\det(A)$ and $\det(A_1, A_2, \ldots, A_n)$ interchangeably, where $A_1, A_2, \ldots, A_n$ are the columns of A. This means we regard the domain of det as either $F^{n \times n}$ or $(F^n)^n$. These two spaces are easily seen to be isomorphic, so this is a very minor abuse of notation.

We begin with one of the most important properties of the determinant, the multiplication property.

Theorem 163 *Let F be a field and let $A, B \in F^{n \times n}$. Then*

$$\det(AB) = \det(A)\det(B).$$

Proof The proof is a calculation using the fact that

$$C_{ij} = \sum_{k=1}^{n} A_{ik} B_{kj}$$

and properties of permutations. We extend the notation $\sigma(\tau)$ to any finite sequence $\tau = (i_1, i_2, \ldots, i_n)$, where each i_k satisfies $1 \leq i_k \leq n$, by defining $\sigma(i_1, i_2, \ldots, i_n) = 0$ if $(i_1, i_2, \ldots, i_n)$ is not a permutation (that is, if $i_k = i_\ell$

for some $k \neq \ell$). Then

$$\begin{aligned}
\det(C) &= \det(C_1, \ldots, C_n) \\
&= \sum_{\tau \in S_n} \sigma(\tau) C_{\tau(1),1} \cdots C_{\tau(n),n} \\
&= \sum_{\tau \in S_n} \sigma(\tau) \left(\sum_{j_1=1}^{n} A_{\tau(1),j_1} B_{j_1,1} \right) \cdots \left(\sum_{j_n=1}^{n} A_{\tau(n),j_n} B_{j_n,n} \right) \\
&= \sum_{\tau \in S_n} \sum_{j_1=1}^{n} \cdots \sum_{j_n=1}^{n} \sigma(\tau) A_{\tau(1),j_1} \cdots A_{\tau(n),j_n} B_{j_1,1} \cdots B_{j_n,n} \\
&= \sum_{j_1=1}^{n} \cdots \sum_{j_n=1}^{n} B_{j_1,1} \cdots B_{j_n,n} \left(\sum_{\tau \in S_n} \sigma(\tau) A_{\tau(1),j_1} \cdots A_{\tau(n),j_n} \right) \\
&= \sum_{j_1=1}^{n} \cdots \sum_{j_n=1}^{n} \det(A_{j_1}, \ldots, A_{j_n}) B_{j_1,1} \cdots B_{j_n,n}.
\end{aligned}$$

Because interchanging two arguments to det changes the sign of the result, we see that

$$\det(A_{j_1}, \ldots, A_{j_n}) = \sigma(j_1, \ldots, j_n)\det(A_1, \ldots, A_n) = \sigma(j_1, \ldots, j_n)\det(A)$$

if $(j_1, \ldots, j_n)$ is a permutation, while

$$\det(A_{j_1}, \ldots, A_{j_n}) = 0$$

if $(j_1, \ldots, j_n)$ is not a permutation. We therefore obtain

$$\begin{aligned}
\det(C) &= \sum_{(j_1,\ldots,j_n) \in S_n} \sigma(j_1, \ldots, j_n)\det(A) B_{j_1,1} \cdots B_{j_n,n} \\
&= \det(A) \sum_{(j_1,\ldots,j_n) \in S_n} \sigma(j_1, \ldots, j_n) B_{j_1,1} \cdots B_{j_n,n} \\
&= \det(A)\det(B).
\end{aligned}$$

QED

We have already seen that $\det(A_1, A_2, \ldots, A_n) = 0$ if $\{A_1, A_2, \ldots, A_n\}$ is linearly dependent. Using the multiplication theorem for determinants, we can easily prove that the converse is also true.

Theorem 164 *Let F be a field and let $\{A_1, A_2, \ldots, A_n\}$ be a linearly independent subset of F^n. Then*

$$\det(A_1, A_2, \ldots, A_n) \neq 0.$$

Proof Let $A = [A_1|A_2|\cdots|A_n]$. If $\{A_1, A_2, \ldots, A_n\}$ is linearly independent, then A is invertible: $AA^{-1} = I$. Therefore,

$$\begin{aligned} & \det\left(AA^{-1}\right) = \det(I) \\ \Rightarrow\ & \det(A)\det\left(A^{-1}\right) = 1, \end{aligned}$$

which implies that $\det(A_1, A_2, \ldots, A_n) = \det(A) \neq 0$.

QED

Phrasing the previous result in terms of matrices, we obtain the following.

Corollary 165 *Let F be a field and suppose $A \in F^{n\times n}$. Then A is nonsingular if and only if* $\det(A) \neq 0$.

Equivalently, we can state that A is singular if and only if $\det(A) = 0$. Obtaining this result was our original motivation for defining the determinant function; we now have a scalar-valued function that can be used to determine if a given matrix is singular or nonsingular. We will see that the determinant function is not particularly useful for testing specific matrices for singularity (there are better ways to determine whether a given matrix is singular or nonsingular); however, the properties of the determinant function make it useful for studying the eigenvalues of matrices, as discussed in the introduction of this chapter.

From the proof of Theorem 164, we also obtain the following result.

Corollary 166 *Let F be a field and let $A \in F^{n\times n}$ be invertible. Then*

$$\det\left(A^{-1}\right) = \frac{1}{\det(A)}.$$

Here is the final fact we will need about the determinant function.

Theorem 167 *Let F be a field and let $A \in F^{n\times n}$. Then*

$$\det\left(A^T\right) = \det(A).$$

Proof By definition, since $\left(A^T\right)_{ij} = A_{ji}$, we have

$$\det\left(A^T\right) = \sum_{\tau\in S_n} \sigma(\tau) A_{1,\tau(1)} A_{2,\tau(2)} \cdots A_{n,\tau(n)}.$$

Since $(\tau(1), \tau(2), \ldots, \tau(n))$ is a permutation of $(1, 2, \ldots, n)$, we can rearrange the factors $A_{1,\tau(1)}, A_{2,\tau(2)}, \ldots, A_{n,\tau(n)}$ so that the column indices are ordered from 1 to n. Since each term is of the form $A_{j,\tau(j)}$, we see that the row index j is always paired with the column index $\tau(j)$; equivalently, we can say that the column index i is always paired with the row index $\tau^{-1}(i)$. Therefore,

$$\det\left(A^T\right) = \sum_{\tau\in S_n} \sigma(\tau) A_{\tau^{-1}(1),1} A_{\tau^{-1}(2),2} \cdots A_{\tau^{-1}(n),n}.$$

We leave it as an exercise to show that $\sigma\left(\tau^{-1}\right) = \sigma(\tau)$, and that $\tau \mapsto \tau^{-1}$ defines a one-to-one correspondence between S_n and itself (see Exercise 4). It then follows that

$$\det\left(A^T\right) = \sum_{\tau^{-1}\in S_n} \sigma\left(\tau^{-1}\right) A_{\tau^{-1}(1),1}A_{\tau^{-1}(2),2}\cdots A_{\tau^{-1}(n),n} = \det(A).$$

QED

Because the columns of A^T are the rows of A, the preceding theorem implies that row operations affect $\det(A)$ in the same fashion as do column operations.

Corollary 168 *Let F be a field and let $A \in F^{n\times n}$.*

1. *If $B \in F^{n\times n}$ is obtained by interchanging two rows of A, then*
$$\det(B) = -\det(A).$$
2. *If $B \in F^{n\times n}$ is obtained by multiplying one row of A by a constant $\lambda \in F$, then*
$$\det(B) = \lambda\det(A).$$
3. *If $B \in F^{n\times n}$ is obtained by adding a multiple of one row of A to another, then*
$$\det(B) = \det(A).$$
In other words, suppose $r_1, r_2, \ldots, r_n$ are the rows of A, λ belongs to F, and B is defined by
$$B = \begin{bmatrix} r_1 \\ \hline \vdots \\ \hline r_i \\ \hline \vdots \\ \hline r_j + \lambda r_i \\ \hline \vdots \\ \hline r_n \end{bmatrix}.$$
Then $\det(B) = \det(A)$.

Exercises

Miscellaneous exercises

1. Prove or give a counterexample: If $A \in \mathbf{R}^{n\times n}$ and all of the diagonal entries of A are zero (that is, $A_{ii} = 0$ for $i = 1, 2, \ldots, n$), then $\det(A) = 0$.
2. Let F be a field and let $A \in F^{n\times n}$. Prove that A^T is singular if and only if A is singular.

3. Let F be a field and let $A \in F^{n\times n}$. Prove that $A^T A$ is singular if and only if A is singular.

4. Let n be a positive integer.

   (a) Show that $\sigma\left(\tau^{-1}\right) = \sigma(\tau)$ for all $\tau \in S_n$.

   (b) Define $f : S_n \to S_n$ by $f(\tau) = \tau^{-1}$. Prove that f is a bijection.

5. Let F be a field, let A be any matrix in $F^{n\times n}$, and let $X \in F^{n\times n}$ be invertible. Prove that $\det(X^{-1}AX) = \det(A)$.

6. Let F be a field and let $A, B \in F^{n\times n}$. Prove that AB is singular if and only if A is singular or B is singular.

7. Let F be a field and suppose $x_1, x_2, \ldots, x_n \in F^n$, $A \in F^{n\times n}$. What is $\det(Ax_1, Ax_2, \ldots, Ax_n)$ in terms of $\det(x_1, x_2, \ldots, x_n)$?

8. Let $A \in F^{m\times n}$, $B \in F^{n\times m}$, where $m > n$. Prove that $\det(AB) = 0$.

9. Let $A \in F^{m\times n}$, $B \in F^{n\times m}$, where $m < n$. Show by example that both $\det(AB) = 0$ and $\det(AB) \neq 0$ are possible.

4.3 Practical computation of $\det(A)$

Since S_n contains $n!$ permutations, the sum defining $\det(A)$ contains $n!$ terms, each the product of n factors. Therefore, computing $\det(A)$ from the definition requires about $n \cdot n!$ multiplications and $n!$ additions, for a total of about $(n+1)!$ operations. Since $n!$ grows so quickly with n, this means that it is impossible to compute $\det(A)$ by the definition if n is at all large. Table 4.3 shows how long it would take to perform $(n+1)!$ arithmetic operations on a three gigahertz computer, assuming the computer performs one arithmetic operation per clock cycle (that is, three billion operations per second).

In spite of these results, it is possible to compute determinants in a reasonable time. We now derive a practical method based on Corollary 168.

Definition 169 *Let F be a field and let A belong to $F^{m\times n}$. We say that A is* diagonal *if $A_{ij} = 0$ for $i \neq j$,* upper triangular *if $A_{ij} = 0$ for $i > j$, and* lower triangular *if $A_{ij} = 0$ for $j > i$. We will say that A is* triangular *if it is either upper or lower triangular.*

The reader should note that a diagonal matrix is a special case of a triangular matrix.

Theorem 170 *Let F be a field and $A \in F^{n\times n}$. If A is triangular, then $\det(A)$ is the product of the diagonal entries of A:*

$$\det(A) = \Pi_{i=1}^{n} A_{ii}.$$

n	operations	time
5	720	$2.40 \cdot 10^{-7}$ sec
10	$3.99 \cdot 10^{7}$	$1.33 \cdot 10^{-2}$ sec
15	$2.09 \cdot 10^{13}$	1.94 hours
20	$5.11 \cdot 10^{19}$	540 years
25	$4.03 \cdot 10^{26}$	4.26 billion years

TABLE 4.1
The number of operations and time required to compute an $n \times n$ determinant on a three gigahertz computer, using the complete expansion of the determinant.

Proof Consider the definition of $\det(A)$:

$$\det(A) = \sum_{\tau \in S_n} \sigma(\tau) A_{\tau(1),1} A_{\tau(2),2} \cdots A_{\tau(n),n}.$$

Since each $\tau \in S_n$ is a bijection, if $\tau(i) > i$ for some i, then there exists j such that $\tau(j) < j$, and conversely. That is, every product $A_{\tau(1),1} A_{\tau(2),2} \cdots A_{\tau(n),n}$ contains at least one entry from below the diagonal of A and one from above, except the product corresponding to the identity permutation: $\tau(i) = i$ for $i = 1, 2, \ldots, n$. If A is triangular, then either all the entries above the diagonal are zero, or all the entries from below the diagonal are. Therefore, every term in the sum defining $\det(A)$ is zero except one: $\sigma(1, 2, \ldots, n) A_{1,1} A_{2,2} \cdots A_{n,n}$. Since $\sigma(1, 2, \ldots, n) = 1$, this gives the desired result.

QED

We now have a practical way to compute $\det(A)$: perform elementary row operations to reduce A to upper triangular form, as in Gaussian elimination, and keep track of how the determinant changes at each step. To facilitate hand calculations, we often denote $\det(A)$ by

$$\begin{vmatrix} A_{11} & A_{12} & \cdots & A_{1n} \\ A_{21} & A_{22} & \cdots & A_{2n} \\ \vdots & \vdots & \ddots & \vdots \\ A_{n1} & A_{n2} & \cdots & A_{nn} \end{vmatrix}.$$

Example 171 *Let $A \in \mathbf{R}^{4 \times 4}$ be defined by*

$$A = \begin{bmatrix} 0 & 1 & 3 & 5 \\ -3 & -2 & 3 & 3 \\ 0 & 8 & 24 & 41 \\ 0 & -1 & 0 & -8 \end{bmatrix}.$$

Then

$$\det(A) = \begin{vmatrix} 0 & 1 & 3 & 5 \\ -3 & -2 & 3 & 3 \\ 0 & 8 & 24 & 41 \\ 0 & -1 & 0 & -8 \end{vmatrix} = -\begin{vmatrix} -3 & -2 & 3 & 3 \\ 0 & 1 & 3 & 5 \\ 0 & 8 & 24 & 41 \\ 0 & -1 & 0 & -8 \end{vmatrix}$$

$$= -\begin{vmatrix} -3 & -2 & 3 & 3 \\ 0 & 1 & 3 & 5 \\ 0 & 0 & 0 & 1 \\ 0 & 0 & 3 & -3 \end{vmatrix}$$

$$= \begin{vmatrix} -3 & -2 & 3 & 3 \\ 0 & 1 & 3 & 5 \\ 0 & 0 & 3 & -3 \\ 0 & 0 & 0 & 1 \end{vmatrix}$$

$$= (-3) \cdot 1 \cdot 3 \cdot 1 = -9.$$

In the above example, we used two types of elementary row operations, interchanging two rows (which changes the sign of the determinant) and adding a multiple of one row to another (which leaves the determinant unchanged). The third type, multiplying a row by a nonzero constant, is never necessary when reducing a matrix to lower triangular form, although it might be used to make arithmetic simpler when doing the calculation by hand.

It can be shown that approximately $(2/3)n^3$ arithmetic operations[3] are required to reduce $A \in F^{n\times n}$ to upper triangular form (see Exercise 4). This means that computing determinants is well within the capabilities of modern computers, even if n is quite large.

There is an interesting point to consider here. We now know that $A \in F^{n\times n}$ is nonsingular if and only if $\det(A) \neq 0$. From Chapter 3, we know that $Ax = b$ has a unique solution for each $b \in F^n$ if and only if A is nonsingular. The following procedure then seems reasonable: Before trying to solve $Ax = b$, compute $\det(A)$ to see if the system is guaranteed to have a unique solution. However, efficient computation of $\det(A)$ requires Gaussian elimination, which is the first part of solving $Ax = b$. If we just proceed with solving $Ax = b$, we will determine whether A is nonsingular or not; there is no reason to compute $\det(A)$.

Moreover, the value of $\det(A)$ is not really helpful in deciding if the solution of $Ax = b$ will succeed. When the computations are performed in finite precision arithmetic on a computer, we rarely obtain a determinant of exactly 0 when A is singular; round-off error typically causes the computed value to be close to zero rather than exactly zero. However, in this context there is no easy way to determine if a given number is "close" to zero, close being a relative term. The determinant is useful in several ways, but not as a practical

[3]To be precise, the number of operations is a polynomial in n, with leading term $(2/3)n^3$. When n is large, the lower degree terms are negligible compared to the leading term.

means of determining in finite precision arithmetic if A is singular or close to singular.

4.3.1 A recursive formula for $\det(A)$

There is another method for computing $\det(A)$, the *method of cofactor expansion*, that is useful in certain circumstances. It requires nearly as many arithmetic operations as does the complete expansion, so it is not practical for computing the determinant of a large matrix, but it is easy to apply for small matrices, and it is particularly useful when the entries of the matrices are variables.

Given a matrix $A \in F^{n\times n}$ and given any indices i, j between 1 and n, we obtain a matrix $A^{(i,j)} \in F^{(n-1)\times(n-1)}$ by deleting the ith row and j column of A. The values $\det\left(A^{(i,j)}\right)$, $i, j = 1, 2, \ldots, n$, are called the *minors* of the matrix A, and the *cofactors* of A are the related numbers $(-1)^{i+j}\det\left(A^{(i,j)}\right)$.

We will need the following technical lemma about permutations.

Lemma 172 *Let n be a positive integer and suppose $\tau = (i_1, i_2, \ldots, i_n)$ is an element of S_n. If $(i_2, i_3, \ldots, i_n)$ is regarded as a permutation of*

$$(1, \ldots, i_1 - 1, i_1 + 1, \ldots, n),$$

then

$$\sigma(i_2, \ldots, i_n) = (-1)^{i_1+1}\sigma(i_1, i_2, \ldots, i_n).$$

Proof Consider the following permutation, defined as the composition of transpositions:

$$(i_1, 1, \ldots, i_1 - 1, i_1 + 1, \ldots, n) = [1, 2][1, 3] \cdots [1, i_1 - 1][1, i_1].$$

Now let $\tau' = (i_2, i_3, \ldots, i_n)$, regarded as a permutation of

$$(1, \ldots, i_1 - 1, i_1 + 1, \ldots, n),$$

be written as a product of transpositions:

$$\tau' = [j_\ell, k_\ell][j_{\ell-1}, k_{\ell-1}] \cdots [j_1, k_1].$$

It follows that

$$\tau = [j_\ell, k_\ell][j_{\ell-1}, k_{\ell-1}] \cdots [j_1, k_1][1, 2][1, 3] \cdots [1, i_1 - 1][1, i_1].$$

Counting the transpositions shows that

$$\sigma(\tau) = \sigma(\tau')(-1)^{i_1-1},$$

and therefore

$$\sigma(\tau') = \sigma(\tau)(-1)^{i_1-1} = \sigma(\tau)(-1)^{i_1+1},$$

as desired.

QED

The following recursive formula for computing $\det(A)$ is called *cofactor expansion along the first column.*

Theorem 173 *Let F be a field and suppose $A \in F^{n\times n}$. Then*

$$\det(A) = \sum_{i=1}^{n}(-1)^{i+1}A_{i,1}\det\left(A^{(i,1)}\right). \tag{4.6}$$

Proof Recalling that $\sigma(i_1, i_2, \ldots, i_n) = 0$ if each i_j is between 1 and n but $(i_1, i_2, \ldots, i_n)$ is not a permutation of $(1, 2, \ldots, n)$, we have

$$\begin{aligned}
&\det(A)\\
= &\sum_{(i_1,i_2,\ldots,i_n)\in S_n} \sigma(i_1, i_2, \ldots, i_n)A_{i_1,1}A_{i_2,2}\cdots A_{i_n,n}\\
= &\sum_{i_1=1}^{n}\sum_{i_2=1}^{n}\cdots\sum_{i_n=1}^{n}\sigma(i_1, i_2, \ldots, i_n)A_{i_1,1}A_{i_2,2}\cdots A_{i_n,n}\\
= &\sum_{i_1=1}^{n}\sum_{\substack{i_2=1\\ i_2\neq i_1}}^{n}\cdots\sum_{\substack{i_n=1\\ i_n\neq i_1}}^{n}\sigma(i_1, i_2, \ldots, i_n)A_{i_1,1}A_{i_2,2}\cdots A_{i_n,n}\\
= &\sum_{i_1=1}^{n}(-1)^{i_1+1}A_{i_1,1}\left(\sum_{\substack{i_2=1\\ i_2\neq i_1}}^{n}\cdots\sum_{\substack{i_n=1\\ i_n\neq i_1}}^{n}\sigma(i_2, \ldots, i_n)A_{i_2,2}\cdots A_{i_n,n}\right).
\end{aligned}$$

We can eliminate the cases that $i_2 = i_1, i_3 = i_1, \ldots, i_n = i_1$ from the summations because $\sigma(i_1, i_2, \ldots, i_n)$ is zero if two arguments are equal. But now, for each i_1,

$$\sum_{\substack{i_2=1\\ i_2\neq i_1}}^{n}\cdots\sum_{\substack{i_n=1\\ i_n\neq i_1}}^{n}\sigma(i_2, \ldots, i_n)A_{i_2,2}\cdots A_{i_n,n} = \det\left(A^{(i_1,1)}\right),$$

and the proof is complete.

QED

As mentioned above, the recursive formula is convenient for hand calculations of small determinants. It is helpful to note that 2×2 determinants are computed as follows:

$$\begin{vmatrix} a & b\\ c & d\end{vmatrix} = ad - bc.$$

(The reader can verify this formula from the complete expansion.)

Example 174 *Let* $A \in \mathbf{R}^{3\times 3}$ *be defined by*

$$A = \begin{bmatrix} 2 & -1 & 3 \\ 4 & 3 & -2 \\ 3 & 1 & 1 \end{bmatrix}.$$

Then

$$\begin{aligned}
\det(A) &= \begin{vmatrix} 2 & -1 & 3 \\ 4 & 3 & -2 \\ 3 & 1 & 1 \end{vmatrix} \\
&= 2\begin{vmatrix} 3 & -2 \\ 1 & 1 \end{vmatrix} - 4\begin{vmatrix} -1 & 3 \\ 1 & 1 \end{vmatrix} + 3\begin{vmatrix} -1 & 3 \\ 3 & -2 \end{vmatrix} \\
&= 2(3\cdot 1 - 1(-2)) - 4((-1)1 - 1\cdot 3) + 3((-1)(-2) - 3\cdot 3) \\
&= 2\cdot 5 - 4(-4) + 3(-7) \\
&= 5.
\end{aligned}$$

As the following theorem shows, we can compute $\det(A)$ using cofactor expansion along any column or row of A. For small examples done by hand, this fact can be used to reduce the work by choosing the row or column containing the most zeros.

Theorem 175 *Let* F *be a field and suppose* $A \in F^{n\times n}$.

1. *For any* $j = 1, 2, \ldots, n$,

$$\det(A) = \sum_{i=1}^{n} (-1)^{i+j} A_{i,j} \det\left(A^{(i,j)}\right).$$

This formula is called cofactor expansion along column j.

2. *For any* $i = 1, 2, \ldots, n$,

$$\det(A) = \sum_{j=1}^{n} (-1)^{i+j} A_{i,j} \det\left(A^{(i,j)}\right).$$

This formula is called cofactor expansion along row i.

Proof Exercise 3.

4.3.2 Cramer's rule

We already know two methods for solving $Ax = b$, where A is a square matrix. One is Gaussian elimination with back substitution, and the other is multiplication by the inverse matrix ($x = A^{-1}b$). There is another method that gives an explicit formula for each component of the solution in terms of certain determinants. To express this formula, we use the following notation:

For any matrix $B \in F^{n\times n}$ and any vector $c \in F^n$, $B_j(c)$ will denote the matrix obtained from B by replacing the jth column by c:

$$B_j(c) = [B_1|\cdots|B_{j-1}|c|B_{j+1}|\cdots|B_n].$$

Theorem 176 (Cramer's rule) *Let F be a field, let $A \in F^{n\times n}$ be a nonsingular matrix, and let $b \in F^n$ be a given vector. Then the unique solution x to $Ax = b$ is defined by*

$$x_i = \frac{\det(A_i(b))}{\det(A)},\ i = 1, 2, \ldots, n.$$

Proof The proof is surprisingly simple and is based on the multiplication theorem for determinants. We know the solution x exists and is unique because A is nonsingular. Let $I \in F^{n\times n}$ be the identity matrix. Then, for each $i = 1, 2, \ldots, n$, $AI_i(x) = A_i(b)$. It follows that

$$\det(A)\det(I_i(x)) = \det(A_i(b))$$

or

$$\det(I_i(x)) = \frac{\det(A_i(b))}{\det(A)}.$$

But it can be shown that $\det(I_i(x)) = x_i$ (see Exercise 6), and the desired result follows.

QED

Exercises

Miscellaneous exercises

1. Compute each of the following determinants two ways: Using cofactor expansion and using row reduction. (The underlying matrices have real entries.)

   (a)
   $$\begin{vmatrix} 3 & 10 & -20 \\ 9 & 28 & -40 \\ -15 & -42 & 25 \end{vmatrix}$$

   (b)
   $$\begin{vmatrix} 1 & -1 & -2 \\ -4 & 4 & 7 \\ 0 & -1 & 4 \end{vmatrix}$$

   (c)
   $$\begin{vmatrix} 0 & 0 & 4 & 0 \\ 3 & 5 & 20 & -6 \\ 0 & -1 & -4 & 0 \\ 0 & 0 & 12 & -2 \end{vmatrix}$$

2. For each matrix, compute its determinant and decide if the matrix is singular or nonsingular.

   (a)
   $$\begin{bmatrix} 1+i & 1-2i \\ 2 & -1-3i \end{bmatrix} \in \mathbf{C}^{2\times 2}$$

   (b)
   $$\begin{bmatrix} 1 & 0 & 1 \\ 1 & 1 & 1 \\ 1 & 0 & 0 \end{bmatrix} \in \mathbf{Z}_2^{3\times 3}$$

   (c)
   $$\begin{bmatrix} 1 & 2 & 1 \\ 2 & 0 & 1 \\ 1 & 1 & 0 \end{bmatrix} \in \mathbf{Z}_3^{3\times 3}$$

3. Prove Theorem 175. (Hint: To prove that cofactor expansion along column j is valid, interchange columns 1 and j and apply Theorem 173. To prove the second part, use the fact that $\det\left(A^T\right) = \det(A)$ and cofactor expansion along the columns of A^T.)

4. Let $A \in F^{n\times n}$ be nonsingular, where F is a field, and let $b \in F^n$.

   (a) Count exactly the number of arithmetic operations required to reduce A to upper triangular form. Show that the result is a polynomial in n whose leading term is $(2/3)n^3$.

   (b) Count exactly the number of arithmetic operations required to reduce the augmented matrix $[A|b]$ to $[I|x]$, where $x = A^{-1}b$. Show that the resulting count is also approximately $(2/3)n^3$.

   In performing these operations counts, row interchanges should be ignored.

5. Let $A \in F^{n\times n}$ be nonsingular, where F is a field, and let $b \in F^n$. How many arithmetic operations are required to solve $Ax = b$ by Cramer's rule? Assume the necessary determinants are computed by row reduction.

6. Let F be a field and let $I \in F^{n\times n}$ be the identity matrix. Prove that, for any $x \in F^n$ and any integer i between 1 and n, $\det(I_i(x)) = x_i$.

7. Suppose $A \in \mathbf{R}^{n\times n}$ is invertible and has integer entries, and assume $\det(A) = \pm 1$. Prove that A^{-1} also has integer entries. (Hint: Use Cramer's rule to solve for each column of A^{-1}.)

8. Use Cramer's rule to derive a formula for A^{-1} in terms of the cofactors of A (and $\det(A)$). Use it to compute the inverse of

$$A = \begin{bmatrix} 0 & -1 & -3 \\ -1 & -2 & -1 \\ 0 & 2 & 7 \end{bmatrix}.$$

9. Use Cramer's rule to solve $Ax = b$, where

$$A = \begin{bmatrix} 0 & -1 & 16 \\ -4 & -5 & 16 \\ 0 & -1 & 12 \end{bmatrix}, \ b = \begin{bmatrix} 15 \\ 3 \\ 11 \end{bmatrix}.$$

10. Use Cramer's rule to find functions u_1 and u_2 satisfying

$$\begin{aligned} \cos(x)u_1(x) + \sin(x)u_2(x) &= 0, \\ -\sin(x)u_1(x) + \cos(x)u_2(x) &= x. \end{aligned}$$

11. Let F be a field, let $x_0, x_1, \ldots, x_n$ be distinct elements of F, and let $V \in F^{(n+1)\times(n+1)}$ be the matrix

$$V = \begin{bmatrix} 1 & x_0 & x_0^2 & \cdots & x_0^n \\ 1 & x_1 & x_1^2 & \cdots & x_1^n \\ \vdots & \vdots & \vdots & \ddots & \vdots \\ 1 & x_n & x_n^2 & \cdots & x_n^n \end{bmatrix}.$$

The matrix V is called a *Vandermonde matrix*, and it appears in applications such as polynomial interpolation (see Section 2.8). The purpose of this exercise is to derive a formula for $\det(V)$ and thereby show that V is nonsingular. We will write $V(x_0, x_1, \ldots, x_n)$ for the Vandermonde matrix defined by the numbers $x_0, x_1, \ldots, x_n$.

(a) Show that

$$\det(V(x_0, x_1, \ldots, x_n)) = (\Pi_{i=1}^n (x_i - x_0)) \det(V(x_1, x_2, \ldots, x_n)).$$

(Hint: Perform one step of Gaussian elimination on V to produce a matrix V', and note that $\det(V') = \det(V)$. Furthermore, notice that $\det(V_1) = \det(V')$, where V_1 is the submatrix of V' consisting of the last n rows and columns of V'. Factor $x_j - x_0$ out of column j, $j = 1, 2, \ldots, n$, of V_1 to obtain a matrix V_2. Then

$$\det(V) = \det(V') = \det(V_1) = (\Pi_{i=1}^n (x_i - x_0)) \det(V_2).$$

Finally, show that $V(x_1, x_2, \ldots, x_n)$ can be obtained from V_2 by a sequence of column operations of type (4.3) of Definition 160 (that is, adding a multiple of one column to another).)

(b) Using the previous result, prove by induction that

$$\det(V(x_0, x_1, \ldots, x_n)) = \Pi_{j=0}^{n-1} \Pi_{i=j+1}^n (x_i - x_j).$$

4.4 A note about polynomials

Up to this point in the book, we have always regarded a polynomial p as defining a function mapping a field F into itself. We have noted the fact that, when the underlying field is finite, the set $\{1, x, x^2, \ldots, x^k\}$ is linearly dependent if k is sufficiently large, which implies that polynomial functions do not have unique representations. For instance, in $\mathcal{P}_2(\mathbf{Z}_2)$, the polynomials 0 and $x^2 + x$ define the same function and hence are equal (since $\mathcal{P}_2(\mathbf{Z}_2)$ is a space of functions by definition).

In discussing eigenvalues, we wish to treat polynomials as algebraic objects, not as functions, and 0 and $x^2 + x$, for example, are regarded as distinct polynomials with coefficients in $\mathbf{Z}_2$. The proper setting for this discussion is the ring of polynomials $F[x]$, where F is a given field. An advantage of this change of viewpoint is that multiplication of polynomials is recognized in the ring structure, whereas it is not a part of vector space theory.

Definition 177 *Let R be a nonempty set on which are defined two operations, called* addition *and* multiplication. *We say that R is a* ring *if and only if these operations satisfy the following properties:*

1. *$\alpha + \beta = \beta + \alpha$ for all $\alpha, \beta \in R$ (commutative property of addition);*
2. *$(\alpha + \beta) + \gamma = \alpha + (\beta + \gamma)$ for all $\alpha, \beta, \gamma \in R$ (associative property of addition);*
3. *there exists an element 0 of R such that $\alpha + 0 = \alpha$ for all $\alpha \in R$ (existence of an additive identity);*
4. *for each $\alpha \in R$, there exists an element $-\alpha \in R$ such that $\alpha + (-\alpha) = 0$ (existence of additive inverses);*
5. *$(\alpha\beta)\gamma = \alpha(\beta\gamma)$ for all $\alpha, \beta, \gamma \in R$ (associative property of multiplication);*
6. *$\alpha(\beta + \gamma) = \alpha\beta + \alpha\gamma$ and $(\beta + \gamma)\alpha = \beta\alpha + \gamma\alpha$ for all $\alpha, \beta, \gamma \in R$ (distributive property of multiplication over addition).*

If, in addition, the following property is satisfied, then R is called a commutative ring.

7. *$\alpha\beta = \beta\alpha$ for all $\alpha, \beta \in R$ (commutative property of multiplication);*

Finally, if R satisfies the following condition, then R is called a ring with unity.

8. *There exists a nonzero element 1 of R such that $\alpha \cdot 1 = \alpha$ for all $\alpha \in R$ (existence of a multiplicative identity);*

The reader will notice the similarity between the definition of a ring and the definition of a field. Indeed, every field is a commutative ring with unity, and a commutative ring with unity requires only the existence of multiplicative inverses to make it a field.

For any field F, we now define the *ring of polynomials* $F[x]$ to be the set of all *polynomials*

$$a_0 + a_1x + \ldots + a_nx^n,$$

where the *coefficients* $a_0, a_1, \ldots, a_n$ are elements of F, n is a nonnegative integer, and $a_n \neq 0$ if $n > 0$. In this expression, x is treated as an undefined symbol and is usually called an *indeterminate*. The *degree* of the polynomial $a_0 + a_1x + \ldots + a_nx^n$ is n if $n > 0$ or if $n = 0$ and $a_0 \neq 0$. The zero polynomial 0 does not have a degree.

Addition and multiplication of polynomials are defined in the usual way:

$$\begin{aligned} & (a_0 + a_1x + \ldots + a_nx^n) + (b_0 + b_1x + \ldots + b_mx^m) \\ = \;& (a_0 + b_0) + (a_1 + b_1)x + \ldots + (a_t + b_t)x^t, \; t = \max\{m, n\}, \\ & (a_0 + a_1x + \ldots + a_nx^n)(b_0 + b_1x + \ldots + b_mx^m) \\ = \;& (a_0b_0) + (a_0b_1 + a_1b_0)x + (a_0b_2 + a_1b_1 + a_2b_0)x^2 + \ldots + (a_nb_m)x^{n+m}. \end{aligned}$$

When necessary, as in the formula for addition, we interpret a_i as 0 when i is greater than the degree of the polynomial.

It is straightforward to verify that $F[x]$ is a commutative ring with unity. The additive identity is the zero polynomial, while the multiplicative identity (the unity) is the constant polynomial 1.

We notice that F is isomorphic to a subset of $F[x]$, namely, the subset consisting of the zero polynomial together with all polynomials of degree zero. It is usual to identify F with this subset of $F[x]$, thus regarding F itself as a subset of $F[x]$.

The ring $F[x]$ has the following additional property: If $p(x), q(x) \in F[x]$ and $p(x)q(x) = 0$, then $p(x) = 0$ or $q(x) = 0$. A commutative ring with unity that satisfies this property is called an *integral domain*. In an integral domain, even though multiplicative inverses do not exist in general, the cancellation law still holds: If $q \neq 0$, then

$$pq = rq \Rightarrow pq - rq = 0 \Rightarrow (p - r)q = 0 \Rightarrow p - r = 0 \Rightarrow p = r.$$

The usual properties of polynomials, including facts about roots and factorization, can be developed in the setting described above. The details are left to Appendix C.

There is a standard construction of a field from an integral domain, which results in the field of quotients of the integral domain. In the case of $F[x]$, the result is the field of rational functions, of which $F[x]$ is (isomorphic to) a subset. The field of rational functions plays no role in what follows, except in one trivial way: $\det(A)$ has been defined for a matrix A with entries in a field. In the next section, we will apply det to matrices whose entries belong to $F[x]$; since $F[x]$ is a subset of a field, it is logically correct to do so.

4.5 Eigenvalues and the characteristic polynomial

We now use the determinant function to investigate the eigenvalues of a matrix $A \in F^{n\times n}$. As mentioned on page 206, we will first study the eigenvalue problem in matrix form, and then translate our results to a general linear operator T.

The scalar $\lambda \in F$ is an eigenvalue of $A \in F^{n\times n}$ if and only if the matrix $\lambda I - A$ is singular, that is, if and only if $r = \lambda$ is a solution to

$$\det(rI - A) = 0.$$

We can view this as an equation in r, which is the main contribution of the determinant function: It reduces the problem of finding the eigenvalues of a matrix to a single equation in a single unknown. Moreover, that equation is a polynomial equation, as the next theorem shows.

Since we typically use the variable x to represent a vector, we will henceforth use r for the indeterminate in a polynomial. Thus polynomials will be written as elements of $F[r]$.

Theorem 178 *Let F be a field and let $A \in F^{n\times n}$. Then $\det(rI - A)$ is a polynomial in the variable r with coefficients from F (that is, $\det(rI - A)$ is an element of $F[r]$). Moreover, $\det(rI - A)$ is a polynomial of degree n with leading coefficient (that is, the coefficient of r^n) equal to* 1.

Proof We use the complete expansion of $\det(rI - A)$:

$$\sum_{\tau\in S_n} \sigma(\tau)(rI_{\tau(1),1} - A_{\tau(1),1})(rI_{\tau(2),2} - A_{\tau(2),2}) \cdots (rI_{\tau(n),n} - A_{\tau(n),n}).$$

This formula shows immediately that $\det(rI - A)$ is a polynomial in r and that the degree is at most n. Since $I_{\tau(j),j} = 0$ unless $\tau(j) = j$, we see that there is exactly one term in the sum of degree n, namely,

$$(r - A_{1,1})(r - A_{2,2}) \cdots (r - A_{n,n}).$$

Moreover, every other term is of degree $n - 2$ or less (since if $\tau(k) \neq k$, then $\tau(j) = k$ must hold for some $j \neq k$, and hence $I_{\tau(i),i} = 0$ for at least two values of i). We can therefore determine the two terms of highest degree in the polynomial:

$$\det(rI - A) = r^n - (A_{1,1} + A_{2,2} + \cdots + A_{n,n})r^{n-1} + \ldots.$$

QED

Definition 179 *Let F be a field and let $A \in F^{n\times n}$. We call*

$$p_A(r) = \det(rI - A)$$

the characteristic polynomial *of A.*

Definition 180 *Let F be a field and let $A \in F^{n\times n}$. The* trace *of A is*

$$\mathrm{tr}(A) = \sum_{i=1}^{n} A_{ii}.$$

Corollary 181 *Let F be a field and let $A \in F^{n\times n}$. Then the characteristic polynomial of A has the form*

$$p_A(r) = r^n - \mathrm{tr}(A)r^{n-1} + \cdots + (-1)^n \det(A).$$

Proof The proof of Theorem 178 shows that

$$p_A(r) = r^n - \mathrm{tr}(A)r^{n-1} + \cdots.$$

Therefore, the only thing left to prove is the value of the constant term, $p_A(0)$. But $p_A(0) = \det(0I - A) = \det(-A) = (-1)^n \det(A)$. Multiplying A by -1 is equivalent to multiplying each column of A by -1, which, by the second property of the determinant function (Definition 160), introduces one factor of -1 for each of the n columns.

QED

We now present several examples.

Example 182 *Let $A \in \mathbf{R}^{2\times 2}$ be defined by*

$$A = \begin{bmatrix} 1 & 2 \\ 2 & 1 \end{bmatrix}.$$

Then

$$\begin{aligned} p_A(r) = \det(rI - A) &= \begin{vmatrix} r-1 & -2 \\ -2 & r-1 \end{vmatrix} \\ &= (r-1)^2 - 4 = r^2 - 2r - 3 = (r+1)(r-3). \end{aligned}$$

Therefore $p_A(r) = 0$ if and only if $r = -1$ or $r = 3$. Thus A has two eigenvalues, -1 and 3.

Example 183 *Let $A \in \mathbf{C}^{2\times 2}$ be defined by*

$$A = \begin{bmatrix} 1+i & 1-2i \\ 2 & -1-3i \end{bmatrix}.$$

Then

$$\begin{aligned} p_A(r) = \det(rI - A) &= \begin{vmatrix} r-1-i & -1+2i \\ -2 & r+1+3i \end{vmatrix} \\ &= (r-1-i)(r+1+3i) + 2(-1+2i) \\ &= r^2 + 2ir = r(r+2i). \end{aligned}$$

The eigenvalues of A are the roots of $p_A(r)$, which are 0 and $-2i$.

Example 184 *Let $A \in \mathbf{Z}_2^{3\times 3}$ be defined by*

$$A = \begin{bmatrix} 1 & 1 & 0 \\ 0 & 1 & 1 \\ 1 & 0 & 1 \end{bmatrix}.$$

Then (recalling that subtraction and addition are the same operation in $\mathbf{Z}_2$)

$$\begin{aligned} p_A(r) &= \det(rI - A) \\ &= \det(rI + A) \\ &= \begin{vmatrix} r+1 & 1 & 0 \\ 0 & r+1 & 1 \\ 1 & 0 & r+1 \end{vmatrix} \\ &= (r+1)\begin{vmatrix} r+1 & 1 \\ 0 & r+1 \end{vmatrix} + \begin{vmatrix} 1 & 0 \\ r+1 & 1 \end{vmatrix} \\ &= (r+1)^3 + 1 \\ &= r^3 + r^2 + r = r(r^2 + r + 1). \end{aligned}$$

The only eigenvalue of A is 0 (note that $r^2 + r + 1$ has no roots in $\mathbf{Z}_2$).

We now recall some facts about polynomials. If $p(r) \in F[r]$, then $\lambda \in F$ is a root of $p(r)$ (that is, $p(\lambda) = 0$) if and only if $r - \lambda$ is a factor of $p(r)$ (that is, $p(r) = (r - \lambda)q(r)$ for some $q(r) \in F[r]$).

Definition 185 *Let $p(r)$ be a polynomial over a field F. If $\lambda \in F$, k is a positive integer, $(r - \lambda)^k$ is a factor of $p(r)$, and $(r - \lambda)^\ell$ is not a factor of $p(r)$ for any $\ell > k$, then λ is called a* root of multiplicity k *of $p(r)$. If $k = 1$, we also say that λ is a* simple root, *while if $k > 1$, then λ is called a* repeated root *or a* multiple root.

Applying the above terminology to eigenvalues, the roots of the characteristic polynomial of a matrix, we say that λ is an eigenvalue of *algebraic multiplicity* k if λ is a root of multiplicity k of $p_A(r)$. We similarly refer to λ as a *simple eigenvalue* if $k = 1$ or *multiple eigenvalue* if $k > 1$.

Having determined the eigenvalues of A by finding the roots of $p_A(r)$, we can find the eigenvectors corresponding to each eigenvalue λ by solving the singular system $(\lambda I - A)x = 0$.

Example 186 *Let A be the matrix from Example 182, and let us find the eigenvector(s) corresponding to the eigenvalue $\lambda = -1$. We must solve the equation $(-I - A)x = 0$, where*

$$-I - A = \begin{bmatrix} -2 & -2 \\ -2 & -2 \end{bmatrix}.$$

It is easy to see that the solution set is $\mathrm{sp}\{(1, -1)\}$. Thus $(1, -1)$ (or any nonzero multiple) is an eigenvector of A corresponding to the eigenvalue -1.

4.5.1 Eigenvalues of real matrix

In many applications, the field of interest is $\mathbf{R}$, which is not algebraically closed. A field is said to be *algebraically closed* if every polynomial with coefficients from that field has a root in that field. Because $\mathbf{R}$ is not algebraically closed, $A \in \mathbf{R}^{n\times n}$ may have no eigenvalues if we restrict ourselves to real numbers.

Example 187 *Let $A \in \mathbf{R}^{2\times 2}$ be defined by*

$$A = \begin{bmatrix} 0 & 1 \\ -1 & 0 \end{bmatrix}.$$

Then

$$p_A(r) = \begin{vmatrix} r & -1 \\ 1 & r \end{vmatrix} = r^2 + 1.$$

The polynomial $p_A(r)$ has no real roots, and hence A has no real eigenvalues.

In contrast to $\mathbf{R}$, the field $\mathbf{C}$ is algebraically closed. This fact is called the *fundamental theorem of algebra*: Every polynomial over $\mathbf{C}$ has a root in $\mathbf{C}$. It follows from the fundamental theorem that every polynomial $p(r) \in \mathbf{C}[r]$ can be factored completely:

$$p(r) = c(r - \lambda_1)^{k_1}(r - \lambda_2)^{k_2} \cdots (r - \lambda_t)^{k_t}.$$

Here $c \in \mathbf{C}$ is a constant and the roots of $p(r)$ are $\lambda_1, \lambda_2, \ldots, \lambda_t$, with multiplicities $k_1, k_2, \ldots, k_t$, respectively. We have $k_1 + k_2 + \cdots + k_t = n$, where n is the degree of $p(r)$. We often say that $p(r)$ has n roots, *counted according to multiplicity*. It follows that if $A \in \mathbf{C}^{n\times n}$, then

$$p_A(r) = (r - \lambda_1)^{k_1}(r - \lambda_2)^{k_2} \cdots (r - \lambda_t)^{k_t},$$

where $k_1 + k_2 + \cdots + k_t = n$, and A has n eigenvalues, counted according to multiplicity.

Eigenvalues and eigenvectors are so useful that, even if our interest is strictly in real numbers, we allow complex eigenvalues and eigenvectors. We therefore make the following special definition for the case of the field of real numbers.

Definition 188 *Let $A \in \mathbf{R}^{n\times n}$. We say that $\lambda \in \mathbf{C}$ is an* eigenvalue *of A and $x \in \mathbf{C}^n$ is a corresponding* eigenvector *if and only if $Ax = \lambda x$ and $x \neq 0$.*

The definition would read the same if A were an element of $\mathbf{C}^{n\times n}$, but this case is already covered by the definition on page 206. Since $\mathbf{R} \subset \mathbf{C}$, the above definition allows for the possibility that $\lambda \in \mathbf{R}$ and $x \in \mathbf{R}^n$.

According to Definition 188, when referring to eigenvalues and eigenvectors, we always regard $A \in \mathbf{R}^{n\times n}$ as an element of $\mathbf{C}^{n\times n}$. Therefore, considering the matrix of Example 187, we would not say that A has no eigenvalues, rather, that the eigenvalues of A are $\pm i$ (the roots of $r^2 + 1$ in $\mathbf{C}$).

When we need to refer to a complex number or vector that is not real, we use the notation $\lambda \in \mathbf{C} \setminus \mathbf{R}$ or $x \in \mathbf{C}^n \setminus \mathbf{R}^n$. For any set X and subset $Y \subset X$, we write $X \setminus Y$ to indicate the elements of X that do not belong to Y:

$$X \setminus Y = \{x \in X \,:\, x \notin Y\}.$$

Theorem 189 *Suppose $A \in \mathbf{R}^{n\times n}$.*

1. *If $\lambda \in \mathbf{C} \setminus \mathbf{R}$ is an eigenvalue of A with corresponding eigenvector x, then $x \in \mathbf{C}^n \setminus \mathbf{R}^n$.*

2. *If $\lambda \in \mathbf{C} \setminus \mathbf{R}$ is an eigenvalue of A and $x \in \mathbf{C}^n$ a corresponding eigenvector, then $\overline{\lambda}$ is also an eigenvalue of A, with corresponding eigenvector $\overline{x}$.*

Proof The first result follows from the simple observation that, if $x \in \mathbf{R}^n$ but $\lambda \in \mathbf{C} \setminus \mathbf{R}$, then $Ax \in \mathbf{R}^n$ but $\lambda x \in \mathbf{C}^n \setminus \mathbf{R}^n$. To prove the second result, we note that, if $A \in \mathbf{R}^{n\times n}$, then $p_A(r)$ has real coefficients. It then follows from the properties of the complex conjugate that

$$p_A(\overline{\lambda}) = \overline{p_A(\lambda)}$$

and hence

$$p_A(\lambda) = 0 \Rightarrow \overline{p_A(\lambda)} = 0 \Rightarrow p_A(\overline{\lambda}) = 0.$$

Therefore, $\overline{\lambda}$ is also an eigenvalue of A. Similarly,

$$Ax = \lambda x \;\Rightarrow\; \overline{Ax} = \overline{\lambda x} \;\Rightarrow\; A\overline{x} = \overline{\lambda}\,\overline{x}.$$

Since $x \neq 0$ implies $\overline{x} \neq 0$, this shows that $\overline{x}$ is an eigenvector corresponding to $\overline{\lambda}$.

QED

Example 190 *Let $A \in \mathbf{R}^{3\times 3}$ be defined by*

$$A = \begin{bmatrix} 1 & 0 & 0 \\ 9 & 1 & 3 \\ 3 & 0 & 2 \end{bmatrix}.$$

We find the eigenvalues of A from the characteristic polynomial:

$$\begin{aligned} p_A(r) = |rI - A| &= \begin{vmatrix} r-1 & 0 & 0 \\ -9 & r-1 & -3 \\ -3 & 0 & r-2 \end{vmatrix} \\ &= (r-1)\begin{vmatrix} r-1 & -3 \\ 0 & r-2 \end{vmatrix} \\ &= (r-1)^2(r-2). \end{aligned}$$

We see that $\lambda_1 = 1$ is an eigenvalue of algebraic multiplicity 2 and $\lambda_2 = 2$ is a simple eigenvalue.

To find the eigenvectors corresponding to λ_1, we solve $(I - A)x = 0$ by row reduction:

$$\left[\begin{array}{rrr|r} 0 & 0 & 0 & 0 \\ -9 & 0 & -3 & 0 \\ -3 & 0 & -1 & 0 \end{array}\right] \rightarrow \left[\begin{array}{rrr|r} 1 & 0 & \frac{1}{3} & 0 \\ 0 & 0 & 0 & 0 \\ 0 & 0 & 0 & 0 \end{array}\right].$$

We obtain $x_1 = -x_3/3$, with x_2 and x_3 free. The general solution is $x = (-\beta/3, \alpha, \beta) = \alpha(0,1,0) + \beta(-1/3,0,1)$. We see that λ_1 has two independent eigenvectors, $(0,1,0)$ and $(-1/3,0,1)$.

We next find the eigenvectors corresponding to $\lambda_2 = 2$ by solving the equation $(2I - A)x = 0$:

$$\left[\begin{array}{rrr|r} 1 & 0 & 0 & 0 \\ -9 & 1 & -3 & 0 \\ -3 & 0 & 0 & 0 \end{array}\right] \rightarrow \left[\begin{array}{rrr|r} 1 & 0 & 0 & 0 \\ 0 & 1 & -3 & 0 \\ 0 & 0 & 0 & 0 \end{array}\right].$$

We see that $x_1 = 0$ and $x_2 = 3x_3$, so $x = (0, 3\alpha, \alpha) = \alpha(0,3,1)$ is the general solution. The eigenvalue λ_2 has only one independent eigenvector, namely, $(0,1,3)$.

The set of all eigenvectors corresponding to an eigenvalue λ is the null space of $\lambda I - A$, except that the zero vector, which belongs to every null space, is not an eigenvector. We use the term *eigenspace* to refer to $\mathcal{N}(\lambda I - A)$ when λ is an eigenvalue of A, and we write $E_\lambda(A) = \mathcal{N}(\lambda I - A)$. We remark that, if A is singular, then 0 is an eigenvalue of A and the corresponding eigenspace is the null space of A. We also note that $\mathcal{N}(\lambda I - A) = \mathcal{N}(A - \lambda I)$, and we will sometimes use the second form.

Definition 191 *Let F be a field and let $A \in F^{n \times n}$. If $\lambda \in F$ is an eigenvalue of A, then the* geometric multiplicity *of λ is the dimension of $E_\lambda(A)$.*

Thus, in Example 190, A has one eigenvalue with geometric multiplicity two, and one with geometric multiplicity one. For both eigenvalues, the algebraic multiplicity equals the geometric multiplicity.

Example 192 *Let $A \in \mathbf{R}^{3\times 3}$ be defined by*

$$A = \begin{bmatrix} 1 & 1 & 1 \\ 0 & 1 & 1 \\ 0 & 0 & 1 \end{bmatrix}.$$

Then

$$rI - A = \begin{bmatrix} r-1 & -1 & -1 \\ 0 & r-1 & -1 \\ 0 & 0 & r-1 \end{bmatrix},$$

and therefore $p_A(r) = \det(rI - A) = (r-1)^3$. *(The reader will recall that the determinant of a triangular matrix is the product of the diagonal entries.) The only eigenvalue is* $\lambda = 1$.

To find the eigenvectors corresponding to $\lambda = 1$*, we solve* $(I - A)x = 0$ *by row reduction. Below is the augmented matrix before and after row reduction:*

$$\left[\begin{array}{ccc|c} 0 & -1 & -1 & 0 \\ 0 & 0 & -1 & 0 \\ 0 & 0 & 0 & 0 \end{array}\right] \rightarrow \left[\begin{array}{ccc|c} 0 & 1 & 0 & 0 \\ 0 & 0 & 1 & 0 \\ 0 & 0 & 0 & 0 \end{array}\right].$$

The transformed equations are $x_2 = 0$ *and* $x_3 = 0$*; the only solutions are of the form* $(\alpha, 0, 0)$*, and* $\{(1, 0, 0)\}$ *is a basis for the eigenspace.*

The preceding example is of interest because the eigenvalue $\lambda = 1$ has algebraic multiplicity 3 but geometric multiplicity only 1. The significance of this will become clear in the next section.

Exercises

Miscellaneous exercises

1. For each of the following real matrices, find the eigenvalues and a basis for each eigenspace.

   (a)
   $$A = \begin{bmatrix} -15 & 0 & 8 \\ 0 & 1 & 0 \\ -28 & 0 & 15 \end{bmatrix}$$

   (b)
   $$A = \begin{bmatrix} -4 & -4 & -5 \\ -6 & -2 & -5 \\ 11 & 7 & 11 \end{bmatrix}$$

2. For each of the following real matrices, find the eigenvalues and a basis for each eigenspace.

   (a)
   $$A = \begin{bmatrix} 6 & -1 & 1 \\ 4 & 1 & 1 \\ -12 & 3 & -1 \end{bmatrix}$$

   (b)
   $$A = \begin{bmatrix} 1 & 2 \\ -2 & 1 \end{bmatrix}$$

3. For each of the following matrices, find the eigenvalues and a basis for each eigenspace.

(a)

$$A = \begin{bmatrix} 2 & 1 \\ 1 & 2 \end{bmatrix} \in \mathbf{Z}_3^{2\times 2}$$

(b)

$$A = \begin{bmatrix} 1 & 1 & 0 \\ 0 & 1 & 1 \\ 1 & 1 & 1 \end{bmatrix} \in \mathbf{Z}_2^{3\times 3}$$

4. For each of the following matrices, find the eigenvalues and a basis for each eigenspace.

   (a)

$$A = \begin{bmatrix} 1+i & 3-4i \\ 3+4i & 1+i \end{bmatrix} \in \mathbf{C}^{2\times 2}$$

   (b)

$$A = \begin{bmatrix} 1 & 1-i & 2 \\ 1+i & 2 & 1-i \\ 2 & 1+i & 3 \end{bmatrix} \in \mathbf{C}^{3\times 3}$$

5. Suppose $A \in \mathbf{R}^{n\times n}$ has a real eigenvalue λ and a corresponding eigenvector $z \in \mathbf{C}^n$. Show that either the real or imaginary part of z is an eigenvector of A.

6. Let $A \in \mathbf{R}^{2\times 2}$ be defined by

$$A = \begin{bmatrix} a & b \\ b & c \end{bmatrix},$$

   where $a, b, c \in \mathbf{R}$. (Notice that A is symmetric, that is, $A^T = A$.)

   (a) Prove that A has only real eigenvalues.

   (b) Under what conditions on a, b, c does A have a multiple eigenvalue?

7. Let $A \in \mathbf{R}^{2\times 2}$ be defined by

$$A = \begin{bmatrix} a & b \\ c & d \end{bmatrix},$$

   where $a, b, c, d \in \mathbf{R}$.

   (a) Under what conditions on a, b, c, d does A have a multiple eigenvalue?

   (b) Assuming A has a multiple eigenvalue λ, under what additional conditions on a, b, c, d is $E_A(\lambda)$ one-dimensional? Two-dimensional?

8. Let $A \in \mathbf{R}^{n\times n}$, where n is odd. Prove that A has a real eigenvalue.

9. Let $q(r) = r^n + c_{n-1}r^{n-1} + \cdots + c_0$ be an arbitrary polynomial with coefficients in a field F, and let

$$A = \begin{bmatrix} 0 & 0 & 0 & \cdots & -c_0 \\ 1 & 0 & 0 & \cdots & -c_1 \\ 0 & 1 & 0 & \cdots & -c_2 \\ \vdots & & \ddots & \ddots & \vdots \\ 0 & 0 & \cdots & 1 & -c_{n-1} \end{bmatrix}.$$

Prove that $p_A(r) = q(r)$. We call A the *companion matrix* of $q(r)$. This shows that for any polynomial, there is a matrix whose characteristic polynomial is a constant multiple of the given polynomial.[4]

10. Let $A \in \mathbf{C}^{n\times n}$, and let the eigenvalues of A be $\lambda_1, \lambda_2, \ldots, \lambda_n$, listed according to multiplicity. Prove that

    (a) $\mathrm{tr}(A) = \lambda_1 + \lambda_2 + \cdots + \lambda_n$;

    (b) $\det(A) = \lambda_1\lambda_2\cdots\lambda_n$.

11. Let F be a field and let A belong to $F^{n\times n}$. Prove that A and A^T have the same characteristic polynomial and hence the same eigenvalues.

12. Let $A \in F^{n\times n}$ be invertible. Show that every eigenvector of A is also an eigenvector of A^{-1}. What is the relationship between the eigenvalues of A and A^{-1}?

13. Let

$$A = \begin{bmatrix} 1 & 2 \\ 3 & 4 \end{bmatrix} \in \mathbf{R}^{2\times 2}.$$

A single elementary row operation reduces A to the upper triangular matrix

$$U = \begin{bmatrix} 1 & 2 \\ 0 & -2 \end{bmatrix}.$$

Show that A and U have different characteristic polynomials and hence different eigenvalues. This shows that the eigenvalues of a matrix are not invariant under elementary row operations.[5]

---

[4]It also proves a deeper result: There is no finite algorithm for computing the eigenvalues of a general matrix. To be more precise, there is no algorithm that will compute the eigenvalues of an arbitrary matrix $A \in \mathbf{C}^{n\times n}$, $n \geq 5$, using a finite number of additions, subtractions, multiplications, divisions, and root extractions. If there were such an algorithm, it could be used to find the roots of any polynomial by first computing the companion matrix of the polynomial. But it is well known that there is no finite algorithm for computing the roots of a general polynomial of degree five or greater.

Even though there is no algorithm for computing the eigenvalues of a matrix exactly in finitely many steps, there are efficient algorithms for computing good approximations to the eigenvalues (see Section 9.9). Roots of a polynomial can be computed by applying such algorithms to the companion matrix of the polynomial. This is done, for example, in the MATLAB® software package.

[5]In the author's experience, students often think eigenvalues *are* invariant under elementary row operations because

14. Let $A \in F^{m \times n}$. The purpose of this exercise is to show that the nonzero eigenvalues of A^TA and AA^T are the same.

    (a) Let $\lambda \neq 0$ be an eigenvalue of A^TA, say $A^TAx = \lambda x$, where $x \in F^n$ is nonzero. Multiply both sides of the equation $A^TAx = \lambda x$ and explain why λ must be an eigenvalue of AA^T.

    (b) Prove the converse: If $\lambda \neq 0$ is a nonzero eigenvalue of AA^T, then λ is also an eigenvalue of A^TA.

    (c) Explain why the reasoning in the previous two parts of this exercise does not apply if $\lambda = 0$.

    (d) Illustrate these results by computing the eigenvalues of A^TA and AA^T, where

$$A = \begin{bmatrix} 1 & 0 & 1 \\ 0 & 1 & 0 \end{bmatrix}.$$

4.6 Diagonalization

In this section, we introduce one of the main applications of eigenvalues and eigenvectors, the diagonalization of matrices. We begin with a preliminary result.

Theorem 193 *Let F be a field, and suppose $A \in F^{n \times n}$. If $x_1, x_2, \ldots, x_m$ are eigenvectors of A corresponding to distinct eigenvalues $\lambda_1, \lambda_2, \ldots, \lambda_m$, respectively, then $\{x_1, x_2, \ldots, x_m\}$ is linearly independent.*

Proof We argue by contradiction and assume that $\{x_1, x_2, \ldots, x_m\}$ is linearly dependent. Then there exists an integer ℓ, $2 \leq \ell \leq m$, such that $\{x_1, x_2, \ldots, x_{\ell-1}\}$ is linearly independent, but $\{x_1, x_2, \ldots, x_\ell\}$ is linearly dependent. We can then write

$$x_\ell = \sum_{i=1}^{\ell-1} \alpha_i x_i \tag{4.7}$$

for some scalars $\alpha_1, \alpha_2, \ldots, \alpha_{\ell-1} \in F$. Multiplying both sides of this equation by A yields

$$Ax_\ell = A\left(\sum_{i=1}^{\ell-1} \alpha_i x_i\right) \Rightarrow Ax_\ell = \sum_{i=1}^{\ell-1} \alpha_i Ax_i,$$

(a) eigenvalues are determined by the determinant of $rI - A$;

(b) the determinant is invariant under the primary elementary row operation (adding a multiple of one row to another).

However, the reader should note that this reasoning shows that eigenvalues are invariant under row operations on $rI - A$, not on A.

which yields

$$\lambda_\ell x_\ell = \sum_{i=1}^{\ell-1} \alpha_i \lambda_i x_i. \tag{4.8}$$

Multiplying both sides of (4.7) by λ_ℓ and subtracting from (4.8), we obtain

$$\sum_{i=1}^{\ell-1} \alpha_i(\lambda_i - \lambda_\ell)x_i = 0.$$

Since $\{x_1, x_2, \ldots, x_{\ell-1}\}$ is linearly independent by assumption, this implies

$$\alpha_i(\lambda_i - \lambda_\ell) = 0, \ i = 1, 2, \ldots, \ell - 1.$$

But we also know that $\lambda_\ell \neq \lambda_i$ for $i = 1, 2, \ldots, \ell - 1$ (the eigenvalues are distinct by assumption), and therefore we obtain $\alpha_1 = \alpha_2 = \ldots = \alpha_{\ell-1} = 0$. This, together with (4.7), implies that $x_\ell = 0$, contradicting that x_ℓ is an eigenvector. This contradiction shows that $\{x_1, x_2, \ldots, x_m\}$ must be linearly independent.

QED

The reader should note that Example 182 shows that the converse of Theorem 193 is false. Linearly independent eigenvectors do not necessarily correspond to distinct eigenvalues.

Corollary 194 *Let F be a field and let $A \in F^{n\times n}$. If $\lambda_1, \ldots, \lambda_m$ are distinct eigenvalues of A and, for each $i = 1, 2, \ldots, m$,*

$$\{x_1^{(i)}, \ldots, x_{k_i}^{(i)}\}$$

is a linearly independent set of eigenvectors of A corresponding to λ_i, then

$$\left\{x_1^{(1)}, \ldots, x_{k_1}^{(1)}, x_1^{(2)}, \ldots, x_{k_2}^{(2)}, \ldots, x_1^{(m)}, \ldots, x_{k_m}^{(m)}\right\}$$

is linearly independent.

Proof Exercise 10.

As we will see below, it is important to know if there is a basis of F^n consisting of eigenvectors of a given matrix $A \in F^{n\times n}$. The following corollary gives an answer to this question.

Corollary 195 *Let F be a field, suppose $A \in F^{n\times n}$, and let $\lambda_1, \lambda_2, \ldots, \lambda_t$ be the distinct eigenvalues of A. There is a basis of F^n consisting of eigenvectors of A if and only if*

$$\dim(E_{\lambda_1}(A)) + \dim(E_{\lambda_2}(A)) + \cdots + \dim(E_{\lambda_t}(A)) = n.$$

Proof Let $k_i = \dim(E_{\lambda_i}(A))$. According to Corollary 194, if $\{x_1^{(i)}, \ldots, x_{k_i}^{(i)}\}$ is a basis for $E_{\lambda_i}(A)$, then

$$\mathcal{X} = \left\{x_1^{(1)}, \ldots, x_{k_1}^{(1)}, x_1^{(2)}, \ldots, x_{k_2}^{(2)}, \ldots, x_1^{(m)}, \ldots, x_{k_m}^{(m)}\right\} \tag{4.9}$$

is linearly independent. Therefore, if $k_1 + k_2 + \cdots + k_t = n$, we have a set of n linearly independent eigenvectors of A, which is necessarily a basis for F^n. Conversely, suppose $k_1 + k_2 + \cdots + k_t < n$. The set $\mathcal{X}$ then spans a proper subspace of F^n, and it is easy to see that this set contains all eigenvectors of A. It follows that there is no set of n linearly independent eigenvectors of A, and therefore no basis of F^n consisting of eigenvectors of A.

QED

Suppose $A \in F^{n\times n}$ has eigenvalues $\lambda_1, \lambda_2, \ldots, \lambda_n$ (not necessarily distinct) and corresponding eigenvectors $x_1, x_2, \ldots, x_n$. We define matrices X and D in $F^{n\times n}$ by

$$X = [x_1|x_2|\ldots|x_n],\ D = \begin{bmatrix} \lambda_1 & 0 & \cdots & 0 \\ 0 & \lambda_2 & \cdots & 0 \\ \vdots & \vdots & \ddots & \vdots \\ 0 & 0 & \cdots & \lambda_n \end{bmatrix}.$$

We then have

$$AX = [Ax_1|Ax_2|\cdots|Ax_n] = [\lambda_1 x_1|\lambda_2 x_2|\cdots|\lambda_n x_n] = XD.$$

We can also work backwards and conclude that if $D \in F^{n\times n}$ is a diagonal matrix and $X \in F^{n\times n}$ satisfies

$$AX = XD,$$

then the columns of X are eigenvectors of A and the diagonal entries of D are the corresponding eigenvalues.

If $\{x_1, x_2, \ldots, x_n\}$ happens to be linearly independent, then X is invertible and

$$X^{-1}AX = D.$$

The following definition describes this situation.

Definition 196 *Let F be a field and suppose $A \in F^{n\times n}$. We say that A is* diagonalizable *if and only if there exists an invertible matrix $X \in F^{n\times n}$ and a diagonal matrix $D \in F^{n\times n}$ such that*

$$X^{-1}AX = D.$$

In general, if A and B are any two matrices in $F^{n\times n}$ and there exists an invertible matrix $X \in F^{n\times n}$ such that $X^{-1}AX = B$, then we say that A is similar *to B and that X defines a* similarity transformation. *Thus A is diagonalizable if and only if it is similar to a diagonal matrix.*

We now proceed to present some facts about similar matrices. We begin by describing the relation of similarity itself.

Theorem 197 *Let F be a field. The relation "is similar to" defines an equivalence relation on $F^{n\times n}$.*

Proof Exercise 11.

According to the preceding theorem, if A is similar to B, then B is also similar to A, and we can simply say that A and B are similar. The following theorem describes properties shared by similar matrices.

Theorem 198 *Let F be a field and $A, B \in F^{n\times n}$ be similar. Then:*

1. $p_A(r) = p_B(r)$;
2. $\det(A) = \det(B)$;
3. $\mathrm{tr}(A) = \mathrm{tr}(B)$;
4. *A and B have the same eigenvalues;*
5. *The algebraic and geometric multiplicities of an eigenvalue λ are the same whether λ is regarded as an eigenvalue of A or B.*

Proof If A and B are similar, then there exists an invertible matrix X in $F^{n\times n}$ such that $A = XBX^{-1}$. We recall that

$$p_A(r) = r^n - \mathrm{tr}(A)r^{n-1} + \cdots + (-1)^n \det(A),$$

and similarly for $p_B(r)$. Conclusions 2–4, as well as the conclusion about the algebraic multiplicities of λ, are therefore implied by $p_A(r) = p_B(r)$. Moreover, $p_A(r) = p_B(r)$ follows from the multiplication theorem for determinants and the fact that $\det\left(X^{-1}\right) = \det(X)^{-1}$:

$$\begin{aligned} p_A(r) = \det(rI - A) &= \det\left(rXX^{-1} - XBX^{-1}\right) \\ &= \det\left(X(rI - B)X^{-1}\right) \\ &= \det(X)\det(rI - B)\det\left(X^{-1}\right) \\ &= \det(rI - B) \\ &= p_B(r). \end{aligned}$$

It remains only to show that the geometric multiplicity of an eigenvalue λ is the same, whether it is regarded as an eigenvalue of A or of B. Let λ be an eigenvalue of A (and hence of B), and let $u_1, \ldots, u_k$ be linearly independent eigenvectors of A corresponding to λ. Notice that

$$Au_j = \lambda u_j \Rightarrow XBX^{-1}u_j = \lambda u_j \Rightarrow B\left(X^{-1}u_j\right) = \lambda\left(X^{-1}u_j\right),$$

which shows that $X^{-1}u_1, \ldots, X^{-1}u_k$ are eigenvalues of B. Since X^{-1} defines an injective operator, $\left\{X^{-1}u_1, \ldots, X^{-1}u_k\right\}$ is linearly independent (cf. the proof of Theorem 93 in Section 3.5), which shows that the geometric multiplicity of λ, regarded as an eigenvalue of B, is at least as great as the geometric multiplicity of λ as an eigenvalue of A. An analogous argument shows that if $u_1, \ldots, u_k$ are linearly independent eigenvectors of B corresponding to λ, then $Xu_1, \ldots, Xu_k$ are linearly independent eigenvectors of A corresponding to λ. This shows that the geometric multiplicity of λ, regarded as an eigenvalue of A, is at least as great as the geometric multiplicity of λ as an eigenvalue of B, and completes the proof.

QED

A matrix A is diagonalizable if and only if there is a basis of F^n consisting of eigenvectors of A. The following theorem leads to an equivalent condition for diagonalizability when F is algebraically closed: that the geometric multiplicity of each eigenvalue of A equals its algebraic multiplicity.

Theorem 199 *Let F be a field and let $A \in F^{n\times n}$. If $\lambda \in F$ is an eigenvalue of A, then the geometric multiplicity of λ is less than or equal to the algebraic multiplicity of λ.*

Proof We argue by induction on the dimension n. If $n = 1$, then the result is obvious, so we assume that the result holds for all matrices of dimension $(n-1)\times(n-1)$. Suppose $\lambda \in F$ is an eigenvalue of $A \in F^{n\times n}$ of algebraic multiplicity $k \geq 1$. Let $x \in F^n$ be an eigenvector of A corresponding to λ. Write $x_1 = x$ and choose $x_2, \ldots, x_n$ so that $\{x_1, x_2, \ldots, x_n\}$ is a basis for F^n, and then define

$$X = [x_1|x_2|\cdots|x_n].$$

Since $\{x_1, x_2, \ldots, x_n\}$ is linearly independent, X is invertible. If $y_1, y_2, \ldots, y_n$ are the rows of X^{-1}, then

$$y_i \cdot x_j = \begin{cases} 1, & i = j, \\ 0, & i \neq j. \end{cases}$$

If $B = X^{-1}AX$, then

$$\begin{aligned} B &= \begin{bmatrix} y_1\cdot Ax_1 & y_1\cdot Ax_2 & \cdots & y_1\cdot Ax_n \\ y_2\cdot Ax_1 & y_2\cdot Ax_2 & \cdots & y_2\cdot Ax_n \\ \vdots & \vdots & \ddots & \vdots \\ y_n\cdot Ax_1 & y_n\cdot Ax_2 & \cdots & y_n\cdot Ax_n \end{bmatrix} \\ &= \begin{bmatrix} y_1\cdot(\lambda x_1) & y_1\cdot Ax_2 & \cdots & y_1\cdot Ax_n \\ y_2\cdot(\lambda x_1) & y_2\cdot Ax_2 & \cdots & y_2\cdot Ax_n \\ \vdots & \vdots & \ddots & \vdots \\ y_n\cdot(\lambda x_1) & y_n\cdot Ax_2 & \cdots & y_n\cdot Ax_n \end{bmatrix} \\ &= \begin{bmatrix} \lambda & y_1\cdot Ax_2 & \cdots & y_1\cdot Ax_n \\ 0 & y_2\cdot Ax_2 & \cdots & y_2\cdot Ax_n \\ \vdots & \vdots & \ddots & \vdots \\ 0 & y_n\cdot Ax_2 & \cdots & y_n\cdot Ax_n \end{bmatrix}. \end{aligned}$$

Write $C = B^{(1,1)}$ (this notation was introduced on page 224); then

$$B = \left[\begin{array}{c|c} \lambda & v \\ \hline 0 & C \end{array}\right],$$

where $v \in F^{n-1}$ is defined by $v = (y_1 \cdot Ax_2, \ldots, y_1 \cdot Ax_n)$.

Computing $\det(rI - B)$ by cofactor expansion down the first column shows that

$$p_A(r) = p_B(r) = (r - \lambda)p_C(r).$$

Since $(r - \lambda)^k$ is a factor of $p_A(r)$, we see that $(r - \lambda)^{k-1}$ is a factor of $p_C(r)$ and hence, by the induction hypothesis, λ is an eigenvalue of C of geometric multiplicity at most $k-1$. One eigenvector of B corresponding to λ is $e_1 = (1, 0, \ldots, 0)$; suppose $z_2, \ldots, z_\ell$ are also eigenvectors of B corresponding to λ such that $\{e_1, z_2, \ldots, z_\ell\}$ is linearly independent. We might as well assume that the first components of $z_2, \ldots, z_\ell$ are all zero, since otherwise we could replace z_2 by $z_2 - (z_2)_1 e_1$ and similarly for $z_3, \ldots, z_\ell$. We write

$$z_j = \left[\begin{array}{c} 0 \\ \hline u_j \end{array}\right], \ j = 2, \ldots, \ell,$$

where each $u_j \in F^{n-1}$. It follows that $\{u_2, \ldots, u_\ell\}$ is linearly independent and $Cu_j = \lambda u_j$ for each j:

$$\begin{aligned} Bz_j = \lambda z_j \quad &\Rightarrow \quad \left[\begin{array}{c|c} \lambda & v \\ \hline 0 & C \end{array}\right] \left[\begin{array}{c} 0 \\ \hline u_j \end{array}\right] = \lambda \left[\begin{array}{c} 0 \\ \hline u_j \end{array}\right] \\ &\Rightarrow \quad \left[\begin{array}{c} v \cdot u_j \\ \hline Cu_j \end{array}\right] = \left[\begin{array}{c} 0 \\ \hline \lambda u_j \end{array}\right] \\ &\Rightarrow \quad Cu_j = \lambda u_j. \end{aligned}$$

The induction hypothesis implies that $\ell - 1 \leq k - 1$, that is, $\ell \leq k$. This shows that the geometric multiplicity of λ, as an eigenvalue of B, is at most its algebraic multiplicity. By Theorem 198, the same is true for λ as an eigenvalue of the matrix A.

QED

The following corollary is the first of many results about eigenvalues and eigenvectors that we obtain by assuming the underlying field to be algebraically closed. The only common algebraically closed field is $\mathbf{C}$, the field of complex numbers, and therefore we will state all such results in terms of $\mathbf{C}$.

Corollary 200 *Let $A \in \mathbf{C}^{n\times n}$. Then A is diagonalizable if and only if the geometric multiplicity of each eigenvalue of A equals its algebraic multiplicity.*

Proof Let $\lambda_1, \ldots, \lambda_m \in \mathbf{C}$ be the distinct eigenvalues of A, let $k_1, \ldots, k_m$ be their algebraic multiplicities and $\overline{k}_1, \ldots, \overline{k}_m$ be their geometric multiplicities. Then, because $\mathbf{C}$ is algebraically closed,

$$k_1 + \cdots + k_m = n.$$

On the other hand, by Corollary 195, A is diagonalizable if and only if

$$\overline{k}_1 + \cdots + \overline{k}_m = n.$$

Since $\overline{k}_i \leq k_i$ for each i, it follows that A is diagonalizable if and only if $\overline{k}_i = k_i$, $i = 1, 2, \ldots, m$.

QED

A matrix $A \in F^{n\times n}$ fails to be diagonalizable if there is no basis of F^n consisting of eigenvectors of A. We have a term to describe such a matrix.

Definition 201 *Let F be a field, and suppose $A \in F^{n\times n}$. We say that A is* defective *if there is no basis of F^n consisting of eigenvectors of A.*

By the preceding corollary, $A \in \mathbf{C}^{n\times n}$ is defective if and only if it has an eigenvalue whose geometric multiplicity is strictly less than its algebraic multiplicity.

Example 202 *Let $A \in \mathbf{R}^{3\times 3}$ be defined by*

$$A = \begin{bmatrix} 1 & 0 & 0 \\ 9 & 1 & 3 \\ 3 & 0 & 2 \end{bmatrix}.$$

We saw in Example 190 that A has two eigenvalues, $\lambda_1 = 1$ and $\lambda_2 = 2$. The first eigenvalue has algebraic multiplicity 2, with eigenvectors $(0,1,0)$ and $(-1/3,0,1)$. Thus the geometric multiplicity is also 2. The second eigenvalue has multiplicity 1, and an eigenvector is $(0,1,3)$. Therefore, each eigenvalue has geometric multiplicity equal to its algebraic multiplicity, and Corollary 200 guarantees that A is diagonalizable.

We define

$$X = \begin{bmatrix} 0 & -\frac{1}{3} & 0 \\ 1 & 0 & 1 \\ 0 & 1 & 3 \end{bmatrix}, \; D = \begin{bmatrix} 1 & 0 & 0 \\ 0 & 1 & 0 \\ 0 & 0 & 2 \end{bmatrix}$$

The reader can verify that

$$AX = XD = \begin{bmatrix} 0 & -\frac{1}{3} & 0 \\ 1 & 0 & 2 \\ 0 & 1 & 6 \end{bmatrix},$$

and that X is invertible (both of these facts are guaranteed by the results derived above).

We end this section with a sufficient condition for A to be diagonalizable. The following theorem is really a corollary of Theorem 193.

Theorem 203 *Let F be a field and $A \in F^{n\times n}$. If A has n distinct eigenvalues, then A is diagonalizable.*

Proof Exercise 12.

It is difficult to overestimate the importance of diagonalization, and applications are given in Sections 4.8, 7.3, and 7.5. Diagonalization is useful because it allows the decoupling of variables, a process which we can quickly illustrate. Suppose we wish to solve $Ax = b$, and A is diagonalizable with $A = XDX^{-1}$. Then

$$\begin{aligned} Ax = b \ &\Rightarrow \ \left(XDX^{-1}\right)x = b \\ &\Rightarrow \ XD(X^{-1}x) = b \\ &\Rightarrow \ D(X^{-1}x) = X^{-1}b \\ &\Rightarrow \ Dy = c, \end{aligned}$$

where $y = X^{-1}x$, $c = X^{-1}b$, and the original system has been reduced to the diagonal system $Dy = c$, which is trivial to solve because the variables are decoupled. Having computed y, we then have $x = Xy$. We can thus solve $Ax = b$ by a three-step process (assuming the decomposition $A = XDX^{-1}$ is already known): compute $c = X^{-1}b$, solve the diagonal system $Dy = c$ to get y, and then compute $x = Xy$.

The process just outlined for solving $Ax = b$ is not a practical algorithm, since diagonalizing A is much more costly than solving the system $Ax = b$ by Gaussian elimination. However, the same idea is applied to more complicated problems involving ordinary and partial differential equations (see Sections 4.8 and 7.5), where decoupling variables is critical to the solution process.

Exercises

Essential exercises

1. Corollary 194 implies that if λ_i, λ_j are distinct eigenvalues of $A \in F^{n\times n}$, then
$$E_{\lambda_i}(A) \cap E_{\lambda_j}(A) = \{0\}.$$
Prove this result.

Miscellaneous exercises

In Exercises 2–8, diagonalize the matrix A if possible, that is, find an invertible matrix X and a diagonal matrix D such that $A = XDX^{-1}$. Recall that $A \in \mathbf{R}^{n\times n}$ is regarded as an element of $\mathbf{C}^{n\times n}$ in this context.

2.
$$A = \begin{bmatrix} 1 & 2 \\ 2 & 1 \end{bmatrix} \in \mathbf{R}^{2\times 2}$$

3.
$$A = \begin{bmatrix} 1 & 2 \\ -1 & 1 \end{bmatrix} \in \mathbf{R}^{2\times 2}$$

4.
$$A = \begin{bmatrix} 0 & 1 & 1 \\ 1 & 0 & 1 \\ 1 & 1 & 0 \end{bmatrix} \in \mathbf{R}^{3\times 3}$$

5.
$$A = \begin{bmatrix} 2 & 0 & 0 \\ -6 & 1 & -6 \\ -3 & 0 & -1 \end{bmatrix} \in \mathbf{R}^{3\times 3}$$

6.
$$A = \begin{bmatrix} 0 & 0 & 1 \\ 1 & 0 & -1 \\ 0 & 1 & 1 \end{bmatrix} \in \mathbf{R}^{3\times 3}$$

7.
$$A = \begin{bmatrix} 1 & 0 \\ 1 & 0 \end{bmatrix} \in \mathbf{Z}_2^{2\times 2}$$

8.
$$A = \begin{bmatrix} 1 & 1 & 0 \\ 1 & 1 & 1 \\ 0 & 1 & 1 \end{bmatrix} \in \mathbf{Z}_2^{3\times 3}$$

9. Show that
$$A = \begin{bmatrix} a & b \\ 0 & c \end{bmatrix}$$
is diagonalizable if $a \neq c$, while
$$B = \begin{bmatrix} a & b \\ 0 & a \end{bmatrix}$$
is not diagonalizable if $b \neq 0$.

10. Prove Corollary 194.

11. Prove Theorem 197.

12. Prove Theorem 203.

13. Let F be a finite field. Prove that F is not algebraically closed. (Hint: Let the elements of F be $\alpha_1, \alpha_2, \ldots, \alpha_q$. Construct a polynomial $p(x)$ of degree q that satisfies $p(\alpha_i) = 1$ for all $i = 1, 2, \ldots, q$.)

14. Let F be a field and suppose $A, B \in F^{n\times n}$ are diagonalizable. Suppose further that the two matrices have the same eigenvectors. Prove that A and B commute, that is, that $AB = BA$. (For the converse of this result, see Exercise 5.4.13.)

15. Let F be a field and suppose $A \in F^{n\times n}$ satisfies $A = XDX^{-1}$, where $D \in F^{n\times n}$ is diagonal and $X \in F^{n\times n}$ is invertible. Find a formula for A^k, where k is a positive integer, in terms of D and X. Explain why this formula provides an efficient way to compute A^k when k is large.

16. Let F be a field and suppose $A \in F^{n\times n}$ is diagonalizable. Using the previous exercise, prove that $p_A(A) = 0$. This result is called the Cayley-Hamilton theorem; in Chapter 5, we will show that it holds for every $A \in F^{n\times n}$. (Note: If $p(r) \in F[r]$, then there exist $c_0, c_1, \ldots, c_k \in F$ such that $p(r) = c_0 + c_1 r + \cdots + c_k r^k$.Then $p(A)$ denotes the matrix $c_0 I + c_1 A + \cdots + c_k A^k$, where $I \in F^{n\times n}$ is the identity matrix. Thus $p_A(A) = 0$ means that $p_A(A)$ is the $n \times n$ zero matrix.)

17. The purpose of this exercise is to prove the following theorem: Let $A \in F^{n\times n}$ be diagonalizable, where F is a field, and let $\lambda \in F$ be an eigenvalue of A. Then $\mathcal{N}((A - \lambda I)^2) = \mathcal{N}(A - \lambda I)$.

    (a) Show that $\mathcal{N}(A - \lambda I) \subset \mathcal{N}((A - \lambda I)^2)$.

    (b) Let $A = XDX^{-1}$, where $D \in F^{n\times n}$ is diagonal, and assume that $X = [X_1|X_2]$, where the columns of X_1 form a basis for $E_\lambda(A)$ and the columns of X_2 are eigenvectors corresponding to eigenvalues of A unequal to λ. Prove that if

    $$\mathcal{N}((A - \lambda I)^2) \not\subset \mathcal{N}(A - \lambda I),$$

    then there exists a vector u of the form $u = X_2 v$ such that

    $$(A - \lambda I)u \neq 0, \ (A - \lambda I)^2 u = 0.$$

    (Hint: Any $u \in F^n$ can be written in the form $u = X_1 w + X_2 v$. Show that if

    $$u \in \mathcal{N}((A - \lambda I)^2) \setminus \mathcal{N}(A - \lambda I)$$

    and $u = X_1 w + X_2 v$, then $X_2 v$ also lies in $\mathcal{N}((A-\lambda I)^2) \setminus \mathcal{N}(A-\lambda I)$.)

    (c) Complete the proof by showing that $u = X_2 v$, $(A - \lambda I)^2 u = 0$ imply that $u = 0$. (Why does this give the desired result?)

18. The purpose of this exercise is to work through the proof of Theorem 199 for a specific matrix $A \in \mathbf{R}^{3\times 3}$. Let

    $$A = \begin{bmatrix} 1 & 0 & 0 \\ -5 & 1 & 1 \\ 0 & 0 & 1 \end{bmatrix}$$

    and notice that $r = 1$, $x = (0, 1, 0)$ form an eigenpair of A.

    (a) Write $x_1 = x$ and extend $\{x_1\}$ to a basis $\{x_1, x_2, x_3\}$ for $\mathbf{R}^3$.

(b) Define $X = [x_1|x_2|x_3]$ and compute the matrix $B = X^{-1}AX$. What are the vector v and the matrix C from the proof of Theorem 199?

(c) Verify that $e_1 = (1, 0, 0)$ is an eigenvector of B corresponding to the eigenvalue $r = 1$.

(d) Find another eigenvector z of B, where the first component of z is zero. Let

$$z = \left[\begin{array}{c} 0 \\ \hline u \end{array}\right],$$

where $u \in \mathbf{R}^2$. Verify that u is an eigenvector of C corresponding to the eigenvalue $r = 1$.

(e) Find another eigenvector $v \in \mathbf{R}^2$ of C, so that $\{u, v\}$ is linearly independent. Write

$$w = \left[\begin{array}{c} 0 \\ \hline v \end{array}\right].$$

Is w another eigenvector of B?

4.7 Eigenvalues of linear operators

We now extend our results about the eigenvalues and eigenvectors of matrices to general linear operators on finite-dimensional spaces. We consider a linear operator $T : V \to V$, where V is a finite-dimensional vector space over a field F. By definition, $\lambda \in F$ and $v \in V$ form an eigenpair of T if $v \neq 0$ and $T(v) = \lambda v$. However, instead of looking for the eigenvalues and eigenvectors of T directly, we can represent T by a matrix and use the results of the previous sections.

If $\mathcal{U} = \{u_1, u_2, \ldots, u_n\}$ is a basis for V, then we can identify $v \in V$ with $x = [v]_{\mathcal{U}} \in F^n$, where

$$v = x_1u_1 + x_2u_2 + \ldots + x_nu_n$$

is the unique representation of v in terms of the given basis. We can then represent T by the matrix $A = [T]_{\mathcal{U},\mathcal{U}} \in F^{n\times n}$, where

$$T(x_1u_1 + \ldots + x_nu_n) = y_1u_1 + \ldots + y_nu_n \Leftrightarrow Ax = y. \tag{4.10}$$

Equivalently, A is defined by

$$T(v) = w \Leftrightarrow A[v]_{\mathcal{U}} = [w]_{\mathcal{U}} \text{ for all } v \in V.$$

The reader should notice that A can be computed from the values of

$$T(u_1), T(u_2), \ldots, T(u_n).$$

Indeed, since $[u_j]_{\mathcal{U}} = e_j$, we have

$$A_j = Ae_j = A[u_j]_{\mathcal{U}} = [T(u_j)]_{\mathcal{U}}.$$

Therefore, if

$$T(u_j) = a_{1j}u_1 + a_{2j}u_2 + \ldots + a_{nj}u_n$$

(that is, $[T(u_j)]_{\mathcal{U}} = (a_{1j}, a_{2j}, \ldots, a_{nj})$), then

$$A_j = \begin{bmatrix} a_{1j} \\ a_{2j} \\ \vdots \\ a_{nj} \end{bmatrix}.$$

Thus

$$A = \begin{bmatrix} a_{11} & a_{12} & \cdots & a_{1n} \\ a_{21} & a_{22} & \cdots & a_{2n} \\ \vdots & \vdots & \ddots & \vdots \\ a_{n1} & a_{n2} & \cdots & a_{nn} \end{bmatrix}.$$

Example 204 *Let $T : \mathcal{P}_2 \rightarrow \mathcal{P}_2$ be defined by $T(p)(x) = p(ax + b)$, where $a, b \in \mathbf{R}$ are constants. Then*

$$\begin{aligned} T(1) &= 1, \\ T(x) &= b + ax, \\ T(x^2) &= (b + ax)^2 = b^2 + 2abx + a^2x^2. \end{aligned}$$

Therefore, using the standard basis $\mathcal{S} = \{1, x, x^2\}$ on $\mathcal{P}_2$, we see that the matrix of T is

$$A = [T]_{\mathcal{S},\mathcal{S}} = \begin{bmatrix} 1 & b & b^2 \\ 0 & a & 2ab \\ 0 & 0 & a^2 \end{bmatrix}.$$

If $\lambda \in F$, $v \in V$ is an eigenpair of T, then

$$T(x) = \lambda x \Rightarrow [T(v)]_{\mathcal{U}} = [\lambda v]_{\mathcal{U}} \Rightarrow [T]_{\mathcal{U},\mathcal{U}}[v]_{\mathcal{U}} = \lambda[v]_{\mathcal{U}}.$$

Thus, with $A = [T]_{\mathcal{U},\mathcal{U}}$, $x = [v]_{\mathcal{U}}$, we have $Ax = \lambda x$.

Conversely, if $A = [T]_{\mathcal{U},\mathcal{U}}$, $x \in F^n$, $\lambda \in F$ satisfy $Ax = \lambda x$, and we define $v \in V$ by

$$v = x_1u_1 + x_2u_2 + \cdots + x_nu_n,$$

then

$$Ax = \lambda x \Rightarrow [T]_{\mathcal{U},\mathcal{U}}[v]_{\mathcal{U}} = \lambda[v]_{\mathcal{U}} \Rightarrow [T(v)]_{\mathcal{U}} = [\lambda v]_{\mathcal{U}} \Rightarrow T(v) = \lambda v.$$

This shows that T and $[T]_{\mathcal{U},\mathcal{U}}$ have the same eigenvalues.

None of the above reasoning depends on the particular basis chosen for

V, so it must be the case that, if $A \in F^{n\times n}$ and $B \in F^{n\times n}$ are matrices representing T with respect to two different bases for V, then A and B have the same eigenvalues. The reader might suspect that A and B must be similar. The following result shows that this is true.

Theorem 205 *Let V be an n-dimensional vector space over a field F, and let $T : V \to V$ be linear. Let $\mathcal{U} = \{u_1, u_2, \ldots, u_n\}$ and $\mathcal{W} = \{w_1, w_2, \ldots, w_n\}$ be bases for V, and let $A = [T]_{\mathcal{U},\mathcal{U}}$, $B = [T]_{\mathcal{W},\mathcal{W}}$. Then A and B are similar.*

Proof Since $\{w_1, w_2, \ldots, w_n\}$ is a basis for V, each vector in the basis

$$\{u_1, u_2, \ldots, u_n\}$$

can be represented as a linear combination of $w_1, w_2, \ldots, w_n$:

$$u_j = X_{1j}w_1 + X_{2j}w_2 + \ldots + X_{nj}w_n, \ j = 1, 2, \ldots, n.$$

Let $X \in F^{n\times n}$ be defined by the scalars X_{ij}, $i, j = 1, 2, \ldots, n$. The matrix X is called the change of coordinates matrix from $\mathcal{U}$ to $\mathcal{W}$. As shown in Exercise 3.6.23, X is invertible and satisfies

$$[v]_{\mathcal{W}} = X[v]_{\mathcal{U}} \text{ for all } v \in V.$$

It follows that for any $v \in V$. we have

$$[T(v)]_{\mathcal{W}} = X[T(v)]_{\mathcal{U}} = XA[v]_{\mathcal{U}} = XAX^{-1}[v]_{\mathcal{W}} \text{ for all } v \in V.$$

However, we also have

$$[T(v)]_{\mathcal{W}} = B[v]_{\mathcal{W}} \text{ for all } v \in V.$$

Therefore,

$$B[v]_{\mathcal{W}} = XAX^{-1}[v]_{\mathcal{W}} \text{ for all } v \in V,$$

that is,

$$By = XAX^{-1}y \text{ for all } y \in \mathbf{R}^n.$$

It follows that $B = XAX^{-1}$, and thus A and B are similar.

QED

The previous theorem, together with Theorem 198, allows us to make the following definitions.

Definition 206 *Let V be a n-dimensional vector space over a field F, and let $T : V \to V$ be linear. Let $\{u_1, u_2, \ldots, u_n\}$ be any basis for V, and let $A \in F^{n\times n}$ be the matrix representing T with respect to this basis.*

1. *The* characteristic polynomial *of T is defined to be $p_A(r)$.*
2. *The* determinant *of T is defined to be* $\det(A)$.

3. *The* trace *of T is defined to be* $\mathrm{tr}(A)$.

Each of these terms is well-defined by the theorems mentioned above.

Example 207 *Let T be the operator defined in Example 204, where a is assumed to be different from 0 and 1. Using the matrix $A = [T]_{\mathcal{S},\mathcal{S}}$ computed in that example, we see that the eigenvalues of T are $\lambda_1 = 1$, $\lambda_2 = a$, $\lambda_3 = a^2$. Since $a \neq 0$, $a \neq 1$ by assumption, we see that T has three distinct eigenvalues. The corresponding eigenvectors are the polynomials*

$$p_1 = 1,\ p_2 = \frac{b}{a-1} + x,\ p_3 = \frac{b^2}{(a-1)^2} + \frac{2b}{a-1}x + x^2.$$

These can be found by computing the eigenvectors of A by the usual method and interpreting the results in terms of the basis $\mathcal{S}$.

Given a linear operator $T : V \rightarrow V$, where V is finite-dimensional, we often wish to choose a basis $\mathcal{U}$ for V so that the matrix $[T]_{\mathcal{U},\mathcal{U}}$ is as simple as possible. The following result is an easy consequence of our discussion in this section.

Theorem 208 *Let V be a finite-dimensional vector space over a field F, and let $T : V \rightarrow V$ be linear. Let $\mathcal{U}$ be a basis for V and define $A = [T]_{\mathcal{U},\mathcal{U}}$. Then A is diagonal if and only if $\mathcal{U}$ consists of eigenvectors of T.*

Proof Write $\mathcal{U} = \{u_1, u_2, \ldots, u_n\}$. We note that A is diagonal if and only if there exists $\lambda_1, \lambda_2, \ldots, \lambda_n \in F$ such that $Ae_j = \lambda_j e_j$ for each j. But

$$\begin{aligned} Ae_j = \lambda_j e_j \ \Leftrightarrow\ [T]_{\mathcal{U},\mathcal{U}}[u_j]_{\mathcal{U}} = \lambda_j [u]_{\mathcal{U}} \ &\Leftrightarrow\ [T(u_j)]_{\mathcal{U}} = [\lambda_j u_j]_{\mathcal{U}} \\ &\Leftrightarrow\ T(u_j) = \lambda_j u_j. \end{aligned}$$

Thus A is diagonal if and only if each vector in the basis $\mathcal{U}$ is an eigenvector of T.

QED

We also point out the following important result:

Theorem 209 *Let V be a nontrivial finite-dimensional vector space over $\mathbf{C}$, and suppose $T : V \rightarrow V$ is linear. Then T has an eigenvalue/eigenvector pair.*

Proof Since V is nontrivial, it has a basis $\mathcal{U}$, and the eigenvalues of T are the eigenvalues of the matrix $[T]_{\mathcal{U},\mathcal{U}} \in \mathbf{C}^{n\times n}$ ($n = \dim(V)$). Every matrix in $\mathbf{C}^{n\times n}$ has an eigenvalue since $\mathbf{C}$ is algebraically closed.

QED

Exercises

Miscellaneous exercises

1. Let $T : \mathbf{R}^3 \to \mathbf{R}^3$ be defined by
$$T(x) = (ax_1 + bx_2, bx_1 + ax_2 + bx_3, bx_2 + ax_3),$$
where $a, b \in \mathbf{R}$ are constants. Prove that there exists a basis $\mathcal{X}$ of $\mathbf{R}^3$ under which $[T]_{\mathcal{X},\mathcal{X}}$ is diagonal.

2. Let $T : \mathbf{R}^3 \to \mathbf{R}^3$ be defined by
$$T(x) = (x_1 + x_2 + x_3, x_2, x_1 + x_3).$$
Find a basis $\mathcal{X}$ for $\mathbf{R}^3$ such that $[T]_{\mathcal{X},\mathcal{X}}$ is diagonal.

3. Let $T : \mathbf{R}^3 \to \mathbf{R}^3$ be defined by
$$T(x) = (x_1 + x_2 + x_3, ax_2, x_1 + x_3),$$
where $a \in \mathbf{R}$ is a constant. For which values of a does there exist a basis $\mathcal{X}$ for $\mathbf{R}^3$ such that $[T]_{\mathcal{X},\mathcal{X}}$ is diagonal?

4. Compute $T(p_i)$, $i = 1, 2, 3$, in Example 207 and verify directly that $T(p_i) = \lambda_i p_i$ for each i.

5. Let T be the operator in Example 204 with $a = 1$ and $b \neq 0$. Show that there is no basis $\mathcal{U}$ for $\mathcal{P}_2$ such that $[T]_{\mathcal{U},\mathcal{U}}$ is diagonalizable.

6. Let $T : \mathcal{P}_n \to \mathcal{P}_n$ be defined by $T(p) = q$, where $q(x) = xp'(x)$. Find the characteristic polynomial and all eigenpairs of T.

7. Let $D : \mathcal{P}_n \to \mathcal{P}_n$ be the differentiation operator: $D(p) = p'$ for all $p \in \mathcal{P}_n$. Find the characteristic polynomial and all eigenpairs of D.

8. Consider the operator T defined in Example 204 and the alternate basis $\mathcal{B} = \{1 - x, 1 + x, 1 - x^2\}$ for $\mathcal{P}_2$. Compute $[T]_{\mathcal{B},\mathcal{B}}$ and verify that the eigenvalues and eigenvectors of T are unchanged when computed using $\mathcal{B}$ in place of $\mathcal{S}$.

9. Let $L : \mathbf{C}^3 \to \mathbf{C}^3$ be defined by
$$L(z) = (z_1, 2z_2, z_1 + z_3).$$
Show that there is no basis $\mathcal{X}$ for $\mathbf{C}^3$ such that $[L]_{\mathcal{X},\mathcal{X}}$ is diagonal.

10. Let $T : \mathbf{C}^3 \to \mathbf{C}^3$ be defined by
$$T(x) = (3x_1 + x_3, 8x_1 + x_2 + 4x_3, -9x_1 + 2x_2 + x_3).$$
Show that there is no basis $\mathcal{X}$ for $\mathbf{C}^3$ such that $[T]_{\mathcal{X},\mathcal{X}}$ is diagonal.

11. Let $T : \mathbf{Z}_2^2 \to \mathbf{Z}_2^2$ be defined by $T(x) = (0, x_1 + x_2)$. Find a basis $\mathcal{X}$ for $\mathbf{Z}_2^2$ such that $[T]_{\mathcal{X},\mathcal{X}}$ is diagonal.

12. Let $T : \mathbf{Z}_3^3 \to \mathbf{Z}_3^3$ be defined by

$$T(x) = (2x_1 + x_2, x_1 + x_2 + x_3, x_2 + 2x_3).$$

Show that there is no basis $\mathcal{U}$ for $\mathbf{Z}_3^3$ such that $[T]_{\mathcal{U},\mathcal{U}}$ is diagonal.

13. Let X be a finite-dimensional vector space over $\mathbf{C}$ with basis

$$\mathcal{X} = \{u_1, \ldots, u_k, v_1, \ldots, v_\ell\}.$$

Let U and V be the subspaces $\text{sp}\{u_1, \ldots, u_k\}$ and $\text{sp}\{v_1, \ldots, v_\ell\}$, respectively. Let $T : X \to X$ be a linear operator with the property $T(u) \in U$ for each $u \in U$. (We say that U is *invariant* under T in this case.)

(a) Prove that there exists an eigenvector of T belonging to U.

(b) Prove that $[T]_{\mathcal{X},\mathcal{X}}$ is a *block upper triangular matrix*, that is, it has the form

$$[T]_{\mathcal{X},\mathcal{X}} = \begin{bmatrix} A & B \\ 0 & C \end{bmatrix} \in \mathbf{C}^{n\times n}, \tag{4.11}$$

where $A \in \mathbf{C}^{k\times k}$, $B \in \mathbf{C}^{k\times \ell}$, $C \in \mathbf{C}^{\ell\times\ell}$, and 0 is the $\ell \times k$ matrix of all zeros.

(c) Show that if V is also invariant under T, then $[T]_{\mathcal{X},\mathcal{X}}$ is *block diagonal*, that is, $B = 0$ in (4.11).

14. Let $A \in \mathbf{C}^{n\times n}$ be an arbitrary matrix. Recall that $B \in \mathbf{C}^{n\times n}$ is called upper triangular if $B_{ij} = 0$ for $i > j$. Prove that A is similar to an upper triangular matrix B. (Hint: Argue by induction on n. Let $\lambda \in \mathbf{C}$, $x_1 \in \mathbf{C}^n$ be an eigenpair of A, and extend $\{x_1\}$ to a basis $\{x_1, x_2, \ldots, x_n\}$ of $\mathbf{C}^n$. Defining $X = [x_1|x_2|\cdots|x_n]$, show that

$$X^{-1}AX = \left[\begin{array}{c|c} \lambda & v \\ \hline 0 & B \end{array}\right],$$

where $v \in \mathbf{C}^{n-1}$ and $B \in \mathbf{C}^{(n-1)\times(n-1)}$. Apply the induction hypothesis to B.)

15. Suppose V is a finite-dimensional vector space over $\mathbf{C}$ and $T : V \to \mathcal{V}$ is linear. Use the previous exercise to prove that there exists a basis $\mathcal{U}$ for V such that $[T]_{\mathcal{U},\mathcal{U}}$ is upper triangular. Why is it necessary (in this exercise and the previous one) to assume that the underlying field is $\mathbf{C}$?

16. Let T be the operator from Exercise 10. Find a basis $\mathcal{X}$ for $\mathbf{C}^3$ such that $[T]_{\mathcal{X},\mathcal{X}}$ is upper triangular (cf. the previous two exercises).

4.8 Systems of linear ODEs

As an application of diagonalization, we will study systems of linear ordinary differential equations (ODEs) with constant coefficients. Here the problem is to find functions $u_1 = u_1(t), \ldots, u_n = u_n(t)$ satisfying

$$\begin{aligned} u_1' &= A_{11}u_1 + A_{12}u_2 + \ldots + A_{1n}u_n, \; u_1(0) = v_1 \\ u_2' &= A_{21}u_1 + A_{22}u_2 + \ldots + A_{2n}u_n, \; u_2(0) = v_2 \\ \vdots \; &\quad \vdots \\ u_n' &= A_{n1}u_1 + A_{n2}u_2 + \ldots + A_{nn}u_n, \; u_n(0) = v_n \end{aligned} \tag{4.12}$$

where $A_{ij} \in \mathbf{R}$, $i, j = 1, 2, \ldots, n$, and $v_1, v_2, \ldots, v_n \in \mathbf{R}$ are given scalars. If we write u for the vector-valued function $u : \mathbf{R} \to \mathbf{R}^n$ defined by

$$u(t) = (u_1(t), u_2(t), \ldots, u_n(t)),$$

and $v = (v_1, v_2, \ldots, v_n)$, then (4.12) can be written in matrix-vector form as

$$u' = Au, \; u(0) = v. \tag{4.13}$$

The problem (4.12) or (4.13) is called an *initial value problem* (IVP), and we will show how to solve the IVP under the assumption that A is diagonalizable. The case in which A is not diagonalizable is discussed in Section 5.5.

Let us suppose there exist an invertible matrix $X \in \mathbf{R}^{n \times n}$ and a diagonal matrix $D \in \mathbf{R}^{n \times n}$ such that $A = XDX^{-1}$. As we have seen, this implies that the columns $x_1, x_2, \ldots, x_n$ of X are eigenvectors of A, and the diagonal entries $\lambda_1, \lambda_2, \ldots, \lambda_n$ are the corresponding eigenvalues. We are assuming for now that A can be diagonalized using only real matrices, that is, that the eigenvalues of A are real. We will discuss below the case in which it is necessary to move into the complex domain in order to diagonalize A.

It is straightforward to use the factorization $A = XDX^{-1}$ to find the general solution of $u' = Au$. We need two facts. First of all, if α is a real number, then the general solution of the scalar differential equation $v' = \alpha v$ is $v(t) = ce^{\alpha t}$, where c is an arbitrary scalar. (Thus the solution space is one-dimensional and is spanned by the function defined by $e^{\alpha t}$.) Second, if B is any constant matrix and $v = v(t)$ is a vector-valued function ($v : \mathbf{R} \to \mathbf{R}^n$), then $(Bv)' = Bv'$. It is easy to prove this from the fact that the derivative operator is linear.

We now reason as follows:

$$\begin{aligned} u' = Au \Leftrightarrow u' = \left(XDX^{-1}\right)u \;&\Leftrightarrow\; X^{-1}u' = D\left(X^{-1}u\right) \\ &\Leftrightarrow\; \left(X^{-1}u\right)' = D\left(X^{-1}u\right) \\ &\Leftrightarrow\; y' = Dy, \end{aligned}$$

where $y = y(t)$ is defined by $y = X^{-1}u$. The system of differential equations appearing in (4.12) is therefore equivalent to the decoupled system

$$\begin{aligned} y_1' &= \lambda_1 y_1, \\ y_2' &= \lambda_2 y_2, \\ &\vdots \\ y_n' &= \lambda_n y_n. \end{aligned}$$

The general solution of $y' = Dy$ is

$$y(t) = \begin{bmatrix} c_1 e^{\lambda_1 t} \\ c_2 e^{\lambda_2 t} \\ \vdots \\ c_n e^{\lambda_n t} \end{bmatrix} = c_1 \begin{bmatrix} e^{\lambda_1 t} \\ 0 \\ \vdots \\ 0 \end{bmatrix} + c_2 \begin{bmatrix} 0 \\ e^{\lambda_2 t} \\ \vdots \\ 0 \end{bmatrix} + \cdots + c_n \begin{bmatrix} 0 \\ 0 \\ \vdots \\ e^{\lambda_n t} \end{bmatrix}.$$

Since $y = X^{-1}x$, we have $x = Xy$, and therefore the general solution of $u' = Au$ is

$$\begin{aligned} x = Xy &= c_1 X \begin{bmatrix} e^{\lambda_1 t} \\ 0 \\ \vdots \\ 0 \end{bmatrix} + c_2 X \begin{bmatrix} 0 \\ e^{\lambda_2 t} \\ \vdots \\ 0 \end{bmatrix} + \cdots + c_n X \begin{bmatrix} 0 \\ 0 \\ \vdots \\ e^{\lambda_n t} \end{bmatrix} \\ &= c_1 e^{\lambda_1 t} x_1 + c_2 e^{\lambda_2 t} x_2 + \cdots + c_n e^{\lambda_n t} x_n, \end{aligned}$$

which expresses the general solution in terms of the eigenvalues and eigenvectors of A. Incidentally, even if A is not diagonalizable, given an eigenpair λ, x of A, the function $u(t) = e^{\lambda t}x$ defines a solution to $u' = Au$, as can be verified directly (see Exercise 2).

We can write the equation $u' = Au$ as a linear operator equation by defining $L : C^1(\mathbf{R};\mathbf{R}^n) \to C(\mathbf{R};\mathbf{R}^n)$ by $L(u) = u' - Au$. In the case that A is diagonalizable with real eigenvalues, we have proved that $\dim(\ker(L)) = n$ and we have found a basis for $\ker(L)$:

$$\left\{ e^{\lambda_1 t} x_1, e^{\lambda_2 t} x_2, \ldots, e^{\lambda_n t} x_n \right\}.$$

Example 210 *Let*

$$A = \begin{bmatrix} 0 & 1 \\ 1 & 0 \end{bmatrix}.$$

The reader can verify that $\lambda_1 = 1$ and $\lambda_2 = -1$ are the eigenvalues of A, and that $x_1 = (1,1)$ and $x_2 = (1,-1)$ are corresponding eigenvectors (thus A is diagonalizable, since the eigenvectors form a basis for $\mathbf{R}^2$). The general solution of $u' = Au$ is

$$u(t) = c_1 e^t \begin{bmatrix} 1 \\ 1 \end{bmatrix} + c_2 e^{-t} \begin{bmatrix} 1 \\ -1 \end{bmatrix} = \begin{bmatrix} c_1 e^t + c_2 e^{-t} \\ c_1 e^t - c_2 e^{-t} \end{bmatrix}.$$

4.8.1 Complex eigenvalues

If A is diagonalizable but has complex eigenvalues, then the above reasoning applies, but the general solution is expressed in terms of complex-valued functions of the form

$$e^{\lambda t}x = e^{(\mu+i\theta)t}(p+iq),$$

where $\lambda = \mu+i\theta$ is an eigenvalue and $x = p+iq$ is a corresponding eigenvector ($p, q \in \mathbf{R}^n$). If A has real entries, it might be undesirable to have complex solutions; fortunately, it is easy to obtain real-valued solutions because complex eigenvalues of a real matrix occur in conjugate pairs. If $\lambda = \mu+i\theta$ is a complex eigenvalue of A with eigenvector $x = p+iq$, then $\overline{\lambda} = \mu - i\theta$, $\overline{x} = p - iq$ is another eigenpair of A. By Euler's formula,

$$e^{(\mu+i\theta)t} = e^{\mu t}\left(\cos\left(\theta t\right) + i\sin\left(\theta t\right)\right)$$

and therefore the solutions to $u' = Au$ corresponding to the eigenvalues $\mu \pm i\theta$ are

$$\begin{aligned} e^{(\mu+i\theta)t}(p+iq) &= e^{\mu t}\left(\cos\left(\theta t\right)p - \sin\left(\theta t\right)q\right) + ie^{\mu t}\left(\sin\left(\theta t\right)p + \cos\left(\theta t\right)q\right), \\ e^{(\mu-i\theta)t}(p-iq) &= e^{\mu t}\left(\cos\left(\theta t\right)p - \sin\left(\theta t\right)q\right) - ie^{\mu t}\left(\sin\left(\theta t\right)p + \cos\left(\theta t\right)q\right). \end{aligned}$$

By linearity, any linear combination of two solutions is another solution, and the reader can verify that

$$\begin{aligned} \frac{1}{2}e^{\lambda t}x + \frac{1}{2}e^{\overline{\lambda}t}\overline{x} &= e^{\mu t}\left(\cos\left(\theta t\right)p - \sin\left(\theta t\right)q\right) \\ -\frac{i}{2}e^{\lambda t}x + \frac{i}{2}e^{\overline{\lambda}t}\overline{x} &= e^{\mu t}\left(\sin\left(\theta t\right)p + \cos\left(\theta t\right)q\right). \end{aligned}$$

The reader should notice that these real-valued solutions are simply the real and imaginary parts of the complex-valued solution $e^{\lambda t}x$.

In forming the general solution of $u' = Au$, if $\lambda_i = \lambda = \mu + i\theta$ and $\lambda_{i+1} = \overline{\lambda} = \mu - i\theta$, then we can replace

$$c_i e^{\lambda_i t}x_i + c_{i+1}e^{\lambda_{i+1}t}x_{i+1}$$

by

$$c_i e^{\mu t}\left(\cos\left(\theta t\right)p - \sin\left(\theta t\right)q\right) + c_{i+1}e^{\mu t}\left(\sin\left(\theta t\right)p + \cos\left(\theta t\right)q\right),$$

where $x_i = p + iq$ is the (complex) eigenvalue corresponding to λ. It can be proved directly that

$$\left\{e^{\mu t}\left(\cos\left(\theta t\right)p - \sin\left(\theta t\right)q\right), e^{\mu t}\left(\sin\left(\theta t\right)p + \cos\left(\theta t\right)q\right)\right\}$$

is a linearly independent set (see Exercise 3).

Example 211 *Let*

$$A = \begin{bmatrix} 0 & 1 \\ -1 & 0 \end{bmatrix}.$$

The eigenvalues of A are $\pm i$, with corresponding eigenvectors $(\mp i, 1)$. With $\lambda = i$, $x = (-i, 1)$, we have

$$e^{\lambda t}x = (\cos(t) + i\sin(t))\begin{bmatrix} -i \\ 1 \end{bmatrix} = \begin{bmatrix} \sin(t) \\ \cos(t) \end{bmatrix} + i\begin{bmatrix} -\cos(t) \\ \sin(t) \end{bmatrix}.$$

The real and imaginary parts of $e^{\lambda t}x$ are real-valued solutions of $u' = Au$, and the general solution is therefore

$$u(t) = c_1\begin{bmatrix} \sin(t) \\ \cos(t) \end{bmatrix} + c_2\begin{bmatrix} -\cos(t) \\ \sin(t) \end{bmatrix} = \begin{bmatrix} c_1\sin(t) - c_2\cos(t) \\ c_1 cos(t) + c_2\sin(t) \end{bmatrix}.$$

If $A \in \mathbf{C}^{n\times n}$, then all of the discussion above still applies, except that it is not necessarily possible to produce real-valued solutions (because complex eigenvalues need not occur in conjugate pairs when $A \in \mathbf{C}^{n\times n}$). In the rest of this section, we allow A to belong to $\mathbf{C}^{n\times n}$.

4.8.2 Solving the initial value problem

With the general solution of $u' = Au$ expressed as

$$u(t) = c_1e^{\lambda_1 t}x_1 + c_2e^{\lambda_2 t}x_2 + \cdots + c_ne^{\lambda_n t}x_n,$$

we have

$$u(0) = c_1x_1 + c_2x_2 + \cdots + c_nx_n = Xc,$$

where $c = (c_1, c_2, \ldots, c_n)$. If we want to solve the IVP (4.13), that is, to find u satisfying $u(0) = v$, we just solve $Xc = v$ to get $c = X^{-1}v$.

Example 212 *Consider the IVP*

$$\begin{aligned} u_1' &= u_2, \ u_1(0) = 1, \\ u_2' &= u_1, \ u_2(0) = 0. \end{aligned}$$

This is equivalent to $u' = Au$, $u(0) = e_1$, where A is the matrix of Example 182. We have already computed the general solution of $u' = Au$:

$$u(t) = c_1e^t\begin{bmatrix} 1 \\ 1 \end{bmatrix} + c_2e^{-t}\begin{bmatrix} 1 \\ -1 \end{bmatrix}.$$

We must find $c \in \mathbf{R}^2$ satisfying

$$Xc = e_1 \Leftrightarrow \begin{bmatrix} 1 & 1 \\ 1 & -1 \end{bmatrix}\begin{bmatrix} c_1 \\ c_2 \end{bmatrix} = \begin{bmatrix} 1 \\ 0 \end{bmatrix}.$$

The solution is easily found: $c_1 = c_2 = 1/2$. Thus the solution of the IVP is

$$u(t) = \frac{1}{2}e^t\begin{bmatrix} 1 \\ 1 \end{bmatrix} + \frac{1}{2}e^{-t}\begin{bmatrix} 1 \\ -1 \end{bmatrix} = \begin{bmatrix} \frac{1}{2}e^t + \frac{1}{2}e^{-t} \\ \frac{1}{2}e^t - \frac{1}{2}e^{-t} \end{bmatrix}.$$

4.8.3 Linear systems in matrix form

We can write the general solution

$$u(t) = c_1 e^{\lambda_1 t} x_1 + c_2 e^{\lambda_2 t} x_2 + \cdots + c_n e^{\lambda_n t} x_n$$

of $u' = Au$ differently as follows:

$$\begin{aligned} u(t) &= c_1 e^{\lambda_1 t} x_1 + c_2 e^{\lambda_2 t} x_2 + \cdots + c_n e^{\lambda_n t} x_n \\ &= [x_1|x_2|\cdots|x_n] \begin{bmatrix} e^{\lambda_1 t} & & & \\ & e^{\lambda_2 t} & & \\ & & \ddots & \\ & & & e^{\lambda_n t} \end{bmatrix} \begin{bmatrix} c_1 \\ c_2 \\ \vdots \\ c_n \end{bmatrix}. \end{aligned}$$

Defining

$$E(t) = \begin{bmatrix} e^{\lambda_1 t} & & & \\ & e^{\lambda_2 t} & & \\ & & \ddots & \\ & & & e^{\lambda_n t} \end{bmatrix} \tag{4.14}$$

this formula simplifies to $u(t) = XE(t)c$. Moreover, with $c = X^{-1}v$, the solution of the IVP (4.13) can be written as

$$u(t) = XE(t)X^{-1}v. \tag{4.15}$$

We will now explore the properties of the matrix-valued functions $E(t)$ and $XE(t)X^{-1}$. We have

$$\begin{aligned} E'(t) &= \begin{bmatrix} \lambda_1 e^{\lambda_1 t} & & & \\ & \lambda_2 e^{\lambda_2 t} & & \\ & & \ddots & \\ & & & \lambda_n e^{\lambda_n t} \end{bmatrix} \\ &= \begin{bmatrix} \lambda_1 & & & \\ & \lambda_2 & & \\ & & \ddots & \\ & & & \lambda_n \end{bmatrix} \begin{bmatrix} e^{\lambda_1 t} & & & \\ & e^{\lambda_2 t} & & \\ & & \ddots & \\ & & & e^{\lambda_n t} \end{bmatrix} = DE(t) \end{aligned}$$

and also $E(0) = I$. Therefore, E solves the matrix IVP

$$E' = DE, \; E(0) = I.$$

If we define $U(t) = XE(t)X^{-1}$, then

$$U' = XE'X^{-1} = XDEX^{-1} = XDX^{-1}XEX^{-1} = AU,$$

since $A = XDX^{-1}$ (the reader should notice the trick of inserting $I = X^{-1}X$ into the middle of the product to produce this simplification). Moreover,

$$U(0) = XE(0)X^{-1} = XIX^{-1} = I,$$

and therefore U satisfies the matrix IVP

$$U' = AU, \; U(0) = I. \tag{4.16}$$

Definition 213 *Let $A \in \mathbf{C}^{n\times n}$ be given. We say that $U \in \mathbf{C}^1(\mathbf{R}; \mathbf{C}^{n\times n})$ is a* fundamental solution *of the differential equation*

$$U' = AU$$

if U is a solution and $U(t)$ is invertible for all $t \in \mathbf{R}$.

If a fundamental solution U satisfies the initial condition $U(0) = I$, then U is called the matrix exponential *of A and we write $U(t) = e^{tA}$.*

It is straightforward to show that both $e^{tD} = E(t)$ and $e^{tA} = XE(t)X^{-1}$ are invertible for all t (see Exercise 4). The scalar exponential function $u(t) = e^{at}$ satisfies

$$u' = au, \; u(0) = 1,$$

which accounts for the terminology and the notation e^{tA} for the matrix exponential.

We have shown that, for a diagonal matrix D, the matrix exponential of D is $E(t)$, where E is defined by (4.14); that is,

$$e^{tD} = \begin{bmatrix} e^{\lambda_1 t} & & & \\ & e^{\lambda_2 t} & & \\ & & \ddots & \\ & & & e^{\lambda_n t} \end{bmatrix}.$$

Also, if A is a diagonalizable matrix and $A = XDX^{-1}$, then $e^{tA} = XE(t)X^{-1}$, that is,

$$e^{tA} = Xe^{tD}X^{-1}.$$

This formula is valid for any $A \in \mathbf{C}^{n\times n}$ that is diagonalizable. The solution to the IVP

$$u' = Au, \; u(0) = v$$

is $u(t) = e^{tA}v$ (cf. (4.15).

Example 214 *Let A be the matrix*

$$A = \begin{bmatrix} 0 & 1 \\ 1 & 0 \end{bmatrix}$$

of Example 210. We have $A = XDX^{-1}$*, where*

$$X = \begin{bmatrix} 1 & 1 \\ 1 & -1 \end{bmatrix}, \ D = \begin{bmatrix} 1 & 0 \\ 0 & -1 \end{bmatrix}.$$

The reader can verify that

$$X^{-1} = \begin{bmatrix} \frac{1}{2} & \frac{1}{2} \\ \frac{1}{2} & -\frac{1}{2} \end{bmatrix}.$$

Therefore,

$$\begin{aligned} e^{tA} = Xe^{tD}X^{-1} &= \begin{bmatrix} 1 & 1 \\ 1 & -1 \end{bmatrix} \begin{bmatrix} e^t & 0 \\ 0 & e^{-t} \end{bmatrix} \begin{bmatrix} \frac{1}{2} & \frac{1}{2} \\ \frac{1}{2} & -\frac{1}{2} \end{bmatrix} \\ &= \begin{bmatrix} \frac{1}{2}\left(e^t + e^{-t}\right) & \frac{1}{2}\left(e^t - e^{-t}\right) \\ \frac{1}{2}\left(e^t - e^{-t}\right) & \frac{1}{2}\left(e^t + e^{-t}\right) \end{bmatrix}. \end{aligned}$$

If $A \in \mathbf{R}^{n\times n}$, then e^{tA} has real entries for all $t \in \mathbf{R}$, even if A has complex eigenvalues (and therefore $X \in \mathbf{C}^{n\times n}$, $e^{tD} \in \mathbf{C}^{n\times n}$). This can be proved using the existence and uniqueness theory of ordinary differential equations, but we will content ourselves with presenting an example.

Example 215 *Let* A *be the matrix*

$$A = \begin{bmatrix} 0 & 1 \\ -1 & 0 \end{bmatrix}$$

of Example 211. We have $A = XDX^{-1}$*, where*

$$X = \begin{bmatrix} -i & i \\ 1 & 1 \end{bmatrix}, \ D = \begin{bmatrix} i & 0 \\ 0 & -i \end{bmatrix}.$$

The reader can verify that

$$X^{-1} = \begin{bmatrix} \frac{i}{2} & \frac{1}{2} \\ -\frac{i}{2} & \frac{1}{2} \end{bmatrix}.$$

Therefore,

$$\begin{aligned} e^{tA} &= Xe^{tD}X^{-1} \\ &= \begin{bmatrix} -i & i \\ 1 & 1 \end{bmatrix} \begin{bmatrix} cos(t) + i\sin(t) & 0 \\ 0 & \cos(t) - i\sin(t) \end{bmatrix} \begin{bmatrix} \frac{i}{2} & \frac{1}{2} \\ -\frac{i}{2} & \frac{1}{2} \end{bmatrix} \\ &= \begin{bmatrix} \cos(t) & \sin(t) \\ -\sin(t) & \cos(t) \end{bmatrix}. \end{aligned}$$

The reader can verify directly that $U(t) = e^{tA}$ *satisfies* $U' = AU$, $U(0) = I$.

The existence and uniqueness theory for ordinary differential equations, as applied to (4.13) (or the matrix problem (4.16)), does not depend on the diagonalizability of A. As a result, it is known the matrix exponential e^{tA} exists for all $A \in \mathbf{C}^{n\times n}$ and also that e^{tA} has real entries if $A \in \mathbf{R}^{n\times n}$. However, if A is not diagonalizable, then we cannot compute e^{tA} by the method described in this section. In Chapter 5, we will develop another decomposition, $A = XJX^{-1}$, that applies to all $A \in \mathbf{C}^{n\times n}$ and produces a matrix J that is as close to diagonal as possible for the given matrix A. (Thus J is diagonal if A is diagonalizable.) The matrix J is called the *Jordan canonical form* of A, and it turns out not to be difficult to compute e^{tJ}, and hence $e^{tA} = Xe^{tJ}X^{-1}$, given that J and X are known. Details are given in Section 5.5.

Exercises

1. Let $\lambda \in \mathbf{C}$. Use Euler's formula to prove directly that
$$\frac{d}{dt}\left[e^{\lambda t}\right] = \lambda e^{\lambda t}.$$
Note that if $f : \mathbf{R} \to \mathbf{C}$, then f can be written as $f(t) = g(t) + ih(t)$, where $g : \mathbf{R} \to \mathbf{R}$ and $h : \mathbf{R} \to \mathbf{R}$ are real-valued. Then, by definition, $f'(t) = g'(t) + ih'(t)$.

2. Let $A \in \mathbf{C}^{n\times n}$ be given, and suppose that $\lambda \in \mathbf{C}$, $x \in \mathbf{C}^n$ form an eigenpair of A. Prove that $u(t) = e^{\lambda t}x$ satisfies $u' = Au$.

3. Let $\mu, \theta \in \mathbf{R}$ and $p, q \in \mathbf{R}^n$, where $\{p, q\}$ is linearly independent. Prove that
$$\left\{e^{\mu t}\left(\cos(\theta t)p - \sin(\theta t)q\right), e^{\mu t}\left(\sin(\theta t)p + \cos(\theta t)q\right)\right\}$$
is a linearly independent subset of $C(\mathbf{R}; \mathbf{R}^n)$.

4. Let $D \in \mathbf{C}^{n\times n}$ be diagonal and $X \in \mathbf{C}^{n\times n}$ be invertible, and define $A = XDX^{-1}$. Prove that e^{tD} and e^{tA} are invertible matrices for all $t \in \mathbf{R}$.

5. Let $A \in \mathbf{R}^{2\times 2}$ be defined by
$$A = \begin{bmatrix} 1 & 4 \\ 9 & 1 \end{bmatrix}.$$
Solve the IVP $u' = Au$, $u(0) = v$, where $v = (1, 2)$.

6. Let $A \in \mathbf{R}^{3\times 3}$ be defined by
$$A = \begin{bmatrix} 1 & 1 & -2 \\ 1 & -2 & 1 \\ -2 & 1 & 1 \end{bmatrix}.$$
Find e^{tA}.

7. Let $A \in \mathbf{R}^{2\times 2}$ be defined by

$$A = \begin{bmatrix} 1 & 2 \\ -2 & 1 \end{bmatrix}.$$

Find e^{tA}.

8. Consider the initial value problem

$$\begin{aligned} u_1' &= u_1 + u_2 + u_3, \ u_1(0) = 1, \\ u_2' &= u_2 + u_3, \ u_2(0) = 0, \\ u_3' &= u_3, \ u_3(0) = 1. \end{aligned}$$

Explain why the methods of this section do not apply to this problem.

4.9 Integer programming

An integer program is a linear program[6] in which the variables must take on integer values. Thus a typical *integer program* (IP) takes the form

$$\begin{aligned} \max \quad & c \cdot x && (4.17a) \\ \text{s.t.} \quad & Ax = b, && (4.17b) \\ & x \geq 0, \ x \in \mathbf{Z}^n && (4.17c) \end{aligned}$$

(or the inequality form in which $Ax = b$ is replaced by $Ax \leq b$). Integer programming problems are difficult to solve efficiently. In this section, we discuss one special case in which the IP can be solved by simply dropping the requirement that $x \in \mathbf{Z}^n$ and solving the resulting *LP*. This is valid because, in the special case to be described below, every basic feasible solution of the LP has integer components.

4.9.1 Totally unimodular matrices

The LP corresponding to (4.17) is

$$\begin{aligned} \max \quad & c \cdot x && (4.18a) \\ \text{s.t.} \quad & Ax = b, && (4.18b) \\ & x \geq 0, && (4.18c) \end{aligned}$$

which is called the *linear programming relaxation* of the IP. We know that if an optimal solution exists, there is an optimal BFS x defined by $x_B = A_B^{-1}b$,

[6]This section depends on the material in Section 3.12 on linear programming.

$x_N = 0$, where B and N are the indices of the basic and nonbasic variables, respectively. We now consider a condition on the matrix A which guarantees that x_B has integer components.

Definition 216 *Let $A \in \mathbf{R}^{m\times n}$. We say that A is* totally unimodular *if the determinant of every square submatrix of A is 0, 1, or -1.*

If $A \in \mathbf{R}^{m\times n}$ is totally unimodular, then each of its entries must be 0, 1, or -1 (since the definition applies square submatrices of all sizes, including 1×1 submatrices).

We have the following simple result.

Theorem 217 *Consider the LP (4.18) and assume that A is totally unimodular and b belongs to $\mathbf{Z}^m$. Then every basic solution x belongs to $\mathbf{Z}^n$.*

Proof This follows from Cramer's rule (see Section 4.3.2). If x is a basic solution, then $x_N = 0$ and $x_B = A_B^{-1}b$. Since A is totally unimodular and A_B is a square, nonsingular submatrix of A, $\det(A_B) = \pm 1$. Then

$$(x_B)_i = \frac{\det\left((A_B)_i(b)\right)}{\det\left(A_B\right)} = \pm\det\left((A_B)_i(b)\right).$$

The determinant of $(A_B)_i(b)$ is computed by multiplying and adding integers, and hence is an integer. Therefore, $(x_B)_i \in \mathbf{Z}$ for all i, which shows that $x \in \mathbf{Z}^n$.

QED

Corollary 218 *If, in the IP (4.17), A is totally unimodular and $b \in \mathbf{Z}^m$, then any optimal BFS of the LP relaxation (4.18) is a solution of the IP.*

Corollary 218 is useful if, in fact, there are interesting IPs in which the constraint matrix is totally unimodular. We will see that this is so. First, we need the following theorem, which allows us to recognize certain totally unimodular matrices.

Theorem 219 *Let A be an $m \times n$ matrix with each entry equal to 0, 1, or -1. Suppose A has at most two nonzeros entries in each column, and that the set of row indices $\{1, 2, \ldots, m\}$ can be partitioned into two subsets S_1 and S_2 such that, for each $j = 1, 2, \ldots, n$,*

1. *if the two nonzeros in column j of A have opposite signs, then their rows indices belong to the same set (S_1 or S_2);*

2. *if the two nonzeros in column j of A have the same sign, then their row indices belong to different sets (one in S_1, the other in S_2).*

Then A is totally unimodular.

Proof We must show that every $k \times k$ submatrix of A has determinant equal to 0, 1, or -1. We argue by induction on k. For $k = 1$, the result is obvious, since each entry of A lies in $\{-1, 0, 1\}$. We assume that the result is true for every submatrix of size $(k-1) \times (k-1)$, and let B be any $k \times k$ submatrix of A. We now consider three cases. First, if B has a column of all zeros, then $\det(B) = 0$ and the desired result holds. Second, if B has a column with a single nonzero (necessarily ± 1), then, by cofactor expansion (Theorem 175), $\det(B)$ equals ± 1 times the determinant of a $(k-1) \times (k-1)$ submatrix of B (and hence of A). By the induction hypothesis, the determinant of the smaller submatrix is 0, 1, or -1, so the result holds in this case as well. If neither the first nor second case holds, then every column of B contains exactly two nonzeros. In this case, let the rows of B correspond to indices $i_1, i_2, \ldots, i_k$, and define $S = \{i_1, i_2, \ldots, i_k\}$, $S_1' = S_1 \cap S$, $S_2' = S_2 \cap S$. Since each column in B contains exactly two nonzeros, it is easy to see that the properties 1 and 2 of S_1 and S_2 also hold for S_1' and S_2'. But these properties imply that for each index j corresponding to a column of B,

$$\sum_{i \in S_1'} A_{ij} - \sum_{i \in S_2'} A_{ij} = 0,$$

which shows that the rows of B are linearly dependent. Thus $\det(B) = 0$ in this case. Thus in every case, $\det(B) \in \{-1, 0, 1\}$ and it follows by induction that A is totally unimodular.

QED

Here are two examples to which the above theorem applies.

Example 220 *Let*

$$A = \begin{bmatrix} 1 & 0 & 1 & 1 & 0 \\ 0 & 1 & 0 & 0 & 1 \\ 1 & 1 & 0 & 0 & 0 \\ 0 & 0 & 1 & 0 & 0 \\ 0 & 0 & 0 & 1 & 1 \end{bmatrix}.$$

We see that A has two nonzero entries in each column. If we define $S_1 = \{1, 2\}$ and $S_2 = \{3, 4, 5\}$, then the hypotheses of Theorem 219 are satisfied because, in each column, there is one nonzero in a row indexed by S_1 and another, of the same sign, in a row indexed by S_2. Thus A is totally unimodular.

Example 221 *Let*

$$A = \begin{bmatrix} -1 & -1 & 0 & 0 & 0 & 0 \\ 1 & 0 & 1 & 0 & 0 & 0 \\ 0 & 0 & -1 & -1 & 1 & 1 \\ 0 & 1 & 0 & 0 & 0 & -1 \\ 0 & 0 & 0 & 1 & -1 & 0 \end{bmatrix}.$$

The matrix A has two nonzeros in each column. If we define $S_1 = \{1,2,3,4,5\}$ and $S_2 = \emptyset$, then A, S_1, S_2 satisfy the hypotheses of Theorem 219: in each column, there are two nonzeros of opposite signs, and they both lie in rows indexed by S_1. Thus A is totally unimodular.

4.9.2 Transportation problems

We now present an application in which the constraint matrix of an IP is totally unimodular. The problem to be described is called a *transportation problem.* We consider a network of n_1 factories, all producing the same product, and n_2 stores that sell the product. The product in question is produced in discrete quantities; examples of such a product might include cars or boxes of laundry detergent. Factory i produces a_i units of the product, while store j wants b_j units. We assume that

$$\sum_{i=1}^{n_1} a_i \ge \sum_{j=1}^{n_2} b_j,$$

which means that the factories can together meet the demand. Finally, we assume that it costs c_{ij} dollars per unit to ship from factory i to store j, and we write x_{ij} for the number of units shipped from factory i to store j. The goal is to choose x_{ij}, $i = 1,2,\ldots,n_1$, $j = 1,2,\ldots,n_2$, to minimize the total shipping cost, which is

$$\sum_{i=1}^{n_1} \sum_{j=1}^{n_2} c_{ij} x_{ij}.$$

The constraints are of two types. First, for each $i = 1,2,\ldots,n_1$, we have the constraint

$$\sum_{j=1}^{n_2} x_{ij} \le a_i, \tag{4.19}$$

which states that we cannot ship more than a_i units from factory i. Second, for each $j = 1,2,\ldots,n_2$, we have the constraint

$$\sum_{i=1}^{n_1} x_{ij} = b_j, \tag{4.20}$$

which states that each store receives the desired number of units. Finally, each x_{ij} must be a nonnegative integer.

To put the resulting IP into the standard equality form, we add slack variables $s_1, s_2, \ldots, s_{n_1}$ to the constraints (4.19). We then write

$$\begin{aligned}
x &= (x_{11}, x_{12}, \ldots, x_{1n_2}, x_{21}, \ldots, x_{n_1,n_2}, s_1, \ldots, s_{n_1}),\\
c &= (c_{11}, c_{12}, \ldots, c_{1n_2}, c_{21}, \ldots, c_{n_1,n_2}, 0, \ldots, 0),\\
d &= (a_1, \ldots, a_{n_1}, b_1, \ldots, b_{n_2}).
\end{aligned}$$

The matrix A is determined by (4.20) and the equality version of (4.19). The matrix has $n_1 + n_2$ rows and $n_1n_2 + n_1$ columns. The first n_1 rows correspond to the factories and the next n_2 rows correspond to the stores. There is one column for each route from a factory to a store, plus n_1 columns for the slack variables. With these definitions, the problem to be solved is the integer program

$$\begin{aligned} \min \quad & c \cdot x \\ \text{s.t.} \quad & Ax = d, \\ & x \geq 0, x \in \mathbf{Z}^{n_1n_2+n_1}. \end{aligned}$$

Let us define $S_1 = \{1, 2, \ldots, n_1\}$ and $S_2 = \{n_1+1, n_1+2, \ldots, n_1+n_2\}$. We claim that, with these definitions of S_1 and S_2, A satisfies the hypotheses of Theorem 219, and therefore A is totally unimodular. To verify this, we note that each entry of A is 0 or 1. A little thought shows that each x_{ij} appears in exactly one of the constraints (4.19) (the one corresponding to the ith factory) and in exactly one of the constraints (4.20) (the one corresponding to the jth store). Thus, in each of the first n_1n_2 columns, there are two nonzeros of the same sign, one in a row indexed by S_1 and the other in a row indexed by S_2. Finally, in the last n_1 column, there is a single nonzero, namely, a 1 in the $i, n_1n_2 + i$ entry. This verifies our claim, and A is totally unimodular.

Since a transportation problem leads to a totally unimodular constraint matrix, the integer program can be solved by finding an optimal BFS by the simplex method applied to the LP relaxation. We now present a specific example.

Example 222 *Consider a network of four factories and six stores. The production at the factories and the demand at the stores (both in numbers of units) are given in the following table:*

Factory	*Production*	*Store*	*Demand*
1	*100*	*1*	*160*
2	*250*	*2*	*150*
3	*200*	*3*	*100*
4	*300*	*4*	*120*
		5	*135*
		6	*145*

The costs of transporting the product from factory i to store j (in dollars per unit) are given in the following table:

$i\backslash j$	*1*	*2*	*3*	*4*	*5*	*6*
1	*1.09*	*1.06*	*1.19*	*1.10*	*1.16*	*0.86*
2	*1.10*	*0.86*	*0.93*	*0.90*	*1.19*	*0.90*
3	*0.91*	*0.84*	*1.03*	*1.00*	*1.02*	*1.14*
4	*1.07*	*1.00*	*0.89*	*1.08*	*0.85*	*0.90*

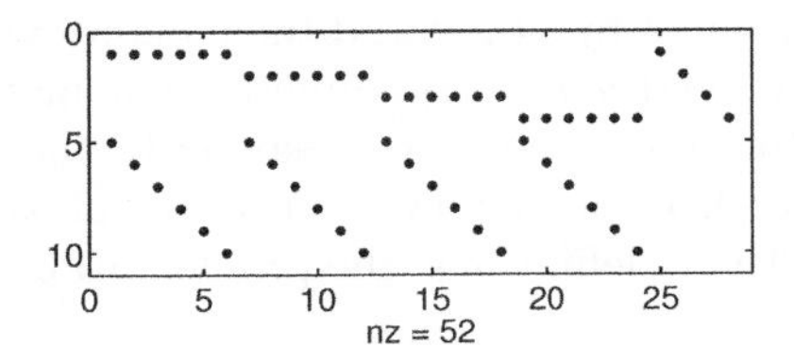

FIGURE 4.2
The pattern of nonzeros in the matrix A from Example 222 (A has 52 nonzeros).

The matrix A describing the constraints (after slack variables have been added to the first n_1 constraints) is $(n_1 + n_2) \times (n_1 n_2 + n_1)$, which in this case is 10×28. All the entries are 0 or 1, and the pattern of ones is shown in Figure 4.2. To complete the specification of the transportation problem as an IP, we define

$$\begin{aligned} c &= (1.09, 1.06, 1.19, 1.10, 1.16, 0.86, 1.10, 0.86, 0.93, \ldots, 0.90, 0, 0, 0, 0), \\ x &= (x_{11}, x_{12}, x_{13}, x_{14}, x_{15}, x_{16}, x_{21}, x_{22}, x_{23}, \ldots, x_{46}, s_1, s_2, s_3, s_4), \\ d &= (100, 250, 200, 300, 160, 150, 100, 120, 135, 145). \end{aligned}$$

We now wish to solve

$$\begin{aligned} \min \quad & c \cdot x \\ s.t. \quad & Ax = d, \\ & x \geq 0, x \in \mathbf{Z}^{28}. \end{aligned}$$

Theorem 219 shows that A is totally unimodular and Corollary 218 shows that the transportation problem described in this example can be solved by simply solving the LP relaxation, that is, by ignoring the constraint that $x \in \mathbf{Z}^{28}$. When we do so, we find the following optimal values of x_{ij}:

$i \backslash j$	1	2	3	4	5	6
1	0	0	0	0	0	100
2	0	110	0	120	0	0
3	160	40	0	0	0	0
4	0	0	100	0	135	45

The value of the slack variables at the optimal BFS are $s_1 = 0$, $s_2 = 20$, $s_3 = 0$, $s_4 = 20$. (All the values of x_{ij} and the slack variables are integers, as expected.) This shows that constraints 1 and 3 are satisfied as equations at the solution and thus all of the production at factories 1 and 3 is used, but 20 units produced at each of factories 2 and 4 are not shipped to any store. This is consistent with the fact that $\sum_{i=1}^{4} a_i = 850$, while $\sum_{j=1}^{6} b_j = 810$.

The optimal value of the objective function is \$712.50.

Exercises

1. Which of the following matrices can be shown to be totally unimodular by Theorem 219?

   (a)
$$\begin{bmatrix} 1 & 0 & 1 & 1 & 0 \\ 0 & 1 & 0 & 0 & 1 \\ 1 & 0 & 0 & 1 & 1 \\ 0 & 1 & 1 & 0 & 0 \end{bmatrix}$$

   (b)
$$\begin{bmatrix} 1 & 0 & 1 & 0 \\ -1 & 1 & 0 & -1 \\ 0 & -1 & 0 & 1 \\ 0 & 0 & -1 & 0 \end{bmatrix}$$

   (c)
$$\begin{bmatrix} 1 & 0 & 0 & 0 & 1 & 0 \\ 0 & 1 & 0 & -1 & 0 & 1 \\ 0 & 0 & 0 & 0 & 0 & 0 \\ 0 & 0 & 1 & -1 & -1 & 1 \end{bmatrix}$$

2. Consider the transportation problem with three factories and two stores. The factories produce 500, 250, and 400 units, respectively, and the stores need 600 and 550 units. The cost per unit of shipping from the first factory are \$14 (to the first store) and \$14 (to the second), the costs from the second factory are \$10 and \$10, respectively, and the costs from the third factory are \$12 and \$13, respectively.

   Formulate the problem of minimizing the total transportation cost as an integer program and solve it with the simplex method (which must produce an integer solution since the constraint matrix is totally unimodular).

3. In Section 3.10, we introduced the adjacency matrix A of a graph G. This matrix is $n \times n$, where G contains n nodes, and A_{ij} is 1 if there is an edge joining nodes i and j, and 0 otherwise. Another way to represent G by matrix is through the *node-edge* incidence matrix B. This matrix is $n \times m$, where n is the number of nodes and m is the number of edges in G. The entry B_{ij} is 1 if node i is one of the endpoints of edge j, and 0 otherwise.

   A graph G is called *bipartite* if its node set V_G can be partitioned into two subsets, V_1 and V_2, and every edge in E_G joins one node in V_1 with one node in V_2. Prove that the node-edge incidence matrix B of a bipartite graph G is totally unimodular. (The constraint matrix for a transportation problem is the node-edge incidence matrix of a bipartite graph, augmented by columns for the slack variables.)

4. A *directed graph*, or *digraph*, is a generalization of a graph in which the edges, now called arcs, are oriented (that is, have a direction). Each arc is said to have a *tail* and a *head*, and is directed from the tail to the head.

   The *node-arc incidence matrix* of a digraph G is the $n \times m$ matrix A, where G has n nodes and m arcs, in which A_{ij} is 1 if node i is the head of arc j, -1 if nodes i is the tail of arc j, and 0 otherwise. Prove that the node-arc incidence matrix of a directed graph is totally unimodular.

5

The Jordan canonical form

In the previous chapter, we saw that a matrix $A \in F^{n\times n}$ is diagonalizable if there is a basis of F^n consisting of eigenvectors of A. A matrix failing to satisfy this condition is called defective.

Given that not all matrices can be diagonalized, we wish to determine the simplest form of a matrix that can be obtained by a similarity transformation. Expressed in terms of a linear operator on a finite-dimensional space, the question is how to choose the basis so that the matrix representing the operator is as simple as possible.

The key to understanding these more general results is the concept of an invariant subspace.

5.1 Invariant subspaces

If $S = E_\lambda(T)$ is an eigenspace of a linear operator $T : X \to X$, then, for every $x \in S$, $T(x) = \lambda x \in S$ (since S, being a subspace, is closed under scalar multiplication). Therefore, T maps each element of S into S itself. There is a name for subspaces having this property.

Definition 223 *Let X be vector space over a field F, let $T : X \to X$ be a linear operator, and let S be a subspace of X. We say that S is* invariant *under T if and only if $T(x) \in S$ for all $x \in S$.*

Similarly, if $A \in F^{n\times n}$ and S is a subspace of F^n, we say that S is invariant under A if and only if $Ax \in S$ for all $x \in S$.

An operator (or matrix) fails to be diagonalizable if it does not have enough independent eigenvectors to form a basis. In such a case, we can understand the structure of the operator by finding its invariant subspaces (some of which may be eigenspaces).

In order to make our discussion as concrete as possible, we will focus on a given matrix $A \in F^{n\times n}$. Our goal is to find an invertible matrix $X \in F^{n\times n}$ such that $J = X^{-1}AX$ is as simple as possible. It is left as a long exercise to rewrite the results of this section in terms of an operator $T : U \to U$ (in which case the goal is to choose a basis for U such that the matrix of T with respect to the basis is as simple as possible).

To see the significance of an invariant subspace, let us suppose that A belongs to $F^{n\times n}$ and S is a subspace of F^n that is invariant under A. We choose a basis $\{x_1, x_2, \dots, x_k\}$ for S and define $X_1 = [x_1|x_2|\cdots|x_k]$. Since S is invariant under A, we see that $Ax_i \in S$ for each $i = 1, 2, \dots, k$, and thus each Ax_i can be expressed as a linear combination of $x_1, x_2, \dots, x_k$. Therefore, there exists scalars B_{ij}, $i, j = 1, 2, \dots, k$, such that

$$Ax_j = \sum_{i=1}^{n} B_{ij}x_i = (X_1B)_j,\ j = 1, 2, \dots, k,$$

where B is the $k \times k$ matrix defined by the scalars B_{ij}. But then, since $Ax_1, Ax_2, \dots, Ax_k$ are the columns of the matrix AX_1, we see that

$$AX_1 = X_1B.$$

We now extend $\{x_1, x_2, \dots, x_k\}$ to a basis $\{x_1, x_2, \dots, x_n\}$ for F^n and define $X_2 = [x_{k+1}|\cdots|x_n]$, $X = [X_1|X_2]$. Then

$$AX = A[X_1|X_2] = [AX_1|AX_2].$$

We know that

$$AX_1 = X_1B = X_1B + X_2 0 = [X_1|X_2]\left[\begin{array}{c} B \\ \hline 0 \end{array}\right].$$

Each vector Ax_j, $j = k + 1, \dots, n$ can be expressed in terms of the basis $\{x_1, x_2, \dots, x_n\}$, which means that there is a $n \times (n - k)$ matrix E such that $AX_2 = XE$. If we write

$$E = \left[\begin{array}{c} C \\ \hline D \end{array}\right],\ C \in F^{k\times(n-k)},\ D \in F^{(n-k)\times(n-k)},$$

then we have

$$AX_2 = XE = [X_1|X_2]\left[\begin{array}{c} C \\ \hline D \end{array}\right] = X_1C + X_2D.$$

The reader should notice the crucial difference between $AX_1 = X_1B$ and $AX_2 = X_1C + X_2D$: Since S is invariant under A, AX_1 can be represented using only the matrix X_1. On the other hand, we have not assumed that $T = \mathrm{sp}\{x_{k+1}, \dots, x_n\}$ is invariant under A, and therefore to represent AX_2 requires both X_1 and X_2.

Putting these relationships into matrix form, we obtain

$$AX = [AX_1|AX_2] = [X_1B|X_1C + X_2D] = [X_1|X_2]\left[\begin{array}{c|c} B & C \\ \hline 0 & D \end{array}\right],$$

or

$$X^{-1}AX = \left[\begin{array}{c|c} B & C \\ \hline 0 & D \end{array}\right].$$

We can draw two conclusions from these results. First, if $\{x_1, \ldots, x_k\}$ spans an invariant subspace, then $X^{-1}AX$ is simpler than a typical matrix (because $X^{-1}AX$ has a block of zeros in the lower left-hand corner). Second, if it happens that $T = \text{sp}\{x_{k+1}, \ldots, x_n\}$ is also invariant under A, then we have $AX_2 = X_2D$ (that is, $C = 0$), and $X^{-1}AX$ is *block diagonal*:

$$X^{-1}AX = \left[\begin{array}{c|c} B & 0 \\ \hline 0 & D \end{array}\right].$$

Example 224 *Let*

$$A = \begin{bmatrix} 2 & 0 & 0 & -1 \\ 0 & 0 & -1 & 2 \\ 2 & 1 & 3 & -2 \\ 1 & 0 & 0 & -1 \end{bmatrix},$$

and define $S = \text{sp}\{x_1, x_2\}$, *where*

$$x_1 = \begin{bmatrix} 0 \\ 1 \\ 0 \\ 0 \end{bmatrix}, \ x_2 = \begin{bmatrix} 0 \\ -1 \\ 1 \\ 0 \end{bmatrix}.$$

We first show that S *is invariant under* A:

$$Ax_1 = \begin{bmatrix} 0 \\ 0 \\ 1 \\ 0 \end{bmatrix} = x_1 + x_2, \ Ax_2 = \begin{bmatrix} 0 \\ -1 \\ 2 \\ 0 \end{bmatrix} = x_1 + 2x_2.$$

If we define $X_1 = [x_1|x_2]$, *then we see that* $AX_1 = X_1B$, *where*

$$B = \begin{bmatrix} 1 & 1 \\ 1 & 2 \end{bmatrix}.$$

Let us suppose we extend $\{x_1, x_2\}$ *to the basis* $\{x_1, x_2, x_3, x_4\}$, *where* x_3, x_4 *are chosen to be* $(1, 0, 0, 0)$ *and* $(0, 0, 0, 1)$, *respectively. We then obtain*

$$X^{-1}AX = \left[\begin{array}{cc|cc} 1 & 1 & 2 & 0 \\ 1 & 2 & 2 & -2 \\ \hline 0 & 0 & 2 & -1 \\ 0 & 0 & 1 & -1 \end{array}\right],$$

where $X = [x_1|x_2|x_3|x_4]$. *In terms of the notation introduced above, we have*

$$C = \begin{bmatrix} 2 & 0 \\ 2 & -2 \end{bmatrix}, \ D = \begin{bmatrix} 2 & -1 \\ 1 & -1 \end{bmatrix},$$

and we see that $T = \text{sp}\{x_3, x_4\}$ *is not invariant under* A.

There is an alternate choice for x_3, x_4 which leads to a block diagonal matrix. If we define $x_3 = (-1, 0, 1, -1)$, $x_4 = (0, 3, -2, -1)$, then

$$X^{-1}AX = \left[\begin{array}{cc|cc} 1 & 1 & 0 & 0 \\ 1 & 2 & 0 & 0 \\ \hline 0 & 0 & 1 & -1 \\ 0 & 0 & -1 & 0 \end{array}\right].$$

With this choice of x_3, x_4, $AX_2 = X_2 D$, where

$$D = \left[\begin{array}{cc} 1 & -1 \\ -1 & 0 \end{array}\right],$$

and $T = \mathrm{sp}\{x_3, x_4\}$ is invariant under A.

5.1.1 Direct sums

The key to the construction described above, resulting in a block diagonal $X^{-1}AX$, is that F^n be the *direct sum* of two invariant subspaces. We have already encountered the (algebraic) sum of two subspaces: If S and T are subspaces of a vector space V, then

$$S + T = \{s + t \, : \, s \in S, t \in T\}.$$

We know that $S + T$ is a subspace of V (see Exercise 2.3.21).

Definition 225 *Let V be a vector space over a field F, and let S, T be subspaces of V. We say that V is the* direct sum *of S and T if and only if each vector $v \in V$ can be written uniquely as $v = s + t$, $s \in S$, $t \in T$.*

More generally, if $S_1, S_2, \ldots, S_t$ are subspaces of V and each $v \in V$ can be written uniquely as $v = v_1 + v_2 + \cdots + v_t$, where $v_i \in S_i$ for $i = 1, 2, \ldots, t$, then we say that V is the direct sum *of $S_1, S_2, \ldots, S_t$.*

We now consider the case that F^n is the direct sum of subspaces $S_1, \ldots, S_t$, and each S_i is invariant under a given $A \in F^{n\times n}$. We assume that the dimension of S_i is m_i, and form an $n \times m_i$ matrix X_i whose columns form a basis for S_i. Then, since S_i is invariant under A, there exists a $m_i \times m_i$ matrix B_i such that

$$AX_i = X_i B_i.$$

Forming the matrix $X = [X_1|X_2|\cdots|X_t]$, we obtain

$$\begin{aligned} AX = [AX_1|AX_2|\cdots|AX_t] &= [X_1B_1|X_2B_2|\cdots|X_tB_t] \\ &= X \left[\begin{array}{c|c|c|c} B_1 & & & \\ \hline & B_2 & & \\ \hline & & \ddots & \\ \hline & & & B_t \end{array}\right]. \end{aligned}$$

Moreover, X is invertible (see Theorem 226 below), and therefore $X^{-1}AX$ is the block diagonal matrix with diagonal blocks $B_1, B_2, \ldots, B_t$.

We can now describe our approach in the following sections: In the remainder of this section and Section 5.2, we show how to express F^n as the direct sum of subspaces that are invariant under a given matrix $A \in F^{n\times n}$. In this way, we will obtain a block diagonal matrix $X^{-1}AX$. In the case that F is algebraically closed, the invariant subspaces are *generalized eigenspaces*, which we define below. In Sections 5.3 and 5.4, we show how to choose the basis for a generalized eigenspace so that the corresponding diagonal block is as simple as possible.

Before we can discuss invariant subspaces further, we need to collect some facts about direct sums.

Theorem 226 *Let V be a finite-dimensional vector space over a field F, and let $S_1, S_2, \ldots, S_t$ be subspaces of V. If V is the direct sum of $S_1, S_2, \ldots, S_t$, then*

1. $V = S_1 + S_2 + \cdots + S_t$;

2. *if* $s_1 + s_2 + \cdots + s_t = 0$, *where* $s_i \in S_i$ *for all* $i = 1, 2, \ldots, t$, *then* $s_1 = s_2 = \ldots = s_t = 0$;

3. $\dim(V) = \dim(S_1) + \dim(S_2) + \cdots + \dim(S_t)$.

Conversely, if any two of these three conditions hold, then V is the direct sum of $S_1, S_2, \ldots, S_t$, and the third condition also holds.

In the case $t = 2$, the second condition is equivalent to $S_1 \cap S_2 = \{0\}$.

Proof Exercise 10.

5.1.2 Eigenspaces and generalized eigenspaces

Let us assume that $A \in F^{n\times n}$ is diagonalizable, and that the distinct eigenvalues of A are $\lambda_1, \lambda_2, \ldots, \lambda_t$. We define

$$m_i = \dim(E_{\lambda_i}(A))$$

and choose a basis $\{x_1^{(i)}, \ldots, x_{m_i}^{(i)}\}$ for each eigenspace $E_{\lambda_i}(A)$. It follows from Corollary 194 of Section 4.6 that

$$\mathcal{X} = \left\{x_1^{(1)}, \ldots, x_{m_1}^{(1)}, x_1^{(2)}, \ldots, x_{m_2}^{(2)}, \ldots, x_1^{(t)}, \ldots, x_{m_t}^{(t)}\right\}$$

is linearly independent. Moreover, $\mathcal{X}$ must form a basis for F^n, since by assumption there is a basis of F^n consisting of eigenvectors of A, and $\mathcal{X}$ contains the maximum number of linearly independent eigenvectors. It follows that F^n is the direct sum of the eigenspaces $E_{\lambda_1}(A), E_{\lambda_2}(A), \ldots, E_{\lambda_t}(A)$, each of which is invariant under A.

We now define

$$X_i = \left[x_1^{(i)}|x_2^{(i)}|\cdots|x_{m_i}^{(i)}\right], \ i = 1, 2, \ldots, t,$$

and $X = [X_1|X_2|\cdots|X_t]$. Since the columns of X_i are eigenvectors of A corresponding to λ_i, we have

$$AX_i = \lambda_i X_i = X_i(\lambda_i I),$$

where I is the $m_i \times m_i$ identity matrix. We then obtain

$$X^{-1}AX = \left[\begin{array}{c|c|c|c} \lambda_1 I & & & \\ \hline & \lambda_2 I & & \\ \hline & & \ddots & \\ \hline & & & \lambda_t I \end{array}\right].$$

This is the best possible case: F^n is the direct sum of invariant subspaces, each of which is an eigenspace of A. In this case, A is similar to a block diagonal matrix, and each diagonal block is a diagonal matrix. Thus A itself is diagonalizable.

Still assuming that A is diagonalizable, let us focus on the first eigenvalue λ_1 (our reasoning will then apply to any eigenvalue, since the eigenvalues can be listed in any order). We define

$$N = E_{\lambda_1}(A) = \mathcal{N}(A - \lambda_1 I), \ R = E_{\lambda_2}(A) + E_{\lambda_3}(A) + \cdots + E_{\lambda_t}(A).$$

Both N and R are invariant under A, and F^n is the direct sum of N and R. This suggests a necessary condition for A to be diagonalizable: Given any eigenvalue λ of A and the corresponding eigenspace $N = \mathcal{N}(A - \lambda I)$ (which is invariant under A), there must exist another subspace R such that R is invariant under A and F^n is the direct sum of N and R. The next few results explore the conditions under which such an R exists, and lead to the concept of a generalized eigenspace.

Theorem 227 *Let F be a field, $A \in F^{n\times n}$, and define $N = \mathcal{N}(A)$. If there exists a subspace R of F^n such that R is invariant under A and F^n is the direct sum of N and R, then $R = \mathrm{col}(A)$.*

Proof Let $k = \dim(N)$, so that $\dim(R) = n - k$ by Theorem 226. We know by the fundamental theorem of linear algebra that $\mathrm{col}(A)$ also satisfies $\dim(\mathrm{col}(A)) = n - k$, so it suffices to prove that $\mathrm{col}(A) \subset R$. Let $x \in \mathrm{col}(A)$; then there exists $y \in F^n$ such that $x = Ay$. Since F^n is the direct sum of N and R, there exists $n \in N$, $r \in R$ such that $y = n + r$. But then

$$x = Ay = A(n + r) = An + Ar = Ar \text{ (since } An = 0).$$

Since $r \in R$ and R is invariant under A by assumption, it follows that Ar, and hence x, belongs to R. This proves that $\mathrm{col}(A) \subset R$; since the two subspaces have the same dimension, they must be equal.

QED

We point out that $\mathrm{col}(A)$ is obviously invariant under A.

Corollary 228 *Let F be a field, $A \in F^{n\times n}$, and define $N = \mathcal{N}(A)$. Then there exists a subspace R of F^n such that R is invariant under A and F^n is the direct sum of N and R if and only if $\mathcal{N}(A) \cap \mathrm{col}(A) = \{0\}$, in which case $R = \mathrm{col}(A)$.*

Proof If there exists such a subspace R, then Theorem 227 implies that $R = \mathrm{col}(A)$, in which case Theorem 226 shows that $\mathcal{N}(A) \cap \mathrm{col}(A) = \{0\}$.

Conversely, suppose $\mathcal{N}(A) \cap \mathrm{col}(A) = \{0\}$. Then Theorem 226 and the fundamental theorem of linear algebra together imply that F^n is the direct sum of $\mathcal{N}(A)$ and $\mathrm{col}(A)$. Since $\mathrm{col}(A)$ is invariant under A, this completes the proof.

QED

We can apply the preceding results to the matrix $A - \lambda I$, where λ is an eigenvalue of A, in place of A, since a subspace is invariant under A if and only if it is invariant under $A - \lambda I$.

Lemma 229 *Let F be a field, and suppose $A \in F^{n\times n}$, $\lambda \in F$. If S is a subspace of F^n, then S is invariant under A if and only if S is invariant under $A - \lambda I$.*

Proof Suppose S is invariant under A. If $x \in S$, then $Ax \in S$ because S is invariant under A and $\lambda x \in S$ because S is a subspace. Therefore,

$$(A - \lambda I)x = Ax - \lambda x \in S$$

because S is a subspace and is therefore close under addition.

The proof of the converse is similar and is left to the reader (see Exercise 12).

QED

If we apply Theorem 227 to the matrix $A - \lambda I$, we see that if there exists a subspace R of F^n such that F^n is the direct sum of $N = \mathcal{N}(A - \lambda I)$ and R, then R must equal $\mathrm{col}(A - \lambda I)$. When A is diagonalizable, we know that R exists, and that it is the direct sum of the other eigenspaces of A. Therefore, $\mathrm{col}(A - \lambda I)$ must equal the direct sum of the eigenspaces of A corresponding to eigenvalues different from λ. Exercise 14 asks the reader to prove this directly.

We have now shown that if A is diagonalizable, then

$$\mathcal{N}(A - \lambda I) \cap \mathrm{col}(A - \lambda I) = \{0\}$$

must hold for each eigenvalue of A. This follows from Corollary 228 and the discussion preceding Theorem 227. The following result gives an equivalent condition that will lead to the concept of a generalized eigenspace.

Theorem 230 *Let F be a field and let $A \in F^{n\times n}$. Then $\mathcal{N}(A)\cap\text{col}(A) = \{0\}$ if and only if $\mathcal{N}(A^2) = \mathcal{N}(A)$.*

Proof We first note that $\mathcal{N}(A) \subset \mathcal{N}(A^2)$ always holds:

$$x \in \mathcal{N}(A) \Rightarrow Ax = 0 \Rightarrow A(Ax) = 0 \Rightarrow A^2x = 0 \Rightarrow x \in \mathcal{N}(A^2).$$

Therefore, $\mathcal{N}(A^2) = \mathcal{N}(A)$ if and only if $\mathcal{N}(A^2) \subset \mathcal{N}(A)$.

Suppose first that $\mathcal{N}(A) \cap \text{col}(A) = \{0\}$. If $x \in \mathcal{N}(A^2)$, then $A(Ax) = 0$, which shows that Ax belongs to both $\text{col}(A)$ and $\mathcal{N}(A)$. By hypothesis, this implies that $Ax = 0$, that is, that $x \in \mathcal{N}(A)$. Therefore, $\mathcal{N}(A^2) \subset \mathcal{N}(A)$.

Conversely, suppose $\mathcal{N}(A^2) \subset \mathcal{N}(A)$, and let $x \in \mathcal{N}(A) \cap \text{col}(A)$. Since $x \in \text{col}(A)$, there exists $y \in F^n$ such that $x = Ay$. But then $x \in \mathcal{N}(A)$ implies $A^2y = 0$, that is, $y \in \mathcal{N}(A^2)$. Since $\mathcal{N}(A^2) \subset \mathcal{N}(A)$ by assumption, it follows that $y \in \mathcal{N}(A)$. But then $x = Ay = 0$, and we have proved that $\mathcal{N}(A) \cap \text{col}(A) = \{0\}$.

QED

Example 231 *Let $A \in \mathbf{R}^{2\times 2}$ be defined by*

$$A = \begin{bmatrix} 1 & 1 \\ 0 & 1 \end{bmatrix}.$$

We can see by inspection that the only eigenvalue of A is 1, and an easy calculation shows that $\mathcal{N}(A - I) = \text{sp}\{(1,0)\}$. On the other hand,

$$A - I = \begin{bmatrix} 0 & 1 \\ 0 & 0 \end{bmatrix},$$

and therefore $\text{col}(A - I)$ also equals $\text{sp}\{(1,0)\}$. Therefore, the condition

$$\mathcal{N}(A - I) \cap \text{col}(A - I) = \{0\}$$

fails, $\mathbf{R}^2$ is not the direct sum of $\mathcal{N}(A - I)$ and $\text{col}(A - I)$, and we cannot find X such that $X^{-1}AX$ is block diagonal.

In terms of the equivalent condition given by Theorem 230, we see that

$$(A - I)^2 = \begin{bmatrix} 0 & 1 \\ 0 & 0 \end{bmatrix}\begin{bmatrix} 0 & 1 \\ 0 & 0 \end{bmatrix} = \begin{bmatrix} 0 & 0 \\ 0 & 0 \end{bmatrix},$$

which implies that $\mathcal{N}((A - I)^2) = \mathbf{R}^2$, which is not equal to $\mathcal{N}(A - I)$. This also shows that we cannot find a basis under which A is block diagonal.

If λ is an eigenvalue of A, and $\mathcal{N}((A-\lambda I)^2) \neq \mathcal{N}(A-\lambda I)$, then Corollary 230 and our reasoning above show that A cannot be diagonalizable. To make sure this reasoning is clear, we will summarize it here. If A is diagonalizable and λ is any eigenvalue of A, then there exists a subspace R (namely, the direct sum of the other eigenspaces of A) such that R is invariant under A and F^n

is the direct sum of $N = \mathcal{N}(A - \lambda I)$ and R. On the other hand, Corollary 230 shows that if $\mathcal{N}((A - \lambda I)^2) \neq \mathcal{N}(A - \lambda I)$, then no such subspace R can exist. Thus $\mathcal{N}((A - \lambda I)^2) \neq \mathcal{N}(A - \lambda I)$ implies that A is not diagonalizable.

In the case that $\mathcal{N}((A - \lambda I)^2) \neq \mathcal{N}(A - \lambda I)$, then

$$E_{\lambda_i}(A) = \mathcal{N}(A - \lambda I) \subset \mathcal{N}((A - \lambda I)^k)$$

for any $k > 1$. For an appropriate value of k, we call $\mathcal{N}((A-\lambda I)^k)$ a generalized eigenspace. In the next section, we show that if $p_A(r)$ can be completely factored as

$$p_A(r) = (r - \lambda_1)^{m_1}(r - \lambda_2)^{m_2} \cdots (r - \lambda_t)^{m_t}, \tag{5.1}$$

then F^n is the direct sum of generalized eigenspaces, each of which is invariant under A, and therefore A is similar to a block diagonal matrix. If F is algebraically closed (practically speaking, if $F = \mathbf{C}$), then the characteristic polynomial of A can always be completely factored, and this shows how (in principle) to find a block diagonal matrix similar to A.

We then show in Sections 5.3 and 5.4 how to choose a basis for a generalized eigenspace to produce the simplest possible corresponding diagonal block, thereby providing a complete answer to the question of how to find the simplest possible matrix similar to $A \in \mathbf{C}^{n \times n}$. In the appendix to Section 5.2, we discuss the situation in which $F \neq \mathbf{C}$ and $p_A(r)$ cannot be factored completely as in (5.1). This discussion is included for the interested reader, but it will not be used in the rest of the book.

Exercises

Miscellaneous exercises

1. Let $A \in \mathbf{R}^{3\times 3}$ be defined by

$$A = \begin{bmatrix} -1 & 1 & -1 \\ 16 & -1 & -2 \\ 32 & -7 & 2 \end{bmatrix}.$$

   (a) Is $S = \text{sp}\{(2, 2, 1), (-1, 2, 0)\}$ invariant under A?

   (b) Is $T = \text{sp}\{(0, 1, 2), (1, 4, 1)\}$ invariant under A?

2. Let $T : \mathcal{P}_2 \to \mathcal{P}_2$ be defined by $T(p)(x) = p(2x + 1)$. Prove that $\mathcal{P}_1$, regarded as a subspace of $\mathcal{P}_2$, is invariant under T.

3. Let $S = \text{sp}\{(1, 3, 1), (1, -1, 0)\}$ and $T = \text{sp}\{(0, 1, 2), (1, -1, 1)\}$ be subspaces of $\mathbf{R}^3$. Is $\mathbf{R}^3$ the direct sum of S and T?

4. Let $S = \text{sp}\{(0, -2, 1, 3), (-4, 1, 3, -5)\}, T = \text{sp}\{(4, -4, 3, 5), (4, -9, 6, 4)\}$ be subspaces of $\mathbf{R}^4$. Is $\mathbf{R}^4$ the direct sum of S and T?

5. Let $A \in \mathbf{R}^{3\times 3}$ be defined by

$$A = \begin{bmatrix} 3 & 0 & -1 \\ -6 & 1 & 3 \\ 2 & 0 & 0 \end{bmatrix},$$

and let $S = \text{sp}\{(0,1,0),(1,0,1)\}$. Prove that S is invariant under A, and find an invertible matrix $X \in \mathbf{R}^{3\times 3}$ such that $X^{-1}AX$ is block upper triangular.

6. Let $A \in \mathbf{R}^{4\times 4}$ be defined by

$$A = \begin{bmatrix} 8 & 0 & 3 & 3 \\ 12 & 2 & 6 & 6 \\ -49 & 5 & -14 & -13 \\ 3 & 3 & 3 & 2 \end{bmatrix},$$

and let $S = \text{sp}\{(1,4,-1,3),(4,7,-19,3)\}$. Show that S is invariant under A, and find an invertible matrix $X \in \mathbf{R}^{4\times 4}$ such that $X^{-1}AX$ is block upper triangular.

7. Let $A \in \mathbf{R}^{4\times 4}$ be defined by

$$A = \begin{bmatrix} 8 & -4 & -1 & 5 \\ 16 & -8 & 3 & 10 \\ 0 & 0 & -1 & 0 \\ -2 & 1 & 6 & -1 \end{bmatrix}.$$

Does there exist a subspace R of $\mathbf{R}^4$ such that $\mathbf{R}^4$ is the direct sum of R and $N = \mathcal{N}(A)$? Answer the question in two ways:

   (a) Determine whether $\mathcal{N}(A^2) = \mathcal{N}(A)$ holds.

   (b) Determine whether $\mathcal{N}(A) \cap \text{col}(A) = \{0\}$ holds.

8. Repeat the previous exercise for

$$A = \begin{bmatrix} -1 & -1 & 1 \\ -2 & 1 & 2 \\ -4 & -1 & 4 \end{bmatrix}.$$

9. Let U be a finite-dimensional vector space over a field F, and let $T : U \to U$ be a linear operator. Let $\mathcal{U} = \{u_1, u_2, \ldots, u_n\}$ be a basis for U and define $A = [T]_{\mathcal{U},\mathcal{U}}$. Suppose $X \in F^{n\times n}$ is an invertible matrix, and define $J = X^{-1}AX$. For each $j = 1, 2, \ldots, n$, define

$$v_j = \sum_{i=1}^{n} X_{ij} u_i.$$

(a) Prove that $\mathcal{V} = \{v_1, v_2, \ldots, v_n\}$ is a basis for U.

(b) Prove that $[T]_{\mathcal{V},\mathcal{V}} = J$.

10. Prove Theorem 226.

11. Let V be a finite-dimensional vector space over a field F, suppose $\{x_1, x_2, \ldots, x_n\}$ is a basis for V, and let k be an integer satisfying $1 \leq k \leq n-1$. Prove that V is the direct sum of $S = \{x_1, \ldots, x_k\}$ and $T = \mathrm{sp}\{x_{k+1}, \ldots, x_n\}$.

12. Complete the proof of Lemma 229.

13. Suppose F is a field, $A \in F^{n\times n}$, N is a subspace of F^n that is invariant under A, and F^n is the direct sum of N and $\mathrm{col}(A)$. Prove that N must be $\mathcal{N}(A)$.

14. Suppose F is a field and $A \in F^{n\times n}$ is diagonalizable. Let $\lambda_1, \lambda_2, \ldots, \lambda_t$ be the distinct eigenvalues of A. Prove that

$$\mathrm{col}(A - \lambda_1 I) = E_{\lambda_2}(A) + \cdots + E_{\lambda_t}(A).$$

15. Is the following a theorem? Let V be a vector space over a field F and let $S_1, S_2, \ldots, S_k$ be subspaces of V. If $V = S_1 + S_2 + \cdots + S_k$ and $S_i \cap S_j = \{0\}$ for all $i, j = 1, 2, \ldots, k$, $i \neq j$, then V is the direct sum of $S_1, S_2, \ldots, S_k$. Prove the result or give a counterexample.

5.2 Generalized eigenspaces

In the next three sections, most of our attention will focus on matrices belonging to $\mathbf{C}^{n\times n}$. Actually, our results apply to any algebraically closed field, but $\mathbf{C}$ is the only such field that concerns us in this book.

In the previous section, we saw that if $\lambda \in F$ is an eigenvalue of $A \in F^{n\times n}$ and $\mathcal{N}((A-\lambda I)^2) = \mathcal{N}(A - \lambda I)$, then we can find $X \in F^{n\times n}$ such that $X^{-1}AX$ has the form

$$\left[\begin{array}{c|c} \lambda I & 0 \\ \hline 0 & D \end{array}\right].$$

On the other hand, we saw in the previous chapter that the diagonalizability of A requires that the algebraic and geometric multiplicity of λ be equal. The reader might suspect that these two conditions ($\mathcal{N}((A-\lambda I)^2) = \mathcal{N}(A-\lambda I)$ and algebraic multiplicity equals geometric multiplicity) are related. The next theorem shows that this is the case. First we need the following result:

Lemma 232 *Let F be a field and let $A \in F^{n\times n}$. If A has the form*

$$A = \begin{bmatrix} B & C \\ 0 & D \end{bmatrix},$$

where $B \in F^{k\times k}$, $C \in F^{k\times(n-k)}$, and $D \in F^{(n-k)\times(n-k)}$, then

$$\det(A) = \det(B)\det(D)$$

and

$$p_A(r) = p_B(r)p_D(r).$$

Proof Exercise 1.

Theorem 233 *Let F be a field, let $\lambda \in F$ be an eigenvalue of $A \in F^{n\times n}$, and suppose $\mathcal{N}((A-\lambda I)^2) = \mathcal{N}(A-\lambda I)$. Then the geometric multiplicity of λ equals its algebraic multiplicity.*

Proof Let m and k be the algebraic and geometric multiplicities, respectively, of λ. We can then find bases $\{x_1,\ldots,x_k\}$ and $\{x_{k+1},\ldots,x_n\}$ for $N = \mathcal{N}(A-\lambda I)$ and $R = \text{col}(A-\lambda I)$, respectively. Moreover, we know from Corollary 228 and Theorem 230 that $\mathcal{N}((A-\lambda I)^2) = \mathcal{N}(A-\lambda I)$ implies that F^n is the direct sum of N and R. Therefore, $\{x_1,\ldots,x_n\}$ is a basis for F^n and, if $X = [x_1|\cdots|x_n]$, then

$$X^{-1}AX = \left[\begin{array}{c|c} \lambda I & 0 \\ \hline 0 & D \end{array}\right], \tag{5.2}$$

where $D \in F^{(n-k)\times(n-k)}$. Since any matrix similar to A has the same characteristic polynomial as does A, we see that

$$p_A(r) = \left|\begin{array}{c|c} (r-\lambda)I & 0 \\ \hline 0 & rI-D \end{array}\right| = (r-\lambda)^k p_D(r).$$

On the other hand, since the algebraic multiplicity of λ is m, we know that

$$p_A(r) = (r-\lambda)^m q(r),$$

where λ is not a root of $q(r)$. To prove that $k = m$, then, it suffices to show that λ is not a root of $p_D(r)$, that is, that λ is not an eigenvalue of D. We will use the notation $X_1 = [x_1|\cdots|x_k]$, $X_2 = [x_{k+1}|\cdots|x_n]$.

Let us argue by contradiction and suppose $u \in F^{n-k}$ satisfies $Du = \lambda u$ and $u \neq 0$. Let us define $x = X_2u$. Equation (5.2) yields

$$A[X_1|X_2] = [X_1|X_2]\left[\begin{array}{c|c} \lambda I & 0 \\ \hline 0 & D \end{array}\right] \Rightarrow AX_1 = \lambda X_1,\ AX_2 = X_2D.$$

But then

$$Ax = AX_2u = X_2Du = X_2(\lambda u) = \lambda(X_2u) = \lambda x.$$

Moreover, since the columns of X_2 are linearly independent and $u \neq 0$, we see that $x \neq 0$ and therefore that x is an eigenvector of A corresponding to the eigenvalue λ. This implies that $x \in N = \mathcal{N}(A - \lambda I)$. On the other hand, every linear combination of the columns of X_2 lies in R, and hence $x \in N \cap R = \{0\}$. This contradicts that $x \neq 0$, and therefore the assumption that λ is an eigenvalue of D.

QED

The converse of Theorem 233 is also true (see Exercise 9).

The following corollary is a direct consequence of the foregoing result.

Corollary 234 *If $A \in \mathbf{C}^{n\times n}$ and, for each eigenvalue λ of A,*

$$\mathcal{N}((A - \lambda I)^2) = \mathcal{N}(A - \lambda I),$$

then A is diagonalizable.

Proof Since $\mathbf{C}$ is algebraically closed, the characteristic polynomial can be factored as

$$p_A(r) = (r - \lambda_1)^{m_1}(r - \lambda_2)^{m_2} \cdots (r - \lambda_t)^{m_t},$$

where $\lambda_1, \lambda_2, \ldots, \lambda_t$ are the distinct eigenvalues of A and $m_1 + m_2 + \cdots + m_t = n$. By the previous theorem, we have $\dim(E_{\lambda_i}) = m_i$ for each $i = 1, 2, \ldots, t$, and therefore

$$\dim(E_{\lambda_1}) + \dim(E_{\lambda_2}) + \cdots + \dim(E_{\lambda_t}) = n.$$

By Corollary 195, there exists a basis for F^n consisting of eigenvectors of A, and therefore A is diagonalizable.

QED

We will now focus on $A \in \mathbf{C}^{n\times n}$. If λ is an eigenvalue of A and

$$\mathcal{N}((A - \lambda I)^2) \neq \mathcal{N}(A - \lambda I),$$

then we cannot write $\mathbf{C}^n$ as the direct sum of $\mathcal{N}(A - \lambda I)$ and another subspace that is invariant under A. In this case, we would like to decompose $\mathbf{C}^n$ into the direct sum of invariant subspaces, one of which contains $\mathcal{N}(A - \lambda I)$ as a subspace. The following theorem shows how to do this.

Theorem 235 *Let F be a field, let $A \in \mathbf{C}^{n\times n}$, and let λ be an eigenvalue of A with algebraic multiplicity m. Then:*

1. *there exists a smallest positive integer k such that*

$$\mathcal{N}((A - \lambda I)^{k+1}) = \mathcal{N}((A - \lambda I)^k);$$

2. *for all $\ell > k$, $\mathcal{N}((A - \lambda I)^\ell) = \mathcal{N}((A - \lambda I)^k)$;*

3. *$N = \mathcal{N}((A-\lambda I)^k)$ and $R = \mathrm{col}((A-\lambda I)^k)$ are invariant under A;*

4. *$\mathbf{C}^n$ is the direct sum of N and R;*

5. $\dim(N) = m$.

Proof

1. Since each $\mathcal{N}((A-\lambda I)^j)$ is a subspace of the finite-dimensional space $\mathbf{C}^n$, the dimension of $\mathcal{N}((A-\lambda I)^j)$ cannot strictly increase as a function of j (since otherwise it would eventually exceed the dimension of $\mathbf{C}^n$). Therefore, there is a smallest positive integer k such that $\mathcal{N}((A-\lambda I)^{k+1}) = \mathcal{N}((A-\lambda I)^k)$.

2. Suppose ℓ is a positive integer with $\ell > k$, say $\ell = k+s$. We can prove by induction on s that $\mathcal{N}((A-\lambda I)^{k+s}) = \mathcal{N}((A-\lambda I)^k)$ for all $s \geq 1$. By definition of k, this is true for $s=1$. Suppose

$$\mathcal{N}((A-\lambda I)^{k+s}) = \mathcal{N}((A-\lambda I)^k).$$

We must show that $\mathcal{N}((A-\lambda I)^{k+s+1}) = \mathcal{N}((A-\lambda I)^{k+s})$. Obviously $\mathcal{N}((A-\lambda I)^{k+s}) \subset \mathcal{N}((A-\lambda I)^{k+s+1})$, so let us assume that x belongs to $\mathcal{N}((A-\lambda I)^{k+s+1})$. Then

$$\begin{aligned}(A-\lambda I)^{k+s+1}x = 0 \quad &\Rightarrow \quad (A-\lambda I)^{k+1}((A-\lambda I)^s x) = 0 \\ &\Rightarrow \quad (A-\lambda I)^k (A-\lambda I)^s x = 0 \\ &\Rightarrow \quad (A-\lambda I)^{k+s} x = 0 \\ &\Rightarrow \quad x \in \mathcal{N}((A-\lambda I)^{k+s}).\end{aligned}$$

(The second step follows from $\mathcal{N}((A-\lambda I)^{k+1}) = \mathcal{N}((A-\lambda I)^k)$.) This completes the proof by induction.

3. Since $(A-\lambda I)^k$ and A commute, it follows that

$$\begin{aligned}& x \in \mathcal{N}((A-\lambda I)^k) \\ \Rightarrow \quad & (A-\lambda I)^k Ax = A(A-\lambda I)^k x = A0 = 0 \\ \Rightarrow \quad & Ax \in \mathcal{N}((A-\lambda I)^k).\end{aligned}$$

Thus $x \in \mathcal{N}((A-\lambda I)^k)$ is invariant under A. A similar proof shows that $\mathrm{col}((A-\lambda I)^k)$ is invariant under A.

4. By the second conclusion, we see that

$$\mathcal{N}((A-\lambda I)^{2k}) = \mathcal{N}((A-\lambda I)^k);$$

therefore, by Corollary 228 and Theorem 230 applied to $(A-\lambda I)^k$, we see that $\mathbf{C}^n$ is the direct sum of $\mathcal{N}((A-\lambda I)^k)$ and $\mathrm{col}((A-\lambda I)^k)$.

5. Let $\{x_1,\ldots,x_\ell\}$, $\{x_{\ell+1},\ldots,x_n\}$ be bases for N and R, respectively. Since $\mathbf{C}^n$ is the direct sum of N and R, $\{x_1,\ldots,x_n\}$ is a basis for $\mathbf{C}^n$ and therefore $X = [x_1|\cdots|x_n]$ is invertible. We write $X_1 = [x_1|\cdots|x_\ell]$ and $X_2 = [x_{\ell+1}|\cdots|x_n]$. The invariance of N and R implies that there exists matrices $B \in \mathbf{C}^{\ell\times\ell}$ and $C \in \mathbf{C}^{(n-\ell)\times(n-\ell)}$ such that $AX_1 = X_1B$ and $AX_2 = X_2C$, and therefore

$$X^{-1}AX = \left[\begin{array}{c|c} B & 0 \\ \hline 0 & C \end{array}\right].$$

It then follows that

$$p_A(r) = p_B(r)p_C(r).$$

Since m is the algebraic multiplicity of λ as an eigenvalue of A, we have $p_A(r) = (r-\lambda)^m q(r)$, where $q(\lambda) \neq 0$.

Notice that

$$Bu = \mu u \Rightarrow AX_1u = X_1Bu = \mu X_1u.$$

It follows that any eigenvalue of B is also an eigenvalue of A, with a corresponding eigenvector lying in N. Exercise 8 asks the reader to show that the only eigenvectors of A lying in $\mathcal{N}((A-\lambda I)^k)$ are eigenvectors corresponding to λ; therefore, we conclude that λ is the only eigenvalue of B. Since $\mathbf{C}$ is algebraically closed, $p_B(r)$ can be factored into the product of linear polynomials, and since λ is the only eigenvalue of B, we obtain

$$p_B(r) = (r-\lambda)^\ell.$$

Next notice that

$$\begin{aligned} Cv = \lambda v \quad &\Rightarrow \quad AX_2v = X_2Cv = \lambda X_2v \\ &\Rightarrow \quad X_2v \in \mathcal{N}(A-\lambda I) \subset N \Rightarrow X_2v = 0, \end{aligned}$$

where the last step follows from the fact that $X_2v \in R$ and $N \cap R = \{0\}$. Since the columns of X_2 form a linearly independent set, it follows that $v = 0$ and hence that λ is not an eigenvalue of C, that is, $p_C(\lambda) \neq 0$.

We now see that

$$p_A(r) = p_B(r)p_C(r) \Rightarrow (r-\lambda)^m q(r) = (r-\lambda)^\ell p_C(r).$$

Since λ is not a root of $q(r)$ or $p_C(r)$, it follows that $\ell = m$ must hold; that is, the geometric multiplicity of λ equals its algebraic multiplicity.

QED

If λ is an eigenvalue of A, $\mathcal{N}((A-\lambda I)^2) \neq \mathcal{N}(A-\lambda I)$, and $k > 1$ is the smallest positive integer such that $\mathcal{N}((A-\lambda I)^{k+1}) = \mathcal{N}((A-\lambda I)^k)$, then we call $\mathcal{N}((A-\lambda I)^k)$ the *generalized eigenspace* of A corresponding to λ, and we write

$$G_\lambda(A) = \mathcal{N}((A-\lambda I)^k).$$

The nonzero elements of $G_\lambda(A)$ are called *generalized eigenvectors.* If it happens that $\mathcal{N}((A-\lambda I)^2) = \mathcal{N}(A-\lambda I)$, then we identify the generalized eigenspace corresponding to λ with the ordinary eigenspace:

$$G_\lambda(A) = \mathcal{N}(A-\lambda I) = E_\lambda(A).$$

We wish to show that, for any $A \in \mathbf{C}^{n\times n}$, $\mathbf{C}^n$ is the direct sum of the generalized eigenspaces of A. We will need the following preliminary results.

Lemma 236 *Let F be a field and let $A \in F^{n\times n}$.*

1. *If $p(r), q(r) \in F[r]$, then $p(A)$ and $q(A)$ commute.*
2. *If S is a subspace of F^n that is invariant under A, then S is also invariant under $p(A)$ for all $p(r) \in F[r]$.*

Proof Exercise 3.

We will mostly apply Lemma 236 to matrices of the form $(A-\lambda I)^k$ (that is, to polynomials of the form $p(r) = (r-\lambda)^k$). However, in the appendix to this section, we will encounter more general polynomials.

Lemma 237 *Let F be a field, and assume $A \in \mathbf{C}^{n\times n}$. If $\lambda_1, \lambda_2, \ldots, \lambda_s$ are distinct eigenvalues of A and $x_1, x_2, \ldots, x_s$ satisfy*

$$x_1 + x_2 + \cdots + x_s = 0, \ x_i \in G_{\lambda_i}(A) \text{ for all } i = 1, 2, \ldots, t,$$

then $x_1 = x_2 = \ldots = x_s = 0$.

Proof We argue by induction on s. There is nothing to prove if $s = 1$. Let us suppose the result holds for some integer s with $1 \leq s < t$ (where t is the number of distinct eigenvalues of A). Assume that $\lambda_1, \lambda_2, \ldots, \lambda_{s+1}$ are distinct eigenvalues of A and that $x_i \in G_{\lambda_i}(A)$ for $i = 1, 2, \ldots, s+1$ satisfy

$$x_1 + x_2 + \cdots + x_{s+1} = 0.$$

Multiply both sides of this equation by the matrix $(A-\lambda_{s+1})^{k_{s+1}}$, where

$$G_{\lambda_{s+1}}(A) = \mathcal{N}((A-\lambda_s I)^{k_{s+1}}).$$

Then

$$(A-\lambda_s I)^{k_{s+1}} x_{s+1} = 0 \text{ since } x_{s+1} \in \mathcal{N}((A-\lambda_s I)^{k_{s+1}})$$

and

$$\tilde{x}_i = (A-\lambda_s I)^{k_{s+1}} x_i \in G_{\lambda_i}(A) \text{ for all } i = 1, 2, \ldots, s$$

since $x_i \in G_{\lambda_i}(A)$ and $G_{\lambda_i}(A)$ is invariant under A, and hence also under $(A - \lambda I)^{k_{s+1}}$. But then we have

$$\tilde{x}_1 + \tilde{x}_2 + \cdots + \tilde{x}_s = 0, \ \tilde{x}_i \in G_{\lambda_i}(A), \ i = 1, 2, \ldots, s.$$

By the induction hypothesis, this implies that

$$\begin{aligned} & \tilde{x}_i = (A - \lambda_s I)^{k_{s+1}} x_i = 0 \text{ for all } i = 1, 2, \ldots, s \\ \Rightarrow \ & x_i \in G_{\lambda_i}(A) \cap G_{\lambda_{s+1}}(A) \text{ for all } i = 1, 2, \ldots, s. \end{aligned}$$

We complete the proof by showing that

$$G_{\lambda_i}(A) \cap G_{\lambda_j}(A) = \{0\} \text{ for all } i \neq j. \tag{5.3}$$

Let $x \in G_{\lambda_i}(A) \cap G_{\lambda_j}(A)$ and assume by way of contradiction that $x \neq 0$. Then there exists an integer $p \geq 0$ such that

$$(A - \lambda_i I)^p x \neq 0 \text{ and } (A - \lambda_i I)^{p+1} x = 0.$$

It follows that $y = (A - \lambda_i I)^p x$ is an eigenvector of A corresponding to λ_i. But

$$(A - \lambda_j I)^{k_j} y = (A - \lambda_j I)^{k_j} (A - \lambda_i I)^p x = (A - \lambda_i I)^p (A - \lambda_j I)^{k_j} x = 0,$$

with the last step following from the fact that $x \in G_{\lambda_j}(A)$. This shows that y is an eigenvector of A corresponding to the eigenvalue λ_i that belongs to $G_{\lambda_j}(A)$. This contradicts the result in Exercise 8, and shows that no such x exists. It follows that (5.3) holds, and the proof is complete.

QED

We can now give the main result.

Theorem 238 *Let $A \in \mathbf{C}^{n \times n}$, and let the characteristic polynomial of A be*

$$p_A(r) = (r - \lambda_1)^{m_1} (r - \lambda_2)^{m_2} \cdots (r - \lambda_t)^{m_t},$$

where $\lambda_1, \lambda_2, \ldots, \lambda_t$ are the distinct eigenvalues of A. Then $\mathbf{C}^n$ is the direct sum of the corresponding generalized eigenspaces

$$G_{\lambda_1}(A), G_{\lambda_2}(A), \ldots, G_{\lambda_t}(A).$$

Proof We know that $m_1 + m_2 + \ldots + m_t = \deg(p_A(r)) = n$, and Theorem 235 implies that $\dim(G_{\lambda_i}(A)) = m_i$. Therefore,

$$\dim(G_{\lambda_1}(A)) + \dim(G_{\lambda_2}(A)) + \cdots + \dim(G_{\lambda_t}(A)) = n.$$

Given this fact and Lemma 237, Theorem 226 implies $\mathbf{C}^n$ is the direct sum of $G_{\lambda_1}(A), G_{\lambda_2}(A), \ldots, G_{\lambda_t}(A)$, as desired.

QED

Let us suppose that $A \in \mathbf{C}^{n \times n}$ and that the characteristic polynomial of A is

$$p_A(r) = (r - \lambda_1)^{m_1}(r - \lambda_2)^{m_2} \cdots (r - \lambda_t)^{m_t},$$

where $\lambda_1, \lambda_2, \ldots, \lambda_t$ are the distinct eigenvalues of A. We can find a basis for each $G_{\lambda_i}(A)$ and form the $n \times m_i$ matrix X_i whose columns are the basis vectors. We then define $X = [X_1|X_2|\cdots|X_t]$. Since each $G_{\lambda_i}(A)$ is invariant under A, there exists a $m_i \times m_i$ matrix B_i such that $AX_i = X_iB_i$, and therefore

$$X^{-1}AX = \begin{bmatrix} B_1 & & & \\ & B_2 & & \\ & & \ddots & \\ & & & B_t \end{bmatrix}.$$

It remains only to choose a basis for each $G_{\lambda_i}(A)$ so that the diagonal block B_i is as simple as possible. This is the subject of the next two sections.

5.2.1 Appendix: Beyond generalized eigenspaces

We now return to the case that F is a general field and A belongs to $F^{n \times n}$. We wish to understand how to express F^n as the direct sum of invariant subspaces in the case that the characteristic polynomial $p_A(r)$ cannot be factored completely into linear factors. The key will be another polynomial related to A, called the *minimal polynomial* of A. We will draw heavily on facts about polynomials that are reviewed in Appendix C.

As usual, we assume that $A \in F^{n \times n}$. Since $F^{n \times n}$ is finite-dimensional, the set $\{I, A, A^2, \ldots, A^s\}$ must be linearly dependent for s sufficiently large. We define the positive integer s by the condition that

$$\{I, A, A^2, \ldots, A^{s-1}\}$$

is linearly independent, but

$$\{I, A, A^2, \ldots, A^s\}$$

is linearly dependent. Then there exist unique scalars $c_0, c_1, \ldots, c_{s-1} \in F$ such that

$$c_0I + c_1A + \cdots + c_{s-1}A^{s-1} + A^s = 0.$$

We define $m_A(r) = c_0 + c_1r + \cdots + c_{s-1}r^{s-1} + r^s$ and call $m_A(r)$ the *minimal polynomial* of A. By definition, $m_A(A) = 0$, which is equivalent to

$$m_A(A)x = 0 \text{ for all } x \in F^n.$$

Also by definition, if $p(r) \in F[r]$ and $p(A) = 0$, then $\deg(p(r)) \geq \deg(m_A(r))$.

Theorem 239 *Let $A \in F^{n \times n}$ be given, and let $m_A(r) \in F[r]$ be the minimal polynomial of A. If $p(r) \in F[r]$ satisfies $p(A) = 0$, then $m_A(r)$ divides $p(r)$, that is, there exists $q(r) \in F[r]$ such that $p(r) = m_A(r)q(r)$.*

Proof By the division algorithm (Theorem 496 in Appendix C), there exist polynomials $q(r), n(r) \in F[r]$ such that

$$p(r) = m_A(r)q(r) + n(r)$$

and either $n(r)$ is the zero polynomial or $\deg(n(r)) < \deg(m_A(r))$. We then have $p(A) = m_A(A)q(A) + n(A)$, which implies that $n(A) = 0$ since both $p(A)$ and $m_A(A)$ are the zero matrix. However, by definition of $m_A(r)$, there is no nonzero polynomial $n(r)$ with degree less than $\deg(m_A(r))$ such that $n(A) = 0$. Therefore $n(r)$ must be the zero polynomial, that is, $p(r) = m_A(r)q(r)$.

QED

We will need the following fact about polynomials and matrices.

Lemma 240 *Let $A \in F^{n\times n}$. If $\lambda \in F$ is an eigenvalue of A, $x \in F^n$ is a corresponding eigenvector, and $p(r) \in F[r]$, then $p(A)x = p(\lambda)x$.*

Proof Exercise 11.

Here is a fundamental fact about the minimal polynomial of A.

Theorem 241 *Let $A \in F^{n\times n}$ be given, and let $m_A(r) \in F[r]$ be the minimal polynomial of A. The roots of $m_A(r)$ are precisely the eigenvalues of A.*

Proof If λ is an eigenvalue of A, then there exists a nonzero vector $x \in F^n$ such that $Ax = \lambda x$, and Lemma 240 implies that $m_A(A)x = m_A(\lambda)x$. But $m_A(A)y = 0$ for all $y \in F^n$, and therefore $m_A(\lambda)x = 0$. Since $x \neq 0$, it follows that $m_A(\lambda) = 0$, and λ is a root of $m_A(r)$.

Conversely, suppose λ is a root of $m_A(r)$. By Corollary 504 in Appendix C, there exists $q(r) \in F[r]$ such that $m_A(r) = (r - \lambda)q(r)$. Now suppose that λ is not an eigenvalue of A, which implies that $(A - \lambda I)x \neq 0$ for all $x \neq 0$. Then, for any $x \in F^n$,

$$m_A(A)x = 0 \Rightarrow (A - \lambda I)q(A)x = 0 \Rightarrow q(A)x = 0$$

(the last step follows from the assumption that $A - \lambda I$ is nonsingular). But then $q(A)x = 0$ for all $x \in \mathbf{R}^n$, which is impossible because $\deg(q(r))$ is strictly less than $\deg(m_A(r))$. This contradiction shows that λ must be an eigenvalue of A.

QED

We recall one more fact about polynomials that will be crucial in our development below. A polynomial $p(r) \in F[r]$ can be factored uniquely as

$$p(r) = cp_1(r)^{k_1} p_2(r)^{k_2} \cdots p_t(r)^{k_t},$$

where $p_1(r), p_2(r), \ldots, p_t(r)$ are distinct irreducible monic polynomials in $F[r]$, $c \in F$, and $k_1, k_2, \ldots, k_t$ are positive integers (see Theorem 508 in Appendix

C). If we apply this theorem to $m_A(r)$, which itself is monic, we see that $c = 1$ and we obtain

$$m_A(r) = p_1(r)^{k_1} p_2(r)^{k_2} \cdots p_t(r)^{k_t}.$$

We will show that F^n is the direct sum of the subspaces

$$\mathcal{N}(p_1(A)^{k_1}), \mathcal{N}(p_2(A)^{k_2}), \ldots, \mathcal{N}(p_t(A)^{k_t}),$$

each of which is invariant under A. We will also show that, in the case that A is diagonalizable, the spaces $\mathcal{N}(p_i(A)^{k_i})$ are the eigenspaces of A, and our results reduce to the usual spectral decomposition of A.

The derivation of the desired results is based on a trick. We define the polynomials

$$q_i(r) = \frac{m_A(r)}{p_i(r)^{k_i}}, \quad i = 1, 2, \ldots, t \tag{5.4}$$

($q_i(r)$ is a polynomial because it is obtained by removing one of the factors of $m_A(r)$). The polynomials $q_1(r), q_2(r), \ldots, q_t(r)$ are relatively prime (any possible common divisor would include $p_i(r)$ for some i, but no $p_i(r)$ divides every $q_j(r)$ since $p_i(r)$ does not divide $q_i(r)$). Therefore, by Theorem 501 of Appendix C, there exist $f_1(r), f_2(r), \ldots, f_t(r) \in F[r]$ such that

$$f_1(r)q_1(r) + f_2(r)q_2(r) + \cdots + f_t(r)q_t(r) = 1. \tag{5.5}$$

Equation (5.5) implies that

$$f_1(A)q_1(A) + f_2(A)q_2(A) + \cdots + f_t(A)q_t(A) = I,$$

where $I \in F^{n\times n}$ is the identity matrix. This in turn implies that

$$f_1(A)q_1(A)x + f_2(A)q_2(A)x + \cdots + f_t(A)q_t(A)x = x \text{ for all } x \in F^n. \tag{5.6}$$

Equation (5.6) shows that

$$F^n = \mathrm{col}(f_1(A)q_1(A)) + \mathrm{col}(f_2(A)q_2(A)) + \cdots + \mathrm{col}(f_t(A)q_t(A)).$$

Throughout the following discussion, the polynomials $p_i(r)$, $q_i(r)$, $f_i(r)$, $i = 1, 2, \ldots, t$ have the meanings assigned above. We will often use the fact that polynomials in A commute (see Lemma 236).

Theorem 242 *For each* $i = 1, 2, \ldots, t$, $\mathrm{col}(f_i(A)q_i(A)) = \mathcal{N}(p_i(A)^{k_i})$.

Proof If $x \in \mathrm{col}(f_i(A)q_i(A))$, then $x = f_i(A)q_i(A)y$ for some $y \in F^n$. It then follows that

$$p_i(A)^{k_1}x = p_i(A)^{k_1}f_i(A)q_i(A)y = f_i(A)p_i(A)^{k_1}q_i(A)y = f_i(A)m_A(A)y = 0$$

(since $m_A(A)$ is the zero matrix). Thus $x \in \mathcal{N}(p_i(A)^{k_1})$.

Conversely, suppose $x \in \mathcal{N}(p_i(A)^{k_i})$. Then

$$x \in \mathcal{N}(f_j(A)q_j(A)) \text{ for all } j \neq i,$$

since each $q_j(A)$, $j \neq i$, contains a factor of $p_i(A)$. But then (5.6) implies that

$$x = f_i(A)q_i(A)x, \tag{5.7}$$

and hence that $x \in \text{col}(f_i(A)q_i(A))$.

Since $x \in \mathcal{N}(p_i(A)^{k_i})$ if and only if $x \in \text{col}(f_i(A)q_i(A))$, we see that $\mathcal{N}(p_i(A)^{k_i}) = \text{col}(f_i(A)q_i(A))$.

QED

The previous theorem, together with (5.7), shows that

$$F^n = \mathcal{N}(p_1(A)^{k_1}) + \mathcal{N}(p_2(A)^{k_2}) + \cdots + \mathcal{N}(p_t(A)^{k_t}). \tag{5.8}$$

Theorem 243 *If $x_i \in \mathcal{N}(p_i(A)^{k_i})$ for $i = 1, 2, \ldots, t$ and*

$$x_1 + x_2 + \cdots + x_t = 0,$$

then $x_1 = x_2 = \ldots = x_t = 0$.

Proof We will prove by induction on k that

$$x_i \in \mathcal{N}(p_i(A)^{k_i}) \text{ for all } i = 1, 2, \ldots, k, \; x_1 + \cdots + x_k = 0 \Rightarrow x_1 = \ldots = x_k = 0.$$

There is nothing to prove for $k = 1$. We now assume that for some k satisfying $1 \leq k < t$, if $y_i \in \mathcal{N}(p_i(A)^{k_i})$ for $i = 1, 2, \ldots, k$ and $y_1 + \cdots + y_k = 0$, then $y_1 = \ldots = y_k = 0$. Suppose $x_i \in \mathcal{N}(p_i(A)^{k_i})$, $i = 1, 2, \ldots, k+1$, are given and satisfy $x_1 + \cdots + x_{k+1} = 0$. Then, multiplying both sides of this last equation by $p_{k+1}(A)^{k_{k+1}}$, we obtain

$$\begin{aligned} & p_{k+1}(A)^{k_{k+1}}x_1 + \cdots + p_{k+1}(A)^{k_{k+1}}x_k + p_{k+1}(A)^{k_{k+1}}x_{k+1} = 0 \\ \Rightarrow \quad & \tilde{x}_1 + \cdots + \tilde{x}_k = 0, \end{aligned}$$

where $\tilde{x}_i = p_{k+1}(A)^{k_{k+1}}x_i$ (since $p_{k+1}(A)^{k_{k+1}}x_{k+1} = 0$). The null space of $p_i(A)^{k_i}$ is invariant under $p_{k+1}(A)^{k_{k+1}}$ (in fact, under any polynomial in A), and therefore

$$\tilde{x}_i \in \mathcal{N}(p_i(A)^{k_i}) \text{ for all } i = 1, 2, \ldots, k.$$

The induction hypothesis then implies that $\tilde{x}_i = 0$ for $i = 1, 2, \ldots, k$. This in turn implies that $x_i \in \mathcal{N}(p_{k+1}(A)^{k_{k+1}}$ for all $i = 1, 2, \ldots, k$, and we are already assuming that $x_i \in \mathcal{N}(p_i(A)^{k_i}$. From (5.6), we see that

$$x_i = f_1(A)q_1(A)x_i + f_2(A)q_2(A)x_i + \cdots + f_t(A)q_t(A)x_i,$$

and every $q_j(A)$, $j = 1, 2, \ldots, t$, contains a factor of $p_i(A)^{k_i}$ or $p_{k+1}(A)^{k_{k+1}}$ (or both). It follows that $x_i = 0$ for all $i = 1, 2, \ldots, k$. But then the equation $x_1 + \cdots + x_{k+1} = 0$ implies that $x_{k+1} = 0$, and the proof is complete.

QED

Corollary 244 *The space F^n is the direct sum of the subspaces*

$$\mathcal{N}(p_1(A)^{k_1}), \mathcal{N}(p_2(A)^{k_2}), \ldots, \mathcal{N}(p_t(A)^{k_t}),$$

each of which is invariant under A.

Proof The previous two theorems, together with Theorem 226, imply that F^n is the direct sum of the given subspaces. Each null space is obviously invariant under A:

$$\begin{aligned} x \in \mathcal{N}(p_i(A)^{k_i}) \ \Rightarrow\ p_i(A)^{k_i}x = 0 \ &\Rightarrow\ p_i(A)^{k_i}Ax = Ap_i(A)^{k_i}x = 0 \\ &\Rightarrow\ Ax \in \mathcal{N}(p_i(A)^{k_i}). \end{aligned}$$

QED

If we now choose a basis for each subspace $\mathcal{N}(p_i(A)^{k_i})$ and form a matrix X whose columns comprise the union of these bases, then $X^{-1}AX$ will be block diagonal. The only remaining question is how to choose a basis for $\mathcal{N}(p_i(A)^{k_i})$ so that the corresponding diagonal block will be as simple as possible. We will not pursue this question here for a general irreducible polynomial $p_i(A)$. In the special case that $p_i(A) = x - \lambda_i$, then λ_i is an eigenvalue of A and there are two subcases to consider, as described above. One possibility is that $\mathcal{N}((A - \lambda_i I)^2) = \mathcal{N}(A - \lambda_i I)$, in which case $k_i = 1$ (see Exercise 12) and

$$\mathcal{N}(p_i(A)^{k_i}) = \mathcal{N}(A - \lambda_i I) = E_{\lambda_i}(A).$$

Then any basis for $\mathcal{N}(p_i(A)^{k_i})$ will do and the corresponding diagonal block will be a diagonal matrix. The second case is that $\mathcal{N}((A-\lambda_i I)^2) \neq \mathcal{N}(A-\lambda_i I)$, and this case will be carefully studied in the next two sections.

If $p_i(A)$ is an irreducible polynomial of degree two or greater, then the results on *cyclic vectors* found in Chapter 7 of Hoffman and Kunze [21] can be used to make the diagonal block relatively simple. However, we will not pursue this development here.

5.2.2 The Cayley-Hamilton theorem

By definition, $m_A(A)$ is the zero matrix for every $A \in F^{n\times n}$, where $m_A(r)$ is the minimal polynomial. The following theorem shows that the characteristic polynomial of A also has this property.

Theorem 245 (The Cayley-Hamilton theorem) *Let F be a field, let $A \in F^{n\times n}$, and let $p_A(r) \in F[r]$ be the characteristic matrix of A. Then $p_A(A)$ is the zero matrix.*

Proof We will argue by induction on n. If $n = 1$, then A can be identified with a scalar, and $p_A(r) = r - A$. It then follows immediately that $p_A(A) = 0$.

Now suppose the result holds for all square matrices of order less than $n > 1$. Let $A \in F^{n\times n}$ be given. We will consider two cases. In the first case, we assume that there exists a subspace S of F^n that has dimension k for some k satisfying $1 \le k < n$. Suppose $\{x_1, \ldots, x_k\}$ is a basis for S, and $\{x_1, \ldots, x_n\}$ is a basis for F^n. We define $X_1 = [x_1|\cdots|x_k]$, $X_2 = [x_{k+1}|\cdots|x_n]$, and $X = [X_1|X_2]$. We then have

$$X^{-1}AX = T = \left[\begin{array}{c|c} B & C \\ \hline 0 & D \end{array}\right],$$

where $B \in F^{k\times k}$, $C \in F^{k\times(n-k)}$, $D \in F^{(n-k)\times(n-k)}$. We know from previous results that $p_A(r) = p_T(r) = p_B(r)p_D(r)$. Since

$$p_A(A) = p_A(XTX^{-1}) = Xp_A(T)X^{-1},$$

it suffices to prove that $p_A(T) = 0$. It is straightforward to prove by induction that

$$T^k = \left[\begin{array}{c|c} B^k & M_k \\ \hline 0 & D^k \end{array}\right]$$

for some matrix $M_k \in F^{k\times(n-k)}$, and hence that

$$p_B(T) = \left[\begin{array}{c|c} p_B(B) & N_k \\ \hline 0 & p_B(D) \end{array}\right],\ p_D(T) = \left[\begin{array}{c|c} p_D(B) & P_k \\ \hline 0 & p_D(D) \end{array}\right],$$

where $N_k, P_k \in F^{k\times(n-k)}$. By the induction hypothesis, $p_B(B) = 0$ and $p_D(D) = 0$. We then obtain

$$\begin{aligned} p_A(T) = p_B(T)p_D(T) &= \left[\begin{array}{c|c} p_B(B) & N_k \\ \hline 0 & p_B(D) \end{array}\right]\left[\begin{array}{c|c} p_D(B) & P_k \\ \hline 0 & p_D(D) \end{array}\right] \\ &= \left[\begin{array}{c|c} 0 & N_k \\ \hline 0 & p_B(D) \end{array}\right]\left[\begin{array}{c|c} p_D(B) & P_k \\ \hline 0 & 0 \end{array}\right] = 0. \end{aligned}$$

This completes the proof in the first case.

In the second case, there is no proper subspace of F^n that is invariant under A. We first prove that, for any nonzero vector x, the set $\{x, Ax, \ldots, A^{n-1}x\}$ is linearly independent. If this set is linearly dependent, there exists k satisfying $1 < k \le n$ such that $\{x, Ax, \ldots, A^{k-2}x\}$ is linearly independent and $\{x, Ax, \ldots, A^{k-1}x\}$ is linearly dependent. It is then easy to show that $S = \text{sp}\{x, Ax, \ldots, A^{k-2}x\}$ is a proper invariant subspace, and this contradiction shows that our claim is correct. Therefore, let x be any nonzero vector in F^n and define $X = [x|Ax|\cdots|A^{n-1}x]$. There exist unique scalars $c_0, c_1, \ldots, c_{n-1} \in F$ such that

$$A^n x + c_{n-1}A^{n-1}x + \cdots + c_1 Ax + c_0 x = 0,$$

that is,

$$\left(A^n + c_{n-1}A^{n-1} + \cdots + c_1 A + c_0 I\right)x = 0.$$

If we now define $X = [x|Ax|\cdots|A^{n-1}x]$, we have

$$\begin{aligned} AX &= [Ax|A^2x|\cdots|A^n x] \\ &= [Ax|A^2x|\cdots|A^{n-1}x| - c_0 x - c_1 Ax - \cdots - c_{n-1}A^{n-1}x] \\ &= X\begin{bmatrix} 0 & 0 & \cdots & 0 & -c_0 \\ 1 & 0 & \cdots & 0 & -c_1 \\ \vdots & \vdots & & \vdots & \vdots \\ 0 & 0 & \cdots & 1 & -c_{n-1} \end{bmatrix} = XV. \end{aligned}$$

Now, $p_A(r) = p_V(r) = r^n + c_{n-1}r^{n-1} + \cdots + c_1 r + c_0$ by Exercise 4.5.9 and the fact that A and V are similar. But then, for all $k \geq 0$,

$$\begin{aligned} & p_A(A)x = p_V(A)x = 0 \\ \Rightarrow\ & p_A(A)A^k x = A^k p_A(A)x = A^k 0 = 0. \end{aligned}$$

Since $\{x, Ax, \ldots, A^{n-1}x\}$ is a basis for F^n, this proves that $p_A(A)$ must be the zero matrix. This completes the proof in the second case.

QED

We have seen above that if $p \in F[r]$ and $p(A) = 0$, then $p(r)$ is a multiple of the minimal polynomial $m_A(r)$. Thus we see that $p_A(r)$ is a multiple of $m_A(r)$. In the case of an algebraically closed field, we can say more.

Corollary 246 *Let $A \in \mathbf{C}^{n\times n}$ have minimal polynomial*

$$m_A(r) = (r-\lambda_1)^{k_1}(r-\lambda_2)^{k_2}\cdots(r-\lambda_t)^{k_t},$$

where $\lambda_1, \lambda_2, \ldots, \lambda_t$ are the distinct eigenvalues of A. Then there exist positive integers $m_1, m_2, \ldots, m_t$, with $m_i \geq k_i$ for all $i = 1, 2, \ldots, t$, such that

$$p_A(r) = (r-\lambda_1)^{m_1}(r-\lambda_2)^{m_2}\cdots(r-\lambda_t)^{m_t}.$$

Proof Since $p_A(A) = 0$, Theorem 239 implies that there exists a polynomial $q(r) \in \mathbf{C}[r]$ such that $p_A(r) = q(r)m_A(r)$. Since $\mathbf{C}$ is algebraically closed, $q(r)$ can be expressed as the product of linear polynomials. By Theorem 241, the only possible roots of $q(r)$ are eigenvalues of A, and therefore

$$q(r) = (r-\lambda_1)^{s_1}(r-\lambda_2)^{s_2}\cdots(r-\lambda_t)^{s_t},$$

where $s_i \geq 0$ for all $i = 1, 2, \ldots, t$. The result now follows.

QED

Exercises

Miscellaneous exercises

1. Prove Lemma 232 as follows: Note that

$$\begin{bmatrix} B & C \\ 0 & D \end{bmatrix} = \begin{bmatrix} I & 0 \\ 0 & D \end{bmatrix} \begin{bmatrix} B & C \\ 0 & I \end{bmatrix}.$$

   Prove, by induction on k, that

$$\begin{vmatrix} I & 0 \\ 0 & D \end{vmatrix} = \det(D).$$

   Then a similar proof yields

$$\begin{vmatrix} B & C \\ 0 & I \end{vmatrix} = \det(B).$$

2. Let A be block upper triangular matrix:

$$A = \begin{bmatrix} B_{11} & B_{12} & \cdots & B_{1t} \\ 0 & B_{22} & \cdots & B_{2t} \\ \vdots & & \ddots & \vdots \\ 0 & \cdots & & B_{tt} \end{bmatrix}.$$

   Prove that:

   (a) The determinant of A is the product of the determinants of the diagonal blocks:

$$\det(A) = \det(B_{11})\det(B_{22}) \cdots \det(B_{tt}).$$

   (b) The set of eigenvalues of A is the union of the sets of eigenvalues of $B_{11}, B_{22}, \ldots, B_{tt}$, and the multiplicity of each eigenvalue λ of A is the sum of the multiplicities of λ as an eigenvalue of $B_{11}, B_{22}, \ldots, B_{tt}$.

3. Prove Lemma 236.

4. Let $A \in \mathbf{R}^{4\times 4}$ be defined by

$$A = \begin{bmatrix} 1 & 1 & 0 & 0 \\ 0 & 1 & 1 & 0 \\ 0 & 0 & 1 & 0 \\ 0 & 0 & 0 & -1 \end{bmatrix}.$$

   (a) Find the characteristic polynomial of A.

(b) Show that $\lambda = 1$ is an eigenvalue of A and find its algebraic multiplicity m.

(c) Find the smallest value of k such that $\mathcal{N}((A-I)^{k+1}) = \mathcal{N}((A-I)^k)$.

(d) Show that $\dim((A-I)^k) = m$.

5. Let $A \in \mathbf{R}^{5\times 5}$ be defined by

$$A = \begin{bmatrix} 0 & 0 & 0 & 0 & -1 \\ 10 & 1 & -1 & 5 & -10 \\ 22 & 0 & 1 & 1 & -8 \\ 6 & 0 & 0 & 1 & -2 \\ 1 & 0 & 0 & 0 & -2 \end{bmatrix}.$$

Find the generalized eigenspaces of A, and verify that $\mathbf{R}^5$ is the direct sum of these subspaces.

6. Let $A \in \mathbf{R}^{3\times 3}$ be defined by

$$A = \begin{bmatrix} -1 & 0 & 0 \\ -6 & 1 & 1 \\ 0 & 0 & 1 \end{bmatrix}.$$

Find an invertible matrix $X \in \mathbf{R}^{3\times 3}$ such that $X^{-1}AX$ is block diagonal. (Hint: First find the generalized eigenspaces of A.)

7. Let $A \in \mathbf{R}^{4\times 4}$ be defined by

$$A = \begin{bmatrix} 2 & 0 & 0 & 0 \\ -3 & 2 & -1 & -2 \\ 1 & 0 & 2 & 5 \\ 0 & 0 & 0 & 1 \end{bmatrix}.$$

(a) Find the eigenvalues of A and their algebraic multiplicities.

(b) For each eigenvalue λ, find the smallest value of k such that $\mathcal{N}((A-\lambda I)^{k+1}) = \mathcal{N}((A-\lambda I)^k)$

8. Let F be a field, let λ be an eigenvalue of A, let k be any positive integer, and define $N = \mathcal{N}((A-\lambda I)^k)$. Prove that if $x \in \mathcal{N}((A-\lambda I)^k)$ is an eigenvector of A, then the eigenvalue corresponding to x is λ.

9. Let F be a field, let $\lambda \in F$ be an eigenvalue of $A \in F^{n\times n}$, and suppose that the algebraic and geometric multiplicities of λ are equal. Prove that $\mathcal{N}((A-\lambda I)^2) = \mathcal{N}(A-\lambda I)$.

10. Let F be a field and let $A \in F^{n\times n}$ be defined by

$$A_{ij} = \begin{cases} 1, & j > i, \\ 0, & j \le i. \end{cases}$$

For example, for $n = 4$,

$$A = \begin{bmatrix} 0 & 1 & 1 & 1 \\ 0 & 0 & 1 & 1 \\ 0 & 0 & 0 & 1 \\ 0 & 0 & 0 & 0 \end{bmatrix}.$$

Prove that A has a single generalized eigenspace (corresponding to the eigenvalue $\lambda = 0$), and show that $k = n$ is the smallest positive integer such that $\mathcal{N}((A - \lambda I)^{k+1}) = \mathcal{N}((A - \lambda I)^k)$.

11. Prove Lemma 240.

12. Let F be a field, assume $A \in F^{n\times n}$, and let $m_A(r)$ be factored into irreducible monic polynomials, as in Appendix 5.2.1 above:

$$m_A(r) = p_1(r)^{k_1} p_2(r)^{k_2} \cdots p_t(r)^{k_t}.$$

   (a) Define ℓ_i to be the smallest positive integer such that

$$\mathcal{N}(p_i(A)^{\ell_i+1}) = \mathcal{N}(p_i(A)^{\ell_i}).$$

   Explain why ℓ_i is well-defined.

   (b) Prove that $\mathcal{N}(p_i(A)^{\ell}) = \mathcal{N}(p_i(A)^{\ell_i})$ for all $\ell \geq \ell_i$.

   (c) Prove that $\ell_i = k_i$. (Hint: Prove that both $\ell_i < k_i$ and $\ell_i > k_i$ lead to contradictions.)

13. Let F be a field and suppose $A \in F^{n\times n}$ has distinct eigenvalues $\lambda_1, \ldots, \lambda_t$. Prove that A is diagonalizable if and only if $m_A(r) = (r - \lambda_1)\cdots(r - \lambda_t)$.

14. Let F be a field and let $A \in F^{n\times n}$. We say that $x \in F^n$ is a *cyclic vector* for A if

$$\mathrm{sp}\{x, Ax, A^2x, \ldots, A^{n-1}x\} = F^n.$$

(This implies, in particular, that $\{x, Ax, A^2x, \ldots, A^{n-1}x\}$ is linearly independent.) Assume that x is a cyclic vector for A.

   (a) Show that there exist unique scalars $c_0, c_1, \ldots, c_{n-1} \in F$ such that

$$A^n x = c_0 x + c_1 Ax + \cdots + c_{n-1}A^{n-1}x.$$

   (b) Define $X = [x|Ax|\cdots|A^{n-1}x]$. Compute the matrix $X^{-1}AX$.

   (c) Prove that the characteristic polynomial of A is

$$p_A(r) = r^n - c_{n-1}r^n - \cdots - c_1 r - c_0.$$

   (Hint: A has the same characteristic polynomial as the matrix $X^{-1}AX$.)

(d) Prove that the minimal polynomial of A is the same as the characteristic polynomial of A. (Hint: Show that $\deg(m_A(r)) < \deg(p_A(r))$ would contradict the linear independence of $\{x, Ax, \ldots, A^{n-1}x\}$.)

15. Let F be a field and let $c_0, c_1, \ldots, c_{n-1}$ be elements of F. Define $A \in F^{n\times n}$ by

$$A = \begin{bmatrix} 0 & 0 & 0 & \cdots & -c_0 \\ 1 & 0 & 0 & \cdots & -c_1 \\ 0 & 1 & 0 & \cdots & -c_2 \\ \vdots & & \ddots & \ddots & \vdots \\ 0 & 0 & \cdots & 1 & -c_{n-1} \end{bmatrix}.$$

Exercise 4.5.9 shows that

$$p_A(r) = r^n + c_{n-1}r^{n-1} + \cdots + c_1 r + c_0.$$

(a) Prove that e_1 is a cyclic vector for A (see the previous exercise).

(b) Prove that $m_A(r) = p_A(r)$. (Hint: We know that $m_A(r)$ is the monic polynomial of smallest degree such that $m_A(A) = 0$. Show that no polynomial $p(r)$ of degree less than $\deg(p_A(r))$ can satisfy $p(A)e_1 = 0$.)

5.3 Nilpotent operators

In the previous section, we saw that if the columns of $X_i \in \mathbf{C}^{n\times m_i}$ for a basis for $N_i = \mathcal{N}\left((A - \lambda_i I)^{k_i}\right)$, then

$$AX_i = X_i B_i,$$

where B_i is an $m_i \times m_i$ matrix. The integers m_i and k_i have the same meaning as in the previous section: m_i is the algebraic multiplicity of λ_i, and k_i is the smallest positive integer such that $\mathcal{N}((A - \lambda_i I)^{k_i+1}) = \mathcal{N}((A - \lambda_i I)^{k_i})$. We want to choose the basis of N_i to make B_i as simple as possible. Doing this requires that we understand the operator defined by $A - \lambda_i I$ acting on N_i.

Since N_i is invariant under $A - \lambda_i I$, we can define an operator $T_i : N_i \to N_i$ by $T(x) = (A - \lambda_i I)x$. This operator has the property that

$$T_i^{k_i}(x) = 0 \text{ for all } x \in N_i,$$

that is, $T_i^{k_i}$ is the zero operator. There is a name for such operators.

Definition 247 *Let V be a vector space over a field F, and let $T : V \to V$ be linear. We say that T is* nilpotent *if and only if $T^k = 0$ for some positive integer k. If k is the smallest integer for which this holds, then we say that T is* nilpotent of index k.

Example 248 *Let $D : \mathcal{P}_2 \to \mathcal{P}_2$ be the derivative operator: $D(p) = p'$. Since the third derivative of any quadratic polynomial is zero, it follows that D is nilpotent of index 3.*

The following fact will be useful.

Theorem 249 *Let V be a vector space over a field F, and let $T : V \to V$ be linear. If $x \in V$ is any vector such that $T^{k-1}(x) \neq 0$ and $T^k(x) = 0$, then*

$$\{x, T(x), T^2(x), \ldots, T^{k-1}(x)\}$$

is linearly independent.

Proof We will argue by contradiction and suppose that

$$\alpha_0 x + \alpha_1 T(x) + \ldots + \alpha_{k-1} T^{k-1}(x) = 0,$$

where $\alpha_0, \alpha_1, \ldots, \alpha_{k-1} \in F$ are not all zero. Let $j \geq 0$ be the smallest index such that $\alpha_j \neq 0$. We can then solve for $T^j(x)$ as follows:

$$T^j(x) = -\sum_{i=j+1}^{k-1} \alpha_j^{-1} \alpha_i T^i(x).$$

We now apply the operator T^{k-j-1} to both sides to obtain

$$T^{k-1}(x) = -\sum_{i=j+1}^{k-1} \alpha_j^{-1} \alpha_i T^{k+i-j-1}(x).$$

Since $k+i-j-1 \geq k$ for all $i = j+1, \ldots, k-1$, it follows that the right-hand side of this equation is zero. But then

$$T^{k-1}(x) = 0,$$

a contradiction.

QED

We will use the following property of nilpotent operators.

Theorem 250 *Let V be a vector space over a field F and let $T : V \to V$ be a nilpotent operator of index k. Suppose $x_0 \in V$ is any vector with the property that $T^{k-1}(x_0) \neq 0$ and define*

$$S = \mathrm{sp}\left\{x_0, T(x_0), \ldots, T^{k-1}(x_0)\right\}.$$

Then S is invariant under T. If $k < \dim(V)$, then there exists a subspace W of V such that W is invariant under T and V is the direct sum of S and W.

Proof Since T transforms the basis of S to

$$\{T(x_0), T^2(x_0), \ldots, T^{k-1}(x_0), 0\} \subset S,$$

we see that S is invariant under T.

To prove the existence of W, we argue by induction on the index k of nilpotency. If $k = 1$, then T is the zero operator and every subspace of V is invariant under T. We can therefore take W to be any subspace such that V is the direct sum of S and W.

We now suppose the result holds for all nilpotent operators of index $k - 1$. The range $R = \mathcal{R}(T)$ is invariant under T and, restricted to R, T is nilpotent of index $k - 1$. We define

$$S_0 = S \cap R.$$

We note that $x_0 \notin R$, since otherwise $T^{k-1}(x_0)$ would be zero. Therefore,

$$S_0 = \text{sp}\left\{T(x_0), T^2(x_0), \ldots, T^{k-1}(x_0)\right\} = \text{sp}\left\{y_0, T(y_0), \ldots, T^{k-2}(y_0)\right\}, \tag{5.9}$$

where $y_0 = T(x_0)$.

By Theorem 249, (5.9) shows that $\dim(S_0) = k - 1$. We can apply the induction hypothesis to produce an invariant subspace W_0 of R with the property that R is the direct sum of S_0 and W_0. We have

$$S = \{x \in V \mid T(x) \in S_0\},$$

and we define

$$W_1 = \{x \in V \mid T(x) \in W_0\}.$$

The desired subspace is not W_1, but we will use W_1 to construct it. We notice for future reference that, since W_0 is invariant under T, $W_0 \subset W_1$.

Since the rest of the proof is fairly delicate, we outline it before proceeding:

- We prove that $V = S + W_1$.
- We then show that $S \cap W_1$ is a one-dimensional space spanned by $T^{k-1}(x_0)$.
- Finally, the desired subset W is the result of removing $T^{k-1}(x_0)$ from a properly chosen basis for W_1.

We first note that $V = S + W_1$. To see this, let x be any vector in V. Then $T(x) \in R = S_0 + W_0$, so there exist $y \in S_0$, $z \in W_0$ such that $T(x) = y + z$. There exist $\alpha_1, \alpha_2, \ldots, \alpha_{k-1} \in F$ such that

$$\begin{aligned} y &= \alpha_1 T(x_0) + \alpha_2 T^2(x_0) + \ldots + \alpha_{k-1} T^{k-1}(x_0) \\ &= T\left(\alpha_1 x_0 + \alpha_2 T(x_0) + \ldots + \alpha_{k-1} T^{k-2}(x_0)\right) \\ &= T(y_1), \end{aligned}$$

where $y_1 = \alpha_1 x_0 + \alpha_2 T(x_0) + \ldots + \alpha_{k-1} T^{k-2}(x_0) \in S$. We then have

$$z = T(x) - y = T(x) - T(y_1) = T(x - y_1),$$

which, by definition of W_1, shows that $x - y_1 \in W_1$ (since $T(x - y_1) \in W_0$). Writing $z_1 = x - y_1$, we obtain $x = y_1 + z_1$, with $y_1 \in S$ and $z_1 \in W_1$, and so $V = S + W_1$, as desired.

We next show that $S \cap W_1 = \mathrm{sp}\{T^{k-1}(x_0)\}$. Since $T^{k-1}(x_0) \in S$ and $T(T^{k-1}(x_0)) = 0 \in W_0$, it follows that $T^{k-1}(x_0) \in W_1$. This shows that $\mathrm{sp}\{T^{k-1}(x_0)\} \subset S \cap W_1$.

On the other hand, suppose $x \in S \cap W_1$. Then $x \in S$ and thus there exist $\alpha_0, \alpha_1, \ldots, \alpha_{k-1} \in F$ such that

$$x = \alpha_0 x_0 + \alpha_1 T(x_0) + \ldots + \alpha_{k-1} T^{k-1}(x_0).$$

Moreover, $T(x) \in W_0$, and, since $T^k(x_0) = 0$,

$$\begin{aligned} T(x) &= T\left(\alpha_0 x_0 + \alpha_1 T(x_0) + \ldots + \alpha_{k-1} T^{k-1}(x_0)\right) \\ &= \alpha_0 T(x_0) + \alpha_1 T^2(x_0) + \ldots + \alpha_{k-2} T^{k-1}(x_0) \in S_0. \end{aligned}$$

But then $T(x) \in S_0 \cap W_0 = \{0\}$, so $T(x) = 0$ and therefore, since

$$\left\{T(x_0), T^2(x_0), \ldots, T^{k-1}(x_0)\right\}$$

is linearly independent, we see that $\alpha_0 = \alpha_1 = \ldots = \alpha_{k-2} = 0$. Thus

$$x = \alpha_{k-1} T^{k-1}(x_0) \in \mathrm{sp}\{T^{k-1}(x_0)\},$$

which is what we wanted to prove.

We are now ready to define the desired subspace W. If the dimension of V is n, then

$$\begin{aligned} & \dim(V) = \dim(S + W_1) = \dim(S) + \dim(W_1) - \dim(S \cap W_1) \\ \Rightarrow \quad & n = k + \dim(W_1) - 1 \\ \Rightarrow \quad & \dim(W_1) = n - k + 1. \end{aligned}$$

Let $\{v_1, v_2, \ldots, v_\ell\}$ be a basis for W_0. Since $T^{k-1}(x_0) \in S_0$ and $S_0 \cap W_0 = \{0\}$, $T^{k-1}(x_0)$ does not belong to W_0. It follows that

$$\{T^{k-1}(x_0), v_1, v_2, \ldots, v_\ell\} \tag{5.10}$$

is linearly independent. Moreover, $W_0 \subset W_1$ implies that the set (5.10) is a linearly independent subset of W_1. Extend this set to a basis

$$\{T^{k-1}(x_0), v_1, v_2, \ldots, v_\ell, v_{\ell+1}, \ldots, v_{n-k}\}$$

of W_1 and define

$$W = \mathrm{sp}\{v_1, v_2, \ldots, v_\ell, v_{\ell+1}, \ldots, v_{n-k}\}.$$

We have $V = S + W_1$, and $T^{k-1}(x_0)$ is in S as well as W_1; it therefore follows that $V = S + W$. Moreover, $S \cap W = \{0\}$, so V is the direct sum of S and W.

Finally, since $\{v_1, v_2, \ldots, v_\ell\}$ is a basis for the invariant subspace W_0 and $v_{\ell+1}, \ldots, v_{n-k} \in W_1$, we have $T(x) \in W_0 \subset W$ for all $x \in W$. Thus W is invariant under T, and the proof is complete.

QED

Using the previous theorem, we can find a special basis for the domain of a nilpotent operator; this basis results in a matrix with a particularly simple form.

Theorem 251 *Let V be a finite-dimensional vector space over F and suppose $T : V \to V$ is a nilpotent linear operator of index k. Then there exist vectors $x_1, x_2, \ldots, x_s \in V$ and integers $r_1, r_2, \ldots, r_s$, with $1 \le r_s \le \cdots \le r_1 = k$, such that*

$$T^{r_i-1}(x_i) \neq 0,\ T^{r_i}(x_i) = 0 \text{ for all } i = 1, 2, \ldots, s,$$

and the vectors

$$\begin{array}{c} x_1, T(x_1), \ldots, T^{r_1-1}(x_1), \\ x_2, T(x_2), \ldots, T^{r_2-1}(x_2), \\ \vdots \quad \vdots \qquad\quad \vdots \\ x_s, T(x_s), \ldots, T^{r_s-1}(x_s) \end{array}$$

form a basis for V.

Proof Since T is nilpotent of index k, there exists a vector $x_1 \in V$ such that $T^{k-1}(x_1) \neq 0$ and $T^k(x_1) = 0$. Let $r_1 = k$ and

$$S_1 = \text{sp}\left\{x_1, T(x_1), \ldots, T^{r_1-1}(x_1)\right\}.$$

By the previous theorem, S_1 is invariant under T and there exists a subspace W_1 of V such that V is the direct sum of S_1 and W_1, and W_1 is also invariant under T.

We note that $\dim(W_1)$ is strictly less than $\dim(V)$ and T, restricted to W_1, is nilpotent of index $r_2 \le r_1 = k$. We choose a vector $x_2 \in W_1$ such that $T^{r_2-1}(x_2) \neq 0$ and $T^{r_2}(x_2) = 0$. We then define

$$S_2 = \text{sp}\left\{x_2, T(x_2), \ldots, T^{r_2-1}(x_2)\right\}.$$

Once again applying the following theorem (now to T restricted to W_1), there exists a subspace W_2 of W_1 such that W_1 is the direct sum of S_2 and W_2 and T is invariant under W_2. We continue in this fashion to find $x_3, \ldots, x_s$; the process ends when $S_s = W_{s-1}$, whereupon the given vectors span the entire space V.

QED

We now describe the matrix of a nilpotent transformation T under the basis of Theorem 251. We begin with the simplest case, in which k, the index of nilpotency of T, equals $\dim(V)$. Then

$$\{x_1, T(x_1), \ldots, T^{k-1}(x_1)\}$$

is a basis for V. We label these basis vectors as

$$u_1 = T^{k-1}(x_1), u_2 = T^{k-2}(x_1), \ldots, u_k = x_1;$$

we then have

$$T(u_1) = 0,\ T(u_j) = u_{j-1},\ j = 2, 3, \ldots, k.$$

In terms of the isomorphism between V and F^k, this means that $T(u_1)$ corresponds to the zero vector and $T(u_j)$ corresponds to e_{j-1} (the $(j-1)$st standard basis vector) for $j > 1$. Therefore, the matrix representing T under the basis $\{u_1, u_2, \ldots, u_k\}$ is

$$J = \begin{bmatrix} 0 & 1 & & & \\ & 0 & 1 & & \\ & & \ddots & \ddots & \\ & & & 0 & 1 \\ & & & & 0 \end{bmatrix}. \tag{5.11}$$

If $s > 1$, then we define a basis $\{u_1, u_2, \ldots, u_n\}$ by

$$\begin{aligned} u_1 &= T^{r_1-1}(x_1), u_2 = T^{r_1-2}(x_1), \ldots, u_{r_1} = x_1, \\ u_{r_1+1} &= T^{r_2-1}(x_2), u_{r_1+2} = T^{r_2-2}(x_2) \ldots, u_{r_1+r_2} = x_2, \\ &\vdots \qquad\qquad \vdots \\ u_{n-r_s+1} &= T^{r_s-1}(x_s), u_{n-r_s+2} = T^{r_s-2}(x_s), \ldots, u_n = x_s. \end{aligned}$$

Then it is not difficult to see that the matrix representing T under the basis $\{u_1, u_2, \ldots, u_n\}$ is

$$A = \begin{bmatrix} J_1 & & & \\ & J_2 & & \\ & & \ddots & \\ & & & J_s \end{bmatrix},$$

where each $J_i \in F^{r_i \times r_i}$ has the form (5.11).

Example 252 *This is a continuation of Example 248. The operator derivative operator D on $\mathcal{P}_2$ is nilpotent of index 3. This is the simple case described above, in which the index of nilpotency equals the dimension of the vector space. Any polynomial p_0 of degree exactly 2 satisfies $D^2(p_0) \neq 0$ and $D^3(p_0) = 0$. For example, if $p_0 = x^2$, then*

$$D(p_0) = 2x,\ D^2(p_0) = 2,\ D^3(p_0) = 0.$$

If we define the basis $\mathcal{B} = \{D^2(p_0), D(p_0), p_0\} = \{2, 2x, x^2\}$, *then*

$$[D]_{\mathcal{B},\mathcal{B}} = \begin{bmatrix} 0 & 1 & 0 \\ 0 & 0 & 1 \\ 0 & 0 & 0 \end{bmatrix}.$$

We could choose a different p_0 *and obtain the same matrix. For instance, with* $p_0 = 1 - x + x^2$, *we have* $D(p_0) = -1 + 2x$, $D^2(p_0) = 2$. *If* $\mathcal{A}$ *is the basis* $\{2, -1 + 2x, 1 - x + x^2\}$, *then* $[D]_{\mathcal{A},\mathcal{A}}$ *is identical to* $[D]_{\mathcal{B},\mathcal{B}}$ *(see Exercise 1).*

Example 253 *Let* $A \in \mathbf{R}^{4\times 4}$ *be defined by*

$$A = \begin{bmatrix} 6 & 2 & 1 & -1 \\ -7 & -1 & -1 & 2 \\ -9 & -7 & -2 & -1 \\ 13 & 3 & 2 & -3 \end{bmatrix}.$$

The reader is invited to verify the following calculations. The characteristic polynomial of A *is* $p_A(r) = r^4 = (r-0)^4$, *which shows that the only eigenvalue of* A *is* $\lambda = 0$, *which has algebraic multiplicity four. We notice that the geometric multiplicity of* $\lambda = 0$ *must be less than four; the only* 4×4 *matrix whose null space has dimension four is the zero matrix.*

A direct calculation shows that A^2 *is the zero matrix, which shows that* A *is nilpotent of index two. We wish to choose as basis* $\{u_1, u_2, u_3, u_4\}$, *as described in Theorem 251, to make the matrix* $X^{-1}AX$ *(where* $X = [u_1|u_2|u_3|u_4]$*) as simple as possible. Before we perform the calculations, we consider the possibilities. Since* A *is nilpotent of index two, there is at least one vector* x_1 *such that* $Ax_1 \neq 0$, $A^2x_1 = 0$. *We would then choose the first two vectors in the basis to be* $u_1 = Ax_1$, $u_2 = x_1$. *Notice that, in this case,* $u_1 = Ax_1$ *is an eigenvector of* A: $Au_1 = A^2x_1 = 0 = 0 \cdot u_1$.

There are now two possibilities for the other two vectors in the basis. There could be a second (independent) vector x_2 *with* $Ax_2 \neq 0$, $A^2x_2 = 0$, *in which case we take* $u_3 = Ax_2$, $u_4 = x_2$. *In this case* u_3 *is a second eigenvector of* A *corresponding to* $\lambda = 0$, *and the geometric multiplicity of* $\lambda = 0$, *that is, the dimension of* $\mathcal{N}(A)$, *is two. The simplified matrix is*

$$X^{-1}AX = \left[\begin{array}{cc|cc} 0 & 1 & & \\ 0 & 0 & & \\ \hline & & 0 & 1 \\ & & 0 & 0 \end{array}\right]. \tag{5.12}$$

The second possibility is that A *has two more independent eigenvectors* x_2, x_3 *corresponding to* $\lambda = 0$. *In this case, the geometric multiplicity of* $\lambda = 0$ *is three* $(\dim(\mathcal{N}(A)) = 3)$, *we take* $u_3 = x_2$, $u_4 = x_3$, *and we obtain*

$$X^{-1}AX = \left[\begin{array}{cc|c|c} 0 & 1 & & \\ 0 & 0 & & \\ \hline & & 0 & \\ \hline & & & 0 \end{array}\right]. \tag{5.13}$$

Notice that we can distinguish the two cases by determining $\dim(\mathcal{N}(A))$. *We find a basis for* $\mathcal{N}(A)$ *by row reducing* $[A|0]$. *The result is*

$$\left[\begin{array}{cccc|c} 1 & 0 & \frac{1}{8} & -\frac{3}{8} & 0 \\ 0 & 1 & \frac{1}{8} & \frac{5}{8} & 0 \\ 0 & 0 & 0 & 0 & 0 \\ 0 & 0 & 0 & 0 & 0 \end{array}\right].$$

This shows that $\dim(\mathcal{N}(A)) = 2$, *and a basis for* $\mathcal{N}(A)$ *is*

$$\{(-1, -1, 8, 0), (3, -5, 0, 8)\}.$$

There must be vectors x_1, x_2 *such that* $u_1 = Ax_1, u_2 = x_1, u_3 = Ax_2, u_4 = x_2$ *is the desired basis. This is the first possibility described above.*

To find x_1, x_2, *we must find two independent vectors in* $\mathcal{N}(A^2)$ *that do not belong to* $\mathcal{N}(A)$. *In this case,* $\mathcal{N}(A^2) = \mathbf{R}^4$, *so we look for two independent vectors in* $\mathbf{R}^4$ *that do not lie in* $\mathcal{N}(A)$. *There are many ways to find two such vectors, which are not unique. For instance, we notice that* $\mathcal{N}(A)$ *is a two-dimensional subspace of* $\mathbf{R}^4$, *so any two independent vectors chosen at random are almost certain to lie outside of* $\mathcal{N}(A)$. *For example, the reader can verify that* $e_1 = (1, 0, 0, 0)$ *and* $e_2 = (0, 1, 0, 0)$ *both lie outside of* $\mathcal{N}(A)$. *We can therefore take* $u_1 = Ae_1 = (6, -7, -9, 13)$, $u_2 = e_1$, $u_3 = Ae_2 = (2, -1, -7, 3)$, $u_4 = (0, 1, 0, 0)$ *to obtain*

$$X = \begin{bmatrix} 6 & 1 & 2 & 0 \\ -7 & 0 & -1 & 1 \\ -9 & 0 & -7 & 0 \\ 13 & 0 & 3 & 0 \end{bmatrix}.$$

The reader can now verify that (5.12) holds with this choice of X. *(Actually, it would be easier to check the equation* $AX = XJ$, *where* J *is the matrix defined by (5.12).)*

A final word about terminology: Every matrix $A \in F^{n \times n}$ defines a linear operator $T : F^n \to F^n$ by $T(x) = Ax$ for all $x \in F^n$. It is therefore natural to refer to a matrix A as nilpotent if the corresponding matrix operator T is nilpotent. If A is nilpotent and k is the index of nilpotency, then k can be characterized as the smallest positive integer such that A^k is the zero matrix.

Exercises

Miscellaneous exercises

1. Let D be the operator of Example 252. Compute directly the matrices of D under the bases $\mathcal{B}$ and $\mathcal{A}$, and verify that they are as given in that example.

2. Suppose X is a vector space over a field F, and suppose $T : X \to X$ is a nilpotent linear operator. Prove that 0 is an eigenvalue of T, and 0 is the only eigenvalue of T.

3. Suppose X is a finite-dimensional vector space over a field F, and suppose $T : X \to X$ is a nilpotent linear operator. Let $I : X \to X$ be the identity operator ($I(x) = x$ for all $x \in X$). Prove the $I + T$ is invertible. (Hint: Argue by contradiction and use the result of the preceding exercise.)

4. This exercise gives an alternate proof of the result of the preceding exercise. Let $T : X \to X$ be a nilpotent linear operator, where X is a vector space over a field F. Let k be the index of nilpotency of T, and let $I : X \to X$ be the identity operator. Prove that $I + T$ is invertible by proving that $S = I - T + T^2 - \cdots + (-1)^{k-1}T^{k-1}$ is the inverse of $I + T$.

5. Let $A \in \mathbf{C}^{n\times n}$ be nilpotent. Prove that the index of nilpotency of A is at most n.

6. Suppose $A \in \mathbf{C}^{n\times n}$ and the only eigenvalue of A is $\lambda = 0$. Prove A is nilpotent of index k for some k satisfying $1 \le k \le n$.

7. (a) Suppose $A \in \mathbf{R}^{n\times n}$ and $A_{ij} > 0$ for all i, j. Prove that A is not nilpotent.

   (b) What can you say if $A \in \mathbf{R}^{n\times n}$ satisfies $A \neq 0$, $A_{ij} \ge 0$ for all i, j? Can A be nilpotent? If so, is there an additional hypothesis that means A cannot be nilpotent?

8. Suppose $A \in \mathbf{R}^{4\times 4}$ and the only eigenvalue of A is $\lambda = 0$. According to the previous exercise, A is nilpotent of index k for some k satisfying $1 \le k \le 4$. Make a list of every possibility for the dimensions of $\mathcal{N}(A), \mathcal{N}(A^2), \ldots, \mathcal{N}(A^{k-1})$ (notice that $\dim(\mathcal{N}(A^k)) = 4$) and the corresponding possibilities for $X^{-1}AX$, where X is chosen according to Theorem 251.

9. Repeat the preceding exercise for $A \in \mathbf{R}^{5\times 5}$.

10. Let A be the nilpotent matrix

$$A = \begin{bmatrix} 1 & -2 & -1 & -4 \\ 1 & -2 & -1 & -4 \\ -1 & 2 & 1 & 4 \\ 0 & 0 & 0 & 0 \end{bmatrix}.$$

    Find a basis for $\mathbf{R}^4$ satisfying the conclusions of Theorem 251 and compute $X^{-1}AX$, where X is the corresponding matrix (the matrix whose columns comprise the basis).

11. Repeat the previous exercise for the matrix

$$A = \begin{bmatrix} 1 & 75 & 76 & -62 \\ 1 & -2 & -1 & -4 \\ -1 & 14 & 13 & -5 \\ 0 & 16 & 16 & -12 \end{bmatrix}.$$

12. Repeat Exercise 10 for the 5×5 matrix

$$A = \begin{bmatrix} -4 & 4 & -12 & 6 & 10 \\ -21 & 21 & -63 & 58 & -27 \\ -1 & 1 & -3 & 6 & -11 \\ 6 & -6 & 18 & -15 & 3 \\ 2 & -2 & 6 & -5 & 1 \end{bmatrix}.$$

13. Suppose X is a finite-dimensional vector space over a field F, and $T : X \to X$ is linear.

    (a) Prove or give a counterexample: If T is nilpotent, then T is singular.

    (b) Prove or give a counterexample: If T is singular, then T is nilpotent.

14. Let X be a finite-dimensional vector space over a field F, and let $S : X \to X, T : X \to X$ be linear operators.

    (a) Prove or give a counterexample: If S and T are nilpotent, then $S + T$ is nilpotent.

    (b) Prove or give a counterexample: If S and T are nilpotent, then ST is nilpotent.

15. Let F be a field and suppose $A \in F^{n \times n}$ is nilpotent. Prove that

$$\det(I + A) = 1.$$

5.4 The Jordan canonical form of a matrix

In Section 5.2, we saw that if $A \in \mathbf{C}^{n \times n}$ and $(\lambda - \lambda_i)^{m_i}$ is a factor of the characteristic polynomial of A, then $N_i = \mathcal{N}((A - \lambda_i I)^{m_i})$ has dimension m_i. There is a smallest integer k_i, with $1 \le k_i \le m_i$, such that

$$\mathcal{N}\left((A - \lambda_i I)^{k_i}\right) = \mathcal{N}\left((A - \lambda_i I)^{m_i}\right).$$

It follows that

$$x \in N_i \Rightarrow (A - \lambda_i I)^{k_i} x = 0,$$

but there exists at least one vector $x \in N_i$ such that

$$(A - \lambda_i I)^{k_i - 1} x \neq 0.$$

In other words, $A - \lambda_i I$, regarded as an operator, is nilpotent of index k_i when restricted to N_i.

We have also seen that $\mathbf{C}^n$ is the direct sum of $N_1, N_2, \ldots, N_t$ and each of these subspaces is invariant under A. If we choose a basis for each N_i and define the matrix $X \in \mathbf{C}^{n \times n}$ whose columns are the vectors in these bases, then

$$X^{-1}AX = \begin{bmatrix} B_1 & & & \\ & B_2 & & \\ & & \ddots & \\ & & & B_t \end{bmatrix}.$$

At the end of Section 5.2, we posed the question of choosing the basis for N_i so that B_i is as simple as possible.

This question is answered by Theorem 251, and we now proceed to show how. Since $A - \lambda_i I$ is nilpotent of index k_i on N_i, there are integers

$$r_{i,1} \geq r_{i,2} \geq \cdots \geq r_{i,s_i},$$

with $r_{i,1} = k_i$ and $r_{i,s_i} \geq 1$, and vectors $x_{i,1}, x_{i,2}, \ldots, x_{i,s_i}$ such that the vectors

$$\begin{array}{c} x_{i,1}, (A - \lambda_i I)x_{i,1}, \ldots, (A - \lambda_i I)^{r_{i,1}-1} x_{i,1}, \\ x_{i,2}, (A - \lambda_i I)x_{i,2}, \ldots, (A - \lambda_i I)^{r_{i,2}-1} x_{i,2}, \\ \vdots \qquad\qquad \vdots \qquad\qquad \vdots \\ x_{i,s_i}, (A - \lambda_i I)x_{i,s_i}, \ldots, (A - \lambda_i I)^{r_{i,s_i}-1} x_{i,s_i} \end{array}$$

form a basis for N_i. We define, for each $j = 1, 2, \ldots, s_i$,

$$X_{i,j} = \left[(A - \lambda_i I)^{r_j - 1} x_{i,j} | (A - \lambda_i I)^{r_j - 2} x_{i,j} | \cdots | x_{i,j} \right] \in \mathbf{C}^{n \times m_i}.$$

The reader should notice that $(A - \lambda_i I)^{r_j - 1} x_{i,j}$ is an eigenvector corresponding to the eigenvalue λ_i. The vectors

$$(A - \lambda_i I)^{r_j - 2} x_{i,j}, (A - \lambda_i I)^{r_j - 3} x_{i,j}, \ldots, x_{i,j}$$

are generalized eigenvectors of A corresponding to λ_i. We refer to

$$(A - \lambda_i I)^{r_j - 1} x_{i,j}, (A - \lambda_i I)^{r_j - 2} x_{i,j}, \ldots, x_{i,j}$$

as an *eigenvector/generalized eigenvector chain.*

As shown at the end of the previous section,

$$(A - \lambda_i I) X_{i,j} = X_{i,j} \hat{J}_{i,j}, \tag{5.14}$$

where

$$\hat{J}_{i,j} = \begin{bmatrix} 0 & 1 & & & \\ & 0 & 1 & & \\ & & \ddots & \ddots & \\ & & & 0 & 1 \\ & & & & 0 \end{bmatrix} \in \mathbf{C}^{r_{i,j} \times r_{i,j}}.$$

Rearranging (5.14), we obtain

$$AX_{i,j} = X_{i,j}J_{i,j},$$

where

$$J_{i,j} = \lambda_i I + \hat{J}_{i,j} = \begin{bmatrix} \lambda_i & 1 & & & \\ & \lambda_i & 1 & & \\ & & \ddots & \ddots & \\ & & & \lambda_i & 1 \\ & & & & \lambda_i \end{bmatrix}.$$

A matrix having the form of $J_{i,j}$ is called a *Jordan block* corresponding to λ_i. We point out that $r_{i,j} = 1$ is possible, in which case $x_{i,j}$ is an eigenvector of A and the Jordan block $J_{i,j}$ is 1×1.

We will temporarily write $u_k = (A - \lambda_i)^{r_j - k} x_{i,j}$ (suppressing the dependence on i and j) to make explicit the defining property of the eigenvector and associated generalized eigenvector. The form of $J_{i,j}$ and the fact that $AX_{i,j} = X_{i,j}J_{i,j}$ shows that

$$\begin{aligned} Au_1 &= \lambda_i u_1, \\ Au_2 &= \lambda_i u_2 + u_1, \\ Au_3 &= \lambda_i u_3 + u_2, \\ &\vdots \\ Au_{r_j} &= \lambda_i u_{r_j} + u_{r_j - 1}. \end{aligned}$$

It is this generalized eigenvalue equation $Au_k = \lambda_i u_k + u_{k-1}$ that makes $J_{i,j}$ as simple as it is.

If we now define

$$X_i = [X_{i,1}|X_{i,2}|\cdots|X_{i,r_i}],$$

then, by Theorem 251, the columns of X_i form a basis for N_i, and

$$AX_i = X_i B_i,$$

where

$$B_i = \begin{bmatrix} J_{i,1} & & & \\ & J_{i,2} & & \\ & & \ddots & \\ & & & J_{i,s_i} \end{bmatrix}.$$

We can then define

$$X = [X_1|X_2|\cdots|X_t] \in \mathbf{C}^{n\times n}$$

to obtain

$$AX = XJ,$$

with

$$J = \begin{bmatrix} B_1 & & & \\ & B_2 & & \\ & & \ddots & \\ & & & B_t \end{bmatrix}.$$

Each of the blocks B_i is made up of Jordan blocks, so J itself is block diagonal, with Jordan blocks on the diagonal.

The matrix J is called the *Jordan canonical form* of the matrix A, and we refer to the factorization $A = XJX^{-1}$ as the *Jordan decomposition* of A. We reiterate that a Jordan block can be 1×1; if every Jordan block of A is 1×1, then A is diagonalizable and the Jordan canonical form of A is a diagonal matrix.

It is important to realize that all the notation used above is necessary, and that the situation really can be this complicated. For any characteristic polynomial

$$p_A(\lambda) = (\lambda - \lambda_1)^{m_1}(\lambda - \lambda_2)^{m_2}\cdots(\lambda - \lambda_t)^{m_t}$$

and any sequence of integers $k_1, k_2, \ldots, k_t$ with $1 \le k_i \le m_i$, and any collection of positive integers $r_{i,j}$, $j = 1,2,\ldots,s_i$, $i = 1,2,\ldots,t$ satisfying $r_{i,j} \le k_i$ for all j and

$$r_{i,1} + r_{i,2} + \ldots + r_{i,s_i} = m_i, \ i = 1,2,\ldots,t,$$

there is a matrix A with the Jordan blocks defined by these numbers.

Example 254 *Let $A \in \mathbf{R}^{8\times 8}$ be defined by*

$$A = \begin{bmatrix} -1 & 0 & 0 & 0 & 0 & 0 & 0 & 0 \\ -1 & 2 & 0 & -1 & -2 & 0 & 1 & -2 \\ -4 & 2 & 0 & -2 & 2 & 1 & 2 & 2 \\ 1 & 11 & -2 & -6 & -13 & -2 & 3 & -13 \\ 1 & 4 & -1 & -2 & -4 & -1 & 1 & -5 \\ -4 & -10 & 3 & 6 & 15 & 2 & -2 & 15 \\ 0 & 10 & -2 & -6 & -11 & -2 & 3 & -11 \\ -1 & -4 & 1 & 2 & 3 & 1 & -1 & 4 \end{bmatrix}.$$

A straightforward calculation shows that

$$p_A(\lambda) = \lambda^8 - 4\lambda^6 + 6\lambda^4 - 4\lambda^2 + 1 = (\lambda+1)^4(\lambda-1)^4,$$

so we write $\lambda_1 = -1$, $m_1 = 4$, $\lambda_2 = 1$, $m_2 = 4$.

The reader is invited to verify the following calculations using a computer algebra system. Considering first $\lambda_1 = -1$, *we have*

$$\begin{aligned}\dim(\mathcal{N}(A+I)) &= 2,\\ \dim(\mathcal{N}((A+I)^2)) &= 3,\\ \dim(\mathcal{N}((A+I)^3)) &= 4,\\ \dim(\mathcal{N}((A+I)^4)) &= 4,\end{aligned}$$

which shows that $k_1 = 3$. *Since* $k_1 > 1$, *we see already that* A *is not diagonalizable. Moreover, we see that there must be a vector* $x_{1,1}$ *such that*

$$(A+I)^2 x_{1,1} \neq 0, \ (A+I)^3 x_{1,1} = 0,$$

that is, a vector $x_{1,1}$ *that belongs to* $\mathcal{N}((A+I)^3)$ *but not to* $\mathcal{N}((A+I)^2)$. *The vector* $x_{1,1} = (-3,-4,1,0,0,1,1,-4)$ *is one such vector. Then*

$$(A+I)^2 x_{1,1}, (A+I)x_{1,1}, x_{1,1}$$

spans a three-dimensional subspace of $\mathcal{N}((A+I)^3)$; $(A+I)^2 x_{1,1}$ *is an eigenvector of* A *corresponding to* λ_1, *and* $(A+I)x_{1,1}, x_{1,1}$ *are associated generalized eigenvectors. Since the dimension of* $\mathcal{N}((A+I)^3)$ *is four, there remains a one-dimensional invariant subspace of* $\mathcal{N}((A+I)^3)$, *which is spanned by an eigenvector* $x_{1,2}$ *independent of the eigenvector* $(A+I)^2 x_{1,1}$. *One such eigenvector is* $x_{1,2} = (1,0,1,0,0,1,1,0)$. *We now define*

$$X_1 = \left[(A+I)^2 x_{1,1} | (A+I)x_{1,1} | x_{1,1} | x_{1,2}\right], \ B_1 = \left[\begin{array}{ccc|c} -1 & 1 & 0 & \\ 0 & -1 & 1 & \\ 0 & 0 & -1 & \\ \hline & & & -1 \end{array}\right],$$

and then $AX_1 = X_1 B_1$.

Turning to the second eigenvalue, $\lambda_2 = 1$, *we find*

$$\begin{aligned}\dim(\mathcal{N}(A-I)) &= 2,\\ \dim(\mathcal{N}((A-I)^2)) &= 4,\\ \dim(\mathcal{N}((A-I)^3)) &= 4,\end{aligned}$$

which shows that $k_2 = 2$. *There must be two independent vectors that belong to* $\mathcal{N}((A-I)^2)$ *but not to* $\mathcal{N}(A-I)$. *Two such vectors are*

$$x_{2,1} = (0,1,-3,2,0,0,-2,0), \ x_{2,2} = (0,1,-3,0,0,2,-6,0).$$

We then find linearly independent vectors

$$\begin{aligned}&(A-I)x_{2,1}, x_{2,1},\\ &(A-I)x_{2,2}, x_{2,2},\end{aligned}$$

and these vectors must span $\mathcal{N}((A-I)^2)$. *We define*

$$X_2 = [(A-I)x_{2,1}|x_{2,1}|(A-I)x_{2,2}|x_{2,2}], \ B_2 = \left[\begin{array}{cc|cc} 1 & 1 & & \\ 0 & 1 & & \\ \hline & & 1 & 1 \\ & & 0 & 1 \end{array}\right],$$

and obtain $AX_2 = X_2B_2$.

Finally, we define

$$X = [X_1|X_2], \ J = \left[\begin{array}{cc} B_1 & 0 \\ 0 & B_2 \end{array}\right] = \left[\begin{array}{ccc|c|cc|cc} -1 & 1 & 0 & & & & & \\ 0 & -1 & 1 & & & & & \\ 0 & 0 & -1 & & & & & \\ \hline & & & -1 & & & & \\ \hline & & & & 1 & 1 & & \\ & & & & 0 & 1 & & \\ \hline & & & & & & 1 & 1 \\ & & & & & & 0 & 1 \end{array}\right].$$

Then X *is invertible,* $A = XJX^{-1}$, *and* J *is the Jordan canonical form of* A. *The matrix* J *is partitioned above to show its four Jordan blocks.*

As a final note, we point out that the minimal polynomial of A *is*

$$m_A(\lambda) = (\lambda+1)^3(\lambda-1)^2.$$

One final point should be made about the Jordan canonical form of a matrix: It cannot be stably computed in finite-precision arithmetic. This means that the round-off error inevitable in finite-precision arithmetic will typically ruin the computed results. This follows from the fact that a small change in the matrix A can result in an abrupt (discontinuous) change in the Jordan canonical form of A.

Example 255 *Consider the* 2×2 *matrix*

$$A = \left[\begin{array}{cc} 1 & 1 \\ 0 & 1 \end{array}\right],$$

which is already in Jordan canonical form. The matrix A *has a single eigenvalue,* $\lambda = 1$, *of algebraic multiplicity* 2. *The only (independent) eigenvector corresponding to* λ *is* $e_1 = (1,0)$, *and the generalized eigenvector is* $e_2 = (0,1)$.

Now consider the matrix

$$B = \left[\begin{array}{cc} 1 & 1 \\ \epsilon^2 & 1 \end{array}\right],$$

where ϵ *is a small positive number. One can compute directly that* B *has two distinct real eigenvalues,*

$$\lambda_1 = 1+\epsilon, \ \lambda_2 = 1-\epsilon,$$

which means that B is similar to a diagonal matrix:

$$X^{-1}BX = \begin{bmatrix} 1+\epsilon & 0 \\ 0 & 1-\epsilon \end{bmatrix}.$$

Moreover, the eigenvectors x_1, x_2 corresponding to λ_1, λ_2, respectively, are

$$x_1 = \begin{bmatrix} 1 \\ \epsilon \end{bmatrix}, \ x_2 = \begin{bmatrix} 1 \\ -\epsilon \end{bmatrix}.$$

Both x_1 and x_2 converge to e_1 as $\epsilon \to 0$, which means that the generalized eigenvector e_2 cannot be seen in the Jordan decomposition of B, no matter how small ϵ is. In this sense, we can say that the Jordan decomposition of A depends discontinuously on the (entries in the) matrix A.

The discontinuous dependence of J on A means that the Jordan decomposition is useful primarily as a theoretical tool, or for small matrices where the necessary computations can be carried out in exact arithmetic. For numerical computations (that is, computations done in finite-precision arithmetic on a computer), a different tool is needed. Often the tool of choice is the *singular value decomposition* (SVD), which is the topic of Chapter 8.

Exercises

Miscellaneous exercises

1. Find the Jordan canonical form J of

$$A = \begin{bmatrix} 1 & 1 & 1 \\ 0 & 1 & 1 \\ 0 & 0 & 1 \end{bmatrix} \in \mathbf{R}^{3\times 3}.$$

   Also find an invertible $X \in \mathbf{R}^{3\times 3}$ such that $A = XJX^{-1}$.

2. Find the Jordan canonical form of

$$A = \begin{bmatrix} 0 & 0 & 2 \\ -2 & 2 & 1 \\ -1 & 0 & 3 \end{bmatrix} \in \mathbf{R}^{3\times 3}.$$

   Also find an invertible $X \in \mathbf{R}^{3\times 3}$ such that $A = XJX^{-1}$.

3. Let $T : \mathcal{P}_3 \to \mathcal{P}_3$ be defined by $T(p) = p'' + p$ for all $p \in \mathcal{P}_3$. Find a basis $\mathcal{X}$ for $\mathcal{P}_3$ such that $[T]_{\mathcal{X},\mathcal{X}}$ is in Jordan canonical form.

4. Let A be a 4×4 matrix. For each of the following characteristic polynomials, how many different Jordan canonical forms are possible? Give each possible form and the corresponding dimensions of $\mathcal{N}((A - \lambda I)^k)$ for each eigenvalue λ and relevant value of k.

(a) $p_A(r) = (r-1)(r-2)(r-3)(r-4)$;
(b) $p_A(r) = (r-1)^2(r-2)(r-3)$;
(c) $p_A(r) = (r-1)^2(r-2)^2$;
(d) $p_A(r) = (r-1)^3(r-2)$;
(e) $p_A(r) = (r-1)^4$.

5. Let $A \in \mathbf{R}^{4\times 4}$ be defined by
$$A = \begin{bmatrix} -3 & 1 & -4 & -4 \\ -17 & 1 & -17 & -38 \\ -4 & -1 & -3 & -14 \\ 4 & 0 & 4 & 10 \end{bmatrix}.$$
Find the Jordan canonical form of A.

6. Let A be a 5×5 matrix. For each of the following characteristic polynomials, how many different Jordan canonical forms are possible? Give each possible form and the corresponding dimensions of $\mathcal{N}((A-\lambda I)^k)$ for each eigenvalue λ and relevant value of k.
   (a) $p_A(r) = (r-1)(r-2)(r-3)(r-4)(r-5)$;
   (b) $p_A(r) = (r-1)^2(r-2)(r-3)(r-4)$;
   (c) $p_A(r) = (r-1)^2(r-2)^2(r-3)$;
   (d) $p_A(r) = (r-1)^3(r-2)(r-3)$;
   (e) $p_A(r) = (r-1)^3(r-2)^2$;
   (f) $p_A(r) = (r-1)^4(r-2)$;
   (g) $p_A(r) = (r-1)^5$.

7. Let $A \in \mathbf{R}^{5\times 5}$ be defined by
$$A = \begin{bmatrix} -7 & 1 & 24 & 4 & 7 \\ -9 & 4 & 21 & 3 & 6 \\ -2 & -1 & 11 & 2 & 3 \\ -7 & 13 & -18 & -6 & -8 \\ 3 & -5 & 6 & 3 & 5 \end{bmatrix}.$$
Find the Jordan canonical form of A.

8. In defining the Jordan canonical form, we list the vectors in an eigenvector/generalized eigenvector chain in the following order:
$$(A-\lambda_i I)^{r_j-1}x_{i,j}, (A-\lambda_i I)^{r_j-2}x_{i,j}, \ldots, x_{i,j}.$$
What is the form of a Jordan block if we list the vectors in the order
$$x_{i,j}, (A-\lambda_i I)x_{i,j}, \ldots, (A-\lambda_i I)^{r_j-1}x_{i,j}$$
instead?

9. Let $\lambda \in \mathbf{C}$ be an eigenvalue of $A \in \mathbf{C}^{n\times n}$. Prove that the number of Jordan blocks corresponding to λ equals the geometric multiplicity of λ.

10. Let $\lambda \in \mathbf{C}$ be given and let

$$J = \begin{bmatrix} \lambda & 1 & & & \\ & \lambda & 1 & & \\ & & \ddots & \ddots & \\ & & & \lambda & 1 \\ & & & & \lambda \end{bmatrix} \in \mathbf{C}^{k\times k}$$

be a corresponding Jordan block. Find a formula for J^m, $m \geq 1$.

11. Let $A \in \mathbf{C}^{n\times n}$ and let J be the Jordan canonical form of A. Using the previous exercise, prove that $p_A(J) = 0$. Since $A = XJX^{-1}$ for some invertible matrix X and $p_A(A) = Xp_A(J)X^{-1}$, this gives another proof of the Cayley-Hamilton theorem. (Hint: Since J is block diagonal, it suffices to prove that $p_A(J_{i,j}) = 0$ for each Jordan block $J_{i,j}$ of J. If $J_{i,j}$ is $k \times k$, then the corresponding eigenvalue λ_i is a root of $p_A(r)$ of multiplicity at least k. It follows that $p_A^{(i)}(\lambda) = 0$ for $i = 0, 1, \ldots, k-1$. Here $p_A^{(i)}(r)$ denotes the ith derivative of $p_A(r)$.)

12. Verify the eigenvalues and eigenvectors of the matrix B in Example 255.

Project: Commuting matrices

We have already seen that if A and B are diagonalizable by a common eigenvector matrix X (that is, if $X^{-1}AX$ and $X^{-1}BX$ are both diagonal), then A and B commute (see Exercise 4.6.14). The purpose of this project is to prove the converse: If A and B are diagonalizable matrices that commute, then a common similarity transformation diagonalizes them. We say that $A, B \in \mathbf{C}^{n\times n}$ are *simultaneously diagonalizable* if there exists an invertible matrix $X \in \mathbf{C}^{n\times n}$ such that $X^{-1}AX$ and $X^{-1}BX$ are both diagonal.

We assume throughout the following exercises that A and B belong to $\mathbf{C}^{n\times n}$.

13. Prove a simpler version of the theorem first: If A and B commute and B has n distinct eigenvalues, then A and B are simultaneously diagonalizable. (The most general form of the theorem will be developed in the remainder of the exercises.)

14. Suppose S is a nontrivial subspace of $\mathbf{C}^n$ that is invariant under A. Prove that S contains an eigenvector of A.

15. Suppose A is diagonalizable, $S \subset \mathbf{C}^n$ is a nontrivial subspace that is invariant under A, and λ is an eigenvalue of A with a corresponding eigenvector belonging to S. Define $T : S \to S$ by $T(x) = Ax$. Prove that $\ker((T - \lambda I)^2) = \ker(T - \lambda I)$. (Here I represents the identity operator on S.)

16. Suppose A is diagonalizable and S is a nontrivial invariant subspace for A. Define T as in the previous exercise. Prove that there is a basis $\mathcal{X}$ of S such that $[T]_{\mathcal{X},\mathcal{X}}$ is diagonal. From this, prove that there exists a basis for S consisting of eigenvectors of A.

17. Suppose A and B commute. Prove that every eigenspace $E_\lambda(A)$ is invariant under B.

18. Use the preceding four exercises to prove that if A and B are diagonalizable and commute, then they are simultaneously diagonalizable.

19. Find a pair of 3×3, nondiagonal matrices that commute and yet are not simultaneously diagonalizable. (Hint: Choose a simple nondiagonalizable matrix A and solve the equation $AB = BA$ for the entries of B.)

5.5 The matrix exponential

One of the simplest initial value problems in ordinary differential equation is

$$u' = au, \ u(0) = u_0, \tag{5.15}$$

where $u = u(t)$ is a scalar-valued function and a, u_0 are given real or complex scalars. The unique solution to (5.15) is

$$u(t) = u_0 e^{at},$$

as is easily verified.

An analogous initial value problem (IVP) occurs when considering a system of linear ODEs with constant coefficients. In Section 4.8, we showed that such a system can be written as $u' = Au$, where $A \in \mathbf{C}^{n\times n}$ and $u : \mathbf{R} \to \mathbf{C}^n$ is the unknown (vector-valued) function. The corresponding IVP is

$$u' = Au, \ u(0) = v, \tag{5.16}$$

where $v \in \mathbf{C}^n$ is a given vector. Defining $L : C^1(a, b; \mathbf{C}^n) \to C(a, b; \mathbf{C}^n)$, the system $u' = Au$ is equivalent to the homogeneous system $L(u) = 0$, and we have seen that $\ker(L)$ has dimension n. If we can find n linearly independent solutions, then we have a basis for $\ker(L)$ (usually called, in differential equations jargon, a fundamental set of solutions) and hence we know the general solution to the system.

The problem of finding the general solution to $u' = Au$ can be expressed in matrix form, where we wish to find a matrix $U = U(t)$ that solves $U' = AU$. By the definition of matrix-matrix multiplication, this equation means that each column of U defines a solution to the system $u' = Au$. If these columns are linearly independent, that is, if U is invertible, then they form a fundamental set of solutions and U itself is called a fundamental solution of the matrix differential equation $U' = AU$. In this case, the general solution of $u' = Au$ is $u(t) = U(t)c$, where $c \in \mathbf{C}^n$ is a constant vector.

It turns out that we can define a matrix-valued function e^{tA} such that the unique solution to (5.16) is

$$u(t) = e^{tA}v.$$

The function e^{tA} satisfies $e^{tA} \in \mathbf{C}^{n\times n}$ for all t and is called the *matrix exponential* of $A \in \mathbf{C}^{n\times n}$. We also refer to e^A (obtained by setting $t = 1$ in the matrix-valued function e^{tA}) as the matrix exponential of A.

The matrix exponential satisfies some, but not all, of the properties of the ordinary scalar-valued exponential function. For example,

$$\frac{d}{dt}\left[e^{tA}\right] = Ae^{tA}$$

holds, but

$$e^{A+B} = e^A e^B$$

fails to hold in general. As we will see, we can compute the matrix exponential e^A (at least in principle) from the Jordan canonical form of A.

5.5.1 Definition of the matrix exponential

The matrix exponential $U(t) = e^{tA}$ is defined to be the unique solution to the IVP

$$U' = AU, \; U(0) = I, \tag{5.17}$$

where I is the $n \times n$ identity matrix. The existence and uniqueness theory for ordinary differential equations guarantees that (5.17) has a unique solution, and therefore e^{tA} is well-defined. (We do not present the existence and uniqueness theory here, but Exercise 8 outlines some of the details.) Moreover, it is straightforward to show that U is a fundamental solution of $U' = AU$, that is, that $U(t)$ is nonsingular for all t (see Exercise 9), and also that $U(t) \in \mathbf{R}^{n\times n}$ for all t if $A \in \mathbf{R}^{n\times n}$.

The next theorem shows that all fundamental solutions are closely related.

Theorem 256 *Let $A \in \mathbf{C}^{n\times n}$. If $U : \mathbf{R} \to \mathbf{C}^{n\times n}$ is any fundamental solution of $U' = AU$, then*

$$U(t)U(0)^{-1} = e^{tA}.$$

Proof Exercise 1.

5.5.2 Computing the matrix exponential

The previous theorem suggests two ways to compute e^{tA}: One is to solve (5.17) directly and the other is to find *any* fundamental solution U and compute

$e^{tA} = U(t)U(0)^{-1}$. In one (important) case, we already know how to find a fundamental solution: If $A \in \mathbf{C}^{n\times n}$ is diagonalizable, then a fundamental solution is

$$U(t) = \left[e^{\lambda_1 t}x_1|e^{\lambda_2 t}x_2|\cdots|e^{\lambda_n t}x_n\right],$$

where $\{x_1, x_2, \ldots, x_n\}$ is a basis of $\mathbf{C}^n$ consisting of eigenvectors of A and $\lambda_1, \lambda_2, \ldots, \lambda_n$ are the corresponding eigenvalues. We write $X = [x_1|x_2|\cdots|x_n]$ and note that $X = U(0)$. Then $U(t) = XE(t)$, where

$$E(t) = \begin{bmatrix} e^{\lambda_1 t} & & & \\ & e^{\lambda_2 t} & & \\ & & \ddots & \\ & & & e^{\lambda_n t} \end{bmatrix}.$$

By Theorem 256, we then obtain

$$e^{tA} = U(t)U(0)^{-1} = XE(t)X^{-1}.$$

Moreover, the matrix $E(t)$ is nothing more than e^{tD}, where D is the diagonal matrix of eigenvalues of A:

$$D = \begin{bmatrix} \lambda_1 & & & \\ & \lambda_2 & & \\ & & \ddots & \\ & & & \lambda_n \end{bmatrix}.$$

To see this, notice that e^{tD} is the solution to the IVP

$$U' = DU, \; U(0) = I,$$

which represents n diagonal systems, each of the form

$$\begin{aligned} u_1' &= \lambda_1 u_1, \; u_1(0) = 0, \\ u_2' &= \lambda_2 u_2, \; u_2(0) = 0, \\ &\vdots \\ u_j' &= \lambda_j u_j, \; u_j(0) = 1, \\ &\vdots \\ u_n' &= \lambda_n u_n, \; u_n(0) = 0. \end{aligned}$$

Since these equations are completely decoupled (each equation involves only one unknown), the solution is obvious: $u_i(t) = 0$, $i \neq j$, and $u_j(t) = e^{\lambda_j t}$. Therefore, e^{tD} is the diagonal matrix with the functions $e^{\lambda_j t}$ on the diagonal, and thus it equals the matrix $E(t)$ defined above.

Therefore, in the case of a diagonalizable matrix $A = XDX^{-1}$, the matrix exponential is

$$e^{tA} = Xe^{tD}X^{-1}. \tag{5.18}$$

This result extends to a general matrix A if the spectral decomposition is replaced by the Jordan decomposition. As we now show, if $A = XJX^{-1}$ is the Jordan decomposition of A, then

$$e^{tA} = Xe^{tJ}X^{-1}. \tag{5.19}$$

We prove (5.19) using a change of variables. We define $V : \mathbf{R} \to \mathbf{C}^{n\times n}$ by $V(t) = X^{-1}U(t)$. Then $U = XV$ and $U' = XV'$ (since X is a constant matrix, that is, it does not depend on t). We then obtain

$$\begin{aligned} & U' = AU \\ \Leftrightarrow\quad & XV' = XJX^{-1}U \\ \Leftrightarrow\quad & XV' = XJV \\ \Leftrightarrow\quad & V' = JV. \end{aligned}$$

Assuming V is the matrix exponential of J, so that $V' = JV$ and $V(0) = I$, we see that $U = XV$ solves $U' = AU$. Moreover, this U is invertible because it is the product of two invertible matrices; thus U is a fundamental solution. It follows that $U(t) = Xe^{tJ}$ and $U(0) = XV(0) = XI = X$. Therefore,

$$e^{tA} = U(t)U(0)^{-1} = Xe^{tJ}X^{-1},$$

which proves (5.19).

It remains only to compute e^{tJ} when J is a matrix in Jordan form. In this case, J is a block diagonal matrix with Jordan blocks on the diagonal:

$$J = \begin{bmatrix} J_{1,1} & & & & \\ & \ddots & & & \\ & & J_{1,s_1} & & \\ & & & \ddots & \\ & & & & J_{t,s_t} \end{bmatrix}$$

(using the notation of Section 5.4). Each $J_{i,j}$ has the form

$$J_{i,j} = \begin{bmatrix} \lambda_i & 1 & & & \\ & \lambda_i & 1 & & \\ & & \ddots & \ddots & \\ & & & \lambda_i & 1 \\ & & & & \lambda_i \end{bmatrix} \in \mathbf{C}^{r_{i,j}\times r_{i,j}},$$

where $r_{i,j} \geq 1$.

The fact that J is block diagonal means that the system $V' = JV$ of differential equations is decoupled into blocks of equations

$$v'_{i,j} = J_{i,j}v_{i,j}, \tag{5.20}$$

where $v_{i,j} = v_{i,j}(t)$ is a vector-valued function containing the components of v corresponding to the columns of J occupied by $J_{i,j}$. It therefore suffices to be able to solve (5.20). To simplify the notation, we will replace $J_{i,j}$ by

$$J = \begin{bmatrix} \lambda & 1 & & & \\ & \lambda & 1 & & \\ & & \ddots & \ddots & \\ & & & \lambda & 1 \\ & & & & \lambda \end{bmatrix} \in \mathbf{C}^{r\times r}$$

and $v_{i,j}$ by v. In other words, we temporarily assume that J consists of a single Jordan block.

We then need to solve the system

$$\begin{aligned} v_1' &= \lambda v_1 + v_2, \\ v_2' &= \lambda v_2 + v_3, \\ &\vdots \\ v_{r-1}' &= \lambda v_{r-1} + v_r, \\ v_r' &= \lambda v_r. \end{aligned} \tag{5.21}$$

Using techniques from elementary ODEs (specifically the method of integrating factors for a first-order, linear ODE), it is not difficult to show that the general solution to this system is

$$\begin{aligned} v_1(t) &= c_1 e^{\lambda t} + c_2 t e^{\lambda t} + c_3 \frac{t^2}{2!} e^{\lambda t} + \ldots + c_r \frac{t^{r-1}}{(r-1)!} e^{\lambda t}, \\ v_2(t) &= c_2 e^{\lambda t} + c_3 t e^{\lambda t} + c_4 \frac{t^2}{2!} e^{\lambda t} + \ldots + c_r \frac{t^{r-2}}{(r-2)!} e^{\lambda t}, \\ &\vdots \\ v_{r-1}(t) &= c_{r-1} e^{\lambda t} + c_r t e^{\lambda t}, \\ v_r(t) &= c_r e^{\lambda t} \end{aligned} \tag{5.22}$$

(see Exercise 2). From this we see that a fundamental solution to $V' = JV$ is

$$V(t) = \begin{bmatrix} e^{\lambda t} & te^{\lambda t} & \frac{t^2}{2!}e^{\lambda t} & \cdots & \frac{t^{r-1}}{(r-1)!}e^{\lambda t} \\ & e^{\lambda t} & te^{\lambda t} & \cdots & \frac{t^{r-2}}{(r-2)!}e^{\lambda t} \\ & & \ddots & \ddots & \vdots \\ & & & \ddots & te^{\lambda t} \\ & & & & e^{\lambda t} \end{bmatrix}. \tag{5.23}$$

Since this matrix satisfies $V(0) = I$, it follows that $V(t) = e^{tJ}$, and we have found the matrix exponential in the case of a single Jordan block.

Returning now to the general case, in which J consists of a number of Jordan block, it is not difficult to see that e^{tJ} is itself block diagonal, each block having the same dimensions as the corresponding Jordan block and having the form (5.23). We illustrate with an example.

Example 257 *Suppose*

$$J = \left[\begin{array}{ccc|c|cc|cc} -1 & 1 & 0 & & & & & \\ 0 & -1 & 1 & & & & & \\ 0 & 0 & -1 & & & & & \\ \hline & & & -1 & & & & \\ \hline & & & & 2 & 1 & 0 & 0 \\ & & & & 0 & 2 & 0 & 0 \\ \hline & & & & & & 2 & 1 \\ & & & & & & 0 & 2 \end{array}\right].$$

Then J consists of four Jordan blocks, namely,

$$J_1 = \begin{bmatrix} -1 & 1 & 0 \\ 0 & -1 & 1 \\ 0 & 0 & -1 \end{bmatrix},\ J_2 = [-1],\ J_3 = \begin{bmatrix} 2 & 1 \\ 0 & 2 \end{bmatrix},\ J_4 = \begin{bmatrix} 2 & 1 \\ 0 & 2 \end{bmatrix}.$$

We have

$$e^{tJ_1} = \begin{bmatrix} e^{-t} & te^{-t} & \frac{t^2}{2!}e^{-t} \\ 0 & e^{-t} & te^{-t} \\ 0 & 0 & e^{-t} \end{bmatrix},\ e^{tJ_2} = \left[e^{-t}\right],$$

$$e^{tJ_3} = \begin{bmatrix} e^{2t} & te^{2t} \\ 0 & e^{2t} \end{bmatrix},\ e^{tJ_4} = \begin{bmatrix} e^{2t} & te^{2t} \\ 0 & e^{2t} \end{bmatrix},$$

and so

$$e^{tJ} = \left[\begin{array}{ccc|c|cc|cc} e^{-t} & te^{-t} & \frac{t^2}{2!}e^{-t} & & & & & \\ 0 & e^{-t} & te^{-t} & & & & & \\ 0 & 0 & e^{-t} & & & & & \\ \hline & & & e^{-t} & & & & \\ \hline & & & & e^{2t} & te^{2t} & & \\ & & & & & e^{2t} & & \\ \hline & & & & & & e^{2t} & te^{2t} \\ & & & & & & 0 & e^{2t} \end{array}\right].$$

Now that we know how to compute the matrix exponential of any matrix J in Jordan form, we can, in theory, compute e^{tA} for any matrix $A \in \mathbf{C}^{n\times n}$ as follows: Find the Jordan decomposition $A = XJX^{-1}$, where J is the Jordan canonical form of A; then

$$e^{tA} = Xe^{tJ}X^{-1}.$$

(Therefore $e^A = Xe^JX^{-1}$, where e^J is obtained by replacing t by 1 in (5.23).) However, we reiterate that, in general, it is not possible to compute the Jordan canonical form a matrix accurately in finite-precision arithmetic. Therefore, this technique for computing e^{tA} is limited to small matrices. The interested reader is referred to the detailed article by Moler and Van Loan [32], which surveys methods for computing the matrix exponential in finite-precision arithmetic.

We close this section with the following result, which was mentioned above.

Theorem 258 *Let $A \in \mathbf{C}^{n\times n}$, $v \in \mathbf{C}^n$ be given. Then the solution to the IVP*

$$u' = Au, \; u(0) = v$$

is $u(t) = e^{tA}v$.

Proof Exercise 7.

Exercises

1. Prove Theorem 256. (Hint: Define $Z(t) = U(t)U(0)^{-1}$ and prove that Z is a solution to (5.17).)

2. Show that (5.22) is the general solution to (5.21). (Hint: The last equation is simple: $v_r' = \lambda v_r$ implies that $v_r(t) = c_re^{\lambda t}$, where c_r is an arbitrary constant. Substitute the formula for v_r into the next-to-last equation and solve for v_{r-1} using the method of integrating factors (see, for example, Zill [46], Section 2.3). Then substitute v_{r-1} into the equation for v_{r-2}' and continue.)

3. Find the matrix exponential e^{tA} for the matrix given in Exercise 5.4.5.

4. Find the matrix exponential e^{tA} for the matrix given in Exercise 5.4.7.

5. Let $A, B \in \mathbf{C}^{n\times n}$.

   (a) Show that if A and B commute, then so do e^{tA} and B (that is, show that if $AB = BA$, then $e^{tA}B = Be^{tA}$). (Hint: Define $U(t) = e^{tA}B - Be^{tA}$ and show that U satisfies the matrix IVP $U' = AU$, $U(0) = 0$.)

   (b) Use the preceding result to show that if A and B commute, the $e^{t(A+B)} = e^{tA}e^{tB}$ holds. (Notice that, in particular, this implies that $e^{A+B} = e^Ae^B$ provided A and B commute.)

   (c) Show by explicit example that $e^{A+B} = e^Ae^B$ does not hold for all matrices A and B.

6. Use the previous exercise to show that, for any $A \in \mathbf{C}^{n\times n}$,

$$\left(e^{tA}\right)^{-1} = e^{-tA}.$$

(Note that e^{tA} is known to be invertible because it is a fundamental solution to $U' = AU$.)

7. Prove Theorem 258.

8. Let $A \in \mathbf{C}^{n\times n}$ be given. The existence and uniqueness theory for ordinary differential equations implies that, given any $t_0 \in \mathbf{R}$ and $v_0 \in \mathbf{C}^n$, the IVP
$$u' = Au, \ u(t_0) = v_0$$
has a unique solution $u : \mathbf{R} \to \mathbf{C}^n$. Moreover, the theory guarantees that u is real-valued (that is, $u : \mathbf{R} \to \mathbf{R}^n$) if $A \in \mathbf{R}^{n\times n}$ and $v_0 \in \mathbf{R}^n$. Use these results to prove that there is a unique solution $U : \mathbf{R} \to \mathbf{C}^{n\times n}$ to the matrix IVP
$$U' = AU, \ U(0) = I,$$
where I is the $n \times n$ identity matrix, and that $U(t) \in \mathbf{R}^{n\times n}$ for all $t \in \mathbf{R}$ if $A \in \mathbf{R}^{n\times n}$.

9. Let $U : \mathbf{R} \to \mathbf{C}^{n\times n}$ be the unique solution to
$$U' = AU, \ U(0) = I,$$
where $A \in \mathbf{C}^{n\times n}$ is given (see the previous exercise). Prove that $U(t)$ is nonsingular for all $t \in \mathbf{R}$. (Hint: If $U(t_0)$ is singular, say $U(t_0)c = 0$ for some $c \in \mathbf{C}^n$, $c \neq 0$, then $u(t) = U(t)c$ solves the IVP $u' = Au$, $u(t_0) = 0$. Use the existence and uniqueness theory from the previous exercise to prove that $U(0)c = 0$, a contradiction.)

5.6 Graphs and eigenvalues

In Section 3.10, we saw that a graph G can be represented by its adjacency matrix A_G, which is a symmetric matrix, all of whose entries are 1 or 0. Since A_G is symmetric, its eigenvalues are real. Although it may seem surprising, these eigenvalues give quite a bit of information about the graph G. The theory that relates properties of graphs to the eigenvalues of the adjacency matrix is called *spectral graph theory.* Spectral graph theory is quite technical, with many specialized results; in this section, we will present a few representative theorems.

5.6.1 Cospectral graphs

We begin with the following definition.

Definition 259 *Let G be a graph. The* eigenvalues *of G are simply the eigenvalues of the adjacency matrix A_G. The collection of eigenvalues of G, together with their multiplicities, is called the* spectrum *of G. Two graphs G and H are called* cospectral *if and only if their spectra are the same.*

We know that similar matrices have the same spectra. Moreover, in Section 3.10, we saw that isomorphic graphs have similar adjacency matrices: If G and H are isomorphic, then there exists a permutation matrix P such that

$$A_G = PA_HP^T.$$

This is a similarity relationship because any permutation matrix P satisfies $P^{-1} = P^T$; see Exercise 6. These considerations yield the following theorem.

Theorem 260 *Isomorphic graphs are cospectral.*

The converse of the preceding theorem does not hold. For instance, the graphs illustrated in Figure 5.1 are cospectral (as the reader is asked to verify in Exercise 7), but they are clearly not isomorphic.

FIGURE 5.1
Two graphs that are cospectral but not isomorphic.

5.6.2 Bipartite graphs and eigenvalues

There are many results in the spectral theory of graphs that apply to special classes of graphs. We will consider two particular classes that are important in applications, bipartite and regular graphs.

A graph G is called *bipartite* if its node set V_G can be partitioned into two subsets, V_1 and V_2, and every edge in E_G joins one node in V_1 with one node in V_2. Bipartite graphs are important in many applications; for instance, in a transportation problem, it may be required to move products from factories to stores. A graph could be defined to represent the transportation network, where edges are the roads from the factories to the stores (with roads from one factory to another, or from one store to another, ignored as unimportant in this problem). The nodes are the factories and the stores, and the graph is bipartite.

An example of a bipartite graph G is shown in Figure 5.2. For this graph, we have $V_G = V_1 \cup V_2$, where $V_1 = \{v_1, v_2, v_3\}$ and $V_2 = \{v_4, v_5, v_6, v_7\}$. The adjacency matrix of G is

$$A_G = \begin{bmatrix} 0 & 0 & 0 & 1 & 0 & 0 & 1 \\ 0 & 0 & 0 & 0 & 1 & 1 & 1 \\ 0 & 0 & 0 & 1 & 0 & 0 & 1 \\ 1 & 0 & 1 & 0 & 0 & 0 & 0 \\ 0 & 1 & 0 & 0 & 0 & 0 & 0 \\ 0 & 1 & 0 & 0 & 0 & 0 & 0 \\ 1 & 1 & 1 & 0 & 0 & 0 & 0 \end{bmatrix}. \tag{5.24}$$

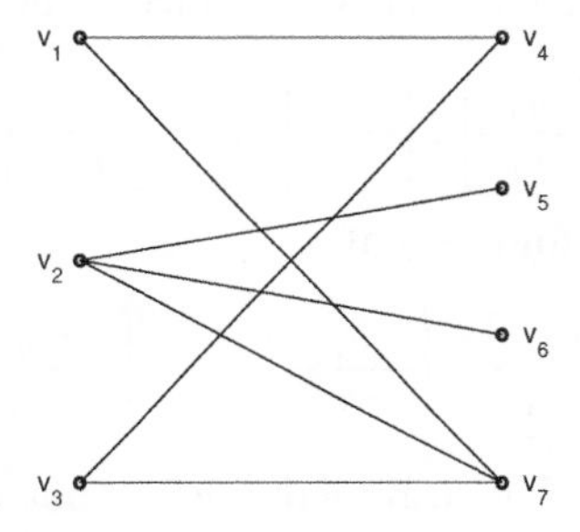

FIGURE 5.2
A bipartite graph.

It is easy to see that if G is a bipartite graph with V_G partitioned into $V_1 \cup V_2$, and if the nodes of G are labeled so that all the nodes in V_1 precede those in V_2 (as in Figure 5.2), then A_G has the form

$$A_G = \left[\begin{array}{c|c} 0 & B \\ \hline B^T & 0 \end{array}\right]. \tag{5.25}$$

If V_1, V_2 contain n_1, n_2 nodes, respectively, then the matrix B is $n_1 \times n_2$.

This structure of the adjacency matrix of a bipartite graph has the following implication for the spectrum.

Theorem 261 *Suppose G is a bipartite graph. Then, if $\lambda > 0$ is an eigenvalue of G of multiplicity m, then $-\lambda$ is also an eigenvalue of G with multiplicity m.*

Proof We can assume that the nodes of G are ordered so that the adjacency matrix A_G is of the form (5.25). The number of nodes in V_G will be denoted by n, and we assume V_G is partitioned as $V_1 \cup V_2$, where V_1, V_2 have n_1, n_2 elements, respectively. Let us assume that λ is a positive eigenvalue of A_G, and let $x \in \mathbf{R}^n$ be a corresponding eigenvector. We can partition x as

$$x = \left[\begin{array}{c} u \\ \hline v \end{array}\right],$$

where $u \in \mathbf{R}^{n_1}$, $v \in \mathbf{R}^{n_2}$. Then

$$A_G x = \lambda x \Rightarrow \left[\begin{array}{c|c} 0 & B \\ \hline B^T & 0 \end{array}\right] \left[\begin{array}{c} u \\ \hline v \end{array}\right] = \lambda \left[\begin{array}{c} u \\ \hline v \end{array}\right] \Rightarrow \begin{array}{c} Bv = \lambda u, \\ B^T u = \lambda v. \end{array}$$

But then $\hat{x} = (u, -v)$ satisfies $A_G \hat{x} = -\lambda \hat{x}$:

$$\left[\begin{array}{c|c} 0 & B \\ \hline B^T & 0 \end{array}\right] \left[\begin{array}{c} u \\ \hline -v \end{array}\right] = \left[\begin{array}{c} -Bv \\ \hline B^T u \end{array}\right] = \left[\begin{array}{c} -\lambda u \\ \hline \lambda v \end{array}\right] = -\lambda \left[\begin{array}{c} u \\ \hline -v \end{array}\right].$$

This shows that if λ is an eigenvalue of A_G, then so is $-\lambda$. The multiplicities of the two eigenvalues must be the same since a set of partitioned vectors of the form

$$\left\{ \left[\begin{array}{c} u_1 \\ \hline v_1 \end{array}\right], \left[\begin{array}{c} u_2 \\ \hline v_2 \end{array}\right], \ldots, \left[\begin{array}{c} u_m \\ \hline v_m \end{array}\right] \right\}$$

is linearly independent if and only if

$$\left\{ \left[\begin{array}{c} u_1 \\ \hline -v_1 \end{array}\right], \left[\begin{array}{c} u_2 \\ \hline -v_2 \end{array}\right], \ldots, \left[\begin{array}{c} u_m \\ \hline -v_m \end{array}\right] \right\}$$

is linearly independent. (This argument shows that the geometric multiplicities of the two eigenvalues are the same; since A_G is symmetric, the geometric and algebraic multiplicities of an eigenvalue are the same.)

QED

As an example, let us consider the bipartite graph G of Figure 5.2, whose adjacency matrix A_G is given by (5.24). A direct calculation shows that the characteristic polynomial of A_G is

$$p_{A_G}(r) = r^7 - 7r^5 + 10r^3 = r^3 \left(r^4 - 7r^2 + 10\right) = r^3 \left(r^2 - 2\right) \left(r^2 - 5\right).$$

Therefore, A_G has simple eigenvalues $-\sqrt{5}, -\sqrt{2}, \sqrt{2}, \sqrt{5}$, and 0 is an eigenvalue of multiplicity 3.

Although we will not prove it here, the converse of Theorem 261 also holds, which means that one can detect the fact that a graph is bipartite from its spectrum.

5.6.3 Regular graphs

The *degree* of a node is the number of edges incident with that node (in other words, the number of edges having that node as an endpoint). A *k-regular* graph is a graph in which every node has degree k.

A *connected graph* is one in which there is a walk between any two nodes in the graph. Referring back to Figure 5.1, the graph on the right is connected, while the graph on the left is not. If a graph is not connected, it can be divided into a number of connected subgraphs, which together contain all the nodes and edges of the graph; each such connected subgraph is called a *connected component* of the graph. Figure 5.3 shows a 3-regular graph with two connected components. If G itself is connected, then it has only one connected component.

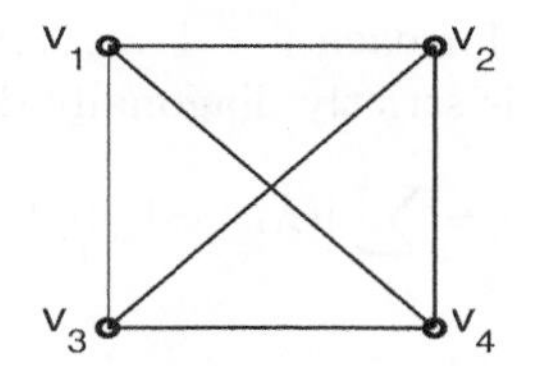

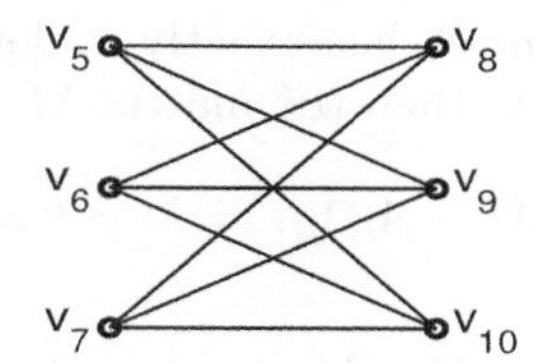

FIGURE 5.3
A 3-regular graph on 10 nodes with two connected components.

The structure of the adjacency matrix A_G is related to the connected components of G in a simple way. Let us assume G has c connected components $G_1, G_2, \ldots, G_c$ (each of which is a graph in its own right). We will label the nodes of G so that the nodes lying in G_1 come first (say $v_1, v_2, \ldots, v_{n_1}$), followed by the nodes lying in G_2 ($v_{n_1+1}, v_{n_1+2}, \ldots, v_{n_1+n_2}$), and so forth. Since no edges exist joining nodes in different connected components, it follows that A_G is *block diagonal*, and each diagonal block is the adjacency matrix of one of the connected components:

$$A_G = \begin{bmatrix} A_{G_1} & 0 & \cdots & 0 \\ 0 & A_{G_2} & \cdots & 0 \\ \vdots & & \ddots & \vdots \\ 0 & \cdots & & A_{G_c} \end{bmatrix}.$$

The following result is implied by Exercise 5.2.2.

Lemma 262 *If G is a graph, then the spectrum of G is the union of the spectra of the connected components of G.*

The following theorem shows the relationship between connectivity of a regular graph and its spectrum.

Theorem 263 *Let G be a k-regular graph. Then the largest eigenvalue of G is k, and the multiplicity of the eigenvalue k is the number of connected components of G. In particular, if k is a simple eigenvalue, then G is connected.*

Proof Let A_G be the adjacency matrix of the k-regular graph G, and assume G has n nodes. By the previous lemma, it suffices to prove the result under the assumption that G is connected; this means that we must show that k is a simple eigenvalue of A_G, and that every other eigenvalue of A_G is strictly smaller in magnitude than k.

Since each node of G is adjacent to exactly k other nodes, each row of A_G has exactly k entries equal to 1, with the rest equal to 0. Therefore, if $e \in \mathbf{R}^n$ is the vector of all ones, it follows that every component of $A_G e$ is k, and hence $A_G e = ke$. Thus k is an eigenvalue of G.

For each node v_i, define S_i to be the set of indices of nodes adjacent to v_i. By assumption, S_i has exactly k elements for each $i = 1, 2, \ldots, n$. If $\lambda \in \mathbf{R}$ satisfies $|\lambda| > k$, then the matrix $\lambda I - A_G$ is strictly diagonally dominant:

$$|(\lambda I - A_G)_{ii}| = |\lambda| > k = \sum_{j \in S_i} 1 = \sum_{j \neq i} |(\lambda I - A_G)ij| .$$

By Exercise 3.6.15, this implies that $\lambda I - A_G$ is nonsingular, and hence λ cannot be an eigenvalue of A_G.

Finally, we wish to show that k is a simple eigenvalue of A_G, that is, that every eigenvector of A_G corresponding to k is a multiple of e. Let x be any eigenvector of A_G corresponding to $\lambda = k$, and let x_i be the largest component of x: $x_i \geq x_j$ for all $j \neq i$. We then have

$$A_G x = kx \;\Rightarrow\; (A_G x)_i = (kx)_i \;\Rightarrow\; \sum_{j \in S_i} x_j = kx_i \;\Rightarrow\; x_i = \frac{1}{k} \sum_{j \in S_i} x_j .$$

(The last equation follows from the fact that v_i is adjacent to exactly k nodes.) But this last equation implies that $x_j = x_i$ for all $j \in S_i$ (since $x_i \geq x_j$ for all $j \in S_i$ and x_i is the average of all such x_j). By a similar argument, we see that $x_k = x_i$ for all nodes v_k adjacent to some v_j, $j \in S_i$. Since G is connected, by repeating this argument, we eventually show that every x_j, $j \neq i$, equals x_i, and hence x is a multiple of e.

QED

5.6.4 Distinct eigenvalues of a graph

Given a graph G, the *distance* $d(u, v)$ between two nodes of G is the length of the shortest walk joining the two nodes, and the *diameter* of the graph is the largest distance between any two nodes (cf. Exercises 3.10.1, 3.10.2). The following theorem shows that the diameter of a graph tells us something about its spectrum.

Theorem 264 *Let G be a connected graph with diameter s. Then G has at least $s + 1$ distinct eigenvalues.*

Proof Since A_G is symmetric, its minimal polynomial (see Appendix 5.2.1 of Section 5.2) is

$$m_{A_G}(r) = (r - \lambda_1)(r - \lambda_2) \cdots (r - \lambda_t),$$

where $\lambda_1, \lambda_2, \ldots, \lambda_t$ are the distinct eigenvalues of A_G. Thus the degree of m_{A_G} is equal to the number of distinct eigenvalues of A_G. Also, as we saw in

Section 5.2.1,

$$m_{A_G}(A_G) = 0,$$

which shows that $\{I, A_G, A_G^2, \ldots, A_G^t\}$ is linearly dependent.

Now, let v_i and v_j be two nodes in G such that $d(v_i, v_j) = s$, the diameter of G. For any $k = 1, 2, \ldots, s$, we can find a node v_{j_k} such that $d(v_i, v_{j_k}) = k$ (each v_{j_k} can be chosen on the walk from v_i to v_j. Since $(A_G^\ell)_{ij_k}$ is the number of walks from v_i to v_{j_k} of length ℓ, it follows that

$$\left(A_G^\ell\right)_{ij_k} = 0 \text{ for all } \ell = 0, 1, \ldots, k-1,$$

while

$$(A_G^s)_{ij_k} \neq 0.$$

From this, it easily follows that A_G^k is not a linear combination of

$$I, A_G, A_G^2, \ldots, A_G^{k-1}$$

for each $k = 1, 2, \ldots, s$, from which it follows that

$$\{I, A_G, A_G^2, \ldots, A_G^s\}$$

is linearly independent. This proves that $t > s$, that is, the number of distinct eigenvalues of A_G is at least $s + 1$.

QED

The results presented in this section are just a small sampling of spectral graph theory. The reader who wishes to learn more can consult the monograph [1] by Biggs.

Exercises

1. Let G be defined by

$$\begin{aligned} V_G &= \{v_1, v_2, v_3, v_4\}, \\ E_G &= \{\{v_1, v_2\}, \{v_2, v_3\}, \{v_3, v_4\}, \{v_1, v_4\}\}. \end{aligned}$$

Find the eigenvalues of G.

2. Let G be defined by

$$\begin{aligned} V_G &= \{v_1, v_2, v_3, v_4, v_5\}, \\ E_G &= \{\{v_1, v_4\}, \{v_2, v_4\}, \{v_2, v_5\}, \{v_3, v_4\}\}. \end{aligned}$$

Find the eigenvalues of G and explain how this example illustrates Theorem 261.

3. Let G be defined by

$$\begin{aligned} V_G &= \{v_1, v_2, v_3, v_4, v_5, v_6, v_7\}, \\ E_G &= \{\{v_1, v_2\}, \{v_2, v_3\}, \{v_3, v_1\}, \{v_4, v_5\}, \{v_5, v_6\}, \{v_6, v_7\}, \{v_4, v_7\}\}. \end{aligned}$$

   Find the eigenvalues of G and explain how this example illustrates Theorem 263.

4. (a) Let G be a bipartite graph with an odd number of vertices. Prove that 0 is an eigenvalue of G.

   (b) Prove by example that 0 need not be an eigenvalue of a bipartite graph with an even number of nodes.

5. Let G be a connected graph with diameter s. Prove by example that the number of distinct eigenvalues can be strictly greater than $s+1$ (cf. Theorem 264).

6. Let P be a permutation matrix. Prove that $P^T = P^{-1}$.

7. Compute the spectrum of each of the graphs in Figure 5.1, and show that the two graphs are cospectral.

6

Orthogonality and best approximation

A variety of problems involving vector spaces and linear operators can only be addressed if we can measure the distance between vectors. For example, if we want to say that y is a good approximation to x, then we need a quantitative measure of the error $y-x$. As another example, when a system $Ax = b$ has no solution, we might want to compute the x that makes Ax as close as possible to b. This also requires that we can measure the distance between vectors.

A *norm* is a real-valued function defined on a vector space that defines the size of the vectors in the space. The standard notation for the norm of a vector u is $\|x\|$; with this notation, the distance between two vectors x and y is denoted by $\|y - x\|$.

In some spaces, we can measure not only the distance between two vectors, but also the angle between them. This is true in Euclidean space, where "angle" has a geometric meaning (at least in two or three dimensions). It turns out to be meaningful to define the angle between two vectors in certain abstract vector spaces. This is defined in terms of an *inner product*, which is a generalization of the Euclidean dot product.

We begin this chapter by defining and exploring norms and inner products. In a departure from the previous chapters, the only field of scalars considered in most of the chapter will be $\mathbf{R}$, the field of real numbers. At the end, we discuss the changes that must be made in order to accommodate the field of complex numbers. In between, we will explore various implications of *orthogonality*, the abstract version of perpendicularity.

6.1 Norms and inner products

The definition of norm is intended to describe the properties that any reasonable measure of size ought to satisfy.

Definition 265 *Let V be a vector space over $\mathbf{R}$. A* norm *on V is a function $\|\cdot\|$ mapping V into $\mathbf{R}$ that satisfies the following properties:*

1. *$\|u\| \geq 0$ for all $u \in V$, and $\|u\| = 0$ if and only if $u = 0$;*
2. *$\|\alpha u\| = |\alpha|\|u\|$ for all $\alpha \in \mathbf{R}$ and all $u \in V$;*

3. $\|u+v\| \le \|u\| + \|v\|$ *for all* $u, v \in V$.

The last property is called the triangle inequality.

A norm is just a special kind of function, and it could be denoted by traditional function notation, for example, as $n(v)$ in place of $\|v\|$, where $n : V \to \mathbf{R}$. However, the notation $\|v\|$ is traditional and convenient.

Each of the properties of a norm expresses a natural property. First, the length of any vector must be a nonnegative number, and only the zero vector has length zero. Second, multiplying a vector by a real scalar α simply stretches or shrinks the vector by a factor $|\alpha|$. Finally, the triangle inequality expresses the limitation illustrated in Figure 6.1: The length of any side of a triangle is at most the sum of the lengths of the other two sides.

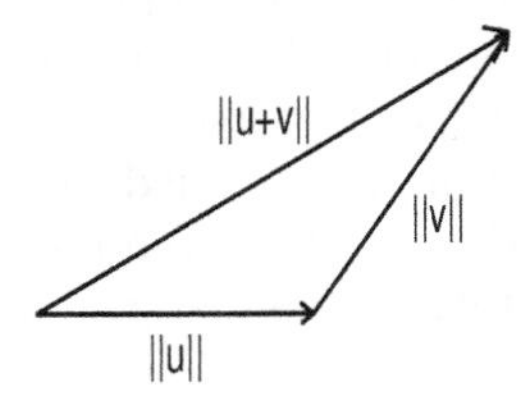

FIGURE 6.1
The triangle inequality: $\|u+v\| \le \|u\| + \|v\|$.

The prototypical norm is the Euclidean norm on $\mathbf{R}^n$:

$$\|x\|_2 = \sqrt{x_1^2 + x_2^2 + \ldots + x_n^2} \text{ for all } x \in \mathbf{R}^n.$$

We denote the Euclidean norm by a subscript of "2" because of the exponent in the definition. There are other norms on $\mathbf{R}^n$, some of which we will describe below, so it is important to distinguish one norm from another. The Euclidean norm is natural because, when we think of a Euclidean vector as an arrow with its tail at the origin and its head at the point with coordinates $(x_1, x_2, \ldots, x_n)$, $\|x\|_2$ is the length of the arrow in the usual sense (the Euclidean distance from the tail to the head).

The Euclidean norm is related to the *dot product* on $\mathbf{R}^n$:

$$x \cdot y = x_1 y_1 + x_2 y_2 + \ldots + x_n y_n.$$

The relationship is $\|x\|_2 = \sqrt{x \cdot x}$. The dot product on $\mathbf{R}^n$ is a special case of an *inner product*, and an inner product always defines a norm in this fashion.

Definition 266 *Let V be a vector space over* $\mathbf{R}$. *An* inner product *on V is a function $\langle \cdot, \cdot \rangle$ mapping $V \times V$ into* $\mathbf{R}$, *and satisfying the following properties:*

$$\begin{aligned} \langle u, v \rangle &= \langle v, u \rangle \text{ for all } u, v \in V, & (6.1) \\ \langle \alpha u + \beta v, w \rangle &= \alpha \langle u, w \rangle + \beta \langle v, w \rangle \text{ for all } \alpha, \beta \in \mathbf{R}, u, v, w \in V, & (6.2) \\ \langle u, u \rangle &\ge 0 \text{ for all } u \in V; \ \langle u, u \rangle = 0 \text{ if and only if } u = 0. & (6.3) \end{aligned}$$

The reader should notice that, together, the first two properties defining an inner product imply

$$\langle u, \alpha v + \beta w\rangle = \alpha \langle u, v\rangle + \beta \langle u, w\rangle \text{ for all } \alpha, \beta \in \mathbf{R}, u, v, w \in V. \tag{6.4}$$

We say that $\langle\cdot,\cdot\rangle$ is *bilinear* (or a *bilinear form*) if it satisfies (6.2) and (6.4). We call $\langle\cdot,\cdot\rangle$ *symmetric* if it satisfies (6.1), and *positive definite* if it satisfies (6.3). Thus, an inner product can be described as a *positive definite symmetric bilinear form.*

Just as we do not use the usual functional notation for a norm, so we have a special notation for an inner product. A function from $V \times V$ into $\mathbf{R}$ normally has a name, such as f, and values of the function are expressed by the notation $f(u, v)$. An inner product has no name as such (which is why we wrote $\langle\cdot,\cdot\rangle$ above), and values of the inner product are written as $\langle u, v\rangle$.

The proof of the following result is left as an exercise.

Lemma 267 *Let V be a vector space over $\mathbf{R}$, and let $\langle\cdot,\cdot\rangle$ be an inner product on V. If either u or v is the zero vector, then $\langle u, v\rangle = 0$.*

Proof Exercise 2.

We wish to show that, given an inner product $\langle\cdot,\cdot\rangle$, we can always define a norm by the formula $\|u\| = \sqrt{\langle u, u\rangle}$. To prove that this defines a norm (in particular, to prove the triangle inequality), it is useful to first derive the following result, which is of independent interest.

Theorem 268 *(The Cauchy-Schwarz inequality) Let V be a vector space over $\mathbf{R}$, and let $\langle\cdot,\cdot\rangle$ be an inner product on V. Then*

$$|\langle u, v\rangle| \le \langle u, u\rangle^{1/2} \langle v, v\rangle^{1/2} \text{ for all } u, v \in V. \tag{6.5}$$

Proof If either u or v is the zero vector, then (6.5) holds because both sides are zero. We can therefore assume that both u and v are nonzero vectors. We first prove the result in the special case that $\langle u, u\rangle = \langle v, v\rangle = 1$. In this case, we have to show that

$$|\langle u, v\rangle| \le 1.$$

By the third property of an inner product, we have $\langle u - v, u - v\rangle \ge 0$. Applying the first two properties of an inner product, we can rewrite $\langle u - v, u - v\rangle$ as follows:

$$\langle u - v, u - v\rangle = \langle u, u\rangle - 2\langle u, v\rangle + \langle v, v\rangle = 2 - 2\langle u, v\rangle.$$

We then obtain

$$2 - 2\langle u, v\rangle \ge 0 \Rightarrow \langle u, v\rangle \le 1.$$

Repeating this reasoning, but starting with the inequality $\langle u + v, u + v\rangle \ge 0$ leads to $-1 \le \langle u, v\rangle$, and thus we see that $|\langle u, v\rangle| \le 1$.

We now consider the general case in which $u, v \in V$ are nonzero vectors. Let us define

$$\hat{u} = \langle u, u \rangle^{-1/2} u, \ \hat{v} = \langle v, v \rangle^{-1/2} v.$$

Then

$$\langle \hat{u}, \hat{u} \rangle = \left\langle \langle u, u \rangle^{-1/2} u, \langle u, u \rangle^{-1/2} u \right\rangle = \langle u, u \rangle^{-1} \langle u, u \rangle = 1,$$

and similarly for $\langle \hat{v}, \hat{v} \rangle$. It follows that $|\langle \hat{u}, \hat{v} \rangle| \le 1$. But

$$\begin{aligned} |\langle \hat{u}, \hat{v} \rangle| &= \left| \left\langle \langle u, u \rangle^{-1/2} u, \langle v, v \rangle^{-1/2} v \right\rangle \right| \\ &= \langle u, u \rangle^{-1/2} \langle v, v \rangle^{-1/2} |\langle u, v \rangle|, \end{aligned}$$

and hence

$$\langle u, u \rangle^{-1/2} \langle v, v \rangle^{-1/2} |\langle u, v \rangle| \le 1,$$

which yields the desired inequality:

$$|\langle u, v \rangle| \le \langle u, u \rangle^{1/2} \langle v, v \rangle^{1/2}.$$

QED

We can now prove that any inner product defines a norm.

Theorem 269 *Let V be a vector space over $\mathbf{R}$, and suppose $\langle \cdot, \cdot \rangle$ is an inner product on V. Then the formula*

$$\|u\| = \sqrt{\langle u, u \rangle} \tag{6.6}$$

defines a norm on V.

Proof We must show that the proposed norm, defined by (6.6), satisfies the three properties of a norm. The first property of a norm follows immediately from the third property of an inner product. The second follows from the second property of an inner product: For any $\alpha \in \mathbf{R}$ and any $u \in V$,

$$\|\alpha u\| = \sqrt{\langle \alpha u, \alpha u \rangle} = \sqrt{\alpha^2 \langle u, u \rangle} = \sqrt{\alpha^2} \sqrt{\langle u, u \rangle} = |\alpha| \|u\|.$$

The third property of a norm, the triangle inequality, follows from the Cauchy-Schwarz inequality:

$$\begin{aligned} \|u+v\|^2 = \langle u+v, u+v \rangle = \langle u, u \rangle + 2 \langle u, v \rangle + \langle v, v \rangle &= \|u\|^2 + 2 \langle u, v \rangle + \|v\|^2 \\ &\le \|u\|^2 + 2\|u\|\|v\| + \|v\|^2 \\ &= (\|u\| + \|v\|)^2. \end{aligned}$$

We then obtain $\|u + v\| \le \|u\| + \|v\|$.

QED

If $\|\cdot\|$ is the norm defined by the inner product $\langle\cdot,\cdot\rangle$, then the Cauchy-Schwarz inequality takes the form

$$|\langle u,v\rangle| \leq \|u\|\|v\| \text{ for all } u,v \in V. \tag{6.7}$$

It is important to know the conditions on u and v that imply equality in the Cauchy-Schwarz inequality ($|\langle u,v\rangle| = \|u\|\|v\|$). Exercise 6.3.15 asks the reader to explore this question.

If V is a vector space and $\|\cdot\|$ is a norm on V, then V together with $\|\cdot\|$ is called a *normed vector space*. Similarly, a vector space together with an inner product is called an *inner product space*. As we see below, a vector space can have many different norms and inner products defined on it.

6.1.1 Examples of norms and inner products

The Euclidean inner product and norm on $\mathbf{R}^n$

The usual dot product on $\mathbf{R}^n$ is defined by

$$x \cdot y = \sum_{i=1}^{n} x_i y_i \text{ for all } x,y \in \mathbf{R}^n.$$

The reader is probably familiar with the dot product on $\mathbf{R}^2$ and $\mathbf{R}^3$, and with the fact that two vectors x and y are perpendicular if and only if $x \cdot y = 0$. The geometric significance of the dot product and inner products in general will be discussed in Section 6.3. The dot product is also called the *Euclidean inner product*. The corresponding norm is the Euclidean norm:

$$\|x\|_2 = \sqrt{\sum_{i=1}^{n} x_i^2} \text{ for all } x \in \mathbf{R}^n.$$

The ℓ^p norms on $\mathbf{R}^n$

For any $p \in \mathbf{R}$, $p \geq 1$, one can define a norm on $\mathbf{R}^n$ by

$$\|x\|_p = \left[\sum_{i=1}^{n} |x_i|^p\right]^{1/p} \text{ for all } x \in \mathbf{R}^n.$$

This is called the ℓ^p norm (read as the "little ell-p norm", to distinguish it from the L^p norm for functions that we will introduce below). The Euclidean norm $\|\cdot\|_2$ is the special case corresponding to $p = 2$ and is therefore also called the ℓ^2 norm.

Besides $p = 2$, the other case that is of practical interest is $p = 1$:

$$\|x\|_1 = \sum_{i=1}^{n} |x_i| \text{ for all } x \in \mathbf{R}^n.$$

A related norm that is often used is the ℓ^∞ norm:

$$\|x\|_\infty = \max\{|x_i| : i = 1, 2, \ldots, n\} \text{ for all } x \in \mathbf{R}^n.$$

It can be shown that, for all $x \in \mathbf{R}^n$,

$$\|x\|_p \to \|x\|_\infty \text{ as } p \to \infty,$$

which explains the notation.

Exercise 3 asks the reader to prove that $\|\cdot\|_1$ and $\|\cdot\|_\infty$ both satisfy the definition of norm. We will not pursue the proof of the fact that $\|\cdot\|_p$, $1 < p < 2$ or $2 < p < \infty$, is also a norm, as these norms are mostly of theoretical interest. We do point out that, of all the norms $\|\cdot\|_p$, $1 \leq p \leq \infty$, only $\|\cdot\|_2$ is defined by an inner product. As we shall see, this makes the Euclidean norm especially useful and easy to use.

The L^2 inner product and norm for functions

The following formula defines an inner product on $C[a, b]$:

$$\langle f, g\rangle_2 = \int_a^b f(x)g(x)\, dx. \tag{6.8}$$

This is called the L^2 inner product (read as the "ell-2 inner product" or the "big ell-2 inner product," if it is necessary to distinguish it from the ℓ^2 inner product). We also refer to this as the $L^2(a, b)$ inner product when we wish to indicate the interval on which the functions are defined.

The only property of an inner product that is not easily verified is that $\langle f, f\rangle_2 = 0$ only if $f = 0$ (that is, f is the zero function). This can be proved from the continuity of f using elementary analysis. We will not go through the proof carefully here, but the idea can be explained concisely: We have

$$\langle f, f\rangle_2 = \int_a^b f(x)^2\, dx.$$

Since the integrand is nonnegative, the integral is also nonnegative. If f were nonzero anywhere on the interval $[a, b]$, by continuity, it would be nonzero on some interval. In this case, $f(x)^2$ would be positive on that interval and then the integral itself would be positive. In other words, if f is not the zero function, then $\langle f, f\rangle_2 > 0$, which is what we want to prove.

Although the definition of the L^2 inner product might seem rather arbitrary, in fact it is the direct generalization of the Euclidean dot product. We can approximate a function $f \in C[a, b]$ by sampling it on a regular grid. A regular grid on the interval $[a, b]$ consists of the gridpoints $x_j = a + j\Delta x$, $j = 0, 1, \ldots, n$, where $\Delta x = (b - a)/n$. Sampling f on this grid leads to the following Euclidean vector:

$$F = (f(x_1), f(x_2), \ldots, f(x_n)) \in \mathbf{R}^n.$$

Given two functions $f, g \in C[a, b]$ and the corresponding vectors $F, G \in \mathbf{R}^n$, we have

$$F \cdot G = \sum_{i=1}^{n} F_i G_i = \sum_{i=1}^{n} f(x_i) g(x_i).$$

The reader will notice the similarity of this expression to the Riemann sum

$$\sum_{i=1}^{n} f(x_i) g(x_i) \Delta x,$$

which is just a scaled version of the dot product $F \cdot G$. Taking the limit as $n \to \infty$ yields

$$\sum_{i=1}^{n} f(x_i) g(x_i) \Delta x \to \int_a^b f(x) g(x)\, dx = \langle f, g \rangle_2 .$$

This explains the naturalness of the L^2 inner product.

The other L^p norms

For any $p \in \mathbf{R}$, $p \geq 1$, we can define the following norm, called the L^p (or $L^p(a, b)$) norm:

$$\|f\|_p = \left[\int_a^b |f(x)|^p \, dx \right]^{1/p} \quad \text{for all } f \in C[a, b].$$

It is not straightforward to show that $\|\cdot\|_p$ is really a norm (except in the case of $p = 2$, when we can use the fact that an inner product always defines a norm), and this is a topic generally covered by a graduate course in analysis. It can be shown that, for all $f \in C[a, b]$, $\|f\|_p \to \|f\|_\infty$ as $p \to \infty$, where

$$\|f\|_\infty = \max\{|f(x)| \,:\, a \leq x \leq b\}.$$

This last norm is called the L^∞ norm.

The similarity between the ℓ^p and L^p norms should be obvious. Only in the case $p = 2$ is the L^p norm defined by an inner product. Analogous to the case of ℓ^p, the most useful of the L^p norms correspond to $p = 1$, $p = 2$, and $p = \infty$.

The spaces $L^p(a, b)$

When using the L^p norms, it is natural to allow any functions for which the norms make sense, that is, any functions for which the integrals are finite. Such functions need not be continuous, and in fact do not even need to be bounded.[1] Roughly speaking, the space $L^p(a, b)$ is the space of all (Lebesgue

---

[1] For example, the function $f(x) = x^{-1/4}$ satisfies

$$\int_0^1 f(x)^2 \, dx < \infty,$$

measurable) functions defined on (a,b) such that

$$\int_a^b |f(x)|^p \, dx < \infty,$$

where the integral in this condition is the Lebesgue integral. Lebesgue measure and integration, and the precise definition of the L^p spaces, form the core of graduate analysis courses, and are beyond the scope of this book. We will merely use these norms (mostly the L^2 inner product and norm) for certain examples. For a development of Lebesgue measure and integration theory, the interested reader can consult the textbooks by Folland [8] or Royden [39].

Exercises

Miscellaneous exercises

1. Let V be a normed vector space over $\mathbf{R}$. Prove that if $u, v \in V$ and v is a nonnegative multiple of u ($v = \alpha u$, $\alpha \geq 0$), then equality holds in the triangle inequality ($\|u+v\| = \|u\| + \|v\|$).

2. Prove Lemma 267.

3. (a) Prove the $\|\cdot\|_1$ defines a norm on $\mathbf{R}^n$.

   (b) Prove the $\|\cdot\|_\infty$ defines a norm on $\mathbf{R}^n$.

4. Prove the following relationships among the common norms on $\mathbf{R}^n$:

   (a) $\|x\|_\infty \leq \|x\|_2 \leq \|x\|_1$ for all $x \in \mathbf{R}^n$.

   (b) $\|x\|_1 \leq \sqrt{n}\|x\|_2$ for all $x \in \mathbf{R}^n$. (Hint: Interpret the sum defining $\|x\|_1$ as a dot product and apply the Cauchy-Schwarz inequality.)

   (c) $\|x\|_2 \leq \sqrt{n}\|x\|_\infty$ for all $x \in \mathbf{R}^n$.

5. Derive relationships among the $L^1(a,b)$, $L^2(a,b)$, and $L^\infty(a,b)$ norms, analogous to the relationships presented in the previous exercises among the ℓ^1, ℓ^2, and ℓ^∞ norms.

6. Define a function $\|\cdot\|$ on $\mathbf{R}^n$ by

   $$\|x\| = |x_1| + |x_2| + \cdots + |x_{n-1}|.$$

   Prove that $\|\cdot\|$ is not a norm on $\mathbf{R}^n$.

7. Define a function $\|\cdot\|$ on $\mathbf{R}^n$ by

   $$\|x\| = \sum_{i=1}^n x_i^2.$$

   Prove that $\|\cdot\|$ is not a norm on $\mathbf{R}^n$.

---

even though this f has an infinite discontinuity at $x = 0$.

8. When we measure distance in $\mathbf{R}^2$ by the Euclidean norm, there is a unique shortest path between any two points in $\mathbf{R}^2$ (namely, the line segment joining those two points). Under the ℓ^1 norm, the distance from $(0,0)$ to $(1,1)$ is 2. Produce two different paths from $(0,0)$ to $(1,1)$, each having length 2.

9. The *unit ball* in a normed vector space V is the set
$$\{v \in V : \|v\| \leq 1\},$$
and the *unit sphere* is the boundary of the unit ball,
$$\{v \in V : \|v\| = 1\}.$$
Draw (on the same coordinate system) the unit spheres in $\mathbf{R}^2$ corresponding to the ℓ^1, ℓ^2, and ℓ^∞ norms. Explain Exercise 4a in terms of your graphs.

10. The $H^1(a,b)$ inner product is defined by
$$\langle f, g \rangle_{H^1} = \int_a^b \{f(x)g(x) + f'(x)g'(x)\}\, dx.$$
   (a) Prove that the H^1 inner product is an inner product on $C^1[a,b]$.
   (b) Suppose ϵ is a small positive number and ω is a large positive number. Prove that $f(x) = \epsilon \sin(\omega x)$ is small when measured in the $L^2(0,1)$ norm, but large when measured in the $H^1(0,1)$ norm. (For a given function u, we could regard $u + f$ as a measurement of u corrupted by high-frequency noise. Whether the error f is regarded as large depends on the chosen norm.)

11. Suppose V is an inner product space and $\|\cdot\|$ is the norm defined by the inner product $\langle \cdot, \cdot \rangle$ on V. Prove that the *parallelogram law* holds:
$$\|u+v\|^2 + \|u-v\|^2 = 2\|u\|^2 + 2\|v\|^2 \text{ for all } u, v \in V.$$
Use this result to prove that neither the ℓ^1 norm nor the ℓ^∞ norm on $\mathbf{R}^n$ is defined by an inner product.[2]

12. Suppose $A \in \mathbf{R}^{n \times n}$ is a nonsingular matrix. Prove that if $\|\cdot\|$ is any norm on $\mathbf{R}^n$, then $\|x\|_A = \|Ax\|$ defines another norm on $\mathbf{R}^n$.

13. Let $\lambda_1, \lambda_2, \ldots, \lambda_n$ be positive real numbers. Prove that
$$\langle x, y \rangle = \sum_{i=1}^{n} \lambda_i x_i y_i$$
defines an inner product on $\mathbf{R}^n$.

[2]The parallelogram law is actually a necessary and sufficient condition for a norm to be defined by an inner product. The proof of sufficiency is tricky and will not be pursued here.

14. Prove that the formula given in the preceding exercise does not define an inner product if one or more of the λ_i satisfies $\lambda_i \leq 0$.

15. Let U and V be vector spaces over $\mathbf{R}$ with norms $\|\cdot\|_U$ and $\|\cdot\|_V$, respectively. Prove that each of the following is a norm on $U \times V$ (see Exercise 2.2.15 for the definition of $U \times V$):

    (a) $\|(u,v)\| = \|u\|_U + \|v\|_V$ for all $(u,v) \in U \times V$;
    (b) $\|(u,v)\| = \sqrt{\|u\|_U^2 + \|v\|_V^2}$ for all $(u,v) \in U \times V$;
    (c) $\|(u,v)\| = \max\{\|u\|_U, \|v\|_V\}$ for all $(u,v) \in U \times V$.

16. Let U and V be vector spaces over $\mathbf{R}$ with inner products $\langle\cdot,\cdot\rangle_U$ and $\langle\cdot,\cdot\rangle_V$, respectively. Prove that

$$\langle (u,v),(w,z)\rangle = \langle u,w\rangle_U + \langle v,z\rangle_V \text{ for all } (u,v),(w,z) \in U \times V$$

defines an inner product on $U \times V$.

6.2 The adjoint of a linear operator

In Section 3.2.5, we introduced the transpose of a matrix. If $A \in F^{m\times n}$, then the transpose A^T of A is the $n \times m$ matrix defined by

$$\left(A^T\right)_{ij} = A_{ji},\ i = 1,2,\ldots,n, j = 1,2,\ldots,m.$$

We can also say that A^T is the matrix whose columns are the rows of A and vice versa. A simple calculation shows why the transpose of a matrix is a natural concept. If $A \in \mathbf{R}^{m\times n}$ and $x \in \mathbf{R}^n$, $y \in \mathbf{R}^m$, then

$$\begin{aligned}
(Ax)\cdot y = \sum_{i=1}^m (Ax)_i y_i = \sum_{i=1}^m \left(\sum_{j=1}^n A_{ij}x_j\right) y_i &= \sum_{i=1}^m \sum_{j=1}^n A_{ij}x_j y_i \\
&= \sum_{j=1}^n \sum_{i=1}^m A_{ij}x_j y_i \\
&= \sum_{j=1}^n \left(\sum_{i=1}^m A_{ij}y_i\right) x_j \\
&= \sum_{j=1}^n \left(A^T y\right)_j x_j \\
&= x \cdot \left(A^T y\right).
\end{aligned}$$

We have proven the following theorem.

Theorem 270 *Let $A \in \mathbf{R}^{m \times n}$. Then*

$$(Ax) \cdot y = x \cdot \left(A^T y\right) \text{ for all } x \in \mathbf{R}^n, y \in \mathbf{R}^m. \tag{6.9}$$

The importance of this result will become apparent in the next few sections. For now, we give a sample application to give the reader a taste of how Theorem 270 is used.

Theorem 271 *Let $A \in \mathbf{R}^{m \times n}$. If b is a nonzero vector in $\mathcal{N}(A^T)$, then $Ax = b$ has no solution. Equivalently,*

$$\mathcal{N}(A^T) \cap \text{col}(A) = \{0\}.$$

Proof Suppose $y \in \mathcal{N}(A^T) \cap \text{col}(A)$. Since $y \in \text{col}(A)$, there exists $u \in \mathbf{R}^n$ such that $y = Au$, and since $y \in \mathcal{N}(A^T)$, we have $A^T y = 0$. We can then reason as follows:

$$\begin{aligned}
A^T y = 0 \quad &\Rightarrow \quad x \cdot (A^T y) = 0 \text{ for all } x \in \mathbf{R}^n \\
&\Rightarrow \quad (Ax) \cdot y \text{ for all } x \in \mathbf{R}^n \\
&\Rightarrow \quad (Ax) \cdot (Au) = 0 \text{ for all } x \in \mathbf{R}^n \\
&\Rightarrow \quad (Au) \cdot (Au) = 0 \text{ (take } x = y) \\
&\Rightarrow \quad Au = 0 \\
&\Rightarrow \quad y = 0.
\end{aligned}$$

The reader will notice how we used a defining property of an inner product ($y \cdot y = 0$ if and only if $y = 0$).

QED

In Section 6.6, we will see an improved version of this theorem, which describes more completely vectors b for which $Ax = b$ has a solution. In addition to results in the spirit of Theorem 271, we will see that the transpose of a matrix is essential in solving best approximation problems (see Section 6.4).

6.2.1 The adjoint of a linear operator

Given $A \in \mathbf{R}^{m \times n}$, the corresponding matrix operator is $T : \mathbf{R}^n \to \mathbf{R}^m$ defined by $T(x) = Ax$ for all $x \in \mathbf{R}^n$. If we define $S : \mathbf{R}^m \to \mathbf{R}^n$ by $S(y) = A^T y$ and write $\langle \cdot, \cdot \rangle$ for the Euclidean dot product, then

$$(Ax) \cdot y \text{ for all } x \in \mathbf{R}^n, y \in \mathbf{R}^m$$

can be written as

$$\langle T(x), y \rangle = \langle x, S(y) \rangle \text{ for all } x \in \mathbf{R}^n, y \in \mathbf{R}^m. \tag{6.10}$$

Equation 6.10 defines a unique operator S that is called the *adjoint* of T. Such an operator exists for any linear operator mapping one finite-dimensional space into another, not just for matrix operators. To prove this, we will need several preliminary results.

The concept of a Gram matrix arises in many contexts.

Definition 272 *Let V be an inner product space over $\mathbf{R}$, and let $\{u_1, \dots, u_n\}$ be a basis for V. Then the matrix $G \in \mathbf{R}^{n\times n}$ defined by*

$$G_{ij} = \langle u_j, u_i \rangle, \ i, j = 1, 2, \dots, n$$

is called the Gram matrix *of the basis $\{u_1, \dots, u_n\}$.*

Theorem 273 *Let V be an inner product space over $\mathbf{R}$, let $\{u_1, \dots, u_n\}$ be a basis for V, and let G be the Gram matrix for this basis. Then G is nonsingular.*

Proof We must show that that $Gx = 0$ only if $x = 0$. Suppose $x \in \mathbf{R}^n$ and $Gx = 0$. Then

$$\begin{aligned} x \cdot Gx = x \cdot 0 \Rightarrow \sum_{i=1}^{n} x_i (Gx)_i = 0 &\Rightarrow \sum_{i=1}^{n}\sum_{j=1}^{n} G_{ij} x_i x_j = 0 \\ &\Rightarrow \sum_{i=1}^{n}\sum_{j=1}^{n} \langle u_j, u_i \rangle x_i x_j = 0 \\ &\Rightarrow \left\langle \sum_{j=1}^{n} x_j u_j, \sum_{i=1}^{n} x_i u_i \right\rangle = 0 \\ &\Rightarrow \langle w, w \rangle = 0, \ w = \sum_{i=1}^{n} x_i u_i \\ &\Rightarrow \sum_{i=1}^{n} x_i u_i = 0 \\ &\Rightarrow x = 0. \end{aligned}$$

The last step follows from the linear independence of the basis $\{u_1, u_2, \dots, u_n\}$. We have therefore shown that $Gx = 0$ implies $x = 0$. Thus G is nonsingular.

QED

The following simple results are frequently useful.

Theorem 274 *Let V be an inner product space over $\mathbf{R}$, and let $x \in V$. Then $\langle x, y \rangle = 0$ for all $y \in V$ if and only if $x = 0$.*

Proof Exercise 7.

Corollary 275 *Let V be an inner product space over $\mathbf{R}$, and let $x, y \in V$. Then*

$$\langle x, v \rangle = \langle y, v \rangle \ \textit{for all} \ v \in V$$

if and only if $x = y$.

Proof Exercise 7.

We can now prove the existence of the adjoint of an arbitrary linear operator.

Theorem 276 *Let X and U be finite-dimensional inner product spaces over* $\mathbf{R}$*, and let $T : X \to U$ be linear. There exists a unique linear operator $S : U \to X$ satisfying*

$$\langle T(x), u \rangle_U = \langle x, S(u) \rangle_X \text{ for all } x \in X, u \in U. \tag{6.11}$$

The operator S is called the adjoint *of T, and is denoted by T^*. Thus T^* is defined by*

$$\langle T(x), u \rangle_U = \langle x, T^*(u) \rangle_X \text{ for all } x \in X, u \in U.$$

Proof Let $\mathcal{X} = \{x_1, x_2, \ldots, x_n\}$ and $\mathcal{U} = \{u_1, u_2, \ldots, u_m\}$ be bases for X and U, respectively. If $S : U \to X$ is linear, then

$$[S(u)]_{\mathcal{X}} = B[u]_{\mathcal{U}},$$

where $B = [S]_{\mathcal{U},\mathcal{X}}$ is the matrix of S with respect to the bases $\mathcal{U}$ and $\mathcal{X}$. Conversely, given any matrix $B \in \mathbf{R}^{n \times m}$, there is a unique linear operator $S : U \to X$ such that $[S]_{\mathcal{U},\mathcal{X}} = B$ (see Exercise 3.3.20). We will prove that there exists a unique matrix $B \in \mathbf{R}^{n \times m}$ such that $S : U \to X$ defined by $[S]_{\mathcal{U},\mathcal{X}} = B$ satisfies (6.11).

In the calculations that follow, we will write $\alpha = [x]_{\mathcal{X}}$ and $\beta = [u]_{\mathcal{U}}$ for generic vectors $x \in X$ and $u \in U$, which means that

$$x = \sum_{i=1}^{n} \alpha_i x_i, \ u = \sum_{j=1}^{m} \beta_j u_j.$$

We first compute $\langle T(x), u \rangle_U$:

$$\begin{aligned} \langle T(x), u \rangle_U &= \left\langle T\left(\sum_{i=1}^{n} \alpha_i x_i\right), \sum_{j=1}^{m} \beta_j u_j \right\rangle_U \\ &= \sum_{i=1}^{n} \sum_{j=1}^{m} \langle T(x_i), u_j \rangle_U \, \alpha_i \beta_j \\ &= \alpha \cdot M\beta, \end{aligned}$$

where $M \in \mathbf{R}^{n \times m}$ is defined by $M_{ij} = \langle T(x_i), u_j \rangle_U$.

Next, if $S : U \to X$ is defined by $[S]_{\mathcal{U},\mathcal{X}} = B \in \mathbf{R}^{n \times m}$, then

$$[S(u)]_{\mathcal{X}} = B[u]_{\mathcal{U}} = B\beta \Rightarrow S(u) = \sum_{k=1}^{n} (B\beta)_k x_k.$$

It then follows that

$$\begin{aligned}\langle x, S(u)\rangle_X &= \left\langle \sum_{i=1}^{n} \alpha_i x_i, \sum_{k=1}^{n} (B\beta)_k x_k \right\rangle_X \\ &= \sum_{i=1}^{n}\sum_{k=1}^{n} \langle x_i, x_k\rangle_X \, \alpha_i (B\beta)_k \\ &= \sum_{i=1}^{n} \alpha_i \left(\sum_{k=1}^{n} \langle x_i, x_k\rangle_X \, (B\beta)_k \right).\end{aligned}$$

Since $\langle x_i, x_k\rangle_X = G_{ik}$, where G is the Gram matrix for the basis $\mathcal{X}$, we have

$$\sum_{k=1}^{n} \langle x_i, x_k\rangle_X \, (B\beta)_k = (GB\beta)_i$$

and therefore

$$\langle x, S(u)\rangle_X = \alpha \cdot (GB\beta) = \alpha \cdot (GB)\beta.$$

We see that

$$\langle T(x), u\rangle_U = \langle x, S(u)\rangle_X \text{ for all } x \in X, u \in U$$

if and only if

$$\alpha \cdot M\beta = \alpha \cdot (GB)\beta \text{ for all } \alpha \in \mathbf{R}^n, \beta \in \mathbf{R}^m.$$

This in turn holds if and only if $GB = M$. Since M is known (it is uniquely determined by the given operator T and the bases $\mathcal{X}$, $\mathcal{U}$) and G is invertible, it follows that there is a unique $B \in \mathbf{R}^{n\times m}$ satisfying the desired condition, namely, $B = G^{-1}M$. Therefore, there is a unique operator $S : U \to X$ satisfying the defining condition for the adjoint of T, and T has a unique adjoint: $T^* = S$.

QED

The proof of Theorem 276 not only proves that the adjoint exists, it shows how to compute it: Given $T : X \to U$, find bases $\mathcal{X}$ and $\mathcal{U}$ for X and U, respectively, compute the Gram matrix G for the basis $\mathcal{X}$ and the matrix M described in the proof, and then compute $B = G^{-1}M$. The adjoint is then defined by $[T^*]_{\mathcal{U},\mathcal{X}} = B$. This means that if $u = \sum_{j=1}^{m} \beta_j u_j$, then

$$T^*(u) = \sum_{i=1}^{n} (B\beta)_i x_i.$$

Example 277 *Let $T : \mathcal{P}_1 \to \mathcal{P}_1$ be defined by $T(p) = p' + p$, where p' is the derivative of the polynomial p. We impose the $L^2(0,1)$ inner product on*

$\mathcal{P}_1$, and we wish to compute the adjoint of T. We choose the standard basis $\mathcal{X} = \{1, x\}$ for $\mathcal{P}_1$ and write $p_1(x) = 1$, $p_2(x) = x$ for the elements in the basis. We then have

$$G_{ij} = \int_0^1 p_i(x)p_j(x)\,dx = \int_0^1 x^{i+j-2}\,dx = \frac{1}{i+j-1}.$$

Therefore,

$$G = \begin{bmatrix} 1 & \frac{1}{2} \\ \frac{1}{2} & \frac{1}{3} \end{bmatrix}.$$

We have

$$T(p_1)(x) = 1,\ T(p_2) = 1 + x,$$

and therefore

$$\begin{aligned} \langle T(p_1), p_1\rangle &= \int_0^1 dx = 1, \\ \langle T(p_1), p_2\rangle &= \int_0^1 x\,dx = \frac{1}{2}, \\ \langle T(p_2), p_1\rangle &= \int_0^1 (1+x)\,dx = \frac{3}{2}, \\ \langle T(p_2), p_2\rangle &= \int_0^1 (1+x)x\,dx = \frac{5}{6}. \end{aligned}$$

We obtain

$$M = \begin{bmatrix} 1 & \frac{1}{2} \\ \frac{3}{2} & \frac{5}{6} \end{bmatrix}.$$

A simple calculation then shows that

$$[T^*]_{\mathcal{X},\mathcal{X}} = B = G^{-1}M = \begin{bmatrix} -5 & -3 \\ 12 & 7 \end{bmatrix},$$

and thus $B\beta = (-5\beta_1 - 3\beta_2, 12\beta_1 + 7\beta_2)$. The final result is

$$q(x) = \beta_1 + \beta_2 x \Rightarrow T^*(q)(x) = (-5\beta_1 - 3\beta_2) + (12\beta_1 + 7\beta_2)x.$$

The following theorems collect some facts about adjoint operators.

Theorem 278 *Let X, U, W be finite-dimensional vector spaces over $\mathbf{R}$, and let $T : X \to U$ and $S : U \to W$ be linear operators. Then:*

1. *$(T^*)^* = T$;*

2. *$(ST)^* = T^*S^*$;*

Proof Exercises 1 and 2.

The corresponding facts for matrices are $(A^T)^T = A$ (which is obvious from the definition of A^T) and $(AB)^T = B^TA^T$ (see Exercise 3).

Theorem 279 *Let X and U be finite-dimensional inner product spaces over $\mathbf{R}$ and assume that $T : X \to U$ is an invertible linear operator. Then T^* is also invertible and*

$$\left(T^*\right)^{-1} = \left(T^{-1}\right)^*.$$

Proof Exercise 4

Since $(T^{-1})^* = (T^*)^{-1}$, we usually write T^{-*} for this operator. The corresponding result for matrices is: If $A \in \mathbf{R}^{n\times n}$ is invertible, then so is A^T, and $(A^T)^{-1} = (A^{-1})^T$ (see Exercise 5). We denote the operator $(A^T)^{-1}$ (or $(A^{-1})^T$) by A^{-T}.

Exercises

Essential exercises

1. Let X and U be finite-dimensional inner product spaces over $\mathbf{R}$, and let $T : X \to U$ be linear. Prove that $\left(T^*\right)^* = T$.

2. Suppose X, U, and W are inner product spaces over $\mathbf{R}$ and $T : X \to U$ and $S : U \to W$ are linear. Prove that $(ST)^* = T^*S^*$.

3. Let $A \in \mathbf{R}^{m\times n}$ and $B \in \mathbf{R}^{n\times p}$. Prove that $(AB)^T = B^TA^T$.

4. Let X and U be finite-dimensional inner product spaces over $\mathbf{R}$ and assume that $T : X \to U$ is an invertible linear operator. Prove that T^* is also invertible, and that

$$\left(T^*\right)^{-1} = \left(T^{-1}\right)^*.$$

5. Suppose $A \in \mathbf{R}^{n\times n}$ is invertible. Prove that A^T is also invertible and that $(A^T)^{-1} = (A^{-1})^T$.

Miscellaneous exercises

6. In the proof of Theorem 276, we used the following fact: If A, B belong to $\mathbf{R}^{m\times n}$ and

$$y \cdot Ax = y \cdot Bx \text{ for all } x \in \mathbf{R}^n, y \in \mathbf{R}^m,$$

then $A = B$. Prove this fact.

7. Prove Theorem 274 and Corollary 275.

8. Let $D : \mathcal{P}_2 \to \mathcal{P}_1$ be defined by $D(p) = p'$. Find D^*, assuming that the $L^2(0,1)$ inner product is imposed on both $\mathcal{P}_1$ and $\mathcal{P}_2$.

9. Let $M : \mathcal{P}_2 \to \mathcal{P}_3$ be defined by $M(p) = q$, where $q(x) = xp(x)$. Find M^*, assuming that the $L^2(0,1)$ inner product is imposed on both $\mathcal{P}_2$ and $\mathcal{P}_3$.

10. Suppose $A \in \mathbf{R}^{n \times n}$ has the following properties: $A^T = A$ and

$$x \cdot Ax > 0 \text{ for all } x \in \mathbf{R}^n, \ x \neq 0.$$

Prove that

$$\langle x, y \rangle_A = x \cdot Ay \text{ for all } x, y \in \mathbf{R}^n$$

defines an inner product on $\mathbf{R}^n$. (Note: We say that A is *symmetric* if $A^T = A$ and *positive definite* if $x \cdot Ax > 0$ for all $x \neq 0$. We will encounter both of these concepts later in the book.)

11. Let X and U be finite-dimensional inner product spaces over $\mathbf{R}$, and suppose $T : X \to U$ is linear. Define $S : \mathcal{R}(T^*) \to \mathcal{R}(T)$ by $S(x) = T(x)$ for all $x \in \mathcal{R}(T^*)$. The goal of this exercise is to prove that S is an isomorphism between $\mathcal{R}(T^*)$ and $\mathcal{R}(T)$.

(a) Prove that S is injective.

(b) The fact that S is injective implies that $\dim(\mathcal{R}(T)) \geq \dim(\mathcal{R}(T^*))$. Explain why, and then use this result to prove that, in fact,

$$\dim(\mathcal{R}(T)) = \dim(\mathcal{R}(T^*)).$$

(Hint: The result applies to any linear operator mapping one finite-dimensional inner product space to another. Apply it to $T^* : U \to X$.)

(c) The previous two parts of the exercise imply that S is also surjective, and hence an isomorphism. Explain why.

The reader should recall that

$$\dim(\mathcal{R}(T)) = \text{rank}(T), \ \dim(\mathcal{R}(T^*)) = \text{rank}(T^*).$$

The fact that $\text{rank}(T) = \text{rank}(T^*)$ is called the *rank theorem*. We will see another proof of the rank theorem in Section 6.6.

12. Use the fundamental theorem of linear algebra and the rank theorem (see the preceding exercise) to prove the following: Let X and U be finite-dimensional inner product spaces over $\mathbf{R}$, and let $T : X \to U$ be linear. Then

(a) T is injective if and only if T^* is surjective;

(b) T is surjective if and only if T^* is injective.

Thus T is bijective if and only if T^* is bijective (cf. Exercise 4).

13. Let X and U be finite-dimensional inner product spaces over $\mathbf{R}$, let $\mathcal{X}$ and $\mathcal{U}$ be bases for X and U, respectively, and let $T : X \to U$ be linear. Find a formula for $[T^*]_{\mathcal{U},\mathcal{X}}$ in terms of $[T]_{\mathcal{X},\mathcal{U}}$ and the Gram matrices for $\mathcal{X}$ and $\mathcal{U}$.

14. Let $f : X \to \mathbf{R}$ be linear, where X is a finite-dimensional inner product space over $\mathbf{R}$. Prove that there exists a unique $u \in X$ such that
$$f(x) = \langle x, u \rangle \text{ for all } x \in X.$$

15. Let $f : \mathcal{P}_2 \to \mathbf{R}$ be defined by $f(p) = p'(0)$. Find $q \in \mathcal{P}_2$ such that $f(p) = \langle p, q \rangle_2$, where $\langle \cdot, \cdot \rangle_2$ is the $L^2(0,1)$ inner product.

6.3 Orthogonal vectors and bases

The reader is probably familiar with the fact that, in $\mathbf{R}^2$, two nonzero vectors x and y are perpendicular if and only if $x \cdot y = 0$. In fact, the law of cosines can be used to show that, for all $x, y \in \mathbf{R}^2$,

$$x \cdot y = \|x\|_2 \|y\|_2 \cos(\theta), \tag{6.12}$$

where θ is the angle between x and y (see Figure 6.2). From (6.12), we see that $\theta = \pi/2$ if and only if $x \cdot y = 0$. The same result holds in $\mathbf{R}^3$ since any two linearly independent vectors x and y determine a plane, which also contains $x - y$. Therefore, the same geometric reasoning holds.

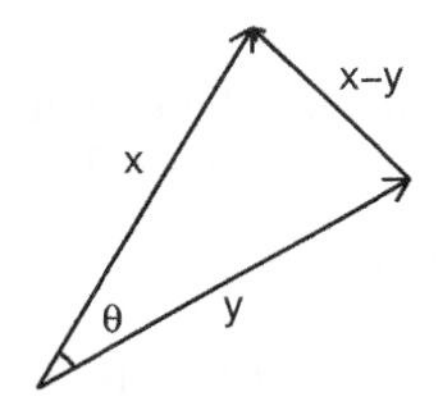

FIGURE 6.2
The angle between two vectors in $\mathbf{R}^2$.

It is not evident how to define the angle between two vectors except in the case of $\mathbf{R}^2$ or $\mathbf{R}^3$, where there is a geometric interpretation of vectors. There is certainly no analogous interpretation to the angle between two functions, for example, and functions form an important class of vectors. However, we can approach this question through the Pythagorean theorem. In the case that $\theta = \pi/2$ in Figure 6.2, we have

$$\|x - y\|_2^2 = \|x\|_2^2 + \|y\|_2^2.$$

Similarly, if the angle between x and y is a right angle, then

$$\|x + y\|_2^2 = \|x\|_2^2 + \|y\|_2^2.$$

(We leave the reader to draw the picture in this case.)

Theorem 280 *Let V be an inner product space over $\mathbf{R}$, and let x, y be vectors in V. If $\langle \cdot, \cdot \rangle$ is the inner product on V and $\|\cdot\|$ is the corresponding norm, then*

$$\|x+y\|^2 = \|x\|^2 + \|y\|^2$$

if and only if $\langle x, y \rangle = 0$. Similarly,

$$\|x-y\|^2 = \|x\|^2 + \|y\|^2$$

if and only if $\langle x, y \rangle = 0$.

Proof We will prove the first result; the proof of the second is entirely similar. We have

$$\|x+y\|^2 = \langle x+y, x+y \rangle = \langle x, x \rangle + 2\langle x, y \rangle + \langle y, y \rangle = \|x\|^2 + 2\langle x, y \rangle + \|y\|^2,$$

from which it immediately follows that $\|x+y\|^2 = \|x\|^2 + \|y\|^2$ if and only if $\langle x, y \rangle = 0$.

QED

Based on the previous theorem, we make the following definition.

Definition 281 *Let V be an inner product space over $\mathbf{R}$.*

1. *We say that vectors $x, y \in V$ are* orthogonal *if and only if $\langle x, y \rangle = 0$.*

2. *We say that $\{u_1, u_2, \ldots, u_k\} \subset V$ is an* orthogonal set *if each u_i is nonzero and $\langle u_i, u_j \rangle = 0$ for all $i \neq j$.*

We use the more general term "orthogonal" in place of the geometric term "perpendicular," since orthogonality applies in situations in which perpendicular has no meaning.

6.3.1 Orthogonal bases

An orthogonal basis is simply a basis that is also an orthogonal set. We are going to show that orthogonal bases are particularly easy to use. We begin with the following result.

Theorem 282 *Let V be an inner product space over $\mathbf{R}$, and let $\{u_1, \ldots, u_k\}$ be an orthogonal subset of V. Then $\{u_1, \ldots, u_k\}$ is linearly independent.*

Proof Suppose $c_1, c_2, \ldots, c_k$ are real numbers and

$$c_1u_1 + c_2u_2 + \ldots + c_ku_k = 0.$$

If we take the inner product of both sides with u_j, we obtain

$$\begin{gathered}\langle c_1u_1 + c_2u_2 + \ldots + c_ku_k, u_j \rangle = \langle 0, u_j \rangle \\ \Rightarrow c_1\langle u_1, u_j \rangle + c_2\langle u_2, u_j \rangle + \ldots + c_k\langle u_k, u_j \rangle = 0\end{gathered}$$

Since $\langle u_i, u_j \rangle = 0$ for $i \neq j$ and $\langle u_j, u_j \rangle = \|u_j\|^2$, we obtain

$$\|u_j\|^2 c_j = 0.$$

The vector u_j is nonzero by definition of orthogonal set, and therefore $\|u_j\|^2$ is nonzero. It follows that $c_j = 0$, and since this holds for all $j = 1, 2, \ldots, k$, we see that $\{u_1, u_2, \ldots, u_k\}$ is linearly independent.

QED

Corollary 283 *Let V be an n-dimensional inner product space over* $\mathbf{R}$. *Then any orthogonal set of n vectors in V is a basis for V*

One of the basic problems we face frequently (often as part of a larger problem) is to represent a given vector as a linear combination of a given basis, that is, to solve

$$\alpha_1 u_1 + \alpha_2 u_2 + \ldots + \alpha_n u_n = v$$

for $\alpha_1, \alpha_2, \ldots, \alpha_n$. In general, this results in an $n \times n$ system of equations that must be solved. However, in the case of an orthogonal basis, the necessary scalars can be computed directly, without solving a system of equations.

Theorem 284 *Let V be an inner product space over* $\mathbf{R}$ *and let $\{u_1, u_2, \ldots, u_n\}$ be an orthogonal basis for V. Then any $v \in V$ can be written*

$$v = \alpha_1 u_1 + \alpha_2 u_2 + \ldots + \alpha_n u_n, \tag{6.13}$$

where

$$\alpha_j = \frac{\langle v, u_j \rangle}{\langle u_j, u_j \rangle}, \quad j = 1, 2, \ldots, n.$$

Proof Since $\{u_1, u_2, \ldots, u_n\}$ is a basis for V, we know that scalars $\alpha_1, \ldots, \alpha_n$ exist that satisfy (6.13). If we take the inner product of both sides of (6.13) with u_j, we obtain

$$\begin{aligned} \langle v, u_j \rangle &= \langle \alpha_1 u_1 + \alpha_2 u_2 + \ldots + \alpha_n u_n, u_j \rangle \\ &= \alpha_1 \langle u_1, u_j \rangle + \alpha_2 \langle u_2, u_j \rangle + \ldots + \alpha_n \langle u_n, u_j \rangle . \end{aligned}$$

Since $\langle u_i, u_j \rangle = 0$ for $i \neq j$, this simplifies to $\langle v, u_j \rangle = \alpha_j \langle u_j, u_j \rangle$, or

$$\alpha_j = \frac{\langle v, u_j \rangle}{\langle u_j, u_j \rangle},$$

as desired.

QED

A vector is said to be *normalized* if its norm is one; such a vector is also called a *unit* vector. We also speak of *normalizing* a vector; this means dividing a nonzero vector by its norm to produce a unit vector:

$$u \neq 0 \Rightarrow \|v\| = 1, \text{ where } v = \|u\|^{-1}u.$$

Even simpler than an orthogonal basis is an orthonormal basis.

Definition 285 *Let V be an inner product space over $\mathbf{R}$. We say that a subset $\{u_1, u_2, \ldots, u_k\}$ of V is an* orthonormal *set if it is orthogonal and each u_i has norm one.*

Corollary 286 *Let V be an inner product space over $\mathbf{R}$, and let $\{u_1, u_2, \ldots, u_n\}$ be an orthonormal basis for V. Then any $v \in V$ can be written as*

$$v = \alpha_1 u_1 + \alpha_2 u_2 + \ldots + \alpha_n u_n,$$

where $\alpha_j = \langle v, u_j \rangle$, $j = 1, 2, \ldots, n$.

Unless otherwise stated, the inner product on $\mathbf{R}^n$ is taken to be the Euclidean dot product. The standard basis $\{e_1, e_2, \ldots, e_n\}$ for $\mathbf{R}^n$ is an orthonormal basis with respect to the dot product. For any $x \in \mathbf{R}^n$, $x \cdot e_j = x_j$, where x_j is the jth component of x. Corollary 286 then leads to the formula

$$x = x_1 e_1 + x_2 e_2 + \ldots + x_n e_n \text{ for all } x \in \mathbf{R}^n,$$

which we knew anyway.

The following example presents an alternate orthogonal basis for $\mathbf{R}^3$.

Example 287 *Consider the following vectors in $\mathbf{R}^3$:*

$$u_1 = \begin{bmatrix} 1 \\ 1 \\ 1 \end{bmatrix}, \ u_2 = \begin{bmatrix} 1 \\ 0 \\ -1 \end{bmatrix}, \ u_3 = \begin{bmatrix} 1 \\ -2 \\ 1 \end{bmatrix}.$$

The inner product on $\mathbf{R}^3$ is the dot product, and the reader can verify that $u_i \cdot u_j = 0$ if $i \neq j$, with

$$u_1 \cdot u_1 = 3, \ u_2 \cdot u_2 = 2, \ u_3 \cdot u_3 = 6. \tag{6.14}$$

Since u_1, u_2, u_3 comprise an orthogonal set of three vectors in a three-dimensional vector space, it follows that these vectors form a orthogonal basis for $\mathbf{R}^3$.

Consider the vector $v = (1, 3, -1)$. According to Theorem 284,

$$v = \frac{v \cdot u_1}{u_1 \cdot u_1} u_1 + \frac{v \cdot u_2}{u_2 \cdot u_2} u_2 + \frac{v \cdot u_3}{u_3 \cdot u_3} u_3.$$

We have

$$v \cdot u_1 = 3, \ v \cdot u_2 = 2, \ v \cdot u_3 = -6,$$

and so

$$v = \frac{3}{3}u_1 + \frac{2}{2}u_2 + \frac{-6}{6}u_3 = u_1 + u_2 - u_3.$$

The reader can verify that this representation is correct.

The inner products (6.14) determine the norms of u_1, u_2, u_3*:*

$$\|u_1\|_2 = \sqrt{3},\ \|u_2\|_2 = \sqrt{2},\ \|u_3\|_2 = \sqrt{6}.$$

Normalizing u_1, u_2, u_3 *yields the orthonormal basis* $\{v_1, v_2, v_3\}$*, where*

$$v_1 = \begin{bmatrix} \frac{1}{\sqrt{3}} \\ \frac{1}{\sqrt{3}} \\ \frac{1}{\sqrt{3}} \end{bmatrix},\ v_2 = \begin{bmatrix} \frac{1}{\sqrt{2}} \\ 0 \\ -\frac{1}{\sqrt{2}} \end{bmatrix},\ v_3 = \begin{bmatrix} \frac{1}{\sqrt{6}} \\ -\frac{2}{\sqrt{6}} \\ \frac{1}{\sqrt{6}} \end{bmatrix}.$$

For hand calculations, it is probably easier to use $\{u_1, u_2, u_3\}$ *and Theorem 284 rather than* $\{v_1, v_2, v_3\}$ *and Corollary 286. On the other hand, for calculations performed on a computer, the orthonormal basis is preferable (a computer is not inconvenienced by "messy" numbers).*

The standard basis for $\mathcal{P}_n$, $\{1, x, x^2, \ldots, x^n\}$, is not orthogonal with respect to the L^2 inner product. For example, regarding elements of $\mathcal{P}_n$ as functions defined on the interval $[0, 1]$, we see that

$$\langle x^i, x^j \rangle_2 = \int_0^1 x^i x^j\, dx = \int_0^1 x^{i+j}\, dx = \frac{1}{i+j+1},\ i, j = 0, 1, \ldots, n.$$

In the following example, we present an orthogonal basis for $\mathcal{P}_2$.

Example 288 *In this example, the vector space will be* $\mathcal{P}_2$ *under the* $L^2(0,1)$ *inner product. We define polynomials*

$$p_1(x) = 1,\ p_2(x) = x - \frac{1}{2},\ p_3(x) = x^2 - x + \frac{1}{6}.$$

The reader can verify that $\langle p_i, p_j \rangle_2 = 0$ *for* $i \neq j$*, while*

$$\begin{aligned} \langle p_1, p_1 \rangle_2 &= \int_0^1 1^2\, dx = 1, \\ \langle p_2, p_2 \rangle_2 &= \int_0^1 \left(x - \frac{1}{2}\right)^2 dx = \frac{1}{12}, \\ \langle p_3, p_3 \rangle_2 &= \int_0^1 \left(x^2 + x - \frac{1}{6}\right)^2 dx = \frac{1}{180}. \end{aligned}$$

Consider the polynomial $q(x) = x^2 + 2x + 3$*. Theorem 284 states that*

$$q(x) = \frac{\langle q, p_1 \rangle_2}{\langle p_1, p_1 \rangle_2} p_1(x) + \frac{\langle q, p_2 \rangle_2}{\langle p_2, p_2 \rangle_2} p_2(x) + \frac{\langle q, p_3 \rangle_2}{\langle p_3, p_3 \rangle_2} p_3(x).$$

Here we have

$$\langle q, p_1\rangle_2 = \int_0^1 (x^2+2x+3)\cdot 1\,dx = \frac{13}{3},$$
$$\langle q, p_2\rangle_2 = \int_0^1 (x^2+2x+3)\left(x-\frac{1}{2}\right)dx = \frac{1}{4},$$
$$\langle q, p_3\rangle_2 = \int_0^1 (x^2+2x+3)\left(x^2-x+\frac{1}{6}\right)dx = \frac{1}{180},$$

so

$$\begin{aligned} q(x) &= \frac{13/3}{1}1 + \frac{1/4}{1/12}\left(x-\frac{1}{2}\right) + \frac{1/180}{1/180}\left(x^2-x+\frac{1}{6}\right) \\ &= \frac{13}{3} + 3\left(x-\frac{1}{2}\right) + \left(x^2-x+\frac{1}{6}\right). \end{aligned}$$

The reader can verify that this equation is correct.

Exercises

Miscellaneous exercises

1. Use the law of cosines to prove (6.12). Referring to Figure 6.2, the law of cosines takes the form

$$\|x-y\|_2^2 = \|x\|_2^2 + \|y\|_2^2 - 2\|x\|_2\|y\|_2\cos(\theta).$$

2. Solve Example 287 without using the fact that $\{u_1, u_2, u_3\}$ is an orthogonal basis: Given the basis $\{u_1, u_2, u_3\}$ and the vector v, express v as a linear combination of u_1, u_2, u_3 by solving

$$\alpha_1 u_1 + \alpha_2 u_2 + \alpha_3 u_3 = v$$

   directly.

3. Solve Example 288 without using the fact that $\{p_1, p_2, p_3\}$ is an orthogonal basis: Given the basis $\{p_1, p_2, p_3\}$ and the polynomial q, express q as a linear combination of p_1, p_2, p_3 by solving

$$\alpha_1 p_1 + \alpha_2 p_2 + \alpha_3 p_3 = q$$

   directly.

4. (a) Show that $\{\sin(\pi x), \sin(2\pi x), \sin(3\pi x)\}$ is an orthogonal basis for a subspace of $L^2(0,1)$.

   (b) More generally, show that, for each positive integer n,

$$\{\sin(\pi x), \sin(2\pi x), \ldots, \sin(n\pi x)\}$$

   is an orthogonal basis for a subspace of $L^2(0,1)$.

5. Consider the following vectors in $\mathbf{R}^4$:

$$u_1 = (1,1,1,1),\ u_2 = (1,-1,1,-1),$$
$$u_3 = (1,0,-1,0),\ u_4 = (0,-1,0,1).$$

   (a) Prove that $\{u_1, u_2, u_3, u_4\}$ is an orthogonal basis for $\mathbf{R}^4$.

   (b) Express $v = (1,2,3,4)$ as a linear combination of u_1, u_2, u_3, u_4.

6. Consider the following quadratic polynomials: $p_1(x) = 1 + x + x^2$, $p_2(x) = 65 - 157x + 65x^2$, $p_3(x) = -21 + 144x - 150x^2$.

   (a) Show that $\{p_1, p_2, p_3\}$ is an orthogonal basis for $\mathcal{P}_2$, regarded as a subspace of $L^2(0,1)$.

   (b) Express each of the standard basis functions 1, x, and x^2 as a linear combination of p_1, p_2, p_3.

7. Consider the functions e^x and e^{-x} to be elements of $C[0,1]$, and regard $C[0,1]$ as an inner product space under the $L^2(0,1)$ inner product. Define $S = \mathrm{sp}\{e^x, e^{-x}\}$. Find an orthogonal basis for S. (Hint: There are many possible solutions. For instance, you could include e^x in the orthogonal basis and find an element of S that is orthogonal to e^x to serve as the second basis function.)

8. Consider the following subspace of $\mathbf{R}^4$: $S = \mathrm{sp}\{(1,1,1,-1), (-2,-2,-1,2)\}$. Find an orthogonal basis for S. (See the hint to the previous exercise.)

9. Notice that $\{x_1, x_2, x_3\}$, where

$$x_1 = (1,0,1,0),\ x_2 = (1,1,-1,1),\ x_3 = (-1,2,1,0),$$

   is an orthogonal subset of $\mathbf{R}^4$. Find a fourth vector $x_4 \in \mathbf{R}^4$ such that $\{x_1, x_2, x_3, x_4\}$ is an orthogonal basis for $\mathbf{R}^4$.

10. Let $\{x_1, x_2, \ldots, x_k\}$, where $1 \leq k < n$, be an orthogonal set in $\mathbf{R}^n$. Explain how to find $x_{k+1} \in \mathbf{R}^n$ such that $\{x_1, x_2, \ldots, x_{k+1}\}$ is orthogonal by solving a system of linear equations.

11. What is the Gram matrix of an orthogonal basis $\{u_1, u_2, \ldots, u_n\}$? What if the basis is orthonormal?

12. Let $\{x_1, x_2, \ldots, x_n\}$ be an orthonormal set in $\mathbf{R}^n$, and define $X = [x_1|x_2|\cdots|x_n]$. Compute X^TX and XX^T. What conclusion do you draw about X^T?

13. Let V be an inner product space over $\mathbf{R}$, and let $\{u_1, \ldots, u_k\}$ be an orthogonal subset of V. Prove that, for all $v \in V$,

$$v \in \mathrm{sp}\{u_1, \ldots, u_k\}$$

if and only if

$$v = \sum_{j=1}^{k} \frac{\langle v, u_j \rangle}{\langle u_j, u_j \rangle} u_j.$$

14. Let V be an inner product space over $\mathbf{R}$, let $\{u_1, \ldots, u_k\}$ be an orthogonal subset of V, and define $S = \mathrm{sp}\{u_1, \ldots, u_k\}$.

    (a) Prove that, for all $v \in V$, $v \notin S$, the vector

    $$v - \sum_{j=1}^{k} \frac{\langle v, u_j \rangle}{\langle u_j, u_j \rangle} u_j$$

    is orthogonal to every vector in S.

    (b) Use the Pythagorean theorem to prove that, if $v \in V$, $v \notin S$, then

    $$\|v\| > \left\| \sum_{j=1}^{k} \frac{\langle v, u_j \rangle}{\langle u_j, u_j \rangle} u_j \right\|.$$

15. Let V be an inner product space over $\mathbf{R}$, and let u, v be vectors in V.

    (a) Assume u and v are nonzero. Prove that $v \in \mathrm{sp}\{u\}$ if and only if $|\langle u, v \rangle| = \|u\|\|v\|$. (Hint: Use the results of the previous two exercises.)

    (b) Using the above result, state precisely conditions on u and v under which equality holds in the Cauchy-Schwarz inequality. (For this part of the exercise, do not assume u and v are necessarily nonzero.)

16. Let V be an inner product space over $\mathbf{R}$. What must be true about $u, v \in V$ for equality to hold in the triangle inequality? (Hint: Use the result in the previous exercise.)

6.4 The projection theorem

Many approximation problems can be formulated abstractly as follows: Given a vector space V, a vector v in V, and a subspace S of V, find the vector $w \in S$ closest to v, in the sense that

$$w \in S, \ \|v - w\| \leq \|v - z\| \text{ for all } z \in S. \tag{6.15}$$

We can write (6.15) as

$$w \in S, \ \|v - w\| = \min\{\|v - z\| : z \in S\}.$$

In this section, we explain how to find w in the case that the norm is defined by an inner product and the subspace is finite-dimensional.[3]

The main result of this section is called the *projection theorem*. The best approximation to v always exists, and it is characterized by the orthogonality condition illustrated in Figure 6.3. This is the content of the following theorem.

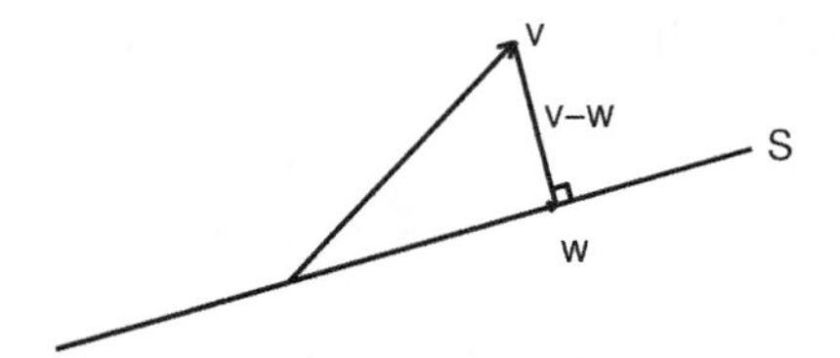

FIGURE 6.3
The best approximation w to v from the subspace S.

Theorem 289 (The projection theorem) *Let V be an inner product space over $\mathbf{R}$, and let S be a finite-dimensional subspace of V.*

1. *For any $v \in V$, there is a unique $w \in S$ satisfying*
$$\|v - w\| = \min\{\|v - z\| \,:\, z \in S\}. \tag{6.16}$$
The vector w is called the best approximation *to v from S or the* projection of v onto S. *It is sometimes denoted by $w = \mathrm{proj}_S v$.*

2. *A vector $w \in S$ is the best approximation to v from S if and only if*
$$\langle v - w, z\rangle = 0 \text{ for all } z \in S. \tag{6.17}$$

3. *If $\{u_1, u_2, \ldots, u_n\}$ is a basis for S, then*
$$\mathrm{proj}_S v = \sum_{i=1}^{n} x_i u_i,$$
where $x = (x_1, x_2, \ldots, x_n) \in \mathbf{R}^n$ is the unique solution to the matrix-vector equation
$$Gx = b.$$
The matrix $G \in \mathbf{R}^{n\times n}$ is the Gram matrix for the basis $\{u_1, u_2, \ldots, u_n\}$ and the vector $b \in \mathbf{R}^n$ is defined as follows:
$$b_i = \langle v, u_i\rangle\,,\ i = 1, 2, \ldots, n. \tag{6.18}$$

[3]The best approximation w exists if the subspace S is infinite-dimensional, provided S is also *complete*, a concept that is beyond the scope of this discussion. See Section 10.3.

Proof We first prove the second conclusion, that w is a solution if and only if it satisfies the orthogonality condition (6.17). Consider any fixed w in S. A vector y lies in S if and only if it can be written as $y = w + tz$ for some $t \in \mathbf{R}$ and some $z \in S$. To see this, we note that if $t \in \mathbf{R}$ and $z \in S$, then $w + tz$ also belongs to S since a subspace is closed under addition and scalar multiplication. On the other hand, given $y \in S$, it can be written as $y = w+tz$ for $t = 1$ and $z = y - w \in S$.

Now consider $\|v - (w + tz)\|^2$:

$$\begin{aligned}\|v - (w + tz)\|^2 &= \|v - w - tz\|^2 \\ &= \langle v - w - tz, v - w - tz\rangle \\ &= \langle v - w, v - w\rangle - 2t\,\langle v - w, z\rangle + t^2\,\langle z, z\rangle \\ &= \|v - w\|^2 - 2t\,\langle v - w, z\rangle + t^2\|z\|^2.\end{aligned}$$

We see that $\|v - (w + tz)\|^2 \geq \|v - w\|^2$ for all $z \in S$ and all $t \in \mathbf{R}$ if and only if

$$t^2\|z\|^2 - 2t\,\langle v - w, z\rangle \geq 0 \text{ for all } z \in S, t \in \mathbf{R}.$$

If we temporarily regard z as fixed, then

$$\phi(t) = t^2\|z\|^2 - 2t\,\langle v - w, z\rangle$$

is a simple quadratic in t. We see that $\phi(0) = 0$ and $\phi(t) \geq 0$ for all $t \in \mathbf{R}$ if and only if $\phi'(t) = 0$ (that is, if and only if ϕ has its minimum at $t = 0$). But

$$\phi'(t) = 2t\|z\|^2 - 2\,\langle v - w, z\rangle$$

and $\phi'(0) = -2\,\langle v - w, z\rangle$. Therefore,

$$\|v - w\|^2 \leq \|v - (w + tz)\|^2 \tag{6.19}$$

holds for all $t \in \mathbf{R}$ if and only if $\langle v - w, z\rangle = 0$, and (6.19) holds for all $t \in \mathbf{R}$ and all $z \in S$ if and only if $\langle v - w, z\rangle = 0$ for all $z \in S$.

We have now shown that (6.17) characterizes any solution to (6.16), but we have not yet shown that a solution exists. We will prove the first and third conclusions of the theorem at the same time. Let us assume that $\{u_1, u_2, \ldots, u_n\}$ is a basis for S. We first make the following observation: $w \in S$ satisfies

$$\langle v - w, z\rangle = 0 \text{ for all } z \in S \tag{6.20}$$

if and only if

$$\langle v - w, u_i\rangle = 0,\ i = 1, 2, \ldots, n \tag{6.21}$$

(see Exercise 10).

We will look for a solution w of (6.17) in the form

$$w = \sum_{j=1}^{n} x_j u_j, \tag{6.22}$$

where $x_1, x_2, \ldots, x_n \in \mathbf{R}$. We have

$$
\begin{aligned}
& \langle v - w, u_i \rangle = 0, \ i = 1, 2, \ldots, n \\
\Leftrightarrow \quad & \left\langle v - \sum_{j=1}^{n} x_j u_j, u_i \right\rangle = 0, \ i = 1, 2, \ldots, n \\
\Leftrightarrow \quad & \langle v, u_i \rangle - \sum_{j=1}^{n} x_j \langle u_j, u_i \rangle = 0 \, i = 1, 2, \ldots, n \\
\Leftrightarrow \quad & \sum_{j=1}^{n} x_j \langle u_j, u_i \rangle = \langle v, u_i \rangle \ i = 1, 2, \ldots, n.
\end{aligned}
$$

Thus w, as defined by (6.22), satisfies (6.17) if and only if x satisfies $Gx = b$, where G is the Gram matrix for $\{u_1, u_2, \ldots, u_n\}$ and b is defined as in (6.18).

It remains only to show that $Gx = b$ has a unique solution. But this follows from the fact that G, being a Gram matrix, is nonsingular by Theorem 273.

QED

Example 290 *Consider the subspace $S = \text{sp}\{u_1, u_2\}$ of $\mathbf{R}^4$, where*

$$u_1 = (1, 2, 1, 1), \ u_2 = (1, 0, 1, 2),$$

and the vector $v = (1, 1, 1, 0) \in \mathbf{R}^4$. We wish to find the projection of v onto S. According to the preceding theorem $w = \text{proj}_S v$ is determined by $w = x_1 u_1 + x_2 u_2$, where $Gx = b$ and

$$G = \begin{bmatrix} u_1 \cdot u_1 & u_2 \cdot u_1 \\ u_1 \cdot u_2 & u_2 \cdot u_2 \end{bmatrix} = \begin{bmatrix} 7 & 4 \\ 4 & 6 \end{bmatrix}, \ b = \begin{bmatrix} v \cdot u_1 \\ v \cdot u_2 \end{bmatrix} = \begin{bmatrix} 4 \\ 2 \end{bmatrix}.$$

Solving $Gx = b$, we obtain $x = (8/13, -1/13)$ and therefore

$$w = \frac{8}{13}(1, 2, 1, 1) - \frac{1}{13}(1, 1, 1, 0) = \left(\frac{7}{16}, \frac{16}{13}, \frac{7}{13}, \frac{6}{13} \right).$$

This vector w is the element of S closest to v.

In the special case that $\{u_1, u_2, \ldots, u_n\}$ is an orthogonal basis, the Gram matrix reduces to a diagonal matrix:

$$G = \begin{bmatrix} \langle u_1, u_1 \rangle & & & \\ & \langle u_2, u_2 \rangle & & \\ & & \ddots & \\ & & & \langle u_n, u_n \rangle \end{bmatrix}.$$

The equation $Gx = b$ can then be solved explicitly to obtain

$$x_i = \frac{\langle v, u_i \rangle}{\langle u_i, u_i \rangle}, \ i = 1, 2, \ldots, n,$$

and the best approximation to v from S is

$$w = \sum_{i=1}^{n} \frac{\langle v, u_i \rangle}{\langle u_i, u_i \rangle} u_i. \tag{6.23}$$

If the basis is not just orthogonal, but orthonormal, this formula simplifies further:

$$w = \sum_{i=1}^{n} \langle v, u_i \rangle \, u_i. \tag{6.24}$$

The reader should notice the similarity of (6.23) and (6.24) to the result of Theorem 284. If $\{u_1, u_2, \ldots, u_k\}$ is an orthogonal basis for S and v is an element of S, then the right-hand side of (6.23) expresses v as a linear combination of $u_1, u_2, \ldots, u_k$; this is the content of Theorem 284. If v does not belong to S, then the same formula (the right-hand side of (6.23)) gives the best approximation to v from S.

We will now present an application of the projection theorem.

6.4.1 Overdetermined linear systems

In this section, we consider a linear system $Ax = y$, where $A \in \mathbf{R}^{m \times n}$, $y \in \mathbf{R}^m$, and $m > n$. By the fundamental theorem of linear algebra, we know that $\text{rank}(A) \leq n < m$, which implies that $\text{col}(A)$ is a proper subspace of $\mathbf{R}^m$. Therefore, for most $y \in \mathbf{R}^m$, the system $Ax = y$ has no solution. In many applications, we nevertheless need to "solve" $Ax = y$ in the sense of finding an approximate solution. A typical example—least-squares line fitting—is presented below.

The usual way to find an approximate solution to $Ax = y$ is to find $x \in \mathbf{R}^n$ to minimize the norm of the *residual* $Ax - y$. Since minimizing $\|Ax - y\|_2$ is equivalent to minimizing $\|Ax - y\|_2^2$, this approach is called the method of *least-squares*. We can formulate this as a best approximation problem by recognizing that $\text{col}(A)$ is a subspace of $\mathbf{R}^m$, and minimizing $\|Ax - y\|_2$ yields the vector in $\text{col}(A)$ closest to y.

By Theorem 289, and in particular the orthogonality condition (6.17), Ax is the best approximation to y from $\text{col}(A)$ if and only if

$$(y - Ax) \cdot w = 0 \text{ for all } w \in \text{col}(A).$$

Since $w \in \text{col}(A)$ if and only if $w = Az$ for some $z \in \mathbf{R}^n$, the orthogonality condition takes the form

$$(y - Ax) \cdot Az \text{ for all } z \in \mathbf{R}^n.$$

Using the basic property (6.9) of A^T, this is equivalent to

$$A^T(y - Ax) \cdot z = 0 \text{ for all } z \in \mathbf{R}^n.$$

By Theorem 274, the only vector in $\mathbf{R}^n$ that is orthogonal to every z in $\mathbf{R}^n$ is the zero vector, so we obtain

$$A^T(y - Ax) = 0,$$

which simplifies to

$$A^T Ax = A^T y.$$

We have thus proved the following theorem.

Theorem 291 *Let $A \in \mathbf{R}^{m\times n}$ and $y \in \mathbf{R}^m$ be given. Then $x \in \mathbf{R}^n$ solves*

$$\min \{\|Az - y\|_2 \, : \, z \in \mathbf{R}^n\}$$

if and only if x satisfies

$$A^T Ax = A^T y. \tag{6.25}$$

The equations represented by $A^T Ax = A^T y$ are called the *normal equations* for the system $Ax = y$. We have not assumed that $A^T A$ is nonsingular, but if it is, then $A^T Ax = A^T y$ has a unique solution, which means that $Ax = y$ has a unique least-squares solution. Even if $A^T A$ is singular, it turns out that $A^T Ax = A^T y$ is consistent for all $y \in \mathbf{R}^m$ (see Exercise 11). In the case that $A^T A$ is singular, $A^T Ax = A^T y$ has infinitely many solutions, and therefore $Ax = y$ has infinitely many least-squares solutions.

We now describe a common application in which the method of least-squares is commonly used. Suppose two variables y and t are thought to be related by the equation $y = c_0 + c_1 t$, where the constants c_0 and c_1 are unknown. We might measure (in a laboratory experiment) data points $(t_1, y_1), (t_2, y_2), \ldots, (t_m, y_m)$ and from these points determine c_0 and c_1. The equations are

$$\begin{aligned} c_0 + c_1 t_1 &= y_1, \\ c_0 + c_1 t_2 &= y_2, \\ \vdots \quad & \quad \vdots \\ c_0 + c_1 t_m &= y_m, \end{aligned}$$

or $Ac = y$, where

$$A = \begin{bmatrix} 1 & t_1 \\ 1 & t_2 \\ \vdots & \vdots \\ 1 & t_m \end{bmatrix}, \; c = \begin{bmatrix} c_0 \\ c_1 \end{bmatrix}, \; y = \begin{bmatrix} y_1 \\ y_2 \\ \vdots \\ y_m \end{bmatrix}.$$

We could find c_0 and c_1 from two data points, but there is a problem: Measured data points are typically noisy, and the measurement errors could cause the computed values of c_0 and c_1 to be poor approximations of the true values.

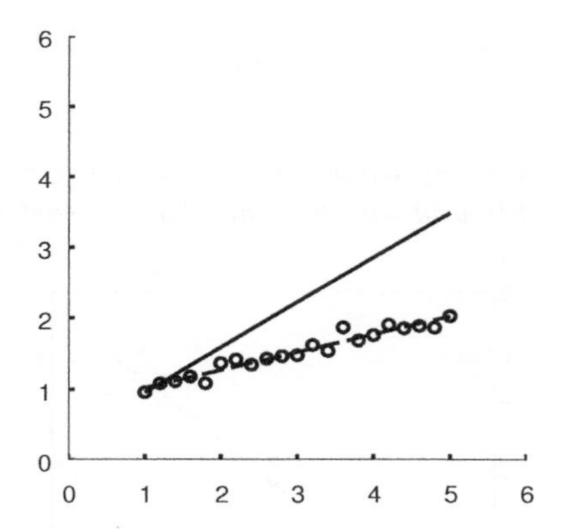

FIGURE 6.4
A collection of data points that nearly lie on a straight line. The dashed line provides a reasonable approximation to the data as a whole, while the solid line is computed using just the first two data points.

On the other hand, if we use more than two data points, then the noise in the data means that the data points are unlikely to lie on a straight line. Thus $Ac = y$ is unlikely to have a solution for $m > 2$. Figure 6.4 illustrates the typical situation.

The usual way of resolving these difficulties is to take c to be the least-squares solution of $Ac = y$. The vector c is found by solving the normal equations $A^TAc = A^Ty$. For this special problem (which is sometimes called the *linear regression* problem), the matrix A^TA and the vector A^Ty take the following special forms:

$$A^TA = \begin{bmatrix} m & \sum_{j=1}^m t_i \\ \sum_{j=1}^m t_i & \sum_{j=1}^m t_i^2 \end{bmatrix}, \; A^Ty = \begin{bmatrix} \sum_{j=1}^m y_i \\ \sum_{j=1}^m t_iy_i \end{bmatrix}.$$

The dashed line in Figure 6.4 was computed by solving the least-squares problem for the given data.

Example 292 *Suppose we believe the variables t and y are related by*

$$y = c_0 + c_1 t,$$

and we measure the following (t, y) pairs:

$$(1, 2.60), \; (2, 4.30), \; (3, 6.15), \; (4, 7.50), \; (5, 8.55).$$

We wish to estimate c_0, c_1 by the method of least-squares. We find c by solving the normal equations $A^TAc = A^Tb$, where

$$A^TA = \begin{bmatrix} 5 & 15 \\ 15 & 55 \end{bmatrix}, \; A^Ty = \begin{bmatrix} 29.10 \\ 102.40 \end{bmatrix}.$$

The result is

$$c = \begin{bmatrix} 1.29 \\ 1.51 \end{bmatrix}.$$

Thus the approximate relationship between t *and* y *is* $y = 1.29 + 1.51t$. *Figure 6.5 shows that data points and line obtained by the least-squares method.*

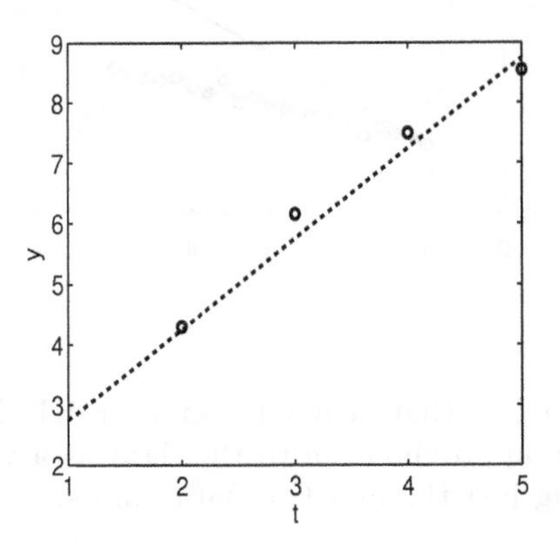

FIGURE 6.5
The data from Example 292 (small circles), together with the approximate line fitting the data, obtained by the method of least-squares.

The least-squares solution to $Ax = y$ uses the orthogonality condition (6.17), but it does not, apparently, depend on the system $Gx = b$ derived in Theorem 289. In fact, however, the system representing the normal equations is precisely the system $Gx = b$. See Exercise 2.

We will present another application of the projection theorem in the next section.

Exercises

Essential exercises

1. Let $A \in \mathbf{R}^{m \times n}$.

   (a) Prove that $\mathcal{N}(A^T A) = \mathcal{N}(A)$.

   (b) Use the preceding result to prove that if A has full rank, then $A^T A$ is invertible.

   (c) Use these results to prove that if A has full rank, then $Ax = y$ has a unique least-squares solution for each $y \in \mathbf{R}^m$, namely, $x = (A^T A)^{-1} A^T y$.

Miscellaneous exercises

2. Let $A \in \mathbf{R}^{m\times n}$ and $y \in \mathbf{R}^m$ be given, and assume that A has full rank. Then $\{A_1, A_2, \ldots, A_n\}$ is a basis for $\mathrm{col}(A)$. Show that $A^T A$ is the Gram matrix for $\{A_1, A_2, \ldots, A_n\}$. (Note that this provides an alternate proof of the fact that $A^T A$ is invertible if A has full rank; cf. Exercise 1.) Similarly, show that $A^T y$ is the vector b mentioned in Theorem 289.

3. Let S be the one-dimensional subspace of $\mathbf{R}^3$ spanned by the vector $(1, 2, -1)$. Find the best approximation from S to $v = (1, 0, 1)$.

4. Let $x_1 = (0, 1, 1)$ and $x_2 = (1, 1, 2)$, and define $S = \mathrm{sp}\{x_1, x_2\} \subset \mathbf{R}^3$. Find the projection of $v = (1, 1, 1)$ onto S.

5. Let $V = C[0, 1]$, regarded as an inner product space under the $L^2(0, 1)$ inner product. Define $S = \mathrm{sp}\{\sin(\pi x), \sin(2\pi x), \sin(3\pi x)\} \subset V$. Find the best approximation to $f(x) = x(1 - x)$ from S.

6. Let $V = C[0, 1]$, regarded as an inner product space under the $L^2(0, 1)$ inner product. Define the following subspaces of V:

$$\begin{aligned} S_1 &= \mathrm{sp}\{\sin(\pi x), \sin(2\pi x), \sin(3\pi x)\}, \\ S_2 &= \mathrm{sp}\{1, \cos(\pi x), \cos(2\pi x)\}. \end{aligned}$$

   Define $f \in V$ by $f(x) = x$. Compute the best approximation to f from both S_1 and S_2. Graph the results. Which is a better approximation?

7. Show that both of the following sets are orthogonal in $C[0, 1]$ under the $L^2(0, 1)$ inner product.
   (a) $\{\sin(\pi x), \sin(2\pi x), \ldots, \sin(n\pi x)\}$;
   (b) $\{1, \cos(\pi x), \cos(2\pi x), \ldots, \cos(n\pi x)\}$.

8. Consider the following data points: $(-2, 14.6)$, $(-1, 4.3)$, $(0, 1.1)$, $(1, 0.3)$, $(2, 1.9)$. Using the method of least-squares, find a quadratic polynomial $p(x) = c_0 + c_1 x + c_2 x^2$ that fits the data as nearly as possible.

9. Consider the following data points: $(0, 3.1)$, $(1, 1.4)$, $(2, 1.0)$, $(3, 2.2)$, $(4, 5.2)$, $(5, 15.0)$. Using the method of least-squares, find the function of the form $f(x) = a_1 e^x + a_2 e^{-x}$ that fits the data as nearly as possible.

10. In the proof of Theorem 289, show that (6.21) implies (6.20).

11. Let $A \in \mathbf{R}^{m\times n}$ and $y \in \mathbf{R}^m$ be given. Prove that the system $A^T A x = A^T y$ always has a solution. (Hint: Use the connection between $A^T A x = A^T y$ and the problem of minimizing $\|Ax - y\|_2$. What does the projection theorem say about the existence of a solution to the latter problem?)

12. In the previous exercise, why does the uniqueness part of the projection theorem not imply that the least-squares solution x to $Ax = y$ is unique?

13. Assume $A \in \mathbf{R}^{m\times n}$. Exercise 6.2.11 implies that $\mathrm{col}(A^T) \subset \mathbf{R}^n$ and $\mathrm{col}(A) \subset \mathbf{R}^m$ are isomorphic. Moreover, that exercise shows that $S : \mathrm{col}(A^T) \to \mathrm{col}(A)$ defined by $S(x) = Ax$ is an isomorphism.

    (a) Show that $R : \mathrm{col}(A) \to \mathrm{col}(A^T)$ defined by $R(y) = A^T y$ is another isomorphism.

    (b) Use these results (that is, the fact that S and R are isomorphisms) to show that, for each $y \in \mathbf{R}^m$, $A^T Ax = A^T y$ has a solution $x \in \mathbf{R}^n$ (cf. Exercise 11; the purpose of this exercise is to give a different proof of the same result).

    (c) Discuss the uniqueness of the solution(s) to $A^T Ax = A^T b$ in terms of these isomorphisms.

14. Let $A \in \mathbf{R}^{m\times n}$ and $y \in \mathbf{R}^m$. Show that there is a unique solution $\overline{x}$ to $A^T Ax = A^T y$ that also belongs to $\mathrm{col}(A^T)$. Here are two ways to approach the proof:

    (a) Use the fact that $\mathrm{col}(A^T)$ and $\mathrm{col}(A)$ are isomorphic—see Exercises 6.2.11 and 13.

    (b) Let $\hat{x} \in \mathbf{R}^n$ be a solution to $A^T Ax = A^T y$, and define $\overline{x}$ to be the projection of $\hat{x}$ onto $\mathrm{col}(A^T)$. Prove that $\overline{x}$ is the desired vector.

15. Let $A \in \mathbf{R}^{m\times n}$, where $m < n$ and $\mathrm{rank}(A) = m$. Let $y \in \mathbf{R}^m$.

    (a) Prove that $Ax = y$ has infinitely many solutions. (Hint: You must prove two conclusions, namely, that $Ax = y$ has a solution and that the solution is not unique. We already know that if a linear system over $\mathbf{R}$ has more than one solution, then it has infinitely many solutions.)

    (b) Prove that AA^T is invertible.

    (c) Let $S = \{x \in \mathbf{R}^n \mid Ax = y\}$, and define

    $$\overline{x} = A^T \left(AA^T\right)^{-1} y.$$

    Prove that $\overline{x} \in S$.

Project: Projection operators

16. Suppose V is a finite-dimensional inner product space over $\mathbf{R}$, S is a finite-dimensional subspace of V, and $P : V \to V$ is defined by $P(v) = \mathrm{proj}_S v$ for all $v \in V$. We call P the *orthogonal projection operator onto* S.

(a) Prove that P is linear. (Hint: Remember that $P(v)$ is characterized by

$$P(v) \in S, \ \langle v - P(v), s \rangle = 0 \text{ for all } s \in S.$$

In other words, $w = P(v)$ is the unique vector in S satisfying $\langle v - w, s \rangle = 0$ for all $s \in S$.)

(b) Prove that $P^2 = P$.

(c) Prove that $P^* = P$. (We say that P is *self-adjoint* or *symmetric* in this case.) (Hint: You have to show that $\langle P(u), v \rangle = \langle u, P(v) \rangle$ for all $u, v \in V$ or, equivalently, $\langle P(u), v \rangle - \langle u, P(v) \rangle = 0$ for all $u, v \in V$. First use the fact that, for all $u \in V$,

$$\langle u - P(u), s \rangle = 0 \text{ for all } s \in S$$

to show that $\langle u, P(v) \rangle = \langle P(u), P(v) \rangle$ for all $u, v \in V$. Then interchange the roles of u and v to obtain the desired result.)

17. Suppose V is a finite-dimensional inner product space over $\mathbf{R}$, and assume that $P : V \to V$ satisfies $P^2 = P$ and $P^* = P$. Prove that there exists a subspace S of V such that $P(v) = \mathrm{proj}_S v$ for all $v \in V$, that is, prove that P is an orthogonal projection operator. (Hint: To do this, you must identify S and then prove that $\langle v - P(v), s \rangle = 0$ for all $s \in S$.)

18. Let $A \in \mathbf{R}^{m \times n}$ have full rank. Then we know from Exercise 1 that $A^T A$ is invertible. Prove that $P : \mathbf{R}^m \to \mathbf{R}^m$ defined by

$$P(y) = A(A^T A)^{-1} A^T y \text{ for all } y \in \mathbf{R}^m$$

is the orthogonal projection operator onto $\mathrm{col}(A)$.

19. Let V be a vector space over a field F, and let $P : V \to V$. If $P^2 = P$ (that is, $P(P(v)) = P(v)$ for all $v \in V$), then we say that P is a *projection operator* (onto $S = \mathcal{R}(P)$). The reader should notice that the orthogonal projection onto a given subspace S is unique, but there may be many nonorthogonal projections onto S. Thus we speak of *the* orthogonal projection operator onto S, but *a* projection operator onto S. Also, we do not need an inner product to define the concept of projection operator (nor does the space V have to be finite-dimensional).

(a) Prove that if P is a projection operator, then so is $I - P$, where $I : V \to V$ represents the identity operator: $I(v) = v$ for all $v \in V$.

(b) Let P be a projection operator, and define

$$S = \mathcal{R}(P), \ T = \mathcal{R}(I - P).$$

i. Prove that $S \cap T = \{0\}$.
ii. Find $\ker(P)$ and $\ker(I - P)$.

iii. Prove that for any $v \in V$, there exist $s \in S$, $t \in T$ such that $v = s + t$. Moreover, prove that this representation is unique.

20. Let $A \in \mathbf{R}^{m\times n}$ have full rank. We know from the preceding results that the matrix $B = I - A(A^TA)^{-1}A^T$ defines a projection operator onto $\text{col}(B)$.

    (a) Prove that $\text{col}(B) = \mathcal{N}(A^T)$.

    (b) Prove that B defines the orthogonal projection operator onto $\mathcal{N}(A^T)$.

    (c) The preceding results imply that given any $y \in \mathbf{R}^m$, there exist unique vectors $r \in \text{col}(A)$ and $z \in \mathcal{N}(A^T)$ such that $y = r + z$. Explain why.

21. Here is an example of a nonorthogonal projection operator: Consider real numbers $x_0, \ldots, x_n$ satisfying $a \le x_0 < \cdots < x_n \le b$. Define $P : V \to V$, where $V = C[a, b]$, by the condition that $p = P(u)$ is the unique polynomial of degree at most n interpolating the function u at the points $x_0, x_1, \ldots, x_n$ (that is, $p(x_i) = u(x_i)$ for $i = 0, 1, \ldots, n$). Prove that P is a projection operator.

22. Let $x_1, x_2, \ldots, x_n$ be linearly independent vectors in $\mathbf{R}^m$ $(n < m)$, and let $S = \text{sp}\{x_1, x_2, \ldots, x_n\}$. Find the matrix representing the orthogonal projection operator onto S. How does result simplify if $\{x_1, x_2, \ldots, x_n\}$ is an orthonormal basis for S?

6.5 The Gram-Schmidt process

It is always possible to transform a given (nonorthogonal) basis into an orthogonal basis for the same space; one way to do this is called the *Gram-Schmidt orthogonalization process.* The basic idea is illustrated by Figure 6.6: Given a (nonorthogonal) basis $\{u, v\}$, it can be replaced by the orthogonal basis $\{u, \hat{v}\}$, where $\hat{v} = v - \text{proj}_u v$. The new set is orthogonal, as guaranteed by the projection theorem, and, since $\hat{v}$ is a linear combination of u and v, it is easy to show that $\text{sp}\{u, \hat{v}\} = \text{sp}\{u, v\}$.

If the basis has three vectors, $\{u, v, w\}$, then finding an orthogonal basis requires two steps: replacing v by $\hat{v} = v - \text{proj}_u v$, and then replacing w by $\hat{w} = w - \text{proj}_{\text{sp}\{u,v\}} w$.

The complete algorithm is described in the following theorem.

Theorem 293 *Let V be an inner product space over a field F, and suppose $\{u_1, u_2, \ldots, u_n\}$ is a basis for V. Let $\{\hat{u}_1, \hat{u}_2, \ldots, \hat{u}_n\}$ be defined by*

$$\begin{aligned}\hat{u}_1 &= u_1,\\ \hat{u}_{k+1} &= u_{k+1} - \text{proj}_{S_k} u_{k+1},\ k = 1, 2, \ldots, n-1,\end{aligned}$$

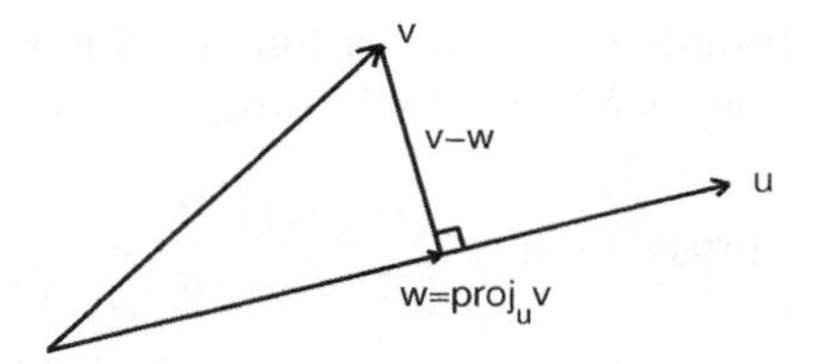

FIGURE 6.6
Orthogonalizing v against u by subtracting off the projection of v onto u.

where

$$S_k = \text{sp}\{u_1, \dots, u_k\},\ k = 1, 2, \dots, n.$$

Then $\{\hat{u}_1, \hat{u}_2, \dots, \hat{u}_n\}$ *is an orthogonal set (and hence linearly independent), and*

$$\text{sp}\{\hat{u}_1, \dots, \hat{u}_k\} = \text{sp}\{u_1, \dots, u_k\} = S_k,\ k = 1, 2, \dots, n.$$

In particular, $\{\hat{u}_1, \hat{u}_2, \dots, \hat{u}_n\}$ *is an orthogonal basis for* V*.*

Proof We will prove by induction on k that $\text{sp}\{\hat{u}_1, \dots, \hat{u}_k\} = \text{sp}\{u_1, \dots, u_k\}$ and $\{\hat{u}_1, \dots, \hat{u}_k\}$ is orthogonal. Both of these statements obviously hold for $k = 1$, since $\hat{u}_1 = u_1$ and a set of one vector is orthogonal by default.

Suppose the results hold for $k = \ell$: $\text{sp}\{\hat{u}_1, \dots, \hat{u}_\ell\} = \text{sp}\{u_1, \dots, u_\ell\}$ and $\{\hat{u}_1, \dots, \hat{u}_\ell\}$ is orthogonal. By the projection theorem, $\text{proj}_{S_\ell} u_{\ell+1}$ is defined by

$$\left\langle u_{\ell+1} - \text{proj}_{S_\ell} u_{\ell+1}, v \right\rangle = 0 \text{ for all } v \in S_\ell.$$

In particular,

$$\left\langle u_{\ell+1} - \text{proj}_{S_\ell} u_{\ell+1}, \hat{u}_i \right\rangle = 0,\ i = 1, 2, \dots, \ell,$$

since, by the induction hypothesis, $\hat{u}_i \in S_\ell$ for $i = 1, 2, \dots, \ell$. This shows that

$$\hat{u}_{\ell+1} = u_{\ell+1} - \text{proj}_{S_\ell} u_{\ell+1}$$

is orthogonal to each vector in $\{\hat{u}_1, \dots, \hat{u}_\ell\}$, and therefore $\{\hat{u}_1, \dots, \hat{u}_{\ell+1}\}$ is orthogonal. Furthermore, since $\text{proj}_{S_\ell} u_{\ell+1} \in S_\ell$, it follows that $\hat{u}_{\ell+1}$ is a linear combination of vectors in $S_{\ell+1}$. Conversely,

$$u_{\ell+1} = \hat{u}_{\ell+1} + \text{proj}_{S_\ell} u_{\ell+1} \in \text{sp}\{u_1, \dots, u_\ell, \hat{u}_{\ell+1}\} = \text{sp}\{\hat{u}_1, \dots, \hat{u}_\ell, \hat{u}_{\ell+1}\}.$$

Therefore,

$$\text{sp}\{\hat{u}_1, \dots, \hat{u}_\ell, \hat{u}_{\ell+1}\} = \text{sp}\{u_1, \dots, u_\ell, u_{\ell+1}\}.$$

This completes the proof.

QED

At each step of the Gram-Schmidt algorithm, we know an orthogonal basis for S_k, and therefore it is easy to compute $\text{proj}_{S_k} u_{k+1}$:

$$\text{proj}_{S_k} u_{k+1} = \sum_{i=1}^{k} \frac{\langle u_{k+1}, \hat{u}_i \rangle}{\langle \hat{u}_i, \hat{u}_i \rangle} \hat{u}_i.$$

Example 294 *Consider the following subspace V of $\mathbf{R}^4$:*

$$V = \text{sp}\{(1,1,1,1), (1,1,1,0), (1,1,0,0)\}.$$

Writing $u_1 = (1,1,1,1)$, $u_2 = (1,1,1,0)$, and $u_3 = (1,1,0,0)$, we can apply the Gram-Schmidt algorithm to find an orthogonal basis for V. We define $\hat{u}_1 = u_1 = (1,1,1,1)$ and $S_1 = \text{sp}\{u_1\}$. Then

$$\text{proj}_{S_1} u_2 = \frac{u_2 \cdot \hat{u}_1}{\hat{u}_1 \cdot \hat{u}_1} \hat{u}_1 = \frac{3}{4}(1,1,1,1) = \left(\frac{3}{4}, \frac{3}{4}, \frac{3}{4}, \frac{3}{4} \right)$$

and

$$\hat{u}_2 = u_2 - \text{proj}_{S_1} u_2 = (1,1,1,0) - \left(\frac{3}{4}, \frac{3}{4}, \frac{3}{4}, \frac{3}{4} \right) = \left(\frac{1}{4}, \frac{1}{4}, \frac{1}{4}, -\frac{3}{4} \right).$$

We define $S_2 = \text{sp}\{\hat{u}_1, \hat{u}_2\}$. Then

$$\begin{aligned}
\text{proj}_{S_2} u_3 &= \frac{u_3 \cdot \hat{u}_1}{\hat{u}_1 \cdot \hat{u}_1} \hat{u}_1 + \frac{u_3 \cdot \hat{u}_2}{\hat{u}_2 \cdot \hat{u}_2} \hat{u}_2 \\
&= \frac{2}{4}(1,1,1,1) + \frac{1/2}{3/4} \left(\frac{1}{4}, \frac{1}{4}, \frac{1}{4}, -\frac{3}{4} \right) \\
&= \left(\frac{1}{2}, \frac{1}{2}, \frac{1}{2}, \frac{1}{2} \right) + \left(\frac{1}{6}, \frac{1}{6}, \frac{1}{6}, -\frac{1}{2} \right) \\
&= \left(\frac{2}{3}, \frac{2}{3}, \frac{2}{3}, 0 \right)
\end{aligned}$$

and

$$\hat{u}_3 = u_3 - \text{proj}_{S_2} u_3 = (1,1,0,0) - \left(\frac{2}{3}, \frac{2}{3}, \frac{2}{3}, 0 \right) = \left(\frac{1}{3}, \frac{1}{3}, -\frac{2}{3}, 0 \right).$$

Example 295 *This is a continuation of the previous example. Suppose we wish to find the best approximation to $u = (1,2,3,4)$ from V. The solution is $\text{proj}_V u$, and the orthogonal basis $\{\hat{u}_1, \hat{u}_2, \hat{u}_3\}$ makes this a direct calculation:*

$$\begin{aligned}
\text{proj}_V u &= \frac{\langle u, \hat{u}_1 \rangle}{\langle \hat{u}_1, \hat{u}_1 \rangle} \hat{u}_1 + \frac{\langle u, \hat{u}_2 \rangle}{\langle \hat{u}_2, \hat{u}_2 \rangle} \hat{u}_2 + \frac{\langle u, \hat{u}_3 \rangle}{\langle \hat{u}_3, \hat{u}_3 \rangle} \hat{u}_3 \\
&= \frac{10}{4}(1,1,1,1) + \frac{-3/2}{3/4} \left(\frac{1}{4}, \frac{1}{4}, \frac{1}{4}, -\frac{3}{4} \right) + \frac{-1}{2/3} \left(\frac{1}{3}, \frac{1}{3}, -\frac{2}{3}, 0 \right) \\
&= \left(\frac{3}{2}, \frac{3}{2}, 3, 4 \right).
\end{aligned}$$

The reader should recall that if we were to compute $\text{proj}_V u$ using the nonorthogonal basis $\{u_1, u_2, u_3\}$, it would be necessary to solve a system of equations.

6.5.1 Least-squares polynomial approximation

We now consider an application of the projection theorem, in which the ability to produce an orthogonal basis is useful. The reader will recall that the projection theorem addresses the problem of finding the best approximation to a given vector from a given subspace.

We will consider approximating a function $f \in C[a,b]$ by a polynomial p from $\mathcal{P}_n$. For example, in designing a calculator or computer, it might be necessary to approximate $f(x) = 2^x$ on the interval $[0,1]$. It is natural to use a polynomial, since any polynomial can be evaluated using only the elementary arithmetic operations of addition and multiplication.[4]

Before we can solve this approximation problem, we have to pose it precisely by stating how the error in $f - p$ is to be measured. There are various ways to proceed, but the problem has the simplest solution if we measure the error in the L^2 norm. Then, since $\mathcal{P}_n$ can be regarded as a subspace of $C[a,b]$, we can apply the projection theorem. Other possibilities are explored in Section 6.8. For definiteness, we will assume that the interval of interest is $[0,1]$ in the following discussion.

We thus wish to minimize $\|f-p\|_2$ over all $p \in \mathcal{P}_n$. This problem can be solved directly using the projection theorem. We use the standard basis $\{1, x, x^2, \ldots, x^n\}$ for $\mathcal{P}_n$ and form $Ga = b$, where

$$G_{ij} = \langle x^j, x^i \rangle_2 = \int_0^1 x^j x^i \, dx = \int_0^1 x^{i+j} \, dx = \frac{1}{i+j+1}, \quad i,j = 0,1,\ldots,n,$$

$$b_i = \langle f, x^i \rangle_2 = \int_0^1 x^i f(x) \, dx, \quad i = 0,1,\ldots,n.$$

The the solution $a = (a_0, a_1, \ldots, a_n)$ gives the weights that represent the best approximation p in terms of the basis:

$$p(x) = a_0 + a_1 x + a_2 x^2 + \ldots + a_n x^n.$$

In this context, it is convenient to index vectors and matrices with $i = 0,1,\ldots,n$ rather than $i = 1,2,\ldots,n+1$.

Example 296 *Let $f(x) = \sin(\pi x)$ and suppose we wish to approximate f by a quadratic polynomial p on $[0,1]$. The matrix G has the simple form given above:*

$$G = \begin{bmatrix} 1 & \frac{1}{2} & \frac{1}{3} \\ \frac{1}{2} & \frac{1}{3} & \frac{1}{4} \\ \frac{1}{3} & \frac{1}{4} & \frac{1}{5} \end{bmatrix}.$$

[4]On first glance, it is necessary to use exponentiation to evaluate x^n, but since n is an integer, this can be, and usually is, done by repeated multiplication.

The right-hand-side vector b is computed as follows:

$$b_0 = \int_0^1 \sin(\pi x)\, dx = \frac{2}{\pi},$$
$$b_1 = \int_0^1 x \sin(\pi x)\, dx = \frac{1}{\pi},$$
$$b_2 = \int_0^1 x^2 \sin(\pi x)\, dx = \frac{\pi^2 - 4}{\pi^3}.$$

Solving $Ga = b$, we obtain

$$a \doteq \begin{bmatrix} -0.050465 \\ 4.1225 \\ -4.1225 \end{bmatrix},$$

and thus the best quadratic approximation is

$$p(x) \doteq -0.050465 + 4.1225x - 4.1225x^2.$$

The function f and the best quadratic approximation are shown in Figure 6.7.

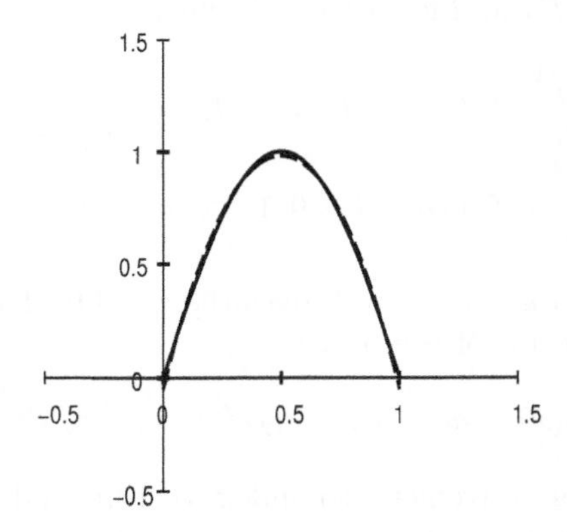

FIGURE 6.7
The function $f(x) = \sin(\pi x)$ (solid curve) and the best quadratic approximation $p(x)$ on $[0, 1]$.

One would expect that f can be approximated more and more accurately if we use polynomials of higher and higher degree. Theoretically, this is true. However, something interesting happens if we repeat the above calculation with $\mathcal{P}_2$ replaced by $\mathcal{P}_n$ for larger n. Figure 6.8 shows the error $|f(x) - p(x)|$, where $p \in \mathcal{P}_n$ for $n = 2, 6, 10, 12$. As one would expect, the error is smaller for $n = 6$ than for $n = 2$, and smaller still for $n = 10$. However, the error actually increases in moving to $n = 12$.

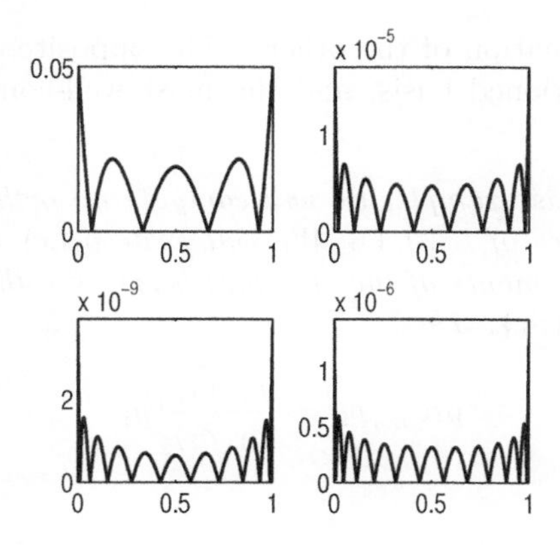

FIGURE 6.8
The error in the best approximation to $f(x) = \sin(\pi x)$ on $[0, 1]$, where the degree of the approximation is $n = 2$ (upper left), $n = 6$ (upper right), $n = 10$ (lower left), and $n = 12$ (lower right).

It can be proved that the error in the best approximation (at least the error as measured in the L^2 norm) cannot increase when n is increased (see Exercise 10). The explanation for the above lies in the nature of the system $Ga = b$. To produce the above results, the linear systems were solved on a computer in floating point (that is, finite-precision) arithmetic. All of the intermediate calculations were subject to round-off error. For many linear systems, round-off error has little effect on the computed solution. Unfortunately, the matrix G in this problem has the property that it becomes increasingly *ill-conditioned* as n increases.

At this point, we will not define the term ill-conditioned precisely, but the general idea is that an ill-conditioned matrix is close to being singular. (The concept of conditioning is covered precisely in Sections 9.5 and 9.6.) When the ill-conditioning is extreme, the matrix is called *numerically singular*, meaning that it cannot be distinguished from a singular matrix in finite-precision arithmetic. The solution to a numerically singular matrix equation can be quite inaccurate.

This is the explanation for the previous example. The best approximation of degree 12 is, in fact, a better approximation to f than is the best approximation of degree 10. However, because the Gram matrix corresponding to $n = 12$ is numerically singular, the computed weights $a_0, a_1, \ldots, a_{12}$ are not very accurate, and so the resulting polynomial approximation is not as good as it should be.

An ill-conditioned Gram matrix results from an ill-conditioned basis, which, informally, is a basis that is barely linearly independent. No element of the basis is a linear combination of the rest, but at least one element is close to being a linear combination of the others. The opposite of an ill-conditioned basis is a well-conditioned basis, and the most well-conditioned basis is an orthogonal basis.

Example 297 *In this example, we will compute an orthogonal basis for* $\mathcal{P}_2$*, regarded as a subspace of* $L^2(0,1)$*. We will write* $p_1(x)=1$, $p_2(x)=x$*, and* $p_3(x)=x^2$ *for the elements of the standard basis. We then begin by defining* $\hat{p}_1=p_1$ *and* $S_1=\text{sp}\{p_1\}$*. Then*

$$\text{proj}_{S_1}p_2=\frac{\langle p_2,\hat{p}_1\rangle_2}{\langle \hat{p}_1,\hat{p}_1\rangle_2}\hat{p}_1.$$

We have

$$\langle p_2,\hat{p}_1\rangle_2=\int_0^1 x\cdot 1\,dx=\frac{1}{2},\ \langle \hat{p}_1,\hat{p}_1\rangle_2=\int_0^1 1^2\,dx=1,$$

and so

$$\text{proj}_{S_1}p_2=\frac{1/2}{1}1=\frac{1}{2}.$$

Therefore,

$$\hat{p}_2(x)=p_2(x)-\text{proj}_{S_1}p_2=x-\frac{1}{2}.$$

Next, we define $S_2=\text{sp}\{p_1,p_2\}$*. We have*

$$\text{proj}_{S_2}p_3=\frac{\langle p_3,\hat{p}_1\rangle_2}{\langle \hat{p}_1,\hat{p}_1\rangle_2}\hat{p}_1+\frac{\langle p_3,\hat{p}_2\rangle_2}{\langle \hat{p}_2,\hat{p}_2\rangle_2}\hat{p}_2.$$

We compute the following inner products:

$$\begin{aligned}\langle p_3,\hat{p}_1\rangle_2&=\int_0^1 x^2\cdot 1\,dx=\frac{1}{3},\\ \langle p_3,\hat{p}_2\rangle_2&=\int_0^1 x^2\left(x-\frac{1}{2}\right)\,dx=\frac{1}{12},\\ \langle \hat{p}_2,\hat{p}_2\rangle_2&=\int_0^1\left(x-\frac{1}{2}\right)^2\,dx=\frac{1}{12}.\end{aligned}$$

We then obtain

$$\text{proj}_{S_2}p_3=\frac{1/3}{1}1+\frac{1/12}{1/12}\left(x-\frac{1}{2}\right)=x-\frac{1}{6},$$

and so

$$\hat{p}_3(x)=p_3(x)-\text{proj}_{S_2}p_3=x^2-x+\frac{1}{6}.$$

Thus an orthonormal basis for $\mathcal{P}_2$*, regarded as a subspace of* $L^2(0,1)$*, is*

$$\left\{1,x-\frac{1}{2},x^2-x+\frac{1}{6}\right\}.$$

Exercise 2 asks the reader to use the orthogonal basis from the previous example to reproduce the results of Example 296.

Although computing an orthogonal basis is time-consuming, it does not have to be done repeatedly. If one wishes to approximate many different functions on a given interval $[a, b]$, one could compute the orthogonal polynomials for that interval once and for all, and then use them to solve each approximation problem as necessary.

Exercises

Miscellaneous exercises

1. Let V be an inner product space over $\mathbf{R}$, and let $u \in V$, $u \neq 0$. The one-dimensional subspace spanned by u can be regarded as a line in V. Find a formula for the projection of an arbitrary vector v onto this line.

2. Repeat Example 296, but use the orthogonal basis for $\mathcal{P}_2$ derived in Example 297 in place of the standard basis. You should find the same best quadratic approximation.

3. Find an orthogonal basis for $\mathcal{P}_3$, regarded as a subspace of $L^2(0, 1)$. (Hint: Rather than starting from scratch, the results of Example 297 can be extended.)

4. Define $f \in C[0, 1]$ by $f(x) = \sin(\pi x)$. Use the results of the previous exercise to approximate f by a cubic polynomial on $[0, 1]$.

5. (a) Find the best cubic approximation, in the $L^2(-1, 1)$ norm, to the function $f(x) = e^x$. Use the standard basis in your computations.
   (b) Find an orthogonal basis for $\mathcal{P}_3$ under the $L^2(-1, 1)$ inner product.
   (c) Repeat the calculations of 5a using the orthogonal basis in place of the standard basis.

6. Let S be the subspace of $\mathbf{R}^4$ spanned by
$$u_1 = (1, 0, 1, -1),\ u_2 = (1, 0, 0, 1),\ u_3 = (1, 2, 1, 1).$$
   (a) Find an orthogonal basis for S.
   (b) Compute $\text{proj}_S x$, where $x = (1, 2, 3, 0)$.

7. Let S be the subspace of $\mathbf{R}^5$ spanned by
$$u_1 = (1, 1, 1, 1, 1),\ u_2 = (1, -1, 1, -1, 1),$$
$$u_3 = (2, 0, 1, 1, 1),\ u_4 = (1, -2, 0, -1, 1).$$
   (a) Find an orthogonal basis for S.
   (b) Compute $\text{proj}_S x$, where $x = (1, 2, 2, 2, 1)$.

8. Let S be the subspace of $C[0,1]$ spanned by e^x and e^{-x}. Regard $C[0,1]$ as an inner product space under the $L^2(0,1)$ inner product.
   (a) Find an orthogonal basis for S.
   (b) Using the orthogonal basis, find the best approximation to $f(x) = x$ from S.

9. Let P be the plane in $\mathbf{R}^3$ defined by the equation $3x - y - z = 0$.
   (a) Find an orthogonal basis for P.
   (b) Find the projection of $u = (1,1,1)$ onto P.

10. Suppose $f \in C[a,b]$ and p_n is the best approximation, in the $L^2(a,b)$ norm, to f from $\mathcal{P}_n$. Prove that, for all $n = 0,1,2,\ldots$,
$$\|f - p_{n+1}\|_2 \leq \|f - p_n\|_2.$$
(Hint: Use the fact that $\mathcal{P}_n \subset \mathcal{P}_{n+1}$.)

11. Consider the $H^1(0,1)$ inner product defined by
$$\langle f, g\rangle = \int_0^1 \{f(x)g(x) + f'(x)g'(x)\}\, dx$$
(see Exercise 6.1.10).
   (a) Compute an orthogonal basis for $\mathcal{P}_2$, regarded as a subspace of $C[0,1]$, under the $H^1(0,1)$ inner product.
   (b) Compute the best approximation to $f(x) = \sin(\pi x)$ from $\mathcal{P}_2$ in the $H^1(0,1)$ norm.
   (c) Compare your results to the best approximation to f in the $L^2(0,1)$ norm (computed in Example 296). In particular, graph the error in f and the error in f' for both approximations. What do the graphs reveal?

12. In Exercise 6.4.7, it was shown that the following sets are orthogonal under the $L^2(0,1)$ inner product:
$$\{\sin(\pi x), \sin(2\pi x), \ldots, \sin(n\pi x)\},$$
$$\{1, \cos(\pi x), \cos(2\pi x), \ldots, \cos(n\pi x)\}.$$
Show that these sets are also orthogonal with respect to the $H^1(0,1)$ inner product (see the previous exercise).

13. Define an inner product on $C[0,1]$ by
$$\langle f, g\rangle = \int_0^1 (1+x)f(x)g(x)\, dx.$$

(a) Show that this formula really does define an inner product on $C[0,1]$.

(b) Find an orthogonal basis for $\mathcal{P}_2$ under the above inner product.

14. Let $\{u_1, u_2, \dots, u_n\}$ be a basis for an inner product space V. Consider the following algorithm: Initially set $\hat{u}_i = u_i$ for all $i = 1, 2, \dots, n$. For each $k = 1, 2, \dots, n-1$, subtract the projection of $\hat{u}_i$ onto u_k from $\hat{u}_i$ for each $i = k+1, \dots, n$:

$$\hat{u}_i \leftarrow \hat{u}_i - \frac{\langle \hat{u}_i, \hat{u}_k \rangle}{\langle \hat{u}_k, \hat{u}_k \rangle} \hat{u}_k, \quad i = k+1, \dots, n.$$

(Notice that the value of $\hat{u}_i$ is updated at each iteration, so this symbol is being used like a variable in a computer program.) Prove that after this algorithm completes, $\{\hat{u}_1, \dots, \hat{u}_n\}$ is an orthogonal basis for V. This approach is called the *modified Gram-Schmidt algorithm.*

15. On page 373, it was stated that the standard basis for $\mathcal{P}_n$ is nearly linearly dependent. The purpose of this exercise is to quantify this statement, when $\mathcal{P}_n$ is regarded as a subspace of $L^2(0,1)$. Compute w_n, the projection of $p_n(x) = x^n$ onto $\mathcal{P}_{n-1}$, for $n = 1, 2, 3, \dots, N$. Then make a table of n versus the relative error in w_n as an approximation to p_n:

$$\frac{\|p_n - w_n\|_2}{\|p_n\|_2}.$$

Take N as large as is practical. The use of a computer algebra system is recommended. What does the table reveal?

6.6 Orthogonal complements

One way to represent a vector space is by a basis. Another way is to decompose it into subspaces that are easy to understand. As with a basis, these subspaces are often chosen to be particularly convenient for the problem at hand. To pursue this idea, we need the following concept.

Definition 298 *Let V be an inner product space over* **R**, *and let S be a nonempty subset of V. The* orthogonal complement *of S is the set*

$$S^\perp = \{u \in V : \langle u, s \rangle = 0 \text{ for all } s \in S\}. \tag{6.26}$$

The symbol $S^\perp$ is usually pronounced "S-perp."

Theorem 299 *Let V be an inner product space over* **R**, *and let S be a nonempty subset of V. Then $S^\perp$ is a subspace of V.*

Proof Exercise 7

Example 300 *Let S be the singleton set $\{u\}$ in $\mathbf{R}^3$, where $u = (1, 2, -1)$. Then $S^\perp$ consists of all vectors $(x, y, z) \in \mathbf{R}^3$ such that $(x, y, z) \cdot (1, 2, -1) = 0$, that is, all vectors (x, y, z) such that*

$$x + 2y - z = 0. \tag{6.27}$$

The reader will recognize (6.27) as the equation of a plane in $\mathbf{R}^3$.

Example 301 *Let V be the vector space $C[0, 1]$ under the $L^2(0, 1)$ inner product, and let*

$$S = \left\{ v \in V : \int_0^1 v(x)\, dx = 0 \right\}.$$

We describe the functions in S as having mean zero.[5] *We wish to determine $S^\perp$.*

We notice that, if $u(x) = C$ for some constant C and $v \in S$, then

$$\langle u, v \rangle_2 = \int_0^1 u(x)v(x)\, dx = \int_0^1 Cv(x)\, dx = C \int_0^1 v(x)\, dx = 0.$$

It follows that every constant function belongs to $S^\perp$. We will show that $S^\perp$ is precisely the set of all constant functions. Given any function $u \in V$, we define the mean *of u to be*

$$\overline{u} = \int_0^1 u(x)\, dx.$$

Then, for any $u \in V$, $u - \overline{u} \in S$:

$$\int_0^1 (u(x) - \overline{u})\, dx = \int_0^1 u(x)\, dx - \int_0^1 \overline{u}\, dx = \int_0^1 u(x)\, dx - \overline{u} = 0.$$

Now suppose $u \in S^\perp$. Then $u - \overline{u} \in S$, and so

$$\langle u, u - \overline{u} \rangle_2 = \int_0^1 u(x) \left(u(x) - \overline{u} \right) dx = 0.$$

We also have

$$\int_0^1 \overline{u} \left(u(x) - \overline{u} \right) dx = \overline{u} \int_0^1 \left(u(x) - \overline{u} \right) dx = 0.$$

---

[5]The mean of v over $[a, b]$ is

$$\frac{1}{b - a} \int_a^b v(x)\, dx.$$

Therefore,

$$\int_0^1 (u(x) - \overline{u})^2 \, dx = \int_0^1 u(x)\,(u(x) - \overline{u}) \, dx - \int_0^1 \overline{u}\,(u(x) - \overline{u}) \, dx = 0 - 0 = 0.$$

But then, since $(u(x) - \overline{u})^2 \geq 0$ *on* $[0,1]$, *the integral of this function is zero if and only if it is identically zero, that is, if and only if* $u(x) = \overline{u}$ *for all* $x \in [0,1]$. *Thus we have shown that if* $u \in S^{\perp}$, *then* u *is a constant function.*

In the previous example, it is easy to show that $\left(S^{\perp}\right)^{\perp} = S$. This is always the case when V is finite-dimensional.[6] In the following preliminary result, we refer to *orthogonal* subspaces; two subspaces are said to be orthogonal if every vector in one is orthogonal to every vector in the other.

Lemma 302 *Let* V *be an inner product space over* $\mathbf{R}$, *and let* S *and* T *be orthogonal subspaces of* V. *Then* $S \cap T = \{0\}$.

Proof Exercise 8.

Theorem 303 *Let* V *be a finite-dimensional inner product space over* $\mathbf{R}$, *and let* S *be a subspace of* V. *Then* $\left(S^{\perp}\right)^{\perp} = S$.

Proof We first show that $S \subset (S^{\perp})^{\perp}$. If $s \in S$ and $y \in S^{\perp}$, then $\langle s, y \rangle = 0$ by definition of $S^{\perp}$. But then s is orthogonal to every element of $S^{\perp}$, so $s \in (S^{\perp})^{\perp}$. Therefore $S \subset (S^{\perp})^{\perp}$.

Now suppose $x \in (S^{\perp})^{\perp}$, and define $s = \text{proj}_S x$. Then, by the projection theorem, $x - s \in S^{\perp}$. On the other hand, since $x \in (S^{\perp})^{\perp}$, $s \in S \subset (S^{\perp})^{\perp}$, and $(S^{\perp})^{\perp}$ is a subspace, it follows that $x - s \in (S^{\perp})^{\perp}$. But the only vector belonging to both $S^{\perp}$ and $(S^{\perp})^{\perp}$ is the zero vector, so $x - s = 0$, which shows that $x = s \in S$. Therefore $(S^{\perp})^{\perp} \subset S$, and we have shown $(S^{\perp})^{\perp} = S$.

QED

We have already introduced the notation $S + T$ for the algebraic sum of two subsets of V, and we know that $S + T$ is a subspace of V if S and T are both subspaces. We will use the notation $S \oplus T$ to represent the algebraic sum of S and T if S and T are orthogonal subspaces.[7]

Lemma 304 *Let* V *be an inner product space over* $\mathbf{R}$, *and let* S *and* T *be orthogonal subspaces of* V. *Then each vector* $v \in S \oplus T$ *can be written uniquely as* $v = s + t$, *where* $s \in S$ *and* $t \in T$. *In other words,* $S \oplus T$ *is a direct sum (see Definition 225 in Section 5.1.1).*

---

[6] If V is infinite-dimensional, as in Example 301, then S must also be *closed* in order for $\left(S^{\perp}\right)^{\perp} = S$ to hold. In Example 301, the subspace S is closed. The concept of a closed set is discussed in Chapter 10.

[7] The notation $S \oplus T$ is used differently by different authors. Some use $S \oplus T$ to mean $S + T$ in the case that this algebraic sum is a direct sum. Others agree with our approach: $S \oplus T$ means $S + T$ in the case that S and T are orthogonal subspaces.

Proof If $v \in S \oplus T$, then, by definition, there exist $s \in S$ and $t \in T$ such that $v = s + t$. Thus the only issue is the uniqueness of s and t. Suppose $s' \in S$ and $t' \in T$ satisfy $v = s' + t'$. We then have

$$s + t = s' + t' \Rightarrow s - s' = t' - t.$$

But $s - s' \in S$ and $t' - t \in T$ because S and T are subspaces, so $s - s' = t' - t$ lies in $S \cap T$. But then, by Lemma 302, $s - s' = t' - t = 0$, that is, $s = s'$ and $t = t'$.

QED

Lemma 305 *Let V be an inner product space over $\mathbf{R}$, and suppose $V = S \oplus T$, where S and T are orthogonal subspaces of V. Then*

$$\dim(V) = \dim(S) + \dim(T).$$

Proof This follows immediately from Theorem 226.

Theorem 306 *Let V be a finite-dimensional inner product space over $\mathbf{R}$, and let S be a subspace of V. Then*

$$V = S \oplus S^{\perp}.$$

Proof Let $v \in V$. Let $s = \text{proj}_S v$; then $t = v - s \in S^{\perp}$ by the projection theorem. Since $v = s + (v - s)$, this shows that every $v \in V$ can be written as $v = s + t$, $s \in S$, $t \in S^{\perp}$. Thus $V \subset S \oplus S^{\perp}$. Since $S \oplus S^{\perp} \subset V$ by definition, we see that $V = S \oplus S^{\perp}$.

QED

This last result gives a way of decomposing a vector space into subspaces. This is particularly useful when solving a linear operator equation in which the operator is singular. We will develop this idea after the next example.

Example 307 *Let S be the subspace of $\mathbf{R}^4$ spanned by the vectors*

$$x_1 = (1, -1, 4, -4), \ x_2 = (1, -1, 3, 0), \ x_3 = (-1, 1, -8, 3).$$

We wish to find a basis for $S^{\perp}$ and show that $\mathbf{R}^4 = S \oplus S^{\perp}$. A vector t belongs to $S^{\perp}$ if and only if

$$\begin{aligned} x_1 \cdot t &= 0, \\ x_2 \cdot t &= 0, \\ x_3 \cdot t &= 0. \end{aligned}$$

These equations are equivalent to the linear system $At = 0$, where the rows of A are the vectors x_1, x_2, x_3, which implies that $S^{\perp} = \mathcal{N}(A)$. A straightforward calculation shows that $S^{\perp} = \mathcal{N}(A) = \text{sp}\{(1, 1, 0, 0)\}$ (in the course of solving

As $= 0$, we also verify that $\text{rank}(A) = 3$*, which means that the rows are linearly independent and hence* $\{x_1, x_2, x_3\}$ *is a basis for* S*).*

If we define $x_4 = (1, 1, 0, 0)$*, then* $\{x_1, x_2, x_3, x_4\}$ *is a basis for* $\mathbf{R}^4$*. This implies that* $\mathbf{R}^4 = S \oplus S^\perp$*, since each vector* $v \in \mathbf{R}^4$ *can be written uniquely as*

$$v = \alpha_1 x_1 + \alpha_2 x_2 + \alpha_3 x_3 + \alpha_4 x_4 = s + t,$$

where $s = \alpha_1 x_1 + \alpha_2 x_2 + \alpha_3 x_3 \in S$ *and* $t = \alpha_4 x_4$*.*

6.6.1 The fundamental theorem of linear algebra revisited

We have already seen that, if X and U are finite-dimensional vector spaces and $T : X \to U$ is linear, then

$$\text{rank}(T) + \text{nullity}(T) = \dim(X).$$

When X and U are inner product spaces, we can say even more. The following theorem describes the relationships among the four subspaces defined by a linear operator T: $\ker(T)$, $\mathcal{R}(T)$, $\ker(T^*)$, and $\mathcal{R}(T^*)$. Because of the importance of these subspaces in understanding the operator T and the operator equation $T(x) = u$, we refer to them as the four *fundamental subspaces* of T.

Theorem 308 *Let* X *and* U *be finite-dimensional inner product spaces over* $\mathbf{R}$*, and let* $T : X \to U$ *be linear. Then*

1. $\ker(T)^\perp = \mathcal{R}(T^*)$ *and* $\mathcal{R}(T^*)^\perp = \ker(T)$*;*
2. $\ker(T^*)^\perp = \mathcal{R}(T)$ *and* $\mathcal{R}(T)^\perp = \ker(T^*)$*.*

Proof It suffices to prove $\mathcal{R}(T^*)^\perp = \ker(T)$, since then

$$\left(\mathcal{R}(T^*)^\perp\right)^\perp = \ker(T)^\perp \Rightarrow \mathcal{R}(T^*) = \ker(T)^\perp.$$

Moreover, the second part of the theorem then follows by applying the first part to T^* in place of T.

Suppose first that $x \in \mathcal{R}(T^*)^\perp$. Then, by definition of range, it follows that

$$\langle x, T^*(u)\rangle_X = 0 \text{ for all } u \in U.$$

But then

$$\langle T(x), u\rangle_U = 0 \text{ for all } u \in U,$$

which implies, by Theorem 274, that $T(x) = 0$. Thus $x \in \ker(T)$.

On the other hand, suppose $x \in \ker(T)$. Then, for any $y \in \mathcal{R}(T^*)$, there exists $u \in U$ such that $y = T^*(u)$, and

$$\langle x, y\rangle_X = \langle x, T^*(u)\rangle_X = \langle T(x), u\rangle_U = \langle 0, u\rangle_U = 0,$$

and hence $x \in \mathcal{R}(T^*)^\perp$. Thus $x \in \mathcal{R}(T^*)^\perp$ if and only if $x \in \ker(T)$, which completes the proof.

QED

The following corollary is called the *rank theorem.*

Corollary 309 *Let X and U be finite-dimensional inner product spaces over* $\mathbf{R}$*, and let $T : X \to U$ be linear. Then*

$$\text{rank}(T^*) = \text{rank}(T).$$

Proof By the fundamental theorem, $\dim(X) = \text{rank}(T) + \text{nullity}(T)$. On the other hand, the preceding theorem, together with Lemma 305, shows that

$$\dim(X) = \dim(\mathcal{R}(T^*)) + \dim(\ker(T)) = \text{rank}(T^*) + \text{nullity}(T).$$

Therefore $\text{rank}(T^*)+\text{nullity}(T) = \text{rank}(T)+\text{nullity}(T)$, or $\text{rank}(T^*) = \text{rank}(T)$, as desired.

QED

Theorem 308 and Corollary 309 shed more light on the fundamental theorem of linear algebra. If $T : X \to U$, then

$$\dim(X) = \text{nullity}(T) + \text{rank}(T)$$

because X is the direct sum of $\ker(T)$, which has dimension $\text{nullity}(T)$, and $\mathcal{R}(T^*)$, which has dimension $\text{rank}(T)$.

Here is an example of the use of Theorem 308: Suppose $u \in U$ and we wish to know whether $T(x) = u$ has a solution. This is equivalent to the question of whether $u \in \mathcal{R}(T)$. Since $\mathcal{R}(T) = \ker(T^*)^{\perp}$, we see that $T(x) = u$ has a solution if and only if

$$T^*(v) = 0 \Rightarrow \langle u, v \rangle_U = 0. \tag{6.28}$$

This characterization is called a *compatibility condition* and is useful in certain applications.

As another application of the previous theorem, suppose we wish to solve $T(x) = y$ and T is singular. Assuming the equation has a solution $\hat{x}$, it has in fact infinitely many solutions; the solution set is $\hat{x} + \ker(T)$. Since

$$X = \mathcal{R}(T^*) \oplus \ker(T),$$

we can write $\hat{x}$ uniquely as $\hat{x} = \overline{x} + z$, where $\overline{x} \in \mathcal{R}(T^*)$ and $z \in \ker(T)$. It can be shown that $\overline{x}$ is the unique solution to $T(x) = u$ that lies in $\mathcal{R}(T^*)$. This gives one way to identify a unique solution to $T(x) = u$, namely, by adding the extra condition that x lie in $\mathcal{R}(T^*)$. The solution $\overline{x}$ has another special property that is explored in the exercises (see Exercises 13ff.).

Finally, we can rephrase Theorem 308 in terms of matrices.

Corollary 310 *Let $A \in \mathbf{R}^{m\times n}$. Then*

1. $\mathcal{N}(A)^{\perp} = \text{col}\left(A^T\right)$ *and* $\text{col}\left(A^T\right)^{\perp} = \mathcal{N}(A)$*;*

2. $\mathcal{N}\left(A^T\right)^\perp = \mathrm{col}(A)$ *and* $\mathrm{col}(A)^\perp = \mathcal{N}\left(A^T\right)$.

As a result, $\mathbf{R}^n = \mathrm{col}(A^T) \oplus \mathcal{N}(A)$ *and* $\mathbf{R}^m = \mathrm{col}(A) \oplus \mathcal{N}(A^T)$.

The rank theorem can also be applied to matrices:

Corollary 311 *Let* $A \in \mathbf{R}^{m\times n}$. *Then* $\mathrm{rank}(A) = \mathrm{rank}\left(A^T\right)$.

The rank of A is the dimension of the column space of A, that is, the number of linearly independent columns of A. Similarly, the rank of A^T is the number of linearly independent columns of A^T. But the columns of A^T are the rows of A, and thus we see that the number of linearly independent rows of A is always the same as the number of linearly independent columns—a fact that is not at all obvious.

Exercises

Miscellaneous exercises

1. Let $S = \mathrm{sp}\{(1,2,1,-1),(1,1,2,0)\}$. Find a basis for $S^\perp$.

2. Let S be a plane in $\mathbf{R}^3$ that passes through the origin. Prove that $S^\perp$ is a line through the origin.

3. Let $A \in \mathbf{R}^{3\times 4}$ be defined by

$$A = \begin{bmatrix} 1 & 2 & -1 & 1 \\ 1 & 3 & 2 & 4 \\ 2 & 8 & 9 & 12 \end{bmatrix}.$$

   Find bases for $\mathrm{col}(A)$ and $\mathcal{N}(A^T)$ and verify that every vector in the first basis is orthogonal to every vector in the second. Also verify that the union of the two bases is a basis for $\mathbf{R}^3$.

4. Let $A \in \mathbf{R}^{4\times 3}$ be defined by

$$A = \begin{bmatrix} 1 & 4 & -4 \\ 1 & 3 & -9 \\ 0 & -1 & -5 \\ 1 & 5 & 1 \end{bmatrix}.$$

   Find bases for $\mathcal{N}(A)$ and $\mathrm{col}(A^T)$ and verify that every vector in the first basis is orthogonal to every vector in the second. Also verify that the union of the two bases is a basis for $\mathbf{R}^3$.

5. Let A be the matrix from the previous exercise.

   (a) Find an orthogonal basis for $\mathbf{R}^3$ that is the union of orthogonal bases for $\mathcal{N}(A)$ and $\mathrm{col}(A^T)$.

(b) Find an orthogonal basis for $\mathbf{R}^4$ that is the union of orthogonal bases for $\text{col}(A)$ and $\mathcal{N}(A^T)$.

6. Let $D : \mathcal{P}_2 \to \mathcal{P}_2$ be defined by $D(p) = p'$. Find bases for the four fundamental subspaces of D, assuming that the $L^2(0,1)$ inner product is imposed on P_2.

7. Prove Theorem 299.

8. Prove Lemma 302.

9. Suppose $A \in \mathbf{R}^{n\times n}$ is symmetric, that is, A satisfies $A^T = A$.

   (a) State Corollary 310 for this special case.

   (b) Formulate the compatibility condition (6.28) for this special case.

10. Let V be vector space over a field F, and let S be a nonempty subset of S. Define $\text{sp}(S)$ to be the set of all finite linear combinations of elements of S; that is,

$$\begin{aligned}\text{sp}(S) &= \{\alpha_1 s_1 + \cdots + \alpha_k s_k : k \text{ is a positive integer,} \\ &\qquad s_1, \ldots, s_k \in S, \alpha_1, \ldots, \alpha_k \in F\}.\end{aligned}$$

   (a) Prove that $\text{sp}(S)$ is a subspace of V.

   (b) Prove that $\text{sp}(S)$ is the smallest subspace of V containing S, in the sense that if T is any subspace of V and $S \subset T$, then $\text{sp}(S) \subset T$.

11. Let V be a finite-dimensional inner product space over $\mathbf{R}$, and let S be a nonempty subset of V. Prove that

$$\left(S^\perp\right)^\perp = \text{sp}(S)$$

(cf. the preceding exercise). (Hint: From Theorem 299, we know that $(S^\perp)^\perp$ is a subspace. First prove that $S \subset (S^\perp)^\perp$. By the preceding exercise, this implies that $\text{sp}(S) \subset (S^\perp)^\perp$. Then prove that $(S^\perp)^\perp \subset \text{sp}(S)$ by showing that for all $x \in (S^\perp)^\perp$, $x = \text{proj}_{\text{sp}(S)} x$.)

12. Let X and U be finite-dimensional inner product spaces over $\mathbf{R}$. Prove the rank theorem directly as follows: Let $x_1, \ldots, x_k$ be vectors in X such that $\{T(x_1), \ldots, T(x_k)\}$ is a basis for $\mathcal{R}(T)$. Prove that

$$\{T^*(T(x_1)), \ldots, T^*(T(x_k))\}$$

is a linearly independent subset of $\mathcal{R}(T^*)$. This shows that

$$\text{rank}(T^*) \geq \text{rank}(T).$$

Since this result holds for all linear operators mapping one finite-dimensional inner product space into another, it holds for $T^* : U \to X$. Explain why this implies that $\text{rank}(T) = \text{rank}(T^*)$.

Project: The pseudoinverse of a matrix

13. Let $A \in \mathbf{R}^{m\times n}$ and $y \in \mathbf{R}^m$ be given. Suppose A is singular and $y \in \text{col}(A)$. Then, by definition, $Ax = y$ has infinitely many solutions and the solution set is $\hat{x} + \mathcal{N}(A)$, where $\hat{x}$ is any one solution.

    (a) Prove that there is a unique solution $\overline{x}$ to $Ax = y$ such that $\overline{x} \in \text{col}(A^T)$.

    (b) Prove that if $x \in \mathbf{R}^n$ is a solution to $Ax = y$ and $x \neq \overline{x}$, then $\|\overline{x}\|_2 < \|x\|_2$.

    We call $\overline{x}$ the *minimum-norm solution* to $Ax = y$.

14. Let $A \in \mathbf{R}^{m\times n}$, $y \in \mathbf{R}^m$. Exercise 6.4.1 shows that the set of all least-squares solutions to $Ax = y$ is $\hat{x}+\mathcal{N}(A)$, where $\hat{x} \in \mathbf{R}^n$ is any one least-squares solution. Exercise 6.4.14 shows that there is a unique vector $\overline{x}$ in $(\hat{x} + \mathcal{N}(A)) \cap \text{col}(A^T)$. Prove that $\overline{x}$ has the smallest Euclidean norm of any element of $\hat{x} + \mathcal{N}(A)$, that is, $\overline{x}$ is the *minimum-norm least-squares solution to* $Ax = y$.

15. Let $A \in \mathbf{R}^{m\times n}$ be given. We define an operator $S : \mathbf{R}^m \to \mathbf{R}^n$ as follows: $\overline{x} = S(y)$ is the minimum-norm least-squares solution of $Ax = y$. We first note that S is well-defined: If $Ax = y$ has a solution $\overline{x}$, that solution is a least-squares solution ($x = \overline{x}$ certainly minimizes $\|Ax-y\|_2$). If there is a unique least-squares solution $\overline{x}$, then $\overline{x}$ certainly has the smallest norm of any least-squares solution (since it is the only one). Therefore, applying the preceding exercises, we see in every case, $Ax = y$ has a unique minimum-norm least-squares solution $\overline{x} = S(y)$.

    (a) Prove that S is a linear operator. It follows that there is a matrix $A^\dagger \in \mathbf{R}^{n\times m}$ such that $S(y) = A^\dagger y$ for all $y \in \mathbf{R}^m$. The matrix $A^\dagger$ is called the *pseudoinverse* of A.

    (b) Find formulas for $A^\dagger$ in each of the following cases:

        i. $A \in \mathbf{R}^{n\times n}$ is nonsingular;

        ii. $A \in \mathbf{R}^{m\times n}$, $m > n$, has full rank;

        iii. $A \in \mathbf{R}^{m\times n}$, $m < n$, has rank m. (Hint: See Exercise 6.4.15.)

16. Find $\mathcal{N}(A^\dagger)$.

17. Prove that $\text{col}(A^\dagger) = \text{col}(A^T)$. (Hint: We already know that $\text{col}(A^\dagger) \subset \text{col}(A^T)$, so it suffices to prove that $\text{col}(A^T) \subset \text{col}(A^\dagger)$.)

18. Prove that $AA^\dagger$ is the matrix defining the orthogonal projection onto $\text{col}(A)$, that is, prove $AA^\dagger y = \text{proj}_{\text{col}(A)} y$ for all $y \in \mathbf{R}^m$.

19. Prove that $A^\dagger A$ is the matrix defining the orthogonal projection onto $\text{col}(A^T)$, that is, prove $A^\dagger Ax = \text{proj}_{\text{col}(A^T)} x$ for all $x \in \mathbf{R}^n$.

20. Prove that the following equations hold for all $A \in \mathbf{R}^{m\times n}$:

    (a) $AA^\dagger A = A$;
    (b) $A^\dagger AA^\dagger = A^\dagger$;
    (c) $A^\dagger A = (A^\dagger A)^T$;
    (d) $AA^\dagger = (AA^\dagger)^T$.

21. Prove that the unique matrix $B \in \mathbf{R}^{n\times m}$ satisfying

$$ABA = A,\ BAB = B,\ BA = (BA)^T,\ AB = (AB)^T$$

    (cf. the previous exercise) is $B = A^\dagger$.

22. Prove or disprove: $(A^\dagger)^\dagger = A$.

6.7 Complex inner product spaces

Up to this point, we have restricted our attention to inner product spaces over the field of real numbers. We now turn our attention to complex inner product spaces. Some of the proofs in this section will just be outlined rather than given in detail, as they are similar to ones already seen in the real case.

One of the key points about an inner product is that it defines a norm by the formula $\|u\| = \sqrt{\langle u,u\rangle}$; if we want this property to extend to inner products over complex vector spaces, then we must require that $\langle u,u\rangle$ be a positive real number for each vector u. However, this implies that something must change in the definition of inner product, since if $\langle u,u\rangle \in \mathbf{R}$ for some u and $\langle\cdot,\cdot\rangle$ is bilinear, then

$$\langle(\alpha+i\beta)u,(\alpha+i\beta)u\rangle = (\alpha+i\beta)^2\langle u,u\rangle,$$

which is complex for most choices of $\alpha+i\beta$.

Because of the issue described above, the properties of symmetry and bilinearity must be adjusted to allow for the field of complex numbers.

Definition 312 *Let V be a vector space over the field $\mathbf{C}$ of complex numbers, and suppose $\langle u,v\rangle$ is a unique complex number for each $u,v\in V$. We say that $\langle\cdot,\cdot\rangle$ is an inner product on V if it satisfies the following properties:*

1. *$\langle u,v\rangle = \overline{\langle v,u\rangle}$ for all $u,v\in V$;*

2. *$\langle\alpha u+\beta v,w\rangle = \alpha\langle u,w\rangle + \beta\langle v,w\rangle$ for all $\alpha,\beta\in\mathbf{C}$ and all $u,v,w\in V$;*

3. *$\langle u,u\rangle \geq 0$ for all $u\in V$ and $\langle u,u\rangle = 0$ if and only if $u=0$.*

The reader should notice that the third property means, in particular, that $\langle u, u\rangle \in \mathbf{R}$ for all $u \in V$. If $u \neq v$, then $\langle u, v\rangle$ can be complex (and not real).

We say that a complex inner product is a *Hermitian* form because it satisfies $\langle u, v\rangle = \overline{\langle v, u\rangle}$. Also, putting together the first two properties of a complex inner product, we obtain

$$\langle u, \alpha v + \beta w\rangle = \overline{\alpha}\,\langle u, v\rangle + \overline{\beta}\,\langle u, w\rangle \text{ for all } \alpha, \beta \in \mathbf{C}, u, v, w \in V.$$

A form $\langle \cdot, \cdot\rangle$ is called *sesquilinear* if it satisfies

$$\langle \alpha u + \beta v, w\rangle = \alpha\,\langle u, w\rangle + \beta\,\langle v, w\rangle \text{ for all } \alpha, \beta \in \mathbf{C}, u, v, w \in V,$$
$$\langle u, \alpha v + \beta w\rangle = \overline{\alpha}\,\langle u, v\rangle + \overline{\beta}\,\langle u, w\rangle \text{ for all } \alpha, \beta \in \mathbf{C}, u, v, w \in V.$$

Therefore, a complex inner product can be described as a *positive definite Hermitian sesquilinear form.*

To show that a complex inner product defines a norm as before, we must verify that the Cauchy-Schwarz inequality holds.

Theorem 313 *Let V be a vector space over $\mathbf{C}$, and let $\langle \cdot, \cdot\rangle$ be an inner product on V. Then*

$$|\langle u, v\rangle| \leq [\langle u, u\rangle]^{1/2}\,[\langle v, v\rangle]^{1/2} \text{ for all } u, v \in V. \tag{6.29}$$

Proof Suppose $u, v \in V$. If $v = 0$, then (6.29) holds because both sides are zero. So we assume that $v \neq 0$, which implies that $\langle v, v\rangle \neq 0$. We then define

$$\lambda = \frac{\langle u, v\rangle}{\langle v, v\rangle}$$

and consider the inequality

$$0 \leq \langle u - \lambda v, u - \lambda v\rangle\,.$$

Expanding the right hand side and simplifying yields

$$0 \leq \langle u, u\rangle - \frac{|\,\langle u, v\rangle\,|^2}{\langle v, v\rangle}$$

or

$$|\,\langle u, v\rangle\,|^2 \leq \langle u, u\rangle\,\langle v, v\rangle\,.$$

Taking the square root of both sides yields the desired result.

QED

We can now define a norm on V as we did in the case of a real inner product:

$$\|v\| = \sqrt{\langle v, v\rangle} \text{ for all } v \in V.$$

The proof that this really is a norm is the same as before, and will be omitted. The reader will recall that the Cauchy-Schwarz inequality is used to show that the triangle inequality holds.

We will refer to a vector space V over $\mathbf{C}$, together with an inner product on V, as a *complex inner product* space.

6.7.1 Examples of complex inner product spaces

Complex Euclidean n-space

We have already encountered the complex version of $\mathbf{R}^n$:

$$\mathbf{C}^n = \{(z_1, z_2, \ldots, z_n) \,:\, z_1, z_2, \ldots, z_n \in \mathbf{C}\}.$$

The *Hermitian dot product* is a generalization of the ordinary dot product:

$$\langle u, v \rangle_{\mathbf{C}^n} = \sum_{i=1}^{n} u_i \overline{v_i} \text{ for all } u, v \in \mathbf{C}^n.$$

It is straightforward to verify that $\langle \cdot, \cdot \rangle_{\mathbf{C}^n}$ satisfies the definition of a complex inner product. To verify the third property, positive definiteness, we use the fact that, for any complex number $\gamma = \alpha + i\beta$,

$$\gamma\overline{\gamma} = (\alpha + i\beta)(\alpha - i\beta) = \alpha^2 + \beta^2 = |\gamma|^2.$$

Using this, we have

$$\langle u, u \rangle_{\mathbf{C}^n} = \sum_{i=1}^{n} u_i \overline{u_i} = \sum_{i=1}^{n} |u_i|^2 \geq 0 \text{ for all } u \in \mathbf{C}^n,$$

with $\langle u, u \rangle_{\mathbf{C}^n} = 0$ only if every component of u is zero.

Complex $L^2(a, b)$

If $[a, b]$ is an interval of real numbers and $f : [a, b] \to \mathbf{C}$, $g : [a, b] \to \mathbf{C}$ are complex-valued functions defined on $[a, b]$, then the complex L^2 inner product of f and g is

$$\langle f, g \rangle_{L^2(a,b)} = \int_a^b f(x)\overline{g(x)}\, dx.$$

The complex L^2 norm is then

$$\|f\|_{L^2(a,b)} = \left[\int_a^b f(x)\overline{f(x)}\, dx\right]^{1/2} = \left[\int_a^b |f(x)|^2\, dx\right]^{1/2}.$$

The space complex $L^2(a, b)$ consists of all complex-valued functions f defined on the interval $[a, b]$ and satisfying

$$\int_a^b |f(x)|^2\, dx < \infty.$$

Certainly all continuous functions defined on $[a, b]$ belong to complex $L^2(a, b)$, as do many discontinuous functions. However, as in the case of real $L^2(a, b)$, a technically correct description of the functions that belong to $L^2(a, b)$ is beyond the scope of this book.

6.7.2 Orthogonality in complex inner product spaces

We define orthogonality as before.

Definition 314 *Let V be a complex inner product space, and let $u, v \in V$. We say that u and v are* orthogonal *if and only if $\langle u, v\rangle = 0$. We call a subset $\{u_1, u_2, \ldots, u_n\}$ of V an* orthogonal set *if and only if each u_j is nonzero and $\langle u_j, u_k\rangle = 0$ for $1 \leq j, k \leq n$, $k \neq j$. Finally, $\{u_1, u_2, \ldots, u_n\} \subset V$ is an* orthonormal set *if and only if it is orthogonal and $\|u_j\| = 1$ for $j = 1, 2, \ldots, n$.*

In some ways, orthogonality in complex spaces is less straightforward than in real spaces. The Pythagorean theorem holds in only one direction.

Theorem 315 *Let V be a complex inner product space, and suppose $u, v \in V$ satisfy $\langle u, v\rangle = 0$. Then*

$$\|u + v\|^2 = \|u\|^2 + \|v\|^2,$$
$$\|u - v\|^2 = \|u\|^2 + \|v\|^2.$$

Proof Exercise 8.

The converse of Theorem 9 is false (see Exercise 9).

The Gram matrix of a basis for a complex inner product space V is unchanged from the real case. If $\{u_1, u_2, \ldots, u_n\}$ is a basis for V, then the Gram matrix $G \in \mathbf{C}^{n\times n}$ is defined by

$$G_{ij} = \langle u_j, u_i\rangle, \ i, j = 1, 2, \ldots, n.$$

The reader should notice that, although $G_{ij} \in \mathbf{C}$ in general, the diagonal entries of G are real: $G_{ii} = \|u_i\|^2$, $i = 1, 2, \ldots, n$. The proof that a Gram matrix is nonsingular is essentially unchanged from the real case.

The projection theorem also extends to complex inner product spaces, although the proof is a little more involved.

Theorem 316 *Let V be an inner product space over $\mathbf{C}$, and let S be a finite-dimensional subspace of V.*

1. *For any $v \in V$, there is a unique $w \in S$ satisfying*

$$\|v - w\| = \min\{\|v - z\| \, : \, z \in S\}.$$

2. *A vector $w \in S$ is the best approximation to v from S if and only if*

$$\langle v - w, z\rangle = 0 \text{ for all } z \in S.$$

3. *If $\{u_1, u_2, \ldots, u_n\}$ is a basis for S, then*

$$\text{proj}_S v = \sum_{i=1}^{n} x_i u_i,$$

where $x = (x_1, x_2, \ldots, x_n) \in \mathbf{R}^n$ *is the unique solution to the matrix-vector equation*

$$Gx = b.$$

The matrix $G \in \mathbf{R}^{n \times n}$ *is the Gram matrix for the basis* $\{u_1, u_2, \ldots, u_n\}$ *and the vector* $b \in \mathbf{R}^n$ *is defined as follows:*

$$b_i = \langle v, u_i \rangle, \ i = 1, 2, \ldots, n. \tag{6.30}$$

Proof Exercise 10

Example 317 *Let* S *be the subspace of* $\mathbf{C}^3$ *defined by* $S = \text{sp}\{v_1, v_2\}$, *where*

$$v_1 = (1 + i, 1 - i, 1), \ v_2 = (2i, 1, i),$$

and let $u = (1, 1, 1)$. *We wish to find the best approximation to* u *from* S. *The solution is* $y = x_1 v_1 + x_2 v_2$, *where* $Gx = b$ *and*

$$\begin{aligned} G &= \begin{bmatrix} \langle v_1, v_1 \rangle_{\mathbf{C}^3} & \langle v_2, v_1 \rangle_{\mathbf{C}^3} \\ \langle v_1, v_2 \rangle_{\mathbf{C}^3} & \langle v_2, v_2 \rangle_{\mathbf{C}^3} \end{bmatrix} = \begin{bmatrix} 5 & 3 + 4i \\ 3 - 4i & 6 \end{bmatrix}, \\ b &= \begin{bmatrix} \langle u, v_1 \rangle_{\mathbf{C}^3} \\ \langle u, v_2 \rangle_{\mathbf{C}^3} \end{bmatrix} = \begin{bmatrix} 3 \\ 1 - 3i \end{bmatrix}. \end{aligned}$$

The solution of $Gx = b$ *is* $x = (3/5 + i, -4/5 - 3/5i)$, *and the best approximation is*

$$y = x_1 v_1 + x_2 v_2 = \left(\frac{4}{5}, \frac{4}{5} - \frac{1}{5} i, \frac{6}{5} + \frac{1}{5} i \right).$$

6.7.3 The adjoint of a linear operator

The concept of adjoint can be extended to complex inner product spaces.

Theorem 318 *Let* V *and* W *be finite-dimensional inner product spaces over* $\mathbf{C}$, *and let* $L : V \to W$ *be linear. Then there exists a unique linear operator* $L^* : W \to V$ *such that*

$$\langle L(v), w \rangle_W = \langle v, L^*(w) \rangle_V \text{ for all } v \in V, w \in W. \tag{6.31}$$

Proof Exercise 11

We now consider $L : \mathbf{C}^n \to \mathbf{C}^m$ defined by $L(x) = Ax$, where $A \in \mathbf{C}^{m\times n}$. For any $x \in \mathbf{C}^n$, $y \in \mathbf{C}^m$,

$$\begin{aligned}
\langle L(x), y\rangle_{\mathbf{C}^m} = \sum_{i=1}^{m}(L(x))_i\overline{y_i} = \sum_{i=1}^{m}(Ax)_i\overline{y_i} &= \sum_{i=1}^{m}\sum_{j=1}^{n} A_{ij}x_j\overline{y_i} \\
&= \sum_{j=1}^{n}\sum_{i=1}^{m} A_{ij}x_j\overline{y_i} \\
&= \sum_{j=1}^{n} x_j \sum_{i=1}^{m} A_{ij}\overline{y_i} \\
&= \sum_{j=1}^{n} x_j \overline{\sum_{i=1}^{m} \overline{A_{ij}} y_i} \\
&= \sum_{j=1}^{n} x_j \overline{\left(\overline{A}^T y\right)_j} \\
&= \left\langle x, \overline{A}^T y\right\rangle_{\mathbf{C}^n}.
\end{aligned}$$

In the above calculation we used the following properties of the complex conjugate:

- $\overline{z_1 + z_2 + \ldots + z_k} = \overline{z_1} + \overline{z_2} + \ldots + \overline{z_k}$ (the conjugate of a sum is the sum of the conjugates);
- $\overline{zw} = \overline{z}\,\overline{w}$ (the conjugate of a product is the product of the conjugates);
- $\overline{\overline{z}} = z$ (the conjugate of a conjugate is the original number).

We now see that, if L is defined by $L(x) = Ax$, then

$$\langle L(x), y\rangle_{\mathbf{C}^m} = \left\langle x, \overline{A}^T y\right\rangle \text{ for all } x \in \mathbf{C}^n, y \in \mathbf{C}^m.$$

It follows that L^* is the operator defined by the matrix $\overline{A}^T$:

$$L^*(y) = \overline{A}^T y \text{ for all } y \in \mathbf{C}^m.$$

We will write $A^* = \overline{A}^T$ and refer to A^* as the *conjugate transpose* of A.

Symmetric and Hermitian matrices

A matrix $A \in \mathbf{R}^{n\times n}$ is called *symmetric* if and only if $A^T = A$, while A belonging to $\mathbf{C}^{n\times n}$ is called *Hermitian* if $A^* = A$. Equivalently, $A \in \mathbf{R}^{n\times n}$ is symmetric if and only if

$$A_{ji} = A_{ij}\ i, j = 1, 2, \ldots, n,$$

and $A \in \mathbf{C}^{n\times n}$ is Hermitian if and only if

$$A_{ji} = \overline{A_{ij}}\ i,j = 1,2,\ldots,n.$$

If $A \in \mathbf{C}^{n\times n}$ is Hermitian, we see, in particular, that

$$A_{ii} = \overline{A_{ii}},\ i = 1,2,\ldots,n,$$

which implies that A_{ii} is real for all i. Thus, while the off-diagonal entries of a Hermitian matrix can be complex, the diagonal entries must be real.

We will see later that symmetric and Hermitian matrices have many special properties. Here is a simple property of Hermitian matrices.

Theorem 319 *Let $A \in \mathbf{C}^{n\times n}$ be Hermitian. Then*

$$\langle Ax, x\rangle_{\mathbf{C}^n} \in \mathbf{R} \text{ for all } x \in \mathbf{C}^n.$$

Proof For any $x \in \mathbf{C}^n$, we have

$$\langle Ax, x\rangle_{\mathbf{C}^n} = \langle x, A^*x\rangle_{\mathbf{C}^n} = \langle x, Ax\rangle_{\mathbf{C}^n}$$

since A is Hermitian. On the other hand, using the properties of a complex inner product, we have

$$\langle Ax, x\rangle_{\mathbf{C}^n} = \overline{\langle x, Ax\rangle_{\mathbf{C}^n}},$$

and hence

$$\langle Ax, x\rangle_{\mathbf{C}^n} = \overline{\langle Ax, x\rangle_{\mathbf{C}^n}}.$$

But a complex number equals its conjugate if and only if it is real.

QED

Exercises

Miscellaneous exercises

1. Let S be the one-dimensional subspace of $\mathbf{C}^3$ spanned by the vector $u = (1-i, 1+i, 1+2i)$. Find the projection of $v = (1, i, 1+i)$ onto S.

2. Find an orthogonal basis for $\mathbf{C}^3$ by applying the Gram-Schmidt process to the basis

$$\{(-i, 1+4i, 0), (-1+4i, -1-1i, i), (2-5i, 3+2i, 1+2i)\}.$$

3. Let

$$S = \left\{e^{-ik\pi x}, e^{-i(k-1)\pi x}, \ldots, e^{-i\pi x}, 1, e^{i\pi x}, \ldots, e^{ik\pi x}\right\}.$$

Prove that S is orthogonal under the (complex) $L^2(-1,1)$ inner product.

4. Let
$$S = \mathrm{sp}\left\{e^{-2\pi ix}, e^{-\pi ix}, 1, e^{\pi ix}, e^{2\pi ix}\right\},$$
regarded as a subspace of complex $L^2(-1,1)$. Find the best approximation to $f(x) = x$ from S.

5. Regard complex $\mathcal{P}_2$ as a subspace of complex $L^2(0,1)$. Find the best approximation to $f(x) = e^{i\pi x}$ from $\mathcal{P}_2$.

6. Let $A \in \mathbf{C}^{3\times 2}$, $b \in \mathbf{C}^3$ be defined by
$$A = \begin{bmatrix} 1 & -2-i \\ 3-i & -5-i \\ 0 & 1-i \end{bmatrix}, \; b = \begin{bmatrix} -2 \\ 2i \\ 1+i \end{bmatrix}.$$
Find the least-squares solution to $Ax = b$.

7. Let $A \in \mathbf{C}^{4\times 2}$, $b \in \mathbf{C}^4$ be defined by
$$A = \begin{bmatrix} 1-i & -4+i \\ 1-i & -3+i \\ -1+i & 4-2i \\ 1-i & -2+i \end{bmatrix}, \; b = \begin{bmatrix} 1 \\ -1 \\ -1+i \\ -2 \end{bmatrix}.$$
Find the least-squares solution to $Ax = b$.

8. Prove Theorem 315.

9. (a) Find two vectors u and v in $\mathbf{C}^2$ such $\|u+v\|^2 = \|u\|^2 + \|v\|^2$ but yet $\langle u, v\rangle_{\mathbf{C}^2} \neq 0$.

   (b) Suppose V is a complex inner product space. What *can* be concluded about $\langle u, v\rangle_{\mathbf{C}^2}$ if $\|u+v\|^2 = \|u\|^2 + \|v\|^2$?

10. Prove Theorem 316. (Hint: Begin by proving the second part, as in the real case. If you mimic the earlier proof, you should conclude that $w \in S$ is the best approximation to v from S if and only if $\mathrm{Re}\,\langle v-w, z\rangle = 0$ for all $z \in S$, where $\mathrm{Re}\gamma$ denotes the real part of the complex number γ (that is, if $\gamma = \alpha + i\beta$, then $\mathrm{Re}\gamma = \alpha$). Then replace z by iz and show that $\mathrm{Im}\,\langle v-w, z\rangle = 0$ for all $z \in S$ also holds. Once the second part of the theorem has been proved, the rest of it follows much as in the real case.)

11. Prove Theorem 318.

12. Let X be a finite-dimensional inner product space over $\mathbf{C}$, and let $f : X \to \mathbf{C}$ be linear. Prove that there exists a unique $u \in X$ such that
$$f(x) = \langle x, u\rangle \text{ for all } x \in X.$$

13. Use Theorem 319 to prove that all the eigenvalues of a Hermitian matrix are real. (Hint: Let $A \in \mathbf{C}^{n\times n}$ be Hermitian, and let $\lambda \in \mathbf{C}$, $x \in \mathbf{C}^n$ satisfy $Ax = \lambda x$, $x \neq 0$. By Theorem 319, $\langle Ax, x\rangle_{\mathbf{C}^n}$ is a real number. Use the fact that $Ax = \lambda x$ and simplify.)

6.8 More on polynomial approximation

The polynomial approximation problem is an interesting application of linear algebra, and in this section we explore it further. We are given a continuous function f and we wish to find the best approximating polynomial $p \in \mathcal{P}_n$ on a given interval $[a,b]$. In the previous sections, it was assumed that the interval was $[0,1]$, but nothing really changes for a different interval except that the calculations are more convenient on some intervals than others.

The previous paragraph does not completely describe the polynomial approximation problem. There remains the choice of norm that defines the best approximation. We have thus far encountered a whole family of norms for functions defined on $[a,b]$, namely, the L^p norms:

$$\|f\|_p = \left[\int_a^b f(x)^p \, dx\right]^{1/p}, \quad 1 \le p < \infty,$$
$$\|f\|_\infty = \max\{|f(x)| \, : \, x \in [a,b]\}.$$

Only one of these norms is defined by an inner product, namely, the L^2 norm, and so the theory derived in the previous sections allows us to solve the problem only in that case (or in the case of some other inner product that we have not encountered yet).

If we solve the best polynomial approximation in the L^2 case, then we can use the projection theorem, and the only remaining question is the choice of basis for $\mathcal{P}_n$. We saw in Example 296 that the standard basis is problematic for numerical reasons—round-off error can ruin the calculation—and so it is preferable to use an orthogonal basis, which can be found by the Gram-Schmidt process. It is also more efficient to use an orthogonal basis, at least if multiple functions are to be approximated and the cost of finding the orthogonal basis (once) can be amortized over several calculations.

One reason for using the L^2 norm to define the best approximation is the convenience of using a norm derived from an inner product. We will see below that solving the best approximation problem in a norm that is not defined by an inner product is more complicated. There is also a statistical reason for preferring the L^2 norm which is beyond the scope of this book. However, in many applications it might be more natural to ask for the best approximation in the L^∞ norm. The L^2 best approximation minimizes the average square error, but the L^∞ best approximation minimizes the largest pointwise error.

The following theorem characterizes the solution to the best polynomial approximation problem posed with respect to the L^∞ norm.

Theorem 320 *Let $f \in C[a,b]$. Then $p \in \mathcal{P}_n$ satisfies*

$$\|f - p\|_\infty = \min\{\|f - q\|_\infty \, : \, q \in \mathcal{P}_n\}$$

if and only if there exist points $x_0, x_1, \ldots, x_{n+1} \in [a, b]$ *such that*

$$|f(x_i) - p(x_i)| = \|f - p\|_\infty, \; i = 0, 1, \ldots, n+1$$

and the sign of $f(x_i) - p(x_i)$ *alternates as* i *increases from* $i = 0$ *to* $i = n$.

The proof of this theorem is beyond the scope of this book.

Example 321 *Let* $f : [-1, 1] \to \mathbf{R}$ *be defined by*

$$f(x) = x \sin(\pi x + 1),$$

(see Figure 6.9) and consider approximating f *by a sixth degree polynomial.*

Let p_∞ *be the best approximation to* f *from* $\mathcal{P}_6$ *in the* L^∞ *norm. Using an algorithm*[8] *from [26], we find*

$$p_\infty(x) = 0.003301 + 0.7585x + 1.610x^2 - 3.444x^3 - 2.405x^4 + 1.856x^5 + 0.7919x^6.$$

The error $f - p_\infty$ *is shown in Figure 6.10, where it can be seen that* p_∞ *satisfies the condition described in Theorem 320.*

We can compare p_∞ *with* p_2*, the best approximation to* f *in the* L^2 *norm. The polynomial* p_2 *can be computed by applying the projection theorem, as in Example 296. The result is*

$$p_2(x) = 0.001311 + 0.7751x + 1.649x^2 - 3.532x^3 - 2.519x^4 + 1.942x^5 + 0.8725x^6.$$

The error $f - p_2$ *is graphed in Figure 6.11.*

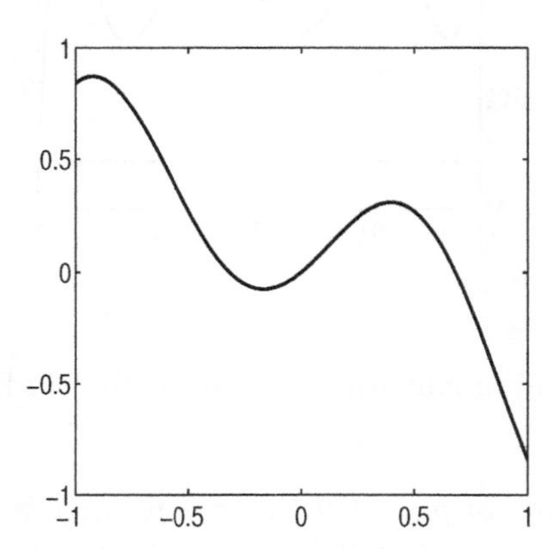

FIGURE 6.9
The function to be approximated in Example 321.

[8]The algorithm used for computing p_∞ can be described roughly as follows: Choose at least $n+2$ points $t_0, t_1, \ldots, t_{n+1}$ in $[-1, 1]$ and choose the polynomial $p_1 \in \mathcal{P}_n$ that minimizes the maximum error $|f(t_i) - p_1(t_i)|$, $i = 0, 1, \ldots, n+1$. Find the point t_{n+2} in $[-1, 1]$ that maximizes $|f(x) - p_1(x)|$ and find the polynomial $p_2 \in \mathcal{P}_n$ that minimizes the maximum error $|f(t_i) - p_2(t_i)|$, $i = 0, 1, \ldots, n+2$. Continue this process to produce a sequence of polynomials $p_1, p_2, p_3, \ldots$. It can be shown that $p_k \to p_\infty$ as $k \to \infty$.

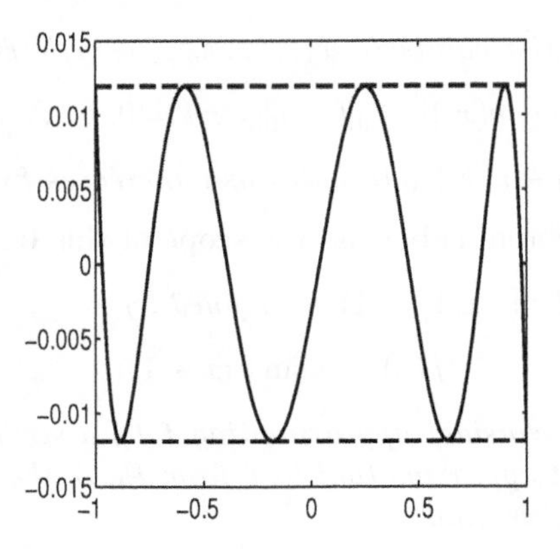

FIGURE 6.10
The error in the best approximation (L^∞ norm) to f in Example 321.

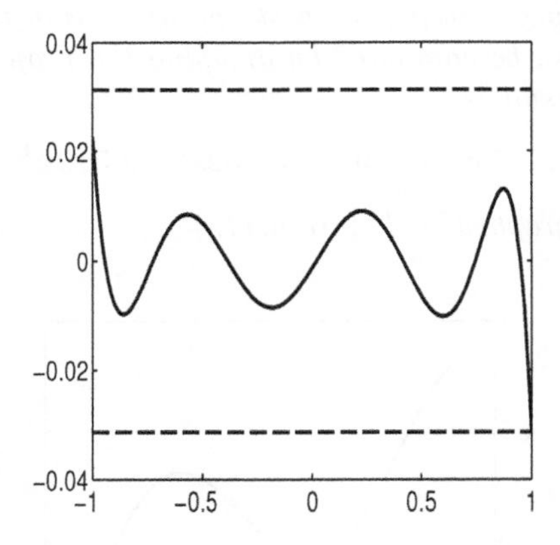

FIGURE 6.11
The error in the best approximation (L^2 norm) to f in Example 321.

The maximum error in p_∞ *(that is,* $\|f - p_\infty\|_\infty$*) is about* 0.01194*, while the maximum error in* p_2 *is about* 0.03146*. Thus we see that* p_2*, while optimal with regard to the* L^2 *norm, is not nearly as good as* p_∞ *when the error is measured in the* L^∞ *norm.*

As the previous example shows, it is possible to compute the best polynomial approximation, in the L^∞ norm, to a given continuous function. However, computing this approximation is much more difficult than applying the projection theorem to compute the best approximation in a norm defined by an inner product.

6.8.1 A weighted L^2 inner product

Figure 6.11 shows that, for the given example, the largest errors in the best L^2 polynomial approximation occur near the endpoints of the interval. This is the typical result for L^2 polynomial approximation. By choosing a different inner product, we can reduce the error near the endpoints and produce an approximation that is much closer to the best approximation in the L^∞ norm.

A *weighted* L^2 inner product on $[-1,1]$ takes the form

$$\langle f, g\rangle_w = \int_{-1}^{1} w(x) f(x) g(x)\, dx,$$

where w is called the weight function. We will not try to describe the exact properties of w that make $\langle \cdot, \cdot \rangle_w$ a valid inner product; instead we define the particular weighted L^2 inner product that is relevant for this discussion:

$$\langle f, g\rangle_c = \int_{-1}^{1} \frac{f(x)g(x)}{\sqrt{1-x^2}}\, dx. \tag{6.32}$$

In this case, the weight function is

$$c(x) = \frac{1}{\sqrt{1-x^2}}. \tag{6.33}$$

It is not clear that $\langle f, g\rangle_c$ is well-defined for all $f, g \in C[-1,1]$, much less for functions that are not continuous, since $c(x) \to \infty$ as $x \to \pm 1$. However, a change of variables shows that, in fact, $\langle f, g\rangle_c$ *is* well-defined for all continuous f, g. We let $x = \cos(\theta)$, $0 \le \theta \le \pi$; then $\sqrt{1-x^2} = \sin(\theta)$ and

$$\langle f, g\rangle_c = \int_{-1}^{1} \frac{f(x)g(x)}{\sqrt{1-x^2}}\, dx = \int_{0}^{\pi} f(\cos(\theta)) g(\cos(\theta))\, d\theta.$$

To solve a polynomial approximation problem with respect to $\langle \cdot, \cdot \rangle_c$, it is helpful to have an orthogonal basis for $\mathcal{P}_n$. We could derive such a basis using the Gram-Schmidt process, but in this case there is a simple formula defining an orthogonal basis. We define the *Chebyshev polynomials* by

$$T_n(x) = \cos(n \arccos(x)),\ n = 0, 1, 2, \ldots. \tag{6.34}$$

At first glance, this formula does not seem to define polynomials at all. However,

$$T_0(x) = \cos(0) = 1,\ T_1(x) = \cos(\arccos(x)) = x,$$

so at least the first two functions are polynomials. Moreover, using the addition and subtraction identities for cosine,

$$\begin{aligned}
&\cos((n+1)\theta) + \cos((n-1)\theta) \\
=\ &\cos(n\theta)\cos(\theta) - \sin(n\theta)\sin(\theta) + \cos(n\theta)\cos(\theta) + \sin(n\theta)\sin(\theta) \\
=\ &2\cos(n\theta)\cos(\theta).
\end{aligned}$$

Therefore,

$$\cos\left((n+1)\theta\right) = 2\cos\left(n\theta\right)\cos\left(\theta\right) - \cos\left((n-1)\theta\right),$$

and setting $\theta = \arccos(x)$, we obtain

$$\begin{aligned} & \cos\left((n+1)\arccos(x)\right) \\ = \; & 2\cos\left(n\arccos(x)\right)\cos\left(\arccos(x)\right) - \cos\left((n-1)\arccos(x)\right), \end{aligned}$$

or

$$T_{n+1}(x) = 2xT_n(x) - T_{n-1}(x).$$

Since T_0 and T_1 are polynomials, this *recurrence relation* shows that T_n is a polynomial for each n. For example,

$$\begin{aligned} T_2(x) &= 2xT_1(x) - T_0(x) = 2x^2 - 1, \\ T_3(x) &= 2xT_2(x) - T_1(x) = 2x(2x^2-1) - x = 4x^3 - 3x, \\ T_4(x) &= 2xT_3(x) - T_2(x) = 2x(4x^3-3x) - (2x^2-1) = 8x^4 - 8x^2 + 1, \end{aligned}$$

and so forth.

Not only is T_n a polynomial for each n, but it is easy to show that these polynomials form an orthogonal set:

$$\begin{aligned} \langle T_n, T_m \rangle_c = \int_{-1}^{1} \frac{T_n(x)T_m(x)}{\sqrt{1-x^2}}\,dx &= \int_0^{\pi} T_n(\cos(\theta))T_m(\cos(\theta))\,d\theta \\ &= \int_0^{\pi} \cos(n\theta)\cos(m\theta)\,d\theta. \end{aligned}$$

Using the trigonometric identity

$$\cos(\alpha)\cos(\beta) = \frac{1}{2}\left(\cos(\alpha+\beta) + \cos(\alpha-\beta)\right),$$

we see that the last integral is zero for $n \neq m$, and so

$$\langle T_n, T_m \rangle_c = 0 \text{ for all } n \neq m.$$

We therefore conclude that $\{T_0, T_1, \ldots, T_n\}$ is an orthogonal basis for $\mathcal{P}_n$.

Example 322 *This is a continuation of Example 321. We compute p_c, the best approximation, in the norm defined by $\langle\cdot,\cdot\rangle_c$, to $f(x) = x\sin(\pi x + 1)$ from $\mathcal{P}_6$. The computation uses the orthogonal basis $\{T_0, T_1, T_2, T_3, T_4, T_5, T_6\}$ described above:*

$$p_c = \sum_{i=0}^{6} \frac{\langle f, T_i \rangle_c}{\langle T_i, T_i \rangle_c} T_i.$$

The result is

$$p_c(x) = 0.001848 + 0.7546x + 1.637x^2 - 3.428x^3 - 2.480x^4 + 1.843x^5 + 0.84237x^6.$$

The error $f - p_c$ is graphed in Figure 6.12. We see that $\|f - p_c\|_\infty \doteq 0.01296$, which is not much more than $\|f - p_\infty\|_\infty$. This is generally true: The least-squares polynomial approximation, defined by $\langle\cdot,\cdot\rangle_c$, to a continuous function f is similar to the best approximation defined by $\|\cdot\|_\infty$, and it is much easier to compute.

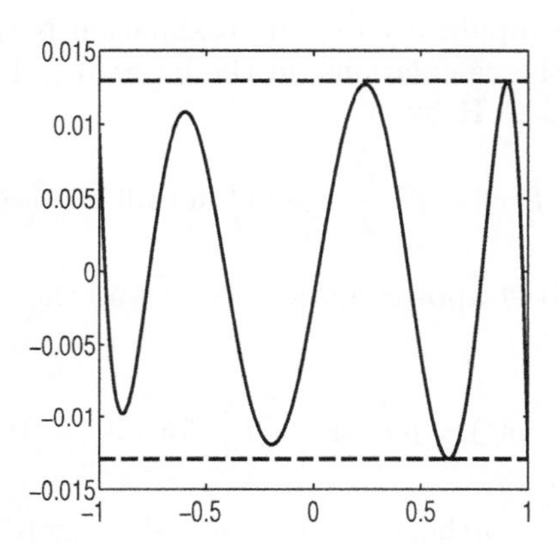

FIGURE 6.12
The error in the best approximation, in the norm defined by $\langle\cdot,\cdot,\rangle_c$, to f in Example 322.

In this section, we have used the interval $[-1,1]$ for convenience, since the Chebyshev polynomials are naturally defined on that interval. We can solve a polynomial approximation problem on another interval $[a,b]$ by utilizing a simple linear change of variables:

$$\begin{aligned} t \in [a,b] \quad &\longleftrightarrow \quad x \in [-1,1], \\ t = a + \frac{b-a}{2}(x+1) \quad &\longleftrightarrow \quad x = -1 + \frac{2}{b-a}(t-a). \end{aligned}$$

The idea is to transform the problem to the interval $[-1,1]$, solve the transformed problem as above, and then transform the answer back to the interval $[a,b]$. The details are left to the exercises.

Exercises

1. Let $f : [-1,1] \to \mathbf{R}$ be defined by $f(t) = e^t$.

   (a) Compute the best quadratic approximation to f in the (unweighted) $L^2(-1,1)$ norm.

   (b) Compute the best quadratic approximation to f in the weighted $L^2(-1,1)$ norm defined by the weight function (6.33).

Which approximation is better when the error is measured in the L^∞ norm?

2. Repeat the previous exercise, but compute cubic approximations.

3. Let $f : [0, \pi] \to \mathbf{R}$ be defined by $f(t) = \sin(t)$. The purpose of this exercise is to compute a cubic approximation to f on $[0, \pi]$. We will perform all of the calculations on the interval $[-1, 1]$. To this end, we define $\tilde{f} : [-1, 1] \to \mathbf{R}$ by

$$\tilde{f}(x) = f\left(\frac{\pi}{2}(x+1)\right) \text{ for all } x \in [-1, 1].$$

We compute a best approximation $\tilde{p}$ to $\tilde{f}$ and then the desired approximation to f is

$$p(t) = \tilde{p}\left(-1 + \frac{2}{\pi}t\right) \text{ for all } t \in [0, \pi].$$

   (a) Compute an orthogonal basis for $\mathcal{P}_3$, regarded as a subspace of $C[-1, 1]$, using the (unweighted) L^2 inner product. Using this basis, compute the best approximation $\tilde{p}$ from $\mathcal{P}_3$ to $\tilde{f}$. Transform $\tilde{p}$ to the interval $[0, \pi]$ and call the result p.

   (b) Compute the best approximation $\tilde{q}$ from $\mathcal{P}_3$ to $\tilde{f}$ with respect to the norm defined by $\langle \cdot, \cdot \rangle_c$. Transform $\tilde{q}$ to $[0, \pi]$ and call the result q.

   (c) Graph the errors $f - p$ and $f - q$ on $[0, \pi]$. Which approximation is better?

4. Let $f \in C[a, b]$ and define $\tilde{f} \in C[-1, 1]$ by

$$\tilde{f}(x) = f\left(a + \frac{b-a}{2}(x+1)\right).$$

Let $\tilde{p}$ be the best approximation, in the (unweighted) $L^2(-1, 1)$ norm, to $\tilde{f}$ from $\mathcal{P}_n$. Define $p \in \mathcal{P}_n$ by

$$p(t) = \tilde{p}\left(-1 + \frac{2}{b-a}(t-a)\right).$$

Prove that p is the best approximation to f, in the $L^2(a, b)$ norm, from $\mathcal{P}_n$.

5. Let V be the set of functions $f \in C[0, \infty)$ satisfying

$$\int_0^\infty f(x)^2 e^{-x}\, dx < \infty.$$

It can be proved that V is a subspace of $C[0, \infty)$. Define an inner product on $C[0, \infty)$ by

$$\langle f, g \rangle = \int_0^{\infty} f(x)g(x)e^{-x}\,dx.$$

(a) Find an orthogonal basis for $\mathcal{P}_2$ (regarded as a subspace of V) with respect to this inner product.

(b) Find the best approximation from $\mathcal{P}_2$, relative to the given inner product, to $f(x) = x^3$.

6.9 The energy inner product and Galerkin's method

An interesting application of orthogonality arises in solving differential equations by the finite element method (FEM). To introduce these ideas, we will study the following boundary value problem (BVP):

$$-\frac{d}{dx}\left(k(x)\frac{du}{dx}\right) = f(x),\ 0 < x < \ell, \tag{6.35a}$$

$$u(0) = 0, \tag{6.35b}$$

$$u(\ell) = 0. \tag{6.35c}$$

The conditions (6.35b–6.35c) are called *Dirichlet boundary conditions*, and (6.35a) is the differential equation. The functions k and f are given, with $k(x) \geq k_0 > 0$ for some constant k_0 and all $x \in (0, \ell)$, and $u = u(x)$ is to be found.

One approach to the BVP (6.35) is Galerkin's method, which is based on using integration by parts to transform the BVP to an equivalent *variational form*. We let $v : [0, \ell] \to \mathbf{R}$ be a *test function* satisfying $v(0) = v(\ell) = 0$. The variational form of the BVP is obtained by multiplying both sides of the differential equation by a test function, integrating both sides over the interval $[0, \ell]$, and applying integration by parts:

$$
\begin{aligned}
& -\frac{d}{dx}\left(k(x)\frac{du}{dx}(x)\right) = f(x),\ 0 < x < \ell \\
\Rightarrow\ & -\frac{d}{dx}\left(k(x)\frac{du}{dx}(x)\right)v(x) = f(x)v(x),\ 0 < x < \ell \\
\Rightarrow\ & -\int_0^{\ell}\frac{d}{dx}\left(k(x)\frac{du}{dx}(x)\right)v(x)\,dx = \int_0^{\ell} f(x)v(x)\,dx \\
\Rightarrow\ & -k(x)\frac{du}{dx}(x)v(x)\Big|_{x=0}^{\ell} + \int_0^{\ell} k(x)\frac{du}{dx}(x)\frac{dv}{dx}(x)\,dx = \int_0^{\ell} f(x)v(x)\,dx \\
\Rightarrow\ & \int_0^{\ell} k(x)\frac{du}{dx}(x)\frac{dv}{dx}(x)\,dx = \int_0^{\ell} f(x)v(x)\,dx.
\end{aligned}
$$

The reader should notice how the boundary conditions $v(0) = v(\ell) = 0$ were used to eliminate the boundary term arising from integration by parts. The final equation holds for all valid test functions; the set of all test functions will be denoted by V. We have now obtained the variational form of the BVP:

$$u \in V, \int_0^\ell k(x)\frac{du}{dx}(x)\frac{dv}{dx}(x)\,dx = \int_0^\ell f(x)v(x)\,dx \text{ for all } v \in V. \tag{6.36}$$

It can be shown that (6.36) is equivalent to (6.35): If u satisfies (6.35), then it also satisfies (6.36) (this is essentially proved by our derivation). Conversely, if u satisfies (6.36) and u is sufficiently smooth, then u also satisfies (6.35). We will not define the space of test functions precisely; for details, the reader is referred to the author's text [14] on finite element methods. For our purposes, it is sufficient to notice that v need only have one derivative, which need not be continuous but only regular enough that the necessary integrals are defined.

The variational form (6.36) is no easier to solve directly than the original BVP (6.35). However, Galerkin's method gives a way to compute approximate solutions. The difficulty with (6.36) is that the space V is infinite-dimensional. In Galerkin's method, we choose a finite-dimensional subspace V_n of V, say

$$V_n = \text{sp}\{\phi_1, \phi_2, \ldots, \phi_n\}.$$

We then replace (6.36) with

$$w_n \in V_n, \int_0^\ell k(x)\frac{dw_n}{dx}(x)\frac{dv}{dx}(x)\,dx = \int_0^\ell f(x)v(x)\,dx \text{ for all } v \in V_n. \tag{6.37}$$

To simplify the notation, we define the bilinear form

$$a(u, v) = \int_0^\ell k(x)\frac{du}{dx}(x)\frac{dv}{dx}(x)\,dx$$

on V, and we also notice that $\int_0^\ell f(x)v(x)\,dx$ is just the $L^2(0, \ell)$ inner product of f and v. It can be shown that $a(\cdot, \cdot)$ defines an inner product, called the *energy inner product*, on V (see Exercise 3). The variational problem (6.36) now takes the form

$$a(u, v) = \langle f, v\rangle_{L^2(0,\ell)} \text{ for all } v \in V, \tag{6.38}$$

while the Galerkin formulation is

$$a(w_n, v) = \langle f, v\rangle_{L^2(0,\ell)} \text{ for all } v \in V_n. \tag{6.39}$$

We will now assume that $u \in V$ solves (6.38) and $w_n \in V_n$ solves (6.39). Restricting v to elements of V_n and subtracting yields

$$a(u, v) - a(w_n, v) = 0 \text{ for all } v \in V_n$$

or

$$a(u - w_n, v) = 0 \text{ for all } v \in V_n. \tag{6.40}$$

Bearing in mind that $a(\cdot, \cdot)$ defines an inner product, (6.40) is precisely the orthogonality condition from the projection theorem (Theorem 289). Therefore, w_n is the best approximation from V_n to the exact solution u in the norm defined by the energy inner product.

This application of the projection theorem is different from the previous examples we have seen. In previous cases, we approximated a known vector (or function); here, we are able to find the best approximation to u even though it is unknown. The price we pay is that we have to use the energy norm, which is a norm defined by the problem itself. We will denote the energy norm by $\|\cdot\|_E$:

$$\|v\|_E = \int_0^\ell k(x) \left(\frac{du}{dx}\right)^2 dx.$$

We already know how to find the solution of (6.39); substituting

$$w_n = \sum_{j=1}^{n} U_j \phi_j$$

and requiring (6.39) to hold for $v = \phi_i$, $i = 1, 2, \ldots, n$ yields the normal equations

$$KU = F,$$

where

$$K_{ij} = a(\phi_j, \phi_i), \; i, j = 1, 2, \ldots, n$$

and

$$F_i = a(u, \phi_i) = \langle f, \phi_i \rangle_{L^2(0,\ell)}, \; i = 1, 2, \ldots, n.$$

In this application, the Gram matrix K is called the *stiffness matrix*[9] and the vector F is called the *load vector*.

For emphasis, we repeat that the computation of the best approximation of an unknown function is possible only because we work with the inner product defined by the variational equation. If we wished to compute the best approximation to u in the L^2 norm, we would have to compute $\langle u, \phi_i \rangle_{L^2(0,\ell)}$, which is not possible. Using the energy inner product, we need instead $a(u, \phi_i)$, which we know from the variational equation: $a(u, \phi_i) = \langle f, \phi_i \rangle_{L^2(0,\ell)}$.

To turn the Galerkin method into a practical numerical method, it remains to choose a family $\{V_n\}$ of finite-dimensional subspaces of V. Such a family of subset must have the following properties:

1. It is possible to approximate u arbitrarily well by taking n large enough: $\|u - w_n\|_E \to 0$ as $n \to \infty$.

[9]This terminology comes from applications in mechanics, in which the function $k(x)$ and the matrix K it defines represent the stiffness of an elastic material.

2. The stiffness matrix K and the load vector F are not too difficult to compute.

3. The normal equations $KU = F$ can be solved efficiently.

The finite element method is based on choosing spaces of *piecewise polynomials*, which were described in Section 2.9.

6.9.1 Piecewise polynomials

Here we will briefly describe the space of continuous piecewise linear functions; the reader is referred to Section 2.9 for details. We begin by establishing a *mesh* on the interval $[0, \ell]$. The mesh consists of the *elements* (subintervals)

$$[x_0, x_1], [x_1, x_2], \ldots, [x_{n-1}, x_n],$$

where $0 = x_0 < x_1 < \cdots < x_n = \ell$ are the *nodes.* For simplicity, we will assume the uniform mesh defined by $h = \ell/n$ and

$$x_i = ih, \ i = 0, 1, \ldots, n.$$

A piecewise linear function is a function $p : [0, \ell] \to \mathbf{R}$ with the property that p is defined by a first-degree polynomial on each element. Only continuous piecewise polynomials are regular enough to belong to the space V of test functions, and a continuous piecewise linear function p is determined entirely by its nodal values $p(x_0), p(x_1), \ldots, p(x_n)$.

The standard basis for the space of all continuous piecewise linear functions (relative to the mesh defined above) is $\{\phi_0, \phi_1, \ldots, \phi_n\}$, which is defined by

$$\phi_j(x_i) = \begin{cases} 1, & i = j, \\ 0, & i \neq j. \end{cases} \tag{6.41}$$

Each ϕ_i therefore has the property that it has value one at exactly one of the nodes and zero at all the rest. A related and crucial property is that ϕ_i is nonzero only on the interval (x_{i-1}, x_{i+1}); it is identically zero on the remainder of $[0, \ell]$. These functions are referred to as *nodal* basis functions; because of the characteristic shape of their graphs, shown in Figure 2.7, $\phi_1, \phi_2, \ldots, \phi_{n-1}$ are sometimes called the *hat functions.*

The defining property (6.41) leads to the conclusion that a continuous piecewise linear function satisfies

$$p(x) = \sum_{i=0}^{n} p(x_i)\phi_i(x).$$

Thus it is simple to represent any continuous piecewise linear function in terms of the nodal basis. For the purpose of solving BVP (6.35), we use the space of piecewise linear functions that satisfy the given boundary conditions:

$$V_n = \text{sp}\{\phi_1, \phi_2, \ldots, \phi_{n-1}\}.$$

Only ϕ_0 is nonzero at $x_0 = 0$, so excluding it from the basis means that every linear combination w of the remaining functions satisfies $w(0) = 0$. The analogous result holds for ϕ_n and $x_n = \ell$.

We now discuss why V_n is a good choice for the approximating subspace in Galerkin's method. Above we specified three properties that should be possessed by the family $\{V_n\}$.

The true solution u should be well-approximated by functions from V_n

We showed in Section 2.9.3 that the piecewise linear interpolant p of a smooth function u satisfies

$$\|u - p\|_{L^\infty(0,\ell)} \leq \frac{M}{8}h^2,$$

where M is a bound for $|u''(x)|$ on $[0, \ell]$. In the present context, we need a bound on the energy norm error in p. For this, we refer the reader to Section 0.4 of [3], where it is shown that

$$\|u - p\|_E \leq Ch, \tag{6.42}$$

where C is a constant depending on the size of the second derivatives of u. In the energy norm, the error in the derivative, rather than the function itself, is measured, and this accounts for an error that is proportional to h rather than h^2.

The reader may wonder why the previous paragraph refers to the interpolant of u rather than the finite element approximation w_n. This is because the analysis is simpler for the interpolant, and we are guaranteed that w_n is the best approximation to u in the energy norm. Therefore, (6.42) immediately implies that

$$\|u - w_n\|_E \leq Ch. \tag{6.43}$$

The condition (6.43) can be proved for many types of approximating subspaces V_n, not just for spaces of piecewise linear functions. It is the other advantages to piecewise linear functions that have led to their prominence in the finite element method.

K and F should be easy to compute

The entries in the stiffness matrix K and the load vector F are defined by integrals of combinations of the basis functions and their derivatives (along with the functions k and f from (6.35)). Since ϕ_i is linear and ϕ_i' is constant on each element, these integrals are about as easy to compute as they could possibly be.

The system $KU = F$ should be easy to solve

We now describe the crucial efficiency advantage in using piecewise linear functions. Since each ϕ_i is nonzero only on (x_{i-1}, x_{i+1}), the same is true for the derivative $d\phi_i/dx$ and so

$$K_{ij} = a(\phi_j, \phi_i) = \int_0^1 k(x)\frac{d\phi_j}{dx}(x)\frac{d\phi_i}{dx}(x)\,dx = 0 \text{ if } |i-j| > 1. \tag{6.44}$$

This follows because, if $|i-j| > 1$, then $[x_{i-1}, x_{i+1}]$ and $[x_{j-1}, x_{j+1}]$ intersect in at most a single point, so the integrand defining $a(\phi_j, \phi_i)$ is identically zero.

Equation (6.44) shows that the stiffness matrix K is *tridiagonal*, that is, every entry in K is zero except those on the diagonal and the first sub- and super-diagonals. This means that the $KU = F$ can be solved very efficiently. A tridiagonal matrix is one type of sparse matrix. A matrix is said to be *sparse* if most of its entries are zero, and *dense* if few or none of its entries are zero. When a matrix is known to be sparse, many operations on that matrix can be done very efficiently by avoiding unnecessary computations. (There is no need to multiply by or add an entry that is known to be zero; the result is already known.) Also, entries known to be zero need not be stored, so there is a savings in memory as well.

For boundary value problems in one dimension, like (6.35), the cost of solving $KU = F$ is probably not very large, even if K is dense. However, in the case of BVPs in two and three dimensions, the cost of storing and solving $KU = F$, when K is dense, can be prohibitive. The sparsity of K is the main reason for the popularity of piecewise polynomials and the finite element method.

Example 323 *We consider the simplest case of (6.35), where k is the constant function* 1. *We will also take $\ell = 1$ and $f(x) = \sin(\pi x)$. The exact solution to (6.35) in this case is*

$$u(x) = \frac{\sin(\pi x)}{\pi^2}.$$

On a uniform mesh with n elements, each element has length $h = 1/n$ and we can easily determine the formulas for the hat functions:

$$\phi_i(x) = \begin{cases} \frac{x-(i-1)h}{h}, & (i-1)h \le x \le ih, \\ -\frac{x-(i+1)h}{h}, & ih \le x \le (i+1)h, \\ 0, & \textit{otherwise}. \end{cases}$$

It then follows that

$$\frac{\phi_i}{dx}(x) = \begin{cases} \frac{1}{h}, & (i-1)h < x < ih, \\ -\frac{1}{h}, & ih < x < (i+1)h, \\ 0, & \textit{otherwise}. \end{cases}$$

We can now compute the stiffness matrix K:

$$
\begin{aligned}
K_{ii} &= \int_0^1 \left(\frac{d\phi_i}{dx}(x)\right)^2 dx = \int_{(i-1)h}^{(i+1)h} \frac{1}{h^2}\, dx = \frac{2}{h}, \\
K_{i+1,i} &= \int_0^1 \frac{d\phi_i}{dx}(x)\frac{d\phi_{i+1}}{dx}\, dx = \int_{ih}^{(i+1)h} \left(-\frac{1}{h}\right)\frac{1}{h}\, dx = -\frac{1}{h}, \\
K_{i,i+1} &= K_{i+1,i}.
\end{aligned}
$$

As explained above, the other entries of K are zero. For this example, the entries on each of the nonzero diagonals of K are constant; this is a consequence of the constant coefficient k in the BVP. The result is

$$
K = \begin{bmatrix}
\frac{2}{h} & -\frac{1}{h} & & & \\
-\frac{1}{h} & \frac{2}{h} & -\frac{1}{h} & & \\
& \ddots & \ddots & \ddots & \\
& & -\frac{1}{h} & \frac{2}{h} & -\frac{1}{h} \\
& & & -\frac{1}{h} & \frac{2}{h}
\end{bmatrix}.
$$

Computing the load vector is a bit harder, since its components are not constant. We have

$$
\begin{aligned}
F_i &= \int_0^1 \phi_i(x)\sin(\pi x)\, dx \\
&= \int_{(i-1)h}^{ih} \frac{x-(i-1)h}{h}\sin(\pi x)\, dx - \int_{ih}^{(i+1)h} \frac{x-(i+1)h}{h}\sin(\pi x)\, dx \\
&= -\frac{2\sin(hi\pi)(1-\cos(h\pi))}{h\pi^2}.
\end{aligned}
$$

We then find the nodal values of the best piecewise linear approximation to the solution by solving $KU = F$ for U. The result, for $n = 10$, is shown in Figure 6.13.

6.9.2 Continuous piecewise quadratic functions

The previous development can be repeated for piecewise polynomials of any degree. We will briefly sketch the construction of a space of continuous piecewise quadratic functions and leave the reader to investigate higher degrees, if desired.

We begin with a uniform mesh on $[0, \ell]$, as before. The elements are

$$[x_0, x_1], [x_1, x_2], \ldots, [x_{n-1}, x_n],$$

where $x_i = ih$, $i = 0, 1, \ldots, n$ and $h = \ell/n$. We will denote by $V_n^{(2)}$ the space of all continuous piecewise quadratic functions, relative to the given

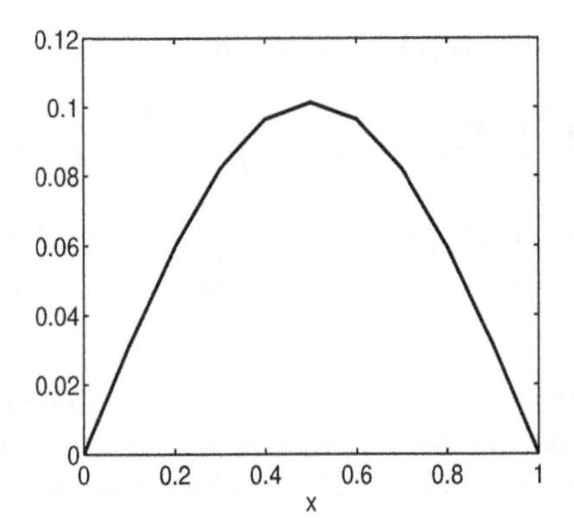

FIGURE 6.13
The piecewise linear approximation to the BVP of Example 323.

mesh and satisfying the boundary conditions. Two nodes per element do not uniquely determine a quadratic piece, so we add the midpoint to each element: $m_i = (x_{i-1}+x_i)/2$. For convenience, we then rename the points as $x_i = ih/2$, $i = 0, 1, \ldots, 2n$. Then x_i is an element endpoint if i is even and an element midpoint if i is odd; the ith element is now denoted $[x_{2i-2}, x_{2i}]$.

Since three nodal values determine a unique quadratic, it follows that any v in $V_n^{(2)}$ is determined completely by the nodal values $v(x_0), v(x_1), \ldots, v(x_{2n})$. Moreover, if we impose the boundary conditions $v(0) = v(\ell) = 0$, then v is determined by $v(x_1), v(x_2), \ldots, v(x_{2n-1})$. This suggests that the dimension of $V_n^{(2)}$ is $2n - 1$. This can be proved by constructing the nodal basis $\{\psi_1, \psi_2, \ldots, \psi_{2n-1}\}$, where $\psi_i \in V_n^{(2)}$ is defined by

$$\psi_i(x_j) = \begin{cases} 1, & i = j, \\ 0 & i \neq j. \end{cases} \tag{6.45}$$

The proof that these functions form a basis for $V_n^{(2n)}$ is essentially the same as in the case of piecewise linear functions; as before, the underlying fact is that the nodal values determine a unique piecewise quadratic function, which in turn depends on the fact that three points determine a unique quadratic.

The defining property of the nodal bases for piecewise linear and piecewise quadratic functions are the same, which makes the piecewise quadratic functions nearly as easy to manipulate as the piecewise linear functions. There is one distinction, however: The function ψ_i has one of two different forms, depending on whether x_i is an endpoint or midpoint of an element. The two possibilities are shown in Figure 6.14. If x_i is an element endpoint (that is, if i is even), then ψ_i is nonzero on two elements, $[x_{i-2}, x_i]$ and $[x_i, x_{i+2}]$. On the other hand, if x_i is an element midpoint (that is, if i is odd), then ψ_i is supported on a single element, $[x_{i-1}, x_{i+1}]$.

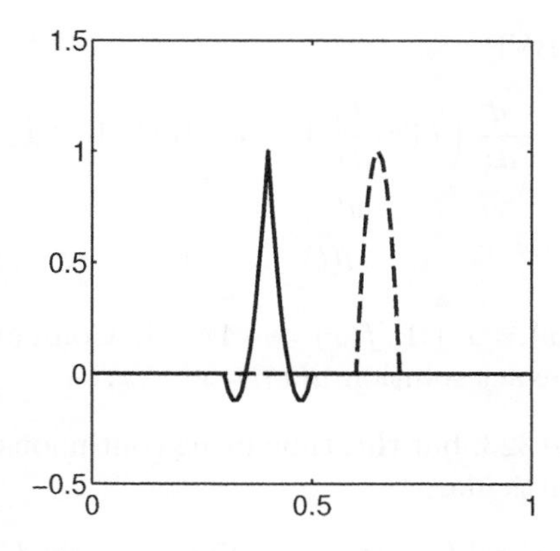

FIGURE 6.14
The two types of nodal basis functions for the space of continuous piecewise quadratic functions. The function on the left (the solid curve) corresponds to an element endpoint, while the one on the right (the dashed curve) corresponds to an element midpoint.

The advantage of using piecewise quadratic functions instead of piecewise linear functions is that, for a given mesh size h, a smooth function can be approximated more accurately using polynomials of higher degree. It can be proved that if u is smooth enough and p is the piecewise quadratic interpolant of u, then

$$\|u - p\|_E \leq Ch^2,$$

where C is a constant depending on the size of the third derivatives of u.

6.9.3 Higher degree finite element spaces

The construction of continuous piecewise polynomials of degree k follows the same pattern as for the quadratic case. A polynomial of degree k is determined by $k+1$ nodal values, so $k-1$ equally-spaced nodes are added to the interior of each element. The advantage of using higher-degree polynomials is that a more accurate approximation can be obtained for a given mesh size h. For details, the reader is referred to [14].

Exercises

1. Apply the finite element method described in this section, with piecewise linear functions on a uniform mesh with ten elements, to approximate a

solution to the BVP

$$\begin{aligned} -\frac{d}{dx}\left(k(x)\frac{du}{dx}\right) &= f(x),\ 0 < x < \ell, \\ u(0) &= 0, \\ u(\ell) &= 0, \end{aligned}$$

where $\ell = 1$, $k(x) = x + 1$, $f(x) = -4x - 1$. Compare your approximate solution to the exact solution $u(x) = x^2 - x$.

2. Repeat Example 323, but this time using continuous piecewise quadratic functions. In particular:

   (a) Find the general form of the stiffness matrix K and the load vector F. (The use of a computer algebra system such as *Mathematica*® or Maple(TM) is recommended for computing the necessary integrals.)

   (b) Find the nodal values of the piecewise quadratic approximation corresponding to $n = 10$ elements.

3. Explain why the energy inner product

$$a(u, v) = \int_0^\ell k(x)\frac{du}{dx}(x)\frac{dv}{dx}(x)\,dx$$

   defines an inner product on

$$V = \{v : [0, \ell] \to \mathbf{R} \mid v(0) = v(\ell) = 0\}.$$

   Note: It is easy to prove that $a(\cdot, \cdot)$ is symmetric, bilinear, and satisfies $a(v, v) \geq 0$ for all $v \in V$. The proof then reduces to showing that $a(v, v) = 0$ only if $v = 0$. A rigorous proof is not possible without a precise definition of V, so an informal argument will suffice.

4. Consider the uniform mesh described in Section 6.9.2.

   (a) Suppose x_i is an element midpoint. Find the formula for ψ_i.

   (b) Suppose x_i is an element endpoint. Find the formula for ψ_i.

5. In this section, we derived the linear system $KU = F$ that produces the Galerkin approximation to the solution of the model problem (6.35). Show how the system changes if the BVP is

$$\begin{aligned} -\frac{d}{dx}\left(k(x)\frac{du}{dx}\right) + p(x)u &= f(x),\ 0 < x < \ell, \\ u(0) &= 0, \\ u(\ell) &= 0. \end{aligned}$$

6. Using the results of the previous exercise, find the approximate solution of

$$\begin{aligned} -\frac{d^2u}{dx^2} + u &= \sin(\pi x),\ 0 < x < 1, \\ u(0) &= 0, \\ u(\ell) &= 0 \end{aligned}$$

on a uniform mesh with ten elements, using piecewise linear functions. Compare to the exact solution (which is a multiple of $\sin(\pi x)$).

7. Let V be the space of continuous piecewise linear functions, relative to a uniform mesh with n elements, on the interval $[0, 1]$, where the functions in V satisfy Dirichlet boundary conditions at the endpoints. Regard V as a subspace of $L^2(0, 1)$. Let $\{\phi_1, \ldots, \phi_{n-1}\}$ be the standard nodal basis for V. Explain how to project $f \in C[0, 1]$ onto V. What is the Gram matrix?

8. Let $u : [0, \ell] \to \mathbf{R}$ be smooth and let $a(\cdot, \cdot)$ be the energy inner product defined by the weight function $k(x) = 1$:

$$a(u, v) = \int_0^\ell u'(x)v'(x)\, dx.$$

Given a (possibly nonuniform) mesh on $[0, \ell]$, prove that the best approximation to u from the space of continuous piecewise linear functions is the same as the piecewise linear interpolant of u. (Hint: Prove that

$$a(u - p, v) = \int_0^1 (u'(x) - p'(x))v'(x)\, dx = 0$$

for all continuous piecewise linear functions v, where p is the piecewise linear interpolant of u. Do this by writing

$$a(u - p, v) = \sum_{i=1}^{n} \int_{x_{i-1}}^{x_i} (u'(x) - p'(x))v'(x)\, dx$$

and showing the integral over each element is zero. Use integration by parts and the fact that p' and v' are constant over $[x_{i-1}, x_i]$.)

Note that this result is generally not true if k is nonconstant.

6.10 Gaussian quadrature

An interesting application of orthogonal polynomials arises in *Gaussian quadrature*. Quadrature refers to numerical integration, that is, to estimating the value of a definite integral using a numerical algorithm. Quadrature forms an important area of numerical analysis, because many functions cannot be integrated analytically. For example, there is no elementary antiderivative for $f(x) = e^{-x^2}$.

6.10.1 The trapezoidal rule and Simpson's rule

It is typical to develop quadrature formulas (also called quadrature rules) on a convenient *reference interval*, since the rules can then be applied to a different interval by a simple change of variables. In this section, we will develop Gaussian quadrature formulas for the interval $[-1, 1]$. The formulas will have the form

$$\int_{-1}^{1} f(x)\,dx \doteq \sum_{i=1}^{n} w_i f(x_i), \tag{6.46}$$

where the *weights* $w_1, w_2, \ldots, w_n$ are arbitrary real numbers and the *nodes* satisfy $-1 \leq x_1 < x_2 < \cdots < x_n \leq 1$.

There are many strategies for constructing quadrature formulas of the form (6.46). One way to proceed is to use interpolating polynomials to approximate the integrand f; the advantage is that polynomials can be integrated easily and exactly. For example, suppose we approximate f on $[-1, 1]$ by a linear polynomial that interpolates f at $x = -1$ and $x = 1$. Such a polynomial is given by

$$\ell(x) = \frac{x-1}{-2} f(-1) + \frac{x+1}{2} f(1)$$

(see Section 2.8). We then have

$$\begin{aligned}
\int_{-1}^{1} f(x)\,dx \doteq \int_{-1}^{1} \ell(x)\,dx &= -\frac{f(-1)}{2}\int_{-1}^{1}(x-1)\,dx + \frac{f(1)}{2}\int_{-1}^{1}(x+1)\,dx \\
&= 1 \cdot f(-1) + 1 \cdot f(1) \\
&= f(-1) + f(1).
\end{aligned}$$

This is called the *trapezoidal rule* (because it estimates the area under $y = f(x)$ by the area of a trapezoid); it is of form (6.46) with $n = 2$, nodes $x_1 = -1$, $x_2 = 1$, and weights $w_1 = 1$, $w_2 = 1$.

The trapezoidal rule is not very accurate, so it is natural to try a higher order polynomial. We can interpolate f at $x_1 = -1$, $x_2 = 0$, and $x_3 = 1$ to obtain the quadratic polynomial

$$\begin{aligned}
q(x) &= \frac{(x-0)(x-1)}{(-1)(-2)} f(-1) + \frac{(x+1)(x-1)}{(1)(-1)} f(0) + \frac{(x+1)(x-0)}{(2)(1)} f(1) \\
&= \frac{x^2 - x}{2} f(-1) - (x^2 - 1) f(0) + \frac{x^2 + x}{2} f(1).
\end{aligned}$$

We then have

$$\begin{aligned}
& \int_{-1}^{1} f(x)\,dx \\
\doteq \; & \int_{-1}^{1} q(x)\,dx \\
= \; & \frac{f(-1)}{2}\int_{-1}^{1}(x^2-x)\,dx - f(0)\int_{-1}^{1}(x^2-1)\,dx + \frac{f(1)}{2}\int_{-1}^{1}(x^2+x)\,dx \\
= \; & \frac{1}{3}f(-1) + \frac{4}{3}f(0) + \frac{1}{3}f(1).
\end{aligned}$$

This approximation formula is called *Simpson's rule*; it is of the form (6.46) with $n = 3$, $x_1 = -1$, $x_2 = 0$, $x_3 = 1$, and $w_1 = 1/3$, $w_2 = 4/3$, $w_3 = 1/3$.

One way to improve upon simple quadrature rules such as the trapezoidal rule or Simpson's rule is to divide $[-1, 1]$ into subintervals and apply the rule on each subinterval. The result would be called, for example, the *composite* trapezoidal rule. (This is probably the form in which the reader has encountered the trapezoidal rule.) Another approach is to continue to increase the order of the interpolating polynomial.

Gaussian quadrature is based on a third approach, which we will introduce with the following observation. Since the trapezoidal rule and Simpson's rule are based on interpolating polynomials, they will give the exact value of $\int_{-1}^{1} f(x)\,dx$ if f happens to be a polynomial whose degree is not too high. For instance, the trapezoidal rule is exact for all $f \in \mathcal{P}_1$ (since the trapezoidal rule is based on a linear interpolating polynomial). Simpson's rule is necessarily exact for all $f \in \mathcal{P}_2$ (since it is based on a quadratic interpolating polynomial), but in fact more is true. By direct calculation, it can be verified that if $f \in \mathcal{P}_3$, then

$$\int_{-1}^{1} f(x)\,dx = \frac{1}{3}f(-1) + \frac{4}{3}f(0) + \frac{1}{3}f(1).$$

In other words, Simpson's rule is exact for all $f \in \mathcal{P}_3$.

6.10.2 Gaussian quadrature

Gaussian quadrature is based on constructing quadrature rules to be exact for polynomials of the highest possible degree. We say that a quadrature rule has *degree of precision* k if it integrates exactly all polynomials of degree k or less. As we will see, the trapezoidal rule and Simpson's rule are not very good by this criterion. The trapezoidal rule uses two nodes to achieve degree of precision 1, while Simpson's rule uses three nodes and achieves degree of precision 3.

A quadrature rule of the form (6.46) with $n = 1$ has two degrees of freedom (one node and one weight), as does a polynomial $f \in \mathcal{P}_1$. Therefore, it seems that it ought to be possible to construct a quadrature rule with one node that is exact for every $f \in \mathcal{P}_1$. Since both integration and (6.46) are linear in f, it suffices that the quadrature rule be exact for $f(x) = 1$ and $f(x) = x$:

$$\sum_{i=1}^{1} w_i \cdot 1 = w_1 = \int_{-1}^{1} 1\,dx = 2,$$
$$\sum_{i=1}^{1} w_i x_i = w_1 x_1 = \int_{-1}^{1} x\,dx = 0.$$

Solving the system

$$\begin{aligned} w_1 &= 2, \\ w_1 x_1 &= 0, \end{aligned}$$

we obtain $w_1 = 2$ and $x_1 = 0$. Thus the quadrature rule

$$\int_{-1}^{1} f(x)\,dx \doteq 2f(0)$$

is exact for all $f \in \mathcal{P}_1$. This is called the *midpoint rule.* The reader should note that the midpoint rule uses only one node to achieve the same degree of precision that the trapezoidal has with two.

For $n = 2$, (6.46) has four degrees of freedom, and $\dim(\mathcal{P}_3) = 4$. Therefore, we want to choose the weights and nodes so that

$$\int_{-1}^{1} f(x)\,dx \doteq \sum_{i=1}^{2} w_i f(x_i) = w_1 f(x_1) + w_2 f(x_2)$$

is exact for $f(x) = 1$, $f(x) = x$, $f(x) = x^2$, and $f(x) = x^3$. The resulting equations are

$$\sum_{i=1}^{2} w_i \cdot 1 = w_1 + w_2 = \int_{-1}^{1} 1\,dx = 2,$$
$$\sum_{i=1}^{2} w_i x_i = w_1 x_1 + w_2 x_2 = \int_{-1}^{1} x\,dx = 0,$$
$$\sum_{i=1}^{2} w_i x_i^2 = w_1 x_1^2 + w_2 x_2^2 = \int_{-1}^{1} x^2\,dx = \frac{2}{3},$$
$$\sum_{i=1}^{2} w_i x_i^3 = w_1 x_1^3 + w_2 x_2^3 = \int_{-1}^{1} x^3\,dx = 0.$$

We then wish to solve the following four equations in the four unknowns

x_1, x_2, w_1, w_2:

$$\begin{aligned} w_1 + w_2 &= 2, \\ w_1x_1 + w_2x_2 &= 0, \\ w_1x_1^2 + w_2x_2^2 &= \frac{2}{3}, \\ w_1x_1^3 + w_2x_2^3 &= 0. \end{aligned}$$

The reader should notice that the system is nonlinear, so it is not obvious that there is a solution or that the solution is unique. With some work, though, we can show that the unique solution[10] is

$$w_1 = 1, w_2 = 1, x_1 = -\frac{1}{\sqrt{3}}, x_2 = \frac{1}{\sqrt{3}}.$$

Thus the Gaussian quadrature rule on two nodes,

$$\int_{-1}^{1} f(x)\,dx \doteq f\left(-\frac{1}{\sqrt{3}}\right) + f\left(\frac{1}{\sqrt{3}}\right),$$

is exact for all $f \in \mathcal{P}_3$. Simpson's rule achieved the same degree of precision, but with three nodes instead of two.

In general, we have $2n$ degrees of freedom to define a quadrature rule with n nodes. Since $\dim(\mathcal{P}_{2n-1}) = 2n$, it seems that we should be able to choose the weights and nodes so that the Gaussian quadrature rule on n nodes has degree of precision $2n-1$. However, the system of nonlinear equations defining the weights and nodes is already difficult to solve algebraically when $n = 2$. One could use Newton's method (see Section 3.8) to find the weights and nodes numerically; however, an estimate of the solution is required to start the algorithm, and the author has found it difficult to find a starting point that will lead to convergence for n much larger than 20. Fortunately, there is a different approach that proves that there is a unique solution for each n and identifies the nodes as the roots of a certain polynomial.

6.10.3 Orthogonal polynomials

Gaussian quadrature rules can be constructed by a two-step process:

1. For any nodes $-1 \le x_1 < x_2 < \cdots < x_n \le 1$, weights $w_1, w_2, \ldots, w_n$ can be chosen so that the resulting quadrature rule (6.46) is exact for all $f \in \mathcal{P}_{n-1}$.

2. For a special choice of nodes, the rule becomes exact for all $f \in \mathcal{P}_{2n-1}$. The roots of a sequence of orthogonal polynomials form the nodes.

[10] Formally, there are two solutions, since the values of x_1 and x_2 can be exchanged to produce a different solution.

Suppose first that the nodes $x_1, \ldots, x_n$ are already chosen. Let $L_1, \ldots, L_n$ be the corresponding Lagrange basis for $\mathcal{P}_{n-1}$:[11]

$$L_i(x) = \frac{(x - x_1) \cdots (x - x_{i-1})(x - x_{i+1}) \cdots (x - x_n)}{(x_i - x_1) \cdots (x_i - x_{i-1})(x_i - x_{i+1}) \cdots (x_i - x_n)}, \ i = 1, 2, \ldots, n.$$

Any $f \in \mathcal{P}_{n-1}$ can be written as

$$f(x) = \sum_{i=1}^{n} f(x_i) L_i(x)$$

(this is the special property of the Lagrange basis). It follows that

$$\int_{-1}^{1} f(x)\,dx = \int_{-1}^{1} \sum_{i=1}^{n} f(x_i) L_i(x)\,dx = \sum_{i=1}^{n} f(x_i) \int_{-1}^{1} L_i(x)\,dx.$$

Therefore, if we define

$$w_i = \int_{-1}^{1} L_i(x)\,dx, \ i = 1, 2, \ldots, n, \tag{6.47}$$

then

$$\int_{-1}^{1} f(x)\,dx \doteq \sum_{i=1}^{n} w_i f(x_i) \tag{6.48}$$

is exact for all $f \in \mathcal{P}_{n-1}$. It can be shown that, for given nodes $x_1, x_2, \ldots, x_n$, (6.47) is the only choice of weights for which (6.48) is exact for all $f \in \mathcal{P}_{n-1}$ (see Exercise 4).

We know from Section 6.5 that it is possible to choose an orthogonal basis $\{p_0, p_1, \ldots, p_n\}$ for $\mathcal{P}_n$, where orthogonality is defined by the $L^2(-1,1)$ inner product. If this basis is chosen by applying the Gram-Schmidt process to the monomial basis $\{1, x, x^2, \ldots, x^n\}$, then p_k is orthogonal to $\mathcal{P}_{k-1}$ for $k = 1, 2, \ldots, n$. We will now show that p_n has n roots in $(-1, 1)$.

Theorem 324 *Suppose p is orthogonal to $\mathcal{P}_{n-1}$ $(n \geq 1)$ under the $L^2(a,b)$ norm. Then p changes signs at least n times on the interval (a,b).*

Proof Suppose p changes signs only m times on (a,b), where $m < n$. Let p change signs at $t_1, t_2, \ldots, t_m$, where $a < t_1 < t_2 < \cdots < t_m < b$, so that p has only one sign on each of the intervals $(a, t_1), (t_1, t_2), \ldots, (t_{m-1}, t_m), (t_m, b)$. Then we can define $q \in \mathcal{P}_m$ by

$$q(x) = \pm(x - t_1)(x - t_2) \cdots (x - t_m),$$

[11] The only difference between the notation here and in Section 2.8 is that here the nodes are indexed starting at $i = 1$ rather than $i = 0$.

where the sign is chosen so that q has the same sign as p throughout the interval $[a,b]$. But then $q \in \mathcal{P}_{n-1}$ and

$$\int_a^b p(x)q(x)\,dx > 0,$$

contradicting that p is orthogonal to $\mathcal{P}_{n-1}$. This contradiction proves the result.

QED

Corollary 325 *Let $\{p_0, p_1, \ldots, p_n\}$ be an orthogonal basis for $\mathcal{P}_n$, where orthogonality is defined by the $L^2(a,b)$ norm and the basis is chosen so that* $\deg(p_k) = k$ *and p_k is orthogonal to $\mathcal{P}_{k-1}$ for $k = 1, 2, \ldots, n$. Then each p_k has exactly k roots in (a,b).*

Proof By the theorem, p_k has at least k roots, and a nonzero polynomial of degree k cannot have more than k roots.

QED

We can now prove that the roots of p_n are the desired quadrature nodes.

Theorem 326 *Let $p_n \in \mathcal{P}_n$ be orthogonal to $\mathcal{P}_{n-1}$ ($n \geq 1$) under the $L^2(-1,1)$ norm, let $x_1, x_2, \ldots, x_n$ be the roots of p_n, and define $w_1, w_2, \ldots, w_n$ by (6.47). Then the quadrature rule (6.48) is exact for all $f \in \mathcal{P}_{2n-1}$.*

Proof We already know that (6.48) is exact for all $f \in \mathcal{P}_{n-1}$. Let $f \in \mathcal{P}_{2n-1}$. Then we can write $f = qp_n + r$, where $q \in \mathcal{P}_{n-1}$ and $r \in \mathcal{P}_{n-1}$ (see Theorem 496 in Appendix C). We have

$$f(x_i) = q(x_i)p_n(x_i) + r(x_i) = r(x_i),\ i = 1, 2, \ldots, n$$

(since $p_n(x_i) = 0$ for all i), and

$$\int_{-1}^{1} q(x_i)p_n(x_i)\,dx = 0$$

since $q \in \mathcal{P}_{n-1}$ and p_n is orthogonal to $\mathcal{P}_{n-1}$. But then

$$\begin{aligned}
\int_{-1}^{1} f(x)\,dx = \int_{-1}^{1} (q(x)p_n(x) + r(x))\,dx &= \int_{-1}^{1} r(x)\,dx \\
&= \sum_{i=1}^{n} w_i r(x_i) \\
&= \sum_{i=1}^{n} w_i f(x_i)
\end{aligned}$$

(using the facts that the quadrature rule is exact for $r \in \mathcal{P}_{n-1}$ and that $f(x_i) = r(x_i)$ for all i). This shows that the quadrature rule is exact for all $f \in \mathcal{P}_{2n-1}$.

QED

If polynomials $p_0, p_1, \ldots$ are defined so that $\{p_0, \ldots, p_k\}$ is an orthogonal basis for $\mathcal{P}_k$ (under the $L^2(-1,1)$ inner product) for each $k = 1, 2, \ldots$, then the polynomials in this sequence are unique up to scalar multiples. Therefore, the roots of the polynomials in this sequence, and hence the nodes for the Gaussian quadrature rules, are uniquely determined.

The first few orthogonal polynomials are

$$\begin{aligned}
p_0(x) &= 1, \\
p_1(x) &= x, \\
p_2(x) &= x^2 - \frac{1}{3}, \\
p_3(x) &= x^3 - \frac{3}{5}x, \\
p_4(x) &= x^4 - \frac{6}{7}x^2 + \frac{3}{35}, \\
p_5(x) &= x^5 - \frac{10}{9}x^3 + \frac{5}{21}x.
\end{aligned}$$

Therefore, if we wish to find, for example, the Gaussian quadrature rule on $n = 5$ nodes, we find the roots of p_5, which are approximately

$$x_1 = -0.9061798,\ x_2 = -0.5384693,\ x_3 = 0,\ x_4 = 0.5384693,\ x_5 = 0.9061798.$$

We can then compute the corresponding weights by (6.47); the results are

$$w_1 = 0.23692689,\ w_2 = 0.47862867,\ w_3 = 0.56888889,\ w_4 = w_2,\ w_5 = w_1.$$

Example 327 *Consider the function $f(x) = \sin(2x^2 + x)$. Using the Gaussian quadrature on $n = 5$ nodes yields*

$$\int_{-1}^{1} f(x)\,dx \doteq \sum_{i=1}^{5} w_i f(x_i) \doteq 0.74180187.$$

We can compare this result with an estimate obtained from Simpson's rule. If we divide the interval $[-1,1]$ into $[-1,0]$ and $[0,1]$ and apply Simpson's rule on each subinterval (the usual way to obtain more precision with Simpson's rule), the calculation also uses five nodes and the result is

$$\int_{-1}^{1} f(x)\,dx \doteq 0.72474582.$$

Using a more sophisticated algorithm, Mathematica® obtains the result

$$\int_{-1}^{1} f(x)\,dx \doteq 0.74140381$$

(correct to the digits shown). Gaussian quadrature produces an error that is about 2.5% of the error in Simpson's rule.

6.10.4 Weighted Gaussian quadrature

In some applications, one must compute many integrals of the form

$$\int_a^b w(x)f(x)\,dx,$$

where w is a fixed positive *weight* function and f can vary. Since

$$\langle f,g\rangle = \int_a^b w(x)f(x)g(x)\,dx \tag{6.49}$$

defines an inner product, the entire development of Gaussian quadrature can be repeated to develop quadrature rules of the form

$$\int_a^b w(x)f(x)\,dx \doteq \sum_{i=1}^{n} w_i f(x_i).$$

The weights in the formula become

$$w_i = \int_{-1}^{1} w(x)L_i(x)\,dx,\ i=1,2,\ldots,n,$$

and the nodes are roots of polynomials that are orthogonal under the weighted inner product (6.49) instead of the ordinary L^2 inner product.

Exercises

1. Use the theory of orthogonal polynomials presented in this section to find the Gaussian quadrature rule with $n=3$ quadrature nodes (on the reference interval $[-1,1]$).

2. Let $w(x)=1+x^2$. Compute the weighted Gaussian quadrature rule

$$\int_{-1}^{1} w(x)f(x)\,dx \doteq \sum_{i=1}^{n} w_i f(x_i)$$

   for $n=1,2,3$.

3. Let $w(x) = 1/\sqrt{1-x^2}$. Compute the weighted Gaussian quadrature rule

$$\int_{-1}^{1} w(x)f(x)\,dx \doteq \sum_{i=1}^{n} w_i f(x_i).$$

   Take $n=3$. (Hint: The orthogonal polynomials for the inner product defined by this weight function are given in Section 6.8. See (6.34) on page 397.)

4. Suppose distinct nodes $x_1,x_2,\ldots,x_n$ are given. Prove that (6.48) is exact for all $f\in\mathcal{P}_{n-1}$ only if the weights $w_1,w_2,\ldots,w_n$ are defined by (6.47).

6.11 The Helmholtz decomposition

If $A \in \mathbf{R}^{m\times n}$ is given, then $\mathbf{R}^m$ can be decomposed as

$$\mathbf{R}^m = \operatorname{col}(A) \oplus \mathcal{N}(A^T).$$

This implies that any $b \in \mathbf{R}^m$ can be written as $b = Ax+y$, where $y \in \mathcal{N}(A^T)$. To find x and y, we multiply by A^T:

$$Ax + y = b \Rightarrow A^TAx + A^Ty = A^Tb \Rightarrow A^TAx = A^Tb$$

(using the fact that $y \in \mathcal{N}(A^T)$). This shows that x is a least-squares solution to $Ax = b$, and can be found by solving the normal equations or by another (possibly more effective) algorithm (see Section 9.8). Once x has been found, we can define y by $y = b - Ax$, and we know from the theory of least-squares problems that $y \in \operatorname{col}(A)^{\perp} = \mathcal{N}(A^T)$. We can also verify this directly:

$$A^Ty = A^T(b - Ax) = A^Tb - A^TAx = 0$$

(since x is defined to be a solution to the normal equations). The reader should notice that y is uniquely determined, but x is unique only if A has full rank. Otherwise, there are infinitely many least-squares solutions to $Ax = b$.

There is an interesting analogy to $\mathbf{R}^m = \operatorname{col}(A) + \mathcal{N}(A^T)$ that arises in vector calculus. We consider *vector fields* defined on a domain Ω in $\mathbf{R}^3$. A vector field on Ω is simply a function $u : \Omega \to \mathbf{R}^3$. We will also refer to *scalar fields* on Ω, which are simply functions of the form $\phi : \Omega \to \mathbf{R}$. The most important operator related to vector fields is the gradient operator ∇, which takes a scalar field ϕ and produces a vector field:

$$\nabla\phi = \begin{bmatrix} \frac{\partial\phi}{\partial x_1} \\ \frac{\partial\phi}{\partial x_2} \\ \frac{\partial\phi}{\partial x_3} \end{bmatrix}.$$

The *Helmholtz decomposition* writes a (more or less) arbitrary vector field u as $u = \nabla\phi + w$, where w lies in the kernel of the adjoint (or transpose) of ∇. Thus the Helmholtz decomposition is analogous to writing $b = Ax + y$, $y \in \mathcal{N}(A^T)$. To see why the Helmholtz decomposition is significant, we must explain some vector calculus.

The Helmholtz decomposition applies to any vector field that has sufficient regularity. In this brief introduction to the topic, we will not try to state precise conditions on the continuity or smoothness required on the vector and scalar fields we manipulate. All of the results we derive in this section are true provided the function involved are sufficiently regular.

6.11.1 The divergence theorem

The *divergence operator*, denoted by $\nabla\cdot$, takes a vector field and produces a scalar field:

$$\nabla \cdot u = \frac{\partial u_1}{\partial x_1} + \frac{\partial u_2}{\partial x_2} + \frac{\partial u_3}{\partial x_3}.$$

The reader should note that the vector field u has components u_1, u_2, u_3:

$$u(x) = \begin{bmatrix} u_1(x) \\ u_2(x) \\ u_3(x) \end{bmatrix} = \begin{bmatrix} u_1(x_1, x_2, x_3) \\ u_2(x_1, x_2, x_3) \\ u_3(x_1, x_2, x_3) \end{bmatrix}.$$

The notation $\operatorname{div} u$ is also used for the divergence of a vector field, but we prefer the notation $\nabla \cdot u$, since formally the divergence of u is the dot product of the gradient operator ∇ with u:

$$\nabla \cdot u = \begin{bmatrix} \frac{\partial}{\partial x_1} \\ \frac{\partial}{\partial x_2} \\ \frac{\partial}{\partial x_3} \end{bmatrix} \cdot \begin{bmatrix} u_1 \\ u_2 \\ u_3 \end{bmatrix} = \frac{\partial u_1}{\partial x_1} + \frac{\partial u_2}{\partial x_2} + \frac{\partial u_3}{\partial x_3}.$$

The significance of the divergence of a vector field is due to the *divergence theorem*: If F is a vector field defined on Ω, $\partial\Omega$ is the boundary of Ω, and n represents the outward-pointing unit normal vector to $\partial\Omega$, then

$$\int_{\Omega} \nabla \cdot F = \int_{\partial\Omega} F \cdot n.$$

The divergence theorem is reminiscent of the fundamental theorem of calculus,

$$\int_a^b f'(x)\,dx = f(x)|_a^b = f(b) - f(a),$$

in that it relates the integral of a derivative over the interior of a domain to the boundary values of the function itself. Indeed, the divergence theorem is the multidimensional version of the fundamental theorem of calculus.

If F is any vector field and S is a surface in $\mathbf{R}^3$, then $\int_S F \cdot n$ is the *flux* (rate of flow) of the vector field across S. For example, if v represents the velocity field of a fluid, then $\int_S v \cdot n$ is the volume of the fluid moving across S, in the direction of n, per unit time. Since $\partial\Omega$ is a closed surface enclosing Ω, $\int_{\partial\Omega} v \cdot n$ represents the net flux out of Ω. Thus, if $\int_{\partial\Omega} v \cdot n > 0$, then there is a net flow out of Ω; this would happen, for example, if v is the velocity of a fluid flowing across Ω and Ω is a region of lower density. A common condition in a model of fluid flow is $\nabla \cdot v = 0$, indicating that the flow is incompressible. More generally, a vector field F is called *solenoidal* or simply *divergence-free* if $\nabla \cdot F = 0$ everywhere.

If ϕ is a scalar field and u is a vector field, then ϕu is another vector field, and

$$\nabla \cdot (\phi u) = \nabla\phi \cdot u + \phi \nabla \cdot u \tag{6.50}$$

(see Exercise 1). Integrating over Ω and applying the divergence theorem, we obtain

$$\int_{\Omega} \nabla \cdot (\phi u) = \int_{\Omega} \nabla\phi \cdot u + \int_{\Omega} \phi \nabla \cdot u$$
$$\Rightarrow \int_{\partial\Omega} \phi u \cdot n = \int_{\Omega} \nabla\phi \cdot u + \int_{\Omega} \phi \nabla \cdot u.$$

If we now assume that $\phi = 0$ on $\partial\Omega$, then

$$\int_{\Omega} \nabla\phi \cdot u + \int_{\Omega} \phi \nabla \cdot u = 0,$$

or

$$\int_{\Omega} \nabla\phi \cdot u = -\int_{\Omega} \phi \nabla \cdot u. \tag{6.51}$$

We can impose a form of the L^2 inner product on the space of vector fields defined on Ω:

$$\langle u, v \rangle = \int_{\Omega} u \cdot v.$$

With this definition, (6.51) can be written as

$$\langle \nabla\phi, u \rangle = \langle \phi, -\nabla \cdot u \rangle.$$

We see that the adjoint of the gradient operator is the negative divergence operator. It is important to keep in mind the boundary conditions $\phi = 0$ on $\partial\Omega$ imposed on the scalar field.

6.11.2 Stokes's theorem

The *curl operator*, denoted by $\nabla\times$, operates on a vector field and produces another vector field according to the formula

$$\nabla \times u = \begin{bmatrix} \frac{\partial u_3}{\partial x_2} - \frac{\partial u_2}{\partial x_3} \\ \frac{\partial u_1}{\partial x_3} - \frac{\partial u_3}{\partial x_1} \\ \frac{\partial u_2}{\partial x_1} - \frac{\partial u_1}{\partial x_2} \end{bmatrix}.$$

The significance of the curl operator is revealed by *Stokes's theorem*.[12] If D is a disk inside of Ω, oriented by a normal vector N (that is, N is perpendicular to D), and F is a vector field on Ω, then

$$\int_{D} (\nabla \times F) \cdot N = \oint_{\partial D} F \cdot t,$$

[12] What follows is a restricted version of Stokes's theorem. See any book on vector calculus, such as [29], for a more general version.

where t is the unit tangent vector to ∂D, pointing in the counterclockwise direction.[13] If we think of D as having a small radius, then $\int_{\partial D} F \cdot t$ measures the tendency of F to rotate around the center of D. We could imagine varying the tilt of D (which is determined by N); the normal vector corresponding to the largest value of $\int_{\partial D} F \cdot t$ lies in the direction of $\nabla \times F$. Therefore, the curl of F determines the axis of maximal rotation of the vector field, and the magnitude of $\nabla \times F$ determines how much F rotates about that axis. If $\nabla \times F = 0$, then F is said to be *irrotational* or *curl-free*.

6.11.3 The Helmholtz decomposition

It is a simple exercise in partial differentiation to verify that, for any scalar field ϕ,

$$\nabla \times \nabla\phi = 0 \tag{6.52}$$

(see Exercise 2). Thus every gradient field is irrotational. If Ω is *simply connected* (which means, loosely speaking, that it has no holes through it), then the converse is true: Every irrotational vector field can be written as the gradient of a scalar field. The Helmholtz decomposition states that every vector field can be written as the sum of an irrotational vector field and a solenoidal vector field:

$$u = \nabla\phi + w, \ \nabla \cdot w = 0.$$

Notice that $\nabla\phi$ lies in the range of the gradient operator and w is to lie in the kernel of the divergence operator. Therefore, this decomposition is exactly analogous to writing $b = Ax + y$, $y \in \mathcal{N}(A^T)$.

We can prove existence of the Helmholtz decomposition directly, although we will have to rely on a standard result from the theory of partial differential equations. If $u = \nabla\phi + w$ holds, where w is divergence-free, then

$$\nabla \cdot u = \nabla \cdot (\nabla\phi + w) = \nabla \cdot (\nabla\phi) + \nabla \cdot w = \nabla \cdot (\nabla\phi).$$

We have

$$\nabla \cdot (\nabla\phi) = \frac{\partial^2\phi}{\partial x_1^2} + \frac{\partial^2\phi}{\partial x_2^2} + \frac{\partial^2\phi}{\partial x_3^2} \tag{6.53}$$

(see Exercise 3). We usually write $\Delta\phi = \nabla \cdot (\nabla\phi)$ and call Δ the *Laplace operator* or the *Laplacian.* We thus want to choose ϕ to satisfy $\Delta\phi = \nabla \cdot u$ in Ω. Like a typical linear differential equation, this partial differential equation has infinitely many solutions and requires side conditions to define a unique solution. It is well-known from the theory of partial differential equations that

$$\begin{aligned} \Delta\phi &= f \text{ in } \Omega, \\ u &= 0 \text{ on } \partial\Omega \end{aligned}$$

[13] The counterclockwise direction is determined by the normal vector and the right-hand rule. If you curl your fingers around the normal vector, with your thumb pointing in the direction of N, your fingers point in the counterclockwise direction.

has a unique solution for any (sufficiently regular) function f. Therefore, we can find ϕ to satisfy

$$\begin{aligned} \Delta\phi &= \nabla\cdot u \text{ in } \Omega, \\ u &= 0 \text{ on } \partial\Omega, \end{aligned}$$

and then define $w = u - \nabla\phi$. Then $u = \nabla\phi + w$ and

$$\nabla\cdot w = \nabla\cdot u - \nabla\cdot(\nabla\phi) = 0,$$

as desired.

Exercises

1. Let Ω be a domain in $\mathbf{R}^3$, and let ϕ, u be a scalar field and a vector field, respectively, defined on Ω. By writing $\nabla\cdot(\phi u)$ explicitly and applying the product rule for differentiation to each term, prove (6.50).

2. Let $\phi:\Omega\to\mathbf{R}$ be a scalar field, where Ω is a domain in $\mathbf{R}^3$. Prove that (6.52) holds.

3. Prove that (6.53) holds for any scalar field ϕ.

4. In this section, we argued that a vector field $u:\Omega\to\mathbf{R}^3$ can be written as $u=\nabla\phi+w$, where $\nabla\cdot w=0$. In fact, it is possible to say more: There exists a vector field $v:\Omega\to\mathbf{R}^3$ such that $w=\nabla\times v$. In other words, we can write a vector field u as $u=\nabla\phi+\nabla\times v$. Without trying to prove that such a v exists, verify:

   (a) $\nabla\cdot(\nabla\times v)=0$. Thus the curl of any vector field is divergence-free.

   (b) $\int_{\partial\Omega}(\nabla\times v)\cdot n=0$. (Hint: Use the divergence theorem.)

   (c) If ϕ, v satisfy $u=\nabla\phi+\nabla\times v$, then

$$\int_{\partial\Omega} u\cdot n = \int_{\partial\Omega} \nabla\phi\cdot n.$$

7

The spectral theory of symmetric matrices

The *spectrum* of a matrix A is defined to be the set of its eigenvalues. For this reason, results about eigenvalues and eigenvectors are referred to as *spectral theory*, and the decomposition $A = XDX^{-1}$ (when it exists) is called the *spectral decomposition* of A.

A matrix $A \in \mathbf{R}^{n\times n}$ is called *symmetric* if $A^T = A$. In this chapter, we explore the spectral theorem for symmetric matrices, which states that a symmetric matrix $A \in \mathbf{R}^{n\times n}$ can always be diagonalized. Moreover, the eigenvalues of a symmetric matrix are always real, and thus there is no need for complex numbers. Finally, the eigenvectors of a symmetric matrix can be chosen to form an orthonormal set, which, as we will see, means that many calculations are easy to perform.

7.1 The spectral theorem for symmetric matrices

The reader will recall that, in the context of spectral theory, a matrix A in $\mathbf{R}^{n\times n}$ is regarded as an element of $\mathbf{C}^{n\times n}$. This is to take advantage of the algebraic closure of $\mathbf{C}$. The first result in this section shows that, for symmetric matrices, it is unnecessary to move into the complex domain.

Theorem 328 *Let $A \in \mathbf{R}^{n\times n}$ be symmetric, and let $\lambda \in \mathbf{C}$ be an eigenvalue of A. Then $\lambda \in \mathbf{R}$ and there exists an eigenvector $x \in \mathbf{R}^n$ corresponding to λ.*

Proof Let $\lambda \in \mathbf{C}$ and $z \in \mathbf{C}^n$ form an eigenpair of A:

$$Az = \lambda z.$$

Without loss of generality, we can assume that $\|z\|_2 = 1$. Using the Hermitian dot product $\langle \cdot, \cdot \rangle_{\mathbf{C}^n}$, we can reason as follows:

$$\begin{aligned} \lambda = \lambda \langle z, z\rangle_{\mathbf{C}^n} = \langle \lambda z, z\rangle_{\mathbf{C}^n} &= \langle Az, z\rangle_{\mathbf{C}^n} \\ &= \langle z, Az\rangle_{\mathbf{C}^n} = \langle z, \lambda z\rangle_{\mathbf{C}^n} = \overline{\lambda} \langle z, z\rangle_{\mathbf{C}^n} = \overline{\lambda}. \end{aligned}$$

The reader will notice the use of the symmetry of A: $\langle Az, z\rangle_{\mathbf{C}^n} = \langle z, Az\rangle_{\mathbf{C}^n}$ (because $A^* = A^T = A$ for a real symmetric matrix). It follows immediately that λ is real, since only a real number can equal its complex conjugate.

The eigenvector z may belong to $\mathbf{C}^n \setminus \mathbf{R}^n$; in this case, we can write $z = x + iy$, where $x, y \in \mathbf{R}^n$ and at least one of x, y is nonzero. Then

$$Az = \lambda z \Rightarrow A(x + iy) = \lambda(x + iy) \Rightarrow Ax + iAy = \lambda x + i\lambda y.$$

Equating the real and imaginary parts of the two sides of the last equation yields

$$\begin{aligned} Ax &= \lambda x, \\ Ay &= \lambda y. \end{aligned}$$

Since x and y cannot both be zero, this shows that x or y (or both) is a real eigenvector of A corresponding to λ.

QED

We know that eigenvectors corresponding to distinct eigenvalues are linearly independent. In the case of a symmetric matrix, we can say even more.

Theorem 329 *Let $A \in \mathbf{R}^{n\times n}$ be symmetric, let $\lambda_1, \lambda_2 \in \mathbf{R}$ be distinct eigenvalues of A, and let $x_1, x_2 \in \mathbf{R}^n$ be eigenvectors corresponding to λ_1, λ_2, respectively. Then x_1 and x_2 are orthogonal.*

Proof We have

$$\lambda_1(x_1 \cdot x_2) = (\lambda_1 x_1) \cdot x_2 = (Ax_1) \cdot x_2 = x_1 \cdot (Ax_2) = x_1 \cdot (\lambda_2 x_2) = \lambda_2(x_1 \cdot x_2),$$

which implies that

$$(\lambda_1 - \lambda_2)(x_1 \cdot x_2) = 0.$$

Since $\lambda_1 - \lambda_2 \neq 0$ by hypothesis, it follows that $x_1 \cdot x_2 = 0$.

QED

To derive the third fundamental spectral property of symmetric matrices, we will use the concept of an orthogonal matrix.

Definition 330 *Let $Q \in \mathbf{R}^{n\times n}$. We say that Q is* orthogonal *if and only if $Q^T = Q^{-1}$.*

We know from Corollary 118 that $Q^T = Q^{-1}$ if and only if $Q^TQ = I$. But if $Q_1, Q_2, \ldots, Q_n$ are the columns of Q, then the condition $Q^TQ = I$ is equivalent to

$$Q_i \cdot Q_j = \begin{cases} 1, & i = j, \\ 0, & i \neq j, \end{cases}$$

that is, to the condition that $\{Q_1, Q_2, \ldots, Q_n\}$ be an orthonormal set.

The proof of the following theorem is similar to that of Theorem 199, specialized to symmetric matrices. We will use the notation of the earlier proof.

Theorem 331 *Let $A \in \mathbf{R}^{n\times n}$ be symmetric and let $\lambda \in \mathbf{R}$ be an eigenvalue of A. Then the geometric multiplicity of λ equals the algebraic multiplicity of λ.*

Proof We argue by induction on the dimension n. If $n = 1$, then the result is obvious, so we assume that the result holds for all matrices of dimension $(n-1)\times(n-1)$. Suppose $\lambda \in \mathbf{R}$ is an eigenvalue of A of algebraic multiplicity $k \geq 1$. Let $x \in \mathbf{R}^n$ be an eigenvector of A corresponding to λ. Finally, write $x_1 = \|x\|^{-1}x$, choose $x_2, \ldots, x_n$ so that $\{x_1, x_2, \ldots, x_n\}$ is an orthonormal basis for $\mathbf{R}^n$, and define

$$X = [x_1|x_2|\cdots|x_n].$$

Since $\{x_1, x_2, \ldots, x_n\}$ is orthonormal, X is orthogonal (and hence invertible).

Define $B = X^TAX$. Then

$$B = \begin{bmatrix} x_1\cdot Ax_1 & x_1\cdot Ax_2 & \cdots & x_1\cdot Ax_n \\ x_2\cdot Ax_1 & x_2\cdot Ax_2 & \cdots & x_2\cdot Ax_n \\ \vdots & \vdots & \ddots & \vdots \\ x_n\cdot Ax_1 & x_n\cdot Ax_2 & \cdots & x_n\cdot Ax_n \end{bmatrix}.$$

We have, for $i \neq 1$,

$$x_i\cdot Ax_1 = x_i\cdot(\lambda x_1) = \lambda(x_i\cdot x_1) = 0$$

and similarly

$$x_1\cdot Ax_i = (Ax_1)\cdot x_i = 0.$$

Therefore, B simplifies to

$$B = \begin{bmatrix} \lambda & 0 & \cdots & 0 \\ 0 & x_2\cdot Ax_2 & \cdots & x_2\cdot Ax_n \\ \vdots & \vdots & \ddots & \vdots \\ 0 & x_n\cdot Ax_2 & \cdots & x_n\cdot Ax_n \end{bmatrix}.$$

Write $C = B^{(1,1)}$; then

$$B = \left[\begin{array}{c|c} \lambda & 0 \\ \hline 0 & C \end{array}\right],$$

where 0 represents the zero vector in $\mathbf{R}^{n-1}$. The above formula for the simplified B shows that C is symmetric (since $x_j\cdot Ax_i = (Ax_j)\cdot x_i = x_i\cdot Ax_j$), and so the induction hypothesis can be applied to C. Since the algebraic multiplicity of λ, as an eigenvalue of C, is $k-1$, it follows that C has $k-1$ eigenvectors $u_2, \ldots, u_k$ corresponding to λ. Moreover, without loss of generality, $\{u_2, \ldots, u_k\}$ can be assumed to be orthonormal (since otherwise we could apply the Gram-Schmidt process to obtain an orthonormal basis for the eigenspace $E_\lambda(C)$). For each $i = 2, \ldots, k$, define $z_i \in \mathbf{R}^n$ by

$$z_i = \left[\begin{array}{c} 0 \\ \hline u_i \end{array}\right].$$

Then

$$Bz_i = \left[\begin{array}{c|c} \lambda & 0 \\ \hline 0 & C \end{array}\right] \left[\begin{array}{c} 0 \\ \hline u_i \end{array}\right] = \left[\begin{array}{c} \lambda \cdot 0 + 0 \cdot u_i \\ \hline \lambda \cdot 0 + Cu_i \end{array}\right] = \left[\begin{array}{c} 0 \\ \hline \lambda u_i \end{array}\right] = \lambda z_i,$$

so each z_i is an eigenvector of B corresponding to λ. Moreover, $\{z_2, \ldots, z_n\}$ is easily seen to be orthonormal.

Finally,

$$Bz_i = \lambda z_i \Rightarrow X^T AXz_i = \lambda z_i \Rightarrow AXz_i = \lambda Xz_i, \; i = 2, \ldots, k,$$

so $Xz_2, \ldots, Xz_k$ are all eigenvectors of A corresponding to λ. We have

$$(Xz_i) \cdot (Xz_j) = z_i \cdot (X^T Xz_j) = z_i \cdot z_j = \begin{cases} 1, & i = j, \\ 0, & i \neq j, \end{cases}$$

using the fact that $X^T X = I$. It follows that $\{Xz_2, \ldots, Xz_k\}$ is an orthonormal set. Since each Xz_i is a linear combination of $x_2, \ldots, x_n$, we see that x_1 is orthogonal to each Xz_i, and thus $\{x_1, Xz_2, \ldots, Xz_k\}$ is an orthonormal (and hence linearly independent) set of eigenvectors of A corresponding to λ. This completes the proof.

QED

As pointed out in the previous proof, we can always find an orthonormal basis for each eigenspace of A. When A is symmetric, vectors from different eigenspaces are orthogonal to one another. Since each eigenspace of a symmetric matrix contains "enough" linearly independent vectors (that is, because the geometric multiplicity of each eigenvalue equals its algebraic multiplicity), the union of the orthonormal bases for the eigenspaces is an orthonormal basis for $\mathbf{R}^n$. We therefore obtain the following corollary.

Corollary 332 *Let $A \in \mathbf{R}^{n\times n}$ be symmetric. Then there exists an orthogonal matrix $X \in \mathbf{R}^{n\times n}$ and a diagonal matrix $D \in \mathbf{R}^{n\times n}$ such that*

$$A = XDX^T.$$

7.1.1 Symmetric positive definite matrices

A special class of symmetric matrices is described by the following definition.

Definition 333 *Let $A \in \mathbf{R}^{n\times n}$ be symmetric. We say that A is symmetric* positive definite *(SPD) if and only if*

$$x \cdot (Ax) > 0 \text{ for all } x \in \mathbf{R}^n, \; x \neq 0.$$

The matrix A is positive semidefinite *if and only if*

$$x \cdot (Ax) \geq 0 \text{ for all } x \in \mathbf{R}^n.$$

The basic property of SPD matrices is that their eigenvalues are not only real, but positive.

Theorem 334 *Let $A \in \mathbf{R}^{n \times n}$ be symmetric. Then A is positive definite if and only if all of the eigenvalues of A are positive, and A is positive semidefinite if and only if all of the eigenvalues of A are nonnegative.*

Proof We will prove the first part of the theorem and leave the second as an exercise. Assume first that A is positive definite, and let $\lambda \in \mathbf{R}$, $x \in \mathbf{R}^n$ be an eigenpair of A. Then $x \cdot (Ax) > 0$. But

$$x \cdot (Ax) = x \cdot (\lambda x) = \lambda(x \cdot x) = \lambda \|x\|_2^2.$$

Since $\|x\| > 0$, it follows that $\lambda > 0$.

On the other hand, suppose the eigenvalues of A are $\lambda_1, \lambda_2, \ldots, \lambda_n$ (listed according to multiplicity), with corresponding eigenvectors $x_1, x_2, \ldots, x_n$, and suppose that each λ_i is positive. Without loss of generality, $\{x_1, x_2, \ldots, x_n\}$ can be taken to be an orthonormal basis for $\mathbf{R}^n$. With $X = [x_1|x_2|\cdots|x_n]$, we have $A = XDX^T$ and X is orthogonal. Therefore, for any $x \in \mathbf{R}^n$, $x \neq 0$, we have

$$x \cdot (Ax) = x \cdot (XDX^T x) = (X^T x) \cdot (DX^T x).$$

This last expression is easily shown to be $\sum_{i=1}^{n} \lambda_i (x_i \cdot x)^2$. Since each λ_i is positive and at least one $x_i \cdot x$ is nonzero (otherwise x would be the zero vector), it follows that $x \cdot (Ax) > 0$, as desired.

QED

SPD matrices occur in many applications and are therefore discussed in some of the optional application sections in the remainder of the text. The following result presents one way that SPD matrices arise.

Theorem 335 *Let X be an inner product space over $\mathbf{R}$ and let $\{x_1, x_2, \ldots, x_n\}$ be a basis for X. Then the Gram matrix G (defined by $G_{ij} = \langle x_j, x_i \rangle_X$) is SPD.*

Proof The symmetry of G follows directly from the symmetry of the inner product. Let $a = (\alpha_1, \alpha_2, \ldots, \alpha_n)$ be any nonzero vector in $\mathbf{R}^n$ and notice that

$$\begin{aligned} a \cdot (Ga) = \sum_{i=1}^{n} \alpha_i (Ga)_i &= \sum_{i=1}^{n} \sum_{j=1}^{n} G_{ij} \alpha_i \alpha_j \\ &= \sum_{i=1}^{n} \sum_{j=1}^{n} \langle x_j, x_i \rangle_X \alpha_i \alpha_j \\ &= \left\langle \sum_{j=1}^{n} \alpha_j x_j, \sum_{i=1}^{n} \alpha_i x_i \right\rangle_X \\ &= \langle y, y \rangle_X , \end{aligned}$$

where $y = \alpha_1 x_1 + \ldots + \alpha_n x_n$. Since not all of $\alpha_1, \ldots, \alpha_n$ are zero, the linear independence of $\{x_1, \ldots, x_n\}$ implies that y is nonzero. Therefore,

$$a \cdot (Ga) = \langle y, y \rangle_X > 0,$$

and the proof is complete.

QED

Corollary 336 *Let $A \in \mathbf{R}^{m \times n}$ be nonsingular. Then $A^T A$ is SPD.*

7.1.2 Hermitian matrices

The reader will recall that $A \in \mathbf{C}^{n \times n}$ is Hermitian if and only if $A^* = A$. Hermitian matrices have many properties in common with symmetric matrices: Their eigenvalues are real, eigenvectors corresponding to distinct eigenvalues are orthogonal, and Hermitian matrices are diagonalizable. The proofs of the following theorems are similar to those for symmetric matrices and are left as exercises.

Theorem 337 *Let $A \in \mathbf{C}^{n \times n}$ be Hermitian, and let λ be an eigenvalue of A. Then $\lambda \in \mathbf{R}$.*

Proof Exercise 10.

While the eigenvalues of a Hermitian matrix are real, the corresponding eigenvectors need not be real.

Theorem 338 *Let $A \in \mathbf{C}^{n \times n}$ be a Hermitian matrix and let $\lambda \in \mathbf{R}$ be an eigenvalue of A. Then the geometric multiplicity of λ equals the algebraic multiplicity of λ.*

Proof Exercise 11.

Theorem 339 *Let $A \in \mathbf{C}^{n \times n}$ be Hermitian, let $\lambda_1, \lambda_2 \in \mathbf{R}$ be distinct eigenvalues of A, and let $x_1, x_2 \in \mathbf{C}^n$ be eigenvectors corresponding to λ_1, λ_2, respectively. Then x_1 and x_2 are orthogonal.*

Proof Exercise 12.

A matrix $U \in \mathbf{C}^{n \times n}$ is called *unitary* if and only if $U^* = U^{-1}$; thus a unitary matrix is the complex analogue of an orthogonal matrix.

Theorem 340 *Let $A \in \mathbf{C}^{n \times n}$ be Hermitian. Then there exists a unitary matrix $X \in \mathbf{C}^{n \times n}$ and a diagonal matrix $D \in \mathbf{R}^{n \times n}$ such that*

$$A = XDX^*.$$

Proof Exercise 13.

The concept of positive definiteness is important for Hermitian matrices as well as for real symmetric matrices.

Definition 341 *Let $A \in \mathbf{C}^{n\times n}$ be Hermitian. We say that A is* positive definite *if*

$$\langle x, Ax\rangle_{\mathbf{C}^n} > 0 \text{ for all } x \in \mathbf{C}^n.$$

The matrix A is positive semidefinite *if and only if*

$$x \cdot (Ax) \geq 0 \text{ for all } x \in \mathbf{C}^n.$$

Theorem 342 *Let $A \in \mathbf{C}^{n\times n}$ be Hermitian. Then A is positive definite if and only if all the eigenvalues of A are positive, and A is positive semidefinite if and only if all of the eigenvalues of A are nonnegative.*

Proof Exercise 14

Theorem 343 *Let X be an inner product space over $\mathbf{C}$ and let $\{x_1, x_2, \ldots, x_n\}$ be a basis for X. Then the Gram matrix G (defined by $G_{ij} = \langle x_j, x_i\rangle_X$) is Hermitian positive definite.*

Proof Exercise 15

Corollary 344 *Let $A \in \mathbf{C}^{m\times n}$ be nonsingular. Then A^*A is Hermitian positive definite.*

Exercises

Essential exercises

1. Let $A \in \mathbf{R}^{m\times n}$. Prove that A^TA is positive semidefinite.

Miscellaneous exercises

2. Prove that a symmetric matrix $A \in \mathbf{R}^{n\times n}$ is positive semidefinite if and only if all the eigenvalues of A are nonnegative.

3. Let $A \in \mathbf{R}^{n\times n}$ be symmetric and positive definite. Define $\langle\cdot,\cdot\rangle$ by

$$\langle x, y\rangle = x \cdot Ay \text{ for all } x, y \in \mathbf{R}^n.$$

   Prove that $\langle\cdot,\cdot\rangle$ is an inner product on $\mathbf{R}^n$.

4. Let $U \in \mathbf{R}^{n\times n}$ be an orthogonal matrix. Prove that multiplication by U preserves norms and dot products:

$$\begin{aligned} \|Ux\|_2 &= \|x\|_2 \text{ for all } x \in \mathbf{R}^n, \\ (Ux)\cdot(Uy) &= x \cdot y \text{ for all } x, y \in \mathbf{R}^n. \end{aligned}$$

5. Suppose multiplication by $A \in \mathbf{R}^{n\times n}$ preserves dot products:

$$(Ax)\cdot(Ay) = x \cdot y \text{ for all } x, y \in \mathbf{R}^n.$$

   Does A have to be orthogonal? Prove or disprove.

6. Suppose multiplication by $A \in \mathbf{R}^{n\times n}$ preserves norms:

$$\|Ax\|_2 = \|x\|_2 \text{ for all } x \in \mathbf{R}^n.$$

Does A have to be orthogonal? Prove or disprove.

7. Let $A \in \mathbf{C}^{n\times n}$ be a Hermitian positive definite matrix. Prove that there exists a Hermitian positive definite matrix $B \in \mathbf{C}^{n\times n}$ such that $B^2 = A$. (The matrix B is called the *square root* of A and is denoted $A^{1/2}$.)

8. Let X be a finite-dimensional inner product space over $\mathbf{C}$, and let $T : X \to X$ be a linear operator. We say that T is self-adjoint if $T^* = T$. Suppose T is self-adjoint. Prove:

   (a) Every eigenvalue of T is real.

   (b) Eigenvectors of T corresponding to distinct eigenvalues are orthogonal.

9. Let X be a finite-dimensional inner product space over $\mathbf{R}$ with basis $\mathcal{X} = \{x_1, \ldots, x_n\}$, and assume that $T : X \to X$ is a self-adjoint linear operator ($T^* = T$). Define $A = [T]_{\mathcal{X},\mathcal{X}}$. Let G be the Gram matrix for the basis $\mathcal{X}$, and define $B \in \mathbf{R}^{n\times n}$ by

$$B = G^{1/2} A G^{-1/2},$$

   where $G^{1/2}$ is the square root of G (see Exercise 7) and $G^{-1/2}$ is the inverse of $G^{1/2}$.

   (a) Prove that B is symmetric. (Hint: First find a relationship between A^T and A involving G.)

   (b) Since A and B are similar, they have the same eigenvalues and there is a simple relationship between their eigenvectors. What is this relationship?

   (c) Use the fact that there is an orthonormal basis of $\mathbf{R}^n$ consisting of eigenvectors of B to prove that there is an orthonormal basis of X consisting of eigenvectors of T.

10. Prove Theorem 337.

11. Prove Theorem 338.

12. Prove Theorem 339.

13. Prove Theorem 340.

14. Prove Theorem 342.

15. Prove Theorem 343.

16. Consider the following *boundary value problem* (BVP):

$$\begin{aligned} -u'' &= f,\ 0 < x < 1, \\ u(0) &= 0, \\ u(1) &= 0. \end{aligned}$$

In this problem, $f = f(x)$ is a given function defined on $[0, 1]$ and the goal is to find a function $u = u(x)$ satisfying the differential equation $u'' = f$ and also the *boundary conditions* $u(0) = 0$ and $u(1) = 0$. One way to (approximately) solve the BVP is to *discretize* it by replacing the exact derivative $-u''$ by the finite difference approximation

$$-u''(x) \doteq \frac{-u(x-h) + 2u(x) - u(x+h)}{h^2},$$

where h is a small positive number. To do this, we first establish a *grid* on $[0, 1]$ by defining $h = 1/n$, $x_i = ih$, $i = 0, 1, \ldots, n$. We know the values $u(x_0) = u(0) = 0$ and $u(x_n) = u(1) = 0$, so we will try to estimate the values $u(x_1), u(x_2), \ldots, u(x_{n-1})$. The estimates will be called $U_1, U_2, \ldots, U_{n-1}$: $U_i \doteq u(x_i)$. From the equations

$$-u''(x_i) = f(x_i),\ i = 1, 2, \ldots, n-1,$$

we obtain the discretized equation

$$\frac{-U_{i-1} + 2U_i - U_{i+1}}{h^2} = f(x_i),\ i = 1, 2, \ldots, n-1.$$

In the equation corresponding to $i = 1$, we use $U_{i-1} = U_0 = 0$, and similarly for $i = n$. The result is a system of $n - 1$ linear equations in $n - 1$ unknowns. In matrix vector form, the equation is $LU = F$, where $U = (U_1, U_2, \ldots, U_{n-1})$, $F = (f(x_0), f(x_1), \ldots, f(x_{n-1}))$, and

$$L = \begin{bmatrix} \frac{2}{h^2} & -\frac{1}{h^2} & & & \\ -\frac{1}{h^2} & \frac{2}{h^2} & -\frac{1}{h^2} & & \\ & \ddots & \ddots & \ddots & \\ & & -\frac{1}{h^2} & \frac{2}{h^2} & -\frac{1}{h^2} \\ & & & -\frac{1}{h^2} & \frac{2}{h^2} \end{bmatrix}$$

(all entries of L not shown are zero).[1] The matrix L is clearly symmetric;

[1]The matrix L is called the *1D discrete Laplacian.* The Laplacian is the differential operator

$$\Delta = \frac{\partial^2}{\partial x^2} + \frac{\partial^2}{\partial y^2}$$

in two variables or

$$\Delta = \frac{\partial^2}{\partial x^2} + \frac{\partial^2}{\partial y^2} + \frac{\partial^2}{\partial z^2}$$

in three; thus the Laplacian reduces to simply the second derivative operator for functions of one variable.

the purpose of this exercise is to show that L is positive definite by producing its eigenvalues and eigenvectors explicitly.

For each n, the function $\phi(x) = \sin(k\pi x)$ satisfies $-\phi'' = (k\pi)^2\phi$, $u(0) = u(1) = 0$. Thus we can regard ϕ as an *eigenfunction* of $-d^2/dx^2$ (subject to the given boundary conditions). The corresponding eigenvalue is $(k\pi)^2$.

(a) Show that for each $k = 1, 2, \ldots, n-1$, the discretized sine wave $U^{(k)}$,
$$U_i^{(k)} = \sin(k\pi x_i),\ i = 1, 2, \ldots, n-1,$$
is an eigenvector of L.

(b) Find the eigenvalue λ_k corresponding to $U^{(k)}$.

(c) Show that $\lambda_k > 0$ for $k = 1, 2, \ldots, n-1$ (which implies that L is positive definite.

The reader should realize that it is quite unusual for the eigenvectors of a matrix to be given explicitly by a simple analytic formula.

7.2 The spectral theorem for normal matrices

Given a matrix A, a similarity relationship $A = XBX^{-1}$ defines the matrix of the linear operator defined by A with respect to an alternate basis. To be precise, let $A \in F^{n\times n}$ be given, where F is a field, and let $T : F^n \to F^n$ be defined by $T(x) = Ax$ for all $x \in F^n$. Then A is the matrix of T with respect to the standard basis for F^n. If $A = XBX^{-1}$ and $\mathcal{X}$ is the basis consisting of the columns of X, then B is the matrix of T with respect to $\mathcal{X}$ (see Exercise 3). Also, for any $x \in F^n$,

$$x = XX^{-1}x \Rightarrow [x]_{\mathcal{X}} = X^{-1}x.$$

If $F = \mathbf{C}$ and A is Hermitian, then we can find X such that $X^{-1}AX$ is diagonal (with real diagonal entries) and X is a unitary matrix ($X^{-1} = X^*$). This means that if we work in the variables defined by the basis $\mathcal{X}$ whose elements are the columns of X, then the matrix becomes diagonal. Moreover, it is simple to move back and forth between the standard basis and the basis $\mathcal{X}$, because the change of variables is multiplication by X^* (there is no need to compute the inverse of X because X is unitary). This is one of the main advantages of a unitary transformation.[2]

This raises the following question: For which matrices can we obtain both of these advantages, that a change of variables yields a diagonal matrix, and

[2] Another advantage is numerical stability, which will be explored in Chapter 9.

that change of variables is defined by a unitary transformation? In other words, which matrices are diagonalizable by a unitary matrix? The following definition describes such matrices.

Definition 345 *Let $A \in \mathbf{C}^{n \times n}$. We say that A is* normal *if and only if*

$$A^* A = AA^*.$$

If A is Hermitian, the A is clearly normal. However, the converse need not be true. For example,

$$A = \begin{bmatrix} 2+i & i \\ i & 2+i \end{bmatrix}$$

is normal but not Hermitian.

We will now derive the properties of normal matrices, culminating with the result that a normal matrix can be diagonalized by a unitary matrix.

Lemma 346 *Let $A \in \mathbf{C}^{n \times n}$ be normal. Then*

$$\|Ax\|_2 = \|A^* x\|_2 \text{ for all } x \in \mathbf{C}^n.$$

Proof We have

$$\langle Ax, Ax \rangle_{\mathbf{C}^n} = \langle x, A^* Ax \rangle_{\mathbf{C}^n} = \langle x, AA^* x \rangle_{\mathbf{C}^n} = \langle A^* x, A^* x \rangle_{\mathbf{C}^n},$$

that is, $\|Ax\|_2^2 = \|A^* x\|_2^2$.

QED

Theorem 347 *Let $A \in \mathbf{C}^{n \times n}$ be normal. If $\lambda \in \mathbf{C}$, $x \in \mathbf{C}^n$ form an eigenpair of A, then $\overline{\lambda}, x$ form an eigenpair of A^*.*

Proof If A is normal, then so is $A - \lambda I$ for any complex number λ. Therefore, by the previous lemma, $\|(A^* - \overline{\lambda} I)x\|_2 = \|(A - \lambda I)x\|_2$ for all $x \in \mathbf{C}^n$, so

$$Ax = \lambda x \Rightarrow \|(A - \lambda I)x\|_2 = 0 \Rightarrow \|(A^* - \overline{\lambda} I)x\|_2 = 0 \Rightarrow A^* x = \overline{\lambda} x.$$

QED

Theorem 348 *Let $A \in \mathbf{C}^{n \times n}$ be normal. Then eigenvectors of A corresponding to distinct eigenvalues are orthogonal.*

Proof Suppose $Ax = \lambda x$, $Ay = \mu y$, with $\lambda \neq \mu$ and $x, y \neq 0$. Then

$$\lambda \langle x, y \rangle_{\mathbf{C}^n} = \langle \lambda x, y \rangle_{\mathbf{C}^n} = \langle Ax, y \rangle_{\mathbf{C}^n} = \langle x, A^* y \rangle_{\mathbf{C}^n} = \langle x, \overline{\mu} y \rangle_{\mathbf{C}^n} = \mu \langle x, y \rangle_{\mathbf{C}^n}.$$

Since $\lambda \neq \mu$, $\lambda \langle x, y \rangle_{\mathbf{C}^n} = \mu \langle x, y \rangle_{\mathbf{C}^n}$ implies that $\langle x, y \rangle_{\mathbf{C}^n} = 0$.

QED

Next, we wish to show that a normal matrix cannot be deficient. We will use the following lemma.

Lemma 349 *Let $A \in \mathbf{C}^{n\times n}$ be normal. Then*

$$\text{col}(A^*) = \text{col}(A) \text{ and } \mathcal{N}(A^*) = \mathcal{N}(A).$$

Proof We know that $\mathbf{C}^n$ is the direct sum of $\mathcal{N}(A)$ and $\text{col}(A^*)$. Let x belong to $\text{col}(A)$, say $x = Ay$, $y \in \mathbf{C}^n$. We can write y as $y = n + z$, $n \in \mathcal{N}(A)$, $z \in \text{col}(A^*)$, and $z \in \text{col}(A^*)$ implies that there exists $w \in \mathbf{C}^n$ such that $z = A^*w$. Then

$$x = Ay = An + Az = AA^*w = A^*Aw \in \text{col}(A^*)$$

(the reader should notice how the normality of A was used). Thus we have shown that $\text{col}(A) \subset \text{col}(A^*)$. The proof of the converse is exactly analogous and is left to the reader.

Now suppose $x \in \mathcal{N}(A)$. We then have

$$\begin{aligned} A^*Ax = 0 \Rightarrow AA^*x = 0 &\Rightarrow \langle x, AA^*x\rangle_{\mathbf{C}^n} = 0 \\ &\Rightarrow \langle A^*x, A^*x\rangle_{\mathbf{C}^n} = 0 \\ &\Rightarrow A^*x = 0 \\ &\Rightarrow x \in \mathcal{N}(A^*), \end{aligned}$$

and therefore $\mathcal{N}(A) \subset \mathcal{N}(A^*)$. Once again, the proof of the converse is similar.

QED

Theorem 350 *Let $A \in \mathbf{C}^{n\times n}$ be normal and let $\lambda \in \mathbf{C}$ be an eigenvalue of A. Then the geometric multiplicity of λ equals the algebraic multiplicity of λ.*

Proof Let λ be an eigenvalue of A. By Theorem 233, it suffices to prove that $\mathcal{N}((A-\lambda I)^2) = \mathcal{N}(A-\lambda I)$, and, according to Theorem 230, this holds if and only if $\text{col}(A-\lambda I) \cap \mathcal{N}(A-\lambda I) = \{0\}$. However, we already know that

$$\text{col}((A-\lambda I)^*) \cap \mathcal{N}(A-\lambda I) = \{0\},$$

and $\text{col}((A-\lambda I)^*) = \text{col}(A-\lambda I)$ by the previous lemma since A, and therefore $A-\lambda I$, is normal. This completes the proof.

QED

Putting together the preceding results, we obtain the spectral theorem for normal matrices.

Theorem 351 *Let $A \in \mathbf{C}^{n\times n}$ be normal. Then there exists a unitary matrix $X \in \mathbf{C}^{n\times n}$ and a diagonal matrix $D \in \mathbf{C}^{n\times n}$ such that*

$$A = XDX^*.$$

The only difference between the spectral theorems for Hermitian and normal matrices is that, in the case of a normal matrix, the eigenvalues need not be real.

The normality of a matrix is necessary as well as sufficient for it to be diagonalizable by a unitary matrix (see Exercise 5).

7.2.1 Outer products and the spectral decomposition

If $A \in \mathbf{C}^{n \times n}$ has the spectral decomposition $A = XDX^*$, where

$$X = [x_1|x_2|\cdots|x_n], \ D = \begin{bmatrix} \lambda_1 & & & \\ & \lambda_2 & & \\ & & \ddots & \\ & & & \lambda_n \end{bmatrix},$$

then, for any $v \in \mathbf{C}^n$,

$$\begin{aligned} Ax = XDX^*v &= XD \begin{bmatrix} \langle v, x_1 \rangle_{\mathbf{C}^n} \\ \langle v, x_2 \rangle_{\mathbf{C}^n} \\ \vdots \\ \langle v, x_n \rangle_{\mathbf{C}^n} \end{bmatrix} \\ &= X \begin{bmatrix} \lambda_1 \langle v, x_1 \rangle_{\mathbf{C}^n} \\ \lambda_2 \langle v, x_2 \rangle_{\mathbf{C}^n} \\ \vdots \\ \lambda_n \langle v, x_n \rangle_{\mathbf{C}^n} \end{bmatrix} \\ &= \sum_{i=1}^{n} \lambda_i \langle v, x_i \rangle_{\mathbf{C}^n} x_i. \end{aligned}$$

This is a simple representation of the operator defined by A in terms of its eigenvalues and eigenvectors. We can express this representation in terms of *outer products*.

Definition 352 *Let U and V be inner product spaces over $\mathbf{R}$ or $\mathbf{C}$. If $u \in U$ and $v \in V$, then the* outer product *of u and v is the operator $u \otimes v : V \to U$ defined by*

$$(u \otimes v)(w) = \langle w, v \rangle u.$$

We see that if there exists an orthonormal basis $\{x_1, x_2, \ldots, x_n\}$ of $\mathbf{C}^n$ consisting of eigenvectors of $A \in \mathbf{C}^{n \times n}$, and if $\lambda_1, \lambda_2, \ldots, \lambda_n$ are the corresponding eigenvalues, then

$$A = \sum_{i=1}^{n} \lambda_i (x_i \otimes x_i).$$

A word about notation

Some authors view $u \in \mathbf{R}^n$ as an $n \times 1$ matrix, in which case u^T is a $1 \times n$ matrix, and

$$u \cdot v = u^T v.$$

Here $u^T v$ denotes the product of the matrices u^T and v. Since u^T is $1 \times n$ and v is $n \times 1$, the product $u^T v$ is a 1×1 matrix, which is identified with the corresponding scalar. We then have $(u \otimes v)w = (uv^T)w$ for any vector w:

$$(uv^T)w = u(v^T w) = (v^T w)u = (v \cdot w)u = (u \otimes v)(w).$$

The reader should notice that, while matrix multiplication is not commutative in general, $v^T w$ is a 1×1 matrix and it is easy to see that $u(v^T w) = (v^T w)u$ is valid. Thus uv^T is the $n \times n$ matrix representing the operator $u \otimes v : \mathbf{R}^n \to \mathbf{R}^n$.

In the complex case, we can represent $u \otimes v$ by the matrix uv^*, where v^* represents the conjugate transpose of the $n \times 1$ matrix v.

Exercises

Miscellaneous exercises

1. Let $A \in \mathbf{R}^{2\times 2}$ be defined by
$$A = \begin{bmatrix} 1 & 1 \\ -1 & 1 \end{bmatrix}.$$
Show that A is normal and find the spectral decomposition of A.

2. Let $A \in \mathbf{R}^{3\times 3}$ be defined by
$$A = \begin{bmatrix} 0 & 1 & 1 \\ -1 & 0 & 1 \\ -1 & -1 & 0 \end{bmatrix}.$$
Show that A is normal and find the spectral decomposition of A.

3. Let $A \in F^{n\times n}$ be given, define $T : F^n \to F^n$ by $T(x) = Ax$ for all $x \in F^n$, and let $\mathcal{X} = \{x_1, \cdots, x_n\}$ be a basis for F^n. Prove that
$$[T]_{\mathcal{X},\mathcal{X}} = X^{-1}AX,$$
where $X = [x_1|\cdots|x_n]$.

4. Let $A \in \mathbf{C}^{n\times n}$ be normal. Prove that $A - \lambda I$ is normal for any $\lambda \in \mathbf{C}$.

5. Let $A \in \mathbf{C}^{n\times n}$. Prove:
   (a) If there exists a unitary matrix $X \in \mathbf{C}^{n\times n}$ and a diagonal matrix $D \in \mathbf{R}^{n\times n}$ such that $A = XDX^*$, then A is Hermitian.
   (b) If there exists a unitary matrix $X \in \mathbf{C}^{n\times n}$ and a diagonal matrix $D \in \mathbf{C}^{n\times n}$ such that $A = XDX^*$, then A is normal.

6. Let $A \in \mathbf{R}^{2\times 2}$ be defined by
$$A = \begin{bmatrix} a & b \\ c & d \end{bmatrix}.$$
Find necessary and sufficient conditions on $a, b, c, d \in \mathbf{R}$ for A to be normal.

7. Let $A \in \mathbf{R}^{n\times n}$. We say that A is *skew-symmetric* if $A^T = -A$.

   (a) Prove that any skew symmetric matrix is normal.

   (b) Prove that a skew symmetric matrix has only purely imaginary eigenvalues.

8. (a) Prove that an orthogonal matrix $A \in \mathbf{R}^{n\times n}$ is normal.

   (b) Prove that a unitary matrix $A \in \mathbf{C}^{n\times n}$ is normal.

9. Let $A \in \mathbf{C}^{n\times n}$ be upper triangular ($A_{ij} = 0$ for $i > j$) and normal. Prove that A is a diagonal matrix.

10. Prove that if $A, B \in \mathbf{C}^{n\times n}$ are normal and commute ($AB = BA$), then AB is normal.

11. Prove that if $A, B \in \mathbf{C}^{n\times n}$ are normal and commute ($AB = BA$), then $A + B$ is normal.

12. Suppose $A \in \mathbf{C}^{n\times n}$ is normal. Prove that there exists a polynomial $p \in \mathbf{C}[r]$ of degree at most $n-1$ such that $A^* = p(A)$. (Hint: Let p be the interpolating polynomial satisfying $p(\lambda_j) = \overline{\lambda}_j$, where $\lambda_1, \lambda_2, \ldots, \lambda_n$ are the eigenvalues of A.)

13. Prove the converse of Lemma 346: If $A \in \mathbf{C}^{n\times n}$ satisfies

$$\|A^*x\|_2 = \|Ax\|_2 \text{ for all } x \in \mathbf{C}^n,$$

    then A is normal.

14. Let V be an inner product space over $\mathbf{R}$ or $\mathbf{C}$, and let u, v be nonzero vectors in V. Find:

    (a) the rank of $u \otimes v$;

    (b) the eigenpairs of $u \otimes v$;

    (c) the characteristic polynomial, determinant, and trace of $u \otimes v$;

    (d) the adjoint of $u \otimes v$.

15. Let $A \in F^{m\times n}$ and $B \in F^{n\times p}$, where F represents $\mathbf{R}$ or $\mathbf{C}$. Find a formula for the product AB in terms of outer products of the columns of A and the rows of B.

16. Let V be an inner product space over $\mathbf{R}$ and let $u \in V$ have norm one. Define $T : V \to V$ by

$$T = I - 2u \otimes u,$$

    where $I : V \to V$ is the identity operator. Prove that T is self-adjoint ($T^* = T$) and orthogonal.

7.3 Optimization and the Hessian matrix

7.3.1 Background

A common problem in applied mathematics is to maximize or minimize a given function $f : \mathbf{R}^n \to \mathbf{R}$, usually called the *objective function.* Stated precisely, the problem (in minimization form) is to find a vector $x^* \in \mathbf{R}^n$ such that $f(x^*) \leq f(x)$ for all other allowable values of x. If all values of $x \in \mathbf{R}^n$ are allowable, then the problem is referred to as an *unconstrained* minimization problem. This problem is written briefly as

$$\min_{x \in \mathbf{R}^n} f(x). \tag{7.1}$$

Sometimes the allowable values of x are limited by *constraints*, such as an *equality constraint* $g(x) = 0$ or an *inequality constraint* $h(x) \geq 0$. The function g can be vector-valued, $g : \mathbf{R}^n \to \mathbf{R}^m$, in which case $g(x) = 0$ represents m equations, $g_1(x) = 0, g_2(x) = 0, \ldots, g_m(x) = 0$. In the same way, $h : \mathbf{R}^n \to \mathbf{R}^p$ can be vector-valued, in which case $h(x) \geq 0$ is interpreted to mean

$$h_1(x) \geq 0, h_2(x) \geq 0, \ldots, h_p(x) \geq 0.$$

The *nonlinear programming problem* (NLP) defined by f, g, and h is to find, among all the values of x satisfying $g(x) = 0$ and $h(x) \geq 0$, one with the smallest value of $f(x)$. An NLP is written in the following form:

$$\begin{aligned} \min \quad & f(x) \\ s.t. \quad & g(x) = 0, \\ & h(x) \geq 0. \end{aligned} \tag{7.2}$$

Any vector that maximizes f minimizes $-f$, and vice versa. Therefore, a maximization problem can be immediately transformed into a minimization problem, and so it is sufficient to develop theory and algorithms for minimization problems only.

In this section, we will study the unconstrained minimization problem (7.1). In the next section, we will discuss some aspects of a special case of (7.2), namely, the case in which there are only equality constraints. Our emphasis is on the role played by linear algebra in the theory of optimization problems.

When studying optimization problems, it is important to understand the difference between local and global solutions. If $f(x^*) \leq f(x)$ for all $x \in \mathbf{R}^n$, then x^* is called a *global minimizer* of f. Nominally, in solving (7.1), we are seeking a global minimizer. However, the theory and algorithms we develop will apply only to the problem of finding a *local minimizer.* The vector x^* is a local minimizer of f if there exists a positive number r such that

$$f(x^*) \leq f(x) \text{ for all } x \text{ satisfying } \|x - x^*\| < r. \tag{7.3}$$

The reason we restrict ourselves to the problem of finding local minimizers is that, in most cases, finding a global minimizer is too difficult. Given a proposed solution x^*, there is no information, computable from f and x^*, that will guarantee that $f(x)$ is not smaller for some value of x far from x^*. However, there is information computable from x^* that can guarantee that $f(x^*) \leq f(x)$ for all nearby values of x.

We will also use the following terminology: x^* is a *strict* local minimizer if

$$f(x^*) < f(x) \text{ for all } x \text{ satisfying } 0 < \|x - x^*\| < r. \tag{7.4}$$

The strict inequality in (7.4) distinguishes it from (7.3). If we merely say that x^* is a local minimizer, this leaves open the possibility that there are nearby values of x for which $f(x)$ is as small as $f(x^*)$.

7.3.2 Optimization of quadratic functions

A quadratic function $q : \mathbf{R}^n \to \mathbf{R}$ has the form

$$q(x) = \frac{1}{2}\sum_{i=1}^{n}\sum_{j=1}^{n} A_{ij}x_ix_j + \sum_{i=1}^{n} b_ix_i + c,$$

where A_{ij}, b_i, c are all real constants. The factor of $1/2$ is included for algebraic convenience; its role will become apparent below. The function q can be expressed in matrix-vector terms as

$$q(x) = \frac{1}{2}x \cdot Ax + b \cdot x + c.$$

Moreover, we might as well assume that A is symmetric, since A can be replaced by a symmetric matrix without changing the value of $x \cdot Ax$ (see Exercise 1).

We wish to minimize q, and we begin by noticing that this is simple when q has a particular form. If A is diagonal and $b = 0$, then

$$q(x) = \frac{1}{2}x \cdot Ax + c = c + \frac{1}{2}\sum_{i=1}^{n} A_{ii}x_i^2.$$

We then see immediately that $x^* = 0$ is the unique global minimizer of q if $A_{ii} > 0$ for $i = 1, 2, \ldots, n$. On the other hand, if any $A_{ii} < 0$, then q has no minimizer, since $q(\alpha e_i) \to -\infty$ as $\alpha \to \infty$. If $A_{ii} \geq 0$ for all i, with at least one $A_{ii} = 0$, then q has a global minimizer at $x = 0$ (with $q(0) = c$), but this minimizer is not unique: $q(\alpha e_i) = c$ for all $\alpha \in \mathbf{R}$.

Since the problem is simple when A is diagonal, it should not be much harder when A is diagonalizable. The matrix A defining q is symmetric, so it is diagonalizable: $A = UDU^T$, where $U \in \mathbf{R}^{n \times n}$ is orthogonal and $D \in \mathbf{R}^{n \times n}$ is diagonal. We then have

$$x \cdot Ax = x \cdot UDU^T x = \left(U^T x\right) \cdot D\left(U^T x\right) = y \cdot Dy,$$

where $y = U^T x$. This change of variables allows us to replace A by the diagonal matrix D.

However, there is another step involved in simplifying q; we also want to eliminate the linear term involving b. In the single variable case, this can be done by completing the square:

$$\begin{aligned}\frac{1}{2}ax^2 + bx + c &= \frac{a}{2}\left(x^2 - 2\left(-\frac{b}{a}\right)x + \left(-\frac{b}{a}\right)^2\right) + c - \frac{a}{2}\left(-\frac{b}{a}\right)^2 \\ &= \frac{a}{2}\left(x - \left(-\frac{b}{a}\right)\right)^2 + c - \frac{b^2}{2a} \\ &= \frac{a}{2}\left(x - x^*\right)^2 + c - \frac{b^2}{2a},\end{aligned}$$

where $x^* = -b/a$. When the matrix A is invertible, we can do the same thing in the multivariable case. For future reference, we remark that A is invertible precisely when all of the eigenvalues of A are nonzero. We define $x^* = -A^{-1}b$, so that $b = -Ax^*$, and then simplify $q(x)$ as follows:

$$\begin{aligned}q(x) &= \frac{1}{2}x \cdot Ax + b \cdot x + c \\ &= \frac{1}{2}x \cdot Ax - (Ax^*) \cdot x + c \\ &= \frac{1}{2}(x - x^*) \cdot A(x - x^*) + c - \frac{1}{2}x^* \cdot Ax^* \\ &= \frac{1}{2}(x - x^*) \cdot A(x - x^*) + \tilde{c},\ \tilde{c} = c - \frac{1}{2}x^* \cdot Ax^*.\end{aligned}$$

If we now define $y = U^T(x - x^*)$, we obtain

$$q(x) = \frac{1}{2}y \cdot Dy + \tilde{c}.$$

We draw the following conclusion: If all the eigenvalues of A are positive,[3] that is, if A is positive definite, then q has a unique global minimizer at $y = 0$, that is, at $x = x^*$. If A has at least one negative eigenvalue, then q is unbounded below and has no minimizer, local or global.

The case that A is singular is more subtle, and we leave the reader to justify the following conclusions (see Exercise 2). If $b \in \mathrm{col}(A)$, so that $Ax = -b$ has a solution x^*, then we can perform the change of variables, exactly as above. If all of the nonzero eigenvalues of A are positive, that is, if A is positive semidefinite, then q has a local minimizer at x^*. However, in this case, the solution to $Ax = -b$ is not unique; rather, the solution set is $x^* + \mathcal{N}(A)$. We therefore conclude that every vector in $x^* + \mathcal{N}(A)$ is a global minimizer of

[3] In this case, 0 is not an eigenvalue of A, A is invertible, and the above change of variables is well-defined.

q. If A has any negative eigenvalues, then q is unbounded below and has no minimizers.

It is also possible that A is singular and $b \notin \text{col}(A)$. In this case, q is unbounded below, even if A is positive semidefinite, because q decreases linearly in some direction. The reader is asked to determine this direction in Exercise 3.

Similar conclusions can be drawn about maximizers if A is *negative definite* or *negative semidefinite*. If A is *indefinite* (that is, has both positive and negative eigenvalues), then q is unbounded above and below, and has neither minimizers nor maximizers. In this case, we say that q has a *saddle point* at $x^* = -A^{-1}b$.

7.3.3 Taylor's theorem

We can extend our results about minimizers of quadratic functions to general nonlinear functions by using the local quadratic approximation to $f : \mathbf{R}^n \to \mathbf{R}$ defined by Taylor's theorem. We begin by reviewing this theorem in one variable.

If $\phi : \mathbf{R} \to \mathbf{R}$ is a function of one variable, then Taylor's theorem with remainder states that

$$\phi(\alpha) = \phi(0) + \phi'(0)\alpha + \frac{\phi''(0)}{2!}\alpha^2 + \ldots + \frac{\phi^{(k)}(0)}{k!}\alpha^k + \frac{\phi^{(k+1)}(c)}{(k+1)!}\alpha^{k+1}, \quad (7.5)$$

where c is some unknown number between 0 and α. The polynomial

$$\phi(0) + \phi'(0)\alpha + \frac{\phi''(0)}{2!}\alpha^2 + \ldots + \frac{\phi^{(k)}(0)}{k!}\alpha^k$$

is the *Taylor polynomial* of ϕ of degree k at $\alpha = 0$, and

$$\frac{\phi^{(k+1)}(c)}{(k+1)!}\alpha^{k+1}$$

is called the *remainder* term. The remainder term is the error when $\phi(\alpha)$ is approximated by the Taylor polynomial. Since the number c is unknown, we often write $O(\alpha^{k+1})$ in place of the remainder term to indicate that the error is bounded by a multiple of α^{k+1}; this assumes that the derivative $\phi^{(k+1)}$ is uniformly bounded over all possible values of c. In the context of optimization, we usually restrict ourselves to the linear and quadratic Taylor approximations, so the equations of interest are

$$\phi(t) = \phi(0) + \phi'(0)\alpha + O(\alpha^2) \quad (7.6)$$

and

$$\phi(t) = \phi(0) + \phi'(0)\alpha + \frac{\phi''(0)}{2}\alpha^2 + O(\alpha^3). \quad (7.7)$$

7.3.4 First- and second-order optimality conditions

When studying the minimization of a function ϕ of one variable, we use Taylor's theorem in the following fashion. If $\alpha = 0$ is a local minimizer of ϕ, then we have

$$\begin{aligned} & \phi(\alpha) \geq \phi(0) \text{ for all } \alpha \text{ near } 0 \\ \Rightarrow\ & \phi(0) + \phi'(0)\alpha + O(\alpha^2) \geq \phi(0) \text{ for all } \alpha \text{ near } 0 \\ \Rightarrow\ & \phi'(0)\alpha + O(\alpha^2) \geq 0 \text{ for all } \alpha \text{ near } 0. \end{aligned}$$

If $\phi'(0)$ were nonzero, then $O(\alpha^2)$ would be negligible compared to $\phi'(0)\alpha$ for all α sufficiently small, which would imply that $\phi'(0)\alpha \geq 0$ for both positive and negative values of α sufficiently small. Since this is not possible, it must be the case that $\phi'(0) = 0$ if $\alpha = 0$ is a local minimizer of ϕ.

The condition $\phi'(0) = 0$ is a well-known *necessary condition* for $\alpha = 0$ to be a local minimizer; however, this condition is not sufficient. It is also satisfied by a function for which $\alpha = 0$ is a local maximizer or a saddle point. For example, $\phi(\alpha) = \alpha^2$, $\phi(\alpha) = -\alpha^2$, and $\phi(\alpha) = \alpha^3$ all satisfy $\phi'(0) = 0$, but $\alpha = 0$ is a minimizer only for the first function.

There is also a second-order necessary condition, which we derive from the quadratic Taylor approximation. If $\alpha = 0$ is a local minimizer of ϕ, then

$$\begin{aligned} & \phi(\alpha) \geq \phi(0) \text{ for all } \alpha \text{ near } 0 \\ \Rightarrow\ & \phi(0) + \phi'(0)\alpha + \frac{\phi''(0)}{2}\alpha^2 + O(\alpha^3) \geq \phi(0) \text{ for all } \alpha \text{ near } 0 \\ \Rightarrow\ & \frac{\phi''(0)}{2}\alpha^2 + O(\alpha^3) \geq 0 \text{ for all } \alpha \text{ near } 0. \end{aligned}$$

To obtain the last inequality, we use the fact that $\phi'(0) = 0$ when $\alpha = 0$ is a local minimizer. If $\phi''(0) \neq 0$, then $O(\alpha^3)$ is negligible compared to $\phi''(0)\alpha^2/2$ for all α sufficiently small. It follows that $\phi''(0) \neq 0$ and

$$\frac{\phi''(0)}{2}\alpha^2 + O(\alpha^3) \geq 0 \text{ for all } \alpha \text{ near } 0$$

both hold if and only if $\phi''(0) \geq 0$. We therefore conclude that if $\alpha = 0$ is a local minimizer of ϕ, then $\phi''(0) \geq 0$. This is the *second-order necessary condition*. Once again, this condition is only necessary, as the example $\phi(\alpha) = \alpha^3$ shows. For this function, $\phi'(0) = 0$ and $\phi''(0) = 0$, so both the first- and second-order necessary conditions are satisfied. However, $\alpha = 0$ is not a local minimizer of ϕ.

A useful sufficient condition can be derived from the above results. Suppose $\phi'(0) = 0$ and $\phi''(0) > 0$. Then the quadratic Taylor approximation yields

$$\phi(\alpha) = \phi(0) + \phi'(0)\alpha + \frac{\phi''(0)}{2}\alpha^2 + O(\alpha^3) = \phi(0) + \frac{\phi''(0)}{2}\alpha^2 + O(\alpha^3).$$

Since $\phi''(0) > 0$, $O(\alpha^3)$ is negligible compared to $\phi''(0)\alpha^2/2$ for all α sufficiently small, so

$$\frac{\phi''(0)}{2}\alpha^2 + O(\alpha^3) > 0 \text{ for all } \alpha \text{ near } 0$$

and hence

$$\phi(\alpha) > \phi(0) \text{ for all } \alpha \text{ near } 0.$$

We therefore obtain the following sufficient conditions: If $\phi'(0) = 0$ and $\phi''(0) > 0$, then $\alpha = 0$ is a strict local minimizer of ϕ.

The necessary conditions and the sufficient conditions presented above are collectively referred to as *optimality* conditions.

To extend the above results from functions of one variable to $f : \mathbf{R}^n \to \mathbf{R}$, we choose any $x^*, p \in \mathbf{R}^n$ and define

$$\phi(\alpha) = f(x^* + \alpha p).$$

The chain rule from multivariable calculus yields the following formulas for the derivatives of ϕ:

$$\phi'(0) = \nabla f(x^*) \cdot p, \tag{7.8}$$

$$\phi''(0) = p \cdot \nabla^2 f(x^*) p. \tag{7.9}$$

The *gradient* of f at x is the vector of partial derivatives of f:

$$\nabla f(x) = \begin{bmatrix} \frac{\partial f}{\partial x_1}(x) \\ \frac{\partial f}{\partial x_2}(x) \\ \vdots \\ \frac{\partial f}{\partial x_n}(x) \end{bmatrix}.$$

The *Hessian* of f at x is the matrix of second partial derivatives of f:

$$\nabla^2 f(x) = \begin{bmatrix} \frac{\partial^2 f}{\partial x_1^2}(x) & \frac{\partial^2 f}{\partial x_2 \partial x_1}(x) & \cdots & \frac{\partial^2 f}{\partial x_n \partial x_1}(x) \\ \frac{\partial^2 f}{\partial x_1 \partial x_2}(x) & \frac{\partial^2 f}{\partial x_2^2}(x) & \cdots & \frac{\partial^2 f}{\partial x_n \partial x_2}(x) \\ \vdots & \vdots & \ddots & \vdots \\ \frac{\partial^2 f}{\partial x_1 \partial x_n}(x) & \frac{\partial^2 f}{\partial x_2 \partial x_n}(x) & \cdots & \frac{\partial^2 f}{\partial x_n^2}(x) \end{bmatrix}.$$

As long as f is sufficiently smooth, $\nabla^2 f(x)$ is always symmetric since the mixed second partial derivatives are equal:

$$\frac{\partial f^2}{\partial x_i \partial x_j}(x) = \frac{\partial f^2}{\partial x_j \partial x_i}(x) \text{ for all } i, j. \tag{7.10}$$

We are not interested in pathological functions f for which (7.10) fails.

We can now extend the optimality conditions for functions of one variable to $f : \mathbf{R}^n \to \mathbf{R}$. If x^* is a local minimizer of f, then

$$\phi'(0) = \nabla f(x^*) \cdot p$$

must equal zero for all directions $p \in \mathbf{R}^n$. By Corollary 274, this implies that

$$\nabla f(x^*) = 0, \tag{7.11}$$

which is the first-order necessary condition for x^* to be a local minimizer of f. Any point x^* that satisfies (7.11) is called a *stationary point* of f.

Similarly, if x^* is a local minimizer of f, then

$$\phi''(0) = p \cdot \nabla^2 f(x^*) p$$

must be nonnegative for all directions $p \in \mathbf{R}^n$. Therefore, the second-order necessary condition for x^* to be a local minimizer of f is that $\nabla^2 f(x^*)$ be positive semidefinite.

Finally, the sufficient conditions for x^* to be a strict local minimizer translate to $\nabla f(x^*) = 0$ and

$$\phi''(0) = p \cdot \nabla^2 f(x^*) p > 0 \text{ for all } p \in \mathbf{R}^n, p \neq 0.$$

If other words, if $\nabla f(x^*) = 0$ and $\nabla^2 f(x^*)$ is positive definite, then x^* is a strict local minimizer of f.

7.3.5 Local quadratic approximations

Taylor's theorem defines a local quadratic approximation to f near any point x^*. Writing $\phi(\alpha) = f(x^* + \alpha p)$, we obtain

$$\begin{aligned}
\phi(\alpha) &\doteq \phi(0) + \phi'(0) + \frac{\phi''(0)}{2}\alpha^2 \\
&= f(x^*) + (\nabla f(x^*) \cdot p)\,\alpha + \left(\frac{p \cdot \nabla^2 f(x^*) p}{2}\right)\alpha^2, \ \alpha \text{ near } 0.
\end{aligned}$$

We can take $\alpha = 1$ and $p = x - x^*$ to obtain

$$f(x) \doteq f(x^*) + \nabla f(x^*) \cdot (x - x^*) + \frac{1}{2}(x - x^*) \cdot \nabla^2 f(x^*)(x - x^*), \ x \text{ near } x^*.$$

We now see that the optimality conditions for x^* to be a *local* minimizer for f are the same as the conditions for x^* to be a *global* minimizer of the local quadratic approximation

$$q(x) = f(x^*) + \nabla f(x^*) \cdot (x - x^*) + \frac{1}{2}(x - x^*) \cdot \nabla^2 f(x^*)(x - x^*).$$

The quadratic q approximates f well near x^*, but may be completely misleading as an approximation to f far from x^*. For this reason, we can only draw local conclusions about the optimality of x^* from q, that is, from $\nabla f(x^*)$ and $\nabla^2 f(x^*)$.

We have seen how linear algebra is crucial in understanding the optimality conditions for unconstrained minimization. In the next section, we show that the same is true for equality-constrained nonlinear programming problems. Linear algebra also plays a critical role in algorithms for both unconstrained and constrained optimization; such algorithms, however, are beyond the scope of this book.

Exercises

1. Suppose $A \in \mathbf{R}^{n\times n}$ and define
$$A_{sym} = \frac{1}{2}\left(A + A^T\right).$$
Prove that A_{sym} is symmetric and that
$$x \cdot Ax = x \cdot A_{sym}x \text{ for all } x \in \mathbf{R}^n.$$

2. Let $A \in \mathbf{R}^{n\times n}$ be symmetric and positive semidefinite, but not positive definite. Define $q : \mathbf{R}^n \to \mathbf{R}$ by
$$q(x) = \frac{1}{2}x \cdot Ax + b \cdot x + c.$$
Prove that if $b \in \text{col}(A)$ and x^* is a solution to $Ax = -b$, then every vector in $x^* + \mathcal{N}(A)$ is a global minimizer of q.

3. Let $A \in \mathbf{R}^{n\times n}$ be symmetric and positive semidefinite, but not positive definite. Define $q : \mathbf{R}^n \to \mathbf{R}$ by
$$q(x) = \frac{1}{2}x \cdot Ax + b \cdot x + c.$$
Suppose $b \notin \text{col}(A)$. Find vectors $x^*, n \in \mathbf{R}^n$ such that
$$q(x^* + \alpha n) \to -\infty \text{ as } \alpha \to \infty.$$

4. Use the chain rule to verify (7.8) and (7.9).

5. Let $q : \mathbf{R}^2 \to \mathbf{R}^2$ be defined by $q(x) = (1/2)x \cdot (Ax) + b \cdot x$, where
$$A = \begin{bmatrix} 1 & 2 \\ 2 & 1 \end{bmatrix},\ b = \begin{bmatrix} -5 \\ -4 \end{bmatrix}.$$
Find all global minimizers of q, if it has any, or explain why none exist.

6. Repeat the preceding exercise with q defined by

$$A = \begin{bmatrix} 2 & 1 \\ 1 & 2 \end{bmatrix}, \; b = \begin{bmatrix} -1 \\ 4 \end{bmatrix}.$$

7. Repeat the preceding exercise with q defined by

$$A = \begin{bmatrix} 1 & 2 \\ 2 & 4 \end{bmatrix}, \; b = \begin{bmatrix} 1 \\ 2 \end{bmatrix}.$$

8. Let $f : \mathbf{R}^2 \to \mathbf{R}$ be defined by

$$f(x) = 100x_2^2 - 200x_1^2 x_2 + 100x_1^4 + x_1^2 - 2x_1 + 1.$$

   Show that $x^* = (1, 1)$ is a stationary point of f. If possible, determine whether it is a local minimizer, a local maximizer, or a saddle point.

7.4 Lagrange multipliers

Linear algebra plays an essential role in the theory of constrained optimization (nonlinear programming). We will consider the equality constrained nonlinear program

$$\min \quad f(x) \tag{7.12a}$$
$$s.t. \quad g(x) = 0, \tag{7.12b}$$

where $f : \mathbf{R}^n \to \mathbf{R}$ and $g : \mathbf{R}^n \to \mathbf{R}^m$ are given. The vector equation $g(x) = 0$ represents m individual constraints. We are interested in the case that $m < n$, in which case the the constraint $g(x) = 0$ defines lower-dimensional subset of $\mathbf{R}^n$ (such as a curve in $\mathbf{R}^2$ or a curve or surface in $\mathbf{R}^3$). We call

$$S = \{x \in \mathbf{R}^n \mid g(x) = 0\}$$

the *feasible* set. Any $x \in S$ is called a *feasible point*, while $x \in \mathbf{R}^n$, $x \notin S$ is *infeasible.*

A point $x^* \in \mathbf{R}^n$ is a local minimizer for (7.12) if x^* is feasible and there exists $r > 0$ such that

$$g(x) = 0, \; \|x - x^*\| < r \; \Rightarrow \; f(x^*) \leq f(x).$$

We wish to derive an optimality condition for x^* to be a local minimizer for (7.12), analogous to the first-order necessary condition $\nabla f(x^*) = 0$ for unconstrained minimization. To do this, we will use the concept of a *feasible* *path* to reduce the problem to one dimension. We say that $x:(-a,a)\to\mathbf{R}^n$ is a feasible path (for (7.12)) through $x^*\in S$ if

$$g(x(\alpha))=0 \text{ for all } \alpha\in(-a,a)$$

and $x(0)=x^*$.

We now suppose that x^* is a local minimizer for (7.12) and that

$$x:(-a,a)\to\mathbf{R}^n$$

is a feasible path through x^*. We define $\phi:(-a,a)\to\mathbf{R}$ by

$$\phi(\alpha)=f(x(\alpha)),\ \ -a<\alpha<a.$$

Then, since x^* is a local minimizer for the nonlinear program, we have in particular that $f(x^*)\le f(x(\alpha))$ for all α sufficiently small. But this implies that $\alpha=0$ is a local minimizer for ϕ, and hence that $\phi'(0)=0$. We have, by the chain rule,

$$\phi'(\alpha)=\nabla f(x(\alpha))\cdot x'(\alpha),$$

where $\nabla f(x)$ is the gradient of f at x and

$$x'(\alpha)=\begin{bmatrix}\frac{dx_1}{d\alpha}(\alpha)\\ \frac{dx_2}{d\alpha}(\alpha)\\ \vdots\\ \frac{dx_n}{d\alpha}(\alpha)\end{bmatrix}.$$

Therefore, $\phi'(0)=0$ is equivalent to

$$\nabla f(x^*)\cdot x'(0)=0. \tag{7.13}$$

This must hold for all feasible paths through x^*.

The condition (7.13) is not a useful optimality condition, because there are typically infinitely many feasible paths through x^*. To make (7.13) practical, we need a usable description of the set

$$Z=\{z\,|\,z=x'(0)\text{ for some feasible path } x \text{ through } x^*\}.$$

We notice that if x is a feasible path through x^*, then

$$\begin{aligned}&g(x(\alpha))=0\text{ for all }\alpha\in(-a,a)\\ \Rightarrow\ &g'(x(\alpha))x'(\alpha)=0\text{ for all }\alpha\in(-a,a)\\ \Rightarrow\ &g'(x^*)x'(0)=0,\end{aligned}$$

where $g'(x)$ is the *Jacobian matrix* of g at x:

$$g'(x)=\begin{bmatrix}\frac{\partial g_1}{\partial x_1}(x) & \frac{\partial g_1}{\partial x_2}(x) & \cdots & \frac{\partial g_1}{\partial x_n}(x)\\ \frac{\partial g_2}{\partial x_1}(x) & \frac{\partial g_2}{\partial x_2}(x) & \cdots & \frac{\partial g_2}{\partial x_n}(x)\\ \vdots & \vdots & \ddots & \vdots\\ \frac{\partial g_m}{\partial x_1}(x) & \frac{\partial g_m}{\partial x_2}(x) & \cdots & \frac{\partial g_m}{\partial x_n}(x)\end{bmatrix}.$$

In optimization theory, it is customary to denote the transpose of the Jacobian $g'(x)$ by $\nabla g(x)$ and call it the *gradient* of g:

$$\nabla g(x) = g'(x)^T.$$

The reader should notice that the columns of $\nabla g(x)$ are the gradients of the components of g:

$$\nabla g(x) = [\nabla g_1(x)|\nabla g_2(x)|\cdots|\nabla g_m(x)].$$

We therefore see that if x is a feasible path through x^* and $z = x'(0)$, then

$$z \in \mathcal{N}\left(\nabla g(x^*)^T\right).$$

If $\nabla g(x^*)$ has full rank, then it can be shown the converse is true: Given any $z \in \mathcal{N}\left(\nabla g(x^*)^T\right)$, there exists a feasible path x through x^* such that $x'(0) = z^*$. (The proof is rather involved and will not be given here; the interested reader can consult [42].) In other words, when $\nabla g(x^*)$ has full rank,

$$Z = \mathcal{N}\left(\nabla g(x^*)^T\right).$$

The necessary condition (7.13) then reduces to

$$\nabla f(x^*) \in \mathcal{N}\left(\nabla g(x^*)^T\right)^{\perp}.$$

But Theorem 308 shows that

$$\mathcal{N}\left(\nabla g(x^*)^T\right)^{\perp} = \operatorname{col}\left(\nabla g(x^*)\right).$$

Thus, if x^* is a local minimizer for (7.12), then

$$\nabla f(x^*) \in \operatorname{col}\left(\nabla g(x^*)\right),$$

that is, there exists $\lambda^* \in \mathbf{R}^m$ such that

$$\nabla f(x^*) = \nabla g(x^*)\lambda^*.$$

The vector λ^* is called a *Lagrange multiplier* for (7.12); its role is to transform the constrained minimization problem to the following system of $n+m$ equations in $n+m$ unknowns:

$$\nabla f(x) = \nabla g(x)\lambda, \tag{7.14a}$$

$$g(x) = 0. \tag{7.14b}$$

If x^* is a local minimizer of (7.12) and $\nabla g(x^*)$ has full rank, then there exists $\lambda^* \mathbf{R}^m$ such that $x = x^*$ and $\lambda = \lambda^*$ together satisfy the above system. Moreover, under these assumptions, the Lagrange multiplier is unique. We say that x^* is a *regular point* of the constraint $g(x) = 0$ if $g(x^*) = 0$ and $\nabla g(x^*)$ has full rank.

Example 353 *We consider the following problem in two variables:*

$$\begin{aligned} \min \quad & f(x) \\ s.t. \quad & g(x) = 0, \end{aligned}$$

where $f : \mathbf{R}^2 \to \mathbf{R}$ *and* $g : \mathbf{R}^2 \to \mathbf{R}$ *are defined by*

$$f(x) = x_1^2 + 2x_2^2, \; g(x) = x_1 + x_2 - 1.$$

In this example, there is a single constraint, and the feasible set is a line in the plane. We have

$$\nabla f(x) = \begin{bmatrix} 2x_1 \\ 4x_2 \end{bmatrix}, \; \nabla g(x) = \begin{bmatrix} 1 \\ 1 \end{bmatrix},$$

and the first-order optimality conditions, $\nabla f(x) = \lambda \nabla g(x)$*,* $g(x) = 0$*, yield three equations in three unknowns:*

$$\begin{aligned} 2x_1 &= \lambda, \\ 4x_2 &= \lambda, \\ x_1 + x_2 &= 1. \end{aligned}$$

It is easy to find the unique solution of this system:

$$x^* = \left(\frac{2}{3}, \frac{1}{3}\right), \; \lambda^* = \frac{4}{3}.$$

Moreover, it is easy to show that $f(x) \to \infty$ *as* $\|x\|_2 \to \infty$ *(*x *feasible), so* x^* *must minimize* f *subject to the constraint* $g(x) = 0$*.*

The system (7.14) is only a necessary condition for x^* to be a minimizer of (7.12), so solutions should be considered as candidates for solutions. Moreover, as with the necessary condition $\nabla f(x^*) = 0$ for unconstrained minimization, (7.14) is also satisfied by maximizers. Nevertheless, this system is useful for designing algorithms because a system of equations is much more tractable than an optimization problem. For instance, one can check whether x^*, λ^* satisfy the system of equations, but checking directly that x^* is a local minimizer requires comparing $f(x^*)$ with $f(x)$ for all (infinitely many) nearby vectors x.

In all but the simplest examples, the system (7.14) is nonlinear, so it is usually not possible to solve it analytically for x^*, λ^*. Therefore, practical algorithms are iterative and approach the exact solution only in the limit. Practical algorithms for nonlinear programming problems are beyond the scope of this book.

Exercises

1. Explain why, if x^* is a local minimizer of (7.12) and a regular point of the constraint $g(x) = 0$, then the Lagrange multiplier λ^* is unique.

2. Let $f : \mathbf{R}^3 \to \mathbf{R}$ and $g : \mathbf{R}^3 \to \mathbf{R}^2$ be defined by
$$f(x) = x_3, \ g(x) = \begin{bmatrix} x_1 + x_2 + x_3 - 12 \\ x_1^2 + x_2^2 - x_3 \end{bmatrix}.$$
Find both the maximizer and minimizer of f subject to the constraint $g(x) = 0$.

3. Let $f : \mathbf{R}^3 \to \mathbf{R}$ be defined by
$$f(x) = (x_1 - 3)^2 + 2(x_2 - 1)^2 + 3(x_3 - 1)^2$$
and $g : \mathbf{R}^3 \to \mathbf{R}^2$ by
$$g(x) = \begin{bmatrix} x_1^2 + x_2^2 + x_3^2 - 1 \\ x_1 + x_2 + x_3 \end{bmatrix}.$$
Find the minimizer and maximizer of f subject to the constraint $g(x) = 0$. (Hint: A computer algebra system might be useful to solve the equations that form the first-order necessary conditions.)

4. Consider the NLP (7.12), where $f : \mathbf{R}^3 \to \mathbf{R}$ and $g : \mathbf{R}^3 \to \mathbf{R}^2$ are defined by
$$f(x) = x_1^2 + (x_2 - 1)^2 + x_3^2, \ g(x) = \begin{bmatrix} x_2^2 - x_3 \\ 2x_2^2 - x_3 \end{bmatrix}.$$
   (a) Show directly that the feasible set is the x_1-axis, that is, the set $\{(x_1, 0, 0) : x_1 \in \mathbf{R}\}$.

   (b) Show that $x^* = (0, 0, 0)$ is the unique (global) solution of the NLP.

   (c) Show that there is no Lagrange multiplier λ^* satisfying
$$\nabla f(x^*) = \nabla g(x^*)\lambda^*.$$
   How can this be, given the discussion in the text?

5. The purpose of this exercise is to prove that, if $A \in \mathbf{R}^{n \times n}$ is symmetric, then
$$\lambda_1 \|x\|_2^2 \le x \cdot Ax \le \lambda_n \|x\|_2^2 \text{ for all } x \in \mathbf{R}^n, \tag{7.15}$$
where λ_1 and λ_n are the smallest and largest eigenvalues of A, respectively. Define $f : \mathbf{R}^n \to \mathbf{R}$ by $f(x) = x \cdot Ax$ and $g : \mathbf{R}^n \to \mathbf{R}$ by $g(x) = x \cdot x - 1$. Prove (7.15) by applying Lagrange multiplier theory to the problem of finding the maximum and minimum values of $f(x)$ subject to the constraint $g(x) = 0$.

6. Consider the following family of nonlinear programs indexed by u in $\mathbf{R}^m$:

$$\begin{aligned} \min \quad & f(x) \\ \text{s.t.} \quad & g(x) = u, \end{aligned}$$

where $f : \mathbf{R}^n \to \mathbf{R}$ and $g : \mathbf{R}^n \to \mathbf{R}^m$. Let $x = x(u)$ be a local minimizer of this NLP, where x is assumed to be a smooth function of u, and define $p(u) = f(x(u))$. Thus $p(u)$ is the optimal value of f for a given u. Assuming that $\nabla g(x(0))$ has full rank, prove that $\nabla p(0) = \lambda$, where λ is the Lagrange multiplier for the NLP when $u = 0$.

7. The purpose of this exercise is to confirm the result of the previous exercise for a specific example. Consider the NLP

$$\begin{aligned} \min \quad & f(x) \\ \text{s.t.} \quad & g(x) = u, \end{aligned}$$

where $f : \mathbf{R}^3 \to \mathbf{R}$ and $g : \mathbf{R}^3 \to \mathbf{R}$ are defined by $f(x) = x_1 + x_2 + x_3$ and $g(x) = x_1^2 + x_2^2 + x_3^2 - 1$. Find the solution $x = x(u)$ of this NLP, and compute $p(u)$, $\nabla p(0)$ explicitly. Also solve the NLP for $u = 0$ and find the Lagrange multiplier λ. Verify that $\nabla p(0) = \lambda$.

7.5 Spectral methods for differential equations

A popular method for solving certain linear differential equations involves finding the *eigenfunctions* of the differential operator defining the equation and expressing the solution as a linear combination of the eigenfunctions. Eigenfunctions are the eigenvectors of a differential operator. The use of eigenfunctions effectively diagonalizes the problem and makes it simple to compute the weights in the linear combination, and thus the solution of the equation. In this section, we will explain this use of eigenvalues and eigenfunctions for a simple model problem. As the reader will see, since the function spaces involved are infinite-dimensional, the differential operators have infinitely many eigenpairs, and the linear combinations become infinite series.

The model problem is the *boundary value problem* (BVP)

$$-\frac{d^2u}{dx^2} = f(x), \; 0 < x < \ell, \tag{7.16a}$$

$$u(0) = 0, \tag{7.16b}$$

$$u(\ell) = 0. \tag{7.16c}$$

Here f is given (known) function defined on $[0, \ell]$, and the goal is to find a function u defined on $[0,\ell]$ and satisfying both the differential equation (7.16a) and the *boundary conditions* (7.16b–7.16c). The particular boundary conditions appearing in the BVP (7.16) are called *Dirichlet* conditions to distinguish them from other types of boundary conditions.

The BVP (7.16) is fairly simple, in that it can be solved by two integrations (using the boundary conditions to determine the two constants of integration). However, it is a model for problems in two or three variables that can be solved by the method of eigenfunctions but not by direct integration, so the method we develop in this section is quite useful. In addition, the eigenfunction method can be used to solve time-dependent problems in one or more spatial variables.

7.5.1 Eigenpairs of the differential operator

We will begin our development of the method of eigenfunctions by computing the eigenpairs of the relevant differential operator. We define the space $C_D^2[0,\ell]$ by

$$C_D^2[0,\ell] = \left\{v \in C^2[0,\ell] \, : \, v(0) = v(\ell) = 0\right\}$$

and $L : C_D^2[0,\ell] \to C[0,\ell]$ by

$$L(v) = -\frac{d^2v}{dx^2}.$$

We wish to find nonzero functions that satisfy $L(v) = \lambda v$ for some scalar λ. (The reader should notice that the boundary conditions have been incorporated into the definition of the domain of L, so that the operator equation $L(v) = \lambda v$ expresses both the differential equation and the boundary conditions.)

If we impose the L^2 inner product on $C[0,\ell]$ (which contains $C_D^2[0,\ell]$ as a subspace), we can draw some useful conclusions about the eigenvalue problem $L(v) = \lambda v$. For any $u, v \in C_D^2[0,\ell]$, we have

$$\begin{aligned}
\langle L(u), v\rangle_{L^2(0,\ell)} &= \int_0^\ell -\frac{d^2u}{dx^2}(x)v(x)\,dx \\
&= -\frac{du}{dx}(x)v(x)\Big|_0^\ell + \int_0^\ell \frac{du}{dx}(x)\frac{dv}{dx}(x)\,dx \\
&= \int_0^\ell \frac{du}{dx}(x)\frac{dv}{dx}(x)\,dx \\
&= u(x)\frac{dv}{dx}(x)\Big|_0^\ell - \int_0^\ell u(x)\frac{d^2v}{dx^2}(x)\,dx \\
&= \langle u, L(v)\rangle_{L^2(0,\ell)}\,.
\end{aligned}$$

The reader should notice the use of integration by parts (twice), and also that the boundary terms canceled because u and v satisfy the boundary conditions.

The above calculation shows that L is symmetric with respect to the $L^2(0,\ell)$ inner product. Even if we extend $C[0,\ell]$, $C^2_D[0,\ell]$ to complex-valued functions and L to operate on these spaces, the symmetry of L still holds, and therefore L can have only real eigenvalues. The proof is exactly as for symmetric (or Hermitian) matrices: If $L(v) = \lambda v$ and $v \neq 0$, then

$$\begin{aligned} \lambda \langle v, v\rangle_{L^2(0,\ell)} = \langle \lambda v, v\rangle_{L^2(0,\ell)} &= \langle L(v), v\rangle_{L^2(0,\ell)} \\ &= \langle v, L(v)\rangle_{L^2(0,\ell)} \\ &= \langle v, \lambda v\rangle_{L^2(0,\ell)} \\ &= \overline{\lambda} \langle v, v\rangle_{L^2(0,\ell)}. \end{aligned}$$

This implies that $\lambda = \overline{\lambda}$, and hence that λ is real. Similarly, eigenfunctions of L corresponding to distinct eigenvalues are orthogonal with respect to the $L^2(0,\ell)$ inner product, as can be shown by extending the proof for symmetric matrices (see Exercise 6).

We also see that

$$\langle L(u), u\rangle_{L^2(0,\ell)} = \int_0^\ell \left(\frac{du}{dx}(x)\right)^2 dx \geq 0 \text{ for all } u \in C^2_D[0,\ell]. \tag{7.17}$$

Applying the boundary conditions and (7.17), it follows that L has only positive eigenvalues (see Exercise 7).

Armed with the knowledge that L has only real and positive eigenvalues, we will now proceed to find all of the eigenvalues and eigenfunctions of L. The equation $L(v) = \lambda v$ can be written as

$$u'' + \lambda u = 0. \tag{7.18}$$

Given that $\lambda > 0$, the general solution of (7.18) is

$$u(x) = c_1 \cos(\sqrt{\lambda}x) + c_2 \sin(\sqrt{\lambda}x)$$

(see Section 3.9). The boundary condition $u(0) = 0$ implies that $c_1 = 0$, so u reduces to $u(x) = c_2 \sin(\sqrt{\lambda}x)$. If u is to be an eigenfunction (and hence nonzero), it must be the case that $c_2 \neq 0$, so we are looking for eigenfunctions of the form $u(x) = \sin(\sqrt{\lambda}x)$. The boundary condition $u(\ell) = 0$ then yields

$$\begin{aligned} \sin(\sqrt{\lambda}\ell) = 0 \quad &\Rightarrow \quad \sqrt{\lambda}\ell = k\pi,\ k = 1,2,3,\ldots \\ &\Rightarrow \quad \lambda = \frac{k^2\pi^2}{\ell^2},\ k = 1,2,3,\ldots. \end{aligned}$$

We have now found all the eigenpairs of L:

$$\lambda_k = \frac{k^2\pi^2}{\ell^2},\ v_k = \sin\left(\frac{k\pi}{\ell}x\right),\ k = 1,2,3,\ldots. \tag{7.19}$$

It can be verified directly that if $m \neq k$, then $\langle v_m, v_k\rangle_{L^2(0,\ell)} = 0$, as guaranteed by the symmetry of L. Also,

$$\langle v_k, v_k\rangle_{L^2(0,\ell)} = \int_0^\ell \sin^2\left(\frac{k\pi}{\ell}x\right) = \frac{\ell}{2},\ k = 1,2,3,\ldots. \tag{7.20}$$

7.5.2 Solving the BVP using eigenfunctions

The idea of the spectral method (or method of eigenfunction expansion) is this: Suppose we wish to solve (7.16) and we can represent the right-hand-side function f as

$$f(x) = \sum_{k=1}^{\infty} c_k v_k(x) = \sum_{k=1}^{\infty} c_k \sin\left(\frac{k\pi}{\ell}x\right). \tag{7.21}$$

Since f is known, we will assume that the weights c_k can be computed (we will discuss how to do this below). Similarly, we assume that the solution u can be written as

$$u(x) = \sum_{k=1}^{\infty} a_k v_k(x) = \sum_{k=1}^{\infty} a_k \sin\left(\frac{n\pi}{\ell}x\right), \tag{7.22}$$

where the coefficients $a_1, a_2, \ldots$ are to be found. The process of finding c_k and then a_k can be explained easily if we proceed in a formal manner (that is, if we do not worry about convergence and related issues). If

$$f = \sum_{k=1}^{\infty} c_k v_k,$$

then, taking the inner product of both sides of this equation with v_k, we obtain

$$\langle f, v_k \rangle_{L^2(0,\ell)} = \left\langle \sum_{j=1}^{\infty} c_j v_j, v_k \right\rangle_{L^2(0,\ell)} = \sum_{j=1}^{\infty} c_j \langle v_j, v_k \rangle_{L^2(0,\ell)} = \frac{\ell}{2} c_k,$$

the last step following from the fact that $\langle v_j, v_k \rangle_{L^2(0,\ell)} = 0$ if $j \neq k$ and $\ell/2$ if $j = k$.[4] It follows that

$$c_k = \frac{2}{\ell} \langle f, v_k \rangle_{L^2(0,\ell)} = \frac{2}{\ell} \int_0^{\ell} f(x) \sin\left(\frac{n\pi}{\ell}x\right) dx, \ k = 1, 2, 3, \ldots. \tag{7.23}$$

Then, with u given by (7.22), we have[5]

$$L(u) = L\left(\sum_{k=1}^{\infty} a_k v_k\right) = \sum_{k=1}^{\infty} a_k L(v_k) = \sum_{k=1}^{\infty} a_k \lambda_k v_k.$$

---

[4] We have already made two unjustified assumptions. The first is that f can be expanded in terms of the eigenfunctions. The second is that

$$\left\langle \sum_{j=1}^{\infty} c_j v_j, v_k \right\rangle_{L^2(0,\ell)} = \sum_{j=1}^{\infty} c_j \langle v_j, v_k \rangle_{L^2(0,\ell)},$$

that is, that the bilinearity of the inner product extends to an infinite sum.

[5] Another unjustified calculation.

Then, equating the series for $L(u)$ and f, we obtain

$$\sum_{k=1}^{\infty} a_k \lambda_k v_k = \sum_{k=1}^{\infty} c_k v_k \Rightarrow a_k \lambda_k = c_k, \; k = 1, 2, 3, \ldots$$

$$\Rightarrow a_k = \frac{c_k}{\lambda_k} = \frac{\ell^2 c_k}{n^2 \pi^2}, \; k = 1, 2, 3, \ldots.$$

Thus we have obtained the values of the coefficients $a_1, a_2, \ldots$, and the solution u has been determined.

Justifying the above calculations is fairly straightforward, although one key step is beyond the scope of this book. If we examine the formula for c_k, we see that it is given by

$$c_k = \frac{\langle f, v_k \rangle_{L^2(0,\ell)}}{\langle v_k, v_k \rangle_{L^2(0,\ell)}},$$

which should look familiar: these are the weights obtained when projecting f onto a subspace spanned by $v_1, v_2, \ldots, v_k$ (cf. Equation (6.23). Therefore, the partial sum

$$\sum_{k=1}^{n} c_k v_k \tag{7.24}$$

is the best approximation (in the $L^2(0, \ell)$ norm) to f from the subspace

$$S_n = \text{sp}\{v_1, v_2, \ldots, v_n\}.$$

As we include more and more eigenfunctions—that is, as the dimension of S_n grows—the best approximation to f from S_n will get closer and closer to f. It is not obvious, however, that the partial sums given by (7.24) will actually converge to f. In fact, however, this is true, as long as we interpret convergence in terms of the L^2 norm. It can be proved that

$$f = \sum_{k=1}^{\infty} c_k v_k$$

for any $f \in L^2(0, \ell)$ in the sense that

$$\left\| f - \sum_{k=1}^{n} c_k v_k \right\| \to 0 \text{ as } n \to \infty$$

(see Chapter 8 of the author's book [13] on partial differential equations). In general, it may not be the case that the infinite series converges to $f(x)$ for every value of $x \in [0, \ell]$. However, if f is sufficiently smooth and satisfies the boundary conditions $f(0) = f(1) = 0$, then, in fact, the series converges to $f(x)$ for each $x \in [0, \ell]$. This need not be true for the function f in this context, as f is only assumed to lie in $C[0,\ell]$, but the corresponding series for the solution u would converge to $u(x)$ for each $x\in[0,\ell]$, since $u\in C_D^2[0,\ell]$ by assumption.

The series (7.21), with the coefficients c_k defined by (7.23), is called a *Fourier sine series.* There are several types of Fourier series, defined by sines and/or cosines, each suitable for different boundary conditions. The interested reader can consult [13] for more details.

Let us now examine the series for u produced by the formal procedure described above:

$$u(x)=\sum_{k=1}^{\infty}a_k v_k,\ a_k=\frac{\ell^2 c_k}{k^2\pi^2}=\frac{2\ell}{k^2\pi^2}\int_0^{\ell}f(x)\sin\left(\frac{k\pi}{\ell}x\right)dx.$$

On the other hand, the Fourier sine series for u is defined by the coefficient

$$\hat{a}_k=\frac{2}{\ell}\int_0^{\ell}u(x)\sin\left(\frac{k\pi}{\ell}x\right)dx,\ k=1,2,3,\ldots.$$

We manipulate this formula by integrating twice:

$$\begin{aligned}
&\frac{2}{\ell}\int_0^{\ell}u(x)\sin\left(\frac{k\pi}{\ell}x\right)dx\\
&=\frac{2}{\ell}\left[-\frac{\ell}{k\pi}u(x)\cos\left(\frac{k\pi}{\ell}x\right)\Big|_0^{\ell}+\frac{\ell}{k\pi}\int_0^{\ell}\frac{du}{dx}(x)\cos\left(\frac{k\pi}{\ell}x\right)dx\right]\\
&=\frac{2}{k\pi}\int_0^{\ell}\frac{du}{dx}(x)\cos\left(\frac{k\pi}{\ell}x\right)dx\\
&=\frac{2}{k\pi}\left[\frac{\ell}{k\pi}\frac{du}{dx}(x)\sin\left(\frac{k\pi}{\ell}x\right)\Big|_0^{\ell}-\frac{\ell}{k\pi}\int_0^{\ell}\frac{d^2u}{dx^2}\sin\left(\frac{k\pi}{\ell}x\right)dx\right]\\
&=\frac{2\ell}{k^2\pi^2}\int_0^{\ell}\left(-\frac{d^2u}{dx^2}\right)\sin\left(\frac{k\pi}{\ell}x\right)dx.
\end{aligned}$$

In the first integration by parts, we use the fact that $u(0)=u(\ell)=0$ to eliminate the boundary term; in the second, we use the fact that $\sin(k\pi x/\ell)$ is zero at $x=0$ and $x=\ell$. Since u solves the differential equation, we have $-du^2/dx^2=f$, and therefore

$$\hat{a}_k=\frac{2\ell}{k^2\pi^2}\int_0^{\ell}f(x)\sin\left(\frac{k\pi}{\ell}x\right)dx=a_k.$$

We see that the series for u computed by the method of eigenfunction expansion is exactly the Fourier sine series of u, which converges to u by the theorem mentioned above.

Example 354 *Consider the BVP*

$$\begin{aligned} -\frac{d^2u}{dx^2} &= x,\ 0 < x < 1, \\ u(0) &= 0, \\ u(1) &= 0 \end{aligned}$$

($\ell = 1$, $f(x) = x$). The Fourier sine coefficients of $f(x) = x$ are

$$c_k = 2\int_0^\ell x \sin(k\pi x)\,dx = \frac{2(-1)^{k+1}}{k\pi},\ k = 1, 2, 3, \ldots.$$

The eigenvalues of $-d/dx^2$ on $[0, 1]$ are $\lambda_k = n^2\pi^2$, so the Fourier sine coefficients of the solution u are

$$a_k = \frac{c_k}{\lambda_k} = \frac{2(-1)^{k+1}}{k^3\pi^3},\ k = 1, 2, 3, \ldots.$$

The solution to the BVP is therefore

$$u(x) = \sum_{k=1}^{\infty} \frac{2(-1)^{k+1}}{k^3\pi^3} \sin(k\pi x).$$

As mentioned above, BVPs such as (7.16) can be solved by two integrations, using the boundary conditions to determine the constants of integration. In this case, we obtain the solution $u(x) = (x - x^3)/6$, which we can compare to the formula for u defined by the infinite series. Figure 7.1 shows the graph of the exact u with four approximations to u defined by the partial Fourier sine series

$$\sum_{k=1}^{n} \frac{2(-1)^{k+1}}{k^3\pi^3} \sin(k\pi x)$$

for $n = 1, 2, 3, 4$. The graphs show that the Fourier sine series converges rapidly to u.

There is a large class of differential equations to which the method of eigenfunction expansion is applicable. However, the eigenpairs can be found explicitly (as in the model problem (7.16)) only for fairly simple differential equations on simple domains. Nevertheless, some of these cases are interesting because they arise in real applications. For example, Laplace's equation is

$$-\Delta u = 0 \text{ in } \Omega,$$

where Δ is the partial differential operator

$$\Delta = \frac{\partial^2}{\partial x^2} + \frac{\partial^2}{\partial y^2}$$

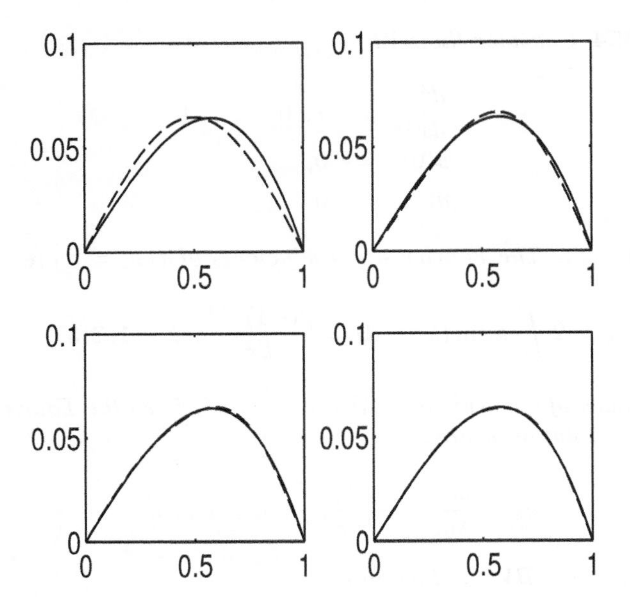

FIGURE 7.1
The solution $u(x) = (x - x^3)/6$ to the BVP of Example 354 and its approximation by partial Fourier sine series. Each plot shows the solution u (solid curve) and the partial Fourier sine series with n terms (dashed line). Upper left: $n = 1$; upper right: $n = 2$; lower left: $n = 3$; lower right: $n = 4$.

in two variables or

$$\Delta = \frac{\partial^2}{\partial x^2} + \frac{\partial^2}{\partial y^2} + \frac{\partial^2}{\partial z^2}$$

in three variables. The set Ω represents a domain in $\mathbf{R}^2$ or $\mathbf{R}^3$. In the two-dimensional case, it is possible to find the eigenvalues and eigenvectors of $-\Delta$ explicitly when Ω is a rectangular or circular domain. The BVP

$$\begin{aligned} -\Delta u &= f \text{ in } \Omega, \\ u &= 0 \text{ on } \partial\Omega \end{aligned}$$

describes the small vertical displacements of a drum whose shape is defined by Ω and which is clamped around the boundary. The function f would represent a small transverse pressure. The eigenvalues of Δ (on Ω and subject to these boundary conditions) are the natural frequencies at which the drum resonates.[6]

[6]An interesting question was posed in 1966 by Kac in the paper *Can one hear the shape of a drum?* [23]. The idea expressed by his title is that the eigenvalues of Δ on Ω are the frequencies we hear. The question, then, is if the collection of these natural frequencies

If Ω forms a more complicated geometric figure than a rectangle or a circle, or the differential operator is more complicated than Δ, then it is probably not possible to find the eigenvalues and eigenfunctions explicitly. However, there are numerical methods to estimate the eigenpairs corresponding to the smallest eigenvalues. As Example 354 suggests, sometimes relatively few eigenpairs can give a good approximation to the solution of a BVP. Therefore, the method of eigenfunction expansion might still be useful in such a case.

Exercises

1. Use the methods of Fourier series to find the solution of the BVP

$$-\frac{d^2u}{dx^2} = 1,\ 0 < x < 1,\ u(0) = u(1) = 0.$$

   Find the solution by integrating twice and compare (graphically) the true solution with several partial sums of the Fourier series.

2. Repeat the previous exercise for the BVP

$$-\frac{d^2u}{dx^2} = e^x,\ 0 < x < 1,\ u(0) = u(1) = 0.$$

3. Consider the operator $M : C_D^2[0,\ell] \to C[0,\ell]$ defined by

$$M(u) = -\frac{d^2u}{dx^2} + u.$$

   (a) Show that M is symmetric with respect to the $L^2(0,1)$ inner product, and has only positive eigenvalues.

   (b) Find the eigenvalues and eigenfunctions of M.

4. Use the results of the preceding exercise to find the solution of

$$-\frac{d^2u}{dx^2} + u = 1 - x,\ 0 < x < 1,\ u(0) = u(1) = 0.$$

5. Define

$$V = \left\{u \in C^2[0,\ell] \,:\, u'(0) = u'(\ell) = 0\right\}$$

   and $K : V \to C[0,\ell]$ by $K(u) = -u''$.

completely determine the shape of the drum. In other words, given an increasing sequence of positive real numbers (tending to infinity), is there one and only domain Ω such that the eigenvalues of Δ on Ω form the given sequence? The question was finally answered in the negative by Gordon, Webb, and Wolpert in 1992 [16].

(a) Show that K is symmetric with respect to the $L^2(0,\ell)$ inner product.

(b) Show that all the eigenvalues of K are nonpositive, and that 0 is an eigenvalue of K.

(c) Find the eigenvalues and eigenfunctions of K.

6. Let $L : C_D^2[0,\ell] \to C[0,\ell]$ be defined by $L(v) = -v''$. Prove that if $L(v_1) = \lambda_1 v_1$, $L(v_2) = \lambda_2 v_2$, $\lambda_1 \neq \lambda_2$, $v_1, v_2 \neq 0$, then $\langle v_1, v_2 \rangle_{L^2(0,\ell)} = 0$. (Hint: Mimic the proof of Theorem 329.)

7. Let $L : C_D^2[0,\ell] \to C[0,\ell]$ be defined by $L(v) = -v''$. Use (7.17) to prove that if $v \in C_D^2[0,\ell]$, $v \neq 0$, $L(v) = \lambda v$, then $\lambda > 0$. (Hint: First use (7.17) and the boundary conditions to prove that $\langle L(u), u \rangle_{L^2(0,\ell)} = 0$ implies $u = 0$.)

8

The singular value decomposition

The Jordan decomposition is one answer to the following question: Given a finite-dimensional vector space X and a linear operator $T : X \to X$, how can we choose a basis for X so the matrix representing T is as simple as possible? As we saw in the previous chapter, it is not always possible to obtain a diagonal matrix; moreover, the basis that results in the Jordan canonical form cannot always be chosen to be orthonormal. Furthermore, the calculation of the Jordan decomposition is problematic in finite-precision arithmetic.

In this chapter, we show that it is always possible to represent a linear transformation by a diagonal matrix using orthonormal bases. To accomplish this, it is necessary (in general) to use different bases for the domain and co-domain of the operator. On the other hand, the result, which is called the *singular value decomposition* (SVD), is applicable even if the domain and co-domain are different (finite-dimensional) vector spaces: $T : X \to U$. The Jordan decomposition only applies to an operator of the form $T : X \to X$.

8.1 Introduction to the SVD

For simplicity, we will develop the SVD first for a square, nonsingular matrix $A \in \mathbf{C}^{n\times n}$. The reader will recall from Section 7.1.2 that A^*A is Hermitian and positive definite in this case. Therefore, there exists an orthonormal basis of $\mathbf{C}^n$ corresponding to eigenvectors of A^*A, and the eigenvalues of A^*A are known to be positive.

Expressed in matrix form, these facts imply that there exist a unitary matrix $V \in \mathbf{C}^{n\times n}$ and a diagonal matrix $D \in \mathbf{R}^{n\times n}$, with positive diagonal entries, such that

$$A^*A = VDV^*.$$

Since the diagonal entries of D are positive, we can write them as $\sigma_1^2, \sigma_2^2, \ldots, \sigma_n^2$. Moreover, by ordering the columns of V (which are the eigenvectors of A^*A) appropriately, we can arrange that $\sigma_1^2 \geq \sigma_2^2 \geq \cdots \geq \sigma_n^2 > 0$. We now write

$v_1, v_2, \ldots, v_n$ for the columns of V:

$$V = [v_1|v_2|\cdots|v_n], \; D = \begin{bmatrix} \sigma_1^2 & & & \\ & \sigma_2^2 & & \\ & & \ddots & \\ & & & \sigma_n^2 \end{bmatrix}.$$

We consider the vectors $Av_1, Av_2, \ldots, Av_n \in \mathbf{C}^n$, which turn out to be orthogonal:

$$\begin{aligned} \langle Av_i, Av_j \rangle_{\mathbf{C}^n} = \langle A^*Av_i, v_j \rangle_{\mathbf{C}^n} &= \left\langle \sigma_i^2 v_i, v_j \right\rangle_{\mathbf{C}^n} \\ &= \sigma_i^2 \left\langle v_i, v_j \right\rangle_{\mathbf{C}^n} \\ &= \begin{cases} \sigma_i^2, & i = j, \\ 0, & i \neq j. \end{cases} \end{aligned}$$

This calculation also shows that $\|Av_i\|_2 = \sigma_i$. We therefore define

$$u_i = \sigma_i^{-1} A v_i, \; i = 1, 2, \ldots, n,$$

and conclude that $\{u_1, u_2, \ldots, u_n\}$ is another orthonormal basis for $\mathbf{C}^n$. If $U = [u_1|u_2|\cdots|u_n]$ and

$$\Sigma = \begin{bmatrix} \sigma_1 & & & \\ & \sigma_2 & & \\ & & \ddots & \\ & & & \sigma_n \end{bmatrix},$$

then $Av_i = \sigma_i u_i$, $i = 1, 2, \ldots, n$, implies that

$$AV = U\Sigma.$$

Since the columns of U are orthonormal, U is unitary, as is V, and we obtain

$$A = U\Sigma V^*. \tag{8.1}$$

This is called the *singular value decomposition* (SVD) of A. The numbers $\sigma_1, \sigma_2, \ldots, \sigma_n$ are called the *singular values* of A, the vectors $v_1, v_2, \ldots, v_n$ are the *right singular vectors*, and $u_1, u_2, \ldots, u_n$ are the *left singular vectors*.

While the singular values of a matrix are unique, the singular vectors and hence the SVD itself are not unique. For instance, if v_i is a right singular vector of A, with corresponding singular σ_i and left singular value u_i, we could multiply both v_i and u_i by -1 (or by any complex number α with $|\alpha| = 1$) and obtain another equally valid right singular vector/left singular vector pair. Exercise 8 asks the reader to explore the question of nonuniqueness of the SVD in more detail. Since the SVD is not unique, we should speak of *an*

SVD instead of *the* SVD. However, we will follow common practice and refer to the SVD of a matrix.[1]

If x is any vector in $\mathbf{C}^n$, then

$$\begin{aligned} Ax = U\Sigma V^*x = \sum_{i=1}^{n}(\Sigma V^*x)_i u_i &= \sum_{i=1}^{n}\sigma_i (V^*x)_i u_u \\ &= \sum_{i=1}^{n}\sigma_i \langle x, v_i\rangle_{\mathbf{C}^n} u_i \\ &= \sum_{i=1}^{n}\sigma_i (u_i \otimes v_i)x \\ &= \left(\sum_{i=1}^{n}\sigma_i (u_i \otimes v_i)\right)x. \end{aligned}$$

Therefore, if $T : \mathbf{C}^n \to \mathbf{C}^n$ is the operator defined by the matrix A, we see that

$$T = \sum_{i=1}^{n}\sigma_i (u_i \otimes v_i).$$

We recall that the operator $u_i \otimes v_i$ is represented by the matrix $u_i v_i^*$.[2] Therefore,

$$A = \sum_{i=1}^{n}\sigma_i u_i v_i^*.$$

Given any $x \in \mathbf{C}^n$, we can express x in terms of the orthonormal basis

$$\{v_1, v_2, \ldots, v_n\}$$

as

$$x = \sum_{j=1}^{n}\langle x, v_j\rangle_{\mathbf{C}^n} v_j.$$

Comparing the formulas

$$\begin{aligned} x &= \sum_{j=1}^{n}\langle x, v_j\rangle_{\mathbf{C}^n} v_j, \\ Ax &= \sum_{i=1}^{n}\sigma_i \langle x, v_i\rangle_{\mathbf{C}^n} u_i, \end{aligned}$$

we see that A acts on x in a particularly simple way: The component of Ax in the u_i direction is just the component of x in the v_i direction, scaled by σ_i.

---

[1]The same ambiguity applies to the spectral decomposition of a diagonalizable matrix.

[2]In the expression $u_i v_i^*$, the vectors u_i, v_i are regarded as $n \times 1$ matrices, and v_i^* is a $1 \times n$ matrix, the conjugate transpose of v_i. It follows that the k, ℓ entry in the matrix $u_i v_i^*$ is $(u_i)_k(\overline{v_i})_\ell$. This was explained in Section 7.2.1.

Example 355 *Let A be the following 2×2 matrix:*

$$A=\begin{bmatrix} 2 & 3 \\ 0 & 2 \end{bmatrix}.$$

This matrix is not diagonalizable; it has a single eigenvalue $\lambda = 2$ with algebraic multiplicity two but geometric multiplicity one. The reader can verify the following calculations.

We have

$$A^TA=\begin{bmatrix} 4 & 6 \\ 6 & 13 \end{bmatrix},$$

so $p_{A^TA} = r^2 - 17r + 16$ and the eigenvalues of A^TA are $\sigma_1^2 = 16$ and $\sigma_2^2 = 1$. Corresponding eigenvectors are

$$v_1=\begin{bmatrix} \frac{1}{\sqrt{5}} \\ \frac{2}{\sqrt{5}} \end{bmatrix},\ v_2=\begin{bmatrix} -\frac{2}{\sqrt{5}} \\ \frac{1}{\sqrt{5}} \end{bmatrix}.$$

We then define

$$u_1=\sigma_1^{-1}Av_1=\begin{bmatrix} \frac{2\sqrt{5}}{5} \\ \frac{\sqrt{5}}{5} \end{bmatrix},\ u_2=\sigma_2^{-1}Av_2=\begin{bmatrix} \frac{-\sqrt{5}}{5} \\ \frac{2\sqrt{5}}{5} \end{bmatrix}.$$

Finally, $A = U\Sigma V^T$, where

$$U=\begin{bmatrix} \frac{2\sqrt{5}}{5} & -\frac{\sqrt{5}}{5} \\ \frac{\sqrt{5}}{5} & \frac{2\sqrt{5}}{5} \end{bmatrix},\ \Sigma=\begin{bmatrix} 4 & 0 \\ 0 & 1 \end{bmatrix},\ V=\begin{bmatrix} \frac{\sqrt{5}}{5} & -\frac{2\sqrt{5}}{5} \\ \frac{2\sqrt{5}}{5} & \frac{\sqrt{5}}{5} \end{bmatrix}.$$

The significance of the SVD can be explained in terms of the previous example. If $T:\mathbf{R}^2\to\mathbf{R}^2$ is the operator defined by $T(x)=Ax$, then there is no choice of basis $\mathcal{B}$ for $\mathbf{R}^2$ such that the matrix $[T]_{\mathcal{B},\mathcal{B}}$ is diagonal. However, if we are willing to use different bases for the domain and co-domain, then we see that $\Sigma = [T]_{\mathcal{V},\mathcal{U}}$ is diagonal for $\mathcal{V}=\{v_1,v_2\}$ and $\mathcal{U}=\{u_1,u_2\}$. Moreover, the bases $\mathcal{V}$ and $\mathcal{U}$ are orthonormal. There are two reasons orthonormal bases are preferred:

1. As we have seen in previous chapters, many calculations are simplified when the basis is orthonormal. This is particularly true for the related calculations of expressing a given vector in terms of the basis and projecting a given vector onto the subspace spanned by the basis.

2. When performing calculations in finite-precision arithmetic, round-off error is reduced when using orthonormal bases (or equivalently, when using orthogonal or unitary matrices). This point will be discussed in Chapter 9, although the reader can consult a book on numerical linear algebra, such as [43] or [19], for a detailed study.

8.1.1 The SVD for singular matrices

If $A \in \mathbf{C}^{n\times n}$ is singular, then A^*A is only positive semidefinite, not positive definite, which implies that A^*A has one or more zero eigenvalues. Let us see how this changes the development of the SVD.

Since A^*A is symmetric and positive semidefinite, it has orthonormal eigenvectors $v_1, v_2, \ldots, v_n$ and corresponding eigenvalues $\sigma_1^2, \sigma_2^2, \ldots, \sigma_n^2 \geq 0$. As before, we order the eigenvalues and eigenvectors so that $\sigma_1 \geq \sigma_2 \geq \cdots \geq \sigma_n$. Now, however, there are one or more zero eigenvalues, which must be the smallest because all of the eigenvalues are nonnegative. We will assume that $\sigma_1, \sigma_2, \ldots, \sigma_r$ are strictly positive and $\sigma_{r+1} = \ldots \sigma_n = 0$. As before, $v_1, v_2, \ldots, v_n$ are the right singular vectors of A and $\sigma_1, \sigma_2, \ldots, \sigma_n$ are the singular values.

We encounter a difference, however, in defining the left singular vectors. We have $A^*Av_i = 0$ for $i = r+1, \ldots, n$, which in turn implies that $Av_i = 0$, $i = r+1, \ldots, n$. We can define

$$u_i = \sigma_i^{-1} A v_i, \ i = 1, 2, \ldots, r,$$

and these vectors are orthonormal, just as before. However, we cannot define $u_{r+1}, \ldots, u_n$ by this formula. It turns out, though, that all we need is for $\{u_1, u_2, \ldots, u_n\}$ to be an orthonormal basis for $\mathbf{C}^n$, and we can always extend a given orthonormal set to an orthonormal basis. (This follows from Theorem 43: Any linearly independent set can be extended to a basis, and we can always orthogonalize using the Gram-Schmidt process.) So we just choose any vectors $u_{r+1}, \ldots, u_n$ in $\mathbf{C}^n$ such that $\{u_1, u_2, \ldots, u_n\}$ is an orthonormal basis. Defining $U = [u_1|u_2|\cdots|u_n]$, $\Sigma = \mathrm{diag}(\sigma_1, \sigma_2, \ldots, \sigma_n)$, and $V = [v_1|v_2|\cdots|v_n]$ as before, we obtain $A = U\Sigma V^*$.

Example 356 *Let*

$$A = \begin{bmatrix} 1 & 0 & 1 \\ 2 & 0 & 2 \\ 1 & 0 & 1 \end{bmatrix}.$$

Then

$$A^TA = \begin{bmatrix} 6 & 0 & 6 \\ 0 & 0 & 0 \\ 6 & 0 & 6 \end{bmatrix},$$

the characteristic polynomial is $p_{A^TA} = r^3 - 12r^2$, and the eigenvalues of A^TA are $\sigma_1^2 = 12$, $\sigma_2^2 = 0$, $\sigma_3^2 = 0$. Corresponding orthonormal eigenvectors are

$$v_1 = \begin{bmatrix} \frac{1}{\sqrt{2}} \\ 0 \\ \frac{1}{\sqrt{2}} \end{bmatrix}, \ v_2 = \begin{bmatrix} \frac{1}{\sqrt{2}} \\ 0 \\ -\frac{1}{\sqrt{2}} \end{bmatrix}, \ v_3 = \begin{bmatrix} 0 \\ 1 \\ 0 \end{bmatrix}.$$

We then have

$$u_1 = \frac{1}{\sigma_1} A v_1 = \frac{1}{\sqrt{12}} \left[\begin{array}{ccc} 1 & 0 & 1 \\ 2 & 0 & 2 \\ 1 & 0 & 1 \end{array} \right] \left[\begin{array}{c} \frac{1}{\sqrt{2}} \\ 0 \\ \frac{1}{\sqrt{2}} \end{array} \right] = \left[\begin{array}{c} \frac{1}{\sqrt{6}} \\ \frac{\sqrt{2}}{\sqrt{3}} \\ \frac{1}{\sqrt{6}} \end{array} \right].$$

We must now choose $u_2, u_3 \in \mathbf{R}^3$ to be any two vectors so that $\{u_1, u_2, u_3\}$ is an orthonormal basis. There are many ways to do this. By inspection,

$$u_2 = \left[\begin{array}{c} \frac{1}{\sqrt{2}} \\ 0 \\ -\frac{1}{\sqrt{2}} \end{array} \right], \ u_3 = \left[\begin{array}{c} \frac{1}{\sqrt{3}} \\ -\frac{\sqrt{2}}{\sqrt{3}} \\ \frac{1}{\sqrt{3}} \end{array} \right]$$

will work. We then obtain $A = U\Sigma V^T$, where

$$U = \left[\begin{array}{ccc} \frac{1}{\sqrt{6}} & \frac{1}{\sqrt{2}} & \frac{1}{\sqrt{3}} \\ \frac{\sqrt{2}}{\sqrt{3}} & 0 & -\frac{\sqrt{2}}{\sqrt{3}} \\ \frac{1}{\sqrt{6}} & -\frac{1}{\sqrt{2}} & \frac{1}{\sqrt{3}} \end{array} \right], \ \Sigma = \left[\begin{array}{ccc} \sqrt{12} & 0 & 0 \\ 0 & 0 & 0 \\ 0 & 0 & 0 \end{array} \right], \ V = \left[\begin{array}{ccc} \frac{1}{\sqrt{2}} & \frac{1}{\sqrt{2}} & 0 \\ 0 & 0 & 1 \\ \frac{1}{\sqrt{2}} & -\frac{1}{\sqrt{2}} & 0 \end{array} \right].$$

Exercises

Miscellaneous exercises

1. Modify Example 356 by choosing u_2, u_3 differently, as follows: Verify that $\{u_1, e_2, e_3\}$, where $e_2 = (0, 1, 0)$, $e_3 = (0, 0, 1)$ are the standard basis vectors, is a basis for $\mathbf{R}^3$. Transform $\{u_1, e_2, e_3\}$ into an orthonormal basis $\{u_1, u_2, u_3\}$ by using Gram-Schmidt.

2. Let
$$A = \left[\begin{array}{cc} 4 & 3 \\ 0 & 5 \end{array} \right].$$
Find the SVD of A in both matrix and outer product form.

3. Let
$$A = \left[\begin{array}{ccc} 2 & 4 & 2 \\ 2 & 4 & 2 \\ 1 & 1 & -3 \end{array} \right].$$
Find the SVD of A in both matrix and outer product form.

4. Let
$$A = \left[\begin{array}{cccc} 3 & 6 & -1 & 4 \\ 0 & 2 & -4 & 2 \\ 0 & 0 & -4 & 4 \\ -3 & 0 & -7 & -2 \end{array} \right].$$
Find the SVD of A and write it in outer product form.

5. Let A be the 2×3 matrix defined as $A = uv^T$, where

$$u = \begin{bmatrix} 1 \\ 2 \end{bmatrix}, \; v = \begin{bmatrix} 1 \\ 0 \\ 1 \end{bmatrix}.$$

   Find the SVD of A.

6. Find a 3×3 matrix whose singular values are $\sigma_1 = 2$, $\sigma_2 = 1$, $\sigma_3 = 1$.

7. Suppose $A \in \mathbf{R}^{n \times n}$ has orthogonal (but not necessarily orthonormal) columns. Find the SVD of A.

8. Let $A \in \mathbf{C}^{n \times n}$ be given. For each of the following descriptions of the singular values of A, show that the SVD of A is not unique, and explain what freedom there exists in choosing the SVD of A.

   (a) A has distinct singular values $\sigma_1 > \sigma_2 > \cdots > \sigma_n$.

   (b) There exist integers $i, i+1, \ldots, i+k$ such that

$$\sigma_i = \sigma_{i+1} = \ldots = \sigma_{i+k},$$

   where $\sigma_1, \sigma_2, \ldots, \sigma_n$ are the singular values of A.

9. Let $A \in \mathbf{R}^{n \times n}$ be a diagonal matrix. Find the SVD of A.

10. Let $A \in \mathbf{C}^{n \times n}$ be a diagonal matrix. Find the SVD of A.

11. Suppose $A \in \mathbf{C}^{n \times n}$ is invertible and $A = U\Sigma V^*$ is the SVD of A. Find the SVD of each of the following matrices:

    (a) A^*

    (b) A^{-1}

    (c) A^{-*}

    Express each SVD in outer product form.

12. Let $A \in \mathbf{R}^{n \times n}$ be symmetric.

    (a) Prove that if A is positive definite, then the SVD is simply the spectral decomposition. That is, suppose $A = VDV^T$, where the columns of V are orthonormal eigenvectors of A and D is a diagonal matrix whose diagonal entries are the corresponding eigenvalues of A. Then the SVD of A is $A = U\Sigma V^T$, where $U = V$ and $\Sigma = D$.

    (b) If A has one or more negative eigenvalues, what changes must be made in the spectral decomposition to produce the SVD?

13. Let $A \in \mathbf{C}^{n \times n}$ be normal, and let $A = XDX^*$ be the spectral decomposition of A. Explain how to find the SVD of A from X and D.

8.2 The SVD for general matrices

It turns out that every matrix $A \in \mathbf{C}^{m\times n}$ has a singular value decomposition, which looks exactly like (8.1). The only difference is the sizes of the matrices.

Theorem 357 *Let $A \in \mathbf{C}^{m\times n}$. Then there exist unitary matrices $U \in \mathbf{C}^{m\times m}$ and $V \in \mathbf{C}^{n\times n}$ and a diagonal matrix $\Sigma \in \mathbf{C}^{m\times n}$, with nonnegative diagonal entries $\sigma_1 \geq \sigma_2 \geq \cdots \geq \sigma_{\min\{m,n\}} \geq 0$, such that*

$$A = U\Sigma V^*.$$

Proof We first assume that $m \geq n$. Let $D \in \mathbf{R}^{n\times n}$, $V \in \mathbf{C}^{n\times n}$ form the spectral decomposition of $A^*A \in \mathbf{C}^{n\times n}$:

$$A^*A = VDV^*.$$

Since A^*A is positive semidefinite, the diagonal entries of D (the eigenvalues of A^*A) are nonnegative, and we can assume that the columns of V (the eigenvectors of A^*A) are ordered so that

$$D = \begin{bmatrix} \sigma_1^2 & & & \\ & \sigma_2^2 & & \\ & & \ddots & \\ & & & \sigma_n^2 \end{bmatrix},$$

where $\sigma_1 \geq \cdots \geq \sigma_r > \sigma_{r+1} = \cdots = \sigma_n = 0$. In this notation, the first r eigenvalues of A^*A are positive ($r = n$ is possible).

We now define $u_1, u_2, \ldots, u_r \in \mathbf{C}^m$ by

$$u_i = \sigma_i^{-1} A v_i, \ i = 1, 2, \ldots, r.$$

It follows from the reasoning on page 464 that $\{u_1, u_2, \ldots, u_r\}$ is orthonormal. If $r < m$, then let $u_{r+1}, \ldots, u_m$ be chosen so that $\{u_1, u_2, \ldots, u_m\}$ is an orthonormal basis for $\mathbf{C}^m$, and define $U = [u_1|u_2|\cdots|u_m] \in \mathbf{C}^{m\times m}$. Since $\{u_1, u_2, \ldots, u_m\}$ is an orthonormal basis, U is unitary. Finally, define $\Sigma \in \mathbf{C}^{m\times n}$ by

$$\Sigma_{ij} = \begin{cases} \sigma_i, & i = j, \\ 0, & i \neq j. \end{cases}$$

It remains only to show that $A = U\Sigma V^*$ or, equivalently, that $AV = U\Sigma$. But the jth column of AV is

$$Av_j = \begin{cases} \sigma_j u_j, & j = 1, \ldots, r, \\ 0, & j = r+1, \ldots, n, \end{cases}$$

while the j column of $U\Sigma$ is

$$\sum_{i=1}^{m} \Sigma_{ij} u_i = \Sigma_{jj} u_j = \begin{cases} \sigma_j u_j, & j = 1, \ldots, r, \\ 0, & j = r+1, \ldots, n, \end{cases}$$

since $\Sigma_{ij} = 0$ for $i \neq j$. This completes the proof in the case that $m \geq n$.

If $n > m$, then the result proved above applies to $A^* \in \mathbf{C}^{n\times m}$, so there exist unitary matrices $U \in \mathbf{C}^{m\times m}$, $V \in \mathbf{C}^{n\times n}$ and a diagonal matrix $\hat{\Sigma} \in \mathbf{C}^{n\times m}$, with diagonal entries $\sigma_1 \geq \cdots \sigma_r > \sigma_{r+1} \geq \cdots \geq \sigma_m = 0$, such that

$$A^* = V\hat{\Sigma}U^*.$$

Taking the adjoint of both sides yields

$$A = U\hat{\Sigma}^T V^*,$$

or, with $\Sigma = \hat{\Sigma}^T$,

$$A = U\Sigma V^*,$$

as desired.

QED

It is helpful to examine carefully the matrices U, Σ, and V in three cases: $m = n$, $m > n$, and $m < n$.

1. If $m = n$, then U, Σ, and V are all square and of the same size.
2. If $m > n$, then $\Sigma \in \mathbf{C}^{m\times n}$ can be partitioned as follows:

$$\Sigma = \left[\begin{array}{c} \Sigma_1 \\ \hline 0 \end{array}\right],$$

where

$$\Sigma_1 = \begin{bmatrix} \sigma_1 & & & \\ & \sigma_2 & & \\ & & \ddots & \\ & & & \sigma_n \end{bmatrix}$$

and 0 is the $(m-n) \times n$ zero matrix. We can also write $U = [U_1|U_2]$, where

$$U_1 = [u_1|\cdots|u_n] \in \mathbf{C}^{m\times n},\ U_2 = [u_{n+1}|\cdots|u_m] \in \mathbf{C}^{m\times(m-n)}.$$

We obtain

$$A = U\Sigma V^* = [U_1|U_2] \left[\begin{array}{c} \Sigma_1 \\ \hline 0 \end{array}\right] V^* = U_1\Sigma_1 V^*.$$

3. If $m < n$, then Σ has the form

$$\Sigma = [\Sigma_1|0],$$

where $\Sigma_1 \in \mathbf{C}^{m\times m}$ has diagonal entries $\sigma_1, \sigma_2, \ldots, \sigma_m$ and 0 is the $m \times (n-m)$ zero matrix. We write $V = [V_1|V_2]$, where

$$V_1 = [v_1|\cdots|v_m] \in \mathbf{C}^{n\times m},\ V_2 = [v_{m+1}|\cdots|v_n] \in \mathbf{C}^{n\times(n-m)},$$

and obtain

$$A = U\Sigma V^* = U\left[\Sigma_1|0\right]\left[\frac{V_1^*}{V_2^*}\right] = U\Sigma_1 V_1^*.$$

In the second and third cases above, $A = U_1\Sigma_1 V^*$ and $A = U\Sigma_1 V_1^*$ are called the *reduced* SVD of A.

In any of the above cases, if $\sigma_1, \ldots, \sigma_r > 0$ and the remaining singular values of A are zero, then we can write

$$A = \sum_{i=1}^{r} \sigma_i u_i \otimes v_i.$$

It is also worth noting that, if A is real, then U and V are also real matrices. (When A is real, not only the eigenvalues but also the eigenvectors of $A^*A = A^TA$ are real. It follows that V and U are real orthogonal matrices.) The SVD eliminates the need to move into the complex domain, as might be necessary when the spectral decomposition is used.

Example 358 *Let A be the following 3×2 matrix:*

$$A = \begin{bmatrix} 2 & 0 \\ 1 & 1 \\ 0 & 2 \end{bmatrix}.$$

Then

$$A^TA = \begin{bmatrix} 5 & 1 \\ 1 & 5 \end{bmatrix},$$

the characteristic polynomial is $p_{A^TA} = r^2 - 10r + 24 = (r-6)(r-4)$, and so we have $\sigma_1^2 = 6$, $\sigma_2^2 = 4$. We easily find the following (orthonormal) eigenvectors of A^TA, which are also the right singular vectors of A:

$$v_1 = \begin{bmatrix} \frac{1}{\sqrt{2}} \\ \frac{1}{\sqrt{2}} \end{bmatrix}, \quad v_2 = \begin{bmatrix} \frac{1}{\sqrt{2}} \\ -\frac{1}{\sqrt{2}} \end{bmatrix}.$$

We can then compute

$$u_1 = \sigma_1^{-1}Av_1 = \begin{bmatrix} \frac{1}{\sqrt{3}} \\ \frac{1}{\sqrt{3}} \\ \frac{1}{\sqrt{3}} \end{bmatrix}, \quad u_2 = \sigma_2^{-1}Av_2 = \begin{bmatrix} \frac{1}{\sqrt{2}} \\ 0 \\ -\frac{1}{\sqrt{2}} \end{bmatrix}.$$

The third left singular vector, u_3, is not determined directly as are u_1 and u_2. Instead, u_3 must be chosen so that $\{u_1, u_2, u_3\}$ is an orthonormal basis for $\mathbf{R}^3$. A suitable choice for u_3 is

$$u_3 = \begin{bmatrix} \frac{1}{\sqrt{6}} \\ -\frac{2}{\sqrt{6}} \\ \frac{1}{\sqrt{6}} \end{bmatrix}.$$

We now have $A = U\Sigma V^T$, *where*

$$U = \begin{bmatrix} \frac{1}{\sqrt{3}} & \frac{1}{\sqrt{2}} & \frac{1}{\sqrt{6}} \\ \frac{1}{\sqrt{3}} & 0 & -\frac{2}{\sqrt{6}} \\ \frac{1}{\sqrt{3}} & -\frac{1}{\sqrt{2}} & \frac{1}{\sqrt{6}} \end{bmatrix}, \ \Sigma = \begin{bmatrix} \sqrt{6} & 0 \\ 0 & 2 \\ 0 & 0 \end{bmatrix}, \ V = \begin{bmatrix} \frac{1}{\sqrt{2}} & \frac{1}{\sqrt{2}} \\ \frac{1}{\sqrt{2}} & -\frac{1}{\sqrt{2}} \end{bmatrix}.$$

As explained above, we can also write $A = U_1\Sigma_1 V$, *where*

$$U = \begin{bmatrix} \frac{1}{\sqrt{3}} & \frac{1}{\sqrt{2}} \\ \frac{1}{\sqrt{3}} & 0 \\ \frac{1}{\sqrt{3}} & -\frac{1}{\sqrt{2}} \end{bmatrix}, \ \Sigma = \begin{bmatrix} \sqrt{6} & 0 \\ 0 & 2 \end{bmatrix}.$$

The SVD of A tells us almost everything we might want to know about a matrix A.

Theorem 359 *Let* $A \in \mathbf{C}^{m\times n}$ *and suppose that* A *has exactly* r *positive singular values. Then* $\text{rank}(A) = r$.

Proof We first note that $u_i \in \text{col}(A)$ for $i = 1, 2, \ldots, r$, since $u_i = A(\sigma_i^{-1}v_i)$. We also have, for any $x \in \mathbf{C}^n$,

$$Ax = \left(\sum_{i=1}^{r} \sigma_i u_i \otimes v_i\right) x = \sum_{i=1}^{r} \sigma_i (u_i \otimes v_i) x = \sum_{i=1}^{r} \sigma_i \langle x, v_i \rangle_{\mathbf{C}^n} u_i.$$

This shows that every element of $\text{col}(A)$ is a linear combination of $u_1, u_2, \ldots, u_r$. Since these vectors form a linearly independent spanning set, it follows that

$$\text{rank}(A) = \dim(\text{col}(A)) = r.$$

QED

In interpreting the following corollary, the reader should notice that the number of singular values of $A \in \mathbf{C}^{m\times n}$ is $\min\{m, n\}$.

Corollary 360 *Let* $A \in \mathbf{C}^{m\times n}$. *Then* A *is singular if and only if the number of nonzero singular values of* A *is less than* n. *If* $m \geq n$, *then* A *is singular if and only if* A *has at least one zero singular value.*

The SVD of A produces orthonormal bases for the most important subspaces associated with A.

Theorem 361 *Let* $A \in \mathbf{C}^{m\times n}$ *and suppose that* A *has exactly* r *positive singular values. Let* $A = U\Sigma V^*$ *be the SVD of* A, *where* $U = [u_1|u_2|\cdots|u_m]$ *and* $V = [v_1|v_2|\cdots|v_n]$. *Then the four fundamental subspaces have the following orthonormal bases:*

1. $\text{col}(A) = \text{sp}\{u_1, \ldots, u_r\}$;
2. $\mathcal{N}(A) = \text{sp}\{v_{r+1}, \ldots, v_n\}$;

3. $\mathrm{col}(A^*) = \mathrm{sp}\{v_1, \ldots, v_r\}$;

4. $\mathcal{N}(A^*) = \mathrm{sp}\{u_{r+1}, \ldots, u_m\}$.

Proof If $\sigma_i = 0$ for $i > r$, then the outer product form of the SVD is

$$A = \sum_{i=1}^{r} \sigma_i u_i v_i^*.$$

We have already used this formula to show that $\{u_1, u_2, \ldots, u_r\}$ is an orthonormal basis for $\mathrm{col}(A)$. If $j > r$, then

$$Av_j = \sum_{i=1}^{r} \sigma_i u_i v_i^* v_j = \sum_{i=1}^{r} \sigma_i \langle v_j, v_i \rangle_{\mathbf{C}^n} u_i = 0,$$

with the last equation following from the fact that $\langle v_j, v_i \rangle_{\mathbf{C}^n} = 0$ for $i < j$. Therefore, we see that $\{v_{r+1}, \ldots, v_n\}$ is an orthonormal set contained in $\mathcal{N}(A)$. Since we know from the fundamental theorem of linear algebra that $\dim(\mathcal{N}(A)) = n - r$, this suffices to show that $\{v_{r+1}, \ldots, v_n\}$ is an orthonormal basis for $\mathcal{N}(A)$.

The SVD of A^* is given by

$$A^* = V\Sigma^T U^* = \sum_{i=1}^{r} \sigma_i v_i \otimes u_i,$$

so the reasoning given above, applied to the SVD of A^*, shows that $\{v_1, v_2, \ldots, v_r\}$ is an orthonormal basis for $\mathrm{col}(A^*)$ and $\{u_{r+1}, \ldots, u_m\}$ is a basis for $\mathcal{N}(A^*)$.

QED

Exercises

Miscellaneous exercises

1. Let A be the 3×2 matrix defined as $A = u_1 v_1^T + u_2 v_2^T$, where

$$u_1 = \begin{bmatrix} 1 \\ 0 \\ 1 \end{bmatrix}, \quad u_2 = \begin{bmatrix} 0 \\ 1 \\ 0 \end{bmatrix}, \quad v_1 = \begin{bmatrix} 1 \\ 2 \end{bmatrix}, \quad v_2 = \begin{bmatrix} -2 \\ 1 \end{bmatrix}.$$

Find the SVD of A and orthonormal bases for the four fundamental subspaces of A.

2. Let

$$A = \begin{bmatrix} 3 & 1 \\ 1 & -1 \\ 1 & -1 \\ -1 & -3 \end{bmatrix}.$$

Find the SVD of A and orthonormal bases for the four fundamental subspace of A.

3. Let A be the matrix from the previous exercise and let $b = (1, 2, 3, 4)$. Find the components of b lying in $\mathrm{col}(A)$ and $\mathcal{N}(A^T)$.

4. Let A be the matrix of Exercise 8.1.3. Find orthonormal bases for the four fundamental subspaces of A.

5. Let A be the matrix of Exercise 8.1.4. Find orthonormal bases for the four fundamental subspaces of A.

6. Let $A \in \mathbf{C}^{m\times n}$ have exactly r nonzero singular values. What are the dimensions of $\mathrm{col}(A)$, $\mathcal{N}(A)$, $\mathrm{col}(A^*)$, and $\mathcal{N}(A^*)$?

7. Let $u \in \mathbf{R}^m$ and $v \in \mathbf{R}^n$ be given, and define $A = uv^T$. (Recall that uv^T is the $m \times n$ matrix whose i, j-entry is $u_i v_j$.) What are the singular values of A? Explain how to compute a singular value decomposition of A.

8. Let $u \in \mathbf{R}^n$ have Euclidean norm one, and define $A = I - 2uu^T$. Find the SVD of A.

9. Let $A \in \mathbf{R}^{m\times n}$ be nonsingular. Compute

$$\min\{\|Ax\|_2 \,:\, x \in \mathbf{R}^n, \|x\|_2 = 1\},$$

and find the vector $x \in \mathbf{R}^n$ that gives the minimum value.

10. Let $A \in \mathbf{C}^{n\times n}$ be arbitrary. Using the SVD of A, show that there exist a unitary matrix Q and a Hermitian positive semidefinite matrix H such that $A = QH$. Alternatively, show that A can be written as $A = GQ$, where G is also Hermitian positive semidefinite and Q is the same unitary matrix. The decompositions $A = QH$ and $A = GQ$ are the two forms of the *polar decomposition* of A.

11. Let $A \in \mathbf{C}^{m\times n}$. Prove that $\|Ax\|_2 \le \sigma_1 \|x\|_2$ for all $x \in \mathbf{C}^n$, where σ_1 is the largest singular value of A.

12. Let $A \in \mathbf{C}^{m\times n}$ be nonsingular. Prove that $\|Ax\|_2 \ge \sigma_n \|x\|_2$ for all $x \in \mathbf{C}^n$, where σ_n is the smallest singular value of A.

13. The pseudoinverse of a matrix was introduced in Exercises 6.6.13ff. Given $A \in \mathbf{C}^{m\times n}$, the pseudoinverse $A^\dagger \in \mathbf{C}^{n\times m}$ is defined by the condition that $x = A^\dagger b$ is the minimum-norm least-squares solution to $Ax = b$.

    (a) Let $\Sigma \in \mathbf{C}^{m\times n}$ be a diagonal matrix. Find $\Sigma^\dagger$.

    (b) Find the pseudoinverse of $A \in \mathbf{C}^{m\times n}$ in terms of the SVD of A.

14. Let $m > n$ and suppose $A \in \mathbf{R}^{m\times n}$ has full rank. Let the SVD of A be $A = U\Sigma V^T$.

(a) Find the SVD of $A(A^TA)^{-1}A^T$.

(b) Prove that $\|A(A^TA)^{-1}A^Tb\|_2 \le \|b\|_2$ for all $b \in \mathbf{R}^m$.

15. The space $\mathbf{C}^{m\times n}$ can be regarded as a vector space. The *Frobenius norm* $\|\cdot\|_F$ on $\mathbf{C}^{m\times n}$ is defined by

$$\|A\|_F = \sqrt{\sum_{i=1}^{m}\sum_{j=1}^{n}|A_{ij}|^2} \text{ for all } A \in \mathbf{C}^{m\times n}.$$

(Note that the Frobenius norm treats A like a Euclidean vector with mn entries.)

(a) Prove that if $U \in \mathbf{C}^{m\times m}$ is unitary, then

$$\|UA\|_F = \|A\|_F \text{ for all } A \in \mathbf{C}^{m\times n}.$$

Similarly, if $V \in \mathbf{C}^{n\times n}$ is unitary, then

$$\|AV\|_F = \|A\|_F \text{ for all } A \in \mathbf{C}^{m\times n}.$$

(b) Let $A \in \mathbf{C}^{m\times n}$ be given, and let r be a positive integer with $r < \text{rank}(A)$. Find the matrix $B \in \mathbf{C}^{m\times n}$ of rank r such that $\|A-B\|_F$ is as small as possible; that is, find $B \in \mathbf{C}^{m\times n}$ to solve

$$\begin{aligned}&\min \|A-B\|_F\\ &\text{s.t. } \text{rank}(B) = r.\end{aligned}$$

(Hint: If $A = U\Sigma V^*$ is the SVD of A, then

$$\|A-B\|_F = \|\Sigma - U^*BV\|_F.)$$

16. Let $A \in \mathbf{C}^{m\times n}$ be given. Find $\|A\|_F$ in terms of the singular values of A. (For the definition of $\|A\|_F$, see the previous exercise.)

8.3 Solving least-squares problems using the SVD

If $A \in \mathbf{R}^{m\times n}$ and $b \in \mathbf{R}^m$, then $x \in \mathbf{R}^n$ is a least-squares solution to $Ax = b$ if and only if x satisfies

$$A^TAx = A^Tb \tag{8.2}$$

(see Theorem 291). If A is nonsingular, then A^TA is square and nonsingular, and hence invertible. In this case, there is a unique least-squares solution. On the other hand, if A is singular, then A^TA is also singular. Since we know from the projection theorem that the system (8.2) has a solution, we can conclude that there are infinitely many least-squares solutions when A is singular.

The SVD can be used to find the set of all least-squares solutions to $Ax = b$ and to distinguish the unique *minimum-norm least-squares solution.* In the following derivation, we use the fact that a unitary or orthogonal matrix U preserves norms: $\|Ux\|_2 = \|x\|_2$ for all x (see Exercise 7.1.4).

Let $A \in \mathbf{R}^{m\times n}$ and $b \in \mathbf{R}^m$, and let $A = U\Sigma V^T$ be the SVD of A. Then

$$\begin{aligned}\|Ax-b\|_2^2 = \left\|U\Sigma V^T x - b\right\|_2^2 &= \left\|U\Sigma V^T x - UU^T b\right\|_2^2 \\ &= \left\|U\left(\Sigma V^T x - U^T b\right)\right\|_2^2 \\ &= \left\|\Sigma V^T x - U^T b\right\|_2^2.\end{aligned}$$

We now define $y = V^T x$ and note that since V is invertible, x is determined once y is known. We assume that A has exactly r positive singular values and write

$$\Sigma_1 = \begin{bmatrix} \sigma_1 & & & \\ & \sigma_2 & & \\ & & \ddots & \\ & & & \sigma_r \end{bmatrix}.$$

Then Σ has one of the following forms:

$$\begin{aligned}\Sigma &= \left[\begin{array}{c|c} \Sigma_1 & 0 \\ \hline 0 & 0 \end{array}\right] \quad (r < \min\{m,n\}), \\ \Sigma &= [\Sigma_1|0] \quad (r = m < n), \\ \Sigma &= \left[\begin{array}{c} \Sigma_1 \\ \hline 0 \end{array}\right] \quad (r = n < m).\end{aligned}$$

We begin with the first case ($r < \min\{m,n\}$), and partition y, U, and $U^T b$ in a compatible way:

$$y = \left[\begin{array}{c} w \\ \hline z \end{array}\right],\ w = \begin{bmatrix} y_1 \\ y_2 \\ \vdots \\ y_r \end{bmatrix},\ z = \begin{bmatrix} y_{r+1} \\ y_{r+2} \\ \vdots \\ y_n \end{bmatrix},$$

$$U = [U_1|U_2],\ U_1 = [u_1|u_2|\cdots|u_r],\ U_2 = [u_{r+1}|u_{r+2}|\cdots|u_m],$$

$$U^T b = \left[\begin{array}{c} U_1^T b \\ \hline U_2^T b \end{array}\right],\ U_1^T b = \begin{bmatrix} u_1\cdot b \\ u_2\cdot b \\ \vdots \\ u_r\cdot b \end{bmatrix},\ U_2^T b = \begin{bmatrix} u_{r+1}\cdot b \\ u_{r+2}\cdot b \\ \vdots \\ u_m\cdot b \end{bmatrix}.$$

We then have

$$\|Ax-b\|_2^2 = \left\|\Sigma y - U^T b\right\|_2^2 = \left\|\left[\begin{array}{c} \Sigma_1 w \\ \hline 0 \end{array}\right] - \left[\begin{array}{c} U_1^T b \\ \hline U_2^T b \end{array}\right]\right\|_2^2 = \left\|\left[\begin{array}{c} \Sigma_1 w - U_1^T b \\ \hline -U_2^T b \end{array}\right]\right\|_2^2.$$

For any partitioned Euclidean vector

$$u = \left[\begin{array}{c} v \\ \hline w \end{array}\right],$$

we have $\|u\|_2^2 = \|v\|_2^2 + \|w\|_2^2$. Therefore,

$$\|Ax - b\|_2^2 = \left\|\Sigma_1 w - U_1^T b\right\|_2^2 + \left\|U_2^T b\right\|_2^2. \tag{8.3}$$

We can draw two conclusions from this formula. First of all, the value of the vector $z = (y_{r+1}, \ldots, y_n)$ does not affect $\|Ax - b\|_2$. This is to be expected, since

$$x = Vy = \sum_{i=1}^{r} y_i v_i + \sum_{i=r+1}^{n} y_i v_i$$

and $\{v_{r+1}, \ldots, v_n\}$ is a basis for the null space of A. Therefore, $y_{r+1}, \ldots, y_n$ just determine the component of x in the null space of A. Second, (8.3) uniquely determines the value of w that minimizes $\|Ax - b\|_2$: Since Σ_1 is nonsingular, the first term $\left\|\Sigma_1 w - U_1^T b\right\|_2^2$ is zero for

$$w = \Sigma_1^{-1} U_1^T b.$$

We then obtain

$$\min_{x \in \mathbf{R}^n} \|Ax - b\|_2 = \left\|U_2^T b\right\|_2.$$

This could have been predicted, since the columns of U_2 form an orthonormal basis for $\mathcal{N}(A^T) = \text{col}(A)^\perp$ and hence $U_2 U_2^T b$ is the component of b orthogonal to $\text{col}(A)$. Moreover, $\|U_2 U_2^T b\|_2 = \|U_2^T b\|_2$. (See Exercise 4.) By choosing an appropriate x, Ax can match the component of b that lies in $\text{col}(A)$, but every Ax is orthogonal to the component of b that lies in $\mathcal{N}(A^T)$.

The desired solution x is given by $x = Vy$. If we write $V_1 = [v_1|v_2|\cdots|v_r]$, $V_2 = [v_{r+1}|v_{r+2}|\cdots|v_n]$, and $V = [V_1|V_2]$, then $x = V_1 w + V_2 z$. We then see that the general least-squares solution of $Ax = b$ is

$$x = V_1 \Sigma_1^{-1} U_1^T b + V_2 z, \; z \in \mathbf{R}^{n-r}.$$

When A is singular, that is, when $\text{rank}(A) = r < n$, then $Ax = b$ has infinitely many least-squares solutions, one for each $z \in \mathbf{R}^{n-r}$. The above formula distinguishes one of the solutions, namely,

$$\hat{x} = V_1 \Sigma_1^{-1} U_1^T b.$$

From Theorem 361, we see that $\hat{x} \in \text{col}(A^T)$ and $V_2 z \in \mathcal{N}(A) = \text{col}(A^T)^\perp$. The Pythagorean theorem implies that, for $x = \hat{x} + V_2 z$,

$$\|x\|_2^2 = \|\hat{x}\|_2^2 + \|V_2 z\|_2^2 \geq \|\hat{x}\|_2^2.$$

Therefore, among all least-squares solutions x, $\hat{x}$ has the smallest norm. For this reason, we refer to

$$\hat{x} = V_1 \Sigma_1^{-1} U_1^T b$$

as the *minimum-norm least-squares solution* to $Ax = b$.

Example 362 *Let $A \in \mathbf{R}^{3\times 3}$, $b \in \mathbf{R}^3$ be defined by*

$$A = \begin{bmatrix} 2 & 1 & 2 \\ 2 & 1 & 2 \\ 1 & 0 & -1 \end{bmatrix}, \quad b = \begin{bmatrix} 1 \\ 2 \\ 3 \end{bmatrix}.$$

We wish to solve the equation $Ax = b$. Since the rows of A are obviously linearly dependent, the matrix is singular. We will find the general solution and the minimum-norm least-squares solution.

The SVD of A is $A = U\Sigma V^T$, where

$$U = \begin{bmatrix} \frac{1}{\sqrt{2}} & 0 & -\frac{1}{\sqrt{2}} \\ \frac{1}{\sqrt{2}} & 0 & \frac{1}{\sqrt{2}} \\ 0 & 1 & 0 \end{bmatrix}, \quad \Sigma = \begin{bmatrix} 3\sqrt{2} & 0 & 0 \\ 0 & \sqrt{2} & 0 \\ 0 & 0 & 0 \end{bmatrix}, \quad V = \begin{bmatrix} \frac{2}{3} & \frac{1}{\sqrt{2}} & \frac{1}{3\sqrt{2}} \\ \frac{1}{3} & 0 & -\frac{2\sqrt{2}}{3} \\ \frac{2}{3} & -\frac{1}{\sqrt{2}} & \frac{1}{3\sqrt{2}} \end{bmatrix}.$$

As expected, A has a zero singular value. In the notation used above, we have

$$\Sigma_1 = \begin{bmatrix} 3\sqrt{2} & 0 \\ 0 & \sqrt{2} \end{bmatrix}, \quad V_1 = \begin{bmatrix} \frac{2}{3} & \frac{1}{\sqrt{2}} \\ \frac{1}{3} & 0 \\ \frac{2}{3} & -\frac{1}{\sqrt{2}} \end{bmatrix}, \quad U_1 = \begin{bmatrix} \frac{1}{\sqrt{2}} & 0 \\ \frac{1}{\sqrt{2}} & 0 \\ 0 & 1 \end{bmatrix},$$

$$V_2 = \begin{bmatrix} \frac{1}{3\sqrt{2}} \\ -\frac{2\sqrt{2}}{3} \\ \frac{1}{3\sqrt{2}} \end{bmatrix}, \quad U_2 = \begin{bmatrix} -\frac{1}{\sqrt{2}} \\ 0 \\ \frac{1}{\sqrt{2}} \end{bmatrix}.$$

The minimum-norm least-squares solution to $Ax = b$ is

$$\hat{x} = V_1 \Sigma_1^{-1} U_1^T b = \begin{bmatrix} \frac{11}{6} \\ \frac{1}{6} \\ -\frac{7}{6} \end{bmatrix}.$$

The reader can check directly that $A^T A\hat{x} = A^T b$ and thus $\hat{x}$ is a least-squares solution to $Ax = b$. The general least-squares solution is

$$\hat{x} + V_2 z = \begin{bmatrix} \frac{11}{6} \\ \frac{1}{6} \\ -\frac{7}{6} \end{bmatrix} + z \begin{bmatrix} \frac{1}{3\sqrt{2}} \\ -\frac{2\sqrt{2}}{3} \\ \frac{1}{3\sqrt{2}} \end{bmatrix},$$

where z can be any real number. Finally, as noted above, the component of b that cannot be matched is

$$U_2 U_2^T b = \begin{bmatrix} -\frac{1}{2} \\ \frac{1}{2} \\ 0 \end{bmatrix}.$$

A direct calculation verifies that $b - A\hat{x} = (-1/2, 1/2, 0)$.

There are still two special cases to consider. The first is the case in which $r = m < n$, where r is the number of nonzero singular values of A (that is, r is the rank of A). Such a matrix A has no zero singular values but is nonetheless singular because it has more columns than rows. Moreover, $\text{col}(A) = \mathbf{R}^m$ (since A has rank m) and therefore $Ax = b$ is consistent for all $b \in \mathbf{R}^m$. Exercise 5 asks the reader to carefully derive the following conclusions. The general solution to $Ax = b$ is

$$x = V_1 \Sigma_1^{-1} U^T b + V_2 z, \tag{8.4}$$

where z is any vector in $\mathbf{R}^{n-m}$, $V_1 = [v_1|\cdots|v_m]$, $V_2 = [v_{m+1}|\cdots|v_n]$, and $\Sigma_1 \in \mathbf{R}^{m\times m}$ is the diagonal matrix whose diagonal entries are $\sigma_1, \sigma_2, \ldots, \sigma_m$. The minimum-norm solution is $\hat{x} = V_1 \Sigma_1^{-1} U^T b$.

Finally, we consider the case in which $A \in \mathbf{R}^{m\times n}$ with $r = n < m$, where $r = \text{rank}(A)$. In this case, A is nonsingular, which implies that $A^T A$ is invertible. The unique least-squares solution is therefore

$$x = \left(A^T A\right)^{-1} A^T b. \tag{8.5}$$

We can write $A = U\Sigma V^T = U_1 \Sigma_1 V^T$, where $U_1 = [u_1|\cdots|u_n]$ and $\Sigma_1 = \text{diag}(\sigma_1, \ldots, \sigma_n) \in \mathbf{R}^{n\times n}$. Direct calculations show that

$$A^T A = V \Sigma_1^2 V^T \text{ and } \left(A^T A\right)^{-1} = V \Sigma_1^{-2} V^T,$$

where Σ_1^{-2} is the diagonal matrix with diagonal entries $\sigma_1^{-2}, \ldots, \sigma_n^{-2}$ (or, equivalently, $\Sigma_1^{-2} = (\Sigma_1^{-1})^2$). Therefore, the unique least-squares solution to $Ax = b$ is

$$x = \left(A^T A\right)^{-1} A^T b = V \Sigma_1^{-2} V^T V \Sigma_1 U_1^T b,$$

which simplifies to

$$x = V \Sigma_1^{-1} U_1^T b. \tag{8.6}$$

Formula (8.6) gives an alternate means of computing the least-squares solution in this case. Instead of forming $A^T A$, $A^T b$ and solving $A^T A x = A^T b$,[3] we can compute the SVD of A and use (8.6). In the next chapter, we will present another method for solving the least-squares problem.

Example 363 *Let*

$$A = \begin{bmatrix} 1 & 2 & 1 \\ 2 & 1 & 1 \end{bmatrix}, \; b = \begin{bmatrix} 1 \\ 3 \end{bmatrix}.$$

[3]The reader should recall that the use of the matrix inverse in formula (8.5), while mathematically correct, should be regarded as a signal that a linear system must be solved. As discussed in Chapter 3, solving the linear system is more efficient than forming and using the inverse matrix.

The SVD of A is $A = U\Sigma V^T$, *where*

$$U = \begin{bmatrix} \frac{1}{\sqrt{2}} & -\frac{1}{\sqrt{2}} \\ \frac{1}{\sqrt{2}} & \frac{1}{\sqrt{2}} \end{bmatrix}, \ \Sigma = \begin{bmatrix} \sqrt{11} & 0 & 0 \\ 0 & 1 & 0 \end{bmatrix}, \ V = \begin{bmatrix} \frac{3}{\sqrt{22}} & \frac{1}{\sqrt{2}} & -\frac{1}{\sqrt{11}} \\ \frac{3}{\sqrt{22}} & -\frac{1}{\sqrt{2}} & -\frac{1}{\sqrt{11}} \\ \frac{2}{\sqrt{22}} & 0 & \frac{3}{\sqrt{11}} \end{bmatrix}.$$

The general solution to $Ax = b$ *is*

$$x = V_1\Sigma_1^{-1}U^Tb + V_2z, \ z \in \mathbf{R},$$

where

$$\Sigma_1 = \begin{bmatrix} \sqrt{11} & 0 \\ 0 & 1 \end{bmatrix}, \ V_1 = \begin{bmatrix} \frac{3}{\sqrt{22}} & \frac{1}{\sqrt{2}} \\ \frac{3}{\sqrt{22}} & -\frac{1}{\sqrt{2}} \\ \frac{2}{\sqrt{22}} & 0 \end{bmatrix}, \ V_2 = \begin{bmatrix} -\frac{1}{\sqrt{11}} \\ -\frac{1}{\sqrt{11}} \\ \frac{3}{\sqrt{11}} \end{bmatrix}.$$

The result is

$$x = \begin{bmatrix} \frac{17}{11} \\ -\frac{5}{11} \\ \frac{4}{11} \end{bmatrix} + z\begin{bmatrix} -\frac{1}{\sqrt{11}} \\ -\frac{1}{\sqrt{11}} \\ \frac{3}{\sqrt{11}} \end{bmatrix}.$$

The minimum-norm solution is $x = (17/11, -5/11, 4/11)$.

Exercises

Miscellaneous exercises

1. Let A be the matrix of Exercise 8.2.1 and let $b = (1, 0, 0)$. Use the SVD of A to find the least squares solution to $Ax = b$.

2. Let A be the matrix of Exercise 8.2.2 and let $b = (1, 2, 2, 1)$. Use the SVD of A to find the least squares solution to $Ax = b$.

3. Let $A \in \mathbf{R}^{4\times 3}$ be defined by

$$A = \begin{bmatrix} 3 & 1 & 1 \\ 2 & 2 & 1 \\ 1 & 3 & 1 \\ 2 & 2 & 1 \end{bmatrix}.$$

   For each of the following vectors b, find the minimum-norm least-squares solution to $Ax = b$.

   (a) $b = (1, 2, -1, 1)$
   (b) $b = (2, 1, 0, 1)$
   (c) $b = (0, -1, 0, 1)$

4. Suppose $A \in \mathbf{R}^{m\times n}$ has SVD $A = U\Sigma V^T$, and we write $U = [U_1|U_2]$, where the columns of U_1 form a basis for col(A) and the columns of U_2 form a basis for $\mathcal{N}(A^T)$. Show that, for $b \in \mathbf{R}^m$, $U_2U_2^Tb$ is the projection of b onto $\mathcal{N}(A^T)$ and $\|U_2U_2^Tb\|_2 = \|U_2^Tb\|_2$.

5. Suppose $A \in \mathbf{R}^{m\times n}$ has rank m and that $A = U\Sigma V^T$ is the SVD of A. Let b be any vector in $\mathbf{R}^m$ and define matrices $V_1 = [v_1|\cdots|v_m]$, $V_2 = [v_{m+1}|\cdots|v_n]$ (where $v_1, v_2, \ldots, v_n$ are the columns of V), and $\Sigma_1 = \mathrm{diag}(\sigma_1, \ldots, \sigma_m)$ (where $\sigma_1, \sigma_2, \ldots, \sigma_m$ are the singular values of A).

   (a) Prove that $x = V_1\Sigma_1^{-1}U^Tb + V_2z$, $z \in \mathbf{R}^{n-m}$, is the general solution to $Ax = b$.

   (b) Prove that $\hat{x} = V_1\Sigma_1^{-1}U^Tb$ has the smallest Euclidean norm of any solution to $Ax = b$.

6. Let $A \in \mathbf{R}^{m\times n}$ with rank$(A) = n < m$. Verify the calculations leading to the formula (8.6) for the unique least-squares solution to $Ax = b$.

7. Let $A \in \mathbf{R}^{m\times n}$ have rank r. Write the formula for the minimum-norm least-squares solution to $Ax = b$ in outer product form.

8. Let $A \in \mathbf{R}^{m\times n}$ be singular. Show that the minimum-norm least-squares solution to $Ax = b$ is the unique least-squares solution lying in col(A^T) (cf. Exercise 6.6.14).

9. Let $A \in \mathbf{R}^{m\times n}$. We say that $B \in \mathbf{R}^{n\times m}$ is a right inverse of A if $AB = I$, where I is the $m \times m$ identity matrix.

   (a) Prove that A has a right inverse if and only if the rows of A are linearly independent. This implies that $m \leq n$.

   (b) With $m \leq n$, we have a reduced SVD of A of the form $A = U\Sigma_1V_1^T$, where $U \in \mathbf{R}^{m\times m}$ is orthogonal, $\Sigma_1 \in \mathbf{R}^{m\times m}$, and $V_1 \in \mathbf{R}^{n\times m}$ has orthonormal columns. Assuming that A has linearly independent rows, find a right inverse of A in terms of the reduced SVD of A.

10. Let $A \in \mathbf{R}^{m\times n}$. We say that $B \in \mathbf{R}^{n\times m}$ is a left inverse of A if $BA = I$, where I is the $n \times n$ identity matrix.

    (a) Prove that A has a right inverse if and only if the columns of A are linearly independent. This implies that $n \leq m$.

    (b) With $n \leq m$, the reduced SVD of A has the form $A = U_1\Sigma_1V^T$, where $V \in \mathbf{R}^{n\times n}$ is orthogonal, $\Sigma_1 \in \mathbf{R}^{n\times n}$, and $U_1 \in \mathbf{R}^{m\times n}$ has orthonormal columns. Assuming A has linearly independent columns, find a left inverse of A in terms of the reduced SVD of A.

8.4 The SVD and linear inverse problems

One of the main applications of the SVD arises in the context of a linear inverse problem, which asks for an estimate of the vector x that satisfies $Ax = b$ for a given $A \in \mathbf{R}^{m\times n}$, $b \in \mathbf{R}^m$. The system $Ax = b$ is regarded as defining an inverse problem when it has the following two properties:

1. The vector b (the data) is likely to be corrupted by noise.

2. The solution x of $Ax = b$, which may or may not be unique, is highly sensitive to errors in the data; that is, a small change in b can lead to a large change in x.

In the context of an inverse problem, the unique solution $x = A^{-1}b$ (if it exists) or the least-squares solution $x = \left(A^TA\right)^{-1}A^Tb$ will be unacceptable, and more sophisticated methods are required to produce an acceptable (approximate) solution x.

To introduce the idea of an inverse problem, we will consider the simplest case in which A is square and nonsingular. We will write b and x for the exact data and solution, respectively, so that $Ax = b$ holds. We will write $\hat{b} = b + \delta b$ for the noisy data, so that δb is the error in the data, and $\hat{x} = x + \delta x$ for the corresponding solution, so that δx is the error in the solution induced by the error in the data.

A square matrix $A \in \mathbf{R}^{n\times n}$ is nonsingular if and only if all of its singular values are strictly positive: $A = U\Sigma V^T$, $\Sigma = \mathrm{diag}(\sigma_1, \sigma_2, \ldots, \sigma_n)$,

$$\sigma_1 \ge \sigma_2 \ge \cdots \ge \sigma_n > 0.$$

In this context, we must distinguish three cases, which can be described qualitatively as follows:

1. The smallest singular value σ_n is much greater than zero, and σ_n is no more than a few orders of magnitude smaller than σ_1.

2. At least one singular value is very close to zero, and there is a definite separation between the largest singular values and those that are almost zero:

   $$\sigma_1 \ge \sigma_2 \ge \cdots \ge \sigma_r \gg \sigma_{r+1} \doteq \cdots \doteq \sigma_n \doteq 0.$$

   In this case, it is also assumed that $\sigma_1, \ldots, \sigma_r$ vary by no more than a few orders of magnitude.

3. The smallest singular value is close to zero when compared to the largest ($\sigma_1 \gg \sigma_n$), but there is no clear separation between large singular values and small ones.

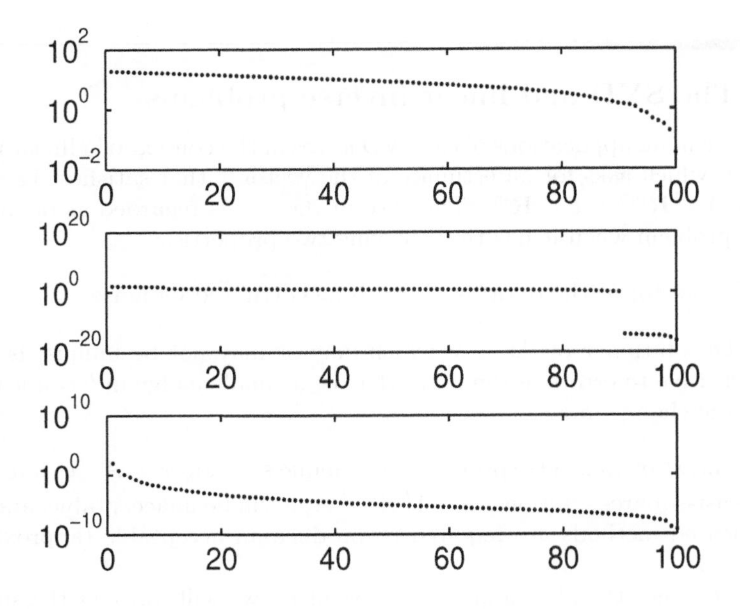

FIGURE 8.1
Singular values for three matrices illustrating the possibilities described on page 483. In the case of the first matrix (top), the singular values vary from about 19 down to about 0.34. The matrix is clearly nonsingular. In the second case (middle), the singular values vary about 10^2 down to 10^{-16}. However, the first 90 singular values vary by only about two orders of magnitude, and the smallest 10 are much smaller than the others. This matrix is numerically singular, with an effective rank of 90. The final case (bottom) shows singular values varying from about 95 down to 10^{-10}. However, there is no separation between large and small singular values, and there is no way to determine the numerical rank.

Figure 8.1 displays a typical collection of singular values for each of the three case.

In the first case, A is definitely nonsingular and the solution x is not sensitive to errors in the data. In other words, small errors in b lead to small errors in x.

In the second case, A can be regarded as numerically singular, which means that in finite-precision computations, it will exhibit the properties of a singular matrix. It is reasonable to set the smallest singular values $(\sigma_{r+1}, \ldots, \sigma_n)$ to zero and regard A as a matrix of rank r. The equation $Ax = b$ should be solved in the least-squares sense, which we learned how to do in the preceding section.

The equation $Ax = b$ is regarded as an inverse problem if it falls into the third case above: The matrix is numerically singular or nearly so, but there is no way to determine the numerical rank of A because there is no separation between large and small singular values. Thus it is not possible to determine which singular values should be regarded as zero. We will now show exactly how this leads to sensitivity to error in the data. We will use the following implication of linearity: If $Ax = b$ and $A(x + \delta x) = b + \delta b$, then, because matrix multiplication defines a linear operator, $A\delta x = \delta b$.

If $A = U\Sigma V^T$ is the SVD of $A \in \mathbf{R}^{n\times n}$ and all the singular values of A are positive, then $A^{-1} = V\Sigma^{-1}U^T$. Using the outer product form of the SVD, we find the following formula for the exact solution of $Ax = b$:

$$x = A^{-1}b = V\Sigma^{-1}U^T b = \sum_{i=1}^{n} \frac{1}{\sigma_i}(v_i \otimes u_i)b = \sum_{i=1}^{n} \frac{u_i \cdot b}{\sigma_i} v_i. \tag{8.7}$$

Similarly,

$$\delta x = A^{-1}\delta b = \sum_{i=1}^{n} \frac{u_i \cdot \delta b}{\sigma_i} v_i. \tag{8.8}$$

The computed solution $x + \delta x$ will be inaccurate if any of the components $(u_i \cdot \delta b)/\sigma_i$ are significant compared to the size $\|x\|_2$ of the exact solution. This can happen even if $u_i \cdot \delta b$ is small for all i, because the smallest singular values are very close to zero. If $u_i \cdot \delta b \gg \sigma_i$, then $(u_i \cdot \delta b)/\sigma_i$ is very large, and this contribution to the error in the computed solution can ruin the final result.

We now present a typical example.

Example 364 *An important class of linear inverse problems consists of Fredholm integral equations of the first kind, which have the form*

$$\int_c^d k(t,s)x(s)\,ds = y(t), \;\; a \le t \le b.$$

In this equation, the kernel[4] $k(t,s)$ *is known, the function* y *is regarded as the data, and the function* x *is to be found. For this example, we will study the particular integral equation*

$$\int_0^1 k(t,s)x(s)\,ds = y(t), \;\; 0 \le t \le 1,$$

where the kernel k *is defined by*

$$k(t,s) = \begin{cases} s(1-t), & 0 \le s \le t, \\ t(1-s), & t \le s \le 1. \end{cases}$$

To transform this integral equation to a (finite-dimensional) linear inverse

[4]The reader should notice that the word "kernel" in this context does not refer to the kernel of a linear operator.

problem $KX = Y$, we must discretize. We define a grid on the interval $[0,1]$ as follows:

$$s_i = \left(i - \frac{1}{2}\right)\Delta s,\ i = 1, 2, \ldots, n,\ \Delta s = \frac{1}{n}.$$

We also write $t_i = s_i$, $i = 1, 2, \ldots, n$. We then have

$$\int_0^1 k(t_i, s)x(s)\,ds \doteq \sum_{j=1}^{n} k(t_i, s_j)x(s_j)\Delta s$$

and so the equations

$$\sum_{j=1}^{n} k(t_i, s_j)x(s_j)\Delta s = y(t_i),\ \ i = 1, 2, \ldots, n$$

should hold, at least approximately. We define the $n \times n$ matrix K by

$$K_{ij} = k(t_i, s_j)\Delta s,\ \ i, j = 1, 2, \ldots, n.$$

We will write $Y \in \mathbf{R}^n$ for the vector defined by $Y_i = y(t_i)$, and we will look for $X \in \mathbf{R}^n$ to satisfy

$$KX = Y,$$

so that $X_j \doteq x(s_j)$ should hold.

For a specific example, we take $n = 100$ and $y(t) = (2t - 5t^4 + 3t^5)/60$. The exact solution to the integral equation is then $x(t) = t^2(1-t)$. The exact data is $Y \in \mathbf{R}^n$ defined by $Y_i = y(t_i)$, $i = 1, 2, \ldots, n$; this vector is shown on the left in Figure 8.2. The matrix K is nonsingular, so there is a unique solution X to $KX = Y$. Figure 8.2 shows the vector X, along with the vector with components $x(t_i)$, $i = 1, 2, \ldots, n$. The two vectors are indistinguishable on this scale, which shows that discretization is adequate for approximating the exact solution.

The results displayed in Figure 8.2 give no indication that there is anything challenging about the system $KX = Y$. However, data from practical problems always contains noise, and noisy data reveals the difficulties inherent in an inverse problem. We therefore add random noise to Y to obtain the noisy data vector $\hat{Y} = Y + \delta Y$ shown on the left in Figure 8.3. The reader should notice that the noise in the data is barely visible on this scale. The computed solution $\hat{X} = X + \delta X$, which satisfies $K\hat{X} = \hat{Y}$, is shown on the right in Figure 8.3. The error in $\hat{Y}$ (that is, $\|\delta Y\|_2/\|Y\|_2$) is about $0.93\,\%$, while the error in $\hat{X}$ is $2268\,\%$. Therefore, the error in $\hat{Y}$ is magnified by a factor of about 2432.

The results for the case of noisy data can be understood by writing $\hat{Y}$ and $\hat{X}$ in terms of the singular vectors of K. If $K = U\Sigma V^T$ is the SVD of K, then $U^T\hat{Y}$ are the components of $\hat{Y}$ in the directions of the left singular vectors of K, while $V^T\hat{X}$ are the corresponding components of $\hat{X}$ in terms of the right singular vectors. These vectors are displayed in Figure 8.4.

The components of $\hat{Y}$ corresponding to the largest singular values are very

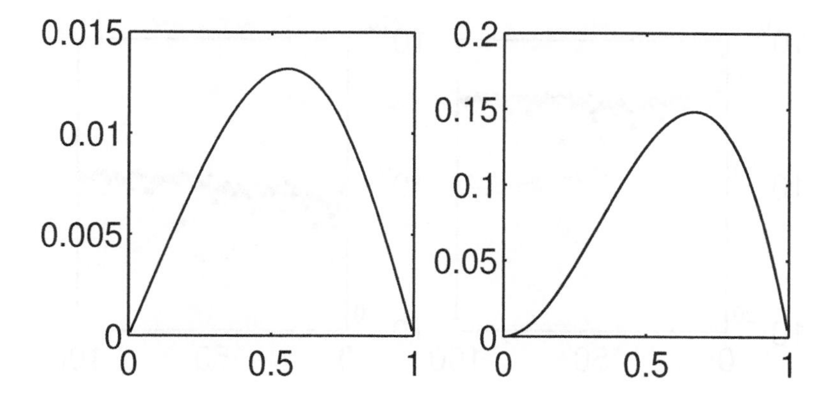

FIGURE 8.2
Left: The exact data Y for the inverse problem $KX = Y$ from Example 364. Right: The solution X computed from the exact data, along with the exact solution. (The two solutions are indistinguishable on this scale.)

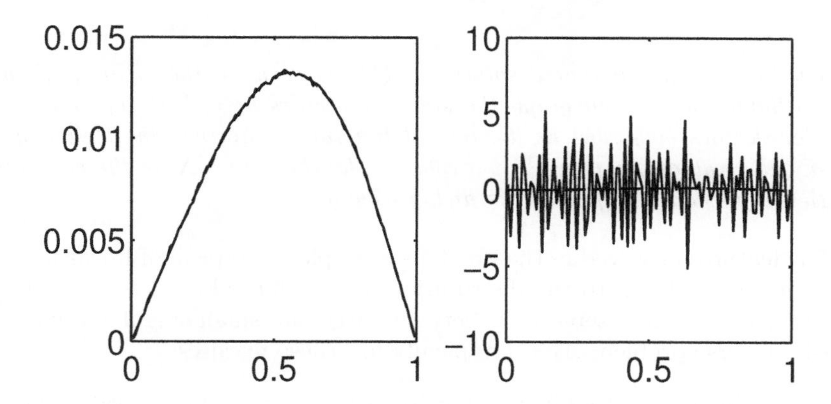

FIGURE 8.3
Left: The noisy data $\hat{Y}$ for the inverse problem $KX = Y$ from Example 364. Right: The computed solution $\hat{X}$. Also graphed is the solution X obtained from noise-free data. The error in $\hat{Y}$ is only about 0.93 %, but the error in $\hat{X}$ is more than 2200 %.

close to the corresponding components of Y. Although the components of $\hat{Y}$ corresponding to the smaller singular values are not close to the corresponding components of Y, these corrupted components are small and contribute little to Y. This explains why $\hat{Y}$ is close to Y.

However, the small noisy components corresponding to the small singular values are divided by those small singular values and are therefore amplified

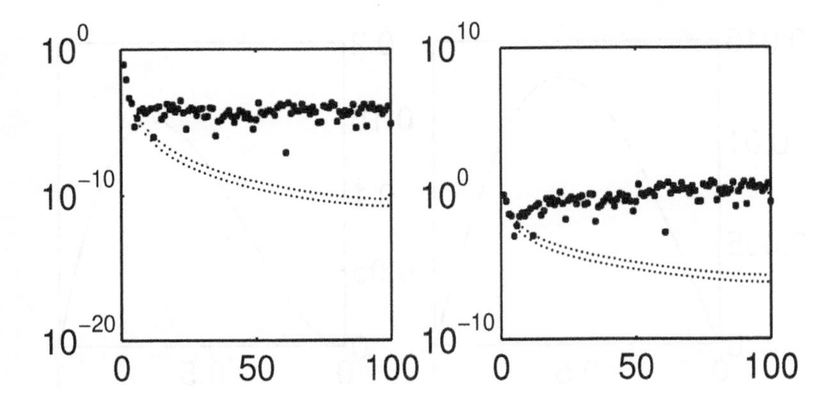

FIGURE 8.4
Left: The components of the exact (small dot) and noisy data (large dot), expressed in terms of the right singular vectors of K. Right: The components of the solutions computed from exact (small dot) and noisy (large dot) data, expressed in terms of the left singular vectors.

enough to ruin the computed solution. This is seen on the right in Figure 8.4. Although we do not graph the singular vectors here, both left and right singular vectors (regarded as discretized functions) increase in frequency as the singular values decrease. This explains why the error δX in the computed solution is not only large but also highly oscillatory.

The features observed in the previous example are typical of linear inverse problems. In addition to the distribution of singular values described above ($\sigma_1 \gg \sigma_n$ but no clear separation between large and small singular values), a realistic inverse problem $Ax = b$ typically has these features:

1. The singular vectors increase in frequency as the singular values decay to zero.

2. The true solution is relatively smooth (nonoscillatory), which means that its components corresponding to small singular vectors are small.

3. The noise, although small, tends to be distributed throughout the singular components, so a significant portion of the noise is high-frequency.

Another fact about inverse problems should be mentioned here. As in Example 364, a (finite-dimensional) linear inverse problem $Ax = b$ usually results from discretizing an infinite-dimensional problem. In computational mathematics, one normally expects a finer discretization to result in a more accurate solution. However, in the case of an inverse problem, a finer discretization means a more accurate resolution of small singular values, which converge to zero as the discretization is refined. Therefore, the results tend to get worse as discretization is refined.

In the above discussion, it was assumed that the matrix A defining the inverse problem was square. There is little change if the matrix is $m \times n$ with $m > n$, in which case it is natural to seek the least-squares solution. The formula producing the least-squares solution in terms of the singular value decomposition is essentially unchanged, as is the explanation for why the solution to an inverse problem is sensitive to noise in the data.

8.4.1 Resolving inverse problems through regularization

The SVD of A shows clearly that the difficulties in solving a linear inverse problem arise from the singular components of the data corresponding to the small singular vectors. This suggests that we might be able to produce a reasonable solution by ignoring or suppressing these singular components. Such an approach is called a *regularization* method. We will briefly introduce two regularization methods here.

8.4.2 The truncated SVD method

We will assume that A belongs to $\mathbf{R}^{m\times n}$, where $m \geq n$, and that the singular values of A are all positive, although they decay to zero in the manner described above. The (unique) least-squares solution to $Ax = \hat{b}$ is then

$$x = \sum_{i=1}^{n} \frac{u_i \cdot \hat{b}}{\sigma_i} v_i.$$

The truncated SVD (TSVD) method simply ignores the singular components corresponding to the smallest singular values. The estimate of the solution is then

$$x_{TSVD}^{(k)} = \sum_{i=1}^{k} \frac{u_i \cdot \hat{b}}{\sigma_i} v_i,$$

where $k \geq 1$ is a positive integer that must be chosen. We will not discuss methods for automatically choosing k, but one common method is trial-and-error, where k is chosen to produce the solution that looks best to an expert in the application area. There are also more quantitative methods, such as L-curve analysis (see [18]).

Example 365 *This is a continuation of Example 364. Using the noisy data from that example, we compute $X_{TSVD}^{(k)}$ for $k = 1, 2, \ldots, 10$. Assuming that it is known that the desired solution is smooth, it is clear that unwanted oscillations begin to appear with $k = 7$, and therefore $k = 6$ is the best choice. Figure 8.5 shows $X_{TSVD}^{(k)}$ for $k = 2, 4, 6, 8$, along with the exact solution. The reader should notice how accurate $X_{TSVD}^{(6)}$ is, despite the fact that it was computed from the same noisy data that produced the solution shown in Figure 8.3.*

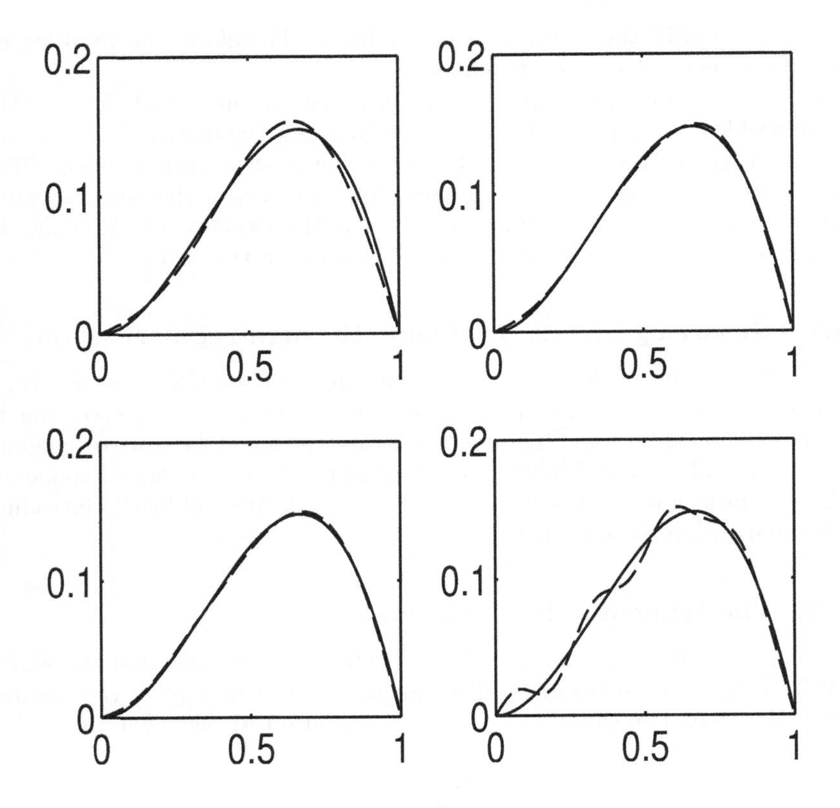

FIGURE 8.5
Four TSVD solutions (dashed curves) to the inverse problem of Example 364: $k = 2$ (upper left), $k = 4$ (upper right), $k = 6$ (lower left), $k = 8$ (lower right). The exact solution (solid curve) is displayed with each TSVD solution.

8.4.3 Tikhonov regularization

A drawback of the TSVD method is that one must compute the SVD of the matrix A. Although we do not discuss practical methods for computing the SVD of a large matrix in this book, it suffices to say that this is quite time-consuming for large matrices and may not be possible in many practical applications. The method of Tikhonov regularization accomplishes much the same thing as TSVD, suppressing the troublesome singular components (although not eliminating them completely), but without the need to compute the SVD.

We recall that the least-squares approach to $Ax = \hat{b}$ is an optimization problem:

$$\min_{x \in \mathbf{R}^n} \|Ax - \hat{b}\|_2^2.$$

Tikhonov regularization is defined by the following related problem:

$$\min_{x\in\mathbf{R}^n} \|Ax-\hat{b}\|_2^2+\epsilon^2\|x\|_2^2. \tag{8.9}$$

The term $\epsilon^2\|x\|_2^2$, where $\epsilon>0$, penalizes large components of x, which often appear as a result of unwanted oscillations in the computed solution. By minimizing the combination of the two terms, $\|Ax-\hat{b}\|_2^2$ and $\epsilon^2\|x\|_2^2$, one tries to balance matching the data on the one hand and avoiding the amplification of noise in the estimate of x on the other. The positive scalar ϵ is called the *regularization parameter* and choosing an appropriate value is a major challenge in this technique (just as is choosing k in the TSVD method). For this brief introduction we will assume that trial-and-error is used. We write x_ϵ for the solution of (8.9) to emphasize the dependence on the regularization parameter ϵ. It is straightforward to show that x_ϵ exists and is unique. In fact, (8.9) is the least-squares formulation of the system

$$\begin{bmatrix} A \\ \epsilon I \end{bmatrix} x = \begin{bmatrix} \hat{b} \\ 0 \end{bmatrix}, \tag{8.10}$$

and the matrix in (8.10) has full rank (see Exercise 1). The normal equations are

$$(A^TA+\epsilon^2 I)x = A^T\hat{b} \tag{8.11}$$

and $A^TA+\epsilon^2 I$ is invertible.

We will give an example of Tikhonov regularization and then explain the connection to the TSVD method.

Example 366 *We continue to study the inverse problem introduced in Example 364. We solve (8.10) by the normal equations approach for several values of the regularization parameter ϵ. In doing so, we notice that if ϵ is too small, then X_ϵ resembles the unregularized solution displayed in Figure 8.3 (large, unwanted oscillations), while if ϵ is too large, then X_ϵ is pushed toward the zero vector.*

Figure 8.6 shows X_ϵ for four values of ϵ. We see that $\epsilon = 0.005$ is a reasonable choice for this problem and yields a good result, roughly as good as we obtained by the TSVD method.

We now use the SVD of A, $A=U\Sigma V^T$, to explain why Tikhonov regularization performs much like the TSVD method. We have

$$\begin{aligned} A^TA+\epsilon^2 I = V\Sigma^TU^TU\Sigma V^T+\epsilon^2VV^T &= V\Sigma^T\Sigma V^T+V(\epsilon^2 I)V^T \\ &= V\left(\Sigma^T\Sigma+\epsilon^2 I\right)V^T, \end{aligned}$$

where $\Sigma^T\Sigma$ is an $n\times n$ diagonal matrix with diagonal entries $\sigma_1^2,\sigma_2^2,\ldots,\sigma_n^2$. Then

$$\begin{aligned} x_\epsilon = \left(A^TA+\epsilon^2 I\right)^{-1}A^T\hat{b} &= V\left(\Sigma^T\Sigma+\epsilon^2 I\right)^{-1}V^TV\Sigma^TU^T\hat{b} \\ &= V\left(\Sigma^T\Sigma+\epsilon^2 I\right)^{-1}\Sigma^TU^T\hat{b}. \end{aligned}$$

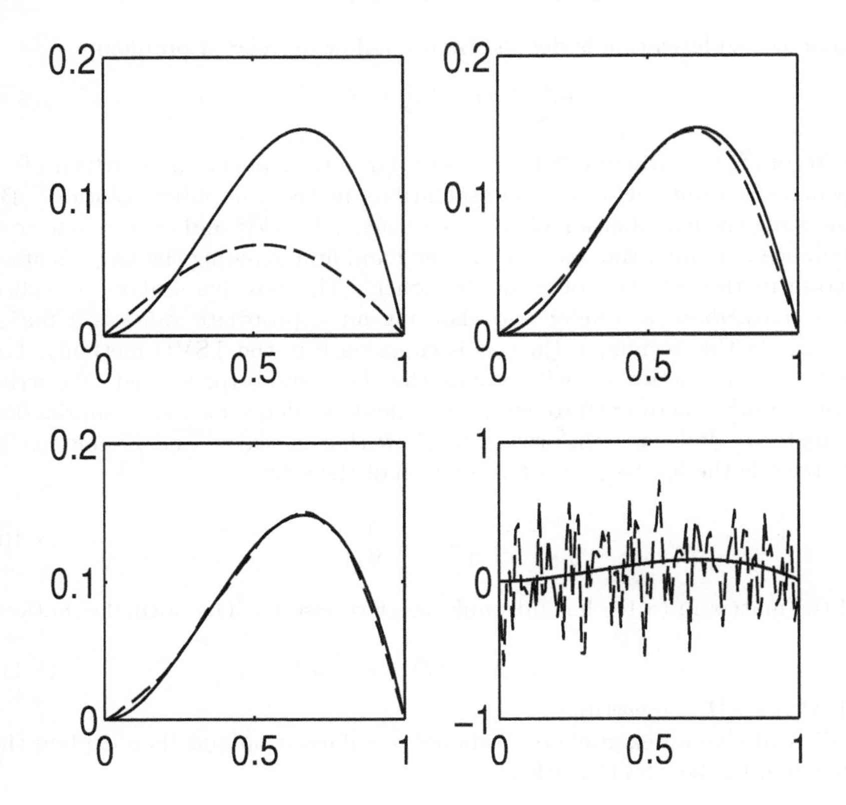

FIGURE 8.6
Four solutions X_ϵ (dashed curves) to the inverse problem of Example 364, obtained by Tikhonov regularization: $\epsilon = 0.1$ (upper left), $\epsilon = 0.01$ (upper right), $\epsilon = 0.005$ (lower left), $\epsilon = 0.0001$ (lower right). The exact solution (solid curve) is displayed with each regularized solution.

The matrix $\left(\Sigma^T\Sigma + \epsilon^2 I\right)^{-1}\Sigma^T$ is diagonal, with diagonal entries

$$\frac{\sigma_i}{\sigma_i^2 + \epsilon^2},\ i = 1, 2, \ldots, n.$$

We therefore obtain

$$x_\epsilon = \sum_{i=1}^{n} \frac{\sigma_i}{\sigma_i^2 + \epsilon^2}(u_i \cdot \hat{b})v_i. \tag{8.12}$$

We can now compare the three solutions to the inverse problem $Ax = \hat{b}$:

1. The unregularized solution:

$$x = \sum_{i=1}^{n} \frac{1}{\sigma_i}(u_i \cdot \hat{b})v_i.$$

2. The TSVD solution

$$x = \sum_{i=1}^{k} \frac{1}{\sigma_i}(u_i \cdot \hat{b})v_i \quad (1 \le k < n).$$

3. The Tikhonov regularized solution:

$$x_\epsilon = \sum_{i=1}^{n} \frac{\sigma_i}{\sigma_i^2 + \epsilon^2}(u_i \cdot b)v_i \quad (\epsilon > 0).$$

We see that the difference between the three methods lies in how they use the components $(u_i \cdot \hat{b})v_i$. The ordinary least-squares solution just multiplies each by $1/\sigma_i$, which, as we have seen, is disastrous if A has small singular values. The TSVD method multiplies each by either $1/\sigma_i$ or 0 and thereby ignores the contributions from small singular components. The Tikhonov regularization method multiplies each by a number lying strictly between $1/\sigma_i$ and 0, namely,

$$\frac{\sigma_i}{\sigma_i^2 + \epsilon^2}. \tag{8.13}$$

We notice that (8.13) is approximately equal to $1/\sigma_i$ if $\sigma_i \gg \epsilon$, while (8.13) is bounded above if $\epsilon \ge \sigma_i$ (see Exercise 2). Therefore, Tikhonov regularization does not allow the singular components corresponding to small singular values to ruin the computed solution, while at the same time, it leaves the components corresponding to large singular values basically unchanged. This explains why Tikhonov regularization can behave similarly to the TSVD method.

Exercises

Note: Exercises 3–4 require the use of a computer package such as MATLAB® that integrates linear algebra capabilities (solving linear systems, computing the SVD) with graphics.

1. (a) Show that (8.9) is equivalent to (8.10).
   (b) Derive the normal equations for (8.10).
   (c) Show that the matrix

   $$\begin{bmatrix} A \\ \epsilon I \end{bmatrix}$$

   appearing in (8.10) has full rank, and explain why this guarantees a unique solution x_ϵ to (8.10) (and hence to (8.9)).

2. (a) Suppose $\epsilon \ge \sigma > 0$. Show that

   $$\frac{\sigma}{\sigma^2 + \epsilon^2} \le \frac{1}{\epsilon}.$$

(b) Suppose $\epsilon \geq \sqrt{\sigma} > 0$. Show that

$$\frac{\sigma}{\sigma^2+\epsilon^2} \leq 1.$$

3. Repeat the analysis in Examples 364–366 for the Fredholm integral equation

$$\int_0^1 k(t,s)x(s)\,ds = y(t),\ 0 \leq t \leq 1,$$

where $k(t,s) = e^{st}$, the exact solution is $x(t) = e^t$, and the exact data is $y(t) = (e^{t+1} - 1)/(t+1)$. Note that you have to generate the noisy data vector from the exact data and a random number generator. (Note: In this problem, the singular values decay to zero very rapidly, so it will not be possible to compute the unregularized solution even in the noise-free case. However, both TSVD and Tikhonov regularization work well.)

4. Repeat the analysis in Examples 364–366 for the Fredholm integral equation

$$\int_0^1 k(t,s)x(s)\,ds = y(t),\ 0 \leq t \leq 1,$$

where $k(t,s) = \sqrt{s^2+t^2}$, $x(t) = t$, and $y(t) = ((1+t^2)^{3/2} - t^3)/3$.

5. Not every Fredholm integral equation of the first kind leads to a true inverse problem. Suppose the kernel has the form

$$k(t,s) = \sum_{m=1}^{N} f_m(t)g_m(s)$$

(such a kernel is called *separable*). Form the matrix K as in Example 364 and show that K can be written as

$$K = \sum_{m=1}^{N} \alpha_m F_m G_m^T,$$

where $F_m, G_m \in \mathbf{R}^n$ for each $m = 1, 2, \ldots, N$ and $\alpha_1, \alpha_2, \ldots, \alpha_m \in \mathbf{R}$. From this, show that K has rank at most N, and therefore the singular values satisfy $\sigma_i = 0$ for $i > N$. Thus such an integral equation falls into the second case discussed on page 483.

8.5 The Smith normal form of a matrix

We have presented several decompositions that allow insight into the nature of a matrix (or the operator it represents): the spectral decomposition (Sections

4.6, 7.1), the Jordan decomposition (Section 5.4), and the singular value decomposition (Sections 8.1, 8.2). These decompositions find application in all areas of mathematics and statistics. In this section, we will present a different decomposition of a matrix that is particularly useful in discrete mathematics. The *Smith normal form* of a matrix reveals the structure of an integer matrix (that is, a matrix whose entries are integers) in such a way as to reveal combinatorial information about the matrix.

To describe the Smith normal form, it is convenient to make the following definition.

Definition 367 *A matrix* $U \in \mathbf{Z}^{n\times n}$ *is called* unimodular *if its determinant is* 1 *or* -1.

Given two integers k, ℓ, we will use the common notation $k|\ell$ to mean that k is a factor of ℓ, that is, that there exists an integer q such that $\ell = kq$.

Theorem 368 *Let* $A \in \mathbf{Z}^{m\times n}$ *be given. There exist unimodular matrices* $U \in \mathbf{Z}^{m\times m}$, $V \in \mathbf{Z}^{n\times n}$ *and a diagonal matrix* $S \in \mathbf{Z}^{m\times n}$ *such that*

$$A = USV,$$

the diagonal entries of S *are* $d_1, d_2, \ldots, d_r, 0, \ldots, 0$, *each* d_i *is a positive integer, and* $d_i|d_{i+1}$ *for* $i = 1, 2, \ldots, r-1$.

The diagonal matrix S is called the *Smith normal form* of A. Since unimodular matrices are invertible, we can define $W = V^{-1}$ and write $A = USW^{-1}$, which shows that S is the matrix representing the operator defined by A, provided the basis defined by the columns of W is used for the domain and the basis defined by the columns of U is used for the co-domain. However, this interpretation is not the focus of the Smith normal form; instead, S is used to deduce combinatorial or number theoretic properties of the matrix A.

The diagonal entries $d_1, d_2, \ldots, d_r$ of the Smith normal form of A are called the *elementary divisors* (or sometimes the *invariant factors*) of A.

8.5.1 An algorithm to compute the Smith normal form

We can prove that the Smith normal form of a matrix exists by defining an algorithm that computes it, and proving that the algorithm works for any integer matrix A. We compute S by multiplying A by a sequence of unimodular matrices:

$$S = P_k P_{k-1} \cdots P_1 A Q_1 Q_2 \cdots Q_\ell,$$

where each P_s belongs to $\mathbf{Z}^{m\times m}$ and each Q_t belongs to $\mathbf{Z}^{n\times n}$. The possible forms for P_s and Q_t are as follows:

1. • P_s is a diagonal matrix; each diagonal entry is 1 except the ith, which is -1. The effect of left-multiplying $B \in \mathbf{Z}^{m\times n}$ by P_s is to multiply the ith row by -1.

- Q_t is a diagonal matrix; each diagonal entry is 1 except the ith, which is -1. The effect of right-multiplying $B \in \mathbf{Z}^{m\times n}$ by Q_t is to multiply the ith column by -1.

2.
   - P_s is equal to the identity matrix, with the 0 in the i,j entry, $j < i$, replaced by a nonzero integer p. The effect of left-multiplying $B \in \mathbf{Z}^{m\times n}$ by P_s is to add p times row j to row i.
   - Q_t is equal to the identity matrix, with the 0 in the i,j entry, $i < j$, replaced by a nonzero integer p. The effect of right-multiplying $B \in \mathbf{Z}^{m\times n}$ by Q_t is to add p times column i to row j.

3.
   - P_s is equal to the identity matrix, with rows i and j interchanged. The effect of left-multiplying $B \in \mathbf{Z}^{m\times n}$ by P_s is to interchange rows i and j of B.
   - Q_t is equal to the identity matrix, with columns i and j interchanged. The effect of right-multiplying $B \in \mathbf{Z}^{m\times n}$ by Q_t is to interchange columns i and j of B.

Replacing B by P_sB (for one of the matrices P_s described above) is called an *elementary row operation*, while replacing B by BQ_t is called an *elementary column operation*. The matrices P_s and Q_t themselves are called *elementary matrices*. Exercise 4 asks the reader to show that each elementary matrix is unimodular.

The algorithm described below indicates how to perform elementary row and column operations on A to obtain the Smith normal form S. Of course, to compute the decomposition $A = USV$, it is necessary to keep track of the matrices $P_1, P_2, \ldots, P_k$ and $Q_1, Q_2, \ldots, Q_\ell$ while computing S from A; then

$$S = P_kP_{k-1}\cdots P_1AQ_1Q_2\cdots Q_\ell \Rightarrow A = USV,$$

where

$$U = P_1^{-1}P_2^{-1}\cdots P_k^{-1},\ V = Q_\ell^{-1}Q_{\ell-1}^{-1}\cdots Q_1^{-1}.$$

The matrices U and V are guaranteed to be unimodular by Exercise 5.

Here is the promised algorithm, taken from [35] (slightly modified to allow for nonsquare and nonsingular matrices). It shows how to perform elementary operations on A to produce S. For simplicity of notation, we will treat A as an array rather than a matrix; in other words, when we refer to A_{ij}, we mean the value of the i,j entry in the transformed matrix at that particular point in the algorithm.

Step 1 Perform row and column interchanges, as necessary, to bring the nonzero entry with the smallest absolute value from the first row and column to the 1, 1 entry. If all entries in the first row and column are zero, go to Step 5.

Step 2 If $A_{11}|A_{1j}$ for each $j = 2, 3, \ldots, n$, go to Step 3. Otherwise, take the smallest value of j such that $A_{11}|A_{j1}$ fails. Then $A_{1j} = qA_{11} + r$, where $0 < r < A_{11}$. Add $-q$ times column 1 to column j. Go to Step 1.

Step 3 If $A_{11}|A_{i1}$ for each $i = 2, 3, \ldots, m$, go to Step 4. Otherwise, take the smallest value of i such that $A_{11}|A_{i1}$ fails. Then $A_{i1} = qA_{11} + r$, where $0 < r < A_{11}$. Add $-q$ times row 1 to row i. Go to Step 1.

Step 4 Now A_{11} divides other entries in the first row and column. Add multiples of column 1 to columns $2, 3, \ldots, n$, as necessary, to zero out those entries. Then apply analogous row operations to zero out entries $2, 3, \ldots, m$ in the first row (but notice that, since the first row has a single nonzero entry, this amounts to replacing $A_{21}, A_{31}, \ldots, A_{m1}$ with zeros).

Step 5 The matrix now has the form

$$A = \left[\begin{array}{c|c} A_{11} & 0 \\ \hline 0 & \tilde{A} \end{array}\right],$$

where $\tilde{A}$ is $(m-1) \times (n-1)$. If $m - 1 = 0$ or $n - 1 = 0$, then A is diagonal; otherwise apply Steps 1 through 4 to the submatrix $\tilde{A}$.

Step 6 The matrix A is now diagonal (and the diagonal entries are integers). If any of the diagonal entries are zero, apply row and column interchanges, if necessary, to bring all nonzero entries before all nonzero entries. Let the diagonal entries of A be $d_1, d_2, \ldots, d_r, 0, \ldots, 0$. Find the first i satisfying $1 \le i \le r - 1$ for which there exists $j > i$ such that $d_i|d_j$ fails. If no such i exists, then stop; the current matrix A is the desired Smith normal form of the original matrix. Otherwise, add row j to row i and go to Step 1.

In this algorithm, Steps 1 through 5 diagonalize the matrix, while the Step 6 ensures that the diagonal entries satisfy the given divisibility property (Step 6 returns to the first five steps if necessary). When the algorithm terminates, A holds S, the desired Smith normal form of the original matrix.

One way to compute U and V is to first compute P and Q as follows: Start with $P = I$ and $Q = I$; each time row operation is performed on A, perform the same operation on P, and each time a column operation is performed on A, perform the same operation on Q. This has the effect of replacing $P = I$ by $P = P_1$, then $P = P_2P_1$, and so forth until $P = P_kP_{k-1} \ldots P_1$. Similarly, $Q = I$ is ultimately replaced by $Q = Q_1Q_2 \ldots Q_\ell$. Having computed P and Q, we then have $U = P^{-1}$ and $V = Q^{-1}$. (This is not the most efficient way to compute U and V; Exercise 6 describes a better approach.)

Example 369 *Let*

$$A = \begin{bmatrix} 8 & 4 & 4 \\ 10 & 5 & 8 \\ 11 & 7 & 7 \end{bmatrix}.$$

We will compute the Smith normal form of A by the above algorithm, and at the same time compute unimodular matrices P and Q such that $S = PAQ$. As *suggested above, we will use the symbols A, P, and Q as arrays and constantly update their values.*

Initially, we have

$$A=\begin{bmatrix} 8 & 4 & 4 \\ 10 & 5 & 8 \\ 11 & 7 & 7 \end{bmatrix},\ P=\begin{bmatrix} 1 & 0 & 0 \\ 0 & 1 & 0 \\ 0 & 0 & 1 \end{bmatrix},\ Q=\begin{bmatrix} 1 & 0 & 0 \\ 0 & 1 & 0 \\ 0 & 0 & 1 \end{bmatrix}.$$

Step 1 calls for us to interchange columns 1 and 2 of A (and we do the same to Q):

$$A=\begin{bmatrix} 4 & 8 & 4 \\ 5 & 10 & 8 \\ 7 & 11 & 7 \end{bmatrix},\ P=\begin{bmatrix} 1 & 0 & 0 \\ 0 & 1 & 0 \\ 0 & 0 & 1 \end{bmatrix},\ Q=\begin{bmatrix} 0 & 1 & 0 \\ 1 & 0 & 0 \\ 0 & 0 & 1 \end{bmatrix}.$$

We now have the smallest number in the first row and column in the $1,1$ *entry, so Step 1 is complete. Moreover,* 4 *divides the other entries in the first row, so there is nothing to do in Step 2. However,* 4 *does not divide* 5 *in the first column, so we add* -1 *times row 1 to row 2 (this is a row operation, so we also perform it on P):*

$$A=\begin{bmatrix} 4 & 8 & 4 \\ 1 & 2 & 4 \\ 7 & 11 & 7 \end{bmatrix},\ P=\begin{bmatrix} 1 & 0 & 0 \\ -1 & 1 & 0 \\ 0 & 0 & 1 \end{bmatrix},\ Q=\begin{bmatrix} 0 & 1 & 0 \\ 1 & 0 & 0 \\ 0 & 0 & 1 \end{bmatrix}.$$

Now we return to Step 1, and since there is now a smaller entry than 4 *in the first column, we interchange rows 1 and 2:*

$$A=\begin{bmatrix} 1 & 2 & 4 \\ 4 & 8 & 4 \\ 7 & 11 & 7 \end{bmatrix},\ P=\begin{bmatrix} -1 & 1 & 0 \\ 1 & 0 & 0 \\ 0 & 0 & 1 \end{bmatrix},\ Q=\begin{bmatrix} 0 & 1 & 0 \\ 1 & 0 & 0 \\ 0 & 0 & 1 \end{bmatrix}.$$

There is nothing to do in Steps 2 and 3, since the new $1,1$ *entry divides all the other entries in the first row and column. We therefore move on to Step 4, and add* -2 *times column 1 to column 2, and* -4 *times column 1 to column 3:*

$$A=\begin{bmatrix} 1 & 0 & 0 \\ 4 & 0 & -12 \\ 7 & -3 & -21 \end{bmatrix},\ P=\begin{bmatrix} -1 & 1 & 0 \\ 1 & 0 & 0 \\ 0 & 0 & 1 \end{bmatrix},\ Q=\begin{bmatrix} 0 & 1 & 0 \\ 1 & -2 & -4 \\ 0 & 0 & 1 \end{bmatrix}.$$

To complete Step 4, we add -4 *times row 1 to row 2 and* -7 *times row 1 to row 3:*

$$A=\begin{bmatrix} 1 & 0 & 0 \\ 0 & 0 & -12 \\ 0 & -3 & -21 \end{bmatrix},\ P=\begin{bmatrix} -1 & 1 & 0 \\ 5 & -4 & 0 \\ 7 & -7 & 1 \end{bmatrix},\ Q=\begin{bmatrix} 0 & 1 & 0 \\ 1 & -2 & -4 \\ 0 & 0 & 1 \end{bmatrix}.$$

We have now put the first row and column in the desired form, and we move on to the 2×2 *submatrix in the lower right-hand corner, returning to Step 1. To put the smallest nonzero entry from the second row and column in the* $2,2$ *entry, we interchange rows 2 and 3 (we are referring to the* $2,2$ *entry and rows* 2 *and* 3 *of the matrix itself; we could also refer to the* $1,1$ *entry and rows* 1 *and* 2 *of the submatrix):*

$$A = \begin{bmatrix} 1 & 0 & 0 \\ 0 & -3 & -21 \\ 0 & 0 & -12 \end{bmatrix}, \; P = \begin{bmatrix} -1 & 1 & 0 \\ 7 & -7 & 1 \\ 5 & -4 & 0 \end{bmatrix}, \; Q = \begin{bmatrix} 0 & 1 & 0 \\ 1 & -2 & -4 \\ 0 & 0 & 1 \end{bmatrix}.$$

The $2,2$ *entry divides the other entries in the second row and column, so there is nothing to do in Steps 2 and 3. For Step 4, we add* -7 *times column 2 to column 3:*

$$A = \begin{bmatrix} 1 & 0 & 0 \\ 0 & -3 & 0 \\ 0 & 0 & -12 \end{bmatrix}, \; P = \begin{bmatrix} -1 & 1 & 0 \\ 7 & -7 & 1 \\ 5 & -4 & 0 \end{bmatrix}, \; Q = \begin{bmatrix} 0 & 1 & -7 \\ 1 & -2 & 10 \\ 0 & 0 & 1 \end{bmatrix}.$$

We have now diagonalized A*, so we move onto Step 6. First we make the diagonal entries nonnegative by multiplying rows 2 and 3 by* -1*:*

$$A = \begin{bmatrix} 1 & 0 & 0 \\ 0 & 3 & 0 \\ 0 & 0 & 12 \end{bmatrix}, \; P = \begin{bmatrix} -1 & 1 & 0 \\ -7 & 7 & -1 \\ -5 & 4 & 0 \end{bmatrix}, \; Q = \begin{bmatrix} 0 & 1 & -7 \\ 1 & -2 & 10 \\ 0 & 0 & 1 \end{bmatrix}.$$

Now it turns out that the diagonal entries of the final matrix have the desired divisibility property, so the algorithm terminates with the desired S*,* P*, and* Q*. The reader can verify that* $S = PAQ$ *and that* $\det(P) = 1$*,* $\det(Q) = -1$*. If we want the decomposition* $A = USV$*, then*

$$U = P^{-1} = \begin{bmatrix} 4 & 0 & -1 \\ 5 & 0 & -1 \\ 7 & -1 & 0 \end{bmatrix}, \; V = Q^{-1} = \begin{bmatrix} 2 & 1 & 4 \\ 1 & 0 & 7 \\ 0 & 0 & 1 \end{bmatrix}.$$

Here is a second example that illustrates more features of the algorithm.

Example 370 *Let*

$$A = \begin{bmatrix} 6 & 4 & 7 \\ 6 & 4 & 4 \\ 3 & 6 & 6 \end{bmatrix}.$$

For this example, we will assume that only the Smith normal form S *is desired (and not* P*,* Q*). For Step 1, we interchange rows 1 and 3:*

$$A = \begin{bmatrix} 3 & 6 & 6 \\ 6 & 4 & 4 \\ 6 & 4 & 7 \end{bmatrix}.$$

There is nothing to do for Steps 2 and 3, since the $1,1$ *entry already divides the remaining entries in the first row and column. We now zero out the entries in the first row, and then first column, to get*

$$A = \begin{bmatrix} 3 & 0 & 0 \\ 6 & -8 & -8 \\ 6 & -8 & -5 \end{bmatrix} \text{ and then } A = \begin{bmatrix} 3 & 0 & 0 \\ 0 & -8 & -8 \\ 0 & -8 & -5 \end{bmatrix}.$$

We now return to Step 1, focusing on the 2×2 submatrix in the lower right-hand corner. There is nothing to do for Steps 1, 2, and 3; for Step 4 we add -1 times column 2 to column 3 to obtain

$$A = \begin{bmatrix} 3 & 0 & 0 \\ 0 & -8 & 0 \\ 0 & -8 & 3 \end{bmatrix},$$

and then add -1 times row 2 to row 3 to get

$$A = \begin{bmatrix} 3 & 0 & 0 \\ 0 & -8 & 0 \\ 0 & 0 & 3 \end{bmatrix}.$$

Finally, we multiply the second row by -1 and we have diagonalized the original matrix:

$$A = \begin{bmatrix} 3 & 0 & 0 \\ 0 & 8 & 0 \\ 0 & 0 & 3 \end{bmatrix}.$$

However, the diagonal entries do not satisfy the divisibility condition and therefore we must invoke Step 6 of the algorithm. Since the $1,1$ entry does not divide the $2,2$ entry, we add row 2 to row 1 to get

$$A = \begin{bmatrix} 3 & 8 & 0 \\ 0 & 8 & 0 \\ 0 & 0 & 3 \end{bmatrix},$$

and then return to Step 1. For brevity, we will now present the sequence of matrices produced by Steps 1 through 5, without explaining each step:

$$\begin{bmatrix} 3 & 8 & 0 \\ 0 & 8 & 0 \\ 0 & 0 & 3 \end{bmatrix} \rightarrow \begin{bmatrix} 3 & 2 & 0 \\ 0 & 8 & 0 \\ 0 & 0 & 3 \end{bmatrix} \rightarrow \begin{bmatrix} 2 & 3 & 0 \\ 8 & 0 & 0 \\ 0 & 0 & 3 \end{bmatrix} \rightarrow \begin{bmatrix} 2 & 1 & 0 \\ 8 & -8 & 0 \\ 0 & 0 & 3 \end{bmatrix}$$
$$\rightarrow \begin{bmatrix} 1 & 2 & 0 \\ -8 & 8 & 0 \\ 0 & 0 & 3 \end{bmatrix} \rightarrow \begin{bmatrix} 1 & 0 & 0 \\ -8 & 24 & 0 \\ 0 & 0 & 3 \end{bmatrix} \rightarrow \begin{bmatrix} 1 & 0 & 0 \\ 0 & 24 & 0 \\ 0 & 0 & 3 \end{bmatrix}.$$

We now once again reach Step 6; we have a diagonal matrix but the $2,2$ entry does not divide the $3,3$ entry. We therefore add the third row to the second *and return to Step 1:*

$$\begin{bmatrix} 1 & 0 & 0 \\ 0 & 24 & 3 \\ 0 & 0 & 3 \end{bmatrix} \to \begin{bmatrix} 1 & 0 & 0 \\ 0 & 3 & 24 \\ 0 & 3 & 0 \end{bmatrix} \to \begin{bmatrix} 1 & 0 & 0 \\ 0 & 3 & 0 \\ 0 & 3 & -24 \end{bmatrix}$$

$$\to \begin{bmatrix} 1 & 0 & 0 \\ 0 & 3 & 0 \\ 0 & 0 & -24 \end{bmatrix} \to \begin{bmatrix} 1 & 0 & 0 \\ 0 & 3 & 0 \\ 0 & 0 & 24 \end{bmatrix}.$$

Now the diagonal entries satisfy the divisibility condition ($1|3, 3|24$)*, and we have computed the Smith normal form of the matrix.*

We will not give a rigorous proof that the above algorithm produces the Smith normal form, but it is not difficult to construct such a proof. If the algorithm terminates, then the final result is the Smith normal form, so the proof consists of proving that the algorithm terminates. Beginning at Step 1, each time the algorithm returns to Step 1, it is guaranteed to reduce the magnitude of the $1,1$ entry; since a positive integer can only be reduced finitely many times before it reaches 1, it is guaranteed that the algorithm returns to Step 1 only finitely many times before going on to Step 4, and then to Step 5. Each time the algorithm reaches Step 5, it has advanced one row and column in diagonalizing the matrix. Thus, the algorithm is guaranteed to diagonalize the matrix in finitely many steps.

The second part of the algorithm consists of ensuring that the divisibility conditions are satisfied by the diagonal entries. We leave it to the reader to argue that this part of the algorithm is also guaranteed to terminate.

8.5.2 Applications of the Smith normal form

We will now present an application of the Smith normal form: determining the p-rank of an integer matrix. Before developing this application, we briefly review the elementary theory of congruence modulo p, which defines the operations in the field $\mathbf{Z}_p$. Throughout this section, p will denote a fixed prime number.

Given any integer k, there exists integers q and r, with $0 \leq r < p$, such that $k = qp + r$. The integer r is called the *congruence class of k modulo p*, and we write $k \equiv r \pmod{p}$. The reader should notice that, if k and p are positive, then r is simply the remainder when k is divided by p. As we saw in Section 2.1, the operations in $\mathbf{Z}_p$ are addition and multiplication modulo p; that is, given $a, b \in \mathbf{Z}_p$, $a + b$ in $\mathbf{Z}_p$ is the congruence class of the ordinary sum $a + b$, and similarly for ab.

In general, for $k, \ell \in \mathbf{Z}$, we say that $k \equiv \ell \pmod{p}$ if p divides $k - \ell$. It is easy to show that, if r is the congruence class of $k \in \mathbf{Z}$, then p divides $k - r$, and hence this is consistent with the earlier definition. Moreover, it is a straightforward exercise to show that $k \equiv \ell \pmod{p}$ if and only if k and ℓ have the same congruence class modulo p.

When working with arbitrary integers in $\mathbf{Z}$, the following properties of congruence modulo n are fundamental.

Theorem 371 *Suppose $a, b, c, d \in \mathbf{Z}$ and $a \equiv c \pmod{p}$, $b \equiv d \pmod{p}$. Then*

$$a + b \equiv c + d \pmod{p}$$

and

$$ab \equiv cd \pmod{p}.$$

Proof Exercise 7.

As an application of the preceding theorem, let us consider two integers $a, b \in \mathbf{Z}_p$ and their congruence classes c, d modulo p, which can be regarded as elements of $\mathbf{Z}_p$. The meaning of the theorem is that we can compute, for example, ab (ordinary product of integers) and then take the congruence class of ab, or we can find the congruence classes c, d of a, b, respectively, and then compute cd in $\mathbf{Z}_p$. Either way, we will get the same result.

The same conclusion applies to a calculation of possibly many additions and subtractions of integers; we can perform the calculation in $\mathbf{Z}$ and then take the congruence class of the result, or take the congruence classes of the initial integers and perform the calculation in $\mathbf{Z}_p$. Either way, the result will be the same. A formal proof by induction of the following theorem is based on these remarks.

Theorem 372 *Let $A \in \mathbf{Z}^{n\times n}$, and let $\tilde{A} \in \mathbf{Z}_p^{n\times n}$ be obtained by replacing each entry of A by its congruence class modulo p. Then the congruence class of $\det(A)$ modulo p is the same as the $\det(\tilde{A})$ in $\mathbf{Z}_p$. In other words,*

$$\det(A) \equiv \det(\tilde{A}) \pmod{p}.$$

Throughout the remainder of this section, given $A \in \mathbf{Z}^{n\times n}$, $\tilde{A}$ will denote the corresponding matrix in $\mathbf{Z}_p^{n\times n}$, as in Theorem 372. We have the following corollary.

Corollary 373 *Let $A \in \mathbf{Z}^{n\times n}$ be given, and let $\tilde{A}$ be the corresponding matrix in $\mathbf{Z}_p^{n\times n}$. Then $\tilde{A}$ is singular if and only if $p|\det(A)$.*

Proof We know that $\tilde{A}$ is singular if and only if $\det(\tilde{A}) = 0$ (where this determinant is computed in $\mathbf{Z}_p$). But

$$\det(A) \equiv \det(\tilde{A}) \pmod{p}$$

by Theorem 372, and $\det(A) \equiv 0 \pmod{p}$ if and only if $p|\det(A)$. This completes the proof.

QED

We notice that if $U \in \mathbf{Z}^{n\times n}$ is unimodular, then so is $\tilde{U} \in \mathbf{Z}_p^{n\times n}$. Given $A \in \mathbf{Z}^{n\times n}$ and the Smith decomposition $A = USV$, we get the Smith decomposition over Z_p: $\tilde{A} = \tilde{U}\tilde{S}\tilde{V}$.

We now make the following definition.

Definition 374 *Let $A \in \mathbf{Z}^{n\times n}$. The* p-rank *of A is the rank of the corresponding matrix $\tilde{A}$ in $\mathbf{Z}_p^{n\times n}$.*

The rank of $A \in \mathbf{Z}^{n\times n}$ or $B \in \mathbf{Z}_p^{n\times n}$ is easily computed from the Smith normal form:

Theorem 375 *1. Let $A \in \mathbf{Z}^{n\times n}$ and let $S \in \mathbf{Z}^{n\times n}$ be the Smith normal form of A, with nonzero diagonal entries $d_1, d_2, \ldots, d_r$. Then the rank of A is r.*

2. Let $B \in \mathbf{Z}_p^{n\times n}$ and let $T \in \mathbf{Z}_p^{n\times n}$ be the Smith normal form over $\mathbf{Z}_p$ of B, with nonzero diagonal entries $e_1, e_2, \ldots, e_s$. Then the rank of B is s.

Proof Exercise 8.

It follows immediately from the preceding theorem that the p-rank of an integer matrix A, for any p, can be computed directly from the Smith normal form.

Corollary 376 *Let $A \in \mathbf{Z}^{n\times n}$ and let $S \in \mathbf{Z}^{n\times n}$ be the Smith normal form of A, with nonzero diagonal entries $d_1, d_2, \ldots, d_r$. Let p be prime and let k be the largest integer such that p does not divide d_k. Then the p-rank of A is k.*

Proof The p-rank of A is just the rank (in $Z_p^{n\times n}$) of $\tilde{A}$. But the Smith normal form of $\tilde{A}$ is just $\tilde{S}$, and the nonzero diagonal entries of $\tilde{S}$ are just the diagonal entries of A that are not divisible by p. The result follows.

QED

Example 377 *Let*

$$A = \begin{bmatrix} 3 & 2 & 10 & 1 & 9 \\ 7 & 6 & 8 & 9 & 5 \\ -100 & -102 & -2 & -204 & 46 \\ -1868 & -1866 & 26 & -3858 & 1010 \\ -27204 & -27202 & 34 & -54734 & 13698 \end{bmatrix}.$$

The Smith normal form of A is

$$S = \begin{bmatrix} 1 & & & & \\ & 2 & & & \\ & & 6 & & \\ & & & 30 & \\ & & & & 0 \end{bmatrix}.$$

From this we see that the rank of A is 4 and, for example, the 5-rank of A is 3. This last result follows from the fact that 5 does not divide 1, 2, or 6, but *does divide* 30*. To check that the* 5*-rank is indeed* 3*, we can compute* $\tilde{A}$*, the corresponding matrix in* $Z_5^{5\times 5}$*, and row-reduce it (using arithmetic in* $\mathbf{Z}_5$*):*

$$\begin{bmatrix} 3 & 2 & 0 & 1 & 4 \\ 2 & 1 & 3 & 4 & 0 \\ 0 & 3 & 3 & 1 & 1 \\ 2 & 4 & 1 & 2 & 0 \\ 1 & 3 & 4 & 1 & 3 \end{bmatrix} \rightarrow \begin{bmatrix} 3 & 2 & 0 & 1 & 4 \\ 0 & 3 & 3 & 0 & 4 \\ 0 & 3 & 3 & 1 & 1 \\ 0 & 1 & 1 & 3 & 4 \\ 0 & 4 & 4 & 4 & 0 \end{bmatrix}$$

$$\rightarrow \begin{bmatrix} 3 & 2 & 0 & 1 & 4 \\ 0 & 3 & 3 & 0 & 4 \\ 0 & 0 & 0 & 1 & 2 \\ 0 & 0 & 0 & 1 & 2 \\ 0 & 0 & 0 & 1 & 2 \end{bmatrix} \rightarrow \begin{bmatrix} 3 & 2 & 0 & 1 & 4 \\ 0 & 3 & 3 & 0 & 4 \\ 0 & 0 & 0 & 1 & 2 \\ 0 & 0 & 0 & 0 & 0 \\ 0 & 0 & 0 & 0 & 0 \end{bmatrix}.$$

The last matrix shows that the rank is 3*.*

The Smith canonical form has many more applications in discrete mathematics. To give just two examples, it can be used to classify equivalent integer programs (see Chapter 7 of [11]), and it is useful in design theory (see, for example, [2] and Section 2.3 of [28]).

Exercises

1. Let
$$A = \begin{bmatrix} 4 & 6 & 5 \\ 6 & 3 & 6 \\ 3 & 6 & 6 \end{bmatrix}.$$
Find the Smith decomposition $A = USV$ of A.

2. Let
$$A = \begin{bmatrix} 5 & 7 & 5 \\ 6 & 6 & 4 \\ 5 & 7 & 3 \end{bmatrix}.$$
Find the Smith decomposition $A = USV$ of A.

3. Let
$$A = \begin{bmatrix} 8 & 4 & 16 \\ 10 & 5 & 20 \\ 11 & 7 & 7 \end{bmatrix}.$$
Find the Smith decomposition $A = USV$ of A.

4. Prove that each elementary matrix P_s described on page 495 is unimodular. (The proofs that the matrices Q_t are unimodular are similar.)

5. (a) Prove that the inverse of a unimodular matrix is unimodular.
   (b) Prove that the product of unimodular matrices is unimodular.

6. Consider the computation of the unimodular matrices U and V in the Smith decomposition $A = USV$. We have

$$U = P^{-1} = (P_k P_{k-1} \cdots P_1)^{-1} = P_1^{-1} P_2^{-1} \cdots P_k^{-1}.$$

We can therefore compute U by beginning with $U = I$ and successively replacing it with $U = P_1^{-1}$, $U = P_1^{-1} P_2^{-1}$, and so forth, until we obtain $U = P_1^{-1} P_2^{-1} \cdots P_k^{-1}$. Explain how to right-multiply a given matrix by P_i^{-1}, when P_i is any of the three possible elementary matrices, and thus how to compute U without first computing P and then P^{-1}.

7. Prove Theorem 371.

8. Prove Theorem 375.

9

Matrix factorizations and numerical linear algebra

The purpose of this chapter is to re-interpret some of the main results of the book in terms of *matrix factorizations*. We have already encountered several factorizations: the spectral decomposition of a square, diagonalizable matrix, the Jordan canonical form of a general square matrix, and the singular value decomposition of a general (not necessarily square) matrix. The factorizations (or decompositions) express concisely the eigenvalue-eigenvector relationships and allow convenient manipulations that take advantage of these relationships.

Apart from eigenvalues and eigenvectors, this book addresses two other concepts: Linear operator equations, which (in the finite-dimensional case) can always be expressed as systems of linear equations, and orthogonality and best approximation. There is a factorization associated with each of these concepts. The *LU factorization* is a concise representation of Gaussian elimination, and the QR factorization orthogonalizes the columns of a matrix (like the Gram-Schmidt process does, although the QR factorization is normally computed by a different algorithm). These and other related factorizations will be explored in this chapter.

In contrast to the preceding chapters, the derivations in this chapter will take into account the issue of stability in the context of finite-precision arithmetic. Practical computer algorithms for the computations of linear algebra comprise the subject of *numerical linear algebra*, to which this chapter is therefore an introduction. Matrix factorizations provide an unifying theme in numerical linear algebra, so it is natural to consider the two topics in tandem.

9.1 The LU factorization

The reader will recall that a square system of linear equations is generally solved by a two-part algorithm, Gaussian elimination with back substitution. The first part of the algorithm, Gaussian elimination, reduces the system $Ax = b$ ($A \in \mathbf{R}^{n\times n}$, $b \in \mathbf{R}^n$) to an upper triangular system $Ux = c$. For example, Gaussian elimination reduces the 4×4 system

$$\begin{aligned} x_1 + 3x_2 + 2x_3 + 3x_4 &= -5, \\ 4x_1 + 15x_2 + 9x_3 - 3x_4 &= 15, \\ 3x_1 + 9x_2 + 5x_3 + 21x_4 &= -38, \\ 2x_1 + 15x_2 + 7x_3 - 42x_4 &= 101 \end{aligned}$$

to

$$\begin{aligned} x_1 + 3x_2 + 2x_3 + 3x_4 &= -5, \\ 3x_2 + x_3 - 15x_4 &= 35, \\ -x_3 + 12x_4 &= -23, \\ -3x_4 &= 6. \end{aligned}$$

In this case, the matrices are

$$A = \begin{bmatrix} 1 & 3 & 2 & 3 \\ 4 & 15 & 9 & -3 \\ 3 & 9 & 5 & 21 \\ 2 & 15 & 7 & -42 \end{bmatrix}, \quad U = \begin{bmatrix} 1 & 3 & 2 & 3 \\ 0 & 3 & 1 & -15 \\ 0 & 0 & -1 & 12 \\ 0 & 0 & 0 & -3 \end{bmatrix}.$$

We begin this section by showing that, if no row interchanges are required when Gaussian elimination is applied, then there is a nonsingular, lower triangular matrix L such that $U = L^{-1}A$. Equivalently, then, we have $A = LU$. The matrix L naturally produced by Gaussian elimination is *unit lower triangular*, that is, a lower triangular matrix with ones on the diagonal. We will say that $A = LU$ is an *LU factorization* of A if L is unit lower triangular and U is upper triangular.

Applying Gaussian elimination to an $n \times n$ matrix A requires $n-1$ steps; at step k, the zeros are introduced below the kth diagonal entry. We will show, in fact, that each such step can be accomplished by multiplication by a lower triangular matrix called an *elementary matrix*. To this end, we will require some notation. Ignoring the effect of Gaussian elimination on the right-hand side b, Gaussian elimination transforms A to U through $n-2$ intermediate matrices. We will write $A^{(1)} = A$, $A^{(2)}$ for the first intermediate matrix, and so on, until $A^{(n)} = U$. The matrix $A^{(k)}$ has the property that $A^{(k)}_{ij} = 0$ for $j < k$ and $i > j$ (that is, all the subdiagonal entries in the first $k-1$ columns are zero). Step k produces $A^{(k+1)}$ from $A^{(k)}$ by adding multiples of row k to rows $k+1, k+2, \ldots, n$. The multipliers in this step are

$$-\frac{A^{(k)}_{ik}}{A^{(k)}_{kk}}, \quad i = k+1, k+2, \ldots, n.$$

To verify this, notice that when row k of $A^{(k)}$ is multiplied by $-A^{(k)}_{ik}/A^{(k)}_{kk}$ and added to row i, the new i,k entry is

$$A_{ik}^{(k+1)} = A_{ik}^{(k)} - \frac{A_{ik}^{(k)}}{A_{kk}^{(k)}} A_{kk}^{(k)} = 0,$$

as desired.

We define, for each $k = 1, 2, \ldots, n-1$,

$$\ell_{ik} = \frac{A_{ik}^{(k)}}{A_{kk}^{(k)}}, \ i = k+1, k+2, \ldots, n. \tag{9.1}$$

We then define the lower triangular matrix L_k to be

$$L_k = \begin{bmatrix} 1 & & & & & & \\ & 1 & & & & & \\ & & \ddots & & & & \\ & & & 1 & & & \\ & & & -\ell_{k+1,k} & \ddots & & \\ & & & \vdots & & \ddots & \\ & & & -\ell_{n,k} & & & 1 \end{bmatrix}. \tag{9.2}$$

The ith row of the product $L_k A^{(k)}$ is a linear combination of the rows of $A^{(k)}$, where the weights are the entries in the ith row of L_k. Since the first k rows of L_k are the first k rows of the identity matrix, it follows that the first k rows of $L_k A^{(k)}$ are identical to the first k rows of $A^{(k)}$. For $i > k$, there are two nonzero entries in row i of L_k, namely, $-\ell_{ik}$ (entry k) and 1 (entry i). The ith row of $L_k A^{(k)}$ is therefore $-\ell_{ik}$ times row k of $A^{(k)}$ plus 1 times row i of $A^{(k)}$. This shows that

$$A^{(k+1)} = L_k A^{(k)}.$$

We now have

$$U = L_{n-1} L_{n-2} \cdots L_1 A.$$

Each L_k is invertible (since its determinant is 1), so

$$A = L_1^{-1} L_2^{-1} \cdots L_{n-1}^{-1} U,$$

or $A = LU$ with

$$L = L_1^{-1} L_2^{-1} \cdots L_{n-1}^{-1}.$$

It remains only to show that L is lower triangular. This follows from the next two theorems.

Theorem 378 *Let $L \in \mathbf{R}^{n \times n}$ be lower triangular and invertible. Then L^{-1} is also lower triangular. Similarly, if $U \in \mathbf{R}^{n \times n}$ is upper triangular and invertible, then U^{-1} is also upper triangular.*

Proof Exercise 3.

Theorem 379 *Let $L_1, L_2 \in \mathbf{R}^{n\times n}$ be two lower triangular matrices. Then L_1L_2 is also lower triangular. Similarly, the product of two upper triangular matrices is upper triangular.*

Proof Exercise 4.

Theorems 378 and 379 together imply that $L = L_1^{-1}L_2^{-1}\cdots L_{n-1}^{-1}$ is lower triangular, and thus we have shown that $A \in \mathbf{R}^{n\times n}$ can be factored as $A = LU$, where L is lower triangular and U is upper triangular, *provided no row interchanges are required when applying Gaussian elimination to A.*

To complete our initial discussion of the LU factorization, we want to show that L can be computed with no additional work when Gaussian elimination is performed, and we also want to derive a condition on the matrix A guaranteeing that no row interchanges are required.

It turns out that the entries in the lower triangle of L are just the multipliers ℓ_{ij} defined by (9.1). This can be shown in two steps, which are left to exercises. Exercise 5 asks you to show that, with L_k defined by (9.2), L_k^{-1} is given by

$$L_k^{-1} = \begin{bmatrix} 1 & & & & & & \\ & 1 & & & & & \\ & & \ddots & & & & \\ & & & 1 & & & \\ & & & \ell_{k+1,k} & \ddots & & \\ & & & \vdots & & \ddots & \\ & & & \ell_{n,k} & & & 1 \end{bmatrix}. \tag{9.3}$$

Next, Exercise 6 outlines a proof that

$$\left(L_k^{-1}L_{k+1}^{-1}\cdots L_{n-1}^{-1}\right)_{ij} = \begin{cases} 0, & j > i, \\ 1, & j = i, \\ \ell_{ij}, & k \le j < i, \\ 0, & j < \min\{k, i\}. \end{cases} \tag{9.4}$$

Putting together the above results, we obtain

$$L = \begin{bmatrix} 1 & & & \\ \ell_{2,1} & 1 & & \\ \ell_{3,1} & \ell_{3,2} & & \\ \vdots & \vdots & \ddots & \\ \ell_{n,1} & \ell_{n,2} & \cdots & 1 \end{bmatrix}.$$

We emphasize that the quantities ℓ_{ij} are the multipliers that must be computed in Gaussian elimination anyway, so there is no cost to assembling the LU factorization of A beyond the cost of Gaussian elimination itself.

Example 380 *Consider the* 3×3 *matrix*

$$A = \begin{bmatrix} 1 & 1 & 0 \\ 2 & 1 & -1 \\ 0 & -3 & -2 \end{bmatrix}.$$

Gaussian elimination proceeds as follows (notice that we operate directly on the matrix A *rather than an augmented matrix):*

$$\begin{bmatrix} 1 & 1 & 0 \\ 2 & 1 & -1 \\ 0 & -3 & -2 \end{bmatrix} \rightarrow \begin{bmatrix} 1 & 1 & 0 \\ 0 & -1 & -1 \\ 0 & -3 & -2 \end{bmatrix}$$
$$\rightarrow \begin{bmatrix} 1 & 1 & 0 \\ 0 & -1 & -1 \\ 0 & 0 & 1 \end{bmatrix}.$$

The first step was to add -2 *times row 1 to row 2 and* 0 *times row 1 to row 3. The second step was to add* -3 *times row 2 to row 3. We therefore have* $\ell_{2,1} = 2$, $\ell_{3,1} = 0$, $\ell_{3,2} = 3$, *and so*

$$L = \begin{bmatrix} 1 & 0 & 0 \\ 2 & 1 & 0 \\ 0 & 3 & 1 \end{bmatrix}, \; U = \begin{bmatrix} 1 & 1 & 0 \\ 0 & -1 & -1 \\ 0 & 0 & 1 \end{bmatrix}.$$

The reader can verify by direct multiplication that $A = LU$.

Not every matrix $A \in \mathbf{R}^{n \times n}$ has an LU factorization. For example, the matrix

$$A = \begin{bmatrix} 0 & 1 \\ 1 & 1 \end{bmatrix}$$

has no LU factorization (see Exercise 7). The key to the existence of the LU factorization is that, at each step $k = 1, 2, \ldots, n-1$, the k,k-entry of the relevant matrix is nonzero. The following theorem provides a sufficient condition for this to be true.

Theorem 381 *Let* $A \in \mathbf{R}^{n \times n}$. *For each* $k = 1, 2, \ldots, n-1$, *let* $M^{(k)} \in \mathbf{R}^{k \times k}$ *be the submatrix extracted from the upper left-hand corner of* A*:*

$$M_{ij}^{(k)} = A_{ij}, \; i, j = 1, 2, \ldots, k.$$

If each $M^{(k)}$, $k = 1, 2, \ldots, n-1$, *is nonsingular, then* A *has a unique LU factorization.*

Proof We argue by induction on n, the size of the matrix. If $n = 1$, then the result is obviously true: Every 1×1 is both lower and upper triangular, and we can take $L = [1]$ and $U = A$. Since the only 1×1 unit lower triangular matrix is $L = [1]$, this is the unique LU factorization of A.

We now assume the result is true for matrices of size $(n-1)\times(n-1)$. Suppose $A \in \mathbf{R}^{n\times n}$ has the property that $M^{(k)}$ is nonsingular for $k = 1,\ldots,n-1$. Then, by the induction hypothesis, $M^{(n-1)}$ has a unique LU factorization $M^{(n-1)} = L'U'$. Moreover, since $M^{(n-1)}$ is nonsingular by hypothesis, so are L' and U'.

An LU factorization of A must have the form

$$\left[\begin{array}{c|c} M^{(n-1)} & a \\ \hline b^T & A_{nn} \end{array}\right] = \left[\begin{array}{c|c} L_1 & 0 \\ \hline u^T & 1 \end{array}\right]\left[\begin{array}{c|c} U_1 & v \\ \hline 0 & \alpha \end{array}\right], \tag{9.5}$$

where $L_1 \in \mathbf{R}^{(n-1)\times(n-1)}$ is unit lower triangular, $U_1 \in \mathbf{R}^{(n-1)\times(n-1)}$ is upper triangular, u, v belong to $\mathbf{R}^{n-1}$, α is a real number, and $a, b \in \mathbf{R}^{n-1}$ are known vectors defined by the entries in the last column and row of A. Equation (9.5) is equivalent to the following four equations:

$$\begin{aligned} M^{(n-1)} &= L_1U_1, \\ L_1v &= a, \\ u^TU_1 &= b^T, \\ u\cdot v + \alpha &= A_{nn}. \end{aligned}$$

Since $M^{(n-1)}$ has a unique LU factorization by the induction hypothesis, we must have $L_1 = L'$, $U_1 = U'$. Moreover, since L' and U' are invertible, the remaining equations are uniquely solvable:

$$v = (L')^{-1}a,\ u = (U')^{-T}b,\ \alpha = A_{nn} - u\cdot v.$$

This shows that A has a unique LU factorization.

QED

9.1.1 Operation counts

When considering practical algorithms, a primary concern is the number of arithmetic operations required. If we have two otherwise satisfactory algorithms, we usually prefer the one requiring fewer operations, as it is expected to execute more quickly on a computer.[1] We will now count the operations required by Gaussian elimination and back substitution.

We will assume that Gaussian elimination is applied to the augmented matrix. The algorithm requires $n-1$ steps. At the kth step, we compute the multipliers $\ell_{ik} = A^{(k)}_{ik}/A^{(k)}_{kk}$, $i = k+1,\ldots,n$ ($n-k$ operations), and then add $-\ell_{ik}$ times row k to row i. Row k has nonzero entries in columns

[1]Depending on the computer architecture used, it may not be true that fewer arithmetic operations leads to a faster execution time. The efficiency of memory access may be more important, and on parallel computers, communication costs between processors may be determinative. However, the first consideration in designing efficient algorithms is normally operation count.

$j = k+1, \ldots, n+1$, so multiplying row k by ℓ_{ik} requires $n-k+1$ operations, as does adding the result to row i (notice that we ignore the entries in column k, because we already know that the entries in rows $k+1, \ldots, n$ will become zero). Thus the row operations for step k require $2(n-k+1)$ operations per row, for $n-k$ rows. Thus the total number of operations for step k is

$$2(n-k)(n-k+1) + n - k = (n-k)(2n-2k+3).$$

The grand total is therefore

$$\begin{aligned}
\sum_{k=1}^{n-1}(n-k)(2n-2k+3) &= \sum_{j=1}^{n-1} j(2j+3) \\
&= 2\sum_{j=1}^{n-1} j^2 + 3\sum_{j=1}^{n-1} j \\
&= 2 \cdot \frac{(n-1)n(2n-1)}{6} + 3 \cdot \frac{(n-1)n}{2} \\
&= \frac{2}{3}n^3 + \frac{1}{2}n^2 - \frac{7}{6}n.
\end{aligned}$$

If we write $[U|c]$ for the augmented matrix after Gaussian elimination, then back substitution can be expressed succinctly as

$$x_k = \left(c_k - \sum_{j=k+1}^{n} U_{kj} x_j \right) / U_{kk}.$$

Computing x_k therefore requires $n-k$ multiplications, $n-k$ subtractions, and 1 division, for a total of $2(n-k)+1$ operations. The total operation count for back substitution is

$$\sum_{k=1}^{n}(2(n-k)+1) = 2\sum_{k=1}^{n}(n-k) + \sum_{k=1}^{n} 1 = 2 \cdot \frac{(n-1)n}{2} + n = n^2.$$

When n is large, the largest power of n dominates in each of the operation counts, so we say that Gaussian elimination requires approximately $2n^3/3$ operations, while back substitution requires approximately $n^2/2$ operations. The entire process of solving $Ax = b$ (Gaussian elimination with back substitution) requires on the order of $2n^3/3$ operations, and we see that the cost of Gaussian elimination dominates. One interpretation of the operation count is that if the size of the system doubles (from n to $2n$ unknowns), then the number of operations increases by a factor of approximately eight, and so should the execution time.[2]

[2]Modern computers use a hierarchical memory system, with one or more levels of fast cache memory. Professionally produced computer code is written to take advantage of

9.1.2 Solving $Ax = b$ using the LU factorization

The LU factorization of A suggests a different way to organize the process of solving $Ax = b$. Instead of performing Gaussian elimination on the augmented matrix $[A|b]$, we operate on A to produce the factorization $A = LU$. The cost of this is essentially the same as calculated above (recall that no additional calculation is required to produce the matrix L; we need only save the multipliers ℓ_{ik} in the lower triangle of L). To be precise, the cost of the LU factorization of A is

$$\frac{2}{3}n^3 - \frac{1}{2}n^2 - \frac{1}{6}n \tag{9.6}$$

(see Exercise 9).

The system $Ax = b$ is equivalent to $LUx = b$, and the solution is $x = U^{-1}L^{-1}b$. We do not actually compute any inverse matrices (this would be inefficient); instead, we solve $LUx = b$ by a two-step process:

1. Solve $Lc = b$ by forward substitution to get $c = L^{-1}b$. Forward substitution is analogous to back substitution; the cost is $n^2 - n$ operations (see Exercise 10).

2. Solve $Ux = c$ to get $x = U^{-1}c = U^{-1}L^{-1}b = A^{-1}b$. The cost was counted above; it is n^2 operations.

The reader should note that, in general, forward substitution has the same operation count as back substitution. However, in this context, L has ones on the diagonal, which eliminates n divisions.

We see that Gaussian elimination (performed on the augmented matrix), together with back substitution, requires

$$\frac{2}{3}n^3 + \frac{1}{2}n^2 - \frac{7}{6}n + n^2 = \frac{2}{3}n^3 + \frac{3}{2}n^2 - \frac{7}{6}n$$

operations, while the LU factorization, together with two triangular solves, requires

$$\frac{2}{3}n^3 - \frac{1}{2}n^2 - \frac{1}{6}n + n^2 - n + n^2 = \frac{2}{3}n^3 + \frac{3}{2}n^2 - \frac{7}{6}n$$

operations. Thus the number of operations required by the two approaches is exactly the same.

However, there is an advantage to using the LU factorization. In many problems, we do not just want to solve $Ax = b$ for a single right-hand-side vector b, but rather $Ax = b_i$ for k vectors $b_1, b_2, \ldots, b_k$. The expensive part of solving $Ax = b$ is Gaussian elimination or, equivalently, finding the LU factorization of A. Since the matrix A is the same for every system, we need

---

cache memory; a side effect is that some problems fit better into cache memory than others, and the running time of most algorithms is not proportional to the number of operations. Therefore, it is not easy to predict the exact performance of a given algorithm on a given computer, and the rule of thumb given above—doubling the problem size increases the execution time by a factor of eight—is only a crude approximation to actual performance.

only compute the LU factorization once, at a cost of about $2n^3/3$ operations, and we can then solve each system $Ax = b_i$ by performing two triangular solves, at a cost of $2n^2 - n$ operations each. The total operation count is approximately

$$\frac{2}{3}n^3 + k(2n^2 - n),$$

which reduces to approximately $2n^3/3$ provided $k \ll n$. If we were to perform Gaussian elimination and back substitution for each right-hand side b_i, the total cost would be approximately $2kn^3/3$, which is much greater.[3]

Exercises

1. Find the LU factorization of

$$A = \begin{bmatrix} 1 & 3 & 2 & 4 \\ 3 & 10 & 5 & 15 \\ -1 & -7 & 3 & -17 \\ -2 & -6 & 1 & -12 \end{bmatrix}.$$

2. Find the LU factorization of

$$A = \begin{bmatrix} 1 & -2 & 2 & 3 \\ -4 & 7 & -4 & -14 \\ -3 & 9 & -17 & 2 \\ 4 & -7 & -1 & -10 \end{bmatrix}.$$

3. Prove Theorem 378. (Hint: Argue by induction on n and use the equation $LL^{-1} = I$.)

4. Prove Theorem 379.

5. Multiply the matrices defined by (9.2) and (9.3), and verify that the result is I. (Hint: It might be easiest to compute the product one row at a time.)

6. Show that, if L_k^{-1} is defined by (9.3), then $L_{n-m}^{-1}L_{n-m-1}^{-1}\cdots L_{n-1}^{-1}$ is defined by (9.4), with $k = n - m$. (Hint: Argue by induction on m. Notice that multiplying any matrix B on the left by L_k^{-1} is equivalent to adding multiples of row k of B to rows $k+1, k+2, \ldots, n$, where the multipliers are $\ell_{k+1,k}, \ell_{k+2,k}, \ldots, \ell_{n,k}$. In the context of this induction proof, the kth row of $B = L_{k+1}^{-1}\cdots L_{n-1}^{-1}$ is simply e_k, the kth standard basis vector. Moreover, the entries of B below B_{kk} are all zero.)

---

[3] If we knew all k right-hand sides at the same time, we could augment A by all k vectors and still need to perform Gaussian elimination once. The operation count would then be the same as for the LU approach. The advantage of the LU approach is that, with the LU factorization, one can solve any system of the form $Ax = b$, even if the right-hand side does not become available until after Gaussian elimination has been performed.

7. (a) Show that there do not exist matrices of the form

$$L = \begin{bmatrix} 1 & 0 \\ \ell & 1 \end{bmatrix}, \; U = \begin{bmatrix} u & v \\ 0 & w \end{bmatrix}$$

such that $LU = A$, where

$$A = \begin{bmatrix} 0 & 1 \\ 1 & 1 \end{bmatrix}.$$

(b) Let

$$A = \begin{bmatrix} 0 & 1 \\ 0 & 1 \end{bmatrix}.$$

Show that A has infinitely many LU factorizations.

8. Prove the converse of Theorem 381: If $A \in \mathbf{R}^{n \times n}$ has a unique LU factorization, then $M^{(1)}, M^{(2)}, \ldots, M^{(n-1)}$ are all nonsingular.

9. Show that the number of arithmetic operations required to compute the LU factorization of an $n \times n$ matrix is given by (9.6).

10. Show that, if L is $n \times n$ and unit lower triangular, then the cost of solving $Lc = b$ is $n^2 - n$ arithmetic operations.

11. Computing A^{-1} is equivalent to solving $Ax = e_j$ for $j = 1, 2, \ldots, n$. What is the operation count for computing A^{-1} by using the LU factorization of A to solve these n systems? Can you use the special nature of the right-hand-side vectors to reduce this operation count?

9.2 Partial pivoting

Based on Theorem 381 (and Exercise 9.1.8), it is easy to produce examples of matrices that have no LU factorization. Such matrices are precisely those for which one or more row interchanges are necessary when performing Gaussian elimination. A row interchange is required when a *pivot*—the diagonal entry used to eliminate nonzeros below it—is zero. A general rule for computing in finite-precision arithmetic is this: If an algorithm is undefined when a certain quantity is zero, it is probably numerically unstable—that is, magnifies round-off error—when that quantity is small. This suggests that Gaussian elimination will be problematic if small but nonzero pivots are encountered, unless row interchanges are performed. In this section, we will demonstrate this statement by explicit example, introduce Gaussian elimination with partial pivoting, which uses row interchanges to make the pivots as large as possible, and derive a matrix factorization similar to the LU factorization but which allows row interchanges. In Section 9.6, we will discuss precisely the concept of numerical stability.

9.2.1 Finite-precision arithmetic

Real numbers can be represented exactly in decimal notation, provided infinitely many digits are allowed. Finite-precision arithmetic implies that only a finite number of digits are available, so that most real numbers are represented only approximately. Modern computers use floating point storage, which is essentially equivalent to scientific notation and which will be summarized below. Modern computers also use binary (base 2) arithmetic instead of decimal (base 10), but most of the issues are the same for any base, so we will restrict ourselves to decimal notation in our explanations.

A floating point system is defined by the base, assumed here to be 10 for convenience, and the numbers t, $e_{min} < 0$, and $e_{max} > 0$. A floating point number has the form

$$\pm 0.d_1 d_2 \cdots d_t \cdot 10^e,$$

where each digit d_i satisfies $0 \leq d_i \leq 9$, with $d_1 \neq 0$, and the exponent e satisfies $e_{min} \leq e \leq e_{max}$. The reader should notice that t is the number of digits in the floating point number. The condition $d_1 \neq 0$ means that the number is *normalized.* It is not difficult to show that any real number that lies between the largest and smallest floating point number can be approximated to within a relative error of

$$\mathbf{u} = 5 \cdot 10^{-t}$$

(see Exercise 3). The number $\mathbf{u}$ is called the *unit round*; it describes the precision of the floating point system. The best possible bound on the relative error in the result of any algorithm (executed in floating point arithmetic) is $\mathbf{u}$. Practically speaking, if one can prove a bound on the relative error that is a small multiple of $\mathbf{u}$, then one has done as well as could be hoped. (But there are subtleties involved in understanding the kinds of error bounds that are possible; these are explored in Section 9.6.)

There are many details about floating point arithmetic that we do not discuss here; for a wealth of details about current floating point systems, specifically the IEEE[4] standard for floating point, the reader can consult Overton [34]. The crucial point to understand for our purposes is that when an algorithm is performed in floating point arithmetic, the result of each elementary operation (addition, subtraction, multiplication, and division) is rounded to the nearest floating point number before the next operation is performed. This means that there can be an enormous number of rounding errors, and the key question is how these propagate to the final computed answer. This is particularly true for an algorithm like Gaussian elimination, which gives an exact answer in finitely many steps when performed in exact arithmetic. In such an algorithm, round-off is the only source of error.

---

[4]Institute of Electrical and Electronics Engineers.

9.2.2 Examples of errors in Gaussian elimination

In this section, we will use a simple floating point system defined by $t = 4$, $e_{min} = -49$, $e_{max} = 49$. We wish to solve several systems of linear equations and examine the effects of round-off error.

Example 382 *The first example is the following system:*

$$\begin{aligned} 10^{-5}x_1 + x_2 &= 1, \\ x_1 + x_2 &= 2. \end{aligned}$$

By Cramer's rule, the exact solution is

$$x_1 = \frac{1}{1-10^{-5}} \doteq 1 + 10^{-5}, \; x_2 = \frac{1 - 2 \cdot 10^{-5}}{1 - 10^{-5}} \doteq 1 - 10^{-5}.$$

In exact arithmetic, the first step of row reduction is as follows:

$$\left[\begin{array}{cc|c} 10^{-5} & 1 & 1 \\ 1 & 1 & 2 \end{array}\right] \rightarrow \left[\begin{array}{cc|c} 10^{-5} & 1 & 1 \\ 0 & 1-10^5 & 2-10^5 \end{array}\right] = \left[\begin{array}{cc|c} 10^{-5} & 1 & 1 \\ 0 & -99999 & -99998 \end{array}\right].$$

However, the numbers -99999 and -99998 require five digits to represent exactly, and our floating point system has only four. Therefore, these numbers are rounded to the nearest floating point number, which is the same in both cases, namely, -10^5. Therefore, row reduction in the given floating point system proceeds as follows:

$$\begin{aligned} \left[\begin{array}{cc|c} 10^{-5} & 1 & 1 \\ 1 & 1 & 2 \end{array}\right] \rightarrow \left[\begin{array}{cc|c} 10^{-5} & 1 & 1 \\ 0 & -10^5 & -10^5 \end{array}\right] &\rightarrow \left[\begin{array}{cc|c} 10^{-5} & 1 & 1 \\ 0 & 1 & 1 \end{array}\right] \\ &\rightarrow \left[\begin{array}{cc|c} 10^{-5} & 0 & 0 \\ 0 & 1 & 1 \end{array}\right] \\ &\rightarrow \left[\begin{array}{cc|c} 1 & 0 & 0 \\ 0 & 1 & 1 \end{array}\right]. \end{aligned}$$

The computed solution is therefore $\hat{x} = (0, 1)$, which is not close to the true solution $x \doteq (1 + 10^{-5}, 1 - 10^{-5})$.

The difficulty in the preceding example is an illustration of the rule of thumb mentioned above: If an algorithm is undefined when a certain quantity is zero, it is probably unstable when that quantity is small. If the coefficient of x_1 in the first equation were exactly zero, then we could not have performed the first step of Gaussian elimination as we did above. Instead, the coefficient is 10^{-5}, which is small (less than the unit round, in this case) compared to the other coefficients in the problem.

Example 383 *We now consider the same system as in the previous example, but we begin by interchanging the two rows:*

$$\left[\begin{array}{cc|c} 10^{-5} & 1 & 1 \\ 1 & 1 & 2 \end{array}\right] \rightarrow \left[\begin{array}{cc|c} 1 & 1 & 2 \\ 10^{-5} & 1 & 1 \end{array}\right] \rightarrow \left[\begin{array}{cc|c} 1 & 1 & 2 \\ 0 & 1-10^{-5} & 1-2\cdot 10^{-5} \end{array}\right].$$

In the given floating point system, $1 - 10^{-5}$ *and* $1 - 2 \cdot 10^{-5}$ *both round to* 1. *We therefore have the following floating point calculation:*

$$\left[\begin{array}{cc|c} 10^{-5} & 1 & 1 \\ 1 & 1 & 2 \end{array}\right] \rightarrow \left[\begin{array}{cc|c} 1 & 1 & 2 \\ 10^{-5} & 1 & 1 \end{array}\right] \rightarrow \left[\begin{array}{cc|c} 1 & 1 & 2 \\ 0 & 1 & 1 \end{array}\right] \rightarrow \left[\begin{array}{cc|c} 1 & 0 & 1 \\ 0 & 1 & 1 \end{array}\right].$$

The computed solution is $\hat{x} = (1, 1)$, *while the exact solution is* $x \doteq (1 + 10^{-5}, 1 - 10^{-5})$. *Gaussian elimination with a row interchange produced an accurate estimate of the solution.*

For most problems, Gaussian elimination with row interchanges—applied when the pivot is not only zero but small—produces satisfactory results. However, this is not true for all problems.

Example 384 *We now consider the system*

$$\begin{aligned} x_1 + 1.001x_2 &= 2, \\ 0.9999x_1 + x_2 &= 2. \end{aligned}$$

Here is row reduction, performed in floating point arithmetic:

$$\left[\begin{array}{cc|c} 1 & 1.001 & 2 \\ 0.9999 & 1 & 2 \end{array}\right] \rightarrow \left[\begin{array}{cc|c} 1 & 1.001 & 2 \\ 0 & -0.001 & 0 \end{array}\right] \rightarrow \left[\begin{array}{cc|c} 1 & 1.001 & 2 \\ 0 & 1 & 0 \end{array}\right] \rightarrow \left[\begin{array}{cc|c} 1 & 0 & 2 \\ 0 & 1 & 0 \end{array}\right].$$

The computed solution is therefore $\hat{x} = (2, 0)$. *The true solution is*

$$x = (20000/8999, -2000/8999),$$

and the true solution, rounded to four digits, is $\overline{x} = (2.222, 0.2222)$. *The computed solution is therefore not very close to the exact solution.*

We are going to see in Section 9.6 that the system studied in this example has a different character than the one in Example 382. In Example 382, the only difficulty was a poor choice of algorithm; Gaussian elimination with a row interchange worked just fine, as shown in Example 383. In this example, the difficulty is intrinsic to the problem, that is, to this particular system. As a hint of this, it it is interesting to note that while $\overline{x}$ *is much closer to* x *than is* $\hat{x}$, *nevertheless* $A\hat{x}$ *is closer to* b *than is* $A\overline{x}$:

$$b - A\hat{x} \doteq \left[\begin{array}{c} 0 \\ 2 \cdot 10^{-4} \end{array}\right], \; b - A\overline{x} \doteq \left[\begin{array}{c} 4.222 \cdot 10^{-4} \\ 4.222 \cdot 10^{-4} \end{array}\right].$$

9.2.3 Partial pivoting

If we decide that we should use row interchanges to avoid small pivots, then we might as well take this idea as far as it will go: At each step, we interchange two rows (if necessary) so that the magnitude of the pivot is as large as possible.

We will write $A^{(1)} = A$ and $A^{(k+1)}$ for the result of the kth step of Gaussian elimination. At the kth step, we compare the entries

$$\left|A_{k,k}^{(k)}\right|, \left|A_{k+1,k}^{(k)}\right|, \ldots, \left|A_{n,k}^{(k)}\right|.$$

If the largest is $|A_{i_k,k}^{(k)}|$, then we interchange rows k and i_k before proceeding. (Of course, it can happen that $i_k = k$, in which case there is no row interchange.)

The use of row interchanges to ensure the largest possible pivot is called *partial pivoting.* A related technique, called *complete pivoting,* uses both row and column interchanges.[5] Complete pivoting requires $(n-k)^2$ comparisons at the kth step, whereas partial pivoting needs only $n-k$. Each comparison is equivalent in cost to an arithmetic operation, and thus complete pivoting is much more expensive. Since, in practice, partial pivoting seems to be as stable as complete pivoting, we will not discuss complete pivoting any further.

In Section 9.6, we will discuss the theory of partial pivoting, specifically, what can be proved about its numerical stability. Here we will summarize the main results: Gaussian elimination with partial pivoting is *not* provably stable; in fact, it is known that the algorithm can be unstable for some problems. However, in practice, it almost always performs in a stable fashion (the same is decidedly not true of Gaussian elimination without partial pivoting, as the examples in Section 9.2.2 show). In fact, the author is aware of only two examples since the advent of electronic computers in which the potential numerical instability of Gaussian elimination has actually been observed on a practical problem (see [45] and [9]).

Here is an example of Gaussian elimination with partial pivoting.

Example 385 *Let*

$$A = \begin{bmatrix} -1.5 & -0.8 & 0.65 & 0.45 \\ 3.0 & 4.0 & 0.5 & 0.0 \\ 6.0 & 2.0 & -1.0 & 4.0 \\ 1.5 & -1.0 & 1.25 & 3.0 \end{bmatrix}, \; b = \begin{bmatrix} 29 \\ -16 \\ 38 \\ 87 \end{bmatrix}.$$

Gaussian elimination with partial pivoting proceeds as follows:

$$\left[\begin{array}{cccc|c} -1.5 & -0.8 & 0.65 & 0.45 & 29.0 \\ 3.0 & 4.0 & 0.5 & 0.0 & -16.0 \\ 6.0 & 2.0 & -1.0 & 4.0 & 38.0 \\ 1.5 & -1.0 & 1.25 & 3.0 & 87.0 \end{array}\right]$$

[5]Row interchanges merely reorder the equations, which has no effect on the solution. A column interchange reorders the variables. These interchanges must be recorded so that, after the algorithm finishes, the variables can be put back in the correct order.

$$\to \left[\begin{array}{cccc|c} 6.0 & 2.0 & -1.0 & 4.0 & 38.0 \\ 3.0 & 4.0 & 0.5 & 0.0 & -16.0 \\ -1.5 & -0.8 & 0.65 & 0.45 & 29.0 \\ 1.5 & -1.0 & 1.25 & 3.0 & 87.0 \end{array}\right] \quad \textit{(Interchange rows 1 and 3)}$$

$$\to \left[\begin{array}{cccc|c} 6.0 & 2.0 & -1.0 & 4.0 & 38.0 \\ 0 & 3.0 & 1.0 & -2.0 & -35 \\ 0 & -0.3 & 0.4 & 1.45 & 38.5 \\ 0 & -1.5 & 1.5 & 2.0 & 77.5 \end{array}\right] \quad \textit{(Zero the entries below the pivot)}$$

$$\to \left[\begin{array}{cccc|c} 6.0 & 2.0 & -1.0 & 4.0 & 38.0 \\ 0 & 3.0 & 1.0 & -2.0 & -35.0 \\ 0 & 0 & 0.5 & 1.25 & 35.0 \\ 0 & 0 & 2.0 & 1.0 & 60.0 \end{array}\right] \quad \textit{(Zero the entries below the pivot)}$$

$$\to \left[\begin{array}{cccc|c} 6.0 & 2.0 & -1.0 & 4.0 & 38.0 \\ 0 & 3.0 & 1.0 & -2.0 & -35.0 \\ 0 & 0 & 2.0 & 1.0 & 60.0 \\ 0 & 0 & 0.5 & 1.25 & 35.0 \end{array}\right] \quad \textit{(Interchange rows 3 and 4)}$$

$$\to \left[\begin{array}{cccc|c} 6.0 & 2.0 & -1.0 & 4.0 & 38.0 \\ 0 & 3.0 & 1.0 & -2.0 & -35.0 \\ 0 & 0 & 2.0 & 1.0 & 60.0 \\ 0 & 0 & 0 & 1.0 & 20.0 \end{array}\right] \quad \textit{(Zero the entry below the pivot)}$$

The reader should notice that, on the second step, no row interchange was necessary; the largest entry was already on the diagonal. After Gaussian elimination with partial pivoting is complete, back substitution proceeds as usual.

A row interchange can be represented as multiplication by a *permutation matrix* P, which is obtained by permuting the rows of the identity matrix I. If $(i_1, i_2, \ldots, i_n)$ is a permutation of $(1, 2, \ldots, n)$, the corresponding permutation matrix P has rows $e_{i_1}, e_{i_2}, \ldots, e_{i_n}$, where $e_1, e_2, \ldots, e_n$ are the standard basis vectors in $\mathbf{R}^n$ (and the rows of the identity matrix). Exercise 6 asks the reader to show that, for any matrix $A \in \mathbf{R}^{n\times m}$ with rows $r_1, r_2, \ldots, r_n \in \mathbf{R}^m$, PA has rows $r_{i_1}, r_{i_2}, \ldots, r_{i_n}$.

In the context of partial pivoting, we are interested in interchanging two rows and leaving the others unchanged. Therefore, we need only consider transpositions, the special permutations interchange two integers and leave the others unchanged (see Section 4.1). We will call a permutation matrix P corresponding to a transposition an *elementary permutation matrix*. We note that an elementary permutation matrix is its own inverse (see Exercise 7). We also point out that right multiplication by a permutation matrix interchanges columns instead of rows (see Exercise 6).

Using elementary permutation matrices to represent row interchanges, Gaussian elimination with partial pivoting takes the form

$$U = L_{n-1}P_{n-1}L_{n-2}P_{n-2}\cdots L_1P_1A, \tag{9.7}$$

where each P_k is an elementary permutation matrix or the identity ($P_k = I$

means that no row interchange was necessary at step k; the largest entry was already on the diagonal). In Example 385, $P_2 = I$ and

$$P_1 = \begin{bmatrix} 0 & 0 & 1 & 0 \\ 0 & 1 & 0 & 0 \\ 1 & 0 & 0 & 0 \\ 0 & 0 & 0 & 1 \end{bmatrix}, \quad P_3 = \begin{bmatrix} 1 & 0 & 0 & 0 \\ 0 & 1 & 0 & 0 \\ 0 & 0 & 0 & 1 \\ 0 & 0 & 1 & 0 \end{bmatrix}.$$

9.2.4 The PLU factorization

The LU factorization can be modified to include row interchanges. Let us suppose that $P_1, P_2, \ldots, P_{n-1}$ are the permutation matrices representing the row interchanges applied during partial pivoting (recall that $P_k = I$ if no row interchange is necessary) and

$$P = P_{n-1}P_{n-2}\cdots P_1.$$

It can be shown that PA has an LU factorization. The factorization $PA = LU$ is sometimes called the PLU factorization of A. Exercise 8 outlines a proof of the existence of the PLU factorization.

As with the ordinary LU factorization, if we need to solve multiple linear systems $Ax = b$ with the same matrix A but different vectors b, then we can compute the PLU factorization of A. Then

$$Ax = b \;\Rightarrow\; PAx = Pb \;\Rightarrow LUx = Pb.$$

Thus each system can be solved by first permuting the entries of b and then performing solving two triangular systems.

Exercises

1. Let
$$A = \begin{bmatrix} -2.30 & 14.40 & 8.00 \\ 1.20 & -3.50 & 9.40 \\ 3.10 & 6.20 & -9.90 \end{bmatrix}, \quad b = \begin{bmatrix} 1.80 \\ -10.50 \\ 22.30 \end{bmatrix}.$$
Solve $Ax = b$ by Gaussian elimination with partial pivoting.

2. Let
$$A = \begin{bmatrix} 8.20 & 11.90 & -16.00 & -8.10 \\ 7.10 & -12.00 & 2.60 & 5.30 \\ 12.90 & -0.20 & -10.60 & 2.20 \\ 6.70 & -1.60 & 14.20 & -9.20 \end{bmatrix}, \quad b = \begin{bmatrix} 7.90 \\ -9.00 \\ 4.10 \\ 8.50 \end{bmatrix}.$$
Solve $Ax = b$ by Gaussian elimination with partial pivoting.

3. Consider a decimal floating point system, as described in this section, determined by the parameters t, e_{min}, and e_{max}.

(a) Let f_{min} be the smallest positive normalized floating point number. What is f_{min}?

(b) Let f_{max} be the largest floating point number. What is f_{max}?

(c) Let x be a real number satisfying $f_{min} \leq x \leq f_{max}$, and let $\mathrm{fl}(x)$ be the floating point number closest to x. Show that

$$\frac{|x - \mathrm{fl}(x)|}{|x|} \leq 5 \cdot 10^{-t}.$$

4. If the linear system from Example 382 is solved by Cramer's rule in floating point arithmetic ($t = 4$, as in Section 9.2.2), what is the computed solution? Answer the same question for the linear system in Example 384.

5. Suppose $A \in \mathbf{R}^{n \times n}$ has an LU decomposition. Prove that the product of the diagonal entries of U equals the product of the eigenvalues of A.

6. Suppose $\{e_1, e_2, \ldots, e_n\}$ is the standard basis for $\mathbf{R}^n$ and $(i_1, i_2, \ldots, i_n)$ is a permutation of $(1, 2, \ldots, n)$. Let P be the permutation matrix with rows $e_{i_1}, e_{i_2}, \ldots, e_{i_n}$.

   (a) Let A be an $n \times m$ matrix. Prove that if the rows of A are $r_1, r_2, \ldots, r_n$, then the rows of PA are $r_{i_1}, r_{i_2}, \ldots, r_{i_n}$.

   (b) Let A be an $m \times n$ matrix with columns $A_1, A_2, \ldots, A_n \in \mathbf{R}^m$. Prove that the columns of AP^T are $A_{i_1}, A_{i_2}, \ldots, A_{i_n}$.

7. Show that an elementary permutation matrix P satisfies $PP = I$. Is this true for a general permutation matrix?

8. Let $A \in \mathbf{R}^{n \times n}$ be reduced to upper triangular form by a sequence of row interchanges and elementary operations:

$$U = L_{n-1}P_{n-1}L_{n-2}P_{n-2} \cdots L_1P_1A.$$

   Define $P = P_{n-1}P_{n-2} \cdots P_1$ and

$$\tilde{L}_k = P_{n-1}P_{n-2} \cdots P_{k+1}L_kP_{k+1}P_{k_2} \cdots P_{n-1}, \ k = 1, 2, \ldots, n-1.$$

   (a) Prove that

$$\tilde{L}_{n-1}\tilde{L}_{n-2} \cdots \tilde{L}_1PA = L_{n-1}P_{n-1}L_{n-2}P_{n-2} \cdots L_1P_1A = U.$$

   (b) Prove that each $\tilde{L}_k$ is an elementary matrix.

   (c) Prove that $PA = LU$ for a certain unit lower triangular matrix L. (What is L?)

9. Let

$$A = \begin{bmatrix} -1.5 & -0.8 & 0.65 & 0.45 \\ 3.0 & 4.0 & 0.5 & 0.0 \\ 6.0 & 2.0 & -1.0 & 4.0 \\ 1.5 & -1.0 & 1.25 & 3.0 \end{bmatrix}, \quad b = \begin{bmatrix} 29 \\ -16 \\ 38 \\ 87 \end{bmatrix}.$$

   Find the PLU factorization of A and use it to solve $Ax = b$.

9.3 The Cholesky factorization

We have seen earlier in this book that a symmetric positive definite (SPD) matrix has many special properties. In this section, we will see that, for SPD matrices, there is an alternative to the PLU factorization that is simpler and less expensive. It is also less susceptible to round-off error, as we will see in Section 9.5.

We begin by showing that when Gaussian elimination is applied to an SPD matrix, partial pivoting is not required. The proof is based on the following preliminary result.

Lemma 386 *Let $A \in \mathbf{R}^{n\times n}$ be symmetric and positive definite. Then all the diagonal entries of A are positive. Moreover, the largest entry in magnitude of A lies on the diagonal.*

Proof By definition, $x \cdot Ax > 0$ for all $x \in \mathbf{R}^n$, $x \neq 0$. Therefore, in particular, $e_i \cdot Ae_i > 0$ (where e_i is the ith standard basis vector). But $e_i \cdot Ae_i = A_{ii}$, so the first conclusion follows. To prove the second conclusion, notice that

$$\begin{aligned}
(e_i - e_j)\cdot A(e_i - e_j) > 0 \ &\Rightarrow\ 2A_{ij} < A_{ii} + A_{jj},\\
(e_i + e_j)\cdot A(e_i + e_j) > 0 \ &\Rightarrow\ 2A_{ij} > -A_{ii} - A_{jj}.
\end{aligned}$$

Putting these inequalities together, we see that $2|A_{ij}| < A_{ii}+A_{jj}$ for all $i \neq j$. This is impossible if A_{ij} is the largest entry in A, and hence the largest entry in A must lie on the diagonal.

QED

The preceding result shows that if A is SPD, then $A_{11} > 0$ and therefore we can take the first step of Gaussian elimination (without a row interchange). The following result shows that, in fact, row interchanges are not necessary at any stage of Gaussian elimination.

Lemma 387 *If a step of Gaussian elimination is applied to an SPD matrix $A \in \mathbf{R}^{n\times n}$, the result has the form*

$$A^{(2)} = \left[\begin{array}{c|c} A_{11} & a^T \\ \hline 0 & B \end{array}\right], \tag{9.8}$$

where $B \in \mathbf{R}^{(n-1)\times(n-1)}$ is also SPD (and $a \in \mathbf{R}^{n-1}$). Moreover,

$$\max\{|B_{ij}| \,:\, i,j = 1,2,\ldots,n-1\} \leq \max\{|A_{ij}| \,:\, i,j=1,2,\ldots,n\}. \tag{9.9}$$

Proof For any $i,j > 1$, we have

$$A^{(2)}_{ij} = A_{ij} - \frac{A_{i1}}{A_{11}}A_{1j}.$$

Therefore,

$$A_{ji}^{(2)} = A_{ji} - \frac{A_{j1}}{A_{11}} A_{1i} = A_{ij} - \frac{A_{1j}}{A_{11}} A_{i1} = A_{ij}^{(2)}$$

(using the symmetry of A). Therefore, the submatrix of $A^{(2)}$ defined by $i, j > 1$ (that is, the matrix B) is symmetric.

We will now show that, for any $x \in \mathbf{R}^{n-1}$, there exists $y_1 \in \mathbf{R}$ such that

$$x \cdot Bx = y \cdot Ay,$$

where

$$y = \left[\begin{array}{c} y_1 \\ \hline x \end{array} \right]. \tag{9.10}$$

With this notation, $x_1 = y_2, \ldots, x_{n-1} = y_n$. We notice that, with $A^{(2)}$ given by (9.8), $a_j = A_{1,j+1}$. Therefore,

$$\begin{aligned} x \cdot Bx &= \sum_{i=2}^{n} \sum_{j=2}^{n} A_{ij}^{(2)} y_i y_j \\ &= \sum_{i=2}^{n} \sum_{j=2}^{n} y_i y_j \left(A_{ij} - \frac{A_{i1}}{A_{11}} A_{1j} \right) \\ &= \sum_{i=2}^{n} \sum_{j=2}^{n} A_{ij} y_i y_j - \frac{1}{A_{11}} \sum_{i=2}^{n} \sum_{j=2}^{n} A_{i1} A_{1j} y_i y_j \\ &= \sum_{i=2}^{n} \sum_{j=2}^{n} A_{ij} y_i y_j - \frac{1}{A_{11}} (a \cdot x)^2. \end{aligned}$$

On the other hand, with y defined by (9.10), we have

$$\begin{aligned} y \cdot Ay &= \left[\begin{array}{c} y_1 \\ \hline x \end{array} \right] \cdot \left[\begin{array}{c|c} A_{11} & a^T \\ \hline a & \tilde{A} \end{array} \right] \left[\begin{array}{c} y_1 \\ \hline x \end{array} \right] \\ &= \left[\begin{array}{c} y_1 \\ \hline x \end{array} \right] \cdot \left[\begin{array}{c} A_{11} y_1 + a \cdot x \\ \hline y_1 a + \tilde{A} x \end{array} \right] \\ &= A_{11} y_1^2 + 2(a \cdot x) y_1 + x \cdot \tilde{A} x \\ &= A_{11} y_1^2 + 2(a \cdot x) y_1 + \sum_{i=2}^{n} \sum_{j=2}^{n} A_{ij} y_i y_j \end{aligned}$$

(where we write $\tilde{A}$ for the lower right-hand corner of A). Comparing these two expressions, we wish to choose y_1 so that

$$A_{11} y_1^2 + 2(a \cdot x) y_1 = -\frac{1}{A_{11}} (a \cdot x)^2.$$

The reader can verify that $y_1 = -(a \cdot x)/A_{11}$ works, and with this value of y, we have

$$x \cdot Bx = y \cdot Ay.$$

It follows that B is positive definite.

Finally, to prove (9.9), it suffices to prove

$$\max\{|B_{ii}| : i = 1, 2, \ldots, n-1\} \leq \max\{|A_{ii}| : i = 1, 2, \ldots, n\}$$

(since the largest entry in an SPD matrix lies on the diagonal). But we know that, for $i > 1$,

$$A_{ii}^{(2)} = A_{ii} - \frac{A_{i1}}{A_{11}} A_{1i} = A_{ii} - \frac{A_{i1}^2}{A_{11}} \leq A_{ii}.$$

This completes the proof.

QED

We can now derive the main result of this section.

Theorem 388 *Let $A \in \mathbf{R}^{n\times n}$ be symmetric and positive definite. Then Gaussian elimination can be performed without partial pivoting; that is, A has an LU factorization. Moreover, the largest entry in any of the intermediate matrices during Gaussian elimination lies in the original matrix A. Finally, in the factorization $A = LU$, all the diagonal entries of U are positive.*

Proof The proof is a straightforward induction argument based on the preceding lemma; we leave the details to the reader. Notice that the diagonal entries of U are the pivots encountered during Gaussian elimination, each of which is a diagonal entry of an SPD matrix. This shows that the diagonal entries of U are positive.

QED

The fact that the entries in the matrix do not grow during Gaussian elimination is significant; it will be discussed in Section 9.6. For now, we just note that in Example 382, which demonstrated the need for partial pivoting, the numerical difficulties arose from a dramatic growth in the entries. This cannot happen with an SPD matrix, even when no row interchanges are applied.

Here is the main consequence of the preceding theorem.

Corollary 389 *Let $A \in \mathbf{R}^{n\times n}$ be a symmetric positive definite matrix. Then A can be factored in the following ways:*

- *$A = LU$, where L is unit lower triangular and U is upper triangular;*
- *$A = LDL^T$, where L is unit lower triangular and D is diagonal with positive diagonal entries;*
- *$A = R^T R$, where R is upper triangular with positive diagonal entries.*

The last factorization is called the Cholesky factorization *of A.*

Proof We have already seen that A has an LU factorization, where the diagonal entries of U are positive. We will derive the other factorizations from $A = LU$. We first define D to be the diagonal matrix whose diagonal entries are the diagonal entries of U. We then define $\tilde{U} = D^{-1}U$ and notice that the entries in the ith row of $\tilde{U}$ are simply the entries of the ith row of U, multiplied by U_{ii}^{-1}. It suffices now to show that $\tilde{U} = L^T$. If we write $A^{(k)}$ for the kth intermediate matrix encountered during Gaussian elimination ($A^{(1)} = A$), then we recall that

$$L_{ik} = \frac{A_{ik}^{(k)}}{A_{kk}^{(k)}},\ i = k+1, \ldots, n.$$

On the other hand, the kth row of U is simply the kth row of $A^{(k)}$ (this is the row used to eliminate entries below it); hence $U_{ki} = A_{ik}^{(k)}$ and

$$\tilde{U}_{ki} = \frac{U_{ki}}{U_{kk}} = \frac{A_{ki}^{(k)}}{A_{kk}^{(k)}} = \frac{A_{ik}^{(k)}}{A_{kk}^{(k)}} = L_{ik}.$$

The last equation follows from the fact that the submatrix of $A^{(k)}$ defined by $i, j \geq k$ is symmetric. We have therefore obtained $A = LDL^T$, as desired.

Finally, since D has positive diagonal entries, the matrix $D^{1/2}$ is well-defined; it is simply the diagonal matrix whose diagonal entries are the square roots of the diagonal entries of D. We define $R = D^{1/2}L^T$. Then R is upper triangular and

$$R^TR = \left(D^{1/2}L^T\right)^T D^{1/2}L^T = LD^{1/2}D^{1/2}L^T = LDL^T = A,$$

as desired.

QED

The Cholesky factorization is often expressed as $A = LL^T$, where L is lower triangular ($L = R^T$). However, the matrix L appearing in this formula is not the same L that appears in $A = LU = LDL^T$.

Having recognized that partial pivoting is not required when row-reducing an SPD matrix, we can compute the Cholesky factorization directly, without applying any row operations at all. The idea is to simply solve the equation $R^TR = A$ for R, taking advantage of the fact that R is upper triangular. The basic equation is

$$A_{ij} = \sum_{k=1}^{n} (R^T)_{ik}R_{kj} = \sum_{k=1}^{n} R_{ki}R_{kj} = \sum_{k=1}^{\min\{i,j\}} R_{ki}R_{kj}.$$

We solve for R one row at a time. First of all,

$$A_{11} = \sum_{k=1}^{1} R_{k1}R_{k1} = R_{11}^2,$$

which yields

$$R_{11} = \sqrt{A_{11}}. \tag{9.11}$$

Since A is SPD, A_{11} is guaranteed to be positive, and therefore we can take its square root to get a positive number R_{11}. We then have

$$A_{1j} = \sum_{k=1}^{1} R_{k1}R_{kj} = R_{11}R_{1j}$$

and thus

$$R_{1j} = \frac{A_{1j}}{R_{11}},\ j = 2, \dots, n. \tag{9.12}$$

Thus we have computed the first row of R. Assuming now that we have computed rows $1, 2, \dots, i-1$, we can compute the ith row as follows. First,

$$A_{ii} = \sum_{k=1}^{i} R_{ki}R_{ki} = \sum_{k=1}^{i-1} R_{ki}^2 + R_{ii}^2,$$

or

$$R_{ii} = \sqrt{A_{ii} - \sum_{k=1}^{i-1} R_{ki}^2}. \tag{9.13}$$

Since we have already shown that the equation $A = R^T R$ has a solution R with a positive diagonal, it follows that the quantity under the square root must be positive when A is SPD. Having computed R_{ii}, we can now compute the rest of the ith row from the equation

$$A_{ij} = \sum_{k=1}^{i} R_{ki}R_{kj} = \sum_{k=1}^{i-1} R_{ki}R_{kj} + R_{ii}R_{ij}.$$

The result is

$$R_{ij} = \frac{A_{ij} - \sum_{k=1}^{i-1} R_{ki}R_{kj}}{R_{ii}},\ j = i+1, \dots, n. \tag{9.14}$$

The reader is asked to verify in Exercise 4 that the Cholesky factorization, computed by this algorithm, requires approximately $n^3/3$ operations. This is to be contrasted with $2n^3/3$ operations for computing the LU factorization of a general matrix A. In other words, the Cholesky factorization takes half as many operations as the LU factorization.

The algorithm presented above is also the most efficient way to check whether a symmetric matrix A is positive definite. It is obviously not possible to check the value of $x \cdot Ax$ for all $x \in \mathbf{R}^n$, and computing the eigenvalues of A is a relatively expensive operation. However, if we simply try to compute the Cholesky factorization $A = R^T R$, one of two things will happen. First, we may encounter a nonpositive number under the square root in (9.11) or (9.13). In this case, we know that A is not positive definite since there is no upper triangular matrix R such that $A = R^T R$. Second, the algorithm may run to completion, producing an upper triangular matrix R (with positive diagonal entries) such that $A = R^T R$. In this case, A is positive definite (see Exercise 6).

In floating point arithmetic, it not sufficient that the quantities under the square roots in (9.11) and (9.13) are positive. We recall the rule of thumb mentioned at the beginning of Section 9.2: If an algorithm is undefined when a certain quantity is zero, then it is probably unstable in floating point arithmetic when that quantity is small. Consequently, we would typically choose a tolerance $\delta > 0$ and call A numerically indefinite if

$$A_{ii} - \sum_{k=1}^{i-1} R_{ki}^2 < \delta$$

for some $i = 1, 2, \ldots, n$. The tolerance δ must be chosen based on the size of the entries in the matrix A, and the reader is referred to Nocedal and Wright [33] for details.

Exercises

1. Let

$$A = \begin{bmatrix} 1 & -3 & 2 \\ -3 & 13 & -10 \\ 2 & -10 & 12 \end{bmatrix}.$$

   (a) Find the Cholesky factorization ($A = R^T R$) of A using the algorithm given in the text.

   (b) Using R, find the factorizations $A = LU$ and $A = LDL^T$.

2. Repeat the preceding exercise for

$$A = \begin{bmatrix} 1 & 1 & 2 & -1 \\ 1 & 5 & -2 & 3 \\ 2 & -2 & 9 & -7 \\ -1 & 3 & -7 & 10 \end{bmatrix}.$$

3. Let

$$A = \begin{bmatrix} 1 & -1 & 3 \\ -1 & 5 & -1 \\ 3 & -1 & 1 \end{bmatrix}.$$

By applying the algorithm given in the text to compute the Cholesky factorization of A, determine whether or not A is positive definite.

4. Count the operations required to compute the Cholesky factorization by (9.11–9.14).

5. Let $A \in \mathbf{R}^{n\times n}$ be SPD. Prove that the Cholesky factorization of A is unique. To be precise, prove that if $A = R_1^T R_1 = R_2^T R_2$ and R_1, R_2 are upper triangular matrices with positive diagonal entries, then $R_1 = R_2$.

6. Suppose $A \in \mathbf{R}^{n\times n}$ is symmetric and there exists an upper triangular matrix $R \in \mathbf{R}^{n\times n}$ with positive diagonal entries such that $A = R^T R$. Prove that A is positive definite.

7. Let $A \in \mathbf{R}^{n\times n}$ be SPD. The factorization $A = LDL^T$ is sometimes preferred over the Cholesky factorization because it avoids the need to compute square roots. Derive an algorithm, analogous to the one presented in the text (see page 527) for the Cholesky factorization, for computing L and D.

9.4 Matrix norms

To study numerical stability carefully, we will need to use norms of matrices (in addition to norms of vectors). The space $\mathbf{R}^{m\times n}$ of $m \times n$ matrices forms a vector space which is isomorphic to $\mathbf{R}^{mn}$. As such, we can generalize any norm defined on Euclidean space to produce a norm on $\mathbf{R}^{n\times n}$. For example, the Euclidean norm

$$\|x\|_2 = \sqrt{\sum_{i=1}^{n} x_i^2}, \ x \in \mathbf{R}^n$$

generalizes to

$$\|A\|_F = \sqrt{\sum_{i=1}^{m}\sum_{j=1}^{n} A_{ij}^2}, \ A \in \mathbf{R}^{m\times n}.$$

This norm is called the *Frobenius norm*, and it satisfies the defining properties of a norm for $\mathbf{R}^{m\times n}$ (see Definition 265). However, the term "matrix norm" means something more than "a norm defined on a space of matrices."

Definition 390 *A* matrix norm $\|\cdot\|$ *on the space* $\mathbf{R}^{n\times n}$ *of square matrices is a norm in the sense of Definition 265 that satisfies the following additional property:*

$$\|AB\| \le \|A\|\|B\| \text{ for all } A, B \in \mathbf{R}^{n\times n}.$$

More generally, a family of matrix norms consists of norms defined on $\mathbf{R}^{m\times n}$

for all positive integers m and n, where each norm is defined by the same symbol $\|\cdot\|$ and collectively they satisfy

$$\|AB\| \leq \|A\|\|B\| \text{ for all } A \in \mathbf{R}^{m\times n}, B \in \mathbf{R}^{n\times p}.$$

It turns out that the Frobenius norm is a matrix norm. This can be shown from the following property of the Frobenius norm, which is of independent interest.

Lemma 391 *If $A \in \mathbf{R}^{m\times n}$ and $x \in \mathbf{R}^n$, then*

$$\|Ax\|_2 \leq \|A\|_F\|x\|_2. \tag{9.15}$$

Proof Let $r_1, r_2, \ldots, r_m \in \mathbf{R}^n$ be the rows of A. Then $(Ax)_i = r_i \cdot x$, and the Cauchy-Schwarz inequality implies that

$$|r_i \cdot x| \leq \|r_i\|_2\|x\|_2.$$

Therefore,

$$(Ax)_i^2 \leq \|r_i\|_2^2\|x\|_2^2 = \left(\sum_{j=1}^n A_{ij}^2\right)\|x\|_2^2,$$

and thus

$$\begin{aligned}
\|Ax\|_2^2 = \sum_{i=1}^m (Ax)_i^2 &\leq \sum_{i=1}^m \left\{\left(\sum_{j=1}^n A_{ij}^2\right)\|x\|_2^2\right\} \\
&= \left(\sum_{i=1}^m\sum_{j=1}^n A_{ij}^2\right)\|x\|_2^2 \\
&= \|A\|_F^2\|x\|_2^2.
\end{aligned}$$

This completes the proof.

QED

Theorem 392 *For any $A \in \mathbf{R}^{m\times n}$, $B \in \mathbf{R}^{n\times p}$,*

$$\|AB\|_F \leq \|A\|_F\|B\|_F.$$

Proof Exercise 2.

The Frobenius norm is thus one example of a matrix norm. As with vector norms, there is not just one matrix norm, but many. In fact, given any norm on $\mathbf{R}^n$,[6] there is a corresponding *induced* matrix norm.

[6] Actually, we need a family of norms, one for each space $\mathbf{R}^n$, $n = 1, 2, \ldots$.

Definition 393 *Let* $\|\cdot\|$ *represent a family of norms defined on* $\mathbf{R}^n$*. The* induced matrix norm *on* $\mathbf{R}^{m\times n}$ *is defined by*

$$\|A\| = \sup\left\{\|Ax\| \,:\, x \in \mathbf{R}^n,\ \|x\| = 1\right\}. \tag{9.16}$$

Notice that the same symbol is used to denote the vector norm and the induced matrix norm.

The norm of $A \in \mathbf{R}^{m\times n}$ is defined to be the *supremum* of $\|Ax\|$, taken over all unit vectors $\|x\|$. The supremum is also called the *least upper bound*, and is defined as follows.

Definition 394 *Let* S *be a nonempty set of real numbers that is bounded above. The* supremum *of* S *is an upper bound* M *of* S *satisfying* $M \le M'$ *for all upper bounds* M' *of* S*. In other words,* $M = \sup S$ *satisfies the two conditions*

$$x \le M \text{ for all } x \in S,$$
$$x \le M' \text{ for all } x \in S \;\Rightarrow\; M \le M'.$$

If S *is unbounded above, then* $\sup S$ *is defined to be* ∞*, while* $\sup S$ *is defined to be* $-\infty$ *if* S *is the empty set.*

Similarly, if S *is bounded below, the* infimum *of* S *is a lower bound* m *of* S *satisfying* $m \ge m'$ *for all lower bounds* m' *of* S*. That is,* $m = \inf S$ *is defined by*

$$m \le x \text{ for all } x \in S,$$
$$m' \le x \text{ for all } x \in S \;\Rightarrow\; m' \le m.$$

If S *is unbounded below, then* $\sup S$ *is defined to be* $-\infty$*, while* $\inf S$ *is defined to be* ∞ *if* S *is the empty set.*

The reader should note that, given any set S of real numbers, both $\inf S$ and $\sup S$ are defined.

The set $S = (0,1) \subset \mathbf{R}$ provides a simple example of the difference between supremum and maximum. The set S has no maximum (no largest element), but $\sup S = 1$. Similarly, S has no minimum, but $\inf S = 0$.

Depending on how the real numbers are constructed, it is either an axiom or a theorem that every nonempty, bounded above set S of real numbers has a (finite) supremum. Moreover, it is easy to show that the supremum is unique. It can be shown that the matrix norm induced by any vector norm is well-defined, that is, that the supremum in (9.16) always exists as a finite number. However, we will not use this fact directly; instead, we will prove directly the existence of the matrix norms induced by the common vector norms by deriving and proving formulas for them. An induced matrix norm is a matrix norm, as will be shown below. The proof that an induced matrix norm defines a norm on $\mathbf{R}^{n\times n}$ is left to an exercise.

Theorem 395 *Let $\|\cdot\|$ be a vector norm on $\mathbf{R}^n$. Then the induced matrix norm $\|\cdot\|$ is a norm (in the sense of Definition 265) on $\mathbf{R}^{n\times n}$.*

Proof Exercise 5.

The following characterization of the induced matrix norm can be considered an alternate definition.

Theorem 396 *Let $\|\cdot\|$ be a norm on $\mathbf{R}^n$ and let the induced matrix norm on $\mathbf{R}^{m\times n}$ be denoted by the same symbol. Then, for all $A\in\mathbf{R}^{m\times n}$,*

$$\|A\| = \sup\left\{\frac{\|Ax\|}{\|x\|} : x\in\mathbf{R}^n,\ x\neq 0\right\}. \tag{9.17}$$

Proof Exercise 3

Corollary 397 *Let $\|\cdot\|$ be a norm on $\mathbf{R}^n$ and let the same symbol denote the induced matrix norm on $\mathbf{R}^{m\times n}$. Then*

$$\|Ax\| \leq \|A\|\|x\| \text{ for all } x\in\mathbf{R}^n.$$

Proof By (9.17), we have

$$\frac{\|Ax\|}{\|x\|} \leq \|A\| \text{ for all } x\in\mathbf{R}^n, x\neq 0,$$

which implies $\|Ax\|\leq\|A\|\|x\|$ for all $x\neq 0$. Since this inequality obviously holds for the zero vector, the result follows.

QED

A second characterization of an induced matrix norm is useful.

Theorem 398 *Let $\|\cdot\|$ denote the matrix norm on $\mathbf{R}^{m\times n}$ induced by a vector norm $\|\cdot\|$. Then*

$$\|A\| = \inf\left\{M>0 : \|Ax\|\leq M\|x\| \text{ for all } x\in\mathbf{R}^n\right\}. \tag{9.18}$$

Proof If $M>0$ satisfies $\|Ax\|\leq M\|x\|$ for all $x\in\mathbf{R}^n$, then this certainly holds for all $x\neq 0$, and so

$$\begin{aligned}
&\frac{\|Ax\|}{\|x\|} \leq M \text{ for all } x\in\mathbf{R}^n, x\neq 0\\
\Rightarrow\ &\sup\left\{\frac{\|Ax\|}{\|x\|} : x\in\mathbf{R}^n,\ x\neq 0\right\} \leq M\\
\Rightarrow\ &\|A\|\leq M.
\end{aligned}$$

Thus $\|A\|$ is a lower bound for the set

$$\left\{M>0 : \|Ax\|\leq M\|x\| \text{ for all } x\in\mathbf{R}^n\right\}.$$

Moreover, by Corollary 397, $\|A\|$ belongs to this set, from which it follows that $\|A\|$ is the infimum of the set.

QED

We can now show that an induced matrix norm is, in fact, a matrix norm.

Corollary 399 *Let $\|\cdot\|$ be a matrix norm induced by a vector norm $\|\cdot\|$. Then*

$$\|AB\| \le \|A\|\|B\| \text{ for all } A \in \mathbf{R}^{m\times n}, B \in \mathbf{R}^{n\times p}.$$

Proof Applying Theorem 397 twice, we have

$$\|ABx\| \le \|A\|\|Bx\| \le \|A\|\|B\|\|x\|.$$

This shows that $M = \|A\|\|B\|$ satisfies $\|ABx\| \le M\|x\|$ for all $x \in \mathbf{R}^n$. It follows from Theorem 398 that $\|AB\| \le \|A\|\|B\|$.

QED

9.4.1 Examples of induced matrix norms

The most useful vector norms are the ℓ^1, ℓ^2 (Euclidean), and ℓ^∞ norms:

$$\begin{aligned}
\|x\|_1 &= \sum_{i=1}^{n} |x_i|, \\
\|x\|_2 &= \sqrt{\sum_{i=1}^{n} |x_i|^2}, \\
\|x\|_\infty &= \max\{|x_i| \,:\, i = 1, 2, \ldots, n\}.
\end{aligned}$$

We will now derive the corresponding induced matrix norms.

Theorem 400 *The matrix norm induced by the ℓ^1 vector norm satisfies*

$$\|A\|_1 = \max\left\{\sum_{i=1}^{m} |A_{ij}| \,:\, j = 1, 2, \ldots, n\right\}.$$

In words, we describe $\|A\|_1$ as the maximum absolute column sum *of A.*

Proof We will prove the result using characterization (9.18) of the induced matrix norm. By definition of $\|\cdot\|_1$, we have

$$\begin{aligned}
\|Ax\|_1 = \sum_{i=1}^{m} |(Ax)_i| &= \sum_{i=1}^{m} \left|\sum_{j=1}^{n} A_{ij}x_j\right| \\
&\le \sum_{i=1}^{m}\sum_{j=1}^{n} |A_{ij}||x_j| \text{ (by the triangle inequality)} \\
&= \sum_{j=1}^{n}\sum_{i=1}^{m} |A_{ij}||x_j| \\
&= \sum_{j=1}^{n}\left(\sum_{i=1}^{m} |A_{ij}|\right)|x_j|.
\end{aligned}$$

Now choose k such that $1 \leq k \leq n$ and

$$\sum_{i=1}^{m} |A_{ik}| = \max\left\{\sum_{i=1}^{m} |A_{ij}| \,:\, j = 1, 2, \ldots, n\right\}.$$

We then obtain

$$\|Ax\|_1 \leq \left(\sum_{i=1}^{m} |A_{ik}|\right) \sum_{j=1}^{n} |x_j| = \left(\sum_{i=1}^{m} |A_{ik}|\right) \|x\|_1.$$

Since this holds for all $x \in \mathbf{R}^n$, (9.18) implies that

$$\|A\|_1 \leq \sum_{i=1}^{m} |A_{ik}|.$$

On the other hand, for the special vector $x = e_k$ (the kth standard basis vector), we have

$$\|Ax\|_1 = \sum_{i=1}^{m} |(Ax)_i| = \sum_{i=1}^{m} |A_{ik}|,$$

which shows that

$$\|A\|_1 \geq \sum_{i=1}^{m} |A_{ik}|.$$

This completes the proof.

QED

The matrix norm induced by the ℓ^∞ norm has a similar formula.

Theorem 401 *The matrix norm induced by the ℓ^∞ vector norm satisfies*

$$\|A\|_\infty = \max\left\{\sum_{j=1}^{n} |A_{ij}| \,:\, i = 1, 2, \ldots, m\right\}.$$

In words, we describe $\|A\|_\infty$ as the maximum absolute row sum *of A.*

Proof Exercise 4.

The reader should note that it is not difficult to compute either $\|A\|_1$ or $\|A\|_\infty$ for a given matrix $A \in \mathbf{R}^{n\times n}$. The ℓ^2 norm, $\|A\|_2$, presents more difficulties. We will need the following result, which is of independent interest.

Theorem 402 *Let $A \in \mathbf{R}^{n\times n}$ be symmetric and let the eigenvalues of A be $\lambda_n \leq \lambda_{n-1} \leq \cdots \leq \lambda_1$. Then*

$$\lambda_n \|x\|_2^2 \leq x \cdot (Ax) \leq \lambda_1 \|x\|_2^2. \tag{9.19}$$

Proof Exercise 6.

We can now derive a formula for $\|A\|_2$.

Theorem 403 *Let $A \in \mathbf{R}^{m \times n}$ and let λ_1 be the largest eigenvalue of $A^T A$. Then $\|A\|_2 = \sqrt{\lambda_1}$.*

Proof For any $x \in \mathbf{R}^n$, we have

$$\|Ax\|_2^2 = (Ax) \cdot (Ax) = x \cdot (A^T Ax) \leq \lambda_1 \|x\|_2^2$$

(applying Theorem 403 to $A^T A$). The matrix $A^T A$ is not only symmetric, but also positive semidefinite, so we know that $\lambda_1 \geq 0$. Therefore, we obtain

$$\|Ax\|_2 \leq \sqrt{\lambda_1} \|x\|_2 \text{ for all } x \in \mathbf{R}^n,$$

which implies that $\|A\|_2 \leq \sqrt{\lambda_1}$. If we choose x to be u_1, an eigenvector of $A^T A$ corresponding to λ_1, then

$$\|Au_1\|_2^2 = (Au_1) \cdot (Au_1) = u_1 \cdot (A^T Au_1) = u_1 \cdot (\lambda_1 u_1) = \lambda_1 \|u_1\|_2^2,$$

and therefore $\|Au_1\|_2 = \sqrt{\lambda_1} \|u_1\|_2$. This shows that $\|A\|_2 \geq \sqrt{\lambda_1}$, and the proof is complete.

QED

The reader might recognize $\sqrt{\lambda_1}$ as the largest singular value of the matrix A (see Section 8.1). We thus have the following result.

Corollary 404 *Let $A \in \mathbf{R}^{m \times n}$. Then*

$$\|A\|_2 = \sigma_1,$$

where σ_1 is the largest singular value of A.

Exercises

Miscellaneous exercises

1. Let $\|\cdot\|$ be any induced matrix norm on $\mathbf{R}^{n \times n}$. Prove that $\rho(A) \leq \|A\|$ for all $A \in \mathbf{R}^{n \times n}$, where $\rho(A)$ is the spectral radius of A:

$$\rho(A) = \max\{|\lambda| : \lambda \text{ is an eigenvalue of } A\}.$$

2. Prove Theorem 392.

3. Prove Theorem 396.

4. Prove Theorem 401.

5. Let $\|\cdot\|$ be a vector norm on $\mathbf{R}^n$. Prove that the induced matrix norm $\|\cdot\|$ satisfies Definition 265 relative to the vector space $\mathbf{R}^{m \times n}$.

6. Prove Theorem 402.

7. Let $A \in \mathbf{R}^{m\times n}$. Prove that $\|A^T\|_2 = \|A\|_2$. (Hint: Use Exercise 4.5.14.)

8. Let $A \in \mathbf{R}^{n\times n}$ be invertible. Show that
$$\|A^{-1}\|_2 = \frac{1}{\sqrt{\lambda_n}} = \frac{1}{\sigma_n},$$
where $\lambda_n > 0$ is the smallest eigenvalue of A^TA and σ_n is the smallest singular value of A. (Note that, since A is nonsingular, A^TA is symmetric and positive definite, and therefore all the eigenvalues of A^TA are positive.) (Hint: Use Exercise 4.5.14.)

9. Let $\|\cdot\|$ denote any norm on $\mathbf{R}^n$ and also the corresponding induced matrix norm. Prove that if $A \in \mathbf{R}^{n\times n}$ is invertible, then
$$\|Ax\| \geq \frac{\|x\|}{\|A^{-1}\|}.$$
(Hint: Write $x = A^{-1}(Ax)$ and compute an upper bound on $\|x\|$ by Corollary 397.)

9.5 The sensitivity of linear systems to errors

We wish to analyze the errors arising when an algorithm such as Gaussian elimination with partial pivoting is performed in floating point arithmetic. We remind the reader that this algorithm produces the exact solution x to a nonsingular system $Ax = b$, provided the calculations are performed in exact arithmetic. Therefore, the only source of error is the necessary rounding that occurs in floating point arithmetic.

We will consider a system $Ax = b$, where $A \in \mathbf{R}^{n\times n}$ is nonsingular, writing $\hat{x}$ for the computed solution and x for the exact solution. It is natural to expect that an error analysis would result in a bound
$$\|\hat{x} - x\| \leq \epsilon, \tag{9.20}$$
where ϵ is a positive number, hopefully small. A better result would bound the relative error:
$$\frac{\|\hat{x} - x\|}{\|x\|} \leq \epsilon. \tag{9.21}$$
A bound on the relative error is much more meaningful than a bound on the absolute error; to say the absolute error is "small" naturally raises the question, "Small compared to what?" In the context of relative error, the concept of "small" is unambiguous.

We will see that, in any inequality of the form (9.20) or (9.21), the bound ϵ depends on the particular matrix A, and that it is not possible to guarantee that ϵ is small for all A. To explain why this is true requires a lengthy digression, in which we consider the effects of errors in A and/or b, rather than round-off error. We will return to the issue of round-off error in the next section.

We begin with the following situation: We wish to solve $Ax = b$ (where $A \in \mathbf{R}^{n\times n}$ is nonsingular) for a given vector b. However, the exact value of b is unavailable; instead, we have an approximation $\hat{b}$ to b (perhaps $\hat{b}$ is obtained from laboratory measurements which contain errors). If we solve $A\hat{x} = \hat{b}$ exactly to get $\hat{x}$, we have

$$A(\hat{x} - x) = A\hat{x} - Ax = \hat{b} - b,$$

or

$$\hat{x} - x = A^{-1}(\hat{b} - b). \tag{9.22}$$

Using any norm $\|\cdot\|$ on $\mathbf{R}^n$ and the induced matrix norm on $\mathbf{R}^{n\times n}$, we have

$$\|\hat{x} - x\| \le \|A^{-1}\|\|\hat{b} - b\|, \tag{9.23}$$

which is a bound on the absolute error in $\hat{x}$. Moreover, this bound is tight in the sense that, for any $b \in \mathbf{R}^n$, there exist $\hat{b}$ such that $\|\hat{x} - x\| = \|A^{-1}\|\|\hat{b} - b\|$ (see Exercise 1).

With a little more work, we can turn (9.23) to a bound on the relative error. We have $b = Ax$, so $\|b\| \le \|A\|\|x\|$ and this, together with (9.23), yields

$$\frac{\|\hat{x} - x\|}{\|A\|\|x\|} \le \|A^{-1}\|\frac{\|\hat{b} - b\|}{\|b\|}$$

or

$$\frac{\|\hat{x} - x\|}{\|x\|} \le \|A\|\|A^{-1}\|\frac{\|\hat{b} - b\|}{\|b\|}. \tag{9.24}$$

This bound is also tight in a certain sense. If A is regarded as fixed, then we can find $b \in \mathbf{R}^n$ and $\hat{b} \in \mathbf{R}^n$ such that (9.24) holds as an equation.

The bound (9.24) provides an answer to the following question: What nonsingular matrices A correspond to linear systems $Ax = b$ that are very sensitive to errors in b? The quantity $\|A\|\|A^{-1}\|$ provides a quantitative measure on the degree to which the relative error in right-hand side can be magnified in the solution. For this reason, we call $\|A\|\|A^{-1}\|$ the *condition number* of the nonsingular matrix A (relative to the norm $\|\cdot\|$) and write $\text{cond}(A) = \|A\|\|A^{-1}\|$. We notice that a matrix does not have a single condition number, but rather one condition number for each matrix norm induced by a vector norm. We will use the following notation for condition numbers defined by the common norms:

$$\begin{aligned}
\text{cond}_1(A) &= \|A\|_1\|A^{-1}\|_1, \\
\text{cond}_2(A) &= \|A\|_2\|A^{-1}\|_2, \\
\text{cond}_\infty(A) &= \|A\|_\infty\|A^{-1}\|_\infty.
\end{aligned}$$

To compute the conditions numbers $\mathrm{cond}_1(A)$ and $\mathrm{cond}_\infty(A)$, the matrix A^{-1} is required. In the case of $\mathrm{cond}_2(A)$, we know that $\|A\|_2 = \sigma_1$, where σ_1 is the largest singular value of A, and

$$\|A^{-1}\|_2 = \frac{1}{\sigma_n},$$

where σ_n is the smallest singular value of A (see Exercise 9.4.8). Therefore,

$$\mathrm{cond}_2(A) = \frac{\sigma_1}{\sigma_n},$$

and thus $\mathrm{cond}_2(A)$ can be computed without explicitly computing A^{-1}. On the other hand, the singular values of A are needed, and computing them is even more expensive than computing A^{-1}. Fortunately, all of these condition numbers can be estimated at a cost much less than that required to compute them exactly.[7] In a practical problem, the order of magnitude of the condition number is usually all that is needed.

We will also need to consider how the solution to $Ax = b$ changes when the matrix A (rather than the vector b) changes. For this, we will need the following results.

Theorem 405 *Suppose $E \in \mathbf{R}^{n\times n}$ and $\|E\| < 1$, where $\|\cdot\|$ is any induced matrix norm. Then $I + E$ is invertible and*

$$\left\|(I+E)^{-1}\right\| \le \frac{1}{1-\|E\|}.$$

Proof To show that $I + E$ is nonsingular, we will argue by contradiction. Suppose there exists $x \neq 0$ such that $(I+E)x = 0$. Without loss of generality, we can assume $\|x\| = 1$. We then have

$$\begin{aligned}(I+E)x = 0 \quad &\Rightarrow \quad x + Ex = 0 \\ &\Rightarrow \quad x = -Ex \\ &\Rightarrow \quad 1 = \|x\| = \|Ex\| \\ &\Rightarrow \quad 1 \le \|E\|\|x\| = \|E\| < 1.\end{aligned}$$

This contradiction shows that $I + E$ must be nonsingular.

Now suppose $B = (I+E)^{-1}$. We then have

$$\begin{aligned}B(I+E) = I \quad &\Rightarrow \quad B + BE = I \\ &\Rightarrow \quad B = I - BE \\ &\Rightarrow \quad \|B\| = \|I - BE\| \le \|I\| + \|BE\| \le 1 + \|B\|\|E\| \\ &\Rightarrow \quad \|B\| - \|B\|\|E\| \le 1 \\ &\Rightarrow \quad \|B\|(1 - \|E\|) \le 1 \\ &\Rightarrow \quad \|B\| \le \frac{1}{1-\|E\|},\end{aligned}$$

[7] Algorithms for inexpensively estimating the condition number of a matrix are beyond the scope of this text. The interested reader can consult Chapter 15 of [19].

as desired.

QED

Corollary 406 *Let $A \in \mathbf{R}^{n\times n}$ be invertible and suppose $E \in \mathbf{R}^{n\times n}$ satisfies*

$$\|E\| < \frac{1}{\|A^{-1}\|},$$

where $\|\cdot\|$ denotes any induced matrix norm. Then $A+E$ is invertible and

$$\|(A+E)^{-1}\| \le \frac{\|A^{-1}\|}{1-\|A^{-1}\|\|E\|}. \tag{9.25}$$

Proof Exercise 2.

We now consider the situation in which the matrix A is perturbed; that is, we wish to solve $Ax = b$ but we have only an approximation $\hat{A}$ to A. Thus we solve $\hat{A}\hat{x} = b$ and we wish to derive a bound on the relative error in $\hat{x}$. We have

$$\begin{aligned}
\hat{A}\hat{x} = Ax \quad &\Rightarrow \quad \hat{A}\hat{x} - Ax = 0 \\
&\Rightarrow \quad \hat{A}\hat{x} - \hat{A}x + \hat{A}x - Ax = 0 \\
&\Rightarrow \quad \hat{A}(\hat{x} - x) = -(\hat{A} - A)x \\
&\Rightarrow \quad \hat{x} - x = -\hat{A}^{-1}(\hat{A} - A)x \\
&\Rightarrow \quad \|\hat{x} - x\| \le \|\hat{A}^{-1}\|\|\hat{A} - A\|\|x\| \\
&\Rightarrow \quad \frac{\|\hat{x} - x\|}{\|x\|} \le \|A\|\|\hat{A}^{-1}\|\frac{\|\hat{A} - A\|}{\|A\|} \\
&\Rightarrow \quad \frac{\|\hat{x} - x\|}{\|x\|} \le \frac{\|A\|\|A^{-1}\|}{1 - \|A^{-1}\|\|\hat{A} - A\|}\frac{\|\hat{A} - A\|}{\|A\|}.
\end{aligned}$$

The only assumption required for the above analysis is that $\hat{A}$ is sufficiently close to A, specifically, that $\|\hat{A} - A\| < \|A^{-1}\|^{-1}$. If $\|\hat{A} - A\| \ll \|A^{-1}\|^{-1}$, then

$$\frac{\|A\|\|A^{-1}\|}{1 - \|A^{-1}\|\|\hat{A} - A\|} \doteq \|A\|\|A^{-1}\| = \text{cond}(A).$$

Thus we see that, at least for small perturbations of A, the magnification of the error in A is controlled by the condition number of A (just as was the magnification of an error in b).

Exercises

Miscellaneous exercises

1. Let $A \in \mathbf{R}^{n\times n}$ be nonsingular.

(a) Suppose $b \in \mathbf{R}^n$ is given. Explain how to produce $\hat{b} \in \mathbf{R}^n$ such that $\|\hat{x} - x\| = \|A^{-1}\|\|\hat{b} - b\|$ (where $Ax = b$ and $A\hat{x} = \hat{b}$).

(b) Show how to produce $b, \hat{b} \in \mathbf{R}^n$ such that

$$\frac{\|\hat{x} - x\|}{\|x\|} = \|A\|\|A^{-1}\|\frac{\|\hat{b} - b\|}{\|b\|}$$

(where $Ax = b$ and $A\hat{x} = \hat{b}$).

2. Prove Corollary 406 by applying Theorem 405 to $I + A^{-1}E$.

3. Suppose $A \in \mathbf{R}^{n\times n}$ is nonsingular and $\hat{A} \in \mathbf{R}^{n\times n}$, $x, \hat{x} \in \mathbf{R}^n$, and $b \in \mathbf{R}^n$ satisfy $Ax = b$, $\hat{A}\hat{x} = b$. Show that

$$\frac{\|\hat{x} - x\|}{\|\hat{x}\|} \le \text{cond}(A)\frac{\|\hat{A} - A\|}{\|A\|}.$$

This is another way to see that the condition number of A determines the sensitivity of the system $Ax = b$ to changes in A. (The reader should notice that $\|\hat{x} - x\|/\|\hat{x}\|$ is a measure of the relative difference in x and $\hat{x}$, although we would prefer to measure the difference relative to $\|x\|$ rather than $\|\hat{x}\|$.)

4. Let $A \in \mathbf{R}^{n\times n}$ be nonsingular, and let $\|\cdot\|$ denote any norm on $\mathbf{R}^n$ and the corresponding induced matrix norm. Prove that

$$\text{cond}(A) = \frac{\max_{x\neq 0}\frac{\|Ax\|}{\|x\|}}{\min_{x\neq 0}\frac{\|Ax\|}{\|x\|}}.$$

5. The purpose of this exercise is to show that the condition number of an invertible matrix A is the reciprocal of the relative distance from A to the nearest singular matrix, which shows directly that cond(A) is large precisely when A is close to a singular matrix. The reader is asked for the complete proof only in the case of the Euclidean vector norm and corresponding matrix norm. Let $A \in \mathbf{R}^{n\times n}$ be invertible, and let $\|\cdot\|$ denote any norm on $\mathbf{R}^n$ and the corresponding induced matrix norm.

(a) Let $B \in \mathbf{R}^{n\times n}$ be any singular matrix. Prove that

$$\|A - B\| \ge \frac{1}{\|A^{-1}\|}.$$

(Hint: By the definition of induced matrix norm, it suffices to find a unit vector x such that $\|(A - B)x\| \ge 1/\|A^{-1}\|$. Prove that any unit vector in $\mathcal{N}(B)$ will do. Exercise 9.4.9 will be useful.)

(b) Using the previous result, show that

$$\inf\left\{\frac{\|A-B\|}{\|A\|} : B \in \mathbf{R}^{n\times n}, \det(B)=0\right\} \geq \frac{1}{\mathrm{cond}(A)}.$$

(c) In the special case of the Euclidean norm on $\mathbf{R}^n$ and induced norm $\|\cdot\|_2$ on $\mathbf{R}^{n\times n}$, construct a singular matrix A' such that

$$\frac{\|A-A'\|_2}{\|A\|_2} = \frac{1}{\mathrm{cond}_2(A)}.$$

(Hint: Construct A' by modifying the SVD of A.)

6. Let $A \in \mathbf{R}^{n\times n}$ be a nonsingular matrix, and consider the problem of computing $y = Ax$ from $x \in \mathbf{R}^n$ (x is regarded as the data and y is regarded as the solution). Suppose we have a noisy value of x, $\hat{x}$, and we compute $\hat{y} = A\hat{x}$. Bound the relative error in $\hat{y}$ in terms of the relative error in $\hat{x}$. What should we regard as the condition number of A with respect to the matrix-vector multiplication problem?

7. Repeat the previous problem for a nonsingular matrix $A \in \mathbf{R}^{m\times n}$, where $m > n$. Use the Euclidean norm in the analysis.

8. Let $x \in \mathbf{R}^n$ be fixed, and consider the problem of computing $y = Ax$, where $A \in \mathbf{R}^{n\times n}$ is regarded as the data and y as the solution. Assume A is limited to nonsingular matrices. Let $\hat{A}$ be a noisy value of A, and define $\hat{y} = \hat{A}x$. Bound the relative error in $\hat{y}$ in terms of the relative error in $\hat{A}$.

9.6 Numerical stability

In certain applications, it is common to encounter equations $Ax = b$ where A is nonsingular but has a very large condition number. Such a matrix is called *ill-conditioned*; a small error in either A or b can result in large errors in the solution x, as we saw in the previous section. It seems reasonable that round-off errors would have an effect similar to errors in A and/or b, and thus that small round-off errors could be magnified into large errors in x when the matrix A is ill-conditioned. If this is true, then it will not be possible to derive a bound on $\|\hat{x}-x\|$ like (9.20) or (9.21) where ϵ is small (at least not a bound that applies to all linear systems).

The concept of ill-conditioning appears in all kinds of problems, not just linear systems. Some problems are intrinsically harder to solve accurately than others, and we cannot expect any numerical algorithm to produce an accurate computed solution to such a problem. The concept of *backward error analysis* was designed to quantify whether an algorithm produces computed solutions that are as good as can be expected. An analysis that produces bounds of the form (9.20) or (9.21) is called a *forward error analysis*. A simple example illustrates the concept of ill-conditioning as well as forward and backward error analysis.

9.6.1 Backward error analysis

Suppose x and y are vectors in $\mathbf{R}^2$ and we wish to compute $x \cdot y$ in floating point arithmetic. To simplify the following analysis, we will assume that x and y are exactly representable in the floating point system being used. We will write $x \odot y$ for the computed dot product and $\mathrm{fl}(\alpha)$ for the floating point representation of a real number α; that is, $\mathrm{fl}(\alpha)$ is the closest floating point number to α. By Exercise 9.2.3, we have

$$\frac{|\mathrm{fl}(\alpha) - \alpha|}{|\alpha|} \leq \mathbf{u},$$

where $\mathbf{u}$ is the unit round of the given floating point system (see page 517). More precisely, we can write

$$\alpha - \mathbf{u}|\alpha| \leq \mathrm{fl}(\alpha) \leq \alpha + \mathbf{u}|\alpha|.$$

Therefore, there exists a real number ϵ such that

$$\mathrm{fl}(\alpha) = \alpha(1+\epsilon) \text{ and } |\epsilon| \leq \mathbf{u}.$$

We make the following assumption about the floating point system: Given two floating point numbers α and β, the computed value of $\alpha+\beta$ is $\mathrm{fl}(\alpha+\beta)$; that is, the algorithm for addition produces the floating point number closest to the exact value. We assume the same to be true for the operations of subtraction, multiplication, and division.[8]

To compute $x \cdot y$ for $x, y \in \mathbf{R}^2$ requires three operations, two multiplications and one addition, and thus there are three rounding errors. The intermediate values are

$$\begin{aligned} \mathrm{fl}(x_1 y_1) &= x_1 y_1 (1+\epsilon_1) \text{ (the computed value of } x_1 y_1), \\ \mathrm{fl}(x_2 y_2) &= x_2 y_2 (1+\epsilon_2) \text{ (the computed value of } x_2 y_2), \end{aligned}$$

and the final result is

$$x \odot y = (x_1 y_1 (1+\epsilon_1) + x_2 y_2 (1+\epsilon_2))(1+\epsilon_3). \tag{9.26}$$

We have $|\epsilon_i| \leq \mathbf{u}$ for $i = 1, 2, 3$. Simplifying, we obtain

$$\begin{aligned} & x \odot y = x_1 y_1 + x_2 y_2 + (\epsilon_1 + \epsilon_3 + \epsilon_1 \epsilon_3) x_1 y_1 + (\epsilon_2 + \epsilon_3 + \epsilon_2 \epsilon_3) x_2 y_2 \\ \Rightarrow \quad & x \odot y = x \cdot y + (\epsilon_1 + \epsilon_3 + \epsilon_1 \epsilon_3) x_1 y_1 + (\epsilon_2 + \epsilon_3 + \epsilon_2 \epsilon_3) x_2 y_2. \end{aligned}$$

[8] The IEEE standard for floating point arithmetic requires this condition; see [34].

We have $|\epsilon_1 + \epsilon_3 + \epsilon_1\epsilon_3| \leq |\epsilon_1| + |\epsilon_3| + |\epsilon_1\epsilon_3|$, which can be just a little larger that $2\mathbf{u}$; a crude estimate is therefore $|\epsilon_1 + \epsilon_3 + \epsilon_1\epsilon_3| \leq 3\mathbf{u}$, and similarly for $|\epsilon_2 + \epsilon_3 + \epsilon_2\epsilon_3|$. Thus

$$|x \odot y - x \cdot y| \leq 3\mathbf{u}(|x_1y_1| + |x_2y_2|)$$

and

$$\frac{|x \odot y - x \cdot y|}{|x \cdot y|} \leq 3\mathbf{u}\frac{|x_1y_1| + |x_2y_2|}{|x \cdot y|}.$$

If $|x_1y_1| + |x_2y_2| = |x \cdot y|$ (for example, if x and y have positive components), then the relative error in $x \odot y$ is bounded by $3\mathbf{u}$, which is perfectly satisfactory. On the other hand, because of cancellation, $|x \cdot y|$ can be much less than $|x_1y_1| + |x_2y_2|$; in fact, the ratio

$$\frac{|x_1y_1| + |x_2y_2|}{|x \cdot y|}$$

can be arbitrarily large. There is therefore no useful bound on the relative error in $x \odot y$ in general. Nevertheless, it is probably clear to the reader that there is no flaw in the *algorithm* used to compute the dot product. The difficulty is simply that, when $|x \cdot y|$ is much less than $|x_1y_1| + |x_2y_2|$, the problem of computing $x \cdot y$ is ill-conditioned.

If we cannot expect to obtain a good bound on the relative error in $x \odot y$, how can we evaluate a proposed numerical algorithm? Another look at the computed value $x \odot y$ provides a clue:

$$\begin{aligned} x \odot y &= (x_1y_1(1+\epsilon_1) + x_2y_2(1+\epsilon_2))(1+\epsilon_3) \\ &= x_1y_1(1+\epsilon_1+\epsilon_3+\epsilon_1\epsilon_3) + x_2y_2(1+\epsilon_2+\epsilon_3+\epsilon_2\epsilon_3) \\ &= x_1\tilde{y}_1 + x_2\tilde{y}_2 \\ &= x \cdot \tilde{y}, \end{aligned}$$

where $\tilde{y}_1 = y_1(1+\epsilon_1+\epsilon_3+\epsilon_1\epsilon_3)$ and $\tilde{y}_2 = y_2(1+\epsilon_2+\epsilon_3+\epsilon_2\epsilon_3)$. We have

$$\frac{|\tilde{y}_i - y_i|}{|y_i|} \leq 3\mathbf{u},\ i = 1, 2,$$

so we see that $\tilde{y}$ is a small perturbation of y. The conclusion is that *the computed solution is the exact solution of a slightly perturbed problem.* A little thought shows that this is the best that can be expected of any algorithm executed in floating point arithmetic; after all, the numbers defining the problem must be rounded to their floating point representations before any computations can be done. If the effect of the errors that accumulate during execution of the algorithm is not much larger than that of the initial rounding errors, then the algorithm is regarded as *numerically stable.* The error in the perturbed problem (in the above case, the error in $\tilde{y}$) is called the *backward error.* Thus numerical stability means that the backward error can be bounded satisfactorily.[9] We also use the phrase *backward stable* to describe an algorithm producing a small backward error.

The error in the computed solution is the *forward error*. While we would like to be able to say that the forward error is small, for an ill-conditioned problem, the forward error can be large (at least in the relative sense) even though the backward error is small.

9.6.2 Analysis of Gaussian elimination with partial pivoting

A backward error analysis of Gaussian elimination with partial pivoting is considerably more complicated than the example given in the preceding section. We will only state the result here and refer the reader to Higham's book [19] for a comprehensive treatment.

Partial pivoting is necessary in Gaussian elimination because otherwise the values in the intermediate matrices

$$A^{(k)} = L_{k-1}P_{k-1}\cdots L_1P_1A$$

will often be large (see Example 382). As we will see below, the stability of Gaussian elimination depends on the actual growth in the entries in these matrices. For this reason, we define the *growth factor* ρ_n by

$$\rho_n = \frac{\max\{|A_{ij}^{(k)}| \,:\, i,j,k = 1,2,\ldots,n\}}{\max\{|A_{ij}| \,:\, i,j = 1,2,\ldots,n\}}.$$

The following result is Theorem 9.5 of [19].[10]

Theorem 407 *Let $A \in \mathbf{R}^{n\times n}$, $b \in \mathbf{R}^n$, and suppose Gaussian elimination with partial pivoting produces a computed solution $\hat{x}$ to $Ax = b$. Then there exists a matrix $\hat{A}$ such that*

$$\hat{A}\hat{x} = b$$

and

$$\frac{\|\hat{A} - A\|_\infty}{\|A\|_\infty} \le \left(\frac{3n^3\mathbf{u}}{1 - 3n\mathbf{u}}\right)\rho_n,$$

where $\mathbf{u}$ is the unit round and ρ_n is the growth factor.

---

[9]In many problems, particularly nonlinear ones, it is difficult or impossible to show that the computed solution is the exact solution to a nearby problem. However, it is almost as good to show that the computed solution is close to the solution of a nearby problem; such a result is called a mixed forward-backward error analysis, and an algorithm satisfying such an analysis is also considered numerically stable. However, we can use the simpler definition of numerical stability in this book.

[10]See page 165 of [19] for a caveat about the derivation of this theorem; the caveat is not important to the interpretation of the theorem. The issue is that Theorem 407 uses bounds on the computed factors L and U that are valid only for the exact L and U. Therefore, Theorem 407 is technically incorrect; the correct version is similar but more complicated. See also the discussion on page 189 of [19].

The interpretation of this result requires consideration of the two factors in the bound on $\|\hat{A} - A\|_\infty / \|A\|_\infty$. First of all, the ratio

$$\frac{3n^3\mathbf{u}}{1 - 3n\mathbf{u}} \tag{9.27}$$

arises from bounding the approximately n^3 rounding errors that occur in the course of the algorithm. The bound results from the worst-case scenario in which all rounding errors have the same sign. In practice, of course, rounding errors are both positive and negative and therefore tend to cancel out.[11] The practical effect is that a bound like (9.27) tends to be quite pessimistic. If we use a long-standing rule of thumb[12] and replace n^3 by $\sqrt{n^3} = n^{3/2}$, then (9.27) is small for any reasonable value of n. For instance, with $n = 10^4$ and $\mathbf{u} \doteq 1.1 \cdot 10^{-16}$ (as in IEEE double precision arithmetic), replacing n^3 by $n^{3/2}$ in (9.27) yields $3.3 \cdot 10^{-10}$. Even this rule of thumb is probably pessimistic.

If we accept that (9.27) (or the quantity it bounds) is small in practice, then it remains only to consider the growth factor ρ_n. If it is small, then Gaussian elimination with partial pivoting can be considered a numerically stable algorithm. However, in fact the only bound possible on ρ_n is $\rho_n \le 2^{n-1}$ (without pivoting, ρ_n can be arbitrarily large, as in Example 382). In fact, if we define $A \in \mathbf{R}^{n \times n}$ by

$$A_{ij} = \begin{cases} 1, & i = j \text{ or } j = n, \\ -1, & i > j, \\ 0, & \text{otherwise}, \end{cases}$$

then $U_{nn} = A_{nn}^{(n)} = 2^{n-1}$, as is easily seen. For instance, in the case $n = 4$, Gaussian elimination proceeds as follows:

$$\begin{bmatrix} 1 & 0 & 0 & 1 \\ -1 & 1 & 0 & 1 \\ -1 & -1 & 1 & 1 \\ -1 & -1 & -1 & 1 \end{bmatrix} \rightarrow \begin{bmatrix} 1 & 0 & 0 & 1 \\ 0 & 1 & 0 & 2 \\ 0 & -1 & 1 & 2 \\ 0 & -1 & -1 & 2 \end{bmatrix}$$
$$\rightarrow \begin{bmatrix} 1 & 0 & 0 & 1 \\ 0 & 1 & 0 & 2 \\ 0 & 0 & 1 & 4 \\ 0 & 0 & -1 & 4 \end{bmatrix} \rightarrow \begin{bmatrix} 1 & 0 & 0 & 1 \\ 0 & 1 & 0 & 2 \\ 0 & 0 & 1 & 4 \\ 0 & 0 & 0 & 8 \end{bmatrix}$$

For this class of matrices, pivoting is never required.

---

[11] In the past, some computers *chopped* real numbers instead of rounding them, meaning that extra digits were simply dropped to produce a floating point approximation. For positive numbers, this means that every rounding error is negative (that is, $\mathrm{fl}(x) \le x$ for all $x > 0$). See page 54 of [19] for an account of dramatic accumulation of "round-off" error when chopping was used instead of rounding.

[12] "It is rather conventional to obtain a "realistic" estimate of the possible overall error due to k round-offs, when k is fairly large, by replacing k by $\sqrt{k}$ in an expression for (or an estimate of) the maximum resultant error." (F. B. Hildebrand, [20])

For several decades, it was felt that examples such as the matrix A described above were contrived and would never occur in practical problems. Indeed, it seems to be the case that matrices that lead to large ρ_n are exceedingly rare. However, at least two practical problems have been reported in the literature in which Gaussian elimination with partial pivoting is unstable due to a large ρ_n (see Wright [45] and Foster [9]).

The above considerations lead to a rather unsatisfactory conclusion: Gaussian elimination with partial pivoting is not provably stable, and examples exist in which it behaves in an unstable fashion. However, such examples are so rare that the algorithm is still used; in fact, it is regarded as the algorithm of choice for solving general systems of the form $Ax = b$. Although there is an alternate algorithm that is always backward stable (the QR algorithm to be described in Section 9.8), it costs twice as many operations as does Gaussian elimination with partial pivoting. This seems a high price to pay when the potential instability of Gaussian elimination is so rarely realized. Also, it is easier to extend Gaussian elimination with partial pivoting to *sparse* systems, that is, systems in which most of the entries in A are zero. Sparse systems are often so large that it is possible to solve them only by taking advantage of the zeros. (We need not store any entry known to be zero and we need not perform an operation when one of the operands is known to be zero. Thus both memory and computation time can be reduced for sparse systems, at least in principle.) For these reasons, Gaussian elimination with partial pivoting is the algorithm of choice for solving $Ax = b$ for a general matrix A.

The situation is better if the matrix A is symmetric and positive definite. In this case, as we saw Theorem 388, the growth factor for Gaussian elimination *without* partial pivoting is $\rho_n = 1$. Therefore, Gaussian elimination applied to an SPD matrix is numerically stable, and pivoting is not required. Normally we would use the Cholesky factorization instead of Gaussian elimination (to save half the arithmetic operations); it is not surprising that the Cholesky factorization approach is also numerically stable. The reader can consult Section 10.1 of [19] for details.

Exercises

1. The purpose of this exercise is to analyze the multiplication of two real numbers for stability.

   (a) Suppose x and y are two real numbers and $\hat{x}$ and $\hat{y}$ are perturbations of x and y, respectively. Obtain bounds on the absolute and relative errors of $\hat{x}\hat{y}$ as an approximation to xy, and show that the problem of computing xy is always well-conditioned.

   (b) Suppose x and y are floating point numbers and that xy is within the range representable by the given floating point system (see Exercise 9.2.3). Perform a backward error analysis on the computation of xy in floating point arithmetic and show that floating point multiplication is backward stable. (Note: This part of the exercise is extremely simple.) You should assume that all floating point operations satisfy the condition described on page 9.6.1: The computed value is the floating point representation of the exact result.

2. Repeat the preceding exercise for addition. You should find that although floating point addition is backward stable, some problems are ill-conditioned. Characterize x and y such that the sum $x+y$ is ill-conditioned.

3. Prove that the growth factor ρ_n is bounded by 2^{n-1} when Gaussian elimination with partial pivoting is applied to an $n \times n$ matrix. (Hint: Notice that the magnitudes of the multipliers—the lower triangular entries in L—are all bounded by 1 when partial pivoting is used.)

4. Consider the problem of matrix-vector multiplication: Given $A \in \mathbf{R}^{n\times n}$, $x \in \mathbf{R}^{n\times n}$, compute $y = Ax$.

   (a) Perform a backward error analysis and show that the result $\hat{y}$ computed in floating point arithmetic can be written as $\hat{y} = (A+\delta A)x$ for some $\delta A \in \mathbf{R}^{n\times n}$.

   (b) Find a bound on $\|\delta A\|$ (in some convenient matrix norm).

   (c) Using the results of Exercise 9.5.8, find a bound on the relative error in $\hat{y}$.

9.7 The sensitivity of the least-squares problem

The reader will recall that the error analysis of Gaussian elimination with partial pivoting consisted of two parts: An analysis of the sensitivity of the solution of $Ax = b$ to errors in A and a backward error analysis of the algorithm to show that the computed solution is the exact solution of a perturbed system $\hat{A}x = b$. In preparation for discussing numerical algorithms for solving the least-squares problem, we will now discuss the sensitivity of the least-squares solution to the overdetermined system $Ax = b$ to errors in A and/or b. As in the case of a square, nonsingular system, we will only use perturbations in A,[13] but we will also consider perturbations in b as a simpler introduction to

[13]There is a simple reason why backward error analysis results in a perturbed system $\hat{A}x = b$ rather than $Ax = \hat{b}$. The point of backward error analysis is to show that all the rounding errors that occur during the algorithm can be regarded as errors in the data defining the problem. To do this, one needs a lot of degrees of freedom in the data. The mn entries in A provide many more degrees of freedom than do the m entries in b.

the analysis. We will only consider the case in which $A \in \mathbf{R}^{m \times n}$ with $m > n$ and $\text{rank}(A) = n$.

The analysis we present below shows that the conditioning of the least-squares problem $Ax = b$ depends on how close b is to $\text{col}(A)$. We will start with the simplest case that $b \in \text{col}(A)$, say $b = Ax$, and we wish to solve $Ax = \hat{b}$ in the least-squares sense. Since A is assumed to have full rank, the solution is

$$\hat{x} = (A^T A)^{-1} A^T \hat{b},$$

so

$$\hat{x} - x = (A^T A)^{-1} A^T \hat{b} - (A^T A)^{-1} A^T b = (A^T A)^{-1} A^T (\hat{b} - b).$$

Therefore,

$$\begin{aligned} & \|\hat{x} - x\| \leq \|(A^T A)^{-1} A^T\| \|\hat{b} - b\| \\ \Rightarrow \quad & \frac{\|\hat{x} - x\|}{\|x\|} \leq \frac{\|(A^T A)^{-1} A^T\| \|\hat{b} - b\|}{\|x\|} = \frac{\|(A^T A)^{-1} A^T\| \|b\|}{\|x\|} \frac{\|\hat{b} - b\|}{\|b\|}. \end{aligned}$$

Since, by assumption, $b = Ax$, we have $\|b\| \leq \|A\| \|x\|$ and thus

$$\frac{\|\hat{x} - x\|}{\|x\|} \leq \frac{\|(A^T A)^{-1} A^T\| \|A\| \|x\|}{\|x\|} \frac{\|\hat{b} - b\|}{\|b\|} = \|(A^T A)^{-1} A^T\| \|A\| \frac{\|\hat{b} - b\|}{\|b\|}.$$

We will now restrict ourselves to the Euclidean norm $\|\cdot\|_2$ and the corresponding induced matrix norm. This is entirely natural, since the least-squares problem itself is defined by orthogonality in the Euclidean dot product. In Exercise 1, we ask the reader to show that

$$\|(A^T A)^{-1} A^T\|_2 = \frac{1}{\sigma_n},$$

where σ_n is the smallest singular value of A ($\sigma_n > 0$ is ensured by the assumption that $\text{rank}(A) = n$). As we saw in Section 9.4.1, $\|A\|_2 = \sigma_1$, where σ_1 is the largest singular value of A. Thus we obtain

$$\frac{\|\hat{x} - x\|_2}{\|x\|_2} \leq \frac{\sigma_1}{\sigma_n} \frac{\|\hat{b} - b\|_2}{\|b\|_2}. \tag{9.28}$$

Because of this result (and related results to be derived below), we extend the concept of condition number to $A \in \mathbf{R}^{m \times n}$, $\text{rank}(A) = n$, defining

$$\text{cond}_2(A) = \frac{\sigma_1}{\sigma_n}.$$

We will now consider the case in which b is to be perturbed, but it is not assumed that b lies in $\text{col}(A)$. If x is the least-squares solution to $Ax = b$, then $b - Ax$ is orthogonal to $\text{col}(A)$, and hence, in particular, $b - Ax$ is orthogonal to Ax. Therefore, since $b = Ax + b - Ax$, it follows from the Pythagorean theorem that

$$\|b\|_2^2 = \|Ax\|_2^2 + \|b - Ax\|_2^2.$$

Moreover, if θ is the angle between b and Ax, then we see that

$$\|Ax\|_2 = \|b\| \cos(\theta), \ \ \|b - Ax\|_2 = \|b\|_2 \sin(\theta). \tag{9.29}$$

We will use these facts below.

If $\hat{x}$ is the least-squares solution to $Ax = \hat{b}$, then

$$\hat{x} = (A^TA)^{-1}A^T\hat{b} \text{ and } \hat{x} - x = (A^TA)^{-1}A^T(\hat{b} - b).$$

It follows that

$$\|\hat{x} - x\|_2 \le \|(A^TA)^{-1}A^T\|_2\|\hat{b} - b\|_2,$$

or

$$\frac{\|\hat{x} - x\|_2}{\|x\|_2} \le \frac{\|(A^TA)^{-1}A^T\|_2\|\hat{b} - b\|_2}{\|x\|_2} = \frac{\|(A^TA)^{-1}A^T\|_2\|b\|_2}{\|x\|_2}\frac{\|\hat{b} - b\|_2}{\|b\|_2}.$$

From (9.29), we know that $\|b\|_2 = \|Ax\|_2/\cos(\theta)$, so we obtain

$$\frac{\|\hat{x} - x\|_2}{\|x\|_2} \le \frac{\|(A^TA)^{-1}A^T\|_2\|Ax\|_2}{\|x\|_2\cos(\theta)}\frac{\|\hat{b} - b\|_2}{\|b\|_2}.$$

Finally, $\|Ax\|_2 \le \|A\|_2\|x\|_2$, and thus

$$\begin{aligned}\frac{\|\hat{x} - x\|_2}{\|x\|_2} &\le \frac{\|(A^TA)^{-1}A^T\|_2\|A\|_2\|x\|_2}{\|x\|_2\cos(\theta)}\frac{\|\hat{b} - b\|_2}{\|b\|_2} \\ &= \frac{\|(A^TA)^{-1}A^T\|_2\|A\|_2}{\cos(\theta)}\frac{\|\hat{b} - b\|_2}{\|b\|_2}.\end{aligned}$$

Since $\|(A^TA)^{-1}A^T\|_2\|A\|_2 = \text{cond}_2(A)$, we see that

$$\frac{\|\hat{x} - x\|_2}{\|x\|_2} \le \frac{\text{cond}_2(A)}{\cos(\theta)}\frac{\|\hat{b} - b\|_2}{\|b\|_2}. \tag{9.30}$$

This shows that the conditioning of the least-squares problem, when b is the data, depends not only on the condition number of A, but also on the angle θ between b and col(A). The conditioning becomes arbitrarily bad as θ approaches $\pi/2$.

When we think about the problem geometrically, this result is not surprising. When b is nearly orthogonal to col(A), the projection of b onto col(A) is very small. The error in $\hat{b}$ can be very small and yet be large compared to $\text{proj}_{\text{col}(A)}b$. More precisely, the error in $\hat{x}$ can be large compared to the error in $\hat{b}$ if $\text{proj}_{\text{col}(A)}(\hat{b} - b)$ is large compared to $\text{proj}_{\text{col}(A)}b$.

Example 408 *Consider the least-squares problem* $Ax = b$*, where*

$$A = \begin{bmatrix} 1 & 0 \\ 1 & 1 \\ 1 & 0 \end{bmatrix}, \ b = \begin{bmatrix} 1.00 \\ 0.00 \\ -1.05 \end{bmatrix}.$$

Since the vector $(1, 0, -1)$ *is orthogonal to* $\mathrm{col}(A)$*, it follows that* b *is nearly orthogonal to* $\mathrm{col}(A)$*. The reader can verify that the least-squares solution to* $Ax = b$ *is*

$$x = \begin{bmatrix} -0.025 \\ 0.025 \end{bmatrix},$$

and that θ *is approximately* $88.6°$*.*

We now consider the perturbed least-squares problem $Ax = \hat{b}$*, where*

$$\hat{b} = \begin{bmatrix} 1.01 \\ 0.005 \\ -1.04 \end{bmatrix}.$$

We have $\|\hat{b} - b\|_2 = 0.015$*, so the relative error in* $\hat{b}$ *is only about* 1%*. However,*

$$\hat{x} = \begin{bmatrix} -0.015 \\ 0.02 \end{bmatrix},$$

and the relative error in $\hat{x}$ *is more than* 31%*.*

We now turn our attention to the least-squares problem in which the matrix A is not known exactly. We continue to write x for the least-squares solution to $Ax = b$, and we now write $\hat{x}$ for the least-squares solution to $\hat{A}x = b$. We will assume that $\mathrm{rank}(\hat{A}) = \mathrm{rank}(A) = n$; this holds for all A sufficiently close to A. It turns out that deriving a useful bound on the relative error in $\hat{x}$ is much more difficult in this case than in the previous cases we have considered. For this reason, we will introduce a different type of analysis, based on first-order perturbation theory.[14] To perform a first-order bound, we write $\hat{A} = A + E$, where $\|E\|/\|A\|$ is assumed to be small and hence $(\|E\|/\|A\|)^2$ is even smaller. We will be satisfied with a bound of the form

$$\frac{\|\hat{x} - x\|}{\|x\|} \le (\cdots) \frac{\|E\|}{\|A\|} + O\left(\frac{\|E\|^2}{\|A\|^2}\right).$$

The factor in parentheses then describes, to the first order, the conditioning of the problem.

We need one preliminary result, which is an extension of Theorem 405.

Lemma 409 *Let* $\|\cdot\|$ *be any matrix norm, and let* $E \in \mathbf{R}^{n \times n}$ *satisfy* $\|E\| < 1$*. Then* $I + E$ *is invertible and*

$$(I + E)^{-1} = I - E + O(\|E\|^2).$$

[14]For an analysis in the spirit of our earlier results, the reader can consult Chapter 20 of [19].

Proof Exercise 2.

Writing $\hat{A} = A + E$, we have

$$\begin{aligned}\hat{A}^T\hat{A} &= (A+E)^T(A+E)\\ &= A^TA + A^TE + E^TA + E^TE\\ &= A^TA(I + (A^TA)^{-1}A^TE + (A^TA)^{-1}E^TA + (A^TA)^{-1}E^TE).\end{aligned}$$

It follows that (assuming E is sufficiently small that $\hat{A}$ has full rank)

$$\begin{aligned}&(\hat{A}^T\hat{A})^{-1}\\ =\ &(I + (A^TA)^{-1}A^TE + (A^TA)^{-1}E^TA + (A^TA)^{-1}E^TE)^{-1}(A^TA)^{-1}\\ =\ &(I - (A^TA)^{-1}A^TE - (A^TA)^{-1}E^TA + O(\|E\|^2))(A^TA)^{-1}\\ =\ &(I - (A^TA)^{-1}A^TE - (A^TA)^{-1}E^TA)(A^TA)^{-1} + O(\|E\|^2).\end{aligned}$$

The reader should notice how we collected the E^TE term with the $O(\|E\|^2)$ term that comes from applying Lemma 409. From now on, we will absorb all terms involving two factors of E into the $O(\|E\|^2)$ error term without comment.

We will now compute $\hat{x}$ to first order:

$$\begin{aligned}\hat{x} &= (\hat{A}^T\hat{A})^{-1}\hat{A}^Tb\\ &= (I - (A^TA)^{-1}A^TE - (A^TA)^{-1}E^TA)(A^TA)^{-1}(A+E)^Tb + O(\|E\|^2)\\ &= (I - (A^TA)^{-1}A^TE - (A^TA)^{-1}E^TA)(x + (A^TA)^{-1}E^Tb) + O(\|E\|^2)\\ &= x + (A^TA)^{-1}E^Tb - (A^TA)^{-1}A^TEx - (A^TA)^{-1}E^TAx + O(\|E\|^2)\\ &= x + (A^TA)^{-1}E^T(b - Ax) - (A^TA)^{-1}A^TEx + O(\|E\|^2).\end{aligned}$$

We therefore obtain

$$\|\hat{x} - x\|_2 \le \|(A^TA)^{-1}E^T(b-Ax)\|_2 + \|(A^TA)^{-1}A^TEx\|_2 + O(\|E\|_2^2)$$

or

$$\frac{\|\hat{x} - x\|_2}{\|x\|_2} \le \frac{\|(A^TA)^{-1}E^T(b-Ax)\|_2}{\|x\|_2} + \frac{\|(A^TA)^{-1}A^TEx\|_2}{\|x\|_2} + O(\|E\|_2^2). \tag{9.31}$$

We now bound each term on the right of (9.31). The second term is simplest:

$$\begin{aligned}\frac{\|(A^TA)^{-1}A^TEx\|_2}{\|x\|_2} &\le \frac{\|(A^TA)^{-1}A^T\|_2\|E\|_2\|x\|_2}{\|x\|_2}\\ &= \|(A^TA)^{-1}A^T\|_2\|E\|_2\\ &= \|(A^TA)^{-1}A^T\|_2\|A\|_2\frac{\|E\|_2}{\|A\|_2}\\ &= \frac{\sigma_1}{\sigma_n}\frac{\|E\|_2}{\|A\|_2}.\end{aligned}$$

For the first term, we have

$$\begin{aligned}\frac{\|(A^TA)^{-1}E^T(b-Ax)\|_2}{\|x\|_2} &\le \frac{\|(A^TA)^{-1}\|_2\|E\|_2\|b-Ax\|_2}{\|x\|_2}\\ &= \frac{\|(A^TA)^{-1}\|_2\|A\|_2^2\|b-Ax\|_2}{\|A\|_2\|x\|_2}\frac{\|E\|_2}{\|A\|_2}\\ &\le \frac{\|(A^TA)^{-1}\|_2\|A\|_2^2\|b-Ax\|_2}{\|Ax\|_2}\frac{\|E\|_2}{\|A\|_2},\end{aligned}$$

the last step following from the fact that $\|Ax\|_2 \le \|A\|_2\|x\|_2$. We now notice that

$$\begin{aligned}&\|(A^TA)^{-1}\|_2 = \sigma_n^{-2},\ \|A\|_2^2 = \sigma_1^2,\\ &\|b-Ax\|_2 = \|b\|_2\sin(\theta),\ \|Ax\|_2 = \|b\|_2\cos(\theta).\end{aligned}$$

Substituting these into the last inequality, we obtain

$$\frac{\|(A^TA)^{-1}E^T(b-Ax)\|_2}{\|x\|_2} \le \frac{\sigma_1^2\sin(\theta)}{\sigma_n^2\cos(\theta)}\frac{\|E\|_2}{\|A\|_2} = \left(\frac{\sigma_1}{\sigma_n}\right)^2\tan(\theta)\frac{\|E\|_2}{\|A\|_2}.$$

Substituting these results into (9.31), we obtain

$$\frac{\|\hat{x}-x\|_2}{\|x\|_2} \le \frac{\sigma_1}{\sigma_n}\frac{\|E\|_2}{\|A\|_2} + \left(\frac{\sigma_1}{\sigma_n}\right)^2\tan(\theta)\frac{\|E\|_2}{\|A\|_2} + O\left(\left(\frac{\|E\|_2}{\|A\|_2}\right)^2\right). \tag{9.32}$$

(Since we are regarding A as constant and examining the effect of varying E, we can replace $O(\|E\|^2)$ by $O(\|E\|^2/\|A\|^2)$. The effect of this is simply to change the constant implicit in the big-oh notation.)

The inequality (9.32) shows that the effect of perturbing A in the least-squares problem $Ax = b$ depends on θ, which describes the size of the residual in the unperturbed problem (that is, it describes how close b is to $\text{col}(A)$). One extreme case is $\theta = 0$, which means that $b \in \text{col}(A)$; in this case, we recover the bound (9.28) that we derived above. The other extreme is that b is close to orthogonal to $\text{col}(A)$, in which case θ is almost $\pi/2$, $\tan(\theta)$ is nearly infinite, and the second term in (9.32) dominates. In this case, the conditioning of the problem is proportional to

$$\left(\frac{\sigma_1}{\sigma_n}\right)^2 = (\text{cond}_2(A))^2.$$

We describe the least-squares problem $Ax = b$ as a *small-residual* problem if b is close to $\text{col}(A)$, while it is a *large-residual* problem if b is nearly orthogonal to $\text{col}(A)$. When $\text{cond}_2(A)$ is large, $(\text{cond}_2(A))^2 \gg \text{cond}_2(A)$, so the conditioning of a large-residual least-squares problem can be much worse than the conditioning of a corresponding small-residual problem. The bound (9.30) shows that the same is true when b rather than A is to be perturbed, but the effect is pronounced when A is perturbed due to the squaring of the condition number. As we will see in the next section, this has important implications for the numerical solution of least-squares problems.

Exercises

1. Let $A \in \mathbf{R}^{m\times n}$ have rank n, and suppose the singular value decomposition of A is $A = U\Sigma V^T$. Find the SVD of $(A^TA)^{-1}A^T$ and use it to prove that
$$\|(A^TA)^{-1}A^T\|_2 = \frac{1}{\sigma_n},$$
where σ_n is the smallest singular value of A.

2. Prove Lemma 409. (Hint: By Theorem 405, we know that $I+E$ is invertible. Use the fact that
$$(I+E)(I-E) = I - E^2 = (I+E)(I+E)^{-1} - E^2,$$
and solve for $(I+E)^{-1}$.)

3. Let $A \in \mathbf{R}^{m\times n}$, $b \in \mathbf{R}^m$ be given, and let x be a least-squares solution of $Ax = b$. Prove that
$$\|Ax\|_2 = \|b\|_2 \cos(\theta), \ \|Ax - b\|_2 = \|b\|_2 \sin(\theta),$$
where θ is the angle between Ax and b.

4. Let $A \in \mathbf{R}^{m\times n}$, $b \in \mathbf{R}^m$ be given, assume $m < n$, rank$(A) = m$, and consider the problem of finding the minimum-norm solution x to $Ax = b$ when b is considered the data. Perform a perturbation analysis on this problem. In other words, suppose $\hat{b}$ is a noisy value of b, and $\hat{x}$ is the minimum-norm solution of $Ax = \hat{b}$. Bound the relative error in $\hat{x}$ in terms of the relative error in $\hat{b}$.

5. Let $A \in \mathbf{R}^{m\times n}$, $b \in \mathbf{R}^m$ be given, assume $m > n$, rank$(A) = r < n$, and consider the problem of finding the minimum-norm solution least-squares solution x to $Ax = b$ when b is considered the data. Perform a perturbation analysis on this problem. In other words, suppose $\hat{b}$ is a noisy value of b, and $\hat{x}$ is the minimum-norm solution of $Ax = \hat{b}$. Bound the relative error in $\hat{x}$ in terms of the relative error in $\hat{b}$.

9.8 The QR factorization

We now turn our attention to the numerical solution of the full-rank least-squares problem. We assume that $A \in \mathbf{R}^{m\times n}$ has rank n and that $b \in \mathbf{R}^m$ is given. We know that we can find the unique least-squares solution of $Ax = b$ by solving the normal equations $A^TAx = A^Tb$, and this appears to be an attractive option. The matrix A^TA is symmetric and positive definite, so

$A^TAx = A^Tb$ can be solved in a numerically stable fashion using the Cholesky factorization.

However, the numerical stability of the Cholesky factorization is not enough to guarantee that the normal equations approach to the least-squares problem is numerically stable. If b is close to $\text{col}(A)$, then the condition number of the least-squares problem is $\text{cond}_2(A) = \sigma_1/\sigma_n$, where σ_1 and σ_n are the largest and smallest singular values of A, respectively. On the other hand, the condition number of the matrix A^TA is $\text{cond}_2(A)^2$, and it is this condition number that is relevant to the numerical solution of $A^TAx = A^Tb$. Therefore, in the case of a small-residual least-squares problem, the normal equations approach squares the condition number of the problem.

If the matrix A is well-conditioned, so that $\text{cond}_2(A)^2$ is not significantly larger than $\text{cond}_2(A)$, then the issue described in the preceding paragraph is not important, and the normal equations method is acceptable. Moreover, if b is far from $\text{col}(A)$, then the condition number of the least-squares problem is controlled by $\text{cond}_2(A)^2$ anyway, so the normal equations approach does not result in any numerical instability. However, in the case of even moderately ill-conditioned, small-residual least-squares problem, the normal equations are to be avoided.

A least-squares problem can also be solved using the singular value decomposition, and this may be the method of choice if the matrix A fails to have full rank. However, computing the SVD of A is expensive, and it is not necessary to know the SVD to solve the full-rank least-squares problem accurately.

In this section, we present a method for solving full-rank least-squares problems based on the QR factorization of a matrix. This factorization has many applications, including the stable calculation of the least-squares solution of an overdetermined system $Ax = b$ (explored in this section), and the calculation of eigenvalues (see Section 9.9). In this section, we will focus our attention on a matrix $A \in \mathbf{R}^{m\times n}$, $m \geq n$, of full rank.

We learned in Section 6.5 that a least-squares problem can be solved by finding an orthonormal basis for the subspace from which the solution is sought. In the case of the overdetermined system $Ax = b$, the subspace is $\text{col}(A)$, and we have a basis $\{A_1, A_2, \ldots, A_n\}$ (where A_j is the jth column of A). We therefore apply the Gram-Schmidt process to obtain an orthonormal basis $\{Q_1, Q_2, \ldots, Q_n\}$ for $\text{col}(A)$:

$$q_1 = A_1,\ Q_1 = \frac{q_1}{\|q_1\|_2}, \tag{9.33a}$$

$$q_2 = A_2 - (A_2 \cdot Q_1)Q_1,\ Q_2 = \frac{q_2}{\|q_2\|_2}, \tag{9.33b}$$

$$\vdots \qquad \vdots$$

$$q_n = A_n - \sum_{i=1}^{n-1} (A_n \cdot Q_i)Q_i,\ Q_n = \frac{q_n}{\|q_n\|_2}. \tag{9.33c}$$

We can solve the above equations for $A_1, A_2, \ldots, A_n$ in terms of $Q_1, Q_2, \ldots, Q_n$; the result is a collection of scalars $r_{11}, r_{12}, r_{22}, \ldots, r_{nn}$ such that

$$A_1 = r_{11}Q_1, \tag{9.34a}$$

$$A_2 = r_{12}Q_1 + r_{22}Q_2, \tag{9.34b}$$

$$\vdots \qquad \vdots$$

$$A_n = r_{1n}Q_1 + r_{2n}Q_2 + \cdots + r_{nn}Q_n \tag{9.34c}$$

(see Exercise 5). If we let the scalars r_{ij} be the nonzero entries in an $n \times n$ upper triangular matrix R_0 and define $Q_0 = [Q_1|Q_2|\cdots|Q_n]$, then we obtain

$$A = Q_0R_0.$$

This is a version of the *QR factorization* of A. The matrix Q_0 is $m \times n$ with orthonormal columns, and R_0 is a nonsingular, $n \times n$ upper triangular matrix. Often we extend Q_0 to an $m \times m$ orthogonal matrix Q by extending the orthonormal set $\{Q_1, \ldots, Q_n\}$ to an orthonormal basis $\{Q_1, \ldots, Q_n, \ldots, Q_m\}$ for $\mathbf{R}^m$ and add an $(m-n) \times n$ block of zeros to the end of the matrix R_0 to produce an $m \times n$ upper triangular matrix R. We then have

$$QR = [Q_0|\tilde{Q}_0]\left[\frac{R_0}{0}\right] = Q_0R_0 + \tilde{Q}_0 0 = Q_0R_0 = A.$$

The factorization $A = QR$ is the QR factorization, and $A = Q_0R_0$ is sometimes called the "economy-sized" QR factorization.

9.8.1 Solving the least-squares problem

Given the (economy-sized) QR factorization of $A \in \mathbf{R}^{m\times n}$, it is easy to solve the overdetermined linear system $Ax = b$ in the least-squares sense. The solution is the unique solution of $A^TAx = A^Tb$, and we have

$$A^TA = (Q_0R_0)^TQ_0R_0 = R_0^TQ_0^TQ_0R_0 = R_0^TR_0,\ A^Tb = (Q_0R_0)^Tb = R_0^TQ_0^Tb$$

(notice that $Q_0^TQ_0 = I$ because Q_0 has orthonormal columns). We then obtain

$$A^TAx = A^Tb \Rightarrow R_0^TR_0x = R_0^TQ_0^Tb \Rightarrow R_0x = Q_0^Tb,$$

the last step following from the fact that R_0^T is invertible. Given $A = Q_0R_0$, then, it is necessary only to form the matrix-vector product Q_0^Tb and solve the upper triangular system $R_0x = Q_0^Tb$ by back substitution.

9.8.2 Computing the QR factorization

We will now show how to compute the QR factorization of A efficiently and in a numerically stable fashion. We warn the reader that *computing the QR fac-*

torization from the Gram-Schmidt algorithm is numerically unstable.[15] Thus the Gram-Schmidt algorithm can be used to show the existence of the QR factorization, as we did above, but it should not be used to compute it.

In practice we use an elimination algorithm to compute R from A. This algorithm differs from Gaussian elimination in that we use orthogonal matrices (called *Householder transformations*) to induce zeros below the diagonal rather than using elementary row operations (or the corresponding elementary matrices). We will now describe the first elimination step, which consists of transforming the first column A_1 of A to αe_1, a multiple of the first standard basis vector. To make the notation a little simpler, we will write $v = A_1$. Gaussian elimination uses a sequence of algebraic transformations (adding multiples of v_1 to the other components of v) to introduce zeros below v_1 (and applies the same transformations to the other columns of A). Thus Gaussian elimination transforms v to $v_1 e_1$.

An orthogonal transformation U must satisfy $\|Uv\|_2 = \|v\|_2$, so we cannot in general transform v to $v_1 e_1$; instead, we can transform v to αe_1, where $\alpha = \pm\|v\|_2$. We will leave the question of the sign to later, and look for an orthogonal matrix U such that $Uv = \alpha e_1$. We can derive such a matrix geometrically from Figure 9.1. Let $x = \alpha e_1 - v$ and define $S = x^{\perp}$. We want

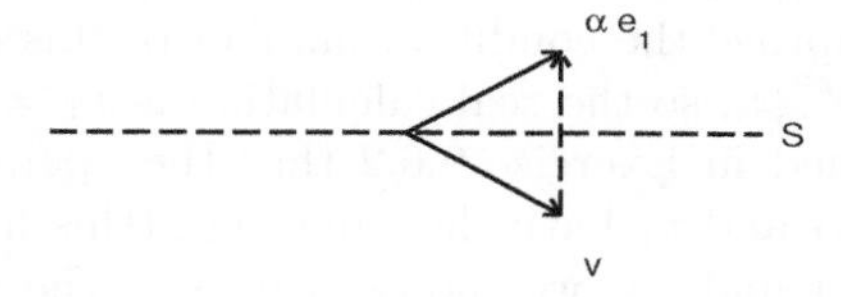

FIGURE 9.1
A given vector v in $\mathbf{R}^m$ and the vector αe_1 ($\alpha = \pm\|v\|_2$). Also shown is the subspace $S = \{x\}^{\perp}$, where $x = \alpha e_1 - v$.

an orthogonal matrix U defining a reflection across the subspace S. This means that $Ux = -x$ and $Uy = y$ for all $y \in S$. We will define $u_1 = x/\|x\|_2$ and let $\{u_2, \ldots, u_m\}$ be an orthonormal basis for S. Then $\{u_1, u_2, \ldots, u_m\}$ is an orthonormal basis for $\mathbf{R}^m$ and we want U to satisfy

$$Uu_1 = -u_1, \; Uu_j = u_j, \; j = 2, 3, \ldots, m.$$

In other words, we know the spectral decomposition of the desired matrix U.

[15]Since the only example of an unstable algorithm that we have encountered so far is Gaussian elimination with partial pivoting, the reader might have formed an erroneous impression about the significance of instability. Gaussian elimination is a rare example of an algorithm that is numerically unstable but is widely used, even by experts, because the numerical instability almost never appears. In almost every other case, numerically instability of an algorithm means that its use should be avoided. Certainly this is true for the Gram-Schmidt algorithm and the QR factorization; the columns of the Q, if computed in floating point arithmetic via the Gram-Schmidt algorithm, will usually be far from orthogonal if n is at all large.

If we define $D \in \mathbf{R}^{m\times m}$ to be the diagonal matrix whose diagonal entries are $-1, 1, \ldots, 1$ and $X = [u_1|u_2|\cdots|u_m]$, then $U = XDX^T$. Exercise 6 asks the reader to verify that

$$U = I - 2u_1 \otimes u_1 = I - 2u_1u_1^T. \tag{9.35}$$

That exercise also asks the reader to verify that U is symmetric and orthogonal.

With U defined by (9.35), $u_1 = x/\|x\|_2$, $x = \alpha e_1 - v$, we can verify that $Uv = \alpha e_1$. There exists $y \in S$ such that

$$v = y - \frac{1}{2}x,\ \alpha e_1 = y + \frac{1}{2}x \tag{9.36}$$

(see Exercise 7). Since $y \in S$, $Uy = y$, and hence

$$Uv = U\left(y - \frac{1}{2}x\right) = Uy - \frac{1}{2}Ux = y - \frac{1}{2}(-x) = y + \frac{1}{2}x = \alpha e_1,$$

as desired.

There is one final point to consider. We must choose the sign for the target vector $\alpha e_1 = \pm\|v\|_2 e_1$. Since we must compute $x = \alpha e_1 - v$, we can choose the sign to improve the condition number of this operation. We have $x_j = v_j$ for $j = 2, 3, \ldots, n$, so the real calculation is $x_1 = \alpha - v_1$. The reader should have determined in Exercise 9.6.2 that the operation $\alpha - v_1$ can be ill-conditioned when α and v_1 have the same sign (this happens when α and v_1 are approximately equal), so we choose α to have the opposite sign of v_1.

We can now show how to compute the QR factorization of a matrix using Householder transformation. We are given a matrix $A \in \mathbf{R}^{m\times n}$ with $\mathrm{rank}(A) = n$. The first step is to define $v = A_1$, the first column of A. We then choose u as described above:

$$u = \frac{x}{\|x\|_2},\ x = \alpha e_1 - v,\ \alpha = -\mathrm{sgn}(v_1)\|v\|_2.$$

Next, we define $Q_1 = I - 2uu^T$, and compute $A^{(2)} = Q_1 A$. The matrix $A^{(2)}$ then has a nonzero $(1,1)$ entry and zeros below the $(1,1)$ entry.

We now suppose that we have applied k steps of the algorithm to obtain $A^{(k+1)} = Q_k Q_{k-1} \cdots Q_1 A$, where

$$A^{(k+1)} = \left[\begin{array}{c|c} R^{(k+1)} & C^{(k+1)} \\ \hline 0 & B^{(k+1)} \end{array}\right]$$

and $R^{(k+1)} \in \mathbf{R}^{k\times k}$ is upper triangular and $B^{(k+1)} \in \mathbf{R}^{(n-k)\times(n-k)}$. We define v to be the first column of $B^{(k+1)}$ ($v \in \mathbf{R}^{n-k}$); $\|v\|_2 \neq 0$ since otherwise the first $k+1$ columns of A are linearly dependent. Next, we define

$$u = \frac{x}{\|x\|_2},\ x = \alpha e_1 - v,\ \alpha = -\mathrm{sgn}(v_1)\|v\|_2$$

(exactly as on the first step, except that now $v, e_1 \in \mathbf{R}^{n-k}$), and

$$\tilde{Q}_{k+1} = I - 2uu^T \in \mathbf{R}^{(n-k)\times(n-k)}.$$

Finally, we define

$$Q_{k+1} = \left[\begin{array}{c|c} I & 0 \\ \hline 0 & \tilde{Q}_{k+1} \end{array}\right].$$

Then

$$\begin{aligned} A^{(k+2)} = Q_{k+1}A^{(k+1)} &= \left[\begin{array}{c|c} I & 0 \\ \hline 0 & \tilde{Q}_{k+1} \end{array}\right] \left[\begin{array}{c|c} R^{(k+1)} & C^{(k+1)} \\ \hline 0 & B^{(k+1)} \end{array}\right] \\ &= \left[\begin{array}{c|c} R^{(k+1)} & C^{(k+1)} \\ \hline 0 & \tilde{Q}_{k+1}B^{(k+1)} \end{array}\right]. \end{aligned}$$

Since $\tilde{Q}_{k+1}$ zeros out the first column of $B^{(k+1)}$ (below its $(1,1)$ entry), we see that $A^{(k+2)}$ has an upper triangular block of size $(k+1)\times(k+1)$. It is easy to verify that Q_{k+1} is orthogonal.

After n steps, we obtain

$$A^{(n+1)} = Q_nQ_{n-1}\cdots Q_1A = \left[\begin{array}{c} R_0 \\ \hline 0 \end{array}\right],$$

where $R_0 \in \mathbf{R}^{n\times n}$ is upper triangular with nonzero diagonal entries. We define

$$Q = (Q_nQ_{n-1}\cdots Q_1)^T = Q_1^TQ_2^T\cdots Q_n^T = Q_1Q_2\cdots Q_n,\ R = \left[\begin{array}{c} R_0 \\ \hline 0 \end{array}\right],$$

and $Q_0 \in \mathbf{R}^{m\times n}$ is defined by the first n columns of Q. Then $A = QR$ is the QR factorization of A and $A = Q_0R_0$ is the economy-sized QR factorization.

Before we show an example, we point out that there is no need to explicitly form the Householder matrices; indeed, it is inefficient to compute $I - 2uu^T$ and then multiply it by another matrix or vector. Instead. if we just record the vector u, we can compute any matrix-vector product efficiently as

$$(I - 2uu^T)y = y - 2(u\cdot y)u.$$

A matrix-matrix product is then computed one column at a time (that is, one matrix-vector product at a time).

Example 410 *Let A be the following 4×3 matrix:*

$$A = \begin{bmatrix} 1 & 1 & 1 \\ 0 & 1 & -1 \\ 1 & 2 & 1 \\ 0 & -1 & 1 \end{bmatrix}.$$

For step 1, we define

$$v = \begin{bmatrix} 1 \\ 0 \\ 1 \\ 0 \end{bmatrix}.$$

Then $\|v\|_2 = \sqrt{2}$ *and*

$$\begin{aligned} x &= -\|v\|_2 e_1 - v = \begin{bmatrix} -\sqrt{2} \\ 0 \\ 0 \\ 0 \end{bmatrix} - \begin{bmatrix} 1 \\ 0 \\ 1 \\ 0 \end{bmatrix} = \begin{bmatrix} -2.4142 \\ 0.0000 \\ -1.0000 \\ 0.0000 \end{bmatrix}, \\ u &= \frac{x}{\|x\|_2} = \begin{bmatrix} -0.92388 \\ 0.0000 \\ -0.38268 \\ 0.0000 \end{bmatrix} \end{aligned}$$

(with all entries rounded to five digits). We can now compute $A^{(2)} = Q_1 A$ *(recall that we do not actually form* $Q_1 = I - 2uu^T$*):*

$$A^{(2)} = \begin{bmatrix} -1.4142 & -2.1213 & -1.4142 \\ 0.0000 & 1.0000 & -1.0000 \\ 0.0000 & 0.70711 & 0.0000 \\ 0.0000 & -1.0000 & 1.0000 \end{bmatrix}.$$

For step 2, we have

$$v = \begin{bmatrix} 1.0000 \\ 0.70711 \\ -1.0000 \end{bmatrix},$$

$\|v\|_2 = 1.5811$*, and*

$$\begin{aligned} x &= -\mathrm{sgn}(v_1)\|v\|_2 e_1 - v = \begin{bmatrix} -2.5811 \\ -0.70711 \\ 1.0000 \end{bmatrix}, \\ u &= \frac{x}{\|x\|_2} = \begin{bmatrix} -0.90345 \\ -0.24750 \\ 0.35002 \end{bmatrix}. \end{aligned}$$

We can then compute $A^{(3)}$ *by applying* $\tilde{Q}_2 = I - 2uu^T$ *to the lower right-hand* 3×3 *submatrix of* $A^{(2)}$*:*

$$A^{(3)} = \begin{bmatrix} -1.4142 & -2.1213 & -1.4142 \\ 0.0000 & -1.5811 & 1.2649 \\ 0.0000 & 0.0000 & 0.62048 \\ 0.0000 & 0.0000 & 0.12251 \end{bmatrix}$$

(actually, the $(3,2)$ entry of $A^{(3)}$ turned out to be $2.7756 \cdot 10^{-17}$ when computed in floating point arithmetic).

Step 3 is the final step. Here is the calculation of the vector u that defines the Householder transformation:

$$v = \begin{bmatrix} 0.62048 \\ 0.12251 \end{bmatrix},$$

$\|v\|_2 = 0.632456$,

$$\begin{aligned} x &= -\text{sgn}(v_1)\|v\|_2 e_1 - v = \begin{bmatrix} -1.2529 \\ -0.12251 \end{bmatrix}, \\ u &= \begin{bmatrix} -0.99525 \\ -0.97318 \end{bmatrix}. \end{aligned}$$

We now compute $A^{(4)}$ by applying $\tilde{Q}_3 = I - 2uu^T$ to the lower right-hand 2×2 submatrix of $A^{(3)}$:

$$A^{(4)} = \begin{bmatrix} -1.4142 & -2.1213 & -1.4142 \\ 0.0000 & -1.5811 & 1.2649 \\ 0.0000 & 0.0000 & -0.63246 \\ 0.0000 & 0.0000 & 0.0000 \end{bmatrix}.$$

We then take $R = A^{(4)}$ and

$$Q = Q_1 Q_2 Q_3 = \begin{bmatrix} -0.70711 & 0.31623 & 0.63246 & -4.1633 \cdot 10^{-17} \\ 0.0000 & -0.63246 & 0.31623 & 0.70711 \\ -0.70711 & -0.31623 & -0.63246 & 4.1633 \cdot 10^{-17} \\ 0.0000 & 0.63246 & -0.31623 & 0.70711 \end{bmatrix}.$$

9.8.3 Backward stability of the Householder QR algorithm

The result of computing the QR factorization by the algorithm described above is quite satisfactory.

Theorem 411 *Let $A \in \mathbf{R}^{m\times n}$, $b \in \mathbf{R}^m$ with $m \geq n$ and $\text{rank}(A) = n$, and suppose the QR factorization of A is computed by the Householder algorithm. If $\hat{Q}\hat{R}$ is the computed QR factorization of A, then there exists an $m \times n$ matrix ΔA such that*

$$(A + \Delta A)x = \hat{Q}\hat{R}.$$

Moreover, ΔA satisfies

$$\|\Delta A\|_F \leq \frac{cmn^{1.5}\mathbf{u}}{1 - cmn\mathbf{u}}\|A\|_F, \tag{9.37}$$

where c represents a small integer constant.

Proof See [19], Theorem 19.4 and the subsequent comments.

We can also give a backward error analysis of the solution of the least-squares problem via the QR algorithm.

Theorem 412 *Let $A \in \mathbf{R}^{m\times n}$, $b \in \mathbf{R}^m$ with $m \geq n$ and* $\text{rank}(A) = n$, *and suppose the system $Ax = b$ is solved in the least-squares sense using the Householder QR factorization of A. Then there exists a matrix $\Delta A \in \mathbf{R}^{m\times n}$ such that the computed solution $\hat{x}$ is the exact least-squares solution to*

$$(A + \Delta A)x = b.$$

Moreover, ΔA satisfies

$$\|\Delta A\|_F \leq \frac{cmn\mathbf{u}}{1 - cmn\mathbf{u}}\|A\|_F, \tag{9.38}$$

where c represents a small integer constant.

Proof See [19], Theorem 20.3 and the subsequent comments.

The reader will notice that Frobenius norm is used to express these results, which show that it is possible to solve the least-squares problem and produce a computed solution that is the exact solution of a slightly perturbed least-squares problem. Thus the Householder QR algorithm for the least-squares problem is backwards stable.

9.8.4 Solving a linear system

If the matrix A is square ($m = n$) and has full rank, then it is invertible and the system $Ax = b$ has a unique solution $x = A^{-1}b$. We can still compute the QR factorization $A = QR$ of A; in this case, both Q and R are invertible and we obtain $A^{-1} = R^{-1}Q^{-1} = R^{-1}Q^T$, $x = R^{-1}Q^Tb$ (of course, in practice we would compute x by solving $Rx = Q^Tb$). If we use this approach to solve $Ax = b$, producing a computed solution $\hat{x}$, then the results of Theorem 412 hold. The significance of this observation is that the bound (9.38) does not contain a growth factor (indeed, using orthogonal transformations, no growth in the entries is possible). *Thus the Householder QR algorithm is a numerically stable method for solving a square, nonsingular system $Ax = b$.* This is to contrasted with Gaussian elimination with partial pivoting, which is numerically stable only if no growth in the entries of the matrix occurs.

In the exercises, the reader is asked to count the number of operations required for the QR factorization and for solving $Ax = b$. The result is that the Householder QR algorithm for solving and $n \times n$ system requires

$$O\left(\frac{4}{3}n^3\right) \tag{9.39}$$

arithmetic operations. This is twice as many operations as required by Gaussian elimination with partial pivoting. We have a choice, then: Use the more efficient algorithm, Gaussian elimination with partial pivoting, which is almost always numerically stable, or the more expensive algorithm, the Householder QR method, which is always numerically stable. In practice, Gaussian elimination is the method of choice, because problems on which it is unstable are so exceedingly rare.

Exercises

1. Find a Householder transformation $Q = I - 2uu^T$ satisfying $Qx = y$, where $x = (1, 2, 1)$ and $y = (2, 1, 1)$.

2. Explain why there is no Householder transformation $Q = I - 2uu^T$ satisfying $Qx = y$, where $x = (1, 2, 1)$ and $y = (1, 0, 1)$.

3. Let
$$A = \begin{bmatrix} 2 & 3 & -2 \\ 3 & 1 & 0 \\ -3 & -3 & 3 \end{bmatrix}.$$
Using the Householder QR algorithm, find the QR factorization of A.

4. Let
$$A = \begin{bmatrix} 1 & -2 & -3 \\ 2 & 1 & 2 \\ 2 & -3 & 1 \\ -1 & -2 & -1 \\ 1 & -3 & 3 \end{bmatrix}.$$
Using the Householder QR algorithm, find the QR factorization of A.

5. Find the values of $r_{11}, r_{12}, r_{22}, \ldots, r_{nn}$ so that (9.34) follows from (9.33).

6. Let $\{u_1, u_2, \ldots, u_m\}$ be an orthonormal basis for $\mathbf{R}^m$, let
$$X = [u_1|u_2|\cdots|u_m],$$
and let $D \in \mathbf{R}^{m\times m}$ be the diagonal matrix whose diagonal entries are $-1, 1, \ldots, 1$. Finally, define $U = XDX^T$.
   (a) Prove that $U = -u_1u_1^T + u_2u_2^T + \ldots + u_mu_m^T$.
   (b) Show that $u_1u_1^T + u_2u_2^T + \ldots + u_mu_m^T = I$, where I is the $m \times m$ identity matrix.
   (c) Show that the formula for U simplifies to (9.35).
   (d) Verify directly that U is symmetric and orthogonal.

7. Let $x \in \mathbf{R}^m$ be given, define $\alpha = \pm\|v\|_2$, $x = \alpha e_1 - v$, $u_1 = x/\|x\|_2$, let $\{u_1, u_2, \ldots, u_m\}$ be an orthonormal basis for $\mathbf{R}^m$, and let
$$S = \{u_1\}^\perp = \mathrm{sp}\{u_2, \ldots, u_m\}.$$
Prove that there exists $y \in S$ such that (9.36) holds.

8. Let $A \in \mathbf{R}^{m\times n}$ have full rank.

   (a) Count the number of operations required to reduce A to an upper triangular matrix $R \in \mathbf{R}^{m\times n}$ using Householder transformations. Remember that the Householder transformations Q_i should not be formed explicitly.

   (b) Count the number of operations required to solve the least-squares problem using the Householder QR method.

   (c) Specialize your results to the case of an $n \times n$ matrix A, and verify (9.39).

   (d) Count the number of operations required to form $Q = Q_1 Q_2 \cdots Q_n$ if it is required. Notice that the product can be formed two ways (left to right or right to left). Be sure to choose the efficient order.

9.9 Eigenvalues and simultaneous iteration

The purpose of this section and the next is to outline an efficient algorithm for accurately estimating the eigenvalues of a square matrix. This algorithm uses the QR factorization and is therefore called the *QR algorithm*. As pointed out in Exercise 4.5.9, there is no finite algorithm for computing all the eigenvalues of a general matrix, so the QR algorithm is necessarily an iterative algorithm that produces estimates converging to the true eigenvalues. The proof of convergence is beyond the scope of this book; we will be content with explaining how the algorithm works and why, intuitively, it should converge.

Before introducing the QR algorithm, we need to present two preliminary topics: a similarity transformation that produces a triangular matrix similar to the given matrix A, and the *power method* for estimating the largest eigenvalue of A.

9.9.1 Reduction to triangular form

As we have seen in earlier chapters, not every matrix is diagonalizable (similar to the diagonal matrix). However, every matrix is similar to a triangular matrix; moreover, the similarity transformation can be chosen to be unitary.

Theorem 413 *If A belongs to $\mathbf{C}^{n\times n}$, then there exists a unitary matrix Q such that $T = Q^*AQ$ is upper triangular.*

Proof We will argue by induction on n. There is nothing to prove in the case $n = 1$ (every 1×1 matrix is upper triangular). We now assume that for every $B \in \mathbf{C}^{(n-1)\times(n-1)}$, there exists a unitary matrix $U \in \mathbf{C}^{(n-1)\times(n-1)}$ such that U^*BU is upper triangular.

Let A be any matrix in $\mathbf{C}^{n\times n}$. Since $\mathbf{C}$ is algebraically closed, there exists an eigenvector λ_1 of A with corresponding eigenvector q_1. We can assume that $\|q_1\|_2 = 1$. Extend $\{q_1\}$ to an orthonormal basis $\{q_1, q_2, \ldots, q_n\}$ for $\mathbf{C}^n$, and define $\hat{Q} = [q_1|q_2|\cdots|q_n]$. We write $\hat{Q}_2 = [q_2|q_3|\cdots|q_n]$, so that $\hat{Q} = [q_1|\hat{Q}_2]$. Then

$$\begin{aligned}
\hat{Q}^*A\hat{Q} = \left[\begin{array}{c} q_1^* \\ \hline \hat{Q}_2^* \end{array}\right] A[q_1|\hat{Q}_2] &= \left[\begin{array}{c} q_1^* \\ \hline \hat{Q}_2^* \end{array}\right] [Aq_1|A\hat{Q}_2] \\
&= \left[\begin{array}{c} q_1^* \\ \hline \hat{Q}_2^* \end{array}\right] [\lambda_1 q_1|A\hat{Q}_2] \\
&= \left[\begin{array}{cc} \lambda_1 q_1^* q_1 & q_1^* A\hat{Q}_2 \\ \hat{Q}_2^* q_1 & \hat{Q}_2^* A\hat{Q}_2 \end{array}\right] \\
&= \left[\begin{array}{cc} \lambda_1 & q_1^* A\hat{Q}_2 \\ 0 & \hat{Q}_2^* A\hat{Q}_2 \end{array}\right].
\end{aligned}$$

The last step follows from the the fact that the columns of $\hat{Q}$ are orthonormal; therefore $q_1^*q_1 = 1$ and $\hat{Q}_2^*q_1 = 0$.

We now apply the induction hypothesis to $B = \hat{Q}_2^*A\hat{Q}_2$, and assume that $U \in \mathbf{C}^{(n-1)\times(n-1)}$ is unitary and $T_2 = U^*BU$ is upper triangular. Define

$$\tilde{Q} = \left[\begin{array}{cc} 1 & 0 \\ 0 & U \end{array}\right].$$

Then $\tilde{Q}$ and $Q = \hat{Q}\tilde{Q}$ are unitary matrices in $\mathbf{C}^{n\times n}$. and

$$\begin{aligned}
Q^*AQ = \tilde{Q}^*\hat{Q}^*A\hat{Q}\tilde{Q} &= \left[\begin{array}{cc} 1 & 0 \\ 0 & U^* \end{array}\right] \left[\begin{array}{cc} \lambda_1 & q_1^*A\hat{Q}_2 \\ 0 & \hat{Q}_2^*A\hat{Q}_2 \end{array}\right] \left[\begin{array}{cc} 1 & 0 \\ 0 & U \end{array}\right] \\
&= \left[\begin{array}{c|c} \lambda_1 & q_1^*A\hat{Q}_2U \\ \hline 0 & U^*\hat{Q}_2^*A\hat{Q}U \end{array}\right] \\
&= \left[\begin{array}{c|c} \lambda_1 & q_1^*A\hat{Q}_2U \\ \hline 0 & T_2 \end{array}\right].
\end{aligned}$$

This shows that Q^*AQ is upper triangular, and completes the proof.

QED

The above result suggests an algorithm for finding all of the eigenvalues of A: Start by finding one eigenpair λ_1, q_1 of A, form the matrix $\hat{Q}$ used in the proof, and compute $\hat{Q}^*A\hat{Q}$. The problem is then reduced to finding the eigenvalues of the $(n-1)\times(n-1)$ matrix in the lower right-hand corner of $\hat{Q}^*A\hat{Q}$. The process of reducing the size of an eigenvalue problem in this fashion is called *deflation*.

To carry out an algorithm based on deflation, we must have a method for finding a single eigenpair of a matrix. The *power method*, to be discussed next, is one such method.

In fact, the QR algorithm is more effective than an algorithm based on the power method and deflation. However, as we will see, the QR method incorporates both of these ideas.

9.9.2 The power method

To explain the power method, we will make the assumption that A is diagonalizable and the eigenvalues $\lambda_1, \lambda_2, \ldots, \lambda_n$ are ordered: $|\lambda_1| \geq |\lambda_2| \geq |\lambda_3| \geq \cdots \geq |\lambda_n|$. Moreover, we assume that $|\lambda_1| > |\lambda_2|$. The assumption that A is diagonalizable is not essential, but the assumption that A has a *dominant eigenvalue* λ_1 is. We take any starting vector v and consider the sequence $v, Av, A^2v, A^3v, \ldots$. We will show that $\{A^kv\}$ converges to an eigenvector of A corresponding to λ_1 (though, as we will see, it is necessary to scale the vectors in the sequence to prevent their components from getting too large or too small).

Since A is diagonalizable by assumption, there exists a basis $\{x_1, x_2, \ldots, x_n\}$ of $\mathbf{R}^n$ consisting of eigenvectors of A: $Ax_i = \lambda_i x_i$ for $i = 1, 2, \ldots, n$. Therefore,

$$v = \alpha_1 x_1 + \alpha_2 x_2 + \ldots + \alpha_n x_n$$

for scalar $\alpha_1, \alpha_2, \ldots, \alpha_n$. It follows that

$$\begin{aligned} A^k v &= \alpha_1 A^k x_1 + \alpha_2 A^k x_2 + \ldots + \alpha_n A^k x_n \\ &= \alpha_1 \lambda_1^k x_1 + \alpha_2 \lambda_2^k x_2 + \ldots + \alpha_n \lambda_n^k x_n \\ &= \lambda_1^k \left(\alpha_1 x_1 + \alpha_2 \left(\frac{\lambda_2}{\lambda_1}\right)^k x_2 + \ldots + \alpha_n \left(\frac{\lambda_n}{\lambda_1}\right)^k x_n \right). \end{aligned}$$

Since $|\lambda_i/\lambda_1| < 1$ for all $i = 2, 3, \ldots, n$, it follows that

$$\left(\frac{\lambda_i}{\lambda_1}\right)^k \to 0 \text{ as } k \to \infty,$$

and therefore that A^kv points increasingly in the direction of v_1 as k increases.

One problem with the above calculation is that, except in the special case that $|\lambda_1| = 1$, $\|A^kv\|$ is converging to zero (if $|\lambda_1| < 1$) or diverging to ∞ (if $|\lambda_1| > 1$). Therefore, although A^kv is converging to the direction of an eigenvector, the calculation is problematic because of very small or very large numbers.[16] For this reason, the power method is implemented in the following manner: Choose a starting vector v and set $v_0 = v$. Given the current estimate v_i of x_1, compute $v_{i+1} = Av_i/\|Av_i\|$.[17]

---

[16]In an actual computation, overflow or underflow could occur. *Overflow* occurs when floating point numbers exceed the range of numbers that can be represented on the computer; *underflow* occurs when numbers become smaller than the smallest representable nonzero numbers. When overflow occurs, the result is an **Inf** (the special floating point "number" representing ∞); when underflow occurs, the result is 0.

[17]Actually, the power method can be performed more efficiently by avoiding the calcu-

Having introduced scaling into the power method makes the convergence less transparent, but the angle between v_i and x_1 is the same as the angle between $A^k v_0$ and x_1, and it suffices to show that this angle converges to zero. The reader is asked to do this in Exercise 3.

The reader may have noticed that the power method fails completely if the v_0 happens to be chosen so that $\alpha_1 = 0$. The possibility of such an occurrence is usually ignored, for two reasons. First of all, if v_0 is chosen at random, then the probability is zero that $\alpha_1 = 0$ (this follows from the fact that $\mathrm{sp}\{x_2, \ldots, x_n\}$ is an $(n-1)$-dimensional subspace of an n-dimensional space and has measure zero in that space). Second, even if $\alpha_1 = 0$ exactly, round-off error will tend to introduce a nonzero x_1 component into $A^k v_0$ as k increases.

The rate of convergence of the power method depends on the ratio $|\lambda_2|/|\lambda_1|$. If this is small (that is, if the dominant eigenvalue is much greater in magnitude the the next largest eigenvalue), then the power method converges rapidly. If the ratio is close to 1, then convergence is quite slow. The power method will fail if $|\lambda_2| = |\lambda_1|$.

9.9.3 Simultaneous iteration

As mentioned above, the power method together with the proof of Theorem 413 suggest an algorithm for computing all of the eigenvalues of an $n \times n$ matrix: Use the power method to find one eigenvalue, use the idea in the proof of Theorem 413 to reduce the problem to finding the eigenvalues of an $(n-1) \times (n-1)$ matrix, and repeat until all the eigenvalues are known (approximately). In fact, this is not a good approach. First of all, the power method may not work at all (if two eigenvalues have the same magnitude) or may converge very slowly (if the two eigenvalues are close in magnitude). Second, one has to apply the deflation procedure with an approximate eigenvalue and the deflation procedure can amplify the error.

On the other hand, there is a power-method like algorithm for computing more than one eigenvalue at a time. We continue to assume that A has eigenvalue $\lambda_1, \lambda_2, \ldots, \lambda_n$ and corresponding eigenvectors $x_1, x_2, \ldots, x_n$, with $|\lambda_1| \geq |\lambda_2| \geq \cdots \geq |\lambda_n|$ (note that λ_1 is no longer assumed to be dominant), and now suppose that $|\lambda_p| > |\lambda_{p+1}|$. The subspace $\mathrm{sp}\{x_1, \ldots, x_p\}$ is invariant under A. Now suppose we begin with orthonormal vectors $v_1, \ldots, v_p$, form the matrix $Q^{(0)} = [v_1|\cdots|v_p]$, and apply A to $Q^{(0)}$ to get $V^{(1)} = AQ^{(0)}$. Reasoning as in our discussion of the power method, the components each column of $V^{(0)}$ lying in the direction of $x_1, \ldots, x_p$ should be magnified compared to those components lying in the direction of $x_{p+1}, \ldots, x_n$. We cannot just compute

lation of $\|Av_i\|$. For instance, one can just divide Av_i by its first component to produce v_{i+1}. In this way, the first component of v_i is always 1. This modification of the power method is not important for our discussion, since we are not interested in the method for its own sake. Rather, understanding the power method aids in making sense of the more complicated QR algorithm.

$A^kQ^{(0)}$ for $k = 1, 2, 3, \ldots$, because then (in addition to the question of overflow or underflow raised above) all the columns would tend toward the dominant eigenvalue (if there is one). However, we can orthogonalize the basis formed by the columns of $V^{(1)}$ by using the QR factorization: $Q^{(1)}R^{(1)} = V^{(1)}$.

Since the QR factorization is just the matrix form of the Gram-Schmidt procedure,[18] it follows that the first column of $Q^{(1)}$ is the (normalized) first column of $V^{(1)}$. This will be true in successive iterations also, so the method we describe here has the effect of applying the power method to the first column of $Q^{(0)}$. However, by computing the QR factorization of $AQ^{(0)}$, we subtract off the contribution of the first column to the other columns, allowing the other components to not be overwhelmed by the dominant eigenvalue (if there is one).

Given $V^{(k)} = Q^{(k)}R^{(k)}$, we continue by computing $V^{(k+1)} = AQ^{(k)}$ and orthogonalizing the columns of $V^{(k+1)}$ by computing the QR factorization: $Q^{(k+1)}R^{(k+1)} = V^{(k+1)}$. For large values of k, the columns of $Q^{(k)}$ should lie nearly in $\text{sp}\{x_1, \ldots, x_p\}$. The fact that the columns of $Q^{(k)}$ span a subspace that is nearly invariant under A implies that there exists a $p \times p$ matrix T_p such that $AQ^{(k)} \doteq Q^{(k)}T_p$.

A problem with the procedure outlined above is that if we do not know the eigenvalues, we probably do not know a value of p such that $|\lambda_p| > |\lambda_{p+1}|$. To get around this problem, we just take $Q^{(0)}$ to be a (square) unitary matrix and apply the iteration described above. The matrices $V^{(k)}$ and $Q^{(k)}$ can be partitioned as $V^{(k)} = [V_1^{(k)}|V_2^{(k)}]$ and $Q^{(k)} = [Q_1^{(k)}|Q_2^{(k)}]$, where $V_1^{(k)}$ and $Q_1^{(k)}$ have p columns, and $R^{(k)}$ is partitioned in a compatible fashion as

$$R^{(k)} = \left[\begin{array}{c|c} R_{11}^{(k)} & R_{12}^{(k)} \\ \hline 0 & R_{22}^{(k)} \end{array}\right],$$

where $R_{11}^{(k)}$ and $R_{22}^{(k)}$ are upper triangular. Then

$$\begin{aligned} V^{(k)} = AQ^{(k)} \quad &\Rightarrow \quad V_1^{(k)} = AQ_1^{(k)}, \\ Q^{(k)}R^{(k)} = V^{(k)} \quad &\Rightarrow \quad Q_1^{(k)}R_{11}^{(k)} = V_1^{(k)}. \end{aligned}$$

Thus, as we apply the iteration to the entire matrix $Q^{(0)}$, we are simultaneously applying it to the first p columns. For large k, we expect

$$AQ_1^{(k)} \doteq Q_1^{(k)}T_{11}, \ T_{11} \in \mathbf{C}^{p \times p}$$

(T_{11} was called T_p above), so we obtain

$$(Q^{(k)})^T AQ^{(k)} \doteq \left[\begin{array}{c|c} T_{11} & T_{12} \\ \hline 0 & T_{22} \end{array}\right].$$

---

[18] We remind the reader that the QR factorization cannot be computed by the Gram-Schmidt procedure, which is numerically unstable. See page 557.

This convergence of $(Q^{(k)})^T AQ^{(k)}$ to block triangular form does not depend on our knowing the value of p.

This algorithm ($V^{(k+1)} = AQ^{(k)}$, $Q^{(k+1)}R^{(k+1)} = V^{(k+1)}$) is called *simultaneous iteration*. In practice, we would not expect a single integer p such that $|\lambda_p| > |\lambda_{p+1}|$, but rather several or many such integers. As a result, we should obtain

$$(Q^{(k)})^T AQ^{(k)} \doteq \begin{bmatrix} T_{11} & T_{12} & \cdots & T_{1q} \\ & T_{22} & \cdots & T_{2q} \\ & & \ddots & \vdots \\ & & & T_{qq} \end{bmatrix},$$

where each T_{ii} is square and corresponds to an invariant subspace determined by eigenvalues of the same magnitude.

The ideal case for simultaneous iteration would be if the eigenvalues of A satisfied $|\lambda_1| > |\lambda_2| > \cdots > |\lambda_n|$. In that case, each block T_{ii} is 1×1 and $\{(Q^{(k)})^T AQ^{(k)}\}$ converges to an upper triangular matrix T. Notice that this reasoning requires that we use complex arithmetic when A has complex eigenvalues. If A is real, we would have to start with a matrix with random complex entries to accomplish this. If A is real with complex eigenvalues, then they occur in conjugate pairs. If every eigenvalue of $A \in \mathbf{R}^{n\times n}$ were complex, then we would have, in the best case,

$$|\lambda_1| = |\lambda_2| > |\lambda_3| = |\lambda_4| > \cdots > |\lambda_{n-1}| = |\lambda_n|.$$

In this case, starting with a real matrix, $AQ^{(k)}$ would converge to a block upper triangular matrix with 2×2 blocks on the diagonal. Each would correspond to a pair of complex conjugate eigenvalues, which could then be easily calculated.

We have not addressed convergence of simultaneous iteration for the case of distinct eigenvalues of the same magnitude, or distinct complex conjugate pairs of eigenvalues having the same magnitude. We will comment on this case below, after we describe the shifted QR algorithm. However, it should be clear, at least intuitively, why simultaneous iteration converges in the two straightforward cases:

1. $A \in \mathbf{R}^{n\times n}$ has n real eigenvalues $\lambda_1, \lambda_2, \ldots, \lambda_n$ satisfying $|\lambda_1| > |\lambda_2| > \cdots > |\lambda_n|$ (or, more generally, $A \in \mathbf{C}^{n\times n}$ has n complex eigenvalues $\lambda_1, \lambda_2, \ldots, \lambda_n$ satisfying $|\lambda_1| > |\lambda_2| > \cdots > |\lambda_n|$).

2. $A \in \mathbf{R}^{n\times n}$ has n distinct eigenvalues, and $|\lambda_k| = |\lambda_\ell|$ only if λ_k, λ_ℓ form a complex conjugate pair.

We now present simple examples of these two cases.

Example 414 *We construct a matrix $A \in \mathbf{R}^{4\times 4}$ with eigenvalues $1, 2, 3, 4$ by choosing a random matrix X and forming $A = XDX^{-1}$, where D is the* *diagonal matrix with diagonal entries* $1, 2, 3, 4$. *The result is*

$$A = \begin{bmatrix} 2.7446 & -2.9753 \cdot 10^{-1} & 1.0301 & 4.5177 \cdot 10^{-1} \\ -5.2738 & 2.1878 & -5.2337 & 1.2273 \\ 2.5710 \cdot 10^{-1} & -2.3636 \cdot 10^{-2} & 2.2620 & -3.0009 \cdot 10^{-1} \\ 4.7824 \cdot 10^{-1} & 2.0855 \cdot 10^{-1} & 2.7770 \cdot 10^{-1} & 2.8056 \end{bmatrix}$$

(rounded to five significant digits). Applying 5 iterations of simultaneous iteration yields $(Q^{(k)})^T AQ^{(k)}$ *equal to*

$$\begin{bmatrix} 4.0121 & 3.5979 \cdot 10^{-1} & 2.1780 & 6.9551 \\ -5.3389 \cdot 10^{-2} & 3.0218 & 3.4115 \cdot 10^{-1} & -1.7069 \cdot 10^{-1} \\ 1.3276 \cdot 10^{-3} & -8.1677 \cdot 10^{-2} & 1.9683 & 6.1465 \cdot 10^{-1} \\ -6.7013 \cdot 10^{-4} & -2.0367 \cdot 10^{-2} & -6.2158 \cdot 10^{-3} & 9.9773 \cdot 10^{-1} \end{bmatrix},$$

while 10 iterations yields

$$\begin{bmatrix} 4.0020 & 1.7106 \cdot 10^{-1} & 2.1629 & 6.9654 \\ -1.2820 \cdot 10^{-2} & 3.0040 & 5.0612 \cdot 10^{-1} & 8.9645 \cdot 10^{-2} \\ 2.7714 \cdot 10^{-5} & -1.1653 \cdot 10^{-2} & 1.9940 & 6.1135 \cdot 10^{-1} \\ -6.4883 \cdot 10^{-7} & -8.1032 \cdot 10^{-5} & 3.2991 \cdot 10^{-5} & 1.0000 \end{bmatrix}.$$

Clearly the diagonal entries are converging to the eigenvalues, while the entries below the diagonal are converging to zero.

Example 415 *For this example, we construct a matrix* $A \in \mathbf{R}^{4\times 4}$ *with two pairs of complex conjugate eigenvalues:* $2 \pm i$, $1 \pm i$. *Similar to the procedure in the preceding example, we choose a random matrix* $X \in \mathbf{R}^{4\times 4}$ *and compute* $A = XTX^{-1}$, *where*

$$T = \begin{bmatrix} 2 & 1 & 0 & 0 \\ -1 & 2 & 0 & 0 \\ 0 & 0 & 1 & 1 \\ 0 & 0 & -1 & 1 \end{bmatrix}$$

(the reader can verify that these 2×2 *blocks correspond to the stated eigenvalues). The resulting matrix is*

$$A = \begin{bmatrix} 1.8212 \cdot 10^{-1} & -2.2639 \cdot 10^{-1} & -1.8658 & 7.5391 \cdot 10^{-1} \\ 1.4853 & 1.6917 & 2.2027 \cdot 10^{-1} & -3.7876 \\ 1.4076 & 4.2241 \cdot 10^{-1} & 3.0387 & -2.4017 \cdot 10^{-2} \\ -9.7814 \cdot 10^{-1} & 1.7585 \cdot 10^{-2} & -7.4759 \cdot 10^{-1} & 1.0876 \end{bmatrix}.$$

Ten iterations of simultaneous iteration yields $(Q^{(k)})^T AQ^{(k)}$ *equal to*

$$\begin{bmatrix} 1.6544 & -1.2406 & -2.7214 & 2.9266 \\ 8.0813 \cdot 10^{-1} & 2.1911 & 2.4948 & 8.9331 \cdot 10^{-1} \\ 4.1741 \cdot 10^{-2} & 9.3636 \cdot 10^{-2} & 7.5971 \cdot 10^{-1} & -1.3004 \\ 4.7960 \cdot 10^{-3} & -4.8750 \cdot 10^{-2} & 8.5991 \cdot 10^{-1} & 1.3948 \end{bmatrix}.$$

The eigenvalues of the upper left-hand 2×2 *block are*

$$1.9227 \pm 9.6465 \cdot 10^{-1} i,$$

while the eigenvalues of the lower right-hand 2×2 *block are*

$$1.0773 \pm 1.0086i.$$

Twenty iterations yields the matrix

$$\begin{bmatrix} 2.4391 & -1.0498 & 1.3972 & 2.7820 \\ 1.1371 & 1.5592 & 3.5011 & 3.5860 \cdot 10^{-1} \\ 5.4666 \cdot 10^{-4} & 9.1691 \cdot 10^{-5} & 1.4361 & -1.2362 \\ 6.1982 \cdot 10^{-5} & -4.9967 \cdot 10^{-4} & 9.6154 \cdot 10^{-1} & 5.6558 \cdot 10^{-1} \end{bmatrix}.$$

Now the eigenvalues of the upper left-hand 2×2 *block are* $1.9991 \pm 1.0001i$, *while the eigenvalues of the lower right-hand block are* $1.0009 \pm 9.9961 \cdot 10^{-1} i$. *The matrix* $(Q^{(k)})^T A Q^{(k)}$ *is converging to a block upper triangular matrix with the correct eigenvalues.*

Simultaneous iteration allows the calculation of all the eigenvalues of a matrix, at least in favorable cases, but it is costly to execute (a QR factorization must be computed at each iteration) and it often converges slowly. In the next section, we present a variant of simultaneous iteration called the QR algorithm that is much more efficient and robust.

Exercises

1. Let
$$A = \begin{bmatrix} 2 & -1 & 0 \\ -1 & 3 & -1 \\ 0 & -1 & 2 \end{bmatrix}.$$
Use the power method to estimate the dominant eigenvalue of A.

2. Let A be the matrix from the previous exercise. Use simultaneous iteration to estimate all the eigenvalues of A.

3. Suppose $A \in \mathbf{C}^{n \times n}$ is diagonalizable, with eigenvalues $\lambda_1, \lambda_2, \ldots, \lambda_n$. Assume $|\lambda_1| > |\lambda_2| \geq \cdots \geq |\lambda_n|$. Let corresponding eigenvectors be $x_1, x_2, \ldots, x_n$, and assume that $v_0 \in \mathbf{C}^n$ satisfies $v_0 = \alpha_1 x_1 + \ldots + \alpha_n x_n$ with $\alpha_1 \neq 0$. Prove that the angle between $A^k v_0$ and $\text{sp}\{x_1\} = E_A(\lambda_1)$ converges to zero as $k \to \infty$.

4. Let $A \in \mathbf{R}^{n \times n}$ have complex conjugate eigenvalues $\lambda = \mu + \theta i$, $\overline{\lambda} = \mu - \theta i$, each of multiplicity one. Prove that there is a two-dimensional subspace of $\mathbf{R}^n$, corresponding to $\lambda, \overline{\lambda}$, that is invariant under A. Specifically, show that there is a basis of $E_\lambda(A) + E_{\overline{\lambda}}(A)$ that consists of two real vectors $x, y \in \mathbf{R}^n$. Prove that $S = \text{sp}\{x, y\}$ (where the span is taken in $\mathbf{R}^n$, that is, S consists of all linear combinations of x and y with real weights) is invariant under A.

5. Let $A \in \mathbf{R}^{n\times n}$. Prove that there exists an orthogonal matrix Q such that Q^TAQ is block upper triangular, with each diagonal block of size 1×1 or 2×2.

9.10 The QR algorithm

If we apply simultaneous iteration to A, we need some way to monitor the progress. Throughout this discussion, we assume that $A \in \mathbf{R}^{n\times n}$ falls into one of the two straightforward cases described above. If the columns of $Q^{(k)}$ form approximate bases for the invariant subspaces of A, then $(Q^{(K)})^TAQ^{(k)}$ should be nearly block upper triangular, with 1×1 or 2×2 diagonal blocks. This suggests that we monitor $(Q^{(k)})^TAQ^{(k)}$ as the iteration proceeds. For simplicity of notation, we will henceforth index the matrices arising in the iteration with subscripts instead of superscripts. We will also assume that the initial matrix for simultaneous iteration is $Q_0 = I$.

We write $A_{i+1} = Q_{i+1}^TAQ_{i+1}$, and recall that the orthogonal matrices Q_i are defined by

$$Q_0 = I,\ Q_{i+1}R_{i+1} = AQ_i,\ i = 0,1,2,\ldots.$$

We now focus directly on computing $A_1, A_2, A_3, \ldots$. Suppose $\hat{A}_0 = A$ and, for each $i = 0,1,2,\ldots$, $\hat{Q}_{i+1}\hat{R}_{i+1} = \hat{A}_i$ is the QR factorization of $\hat{A}_i$ and $\hat{A}_{i+1} = \hat{Q}_{i+1}^T\hat{A}_i\hat{Q}_{i+1}$. We will show that, for each i,

$$A_i = \hat{A}_i,\ R_i = \hat{R}_i,\ Q_i = \hat{Q}_1\hat{Q}_2\cdots\hat{Q}_i. \tag{9.40}$$

The proof of this assertion relies on uniqueness of the QR factorization of a given matrix. In fact, there is more than one QR factorization of A, but if A is nonsingular, then there is a unique factorization in which all the diagonal entries of R are positive. If this uniqueness does not hold (for instance, because A is singular), then (9.40) is not necessarily true, but we could perform the QR factorizations in such a way as to make (9.40) true.

We thus wish to compare the following two iterations:

$$Q_0 = I,\ A_i = Q_i^TAQ_i,\ Q_{i+1}R_{i+1} = AQ_i,\ i = 0,1,2,\ldots,$$
$$\hat{A}_0 = A, \hat{Q}_0 = I,\ \hat{Q}_{i+1}\hat{R}_{i+1} = \hat{A}_i,\ \hat{A}_{i+1} = \hat{Q}_{i+1}^T\hat{A}_i\hat{Q}_{i+1},\ i = 0,1,2,\ldots.$$

We argue by induction on i. We have $Q_1R_1 = AQ_0 = A$ and $\hat{Q}_1\hat{R}_1 = \hat{A}_0 = A$, so $\hat{Q}_0 = Q_0$ and $\hat{R}_0 = R_0$. Furthermore, $\hat{A}_1 = \hat{Q}_1^T\hat{A}_0\hat{Q}_1 = Q_1^TAQ_1 = A_1$. Thus (9.40) holds for $i = 1$.

We now assume that (9.40) holds for some $i \geq 1$. Then

$$\begin{aligned} Q_{i+1}R_{i+1} = AQ_i &\Rightarrow Q_{i+1}R_{i+1} = A\hat{Q}_1 \cdots \hat{Q}_i \\ &\Rightarrow \hat{Q}_i^T \cdots \hat{Q}_1^T Q_{i+1}R_{i+1} = \hat{Q}_i^T \cdots \hat{Q}_1^T A\hat{Q}_1 \cdots \hat{Q}_i \\ &\Rightarrow \hat{Q}_i^T \cdots \hat{Q}_1^T Q_{i+1}R_{i+1} = Q_i^T AQ_i = A_i = \hat{A}_i \\ &\Rightarrow \hat{Q}_i^T \cdots \hat{Q}_1^T Q_{i+1}R_{i+1} = \hat{Q}_{i+1}\hat{R}_{i+1} \\ &\Rightarrow Q_{i+1}R_{i+1} = \hat{Q}_1 \cdots \hat{Q}_i\hat{Q}_{i+1}\hat{R}_{i+1}. \end{aligned}$$

It follows that $Q_{i+1}R_{i+1}$ and $\hat{Q}_1 \cdots \hat{Q}_i\hat{Q}_{i+1}\hat{R}_{i+1}$ are QR factorizations of the same matrix, and hence

$$Q_{i+1} = \hat{Q}_1 \cdots \hat{Q}_i\hat{Q}_{i+1},\ R_{i+1} = \hat{R}_{i+1}$$

(modulo the comments made above about the uniqueness of the QR factorization). It is then straightforward to show that $\hat{A}_{i+1} = A_{i+1}$.

We can therefore, if we wish, compute $\hat{A}_i$, $\hat{Q}_i$, and $\hat{R}_i$ instead of A_i, Q_i, and R_i. Moreover, using the fact that $\hat{Q}_{i+1}^T\hat{A}_i = \hat{R}_{i+1}$, we obtain

$$\hat{A}_{i+1} = \hat{Q}_{i+1}^T A_i \hat{Q}_{i+1} = \hat{R}_{i+1}\hat{Q}_{i+1}.$$

This results in the *QR algorithm:*

$$\hat{A}_0 = A,\ \hat{Q}_{i+1}\hat{R}_{i+1} = \hat{A}_i,\ \hat{A}_{i+1} = \hat{R}_{i+1}\hat{Q}_{i+1},\ i = 0, 1, 2, \ldots.$$

This version is preferred over performing simultaneous iteration explicitly, so we will now change notation and write the QR algorithm as follows:

$$A_0 = A,\ Q_{i+1}R_{i+1} = A_i,\ A_{i+1} = R_{i+1}Q_{i+1},\ i = 0, 1, 2, \ldots. \tag{9.41}$$

This algorithm produces a sequence $A_0, A_1, A_2, \ldots$ of matrices, each of which is similar to A (with an orthogonal similarity transformation):

$$A_i = (Q_1 \cdots Q_i)^T A(Q_1 \cdots Q_i),\ i = 0, 1, \ldots.$$

Moreover, at least in the two cases described above (before Example 414), we expect $\{A_i\}$ to converge to a block upper triangular matrix, with 1×1 or 2×2 blocks on the diagonal, revealing the eigenvalues of A.

9.10.1 A practical QR algorithm

As presented above, the QR algorithm is impractical because of its computational cost. Each iteration requires a QR factorization, which costs on the order of $(4/3)n^3$ arithmetic operations. Moreover, the rate of convergence of the QR algorithm is controlled by the ratios $|\lambda_{k+1}|/|\lambda_k|$. If A has two eigenvalues that are nearly equal in magnitude, the convergence will be very slow, while if A has two distinct eigenvalues with the same magnitude, the algorithm will not converge at all.

9.10.1.1 Reduction to upper Hessenberg form

We will now show how both the cost per iteration and the number of iterations can be greatly reduced. First of all, we will show that if A is almost upper triangular in the sense that $A_{ij} = 0$ for $i > j+1$, then all of the matrices $A_1, A_2, \ldots$ satisfy the same property, and the QR factorization of A_i is inexpensive. We say that A is *upper Hessenberg* if $A_{ij} = 0$ for all $i > j+1$. This means that A can have nonzeros on and above the diagonal, and also on the first subdiagonal, but not below the first subdiagonal. Thus an upper Hessenberg matrix has the following form:

$$\begin{bmatrix} \times & \times & \times & \times & \times \\ \times & \times & \times & \times & \times \\ 0 & \times & \times & \times & \times \\ 0 & 0 & \times & \times & \times \\ 0 & 0 & 0 & \times & \times \end{bmatrix}.$$

By using special orthogonal matrices, the QR factorization of an upper Hessenberg matrix can be computed in approximately $3n^2$ arithmetic operations (as opposed to $(4/3)n^3$ operations for the QR factorization of a general matrix). The algorithm that accomplishes this introduces zeros below the diagonal one column at a time (just as does the algorithm described in Section 9.8 for computing the QR factorization of a general matrix).

Let us suppose H is an $n \times n$ matrix that has been partially reduced from upper Hessenberg to upper triangular form. Specifically, we have found orthogonal matrices $P_1, \ldots, P_{j-1}$ such that

$$P_{j-1} \cdots P_1 H = \left[\begin{array}{c|c} R_{j-1} & C_{j-1} \\ \hline 0 & H_{j-1} \end{array}\right],$$

where $R_{j-1} \in \mathbf{R}^{(j-1)\times(j-1)}$ is upper triangular and $H_{j-1} \in \mathbf{R}^{(n-j+1)\times(n-j+1)}$ is upper Hessenberg. We will find an orthogonal matrix P_j of the form

$$P_j = \left[\begin{array}{c|cc|c} I & & & \\ \hline & c & s & \\ & -s & c & \\ \hline & & & I \end{array}\right]$$

that advances the factorization by one more column, that is, that introduces a zero in the $j+1, j$ entry. The numbers c, s must satisfy $c^2 + s^2 = 1$. A matrix of the form P_j is called a *plane rotation* because it is equivalent to the rotation matrix

$$\tilde{R} = \begin{bmatrix} c & s \\ -s & c \end{bmatrix} = \begin{bmatrix} \cos(\theta) & \sin(\theta) \\ -\sin(\theta) & \cos(\theta) \end{bmatrix}$$

(see Exercise 3.2.2) acting in the plane determined by coordinates j and $j+1$. The reader can verify that P_j is orthogonal. Moreover, multiplying

$P_{j-1} \cdots P_1 H$ on the left by P_j affects only rows j and $j+1$. We write $\tilde{H}$ for the 2×2 submatrix of H_{j-1} determined by the first two rows and columns:

$$\tilde{H} = \begin{bmatrix} \alpha & \gamma \\ \beta & \delta \end{bmatrix}.$$

We want to choose c and s so that

$$\tilde{R}\tilde{H} = \begin{bmatrix} \eta & \tilde{\gamma} \\ 0 & \tilde{\delta} \end{bmatrix},$$

where $\eta = \sqrt{\alpha^2 + \beta^2}$ must hold (because $\tilde{R}$ is orthogonal). The reader can verify that $c = \alpha/\eta$ and $s = \beta/\eta$ are the desired numbers. With these values of c and s, we obtain

$$P_j P_{j-1} \cdots P_1 H = \left[\begin{array}{c|c} R_j & C_j \\ \hline 0 & H_j \end{array}\right],$$

where now $R_j \in \mathbf{R}^{j \times j}$ is upper triangular and $H_j \in \mathbf{R}^{(n-j)\times(n-j)}$ is upper Hessenberg. After $n-1$ such steps, we obtain an upper triangular matrix $R = P_{n-1} \cdots P_1 H$. Then

$$H = P_1^T \cdots P_{n-1}^T R = QR, \; Q = P_1^T \cdots P_{n-1}^T.$$

This is the QR factorization of H. Exercise 3 asks the reader to show that computing the QR factorization of H in this manner requires $3n^2 + 2n - 5$ arithmetic operations, plus $n-1$ square roots.

Now suppose we apply the QR algorithm to compute the eigenvalues of an upper Hessenberg matrix A. We have $A_0 = A$ (so A_0 is upper Hessenberg), and

$$A_{i+1} = R_{i+1} Q_{i+1}, \text{ where } Q_{i+1} R_{i+1} = A_i.$$

Assuming (by way of induction) that A_i is upper Hessenberg, we can compute the QR factorization of A_i using plane rotations, as described above. We then have

$$A_{i+1} = R_{i+1} P_1^T \cdots P_{n-1}^T.$$

Since R_{i+1} is upper triangular and multiplying by P_j^T on the right combines columns $j, j+1$ of R_{j+1} to form columns $j, j+1$, it follows that A_{i+1} is upper Hessenberg.

We therefore see that the QR algorithm can be used to compute the eigenvalues of an upper Hessenberg matrix at a cost of only $3n^2$ operations per iteration, rather then the $(4/3)n^3$ operations required for a general matrix. This observation is relevant because, for a given matrix $A \in \mathbf{R}^{n \times n}$, there exists an orthogonal matrix $\tilde{Q}$, computable in finitely many operations, such that $\tilde{Q}^T A \tilde{Q}$ is upper Hessenberg.[19] We now describe briefly how to find such

[19]The reader should recall that there also exists an orthogonal matrix Q such that $Q^T AQ$ is upper triangular (Theorem 413). However, it is not possible to compute such a Q in finitely many operations.

a $\tilde{Q}$. Let $x_1 \in \mathbf{R}^{n-1}$ be the first column of A, excluding the first entry. We can find a Householder transformation $T_1 \in \mathbf{R}^{(n-1)\times(n-1)}$ such that $T_1 x_1$ is a multiple of $e_1 \in \mathbf{R}^{(n-1)}$. We define

$$\tilde{Q}_1 = \left[\begin{array}{c|c} 1 & 0 \\ \hline 0 & T_1 \end{array}\right].$$

Then $\tilde{Q}_1 A$ has zeros in the first column below the 2, 1 entry; in other words, the first column of $\tilde{Q}_1 A$ is consistent with upper Hessenberg form. Moreover, multiplying $\tilde{Q}_1 A$ on the right by $\tilde{Q}_1^T$ leaves the first column unchanged; hence $\tilde{Q}_1 A \tilde{Q}_1^T$ has a first column consistent with upper Hessenberg form. We now move on to the second column, putting it in upper Hessenberg form by multiplying on the left by

$$\tilde{Q}_2 = \left[\begin{array}{c|c} I & 0 \\ \hline 0 & T_2 \end{array}\right],$$

where I is the 2×2 identity matrix and $T_2 \in \mathbf{R}^{(n-2)\times(n-2)}$ is a Householder transformation. Multiplying $\tilde{Q}_2\tilde{Q}_1 A \tilde{Q}_1^T$ on the right by $\tilde{Q}_2^T$ leaves the first two columns unchanged, and so $\tilde{Q}_2\tilde{Q}_1 A \tilde{Q}_1^T \tilde{Q}_2^T$ has the first two columns in upper Hessenberg form. Continuing in this fashion, we see that

$$\tilde{Q}_{n-2}\cdots\tilde{Q}_1 A \tilde{Q}_1^T \cdots \tilde{Q}_{n-1}^T$$

is in upper Hessenberg form. We define

$$\tilde{Q} = \left(\tilde{Q}_{n-2}\cdots\tilde{Q}_1\right)^T = \tilde{Q}_1\cdots\tilde{Q}_{n-2}.$$

The QR algorithm is now modified to

$$A_0 = \tilde{Q}^T A \tilde{Q},\ Q_{i+1}R_{i+1} = A_i,\ A_{i+1} = R_{i+1}Q_{i+1},\ i = 0, 1, 2, \ldots.$$

The matrix $\tilde{Q}$ is determined as described above, and each A_i is upper Hessenberg.

Finally, we note that if A happens to be symmetric, then the above method produces a sequence of tridiagonal matrices, that is, matrices whose only nonzeros lie on the diagonal and the first sub- and super-diagonals. (This is true because an orthogonal similarity transformation preserves symmetry, and a symmetric upper Hessenberg matrix is tridiagonal.) The cost of each iteration then reduces to a small multiple of n operations. Exercise 4 asks the reader for an exact operation count.

9.10.1.2 The explicitly shifted QR algorithm

The use of upper Hessenberg matrices makes the cost of each iteration affordable. However, the QR method can converge very slowly and still require an unacceptable amount of time. Fortunately, there is a way to speed up the convergence of the QR method. We will explain this method in the simplest case, namely, that $A \in \mathbf{R}^{n\times n}$ is real with real eigenvalues.

As noted above, the rate of convergence of the QR algorithm depends on the ratios $|\lambda_{k+1}|/|\lambda_k|$; a ratio close to 1 means slow convergence. Let us suppose we have an estimate of λ_{k+1}: $\mu \doteq \lambda_{k+1}$. If μ is closer to λ_{k+1} than to λ_k, then

$$\frac{|\lambda_{k+1}-\mu|}{|\lambda_k-\mu|} < \frac{|\lambda_{k+1}|}{|\lambda_k|} \tag{9.42}$$

(see Exercise 5). We can take advantage of this fact because it is easy to shift the eigenvalues of A (or A_i) by μ: If the eigenvalues of A_i are $\lambda_1, \lambda_2, \ldots, \lambda_n$, then the eigenvalues of $A-\mu I$ are$\lambda_1-\mu, \lambda_2-\mu, \ldots, \lambda_n-\mu$. The *explicitly shifted QR algorithm* takes the form

$$A_0 = A,\ Q_{i+1}R_{i+1} = A_i - \mu I,\ A_{i+1} = R_{i+1}Q_{i+1} + \mu I.$$

By adding back the shift μI, we obtain (since $R_{i+1} = Q_{i+1}^T(A_i - \mu I)$)

$$\begin{aligned} A_{i+1} = R_{i+1}Q_{i+1} + \mu I &= Q_{i+1}^T(A_i - \mu I)Q_{i+1} + \mu I \\ &= Q_{i+1}^T A_i Q_{i+1} - \mu I + \mu I \\ &= Q_{i+1}^T A_i Q_{i+1}. \end{aligned}$$

Thus we still obtain a sequence of matrices similar to A. However, it is important to note that the similarity transformations have changed—Q_i is now determined by $A_i - \mu I$ instead of by A_i, so the eigenvector corresponding to the eigenvalue closest to μ should be better determined. The reader is encouraged to work out the corresponding simultaneous iteration approach, which should make this clearer.

In practice, we do not use a constant μ. Rather, since our estimate of λ_{k+1} improves with each iteration, we update μ each time. One possibility is to use $\mu_i = A_{nn}^{(i)}$ as an estimate of λ_n (since, when A has real eigenvalues, the n,n entry of A_i is expected to converge to the smallest eigenvalue of A). As a result of this improve value of μ_i, the iteration converges rapidly to an accurate estimate of λ_n. Once λ_n is determined accurately, we move on to estimate λ_{n-1}. When $A_0, A_1, A_2, \ldots$ are in upper Hessenberg form, it is easy to tell when $A_{nn}^{(i)}$ is close to λ_n: the $n, n-1$ entry of A_i becomes negligible.

One remark is in order before we do an example: If we begin shifting immediately, before $A_{nn}^{(i)}$ is close to λ_n, the initial value of μ could (and probably is) closest to some λ_j, $j \neq n$. The only consequence is that we cannot expect the eigenvalues to be ordered from largest to smallest on the diagonal of A_i.

Example 416 *Consider the* 4×4 *upper Hessenberg matrix*

$$A = \begin{bmatrix} 1.1203\cdot 10^{1} & 1.6772 & -1.4286 & 1.5319 \\ 6.6762\cdot 10^{-1} & 8.7199 & -5.0941\cdot 10^{-1} & -1.0088\cdot 10^{-1} \\ 0 & -9.4267\cdot 10^{-1} & 1.0068\cdot 10^{1} & -7.5483\cdot 10^{-1} \\ 0 & 0 & -3.7444\cdot 10^{-1} & 9.3195 \end{bmatrix}.$$

The eigenvalues of A are

$$\lambda_1 = 11.825,\ \lambda_2 = 10.097,\ \lambda_3 = 9.2276,\ \lambda_4 = 8.1608.$$

The reader should notice that these eigenvalues are not well separated, so we should expect the (unshifted) QR algorithm to converge slowly. Twelve iterations produce

$$A_{12} = \begin{bmatrix} 1.1750\cdot 10^{1} & 1.8648 & 2.0311\cdot 10^{-1} & 1.3432 \\ 6.6748\cdot 10^{-2} & 1.0193\cdot 10^{1} & 1.7329\cdot 10^{-1} & 3.8889\cdot 10^{-1} \\ 0 & -4.5645\cdot 10^{-1} & 8.5628 & -7.6665\cdot 10^{-1} \\ 0 & 0 & -4.1556\cdot 10^{-1} & 8.8053 \end{bmatrix}.$$

Before we apply the shifted algorithm, we notice that $A_{4,4}$ is close to λ_3; thus we should expect to find an accurate estimate of λ_3 first if we begin with the shift $\mu = A_{4,4}$. Indeed, here is the result of four iterations of the explicitly shifted QR algorithm:

$$\begin{bmatrix} 1.1813\cdot 10^{1} & 1.0211 & -2.0002 & 4.3441\cdot 10^{-1} \\ 3.1155\cdot 10^{-2} & 8.0975 & 3.8661\cdot 10^{-1} & -1.3384\cdot 10^{-1} \\ 0 & -3.5881\cdot 10^{-1} & 1.0173\cdot 10^{1} & 8.0261\cdot 10^{-2} \\ 0 & 0 & -2.5041\cdot 10^{-17} & 9.2276 \end{bmatrix}.$$

The size of the 4, 3 entry shows that the 4, 4 entry is an accurate eigenvalue; indeed, it is λ_3 correct to five digits. We now deflate the problem by eliminating the last row and column. Four iterations of the shifted algorithm on the resulting 3×3 matrix yields

$$\begin{bmatrix} 1.1814\cdot 10^{1} & 6.8371\cdot 10^{-1} & -2.1483 \\ 5.6721\cdot 10^{-2} & 8.1714 & 7.2672\cdot 10^{-1} \\ 0 & -5.3572\cdot 10^{-21} & 1.0097\cdot 10^{1} \end{bmatrix}.$$

This time we succeeded in estimating λ_2, correct to the digits shown. Finally, we deflate and apply four more iterations to obtain

$$\begin{bmatrix} 1.1825\cdot 10^{1} & 6.2699\cdot 10^{-1} \\ -2.7511\cdot 10^{-40} & 8.1608 \end{bmatrix}.$$

The diagonal entries are accurate estimates of λ_1 and λ_4, respectively.

Although the ordinary QR algorithm cannot estimate distinct eigenvalues λ_{i+1}, λ_i with $|\lambda_{i+1}| = |\lambda_i|$ (for example, $\lambda_{i+1} = -\lambda_i$), the shifted QR algorithm has no difficulty with this situation. The ratio $|\lambda_{i+1} - \mu|/|\lambda_i - \mu|$ can be much less than 1 even though $|\lambda_{i+1}|/|\lambda_i| = 1$.

A modification of the procedure described above makes the convergence even faster. It has been found that μ equal to the eigenvalue of

$$\begin{bmatrix} A^{(i)}_{n-1,n-1} & A^{(i)}_{n-1,n} \\ A^{(i)}_{n,n-1} & A^{(i)}_{n,n} \end{bmatrix} \tag{9.43}$$

closest to $A_{n,n}^{(i)}$ works even better than $\mu = A_{n,n}^{(i)}$. We will not pursue this enhancement here, but it is related to the implicitly shifted QR algorithm presented in the next section.

9.10.1.3 The implicitly shifted QR algorithm

In the case that $A \in \mathbf{R}^{n\times n}$ has complex eigenvalues, we could execute the explicitly shifted QR algorithm in complex arithmetic (provided we use the enhancement described in the last paragraph above). However, it would be more efficient to work with real numbers and produce a sequence $\{A_i\}$ converging to a block upper triangular matrix T, where complex conjugate eigenvalues correspond to 2×2 blocks on the diagonal. This saves the expense of arithmetic with complex numbers, and it costs little to determine the complex conjugate eigenvalues from the 2×2 blocks.

Suppose that, at a certain step, the matrix (9.43) has complex conjugate eigenvalues μ, $\overline{\mu}$. We could perform QR iterations with shifts μ and $\overline{\mu}$, as follows:

$$\begin{aligned} Q_{i+1}R_{i+1} = A_i - \mu I,\ A_{i+1} = R_{i+1}Q_{i+1} + \mu I, \\ Q_{i+2}R_{i+2} = A_{i+1} - \overline{\mu} I,\ A_{i+2} = R_{i+2}Q_{i+2} + \overline{\mu} I. \end{aligned}$$

As written, these steps require complex arithmetic. However, we have

$$\begin{aligned} Q_{i+1}Q_{i+2}R_{i+2}R_{i+1} &= Q_{i+1}(A_{i+1} - \overline{\mu}I)R_{i+1} \\ &= Q_{i+1}(R_{i+1}Q_{i+1} + \mu I - \overline{\mu}I)R_{i+1} \\ &= Q_{i+1}R_{i+1}Q_{i+1}R_{i+1} + (\mu - \overline{\mu})Q_{i+1}R_{i+1} \\ &= (A_i - \mu I)^2 + (\mu - \overline{\mu})(A_i - \mu I) \\ &= (A_i - \mu I)(A_i - \overline{\mu}I) \\ &= A_i^2 - (\mu + \overline{\mu})I + \mu\overline{\mu}I. \end{aligned}$$

Since $\mu + \overline{\mu}, \mu\overline{\mu} \in \mathbf{R}$, we see that $(A_i - \mu I)(A_i - \overline{\mu}I)$ is a real matrix, and $(Q_{i+1}Q_{i+2})(R_{i+2}R_{i+1})$ is a QR factorization of it. (The reader should notice that the product of unitary matrices is unitary and the product of upper triangular matrices is upper triangular.) Therefore, Q_{i+1} and Q_{i+2} can be chosen so that $\hat{Q}_{i+1} = Q_{i+1}Q_{i+2}$ is real and therefore orthogonal. But then

$$A_{i+2} = Q_{i+2}^T A_{i+1} Q_{i+2} = Q_{i+2}^T Q_{i+1}^T A_i Q_{i+1} Q_{i+2} = \hat{Q}_{i+1}^T A_i \hat{Q}_{i+1}.$$

This shows that we can compute the matrix A_{i+2} without moving into the complex domain:

$$\hat{Q}_{i+1}\hat{R}_{i+1} = A_i^2 - (\mu + \overline{\mu})I + \mu\overline{\mu}I,\ A_{i+2} = \hat{Q}_{i+1}^T A_i \hat{Q}_{i+1}.$$

This computation is not practical, since the calculation of $A_i^2 - (\mu+\overline{\mu})I + \mu\overline{\mu}I$ requires operations on the order of n^3 rather than n^2. However, it is possible to produce A_{i+2} indirectly using real arithmetic and only about $10n^2$ operations. The details can be found in Golub and Van Loan [15].

Exercises

1. Use Householder transformations, as described in this section, to reduce

$$A = \begin{bmatrix} -2 & -1 & 0 & 0 \\ 2 & -2 & -1 & 3 \\ -2 & -2 & 2 & -1 \\ 3 & 1 & 1 & 2 \end{bmatrix}$$

to upper Hessenberg form via a sequence of similarity transformations.

2. Let $H \in \mathbf{R}^{4\times 4}$ be the following upper Hessenberg matrix:

$$H = \begin{bmatrix} -3 & -2 & 0 & -3 \\ 3 & 2 & -1 & -2 \\ 0 & 1 & -3 & -2 \\ 0 & 0 & -2 & -1 \end{bmatrix}.$$

Compute the QR factorization of A using plane rotations.

3. Let H be an $n \times n$ upper Hessenberg matrix. Show that the algorithm described in this section (using plane rotations) requires $3n^2 + 2n - 5$ arithmetic operations, plus the calculation of $n - 1$ square roots, to find the QR factorization of H. (Note: This count does not include the cost of computing $Q = P_1^T \cdots P_{n-1}^T$ explicitly. In most applications, it is preferable to leave Q in factored form.)

4. (a) Count the number of operations required to find the QR factorization of a symmetric tridiagonal matrix.

   (b) Count the number of operations for each iteration of the QR algorithm, applied to a symmetric tridiagonal matrix.

5. Prove that (9.42) holds if and only if the relative error in μ as an estimate of λ_{k+1} is less than the relative error in μ as an estimate of λ_k.

10

Analysis in vector spaces

In this chapter, we will give a brief introduction to analysis—questions of convergence, continuity, and so forth—in vector spaces. Our treatment is necessarily brief; we wish to make two basic points. First of all, all norms on a given finite-dimensional vector space are equivalent, in the sense that they define the same notion of convergence. We learned early in this book that all finite-dimensional vector spaces are isomorphic in the algebraic sense; the results of this chapter (Section 10.1) show that this is true in the analytic sense as well. Second, infinite-dimensional vector spaces are much more complicated. In particular, it is possible to define nonequivalent norms on an infinite-dimensional vector spaces. We will find it useful to define notions of convergence that do not depend on a norm.

This chapter assumes that the reader is familiar with basic analysis of the space of real numbers; the results we use are summarized in Appendix D.

10.1 Analysis in $\mathbf{R}^n$

Analysis in $\mathbf{R}^n$ is based on extending the concepts and results from $\mathbf{R}$ by substituting a norm $\|\cdot\|$ for the absolute value function $|\cdot|$. In $\mathbf{R}$, x and y are close if $|x-y|$ is small; similarly, $x, y \in \mathbf{R}^n$ are regarded as close if $\|x-y\|$ is small. Analysis in $\mathbf{R}^n$ is complicated somewhat by the fact that many different norms can be defined on $\mathbf{R}^n$. However, one of the main results we will present is that all norms on $\mathbf{R}^n$ are equivalent; for example, if $\{x_k\}$ converges to x in one norm, then it does so in any other norm. We need quite a few preliminary results before we get to this point, however, so we begin by defining convergence and related concepts in terms of the ℓ^∞ norm,[1]

$$\|x\|_\infty = \max\{|x_i| \,:\, i = 1, 2, \ldots, n\}.$$

[1]We could also perform the following development by starting with another familiar norm, such as the Euclidean norm $\|\cdot\|_2$ or the ℓ^1 norm $\|\cdot\|_1$.

10.1.1 Convergence and continuity in $\mathbf{R}^n$

Definition 417 *Let $\{x_k\}$ be a sequence of vectors in $\mathbf{R}^n$, and let $x \in \mathbf{R}^n$. We say that $\{x_k\}$* converges *to x if and only if for every $\epsilon > 0$, there exists a positive integer N such that*

$$k \geq N \;\Rightarrow\; \|x_k - x\|_\infty < \epsilon.$$

If $\{x_k\}$ converges to x, we write $x_k \to x$ or

$$\lim_{k\to\infty} x_k = x.$$

If $\{x_k\}$ does not converge to any vector, then we say that the sequence diverges.

If $\{x_k\}$ is a sequence of vectors, then $\{\|x_k\|_\infty\}$ is a sequence of real numbers, as is $\{\|x_k - x\|_\infty\}$. The reader should notice that $x_k \to 0$ if and only if $\|x_k\|_\infty \to 0$, while $x_k \to x$ if and only if $\|x_k - x\|_\infty \to 0$.

In the proof of the following theorem and at other points in this section, we will have to refer to the components of vectors in a sequence. When necessary, we can denote the terms of the sequence as $x^{(k)}$, reserving subscripts to indicate components of vectors.

Theorem 418 *Let $\{x^{(k)}\}$ be a sequence of vectors in $\mathbf{R}^n$, and let x be a given vector in $\mathbf{R}^n$. Then $x^{(k)} \to x$ if and only if for each $i = 1, 2, \ldots, n$, $x_i^{(k)} \to x_i$.*

Proof The "only if" direction of the proof is straightforward and is left as an exercise (see Exercise 1). We now suppose that for each $i = 1, 2, \ldots, n$, $x_i^{(k)} \to x_i$ as $k \to \infty$. Let $\epsilon > 0$ be given. Then, for each i, there exists a positive integer N_i such that

$$k \geq N_i \;\Rightarrow\; |x_i^{(k)} - x_i| < \epsilon.$$

Let $N = \max\{N_1, N_2, \ldots, N_n\}$. Then, if $k \geq N$, it follows that $|x_i^{(k)} - x_i| < \epsilon$ for all $i = 1, 2, \ldots, n$ (since $k \geq N \geq N_i$). But then

$$k \geq N \;\Rightarrow\; \|x^{(k)} - x\|_\infty < \epsilon,$$

and we have shown that $x^{(k)} \to x$ as $k \to \infty$.

QED

In $\mathbf{R}$, we often refer to intervals of the form $(x - \epsilon, x + \epsilon)$. The analogous construction in $\mathbf{R}^n$ is the *open ball*[2] $B_{\epsilon,\infty}(x)$ of radius ϵ centered at x,

$$B_{\epsilon,\infty}(x) = \{y \in \mathbf{R}^n \,:\, \|y - x\|_\infty < \epsilon\}.$$

Using open balls, we can define concepts like accumulation point and open set.

[2]Soon we will define $B_\epsilon(x)$ for an arbitrary norm. If we use the Euclidean norm $\|\cdot\|_2$ in $\mathbf{R}^3$, then $B_{\epsilon,2}(x)$ is a spherical ball of radius ϵ, centered at x. We use the same terminology for any other norm, even if (as in the ℓ_∞ norm) the "ball" is not round.

Definition 419 *Let S be a subset of $\mathbf{R}^n$. We say that $y \in \mathbf{R}^n$ is an* accumulation point *of S if for every $\epsilon > 0$, the open ball $B_{\epsilon,\infty}(x)$ contains infinitely many points of S.*

Definition 420 *Let S be a subset of $\mathbf{R}^n$. We say that S is* open *if for each $x \in S$, there exists $\epsilon > 0$ such that $B_{\epsilon,\infty}(x) \subset S$.*

On the other hand, we say that S is closed *if $\mathbf{R}^n \setminus S$ is open.*

We can now define the notion of convergence of function values.

Definition 421 *Let S be a subset of $\mathbf{R}^n$, let $f : S \to \mathbf{R}$, and suppose y is an accumulation point of S. We say that $f(x)$ converges to $L \in \mathbf{R}$ as $x \to y$ if for every $\epsilon > 0$, there exists $\delta > 0$ such that*

$$x \in S, \|x - y\|_\infty < \delta, x \neq y \;\Rightarrow\; |f(x) - L| < \epsilon.$$

If $f(x)$ converges to L as $x \to y$, we write

$$\lim_{x \to y} f(x) = L$$

or $f(x) \to L$ as $x \to y$. If there is no real number L such that $f(x) \to L$ as $x \to y$, then we say that $f(x)$ diverges *as $x \to y$.*

We define continuity of a function as follows.

Definition 422 *Let S be a subset of $\mathbf{R}^n$, and let $f : S \to \mathbf{R}$ be a function. We say that f is* continuous *at $x \in S$ if for any $\epsilon > 0$, there exists $\delta > 0$ such that*

$$y \in S, \|y - x\|_\infty < \delta \;\Rightarrow\; |f(y) - f(x)| < \epsilon.$$

We say that f is continuous on *S, or simply* continuous, *if it is continuous at every $x \in S$.*

It turns out that every norm on $\mathbf{R}^n$ is continuous. We will prove this using the following preliminary results.

Lemma 423 *Let $\|\cdot\|$ be any norm on $\mathbf{R}^n$. Then there exists a constant $M > 0$ such that*

$$\|x\| \leq M\|x\|_\infty \text{ for all } x \in \mathbf{R}^n.$$

Proof We have

$$x = \sum_{i=1}^{n} x_i e_i,$$

where $\{e_1, e_2, \ldots, e_n\}$ is the standard basis on $\mathbf{R}^n$. By the triangle inequality, we have

$$\|x\| \leq \sum_{i=1}^{\infty} \|x_i e_i\| = \sum_{i=1}^{\infty} |x_i|\|e_i\| \leq \|x\|_\infty \sum_{i=1}^{\infty} \|e_i\|.$$

Thus the desired result holds with $M = \sum_{i=1}^{\infty} \|e_i\|$.

QED

Lemma 424 (Reverse triangle inequality) *Let V be a vector space over $\mathbf{R}$, and let $\|\cdot\|$ be any norm on V. Then*

$$|\|x\| - \|y\|| \le \|x - y\| \text{ for all } x, y \in V.$$

Proof We have $\|x\| = \|x-y+y\| \le \|x-y\| + \|y\|$ by the triangle inequality, which yields $\|x\| - \|y\| \le \|x - y\|$. Interchanging the roles of x and y yields $\|y\| - \|x\| \le \|y - x\| = \|x - y\|$. Putting together these results yields the desired inequality.

QED

Theorem 425 *Let $\|\cdot\|$ be any norm on $\mathbf{R}^n$. Then $\|\cdot\|$ is a continuous function.*

Proof Let x be any vector in $\mathbf{R}^n$, and let $\epsilon > 0$ be given. Applying the previous two lemmas, we have

$$|\|y\| - \|x\|| \le \|y - x\| \le M\|y - x\|_\infty.$$

Let $\delta = \epsilon/M$; then it follows that $\|y - x\|_\infty < \delta$ implies that $|\|y\| - \|x\|| < \epsilon$, as desired.

QED

The preceding result will be used to show that all norms on $\mathbf{R}^n$ are equivalent.

Corollary 426 *Let $y \in \mathbf{R}^n$ be given. Then $\|x\|_\infty \to \|y\|_\infty$ as $x \to y$.*

10.1.2 Compactness

The Bolzano-Weierstrass theorem holds in $\mathbf{R}^n$ as well as in $\mathbf{R}$.

Definition 427 *Let S be a subset of $\mathbf{R}^n$. We say that S is* bounded *if there exists $R > 0$ such that $\|x\|_\infty \le R$ for all $x \in S$.*

The reader should notice that since $\mathbf{R}^n$ $(n > 1)$ is not ordered, there is no concept of "bounded above" or "bounded below", as there is in $\mathbf{R}$.

Theorem 428 *Let S be a nonempty, closed, and bounded subset of $\mathbf{R}^n$, and let $\{x^{(k)}\}$ be a sequence of vectors in S. Then there exists a subsequence $\{x^{(k_j)}\}$ that converges to a vector $x \in S$.*

Proof Suppose $\|x\|_\infty \le R$ for all $x \in S$. For each $i = 1, 2, \ldots, n$, the sequence of vectors defines a sequence of real numbers $\{x_i^{(k)}\}$. Let us consider the sequence $\{x_1^{(k)}\}$. By definition of the ℓ^∞ norm, we see that $\{x_1^{(k)}\}$ belongs to the closed and bounded interval $[-R, R]$, and hence there exists a subsequence $\{x_1^{(k_j)}\}$ that converges to some $x_1 \in [-R, R]$. Next, we consider the sequence $\{x_2^{(k_j)}\}$, which also belongs to the interval $[-R, R]$. It follows that there exists a subsequence $\{x_2^{(k_{j_p})}\}$ of $\{x_2^{(k_j)}\}$ that converges to some $x_2 \in [-R, R]$. Since $\{x_1^{(k_j)}\}$ converges to x_1, so does every subsequence of $\{x_1^{(k_j)}\}$. Therefore,

$$x_1^{(k_{j_p})} \to x_1 \text{ as } p \to \infty,$$
$$x_2^{(k_{j_p})} \to x_2 \text{ as } p \to \infty.$$

We will now relabel the indices so that $\{x_2^{(k_{j_p})}\}$ becomes $\{x_2^{(k_j)}\}$ (otherwise, by the time we arrived at the end of this proof, we would have n subscripts on the superscript k). We now repeat the above argument to obtain a subsequence of $\{x_3^{(k_j)}\}$, again denoted by $\{x_3^{(k_j)}\}$, converging to $x_3 \in [-R, R]$. Repeating this argument a finite number of times, we obtain $\{k_j\}$ and $x_1, \ldots, x_n$ such that $x_i^{(k_j)} \to x_i$ as $j \to \infty$ for each $i = 1, 2, \ldots, n$. By Theorem 418, it follows that $x^{k_j} \to x = (x_1, x_2, \ldots, x_n)$ as $j \to \infty$.

QED

The reader should notice the key step in the above proof that used the finite dimensionality of $\mathbf{R}^n$. The process of extracting a subsequence of a given sequence was repeated only a finite number of times. The principle being used is that "a subsequence of a subsequence is a subsequence of the original sequence." This can be iterated, for example as "a subsequence of a subsequence of a subsequence is a subsequence of the original sequence," but only a finite number of times.

We now see that a closed and bounded subset of $\mathbf{R}^n$ is sequentially compact. This allows us to prove Weierstrass's theorem (Theorem 534) in $\mathbf{R}^n$.

Theorem 429 *Let S be a closed and bounded subset of $\mathbf{R}^n$, and let $f : S \to \mathbf{R}$ be continuous. Then there exist $m_1, m_2 \in S$ such that*

$$f(m_1) \le f(x) \le f(m_2) \text{ for all } x \in S.$$

In short, we say that f attains its maximum and minimum values on S.

Proof If $M = \sup\{f(x) : x \in S\}$, then there exists a sequence $\{x_k\}$ in S such that $f(x_k) \to M$. Since S is closed and bounded, there exists a subsequence $\{x_{k_j}\}$ of $\{x_k\}$ and a vector $m_2 \in S$ such that $x_{k_j} \to m_2$. But then $f(x_{k_j}) \to f(m_2)$ (since f is continuous) and $f(x_{k_j}) \to M$ (since a subsequence of a convergent sequence converges to the same limit). Thus $f(m_2) = M$, and hence $f(m_2) \ge f(x)$ for all $x \in S$.

The proof that m_1 exists is similar.

QED

10.1.3 Completeness of $\mathbf{R}^n$

The concept of Cauchy sequences and completeness are fundamental in the analysis of $\mathbf{R}^n$.

Definition 430 *Let $\{x_k\}$ be a sequence of vectors in $\mathbf{R}^n$. We say that $\{x_k\}$ is a* Cauchy sequence *if for every $\epsilon > 0$, there exists a positive integer N such that*

$$m, n \geq N \Rightarrow \|x_n - x_m\|_\infty < \epsilon.$$

It is straightforward to show that $\mathbf{R}^n$ is complete.

Theorem 431 *The space $\mathbf{R}^n$ is complete.*

Proof Let $\{x^{(k)}\}$ be a Cauchy sequence in $\mathbf{R}^n$, and let $\epsilon > 0$ be given. Then, by definition, there exists a positive integer N such that

$$m, n \geq N \Rightarrow \|x^{(m)} - x^{(n)}\|_\infty < \epsilon.$$

But then the definition of $\|\cdot\|_\infty$ implies that, for each $i = 1, 2, \ldots, n$,

$$m, n \geq N \Rightarrow |x_i^{(m)} - x_i^{(n)}| < \epsilon,$$

which shows that $\{x_i^{(k)}\}$ is a Cauchy sequence of real numbers. Since $\mathbf{R}$ is complete, there exist numbers $x_1, x_2, \ldots, x_n \in \mathbf{R}$ such that

$$x_i^{(k)} \to x_i \text{ as } k \to \infty,\ i = 1, 2, \ldots, n.$$

Defining $x = (x_1, x_2, \ldots, x_n)$, it follows from Theorem 418 that $x^{(k)} \to x$. This shows that $\mathbf{R}^n$ is complete.

QED

10.1.4 Equivalence of norms on $\mathbf{R}^n$

We now have the tools we need to prove the fundamental result that all norms on $\mathbf{R}^n$ are equivalent. Before we do this, of course, we must define the meaning of equivalence of norms.

Definition 432 *Let X be a vector space over $\mathbf{R}$, and let $\|\cdot\|$ and $\|\cdot\|_*$ be two norms on X. We say that $\|\cdot\|_*$ is equivalent to $\|\cdot\|$ if there exist $c_1, c_2 > 0$ such that*

$$c_1\|x\| \leq \|x\|_* \leq c_2\|x\| \text{ for all } x \in X. \tag{10.1}$$

We notice that if (10.1) holds, then

$$c_2^{-1}\|x\|_* \leq \|x\| \leq c_1^{-1}\|x\|_* \text{ for all } x \in X,$$

so we can simply say that $\|\cdot\|$ and $\|\cdot\|_*$ are equivalent. In fact, equivalence of norms is an equivalence relation; see Exercise 2.

We have been working towards the following fundamental theorem.

Theorem 433 *Let $\|\cdot\|$ be any norm on $\mathbf{R}^n$. Then $\|\cdot\|$ is equivalent to $\|\cdot\|_\infty$.*

Proof Consider the unit sphere S in $\mathbf{R}^n$ (relative to the ℓ^∞ norm):

$$S = \{x \in \mathbf{R}^n : \|x\|_\infty = 1\}.$$

The set S is bounded by definition. If $\{x_k\}$ is a sequence of vectors in S that converges to $x \in \mathbf{R}^n$, then, by Corollary 425,

$$\lim_{k\to\infty} \|x_k\|_\infty = \|x\|_\infty.$$

But $\|x_k\|_\infty = 1$ for all k, so it follows that $\|x\|_\infty = 1$. Thus $x \in S$, and we have shown that S is closed.

Since S is closed and bounded and $\|\cdot\|$ is continuous, Theorem 429 implies that there exist constants c_1 and c_2 such that

$$c_1 \le \|x\| \le c_2 \text{ for all } x \in S.$$

Moreover, since $c_1 = \|y_1\|$, $c_2 = \|y_2\|$ for some $y_1, y_2 \in S$ and S does not contain the zero vector, it follows that c_1 and c_2 are positive constants.

If now y is any nonzero vector in $\mathbf{R}^n$, then $y = \|y\|_\infty x$, where $x = y/\|y\|_\infty$. The vector x lies in S, so

$$c_1 \le \|x\| \le c_2.$$

Multiplying through by $\|y\|_\infty$ yields

$$c_1\|y\|_\infty \le \|y\|_\infty\|x\| \le c_2\|y\|_\infty.$$

But

$$\|y\|_\infty\|x\| = \|\|y\|_\infty x\| = \|y\|,$$

and hence we have

$$c_1\|y\|_\infty \le \|y\| \le c_2\|y\|_\infty,$$

as desired.

QED

Corollary 434 *Any two norms on $\mathbf{R}^n$ are equivalent.*

Proof This follows from the preceding theorem and Exercise 2.

The preceding result is quite strong; it means that regardless of the norm we choose to impose on $\mathbf{R}^n$, any analytical relationship is unchanged. To be specific, suppose $\|\cdot\|$ and $\|\cdot\|_*$ are two norms on $\mathbf{R}^n$. We define, for example, $x_k \to x$ under $\|\cdot\|$ just as in Definition 417, except that $\|\cdot\|_\infty$ is replaced by $\|\cdot\|$. The same is true for the other terms defined in this section in terms of $\|\cdot\|_\infty$—any could be defined in terms of $\|\cdot\|$ instead. The following results are left as exercises for the reader (see Exercises 3–12).

1. If $\{x_k\}$ is a sequence of vectors in $\mathbf{R}^n$, then $\{x_k\}$ converges to $x \in \mathbf{R}^n$ under $\|\cdot\|$ if and only if $x_k \to x$ under $\|\cdot\|_*$.

2. If S is a subset of $\mathbf{R}^n$, then S is open with respect to the norm $\|\cdot\|$ if and only if S is open with respect to the norm $\|\cdot\|_*$.

3. If S is a subset of $\mathbf{R}^n$, then S is closed with respect to the norm $\|\cdot\|$ if and only if S is closed with respect to the norm $\|\cdot\|_*$.

4. If S is a subset of $\mathbf{R}^n$, then $x \in \mathbf{R}^n$ is an accumulation point of S under $\|\cdot\|$ if and only if x is an accumulation point of S under $\|\cdot\|_*$.

5. If S is a subset of $\mathbf{R}^n$, $f : S \to \mathbf{R}$ is a function, and $y \in \mathbf{R}^n$, then $\lim_{x\to y} f(x) = L$ under $\|\cdot\|$ if and only if $\lim_{x\to y} f(x) = L$ under $\|\cdot\|_*$.

6. If S is a subset of $\mathbf{R}^n$, $f : S \to \mathbf{R}$ is a function, and $y \in \mathbf{R}^n$, then f is continuous at y under $\|\cdot\|$ if and only if f is continuous at y under $\|\cdot\|_*$.

7. Let S be a subset of $\mathbf{R}^n$. Then S is bounded under $\|\cdot\|$ if and only if S is bounded under $\|\cdot\|_*$.

8. Let S be a subset of $\mathbf{R}^n$. Then S is sequentially compact under $\|\cdot\|$ if and only if S is sequentially compact under $\|\cdot\|_*$.

9. Let $\{x_k\}$ be a sequence of vectors in $\mathbf{R}^n$. Then $\{x_k\}$ is Cauchy under $\|\cdot\|$ if and only if $\{x_k\}$ is Cauchy under $\|\cdot\|_*$.

10. The space $\mathbf{R}^n$ is complete under $\|\cdot\|$ if and only if it is complete under $\|\cdot\|_*$.

These results show that it is unnecessary to identify the norm when describing an analytical relationship in $\mathbf{R}^n$. For instance, we can simply say that $x_k \to x$ or y is an accumulation point of S, with no need to add the phrase "under $\|\cdot\|$."

This strong property of $\mathbf{R}^n$ is not shared by infinite dimensional vector spaces. For example, if X is infinite-dimensional, then $f : X \to \mathbf{R}$ can be continuous under one norm and not under another, or $\{x_k\}$ can converge under one norm and not under another. We will see specific examples later in this chapter.

Exercises

1. Let $\{x^{(k)}\}$ be a sequence in $\mathbf{R}^n$ and suppose $x^{(k)} \to x \in \mathbf{R}^n$. Let i be an integer, $1 \le i \le n$. Prove that the sequence $\{x_i^{(k)}\}$ of real numbers converges to the real number x_i.

2. Use the remarks following Definition 432 to prove that equivalence of norms is an equivalence relation.

3. Let $\|\cdot\|$ and $\|\cdot\|_*$ be two norms on $\mathbf{R}^n$. Prove that if $\{x_k\}$ is a sequence of vectors in $\mathbf{R}^n$ and $x \in \mathbf{R}^n$, then $x_k \to x$ under $\|\cdot\|$ if and only if $x_k \to x$ under $\|\cdot\|_*$.

4. Let $\|\cdot\|$ and $\|\cdot\|_*$ be two norms on $\mathbf{R}^n$, and let S be a nonempty subset of $\mathbf{R}^n$. Prove that S is open under $\|\cdot\|$ if and only if S is open under $\|\cdot\|_*$.

5. Let $\|\cdot\|$ and $\|\cdot\|_*$ be two norms on $\mathbf{R}^n$, and let S be a nonempty subset of $\mathbf{R}^n$. Prove that S is closed under $\|\cdot\|$ if and only if S is closed under $\|\cdot\|_*$.

6. Let $\|\cdot\|$ and $\|\cdot\|_*$ be two norms on $\mathbf{R}^n$, and let S be a nonempty subset of $\mathbf{R}^n$. Prove that x is an accumulation point of S under $\|\cdot\|$ if and only if x is an accumulation point of S under $\|\cdot\|_*$.

7. Let $\|\cdot\|$ and $\|\cdot\|_*$ be two norms on $\mathbf{R}^n$, let S be a nonempty subset of $\mathbf{R}^n$, let $f : S \to \mathbf{R}^n$ be a function, and let y be an accumulation point of S. Prove that $\lim_{x\to y} f(x) = L$ under $\|\cdot\|$ if and only if $\lim_{x\to y} f(x) = L$ under $\|\cdot\|_*$.

8. Let $\|\cdot\|$ and $\|\cdot\|_*$ be two norms on $\mathbf{R}^n$, let S be a nonempty subset of $\mathbf{R}^n$, let $f : S \to \mathbf{R}^n$ be a function, and let y be a point in S. Prove that f is continuous at y under $\|\cdot\|$ if and only if f is continuous at y under $\|\cdot\|_*$.

9. Let $\|\cdot\|$ and $\|\cdot\|_*$ be two norms on $\mathbf{R}^n$, and let S be a nonempty subset of $\mathbf{R}^n$. Prove that S is bounded under $\|\cdot\|$ if and only if S is bounded under $\|\cdot\|_*$.

10. Let $\|\cdot\|$ and $\|\cdot\|_*$ be two norms on $\mathbf{R}^n$, and let S be a nonempty subset of $\mathbf{R}^n$. Prove that S is sequentially compact under $\|\cdot\|$ if and only if S is sequentially compact under $\|\cdot\|_*$.

11. Let $\|\cdot\|$ and $\|\cdot\|_*$ be two norms on $\mathbf{R}^n$, and let $\{x_k\}$ be a sequence in $\mathbf{R}^n$. Prove that $\{x_k\}$ is Cauchy under $\|\cdot\|$ if and only if $\{x_k\}$ is Cauchy under $\|\cdot\|_*$.

12. Let $\|\cdot\|$ and $\|\cdot\|_*$ be two norms on $\mathbf{R}^n$. Prove that $\mathbf{R}^n$ is complete under $\|\cdot\|$ if and only if $\mathbf{R}^n$ is complete under $\|\cdot\|_*$.

13. Let X be a vector space with norm $\|\cdot\|$, and suppose $\{x_k\}$ is a sequence in X converging to $x \in X$ under $\|\cdot\|$ (that is, $\|x_k - x\| \to 0$ as $k \to \infty$). Then $\{x_k\}$ is a Cauchy sequence.

10.2 Infinite-dimensional vector spaces

We have already seen that the space $\mathcal{P}$ of all polynomials is infinite-dimensional, and therefore so are spaces of functions for which $\mathcal{P}$ can be regarded as a subspace. We will now describe another infinite-dimensional space that is in some ways simpler than function spaces.

The space ℓ^2 is defined to be the space of all infinite sequences $\{x_i\}_{i=1}^{\infty}$ of real numbers such that

$$\sum_{i=1}^{\infty} x_i^2 < \infty.$$

We will write $x = \{x_i\}$ for a given element $\{x_i\}$ in ℓ^2.

If ℓ^2 is to be a vector space, it must be closed under addition and scalar multiplication. This can be proved using the following elementary but extremely useful lemma.

Lemma 435 *Let a and b be two real numbers. Then*

$$|2ab| \leq a^2 + b^2.$$

Proof We have $(|a| - |b|)^2 \geq 0$, and this inequality can be rearranged to give $2|a||b| \leq |a|^2 + |b|^2$, which is equivalent to the desired result.

QED

We now defined addition and scalar multiplication in ℓ^2 in the obvious way:

$$\begin{aligned} \{x_i\} + \{y_i\} &= \{x_i + y_i\}, \\ \alpha\{x_i\} &= \{\alpha x_i\}. \end{aligned}$$

Theorem 436 *The space ℓ^2 is a vector space over* **R**.

Proof We must verify that ℓ^2 is closed under addition and scalar multiplication. If x and y belong to ℓ^2, then

$$\sum_{i=1}^{\infty} x_i^2 < \infty, \ \sum_{i=1}^{\infty} y_i^2 < \infty.$$

For each i, the preceding lemma implies that $(x_i + y_i)^2 \leq 2x_i^2 + 2y_i^2$, which in turn implies that

$$\sum_{i=1}^{\infty} (x_i + y_i)^2 \leq 2\sum_{i=1}^{\infty} x_i^2 + 2\sum_{i=1}^{\infty} y_i^2 < \infty.$$

Therefore, $x + y = \{x_i + y_i\} \in \ell^2$. Scalar multiplication is simpler: If $x \in \ell^2$, then $\alpha x = \{\alpha x_i\}$ and

$$\sum_{i=1}^{\infty}(\alpha x_i)^2 = \sum_{i=1}^{\infty}\alpha^2 x_i^2 = \alpha^2\sum_{i=1}^{\infty}x_i^2 < \infty.$$

Therefore, $\alpha x \in \ell^2$.

It is easy to verify that the various algebraic properties of a vector space hold, where the zero vector is the zero sequence (that is, the sequence whose terms are all zero), and the additive inverse of $x = \{x_i\}$ is $-x = \{-x_i\}$. The details are left to the reader.

QED

The reader will recall that the symbol ℓ^2 is also used to describe the Euclidean norm on $\mathbf{R}^n$. It should be apparent that the space ℓ^2 is the natural infinite-dimensional generalization of $\mathbf{R}^n$ under the Euclidean norm.

We can define an inner product on ℓ^2 by

$$\langle x, y\rangle_{\ell^2} = \sum_{i=1}^{\infty} x_i y_i. \tag{10.2}$$

If x and y belong to ℓ^2, then, by Lemma 435, we have

$$\sum_{i=1}^{\infty}|x_i y_i| \le \frac{1}{2}\sum_{i=1}^{\infty}x_i^2 + \frac{1}{2}\sum_{i=1}^{\infty}y_i^2 < \infty \Rightarrow \sum_{i=1}^{\infty}x_i y_i < \infty.$$

This shows that $\langle x, y\rangle_{\ell^2}$ is well-defined. We will leave it to the reader to verify that $\langle\cdot,\cdot\rangle_{\ell^2}$ defines an inner product on ℓ^2. The corresponding norm is

$$\|x\|_{\ell^2} = \left[\sum_{i=1}^{\infty}x_i^2\right]^{1/2}.$$

We remark that $\|\cdot\|_{\ell^2}$ is continuous as a real-valued function on ℓ^2, when continuity is defined in terms of the norm $\|\cdot\|_{\ell^2}$ itself. This is a special case of the following result.

Theorem 437 *Let X be a vector space over $\mathbf{R}$, and let $\|\cdot\|$ be a norm on X. Then $\|\cdot\|$ is a continuous function (where continuity is defined as in Definition 421 but using the norm $\|\cdot\|$).*

Proof Exercise 1.

We will now show that ℓ^2 is infinite-dimensional. We will write e_k for the sequence whose terms are all zero, except for the kth term, which is one.

Lemma 438 *For any positive integer n, the subset $\{e_1, e_2, \ldots, e_n\}$ of ℓ^2 is linearly independent.*

Proof We will write $\{x_i\}$ for the sequence $\alpha_1 e_1 + \alpha_2 e_2 + \ldots + \alpha_n e_n$, where $\alpha_1, \alpha_2, \ldots, \alpha_n$ are scalars. We see that

$$x_i = \begin{cases} \alpha_i, & i = 1, 2, \ldots, n, \\ 0, & i > n. \end{cases}$$

It follows that $\{x_i\} = 0$ implies that $\alpha_i = 0$ for $i = 1, 2, \ldots, n$. This shows that $\{e_1, e_2, \ldots, e_n\}$ is linearly independent.

QED

Corollary 439 *The space ℓ^2 is infinite-dimensional.*

Proof For each positive integer n, ℓ^2 contains a linearly independent subset of n elements. It follows from Theorem 34 that ℓ^2 cannot be finite-dimensional.

QED

We will now show that the Bolzano-Weierstrass theorem fails in ℓ^2. To be more specific, we will produce a closed and bounded subset of ℓ^2 and a sequence in that subset that has no convergent subsequence. The subset will be the closure of the unit ball:

$$\overline{B} = \overline{B_1(0)}.$$

The sequence $\{e_k\}$ belongs to $\overline{B}$. For any $k \neq j$,

$$\|e_k - e_j\|_{\ell^2} = \left[\sum_{i=1}^{\infty} ((e_k)_i - (e_j)_i)^2\right]^{1/2} = \sqrt{1+1} = \sqrt{2}$$

(note that the sequence $e_k - e_j$ has exactly two nonzero terms, each of which has magnitude one). This shows that $\{e_k\}$ cannot be Cauchy, nor can any of its subsequences. Therefore Exercise 10.1.13 shows that $\{e_k\}$ has no convergent subsequences, and hence $\overline{B}$ is not sequentially compact.

10.2.1 Banach and Hilbert spaces

Given a normed linear space X with norm $\|\cdot\|$, the concept of a Cauchy sequence is defined just as in the previous section: $\{x_k\}$ is Cauchy if and only if for all $\epsilon > 0$, there exists a positive integer N such that

$$n, m \geq N \Rightarrow \|x_n - x_m\| < \epsilon.$$

Definition 440 *Let X be a normed linear space. We say that X is* complete *if and only if every Cauchy sequence in X converges to an element of X. A complete normed linear space is called a* Banach space.

If X is complete and the norm on X is defined by an inner product, then X is called a Hilbert space. *In other words, a Hilbert space is a complete inner product space.*

The reader should note that a Hilbert space is a special kind of Banach space.

We have already seen that every finite-dimensional vector space is complete, so $\mathbf{R}^n$ is a Hilbert space under the Euclidean dot product or a Banach space under the ℓ^1 or ℓ^∞ norms (or any other norm). We will now present an infinite-dimensional a Banach space, and an alternate norm on that space under which it is not complete. A key conclusion to draw from this example is that, on an infinite-dimensional space, not all norms are equivalent. (If all norms were equivalent, then the space would either be complete under all of them or complete under none of them.)

The space $C[a,b]$

The reader will recall that $C[a,b]$ is the space of all continuous, real-valued functions defined on the interval $[a,b]$. This space is infinite-dimensional; for example, the space $\mathcal{P}$ of all polynomials can be regarded as a subspace of $C[a,b]$, which shows that $C[a,b]$ is not finite-dimensional.

We will now discuss two norms that can be defined on $C[a,b]$. The first is the L^∞ norm:

$$\|f\|_\infty = \max\{|f(x)| \,:\, a \leq x \leq b\}.$$

Convergence in the L^∞ norm is related to uniform convergence.

Definition 441 *Let S be any set and suppose that*

$$f : S \to \mathbf{R},\ f_k : S \to \mathbf{R},\ k = 1, 2, \ldots,$$

are real-valued functions defined on S. We say that $\{f_k\}$ converges to f uniformly on S if and only if for all $\epsilon > 0$, there exists a positive integer N such that

$$k \geq N \ \Rightarrow\ |f_k(x) - f(x)| < \epsilon \text{ for all } x \in S.$$

Notice that if $\{f_k\}$ converges uniformly to f on S, then

$$\lim_{k\to\infty} f_k(x) = f(x) \text{ for all } x \in S. \tag{10.3}$$

With no further conditions, (10.3) defines *pointwise convergence* ((10.3) states that $\{f_k(x)\}$ converges at each point of S). Uniform convergence implies something stronger: not only does $\{f_k(x)\}$ converge for each $x \in S$, but these sequences converge at a uniform rate.

The L^∞ norm is sometimes described as the norm of uniform convergence: If $\{f_k\}$ is a sequence of functions in $C[a,b]$ and $\{f_k\}$ converges to a function $f[a,b] \to f$ in the sense that $\|f_k - f\|_\infty \to 0$, then $f_k \to f$ uniformly (see Exercise 2). A fundamental theorem of analysis states that if a sequence of continuous functions converges uniformly, then the limit function must also be continuous. Accepting this result, we can prove the following theorem.

Theorem 442 *$C[a,b]$ is complete under the L^∞ norm.*

Proof Let $\{f_k\}$ be a Cauchy sequence in $C[a,b]$. Therefore, if $\epsilon > 0$ is given, there exists a positive integer N such that

$$m,n \geq N \Rightarrow \|f_n - f_m\|_\infty < \epsilon.$$

For any $x \in [a,b]$, we have $|f_n(x) - f_m(x)| \leq \|f_n - f_m\|_\infty$, so

$$m,n \geq N \Rightarrow |f_n(x) - f_m(x)| < \epsilon.$$

But this implies that $\{f_k(x)\}$ is a Cauchy sequence in $\mathbf{R}$ and hence is convergent (because $\mathbf{R}$ is complete). This is true for each $x \in [a,b]$, so we can define $f : [a,b] \to \mathbf{R}$ by

$$f(x) = \lim_{k\to\infty} f_k(x), \ x \in [a,b].$$

We can then show that $\{f_k\}$ converges to f in the L^∞ norm (see Exercise 3). Since the uniform limit of a sequence of continuous functions is continuous, it follows that f is continuous, that is, $f \in C[a,b]$. Thus $f_k \to f$ in the L^∞ norm, and we have shown that $C[a,b]$ is complete.

QED

The preceding theorem shows that $C[a,b]$, under the L^∞ norm, is a Banach space.

The space $L^2(a,b)$

We will now consider the space $C[a,b]$ under a different norm, namely, under the $L^2(a,b)$ norm introduced in Section 6.1:

$$\|f\|_{L^2(a,b)} = \left[\int_a^b f(x)^2 \, dx\right]^{1/2}.$$

We will immediately show that the L^2 norm is not equivalent to the L^∞ norm by showing that $C[a,b]$ is not complete under the L^2 norm.

Example 443 *We define $f_k \in C[0,1]$ by*

$$f_k(x) = \begin{cases} 0, & 0 \leq x \leq \frac{1}{2} - \frac{1}{k+1}, \\ \frac{k+1}{2}\left(x - \frac{1}{2} + \frac{1}{k+1}\right), & \frac{1}{2} - \frac{1}{k+1} < x < \frac{1}{2} + \frac{1}{k+1}, \\ 1, & \frac{1}{2} + \frac{1}{k+1} \leq x \leq 1. \end{cases}$$

For any positive integers m,n with $m > n$, f_m and f_n agree except on the interval $[1/2 - 1/(n+1), 1/2 + 1/(n+1)]$, where the difference is certainly less than 1. Therefore,

$$\|f_m - f_n\|_{L^2(0,1)} = \left[\int_{\frac{1}{2}-\frac{1}{n+1}}^{\frac{1}{2}+\frac{1}{n+1}} (f_m(x) - f_n(x))^2 \, dx\right]^{1/2} \leq \sqrt{\frac{2}{n+1}}.$$

From this it follows that $\{f_k\}$ *is a Cauchy sequence. Moreover,* $\{f_k\}$ *converges to* $f : [0,1] \to \mathbf{R}$ *defined by*

$$f(x) = \begin{cases} 0, & 0 \le x \le \frac{1}{2}, \\ 1, & \frac{1}{2} < x \le 1 \end{cases}$$

(it is easy to show that $\|f_k - f\|_{L^2(0,1)} \to 0$*).*

The significance of Example 443 is that f is not continuous and hence does not belong to $C[0,1]$. Therefore $C[0,1]$ is not complete under the $L^2(0,1)$ norm, since $\{f_k\}$ is a Cauchy sequence that has no limit point within the space. This is in some ways analogous to a sequence of rational numbers converging to the irrational number $\sqrt{2}$, which shows that $\mathbf{Q}$ is not complete. However, in this case, we have a space that is complete under one norm but not under another. This is possible for an infinite-dimensional space but not for a finite-dimensional space.

Given a space that is not complete, it is always possible to complete that space by adding in limits for all Cauchy sequences. The larger space constructed in this way is called the *completion* of the original space. This is one way to construct the real numbers from the space of rational numbers, for example, and $\mathbf{R}$ can be regarded as the completion of $\mathbf{Q}$.

To define a completion of a normed vector space X precisely, we use the concept of equivalence classes: We define two Cauchy sequences $\{x_k\}$ and $\{y_k\}$ in the original space to be equivalent if $\|x_k - y_k\|_X \to 0$. It can be shown that this is an equivalence relation. The completion $\hat{X}$ is defined to be the space of all equivalence classes of Cauchy sequences. We then show that X can be regarded as a subspace of $\hat{X}$, and extend the vector space operations (addition and scalar multiplication) to $\hat{X}$. The completion $\hat{X}$ can then be shown to be complete.

The completion of $C[a,b]$ under the L^2 norm is called $L^2(a,b)$. As described above, the space $L^2(a,b)$ consists of equivalence classes of Cauchy sequences in $C[a,b]$. Informally, however, $L^2(a,b)$ is regarded as the space of all functions that are *square-integrable*:

$$L^2(a,b) = \left\{ f : (a,b) \to \mathbf{R} \mid \int_a^b f(x)^2 \, dx < \infty \right\}.$$

In this informal view of $L^2(a,b)$, f need not be continuous to belong to $L^2(a,b)$. For example, the function f in Example 443 has a discontinuity at $x = 1/2$ and yet belongs to $L^2(0,1)$. (To be precise, there is an equivalence class of Cauchy sequences in $C[0,1]$ that have a common limit of f; in this sense we can regard f as belonging to $L^2(0,1)$.)[3]

[3]There is another way to construct the space of square-integrable functions: $L^2(a,b)$ is the space of all measurable real-valued functions f defined on the interval $[a,b]$ such that

$$\int_a^b f(x)^2 \, dx < \infty.$$

The space $L^2(a, b)$ and related Hilbert spaces are critical in the modern theory of differential equations (particularly partial differential equations), where vector space methods are central.

Exercises

1. Prove Theorem 437.

2. Suppose f belongs to $C[a, b]$, $\{f_k\}$ is a sequence in $C[a, b]$, and

$$\|f_k - f\|_\infty \to 0 \text{ as } k \to \infty.$$

   Prove that $\{f_k\}$ converges uniformly to f on $[a, b]$.

3. Suppose $\{f_k\}$ is a Cauchy sequence in $C[a, b]$ (under the L^∞ norm) that converges pointwise to $f : [a, b] \to \mathbf{R}$. Prove that $f_k \to f$ in the L^∞ norm (that is, that $\|f_k - f\|_\infty \to 0$ as $k \to \infty$).

4. Let $f_k : [0, 1] \to \mathbf{R}$ be defined by $f_k(x) = x^k$. Prove that $\{f_k\}$ is Cauchy under the $L^2(0, 1)$ norm but not under the $C[0, 1]$ norm.

5. Recall that

$$C^1[a, b] = \{f : [a, b] \to \mathbf{R} \,:\, f, f' \text{ are continuous}\}.$$

   The $C^1[a, b]$ norm is defined by

$$\begin{aligned}\|f\|_{C^1[a,b]} &= \max\{\max\{|f(x)| \,:\, x \in [a, b]\}, \max\{|f'(x)| \,:\, x \in [a, b]\}\} \\ &= \max\{\|f\|_\infty, \|f'\|_\infty\}.\end{aligned}$$

   It can be shown that $C^1[a, b]$, under $\|\cdot\|_{C^1[a,b]}$, is complete. Prove that $C^1[a, b]$ is not complete under $\|\cdot\|_\infty$ by finding a sequence $\{f_k\}$ in $C^1[a, b]$ that is Cauchy under $\|\cdot\|_\infty$ but converges to a function that does not belong to $C^1[a, b]$.

10.3 Functional analysis

"Functional analysis" is the name given to the study of infinite-dimensional linear algebra. This name comes from the fact that we can define alternate

In this definition, the measure and integral used are the Lebesgue measure and integral (see [8] or [39]), and two functions are regarded as equal if they differ only on a set of measure zero. This alternate method is more intuitive, but the reader should note that in a precise development, one does not avoid using equivalence classes in this alternate construction. The elements of $L^2(a, b)$ are actually equivalence classes comprised of functions that differ only on a set of measure zero.

The question as to whether the two methods give the same result was a famous problem in analysis in the twentieth century; it was solved in by Meyers and Serrin in the paper [31].

notions of convergence on an infinite-dimensional vector space by considering the linear functionals defined on that space. We will give a brief introduction to functional analysis without, however, trying to prove all of our assertions.

A linear functional on a normed vector space V is simply a linear function $f : V \to \mathbf{R}$. We call it a "functional" rather than simply a "function" simply because, in many applications, V itself is a space of functions. Having a different word for functionals leads to less confusion.

If V is finite-dimensional, then every linear functional on V is continuous. However, this is not so in infinite dimensions, as the following example shows.

Example 444 *Let $V = C[0,1]$ under the $L^2(0,1)$ norm, and let $f : V \to \mathbf{R}$ be defined by $f(v) = v(1)$ for all $v \in V$ (that is, f evaluates the function $v = v(x)$ at $x = 1$). The function f is linear:*

$$f(u+v) = (u+v)(1) = u(1) + v(1) = f(u) + f(v) \text{ for all } u, v \in V,$$
$$f(\alpha v) = (\alpha v)(1) = \alpha v(1) = \alpha f(v) \text{ for all } v \in V, \alpha \in \mathbf{R}.$$

Now, suppose f is continuous. If $\{v_k\}$ is a sequence in V and $v_k \to v \in V$, then $f(v_k) \to f(v)$ must hold. But consider the sequence $\{v_k\}$ defined by $v_k(x) = x^k$. We have

$$\|v_k\|_{L^2(0,1)} = \left[\int_0^1 (x^k)^2\, dx\right]^{1/2} = \frac{1}{\sqrt{2k+1}},$$

which shows that $\|v_k\|_{L^2(0,1)} \to 0$ as $k \to \infty$, and hence that $v_k \to v$, where v is the zero function. But $f(v_k) = 1^k = 1$ for all k, while $f(v) = v(1) = 0$. Therefore $\{f(v_k)\}$ does not converge to $f(v)$, which shows that f is not continuous.

We want to exclude functionals like the one in the preceding example from consideration, so we restrict our attention to continuous linear functionals.

Definition 445 *Let V be a normed vector space over $\mathbf{R}$. The* dual space *V^* of V is the space of continuous linear functionals defined on V:*

$$V^* = \{f : V \to \mathbf{R} \,|\, f \text{ is linear and continuous}\}.$$

Before we can explore the properties of V^*, we develop an equivalent condition for the continuity of a linear functional.

Definition 446 *Let V be a normed vector space over $\mathbf{R}$, and let $f : V \to \mathbf{R}$ be linear. We say that f is* bounded *if and only if there exists a positive number M such that*

$$|f(v)| < M \text{ for all } v \in V, \|v\| \leq 1.$$

In most contexts, if we say that a real-valued function is bounded, we mean that it is bounded on its entire domain. Linear functionals (with the exception of the zero functional) are never bounded in this sense; instead, boundedness for a linear functional means that the functional is bounded on the unit ball $B_1(0)$.

Theorem 447 *Let V be a normed vector space, and let $f : V \to \mathbf{R}$ be linear. Then f is continuous if and only if it is bounded.*

Proof Suppose first that f is continuous. Then it is continuous at $v = 0$, which means that given any $\epsilon > 0$, there exists $\delta > 0$ such that

$$\|u - v\| < \delta \Rightarrow |f(u) - f(v)| < \epsilon.$$

Since v is the zero vector and $f(v) = 0$, this reduces to

$$\|u\| < \delta \Rightarrow |f(u)| < \epsilon.$$

Let us suppose $\epsilon > 0$ is given and fixed, and δ is chosen according to the definition of continuity at zero. Let δ' be any number satisfying $0 < \delta' < \delta$. Then, for any $u \in B_1(0)$, $\delta' u$ satisfies $\|\delta' u\| < \delta$ and hence $|f(\delta' u)| < \epsilon$. By linearity, this yields

$$|f(u)| < M = \frac{\epsilon}{\delta'}.$$

Since this holds for all $u \in V$, we see that f is bounded.

Conversely, suppose f is bounded:

$$\|u\| \leq 1 \Rightarrow |f(u)| < M.$$

Let v be any vector in V and let $\epsilon > 0$ be given. Let $\delta = \epsilon/M$. Let $u \in V$ satisfy $\|u - v\| < \delta$ and define $w = u - v$. Then

$$\left| f\left(\frac{w}{\|w\|}\right) \right| < M,$$

which by linearity implies that

$$|f(w)| < M\|w\| < M\delta = \epsilon.$$

But then

$$\|u - v\| < \delta \Rightarrow |f(u) - f(v)| = |f(u - v)| < \epsilon,$$

and thus f is continuous at v. Since this holds for all $v \in V$, we see that f is continuous.

QED

Like any other function space, V^* is a vector space. The following results allow us to define a norm on V^*.

Lemma 448 *Let V be a normed vector space over $\mathbf{R}$ and let $f \in V^*$. Then*

$$\begin{aligned} & \sup\{|f(v)| : v \in V, \|v\| \leq 1\} \\ = \; & \inf\{M > 0 : |f(v)| \leq M \text{ for all } v \in V, \|v\| \leq 1\}. \end{aligned}$$

Proof Exercise 1.

Theorem 449 *Let V be a normed vector space. For each $f \in V^*$, define*

$$\|f\|_{V^*} = \sup\{|f(v)| \,:\, v \in V, \|v\|_V \leq 1\}.$$

Then $\|\cdot\|_{V^*}$ defines a norm on V^*.

Proof The previous lemma guarantees that $\|f\|_{V^*}$ is well-defined for all f in V^*. The definition of $\|\cdot\|_{V^*}$ shows that $\|f\|_{V^*} \geq 0$ for all $f \in V^*$. Moreover, if $f \in V^*$, $f \neq 0$, then there exists $v \in V$ such that $f(v) \neq 0$. Then $f(w)$ is also nonzero, where $w = v/\|v\|_V$, and

$$\sup\{|f(v)| \,:\, v \in V, \|v\|_V \leq 1\} \geq |f(w)| > 0.$$

It follows that $\|f\|_{V^*} = 0$ if and only if $f = 0$.

The second property of a norm is straightforward to verify:

$$\begin{aligned}
\|\alpha f\|_{V^*} &= \sup\{|(\alpha f)(v)\| \,:\, v \in V, \|v\|_V \leq 1\} \\
&= \sup\{|\alpha(f(v))\| \,:\, v \in V, \|v\|_V \leq 1\} \\
&= \sup\{|\alpha||f(v)\| \,:\, v \in V, \|v\|_V \leq 1\} \\
&= |\alpha| \sup\{|f(v)\| \,:\, v \in V, \|v\|_V \leq 1\} \\
&= |\alpha|\|f\|_{V^*}.
\end{aligned}$$

Finally, we can prove the triangle inequality as follows:

$$\begin{aligned}
\|f+g\|_{V^*} &= \sup\{|(f+g)(v)| \,:\, v \in V, \|v\| \leq 1\} \\
&\leq \sup\{|f(v)| + |g(v)| \,:\, v \in V, \|v\| \leq 1\} \\
&\leq \sup\{|f(v)| \,:\, v \in V, \|v\| \leq 1\} + \sup\{|g(v)| \,:\, v \in V, \|v\| \leq 1\} \\
&= \|f\|_{V^*} + \|g\|_{V^*}.
\end{aligned}$$

For a justification of the second step in this reasoning, see Exercise 2.

QED

The space V^* is always complete, even if V is not. The proof of this fact is lengthy but elementary.

Theorem 450 *Let V be a normed vector space. Then V^*, under the norm defined in Theorem 449, is complete.*

Proof Exercise 3.

Although boundedness and continuity are equivalent for linear functionals, the concept of boundedness is usually more useful. The reason is the following result.

Theorem 451 *Let V be a normed vector space and let f belong to V^*. Then*

$$|f(v)| \leq \|f\|_{V^*}\|v\|_V \text{ for all } v \in V.$$

Proof Exercise 4.

Example 452 *In Section 6.1, we briefly introduced the space $L^1(a,b)$ of integrable functions on the interval $[a,b]$. To be precise, $L^1(a,b)$ is the space of all measurable real-valued functions f defined on the interval $[a,b]$ such that*

$$\int_a^b |f(x)|\,dx < \infty.$$

In this definition, the measure and integral used are the Lebesgue measure and integral (see [8] or [39]), and two functions are regarded as equal if they differ only on a set of measure zero.[4]

Suppose now that $g:[a,b]\to\mathbf{R}$ is bounded and measurable. We then have

$$\begin{aligned}\left|\int_a^b f(x)g(x)\,dx\right| &\le \int_a^b |f(x)g(x)|\,dx\\ &\le \max\{|g(x)| : a\le x\le b\}\int_a^b |f(x)|\,dx\\ &= \|g\|_\infty \|f\|_{L^1(a,b)}.\end{aligned}$$

We can therefore define a linear functional $\ell : L^1(a,b)\to\mathbf{R}$ by

$$\ell(f) = \int_a^b f(x)g(x)\,dx.$$

We have seen that $|\ell(f)| \le \|g\|_\infty\|f\|_{L^1(a,b)}$, and therefore ℓ is bounded (continuous). It follows that $\ell\in(L^1(a,b))^*$.

In fact, every bounded linear functional on $L^1(a,b)$ is defined in this fashion. To be precise, given $\ell\in(L^1(a,b))^*$, there exists $g\in L^\infty(a,b)$ such that

$$\ell(f) = \int_a^b f(x)g(x)\,dx.$$

The space $L^\infty(a,b)$ is the space of all measurable real-valued functions defined on $[a,b]$ that are bounded almost everywhere (that is, except on a set of Lebesgue measure zero). In other words, the dual of $L^1(a,b)$ is $L^\infty(a,b)$. The proof of this fact is beyond the scope of this book.

10.3.1 The dual of a Hilbert space

The reader will recall that a Hilbert space is a complete inner product space. Such spaces have the special property that the dual of a Hilbert space H is

[4]In the previous section, we described two ways to construct $L^2(a,b)$. There are also two ways to construct $L^1(a,b)$ and, in general, $L^p(a,b)$ for any p satisfying $1\le p<\infty$. Here we have chosen the more intuitive description.

isomorphic to H itself. This fact is a consequence of the Riesz Representation theorem to be proved below. We will need some preliminary results. We begin by exploring the concept of orthogonality in a Hilbert space.

One of the most useful consequences of orthogonality is the projection theorem, which allows us to find the best approximation to a given vector from a finite-dimensional subspace. It turns out that the same result holds for an infinite-dimensional subspace, provided the subspace is closed. The proof of this fact is based on the parallelogram law

$$\|u+v\|^2 + \|u-v\|^2 = 2\|u\|^2 + 2\|v\|^2 \text{ for all } u, v \in H,$$

which holds in any inner product space H (see Exercise 6.1.11).

Theorem 453 (The projection theorem) *Let H be a Hilbert space over $\mathbf{R}$, and let S be a closed subspace of H.*

1. *For any $v \in H$, there is a unique best approximation to v from S, that is, a unique $w \in S$ satisfying*

$$\|v-w\| = \min\{\|v-z\| : z \in S\}. \tag{10.4}$$

2. *A vector $w \in S$ is the best approximation to v from S if and only if*

$$\langle v-w, z\rangle = 0 \text{ for all } z \in S. \tag{10.5}$$

Proof The reader may recall from Section 6.4 that when S is finite-dimensional, we can prove the existence of the best approximation to v from S directly, by using a basis for S. When S is infinite-dimensional, we have to adopt a less direct approach. Since $\|v-z\|_H \geq 0$ for all $z \in S$,

$$d = \inf\{\|v-z\|_H : z \in S\}$$

is a nonnegative number. There exists a sequence $\{z_k\}$ in S such that

$$\lim_{k\to\infty} \|v-z_k\|_H = d$$

(the sequence $\{z_k\}$ is called a minimizing sequence). We will prove that $\{z_k\}$ is a Cauchy sequence. Applying the parallelogram law, we have

$$\begin{aligned}\|z_m - z_n\|_H^2 &= \|(z_m-v)-(z_n-v)\|_H^2 \\ &= 2\|z_m-v\|_H^2 + 2\|z_n-v\|_H^2 - \|(z_m-v)+(z_n-v)\|_H^2.\end{aligned}$$

Now notice that

$$\begin{aligned}\|(z_m-v)+(z_n-v)\|_H = \|z_m+z_n-2v\|_H &= \left\|2\left(\frac{z_m+z_n}{2}-v\right)\right\|_H \\ &= 2\left\|\frac{z_m+z_n}{2}-v\right\|_H.\end{aligned}$$

Since S is a subspace, $(z_m + z_n)/2 \in S$, and therefore

$$\left\| \frac{z_m + z_n}{2} - v \right\|_H \geq d.$$

But then

$$\begin{aligned} \|z_m - z_n\|_H^2 &= \|(z_m - v) - (z_n - v)\|_H^2 \\ &= 2\|z_m - v\|_H^2 + 2\|z_n - v\|_H^2 - \|(z_m - v) + (z_n - v)\|_H^2 \\ &\leq 2\|z_m - v\|_H^2 + 2\|z_n - v\|_H^2 - 4d^2, \end{aligned}$$

and since $\|z_m - v\|_H, \|z_n - v\|_H \to d$, we see that $\|z_m - z_n\|_H \to 0$ as $m, n \to \infty$. From this it is easy to prove that $\{z_k\}$ is a Cauchy sequence. Since H is complete, there exists $w \in H$ such that $z_k \to w$. Since $\{z_k\}$ belongs to S and S is closed, it follows that $w \in S$. Finally, the continuity of the norm implies that $\|z_k - v\|_H \to \|w - v\|_H$. But we already know that $\|z_k - v\|_H \to d$, and so $\|w - v\|_H = d$. Thus w is a best approximation to v from S.

Condition (10.5) can be proved exactly as in Section 6.4. The uniqueness of w can be derived from (10.5); if $u \in S$ is another best approximation to v, then we have

$$\langle w - v, z \rangle_H = 0, \ \langle u - v, z \rangle_H = 0 \text{ for all } z \in S.$$

Subtracting yields

$$\langle w - u, z \rangle_H = 0 \text{ for all } z \in S.$$

Since $w - u \in S$, it follows that $w - u = 0$, that is, $w = u$. This completes the proof.

QED

As in a finite-dimensional space, we define the orthogonal complement $S^\perp$ of a subspace S by

$$S^\perp = \{v \in H : \langle v, u \rangle_H = 0 \text{ for all } u \in S\}.$$

The reader will recall the fundamental result that if S is a subspace of a finite-dimensional inner product space, then $(S^\perp)^\perp = S$. The same result holds in a Hilbert space, provided S is not only a subspace but also a closed set.

Theorem 454 *Let H be a Hilbert space and let S be a closed subspace of H. Then $(S^\perp)^\perp = S$.*

Proof The proof is the same as that of Theorem 303. Notice that S must be closed so that we can apply the projection theorem.

QED

We need one more preliminary result before we can prove the Riesz representation theorem. An infinite-dimensional subspace S might have the property that $S^\perp$ is finite-dimensional. In such a case we speak of the *co-dimension* of S, which is just the dimension of $S^\perp$. The following lemma will allow us to show that one vector is enough to define a bounded linear functional on a Hilbert space.

Lemma 455 *Let H be a Hilbert space, and let $f \in H^*$, $f \neq 0$. Then* $\ker(f)$ *is a closed subspace with co-dimension one.*

Proof We already know that $\ker(f)$ is a subspace. If $\{v_k\}$ is a sequence in $\ker(f)$ and $v_k \to v \in H$, then, by continuity of f, we have

$$f(v) = \lim_{k\to\infty} f(v_k) = \lim_{k\to\infty} 0 = 0.$$

Therefore $f(v) = 0$, which implies that $v \in \ker(f)$. Thus $\ker(f)$ is closed.

If u and w are nonzero vectors in $\ker(f)^\perp$, then $f(u)$ and $f(w)$ are nonzero. It follows that $f(u) - \alpha f(w) = 0$, where

$$\alpha = \frac{f(u)}{f(w)}.$$

But then the linearity of f implies that $f(u - \alpha w) = 0$, whence $u - \alpha w \in \ker(f)$. But $u - \alpha w$ also belongs to $\ker(f)^\perp$, and the only vector belonging to both $\ker(f)$ and $\ker(f)^\perp$ is the zero vector. Thus $u - \alpha w = 0$, that is, $u = \alpha w$. Since f is not the zero functional, $\ker(f)^\perp$ contains at least one nonzero vector w; however, the above reasoning shows that every other nonzero vector in $\ker(f)^\perp$ is a multiple of w. Thus $\ker(f)^\perp$ is one-dimensional, that is, $\ker(f)$ has co-dimension one.

QED

Theorem 456 (Riesz representation theorem) *Let H be a Hilbert space over* $\mathbf{R}$. *If $f \in H^*$, then there exists a unique vector u in H such that*

$$f(v) = \langle v, u\rangle_H \text{ for all } v \in H.$$

Moreover, $\|u\|_H = \|f\|_{H^*}$.

Proof We will prove existence, and leave uniqueness as an exercise (see 6).

If f is the zero functional, then v can be taken to be the zero vector in H, and the conclusion of the theorem obviously holds. We will therefore assume that f is nonzero and take w to be any nonzero vector in $\ker(f)^\perp$. Next, we define $u \in \ker(f)$ by

$$u = \frac{f(w)}{\|w\|_H^2} w.$$

Then

$$\langle w, u\rangle_H = \left\langle w, \frac{f(w)}{\|w\|_H^2} w \right\rangle_H = \frac{f(w)}{\|w\|_H^2} \langle w, w\rangle_H = f(w).$$

Therefore, $f(w) = \langle w, u \rangle_H$ and hence, by linearity, $f(\beta w) = \langle \beta w, u \rangle_H$ for all $\beta \in \mathbf{R}$. Since $\ker(f)^{\perp} = \text{sp}\{w\}$, it follows that $f(x) = \langle x, u \rangle_H$ for all x in $\ker(f)^{\perp}$.

Every vector $v \in H$ can be written as

$$v = x + y, \ x \in \ker(f)^{\perp}, y \in \ker(f).$$

It follows that

$$f(v) = f(x + y) = f(x) + f(y) = f(x) = \langle x, u \rangle_H$$

($f(y) = 0$ since $y \in \ker(f)$), while

$$\langle v, u \rangle_H = \langle x + y, u \rangle_H = \langle x, u \rangle_H + \langle y, u \rangle_H = \langle x, u \rangle_H$$

($\langle y, u \rangle_H = 0$ since $y \in \ker(f)$, $u \in \ker(f)^{\perp}$). Thus we see that $f(v) = \langle v, u \rangle_H$ for all $v \in H$.

Finally, by the Cauchy-Schwarz inequality,

$$|f(v)| = |\langle v, u \rangle_H| \leq \|v\|_H \|u\|_H \text{ for all } v \in H,$$

which shows that $\|f\|_{H^*} \leq \|u\|_H$. On the other hand,

$$|f(u)| = |\langle u, u \rangle_H| = \|u\|_H \|u\|_H,$$

which shows that $\|f\|_{H^*} \geq \|u\|_H$. Thus we see that $\|f\|_{H^*} = \|u\|_H$.

QED

Exercises

1. Prove Lemma 448.

2. Let S be any set and let $f : S \to \mathbf{R}$, $g : S \to \mathbf{R}$ be given functions. Prove that

$$\sup\{f(x) + g(x) : x \in S\} \leq \sup\{f(x) : x \in S\} + \sup\{g(x) : x \in S\}.$$

3. Prove Theorem 450. (Hint: For a given Cauchy sequence $\{f_k\}$ in V^*, show that $\{f_k(v)\}$ is a Cauchy sequence in $\mathbf{R}$ for each $v \in V$. Define $f : V \to \mathbf{R}$ by $f(v) = \lim_{k\to\infty} f_k(v)$ and show that f is linear and bounded. Finally, show that $\|f_k - f\|_{V^*} \to 0$ as $k \to \infty$.

4. Prove Theorem 451.

5. Suppose H is a Hilbert space and S is a subspace of H that fails to be closed. What is $(S^{\perp})^{\perp}$ in this case?

6. Prove that the vector $u \in H$ satisfying the conclusions of the Riesz representation theorem is unique.

7. Let X, U be Hilbert spaces, and let $T : X \to U$ be linear. We say that T is *bounded* if and only if there exists $M > 0$ such that

$$\|T(x)\|_U \le M\|x\|_X \text{ for all } x \in X.$$

Prove that T is continuous if and only if T is bounded. (Hint: The proof is similar to the proof that a linear functional is continuous if and only if it is bounded.)

8. Let X, U be Hilbert spaces, and let $T : X \to U$ be linear and bounded. Use the Riesz representation theorem to prove that there exists a unique bounded linear operator $T^* : U \to X$ such that

$$\langle T(x), u\rangle_U = \langle x, T^*(u)\rangle_X \text{ for all } x \in X, u \in U.$$

The operator T^* is called the *adjoint* of T.

10.4 Weak convergence

We will now explain the main reason that it is useful to study linear functionals on infinite-dimensional vector spaces. A norm on a vector space defines a topology, that is, a collection of open sets. The reader will recall that convergence, continuity, compactness, and other basic notions can be defined in terms of open sets if desired, so reference to the norm topology implies, among other things, the notion of convergence defined by the norm.

We have seen several examples of infinite-dimensional vector spaces, including $\mathcal{P}$ (the space of all polynomials), $C[a, b]$ (for which $\mathcal{P}$ can be regarded as a subspace), and the space ℓ^2 of square-summable sequences. In Section 10.2, we showed that the unit ball in ℓ^2 is not sequentially compact in the norm topology. In fact, the unit ball in an infinite-dimensional vector space is never sequentially compact with respect to the norm topology. We will prove this result for inner product spaces, where the proof is simpler, but it holds in general normed vector spaces as well.

Theorem 457 *Let H be an inner product space over $\mathbf{R}$, and let $\overline{B}$ be the closed unit ball in H. Then $\overline{B}$ is sequentially compact (with respect to the norm topology) if and only if H is finite-dimensional.*

Proof We have already seen that if H is finite-dimensional, then it is isomorphic to $\mathbf{R}^n$, and the closed unit ball in $\mathbf{R}^n$ is sequentially compact (since it is a closed and bounded set). Therefore, if H is finite-dimensional, then $\overline{B}$ is sequentially compact.

To prove the converse, we assume that H is infinite-dimensional and prove that $\overline{B}$ is not sequentially compact by exhibiting a sequence in $\overline{B}$ that has no convergent subsequence. We will do this by constructing a sequence $\{x_k\}$ such that $\|x_k\|_H = 1$ for all k and $\langle x_k, x_\ell\rangle_H = 0$ for all $k \neq \ell$. From this it follows that $\|x_k - x_\ell\|_H = \sqrt{2}$ for all k, ℓ, and hence no subsequence of $\{x_k\}$ can be Cauchy. Therefore, no subsequence of $\{x_k\}$ is convergent.

We will use induction to construct an orthogonal set $\{x_1, x_2, x_3, \ldots\}$ with $\|x_k\|_H = 1$ for all $k = 1, 2, 3 \ldots$. We begin by choosing any nonzero vector with norm one and calling it x_1. Given an orthonormal set $\{x_1, x_2, \ldots, x_n\}$, we choose any vector $y \in H$, $y \notin S_n = \text{sp}\{x_1, x_2, \ldots, x_n\}$. We then define $w = y - \text{proj}_{S_n} y$ and $x_{n+1} = w/\|w\|_H$ (the definition of w involves projection onto a finite-dimensional subspace, as in original version of the projection theorem presented in Section 6.4). Then $\{x_1, x_2, \ldots, x_n, x_{n+1}\}$ is also orthonormal (cf. the Gram-Schmidt process in Section 6.5). Since H is infinite-dimensional, it is always possible to choose $y \notin S_n$, regardless of how large n is. Thus the desired sequence $\{x_k\}$ exists, and the proof is complete.

QED

We emphasize that only the proof above depends on the existence of an inner product; the result is equally valid in a normed linear space where the norm is not defined by an inner product.

Corollary 458 *If H is an infinite-dimensional Hilbert space and S is a subset of H with a nonempty interior, then S is not sequentially compact.*

Proof Exercise 1

Compactness is such a useful property that it is worth asking whether we could change the topology so as to make the closed unit ball compact. This is possible. We will sketch this development, although providing all of the details is beyond the scope of this book.

Definition 459 *Let H be a Hilbert space over* **R**. *The* weak topology *on H is the weakest topology (that is, the topology with the fewest open sets) such that each $f \in H^*$ is still continuous. If a sequence in H converges with respect to the weak topology, then it is said to* converge weakly *or to be* weakly convergent.

We have not carefully defined a topology, so the reader cannot be expected to apply the above definition precisely without further study. However, there is a simple characterization of convergence of sequences in the weak topology.

Theorem 460 *Let H be a Hilbert space over* **R**, *and let $\{x_k\}$ be a sequence in H. Then $\{x_k\}$ converges weakly to $x \in H$ if and only if*

$$f(x_k) \to f(x) \text{ for all } f \in H^*.$$

By the Riesz representation theorem, we can equivalently say that $\{x_k\}$ converges weakly to $x \in H$ if and only if

$$\langle x_k, u\rangle_H \to \langle x, u\rangle_H \text{ for all } u \in H.$$

Example 461 *Consider the Hilbert space ℓ^2 introduced in Section 10.2, and the sequence $\{e_k\}$ defined there. Recall that each element of ℓ^2 is a square-summable sequence, and e_k is the sequence whose terms are all zero, except the kth term, which is one. We have already seen that $\{e_k\}$ is not convergent in the norm topology. However, given any $u \in \ell^2$, we have $\langle e_k, u \rangle_{\ell^2} = u_k$ (where u_k is the kth term in the sequence u) and $u_k \to 0$ since*

$$\sum_{k=1}^{\infty} u_k^2 < \infty.$$

It follows that $\{e_k\}$ converges weakly to the zero vector in ℓ^2.

The sequence $\{x_k\}$ converges to x under the norm topology if and only if

$$\|x_k - x\| \to 0 \text{ as } k \to \infty.$$

When we need to distinguish convergence in the norm topology from weak convergence, we say $x_k \to x$ *strongly* or $x_k \to x$ *in norm*.

Convergence of sequences does not completely characterize the weak topology; if we want to completely describe a topology in terms of convergence, we need the concept of a *net*, which is a generalization of the concept of a sequence. Nets are beyond the scope of this brief introduction to the weak topology. To avoid introducing the additional background needed for more generality, we will restrict some of our results to refer only to sequences. For instance, at certain points we will refer to a set S as *sequentially closed*, meaning that if $\{x_k\} \subset S$ converges to x (in the topology under consideration), then $x \in S$.

As suggested above, the weak topology is useful because sets that are not compact in the norm topology can be compact in the weak topology.

Theorem 462 *Let H be a Hilbert space over $\mathbf{R}$. Then the closed unit ball $\overline{B}$ is sequentially compact in the weak topology.*

The proof of this result is beyond the scope of this book.

Corollary 463 *Let H be a Hilbert space over $\mathbf{R}$, and let S be a closed and bounded subset of H. If $\{x_k\}$ is a sequence in S, then there exists a subsequence $\{x_{k_j}\}$ and a vector $x \in H$ such that $x_{k_j} \to x$ weakly.*

Proof Exercise 2.

Here are a couple of elementary facts about weak convergence.

Theorem 464 *Let H be a Hilbert space over $\mathbf{R}$, and suppose $\{x_k\}$ is a sequence in H converging strongly to $x \in H$. Then $x_k \to x$ weakly.*

Proof Exercise 3

Theorem 465 *Let H be a Hilbert space over $\mathbf{R}$, and suppose $\{x_k\}$ is a sequence in H. If $x_k \to x \in H$ weakly and $\|x_k\| \to \|x\|$, then $x_k \to x$ strongly.*

Proof The result is proved by the following calculation:

$$\begin{aligned}\|x_k - x\|^2 = \langle x_k - x, x_k - x\rangle &= \langle x_k, x_k\rangle - 2\langle x_k, x\rangle + \langle x, x\rangle \\ &= \|x_k\|^2 - 2\langle x_k, x\rangle + \|x\|^2 \\ &\to \|x\|^2 - 2\langle x, x\rangle + \|x\|^2 = 0.\end{aligned}$$

QED

From the preceding result, we see that if $x_k \to x$ weakly but not strongly, then $\|x_k\| \not\to \|x\|$. However, there is a general relationship between the norms of the terms x_k and $\|x\|$. To describe this relationship, we need the following concept.

Definition 466 *Let $\{\alpha_k\}$ be a sequence of real numbers. The* limit inferior *of $\{\alpha_k\}$ is defined by*

$$\liminf_{k\to\infty} \alpha_k = \lim_{k\to\infty} \inf\{\alpha_\ell : \ell \ge k\}.$$

Similarly, we define the limit superior *of $\{\alpha_k\}$ as*

$$\limsup_{k\to\infty} \alpha_k = \lim_{k\to\infty} \sup\{\alpha_\ell : \ell \ge k\}.$$

It is possible that $\inf\{\alpha_k : k \ge 1\} = -\infty$; in this case, there is a subsequence of $\{a_k\}$ converging to $-\infty$, $\inf\{x_\ell : \ell \ge k\} = -\infty$ for all k, and we obtain $\liminf_{k\to\infty} \alpha_k = -\infty$. Otherwise, $\{\inf\{x_\ell : \ell \ge k\}\}$ is a monotonically increasing sequence of real numbers, and $\liminf_{k\to\infty} \alpha_k$ exists as a real number or ∞. Thus $\liminf_{k\to\infty} \alpha_k$ always exists, although it may equal $\pm\infty$. The same is true of $\limsup_{k\to\infty} \alpha_k$.

Theorem 467 *Let $\{\alpha_k\}$ be a sequence of real numbers.*

1. *There exists a subsequence $\{\alpha_{k_j}\}$ such that*

$$\lim_{j\to\infty} \alpha_{k_j} = \liminf_{k\to\infty} \alpha_k.$$

2. *There exists a subsequence $\{\alpha_{k_j}\}$ such that*

$$\lim_{j\to\infty} \alpha_{k_j} = \limsup_{k\to\infty} \alpha_k.$$

3. *If $\{\alpha_{k_j}\}$ is any convergent subsequence of $\{\alpha_k\}$, then*

$$\liminf_{k\to\infty} \alpha_k \le \lim_{j\to\infty} \alpha_{k_j} \le \limsup_{k\to\infty} \alpha_k.$$

4. *If $\lim_{k\to\infty} \alpha_k$ exists, then*

$$\liminf_{k\to\infty} \alpha_k = \limsup_{k\to\infty} \alpha_k = \lim_{k\to\infty} \alpha_k.$$

Proof Exercise 4.

Here is the relationship between the norm and weakly convergent sequences.

Theorem 468 *Let H be a Hilbert space over* $\mathbf{R}$*, and let $\{x_k\}$ be a sequence in H converging weakly to $x \in H$. Then*

$$\|x\| \leq \liminf_{k\to\infty} \|x_k\|.$$

Proof It is obvious that $\liminf_{k\to\infty} \|x_k\| \geq 0$; hence the result is obvious for $x = 0$. Hence we will assume that x is not the zero vector.

Let $\{x_{k_j}\}$ be a subsequence of $\{x_k\}$ such that

$$\lim_{j\to\infty} \|x_{k_j}\| = \liminf_{k\to\infty} \|x_k\|.$$

We have

$$\langle x_{k_j}, x\rangle \leq \|x_{k_j}\|\|x\|$$

and thus

$$\|x\|^2 = \lim_{j\to\infty} \langle x_{k_j}, x\rangle \leq \lim_{j\to\infty} \|x_{k_j}\|\|x\| = \|x\| \liminf_{k\to\infty} \|x_k\|.$$

Since $\|x\| \neq 0$ by assumption, this yields the desired result.

QED

We have terminology to describe the preceding result.

Definition 469 *Let X be a normed vector space and let $f : X \to$* $\mathbf{R}$.

1. *We say that f is* lower semicontinuous *at x if*

$$f(x) \leq \liminf_{k\to\infty} f(x_k)$$

for all sequences $x_k \to x$.

2. *We say that f is* upper semicontinuous *at x if*

$$f(x) \geq \limsup_{k\to\infty} f(x_k)$$

for all sequences $x_k \to x$.

Two points must be made about these definitions. First of all, continuity or semicontinuity of a function is relative to an underlying topology. In the case of the above definition, the topology in question is the one that defines convergence of $\{x_k\}$ to x. Thus a function could be lower semicontinuous under one topology but not under another. Theorem 468 says (essentially) that the norm is *weakly lower semicontinuous.* Example 461 shows that the norm need not be weakly continuous; in that example, $\epsilon_k \to 0$ weakly, but $\|e_k\| \not\to 0$ (in fact, $\|e_k\| = 1$ for all k).

The second point about Definition 469 is that it defines semicontinuity in terms of sequences. In a topology like the weak topology, which is not completely described by convergence of sequences, the correct definition of, for example, lower semicontinuity requires that

$$f(x) \leq \liminf_{y \to x} f(y).$$

However, we cannot define this type of limit inferior without a better understanding of topologies that are not defined by norms. Thus Definition 469 really defines what one might call *sequential semicontinuity*. We will not use the additional modifier in the remainder of this brief discussion.

Here is a consequence of the weak lower semicontinuity of the norm.

Theorem 470 *Let H be a Hilbert space over $\mathbf{R}$, and let S be a closed and bounded subset of H. If S is also closed with respect to the weak topology, then there exists $\overline{x} \in S$ such that*

$$\|\overline{x}\| = \inf\{\|x\| \, : \, x \in S\}.$$

Proof Let $\{x_k\} \subset S$ be a minimizing sequence:

$$\lim_{k \to \infty} \|x_k\| = \inf\{\|x\| \, : \, x \in S\}.$$

Since S is closed and bounded (and closed in the weak topology), it is weakly sequentially compact, and thus there exists a subsequence $\{x_{k_j}\}$ and $\overline{x} \in S$ such that $x_{k_j} \to \overline{x}$ weakly. But then we have

$$\|\overline{x}\| \leq \liminf_{j \to \infty} \|x_{k_j}\| = \lim_{j \to \infty} \|x_{k_j}\| = \inf\{\|x\| \, : \, x \in S\},$$

as desired.

QED

The reader will recall that one of the main uses of compactness is to prove the existence of maximizers and minimizers by Weierstrass's theorem: A continuous function attains its maximum and minimum over a sequentially compact set. In infinite dimensions, one rarely works over a compact set, at least if compactness is defined by the norm topology (recall that by Corollary 458, no infinite-dimensional subset with a nonempty interior is compact in the norm topology). Weak compactness is often available; however, functions that are continuous with respect to the norm topology are often not continuous with respect to the weak topology. In short, under a weaker topology, more sets are compact but fewer functions are continuous. Nevertheless, as the proof of the previous theorem shows, weak lower semicontinuity is an adequate substitute for continuity.

We will end our brief introduction to analysis in infinite-dimensional space with one class of problems for which we have both weak sequential compactness and weak lower semicontinuity.

10.4.1 Convexity

We begin with the following definitions.

Definition 471 *Let V be a vector space over $\mathbf{R}$, and let C be a subset of V. We say that C is* convex *if and only if*

$$x, y \in C,\ \alpha \in [0,1] \ \Rightarrow\ (1-\alpha)x + \alpha y \in C.$$

We refer to $(1-\alpha)x+\alpha y$ ($\alpha \in [0,1]$) as a convex combination *of x and y; the set of all convex combinations of x and y form the* line segment *with endpoints x and y.*

Examples of convex sets include subspaces (which include all linear combinations, not just convex combinations) and balls $B_r(x)$ in any normed linear space (see Exercise 5).

Definition 472 *Let C be a convex subset of a vector space V over $\mathbf{R}$. We say that $f : C \to \mathbf{R}$ is a* convex *function if*

$$f((1-\alpha)x + \alpha y) \le (1-\alpha)f(x) + \alpha f(y) \text{ for all } x, y \in C, \alpha \in [0,1].$$

The characteristic feature of a convex function is that its graph lies on or below any chord joining two points on the graph (see Figure 10.1).

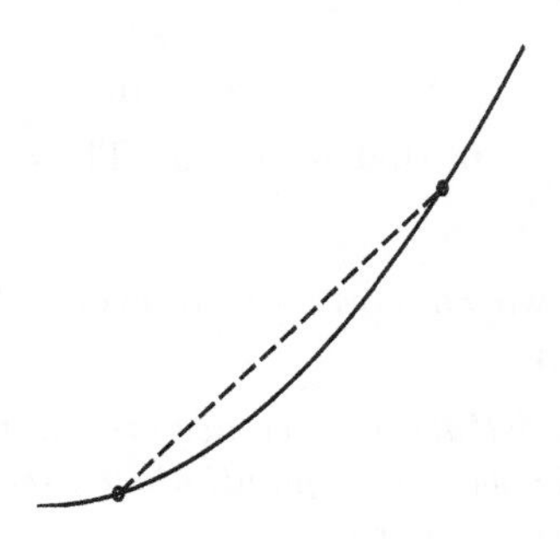

FIGURE 10.1
The graph of a convex function.

We now proceed to develop a few of the main properties of convex sets and functions in Hilbert spaces. We will see the fundamental role played by weak convergence. The first fact is that the projection theorem extends to any closed convex subset, not just to closed subspaces.

Theorem 473 *Let H be a Hilbert space over $\mathbf{R}$, and let C be a nonempty, closed, convex subset of H. For any $x \in H$, there exists a unique $\overline{x} \in C$ such that*

$$\|x - \overline{x}\| = \inf\{\|x - z\| \,:\, z \in C\}.$$

Moreover, $\overline{x}$ is the unique vector in C satisfying

$$\langle x - \overline{x}, z - \overline{x} \rangle \leq 0 \text{ for all } z \in C. \tag{10.6}$$

Proof The existence of a minimizer $\overline{x}$ is proved exactly as in the proof of Theorem 453 (choose a minimizing sequence, use the parallelogram law to prove that it is a Cauchy sequence and thus converges to some $\overline{x}$, and use the fact that C is closed to argue that $\overline{x} \in C$). We then have

$$(1-\alpha)\overline{x} + \alpha z = \overline{x} + \alpha(z - \overline{x}) \in C \text{ for all } \alpha \in [0,1].$$

If we define $\phi : [0,1] \to \mathbf{R}$ by

$$\phi(\alpha) = \|x - (\overline{x} + \alpha(z - \overline{x}))\|^2 = \|x - \overline{x}\|^2 - 2\alpha \langle x - \overline{x}, z - \overline{x} \rangle + \alpha^2 \|z - \overline{x}\|^2,$$

then $\phi(\alpha) \geq \phi(0)$ for all $\alpha \in [0,1]$. It follows that $\phi'(\alpha) \geq 0$. But

$$\phi'(0) = -2\alpha \langle x - \overline{x}, z - \overline{x} \rangle,$$

and thus we see that (10.6) must hold.

Finally, if $\hat{x} \in C$ also satisfies

$$\|x - \hat{x}\| = \inf\{\|x - z\| : z \in C\},$$

then, by the preceding argument, $\hat{x}$ also satisfies (10.6) (with $\hat{x}$ in place of $\overline{x}$):

$$\langle x - \hat{x}, z - \hat{x} \rangle \leq 0 \text{ for all } z \in C. \tag{10.7}$$

Substituting $\hat{x}$ for z in (10.6), $\overline{x}$ for z in (10.7), adding, and manipulating yields

$$\langle \hat{x} - \overline{x}, \hat{x} - \overline{x} \rangle \leq 0.$$

This implies that $\|\hat{x} - \overline{x}\| = 0$, that is, $\hat{x} = \overline{x}$. Thus $\overline{x}$ is unique.

QED

Using Theorem 473, we can prove that every closed convex set is also weakly sequentially closed.

Theorem 474 *Let H be a Hilbert space over $\mathbf{R}$, and let C be a closed convex subset of H. Then C is weakly sequentially closed; that is, if $\{x_k\} \subset C$ converges weakly to $x \in H$, then $x \in C$.*

Proof Suppose $x_k \to x$ weakly, and let $\overline{x}$ be the vector in C closest to x; it suffices to show that $\overline{x} = x$. By (10.6),

$$\begin{aligned} & \langle z - \overline{x}, x - \overline{x} \rangle \leq 0 \text{ for all } z \in C \\ \Rightarrow\ & \langle x_k - \overline{x}, x - \overline{x} \rangle \leq 0 \text{ for all } k = 1, 2, 3, \ldots \\ \Rightarrow\ & \lim_{k \to \infty} \langle x_k - \overline{x}, x - \overline{x} \rangle \leq 0 \\ \Rightarrow\ & \langle x - \overline{x}, x - \overline{x} \rangle \leq 0 \\ \Rightarrow\ & \|x - \overline{x}\| \leq 0. \end{aligned}$$

This gives the desired result.

QED

The reader is doubtless accustomed to using the graph of $f : \mathbf{R} \to \mathbf{R}$ to study properties of the function. Many properties of a function defined on vector spaces can be most easily derived using the epigraph of the function.

Definition 475 *Let V be a vector space over $\mathbf{R}$, let S be a subset of V, and suppose $f : S \to \mathbf{R}$. The* epigraph *of f is the following subset of $S \times \mathbf{R}$:*

$$\mathrm{epi}(f) = \{(x, r) \in V \times \mathbf{R} \, : \, f(x) \leq r\}.$$

We will use two facts about epigraphs.

Theorem 476 *Let V be a vector space over $\mathbf{R}$, and let C be a convex subset of V. A function $f : C \to \mathbf{R}$ is convex if and only if $\mathrm{epi}(f)$ is a convex set.*

Proof Suppose first that f is convex, and let $(x, r), (y, s)$ be points in $\mathrm{epi}(f)$. By definition, $f(x) \leq r$, $f(y) \leq s$, and thus if $\alpha \in [0, 1]$, then

$$f((1 - \alpha)x + \alpha y) \leq (1 - \alpha)f(x) + \alpha f(y) \leq (1 - \alpha)r + \alpha s.$$

It follows that $(1-\alpha)(x, r)+\alpha(y, s) = ((1-\alpha)x+\alpha y, (1-\alpha)r+\alpha s) \in \mathrm{epi}(f)$. Therefore, $\mathrm{epi}(f)$ is convex.

Conversely, suppose $\mathrm{epi}(f)$ is convex. Let $x, y \in C$ and suppose $\alpha \in [0, 1]$. Then $(x, f(x)), (y, f(y)) \in \mathrm{epi}(f)$ and, since $\mathrm{epi}(f)$ is convex, we have

$$(1 - \alpha)(x, f(x)) + \alpha(y, f(y)) = ((1 - \alpha)x + \alpha y, (1 - \alpha)f(x) + \alpha f(y)) \in \mathrm{epi}(f).$$

By definition of $\mathrm{epi}(f)$, this implies that

$$f((1 - \alpha)x + \alpha y) \leq (1 - \alpha)f(x) + \alpha f(y),$$

and hence that f is convex.

QED

The next result says that $\mathrm{epi}(f)$ is closed in a given topology if and only if f is lower semicontinuous with respect to that topology.

Theorem 477 *Let H be a Hilbert space over $\mathbf{R}$, let S be a closed subset of H, and let $f : S \to \mathbf{R}$. Then f is lower semicontinuous with respect to a given topology if and only if $\mathrm{epi}(f)$ is sequentially closed with respect to that topology.*[5]

Proof Suppose first that f is lower semicontinuous. Let $\{(x_k, r_k)\}$ be a sequence in $\mathrm{epi}(f)$ with $(x_k, r_k) \to (x, r)$. Then, since $f(x_k) \leq r_k$ for all k, we have

$$f(x) \leq \liminf_{k \to \infty} f(x_k) \leq \lim_{k \to \infty} r_k = r.$$

[5]Actually, $\mathrm{epi}(f)$ is a subset of $H \times \mathbf{R}$, so the topology in question is the *product topology* determined by the given topology on H and the standard topology on $\mathbf{R}$.

This shows that $(x, r) \in \text{epi}(f)$, and hence $\text{epi}(f)$ is sequentially closed.

Conversely, suppose $\text{epi}(f)$ is sequentially closed. Let $\{x_k\} \subset S$ converge to $x \in S$. Then $(x_k, f(x_k)) \in \text{epi}(f)$ for all k. Moreover, there is a subsequence $\{x_{k_j}\}$ such that

$$r = \liminf_{k\to\infty} f(x_k) = \lim_{j\to\infty} f(x_{k_j}).$$

It follows that $(x_{k_j}, f(x_{k_j})) \to (x, r)$, and $(x, r) \in \text{epi}(f)$ since $\text{epi}(f)$ is sequentially closed. But then, by definition of $\text{epi}(f)$,

$$f(x) \leq r = \liminf_{k\to\infty} f(x_k),$$

and thus f is lower semicontinuous.

QED

Putting together the results given above, we have the following fundamental theorem.

Theorem 478 *Let H be a Hilbert space over $\mathbf{R}$, let C be a closed and bounded convex subset of H, and let $f : C \to \mathbf{R}$ be convex and lower semicontinuous (with respect to the norm topology). Then there exists $\overline{x} \in C$ such that*

$$f(\overline{x}) = \inf\{f(x) : x \in C\}.$$

Proof Since f is convex and lower semicontinuous with respect to the norm topology, it follows from Theorems 476 and 477 that $\text{epi}(f)$ is a closed convex subset of $H \times \mathbf{R}$. But then, by Theorem 474, $\text{epi}(f)$ is weakly sequentially closed, and hence f is weakly lower semicontinuous by Theorem 477.

Now let $\{x_k\} \subset C$ be a minimizing sequence for f:

$$\lim_{k\to\infty} f(x_k) = \inf\{f(x) : x \in C\}.$$

Since C is closed and convex, it is weakly closed by Theorem 474 and hence weakly sequentially compact by Corollary 463. Hence there exists a subsequence $\{x_{k_j}\}$ and a vector $\overline{x} \in C$ such that $x_{k_j} \to \overline{x}$ weakly. But then the weak lower semicontinuity of f implies that

$$f(\overline{x}) \leq \liminf_{j\to\infty} f(x_{k_j}) = \lim_{k\to\infty} f(x_k) = \inf\{f(x) : x \in C\}.$$

This completes the proof.

QED

This concludes our brief introduction to convexity, which was intended to demonstrate the utility of the weak topology. Although Weierstrass's theorem does not apply to a closed and bounded subset of an infinite-dimensional space, at least for convex functions we can recover a version of Weierstrass's theorem by appealing to the weak topology.

Exercises

1. Prove Corollary 458.

2. Prove Corollary 463.

3. Prove Theorem 464. (Hint: Use the Cauchy-Schwarz inequality to prove that $|\langle x_k, y\rangle - \langle x, y\rangle| \to 0$ for each $y \in H$.)

4. Prove Theorem 467.

5. Let V be a normed linear space over $\mathbf{R}$, let x be any vector in V, and let $S = B_r(x) = \{y \in V : \|y - x\| < r\}$ for any $r > 0$. Prove that S is convex.

6. Let $f : \mathbf{R}^n \to \mathbf{R}$ be a convex function, and suppose f is continuously differentiable. Prove that
$$f(x) \geq f(y) + \nabla f(y) \cdot (x - y) \text{ for all } x, y \in \mathbf{R}^n.$$

7. Use the results of the previous exercise to prove that if $f : \mathbf{R}^n \to \mathbf{R}$ is convex and continuously differentiable, then
$$(\nabla f(x) - \nabla f(y)) \cdot (x - y) \geq 0 \text{ for all } x, y \in \mathbf{R}^n.$$

8. Prove the converse of Exercise 6: If $f : \mathbf{R}^n \to \mathbf{R}$ is continuously differentiable and satisfies
$$f(x) \geq f(y) + \nabla f(y) \cdot (x - y) \text{ for all } x, y \in \mathbf{R}^n,$$
then f is convex.

9. Let H be a Hilbert space over $\mathbf{R}$, let S be a subset of H, and let $f : S \to \mathbf{R}$ be continuous. Prove that f is both lower and upper semicontinuous.

10. Let H be a Hilbert space, and let $\{u_k\}$ be an orthonormal sequence in H:
$$\langle u_j, u_k\rangle = \begin{cases} 1, & j = k, \\ 0, & j \neq k. \end{cases}$$
    (a) Prove *Bessel's inequality*: For all $x \in H$,
$$\sum_{k=1}^{\infty} |\langle x, u_k\rangle|^2 < \infty.$$
    (Hint: Use the fact that the partial sum $\sum_{k=1}^{n} |\langle x, u_k\rangle|^2$ is the square of the norm of $\text{proj}_{S_n} x$, where $S_n = \text{sp}\{u_1, \ldots, u_n\}$.)
    (b) Using Bessel's inequality, prove that $\{u_k\}$ converges weakly to the zero vector.

11. Let X, U be Hilbert spaces, and let $T : X \to U$ be linear and bounded. We know from Exercise 10.3.7 that T is continuous. Prove that T is also weakly continuous, meaning that

$$\{x_k\} \subset X, \; x_k \to x \text{ weakly} \;\Rightarrow\; T(x_k) \to T(x) \text{ weakly}.$$

(Hint: Make use of the existence of T^*—see Exercise 10.3.8.)

A

The Euclidean algorithm

The Euclidean algorithm is a method for computing the greatest common divisor (gcd) of two integers. As we show below, we can compute multiplicative inverses in $\mathbf{Z}_p$ using the Euclidean algorithm.

The Euclidean algorithm is based on the following theorem, which is usually called the *division algorithm* (even though it is a theorem, not an algorithm).

Theorem 479 *Let a and b be integers, with $a > 0$. Then there exist integers q (the* quotient*) and r (the* remainder*) such that*

$$b = qa + r$$

and $0 \leq r < a$.

The proof of this theorem can be found in any book on elementary number theory, such as [38].

Let us suppose we wish to find $\gcd(a, b)$, where $b > a > 0$. For the sake of convenience, we write $r_0 = b$, $r_1 = a$. We can apply the division algorithm to obtain

$$r_0 = q_1 r_1 + r_2, \ 0 \leq r_2 < r_1.$$

From this equation, it is easy to see that a positive integer k divides both r_0 and r_1 if and only if k divides both r_1 and r_2. This implies

$$\gcd(a, b) = \gcd(r_1, r_0) = \gcd(r_2, r_1).$$

Moreover, $r_2 < r_1$ and $r_1 < r_0$, and therefore the problem of computing the gcd has been simplified.

We now apply the division algorithm again to write

$$r_1 = q_2 r_2 + r_3, \ 0 \leq r_3 < r_2.$$

Reasoning as above, we have $\gcd(r_2, r_1) = \gcd(r_3, r_2)$. We continue in this fashion to produce a strictly decreasing sequence of nonnegative integers $r_0, r_1, \ldots$. After a finite number of steps, we must obtain $r_{j+1} = 0$, which implies that

$$r_{j-1} = q_j r_j$$

and hence $\gcd(a, b) = \gcd(r_{j-1}, r_j) = r_j$. This is the Euclidean algorithm for finding $\gcd(a, b)$.

Example 480 *Let use find* gcd(56, 80). *We compute as follows:*

$$\begin{aligned} 80 = 1 \cdot 56 + 24 &\Rightarrow \gcd(56,80) = \gcd(24,56), \\ 56 = 2 \cdot 24 + 8 &\Rightarrow \gcd(56,80) = \gcd(8,24), \\ 24 = 3 \cdot 8 + 0 &\Rightarrow \gcd(56,80) = 8. \end{aligned}$$

Therefore the greatest common divisor of 56 *and* 80 *is* 8.

A.0.1 Computing multiplicative inverses in $\mathbf{Z}_p$

Applying the Euclidean algorithm results in a sequence of equations

$$r_0 = q_1 r_1 + r_2, \tag{A.1a}$$
$$r_1 = q_2 r_2 + r_3, \tag{A.1b}$$
$$\vdots \qquad \vdots$$
$$r_{j-2} = q_{j-1} r_{j-1} + r_j, \tag{A.1c}$$
$$r_{j-1} = q_j r_j, \tag{A.1d}$$

where $r_0 = b$, $r_1 = a$, and $r_j = \gcd(a, b)$. We can use these relationships to compute the multiplicative inverse of an element of $\mathbf{Z}_p$.

Let us assume p is a prime number and a is an integer satisfying $1 \leq a \leq p - 1$. We can regard a as an element of $\mathbf{Z}_p$, and we wish to find $x \in \mathbf{Z}_p$ such that $ax = 1$ in $\mathbf{Z}_p$, that is, $ax = py + 1$, where $y \in \mathbf{Z}$. This in turn is equivalent to $ax - py = 1$, where x and y are integers and $0 \leq x \leq p - 1$.

To find such integers x and y, apply the Euclidean algorithm to $r_0 = p$ and $r_1 = a$. In this case, we know that $\gcd(r_1, r_0) = r_j = 1$ (since p is prime), and Equation (A.1c) yields

$$1 = r_{j-2} - q_{j-1} r_{j-1},$$

or

$$1 = r_{j-2} y_j + r_{j-1} x_j \ (y_j = 1, \ x_j = -q_{j-1}).$$

Next, we notice that

$$r_{j-1} = r_{j-3} - q_{j-2} r_{j-2},$$

and therefore

$$1 = r_{j-2} y_j + (r_{j-3} - q_{j-2} r_{j-2}) x_j,$$

which can be rearranged to yield

$$1 = r_{j-3} y_{j-1} + r_{j-2} x_{j-1}$$

for integers y_{j-1} and x_{j-1}. We can continue this process to eventually obtain

$$1 = r_0 y_2 + r_1 x_2 = p y_2 + a x_2.$$

The integer x_2 can be written (by the division algorithm) as $x_2 = kp + x$, where $0 \leq x \leq p-1$, and then we obtain

$$\begin{aligned} ax_2 + py_2 = 1 \quad &\Rightarrow \quad a(kp+x) + py_2 = 1 \\ &\Rightarrow \quad ax + p(y_2 + ka) = 1 \\ &\Rightarrow \quad ax - py = 1, \end{aligned}$$

where $y = -(y_2 + ka)$. The integer x (regarded as an element of $\mathbf{Z}_p$) is the multiplicative inverse of a in $\mathbf{Z}_p$.

Example 481 *Consider $\mathbf{Z}_p$ with $p = 101$. We will find the multiplicative inverse of $a = 90$. First we apply the Euclidean algorithm:*

$$\begin{aligned} 101 &= 1 \cdot 90 + 11, \\ 90 &= 8 \cdot 11 + 2, \\ 11 &= 5 \cdot 2 + 1. \end{aligned}$$

Now we work backwards:

$$\begin{aligned} 1 = 11 - 5 \cdot 2 = 11 - 5 \cdot (90 - 8 \cdot 11) &= -5 \cdot 90 + 41 \cdot 11 \\ &= -5 \cdot 90 + 41 \cdot (101 - 90) \\ &= 41 \cdot 101 - 46 \cdot 90. \end{aligned}$$

We have $-46 = -1 \cdot 101 + 55$, and therefore

$$1 = 41 \cdot 101 + (-101 + 55) \cdot 90 = 90 \cdot 55 - 101 \cdot 49 \Rightarrow 90 \cdot 55 = 101 \cdot 49 + 1.$$

This shows that $90 \cdot 55$ is congruent to 1 modulo 101, that is, $90 \cdot 55 = 1$ in $\mathbf{Z}_{101}$.

A.0.2 Related results

We will need the following result.

Theorem 482 *Let a, b be positive integers. Then there exist integers x, y such that $ax + by = \gcd(a, b)$.*

Proof The proof for the case $\gcd(a, b) = 1$ is given above (before Example 481), and the proof in the general case is no different.

QED

If a, b are positive integers, and $\gcd(a, b) = 1$, then a and b are said to be *relatively prime*. In the case that a and b are relatively prime, there exist integers x, y such that

$$ax + by = 1.$$

We can extend the concept of greatest common divisor to any set of positive integers. If $a_1, a_2, \ldots, a_k$ are positive integers, then $\gcd(a_1, a_2, \ldots, a_k)$ is the largest positive integer d such that $d|a_i$ for all $i = 1, 2, \ldots, k$. We then say that $a_1, a_2, \ldots, a_k$ are relatively prime if and only if $\gcd(a_1, a_2, \ldots, a_k) = 1$.

Corollary 483 *Let $a_1, a_2, \ldots, a_k$ be positive integers, where $k \geq 2$. Then there exist integers $x_1, x_2, \ldots, x_k$ such that*

$$a_1 x_1 + a_2 x_2 + \cdots + a_k x_k = \gcd(a_1, a_2, \ldots, a_k).$$

Proof We argue by induction on k. For $k = 2$, the result is Theorem 482. Let us suppose the result holds for any $k-1$ positive integers, and let $a_1, a_2, \ldots, a_k$ be positive integers. By the induction hypothesis, there exist integers $x_1', x_2', \ldots, x_{k-1}'$ such that

$$a_1 x_1' + a_2 x_2' + \cdots + a_{k-1} x_{k-1}' = \gcd(a_1, \ldots, a_{k-1}).$$

But then, applying Theorem 482, we find integers x_1'' and x_k such that

$$\gcd(a_1, \ldots, a_{k-1}) x_1'' + a_k x_k = \gcd(\gcd(a_1, \ldots, a_{k-1}), a_k).$$

It is easy to show that

$$\gcd(\gcd(a_1, \ldots, a_{k-1}), a_k) = \gcd(a_1, a_2, \ldots, a_k),$$

and

$$\gcd(a_1, \ldots, a_{k-1}) x_1'' + a_k x_k = a_1 x_1 + a_2 x_2 + \cdots + a_k x_k,$$

where $x_i = x_1'' x_i'$ for $i = 1, 2, \ldots, k-1$. This completes the proof.

QED

B

Permutations

In this appendix, we will define permutations, transpositions, and the signature of a permutation. We will then prove the two theorems about permutations that are used in Section 4.1. Throughout this discussion, we write $\mathbf{n}$ for the set of the first n positive integers: $\mathbf{n} = \{1, 2, \ldots, n\}$.

Definition 484 *A* permutation *is a bijection of* $\mathbf{n}$ *onto itself. We will often denote* $\tau : \mathbf{n} \to \mathbf{n}$ *by listing its values in a finite sequence:* $\tau = (\tau(1), \tau(2), \ldots, \tau(n))$. *The set of all permutations of* $\mathbf{n}$ *is denoted by* S_n.[1]

Therefore, $\tau = (i_1, i_2, \ldots, i_n)$ means $\tau(j) = i_j$.

Definition 485 *A* transposition *of* $\mathbf{n}$ *is a permutation that interchanges two elements of* $\mathbf{n}$. *In other words, a permutation* $\tau \in S_n$ *is a transposition if there exist integers* $i, j \in \mathbf{n}$, $i \neq j$, *such that*

$$\tau(k) = \begin{cases} j, & k = i, \\ i, & k = j, \\ k, & \text{otherwise.} \end{cases}$$

The transposition τ *defined by* $i \neq j$ *will be denoted* $[i, j]$, *and we will write*

$$[i, j](k) = \begin{cases} j, & k = i, \\ i, & k = j, \\ k, & \text{otherwise} \end{cases}$$

when convenient.

We will use product notation to denote the composition of permutations in general and transpositions in particular. For example, if $n = 5$, then

$$\begin{aligned} [2, 4] &= (1, 4, 3, 2, 5), \\ [1, 2] &= (2, 1, 3, 4, 5), \\ [1, 2][2, 4] &= (2, 4, 3, 1, 5). \end{aligned}$$

To verify this last result, one could compute as follows:

$$[1, 2][2, 4](1) = [1, 2](1) = 2, \; [1, 2][2, 4](2) = [1, 2](4) = 4, \; \ldots.$$

[1] S_n is a group called the *symmetric group on n symbols.*

Alternatively, we have

$$[1,2][2,4] = [1,2](1,4,3,2,5) = (2,4,3,1,5).$$

We now prove the first of the theorems that we need.

Theorem 486 *Let n be a positive integer and let τ be a permutation of* $\mathbf{n}$. *Then τ can be written as a product of transpositions.*

Proof Suppose $\tau = (i_1, i_2, \ldots, i_n)$. We will show how to τ as the product of n permutations $\tau^{(1)}, \tau^{(2)}, \ldots, \tau^{(n)}$, each of which is a transposition or the identity permutation. In the following derivation, we will write

$$\tau^{(j)}\tau^{(j-1)}\cdots\tau^{(1)} = (i_1^{(j)}, i_2^{(j)}, \ldots, i_n^{(j)}).$$

We will show that we can define $\tau^{(1)}, \tau^{(2)}, \ldots, \tau^{(n)}$ so that

$$\tau = \tau^{(n)}\tau^{(n-1)}\cdots\tau^{(1)}.$$

We begin by defining $\tau^{(1)} = [1, i_1]$, with the understanding that this indicates the identity permutation if $i_1 = 1$. We then have

$$\tau^{(1)} = (i_1, 2, \ldots, i_1 - 1, 1, i_1 + 1, \ldots, n-1, n),$$

which we denote by

$$(i_1^{(1)}, i_2^{(1)}, \ldots, i_n^{(1)}).$$

We note that $i_1^{(1)} = i_1$. Next, we define $\tau^{(2)} = [i_2^{(1)}, i_2]$ and denote

$$\tau^{(2)}\tau^{(1)} = (i_1^{(2)}, i_2^{(2)}, \ldots, i_n^{(2)}).$$

Since $i_2^{(1)} \neq i_1$, $i_2 \neq i_1$, we still have $i_1^{(2)} = i_1$, and now $i_2^{(2)} = i_2$. We continue in this fashion, defining $\tau^{(3)} = [i_3^{(2)}, i_3]$, $\tau^{(4)} = [i_4^{(3)}, i_4]$, and so on. At each step, we have $i_k^{(\ell)} = i_k$ for $k \leq \ell$. The result is

$$\tau^{(n)}\tau^{(n-1)}\cdots\tau^{(1)} = (i_1^{(n)}, i_2^{(n)}, \ldots, i_n^{(n)}),$$

and since $i_k^{(n)} = i_k$ for all $k \leq n$, we see that

$$\tau^{(n)}\tau^{(n-1)}\cdots\tau^{(1)} = \tau,$$

as desired.

QED

We will need two more concepts to prove our second main theorem. First of all, we will refer to a transposition of the form $[i, i+1]$ as an *adjacent transposition.* Second, given a transposition τ, a pair (i, j) of integers is called an *inversion pair* for τ if $i < j$ and $\tau(i) > \tau(j)$.

Definition 487 *Let $\tau \in S_n$. We define $N(\tau)$ to be the number of inversion pairs of τ.*

For example, $N((1,3,2)) = 1$, $N((4,2,1,3)) = 4$, and $N((4,3,2,1)) = 6$.

We will use the following lemma.

Lemma 488 *Let $[i,j]$ be a transposition of $\mathbf{n}$. Then $[i,j]$ can be written as the product of an odd number of adjacent transpositions.*

Proof Since $[i,j] = [j,i]$, we can assume that $i < j$. The proof consists of showing that

$$[i,j] = [i,i+1][i+1,i+2]\cdots[j-1,j][j-2,j-1]\cdots[i,i+1]. \tag{B.1}$$

The above product consists of $j - i + j - i - 1 = 2(j-i) - 1$ adjacent transpositions, and $2(j-i) - 1$ is odd for all i, j.

Let τ be the permutation defined by the right side of (B.1). If $k < i$ or $k > j$, then clearly k is fixed by τ, that is, $\tau(k) = k$. We have

$$\begin{aligned}
\tau(i) &= [i,i+1][i+1,i+2]\cdots[j-1,j][j-2,j-1]\cdots[i,i+1](i)\\
&= [i,i+1][i+1,i+2]\cdots[j-1,j][j-2,j-1]\cdots[i+1,i+2](i+1)\\
&\;\;\vdots\\
&= [i,i+1][i+1,i+2]\cdots[j-1,j](j-1)\\
&= [i,i+1][i+1,i+2]\cdots[j-2,j-1](j)\\
&= j
\end{aligned}$$

(since the last $j-i-1$ factors fix j). Next, since the first $j-i-1$ factors fix j, we have

$$\begin{aligned}
\tau(j) &= [i,i+1][i+1,i+2]\cdots[j-1,j][j-2,j-1]\cdots[i,i+1](j)\\
&= [i,i+1][i+1,i+2]\cdots[j-1,j](j)\\
&= [i,i+1][i+1,i+2]\cdots[j-2,j-1](j-1)\\
&\;\;\vdots\\
&= [i,i+1](i+1)\\
&= i.
\end{aligned}$$

Finally, if $i < k < j$, then one factor of τ maps k to $k-1$, another maps $k-1$ back to k, and the rest have no effect; we leave the verification to the reader. Therefore $\tau(k) = k$.

Thus we have shown that

$$\tau(k) = \begin{cases} j, & k = i,\\ i, & k = j,\\ k, & \text{otherwise,}\end{cases}$$

that is, $\tau = [i,j]$.

QED

We are now ready to prove the main theorem.

Theorem 489 *Let $\tau \in S_n$. Then every factorization of τ into a product of transpositions has an even number of factors, or every factorization of τ into a product of transpositions has an odd number of factors.*

Proof Suppose $\mu_1, \mu_2, \ldots, \mu_k$ and $\nu_1, \nu_2, \ldots, \nu_\ell$ are transpositions such that

$$\tau = \mu_1\mu_2 \cdots \mu_k = \nu_1\nu_2 \cdots \nu_\ell.$$

It suffices to prove that $k - \ell$ is even, since then k and ℓ are either both even or both odd. If we replace each μ_i by a product of an odd number of adjacent transpositions, we obtain a factorization of τ into k' adjacent transpositions, and the parity of k' is the same as the parity of k (k' is the sum of k odd numbers). Therefore, without loss of generality, we can assume that each μ_i is an adjacent transposition, and similarly for ν_j.

Now notice that multiplying any permutation τ' by an adjacent transposition $[i, i+1]$ changes $N(\tau')$ by exactly 1: $N([i, i+1]\tau') = N(\tau') - 1$ if $(i, i+1)$ is an inversion pair for τ' or $N([i, i+1]\tau') = N(\tau') + 1$ if $(i, i+1)$ is not an inversion pair for τ'. (Note: This is the reason we had to factor τ into adjacent transpositions; the same result does not hold for a general transposition. For an example, consider $[1, 4](4, 3, 2, 1)$.) Therefore, multiplying a permutation τ' by an adjacent transposition changes the parity of $N(\tau')$.

Let ι be the identity permutation, and notice that $N(\iota) = 0$. Therefore, $N(\mu_k) = N(\mu_k\iota)$ is odd, $N(\mu_{k-1}\mu_k)$ is even, and so forth. We conclude that the parity of $N(\tau) = N(\mu_1\mu_2 \cdots \mu_k)$ is the parity of k. By the same reasoning, the parity of $N(\tau)$ must be the parity of ℓ. Therefore, k and ℓ are either both even or both odd, and the proof is complete.

QED

We have now shown that the concept of the parity of a permutation is well-defined.

Definition 490 *Let $\tau \in S_n$. We say that τ is* even *if it can be factored into the product of an even number of transpositions and* odd *if it can be factored into the product of an odd number of transpositions.*

Finally, we can define the signature of a permutation.

Definition 491 *Let $\tau \in S_n$. We define the* signature $\sigma(\tau)$ *of τ as follows:*

$$\sigma(\tau) = \begin{cases} 1, & \text{if } \tau \text{ is even,} \\ -1, & \text{if } \tau \text{ is odd.} \end{cases}$$

The proof of Theorem 489 shows that $\sigma(\tau) = (-1)^{N(\tau)}$.

C

Polynomials

C.1 Rings of polynomials

In this appendix, we develop the elementary properties of polynomials that are used in the text.

Definition 492 *Let F be a field. Then the* ring of polynomials $F[x]$ *is the set of all expressions of the form*

$$a_0 + a_1x + a_2x^2 + \ldots,$$

where $a_0, a_1, \ldots \in F$ and there exists $M \geq 0$ such that $a_k = 0$ for all $k \geq M$. Each such expression is called a polynomial *in the* indeterminate x. *We say that the* degree *of $a_0 + a_1x + a_2x^2 + \ldots$ is undefined if all of the* coefficients *$a_0, a_1, \ldots$ are 0; otherwise, the degree is the largest integer n such that $a_n \neq 0$.*

Addition in $F[x]$ is defined by

$$(a_0 + a_1x + \ldots) + (b_0 + b_1x + \ldots) = (a_0 + b_0) + (a_1 + b_1)x + \ldots,$$

and multiplication by

$$(a_0 + a_1x + \ldots)(b_0 + b_1x + \ldots) = (c_0 + c_1x + \ldots),$$

where

$$c_k = \sum_{i+j=k} a_ib_j.$$

Some comments about this definition are in order. The indeterminate x is regarded as a pure symbol; therefore, in the expression $a_0 + a_1x + a_2x^2 + \ldots$, the "+" should not be regarded as addition nor should x^2 be regarded as x times x, at least not until such interpretations have been justified. If the introduction of the undefined indeterminate x is regarded as inelegant, then a polynomial could be defined simply as the sequence of its coefficients. From this viewpoint, we would write $(a_0, a_1, a_2, \ldots)$ instead of $a_0 + a_1x + a_2x^2 + \ldots$.

We will leave it as an exercise for the reader to prove that $F[x]$, as defined above, is a commutative ring with unity, where the unit (multiplicative identity) is the polynomial $1 + 0x + 0x^2 + \ldots$ and the additive identity in $F[x]$ is $0 + 0x + 0x^2 + \ldots$. Part of this exercise involves verifying that addition and multiplication are well-defined (that is, yield elements of $F[x]$).

We now argue that $a_0+a_1x+a_2x^2+\ldots$ can be regarded the sum of simpler elements of $F[x]$, namely $a_0, a_1x, a_2x^2, \ldots$ and that each of these elements can be regarded as an element of F times a power of x. We begin by noticing that F is isomorphic to a subset of $F[x]$, namely

$$F' = \{a + 0x + 0x^2 + \ldots : a \in F\}.$$

The isomorphism is $\phi : F \to F'$, $\phi(a) = a + 0x + 0x^2 + \ldots$. The mapping ϕ is not only a bijection, but it preserves the field operations of addition and multiplication, as is easily verified:

$$\begin{aligned} \phi(a+b) &= \phi(a) + \phi(b) \text{ for all } a, b \in F, \\ \phi(ab) &= \phi(a)\phi(b) \text{ for all } a, b \in F. \end{aligned}$$

We will identify F with F', which means that F can be regarded as a subset of $F[x]$ and we will henceforth write simply a for $a + 0x + 0x^2 + \ldots$.

Next, we define the polynomial x to be $0+1x+0x^2+\ldots$ and we note that

$$\begin{aligned} x \cdot x &= 0 + 0x + 1x^2 + 0x^3 + \ldots, \\ x \cdot x \cdot x &= 0 + 0x + 0x^2 + 1x^3 + 0x^4 + \ldots, \end{aligned}$$

and so forth. Therefore, if we write x^i for

$$0 + 0x + \ldots + 0x^{i-1} + 1x^i + 0x^{i+1} + \ldots,$$

then we can interpret x^i in the usual way (as the product of i factors of x). It is straightforward to verify that

$$\begin{aligned} &(a + 0x + 0x^2 + \ldots)(0 + 0x + \ldots + 0x^{i-1} + 1x^i + 0x^{i+1} + \ldots) \\ = \;& 0 + 0x + \ldots + 0x^{i-1} + ax^i + 0x^{i+1} + \ldots, \end{aligned}$$

so ax^i can be interpreted as a times x^i. Finally, it is also straightforward to verify that

$$a_0 + a_1x + a_2x^2 + \ldots$$

is the sum of the polynomials $a_0, a_1x, a_2x^2, \ldots$.

Thus $a_0 + a_1x + a_2x^2 + \ldots$ can be interpreted in the usual fashion, and there is no reason to write terms with coefficient zero. Therefore, we will write, for example, $2 + x - x^3$ instead of

$$2 + 1x + 0x^2 + (-1)x^3 + 0x^4 + \ldots.$$

We will represent arbitrary elements of $F[x]$ as $p(x)$ or $q(x)$, and we write $\deg(p(x))$ for the degree of a nonzero element of $F[x]$. Finally, we will write $x^0 = 1$ when convenient.

We will now state the precise results we use in the text. Proofs of the more obvious results will be left to the reader.

Lemma 493 *Let $p(x), q(x)$ be nonzero elements of $F[x]$, where F is a field. Then*

$$\deg(p(x)q(x)) = \deg(p(x)) + \deg(q(x))$$

and, if $p(x) + q(x)$ is nonzero,

$$\deg(p(x) + q(x)) \leq \max\{\deg(p(x)), \deg(q(x))\}.$$

Proof Exercise.

The following result implies that $F[x]$ is an integral domain (see Section 4.4).

Lemma 494 *Let F be a field. If $p(x), q(x) \in F[x]$ and $p(x)q(x) = 0$, then $p(x) = 0$ or $q(x) = 0$. Moreover, if $p(x), q(x), r(x) \in F[x]$ and $r(x) \neq 0$, then $p(x)r(x) = q(x)r(x)$ implies $p(x) = q(x)$.*

Proof Exercise.

We will need the following lemma.

Lemma 495 *If F is a field and $f(x), g(x)$ are nonzero elements of $F[x]$ with $\deg(g(x)) \leq \deg(f(x))$, then there exists $q(x) \in F[x]$ such that either*

$$f(x) = q(x)g(x)$$

or

$$\deg(f(x) - q(x)g(x)) < \deg(f(x)).$$

Proof Let $m = \deg(f(x))$, $n = \deg(g(x))$, and let $a_m x^m$, $b_n x^n$ be the leading terms (that is, nonzero terms of the highest degree)) of $f(x)$ and $g(x)$, respectively. Define $q(x) = a_m b_n^{-1} x^{m-n}$. Then $\deg(q(x)g(x)) = m$ and the leading term of $q(x)g(x)$ is $a_n x^n$. The result now follows.

QED

We can now prove our first major result. The following theorem is called the *division algorithm* for polynomials.

Theorem 496 *Let F be a field and $f(x), g(x)$ elements of $F[x]$, with $g(x)$ nonzero. Then there exist elements $q(x), r(x) \in F[x]$ such that*

$$f(x) = q(x)g(x) + r(x),$$

where $r(x) = 0$ or $\deg(r(x)) < \deg(g(x))$. Moreover, $q(x)$ and $r(x)$ are unique.

Proof We begin by establishing the existence of $q(x)$ and $r(x)$. We first note that if $\deg(g(x)) = 0$, then $g \in F$, $g \neq 0$ implies that g^{-1} exists. We then obtain $f(x) = q(x)g(x) + r(x)$, where $q(x) = g^{-1}f(x)$ and $r(x) = 0$. Next, if $\deg(f(x)) < \deg(g(x))$, we have $f(x) = q(x)g(x) + r(x)$, where $q(x) = 0$ and $r(x) = f(x)$. For the final simple case, if $\deg(f(x)) = \deg(g(x))$, then by Lemma 495, there exists $q(x) \in F[x]$ such that $r(x) = f(x) - q(x)g(x)$ satisfies either $r(x) = 0$ or $\deg(r(x)) < \deg(f(x)) = \deg(g(x))$. Thus the result holds in this case as well.

We can now assume that $\deg(g(x)) \geq 1$ and $\deg(f(x)) \geq \deg(g(x))$. We complete the proof by induction on the degree of $f(x)$. The base case is $\deg(f(x)) = \deg(g(x))$, and we have already proved the result in this case. We now assume the result holds for $\deg(f(x)) < n$, where $n > \deg(g(x))$. Let $f(x) \in F[x]$, $\deg(f(x)) = n$. By Lemma 495, there exists $q_1(x) \in F[x]$ such that $f_1(x) = f(x) - q_1(x)g(x)$ satisfies $f_1(x) = 0$ or $\deg(f_1(x)) < \deg(f(x))$. If $f_1(x) = 0$, then the result holds with $q(x) = q_1(x)$ and $r(x) = 0$. Otherwise, the induction hypothesis implies that there exist $q_2(x), r(x) \in F[x]$ such that

$$f_1(x) = q_2(x)g(x) + r(x)$$

and either $r(x) = 0$ or $\deg(r(x)) < \deg(g(x))$. But then

$$f(x) - q_1(x)g(x) = q_2(x)g(x) + r(x) \Rightarrow f(x) = (q_1(x) + q_2(x))g(x) + r(x).$$

In this case the result holds with $q(x) = q_1(x) + q_2(x)$ and the given $r(x)$. We have now proved by induction that $q(x)$ and $r(x)$ exist for all $f(x)$ with $\deg(f(x)) \geq \deg(g(x)) \geq 1$, and the existence proof is complete.

It remains to prove uniqueness. Assume there exist $q_1(x), q_2(x), r_1(x), r_2(x)$ in $F[x]$ such that

$$f(x) = q_1(x)g(x) + r_1(x) = q_2(x)g(x) + r_2(x),$$

and either $r_1(x) = 0$ or $\deg(r_1) < \deg(g(x))$, and similarly for $r_2(x)$. We then have

$$(q_1(x) - q_2(x))g(x) = r_2(x) - r_1(x). \tag{C.1}$$

If $q_1(x) - q_2(x)$ and $r_2(x) - r_1(x)$ are both nonzero, then we have

$$\deg(r_2(x) - r_1(x)) = \deg((q_1(x) - q_2(x))g) = \deg(q_1(x) - q_2(x)) + \deg(g(x)),$$

which is a contradiction since

$$\deg(r_2(x) - r_1(x)) \leq \max\{\deg(r_1(x)), \deg(r_2(x))\} < \deg(g(x)).$$

Therefore at least one of $q_1(x) - q_2(x)$, $r_2(x) - r_1(x)$ is zero. It is easy to show that if one of $q_1(x) - q_2(x)$, $r_2(x) - r_1(x)$ is zero, then (C.1) implies that the other is also. This proves uniqueness, and the proof is complete.

QED

The reader will notice the similarity of the division algorithm for polynomials and the division algorithm for integers (Theorem 479). We can define greatest common divisor for polynomials just as for integers, and use the Euclidean algorithm to compute it. We also obtain Theorem 499 below, which expresses the gcd of $p(x)$ and $q(x)$ as a combination of $p(x)$ and $q(x)$, and which should be compared to Theorem 482.

Given $p(x), d(x) \in F[x]$, we say that $d(x)$ divides $p(x)$ (or is a divisor of $p(x)$) if there exists $s(x) \in F[x]$ such that $p(x) = s(x)d(x)$.

Definition 497 *Let F be a field and let $p(x), q(x) \in F[x]$ be nonzero. We say that $d(x) \in F[x]$ is a* greatest common divisor *of $p(x)$ and $q(x)$ if $d(x)$ is a common divisor of $p(x), q(x)$ and if, given any common divisor $r(x)$ of $p(x), q(x)$, $d(x)$ divides $r(x)$.*

It is easy to see that a greatest common divisor of $p(x), q(x)$ (if it exists) is not unique; given one greatest common divisor $d(x)$ and any nonzero $c \in F[x]$, the polynomial $cd(x)$ is also a greatest common divisor of $p(x), q(x)$. In order to impose uniqueness, we introduce the concept of a monic polynomial.

Definition 498 *Let F be a field and $p(x)$ an element of $F[x]$. If $\deg(p(x)) = n$ and $p(x) = a_0 + a_1x + \cdots + a_nx^n$, then we call a_n the* leading coefficient *of $p(x)$, and we say that $p(x)$ is* monic *if $a_n = 1$.*

It is easy to see that if $p(x), q(x) \in F[x]$ has a greatest common divisor, then it has a unique monic greatest common divisor; thus, when we refer to *the* gcd of $p(x)$ and $q(x)$, we mean the monic gcd. The following theorem, in part, asserts the existence of a gcd.

Theorem 499 *Let F be a field and let $p(x), q(x)$ be nonzero elements of $F[x]$. Then $p(x)$ and $q(x)$ have a unique monic greatest common divisor $d(x) \in F[x]$. Moreover, there exist polynomials $u(x), v(x) \in F[x]$ such that*

$$d(x) = u(x)p(x) + v(x)q(x).$$

The theorem can be proved using the Euclidean algorithm, exactly as the analogous results were proved for positive integers in Appendix A. We will illustrate with an example.

Example 500 *Let $p(x), q(x) \in \mathbf{R}[x]$ be defined by*

$$p(x) = x^4 - x^3 + 2x^2 - x - 1, \ q(x) = x^3 - x^2 + x - 1.$$

We first apply the Euclidean algorithm to find the gcd of $p(x)$ and $q(x)$:

$$\begin{aligned} x^4 - x^3 + 2x^2 - x - 1 &= x(x^3 - x^2 + x - 1) + x^2 - 1, \\ x^2 - x^2 + x - 1 &= (x-1)(x^2 - 1) + 2x - 2, \\ x^2 - 1 &= \frac{1}{2}x(2x-2) + x - 1, \\ 2x - 2 &= 2(x-1). \end{aligned}$$

We see that the gcd of $p(x)$ and $q(x)$ is $x - 1$. We can now work backwards *to find* $u(x), v(x) \in \mathbf{R}[x]$ *such that* $x - 1 = u(x)p(x) + v(x)q(x)$*:*

$$\begin{aligned}
x - 1 &= x^2 - 1 - \frac{1}{2}x(2x-2) \\
&= x^2 - 1 - \frac{1}{2}x\left(x^3 - x^2 + x - 1 - (x-1)(x^2-1)\right) \\
&= \left(1 + \frac{1}{2}x(x-1)\right)(x^2-1) - \frac{1}{2}x(x^3 - x^2 + x - 1) \\
&= \left(\frac{1}{2}x^2 - \frac{1}{2}x + 1\right)(x^4 - x^3 + 2x^2 - x - 1 - x(x^3 - x^2 + x - 1)) - \\
&\quad \frac{1}{2}x(x^3 - x^2 + x - 1) \\
&= \left(\frac{1}{2}x^2 - \frac{1}{2}x + 1\right)(x^4 - x^3 + 2x^2 - x - 1) + \\
&\quad \left(-\frac{1}{2}x - x\left(\frac{1}{2}x^2 - \frac{1}{2}x + 1\right)\right)(x^3 - x^2 + x - 1) \\
&= \left(\frac{1}{2}x^2 - \frac{1}{2}x + 1\right)(x^4 - x^3 + 2x^2 - x - 1) + \\
&\quad \left(-\frac{1}{2}x^3 + \frac{1}{2}x^2 - \frac{3}{2}x\right)(x^3 - x^2 + x - 1).
\end{aligned}$$

We can extend the previous theorem to more than two polynomials. We say that $d(x)$ is the greatest common divisor of $p_1(x), p_2(x), \ldots, p_k(x)$ if $d(x)$ divides each $p_i(x)$ and if any polynomial $e(x)$ that divides every $p_i(x)$ also divides $d(x)$.

Theorem 501 *Let F be a field and $p_1(x), p_2(x), \ldots, p_k(x)$ be nonzero elements of $F[x]$. There exists a unique monic greatest common divisor $d(x)$ of $p_1(x), p_2(x), \ldots, p_k(x)$. Moreover, there exist $u_1(x), u_2(x), \ldots, u_k(x) \in F[x]$ such that*

$$d(x) = u_1(x)p_1(x) + u_2(x)p_2(x) + \cdots + u_k(x)p_k(x).$$

The proof of this theorem is similar to the proof of Corollary 483, the analogous result for integers.

C.2 Polynomial functions

Each polynomial

$$p(x) = a_0 + a_1x + \ldots + a_nx^n \in F[x]$$

defines a polynomial function $p: F \to F$ by the rule

$$p(t) = a_0 + a_1 t + \ldots + a_n t^n.$$

Notice that we use the same symbol p to denote both the polynomial and the polynomial function; the context should resolve the ambiguity.

The most important question about polynomials concerns their roots.

Definition 502 *Let F be a field and $p(x) \in F[x]$. We say that $\alpha \in F$ is a* root *of $p(x)$ if and only if $p(\alpha) = 0$.*

Given $p(x) \in F[x]$, we wish to know how many roots $p(x)$ has. The following result, called the *remainder theorem*, sheds some light on this subject.

Theorem 503 *Let F be a field and let $p(x) \in F[x]$ and $\alpha \in F$ be given. Then there exists $q(x) \in F[x]$ such that*

$$p(x) = q(x)(x - \alpha) + p(\alpha),$$

where $p(\alpha)$ is the value of the polynomial function p at α.

Proof By Theorem 496, there exist $q(x), r(x) \in F[x]$ such that

$$p(x) = q(x)(x - \alpha) + r(x),$$

where $r(x) = 0$ or $\deg(r(x)) < \deg(x - \alpha) = 1$. We see that in either case, $r(x) = r \in F$. The two polynomials $p(x)$ and $q(x)(x - \alpha) + r$, being equal, define the same polynomial function, and if we evaluate both at $x = \alpha$, we obtain

$$p(\alpha) = q(\alpha)(\alpha - \alpha) + r,$$

or $p(\alpha) = r$, as desired.

QED

Corollary 504 *Let F be a field and let $p(x) \in F[x]$ be given. Then $\alpha \in F$ is a root of $p(x)$ if and only if there exists $q(x) \in F[x]$ such that $p(x) = q(x)(x-\alpha)$.*

Using Corollary 504, we obtain the following fundamental fact about roots of polynomials.

Theorem 505 *Let F be a field and let $p(x) \in F[x]$ be nonzero with degree n. Then $p(x)$ has at most n distinct roots in F.*

Proof Exercise.

The preceding theorem gives an upper bound on the number of roots of a polynomial, but it does not guarantee that a given polynomial has any roots. In fact, a polynomial need not have any roots over a given field, a fact that is surely familiar to the reader. For instance, if the field is **R**, then $x^2 + 1 \in \mathbf{R}[x]$ has no roots. On the other hand, if we replace **R** by **C** (an *extension field* of **R**), then $x^2 + 1$ has two roots, $\pm i$. This is an example of the following theorem.

Theorem 506 (The fundamental theorem of algebra) *Let $p(x) \in \mathbf{C}[x]$. Then $p(x)$ has a root.*

There is no elementary proof of this theorem, which was first proved by the famous German mathematician Gauss. The reader can consult [30] for a fairly straightforward proof based on results from multivariable calculus. We can combine the fundamental theorem and Corollary 504 to obtain our final result.

Theorem 507 *Let $p(x) \in \mathbf{C}[x]$ have degree n. Then $p(x)$ can be factored as*

$$p(x) = \gamma(x - \alpha_1)^{k_1}(x - \alpha_2)^{k_2} \cdots (x - \alpha_k)^{k_k}, \tag{C.2}$$

where $\gamma \in \mathbf{C}$, $\alpha_1, \ldots, \alpha_k$ are distinct elements of $\mathbf{C}$, and $k_1, k_2, \ldots, k_k$ are positive integers satisfying $k_1 + k_2 + \ldots + k_k = n$. Moreover, this factorization is unique up to the order of the factors.

Proof The proof is a straightforward induction argument that is left to the reader.

QED

The number k_i in (C.2) is called the *multiplicity* of the root α_i, and we usually express Theorem 507 by saying that every polynomial with coefficients in $\mathbf{C}$ has n complex roots, counted according to multiplicity. Since $\mathbf{R}[x] \subset \mathbf{C}[x]$, this result applies to polynomials with real coefficients as well, but it means that we might have to go into the complex domain to find all the roots of a given $p(x) \in \mathbf{R}[x]$, another well-known fact.

C.2.1 Factorization of polynomials

A polynomial $p(x) \in F[x]$ of positive degree is called *irreducible* if it cannot be written as the product of two polynomials of positive degree. It should be noticed that every polynomial of degree 1 is irreducible, since the product of polynomials of positive degree is at least 2. Irreducible polynomials are analogous to prime numbers, and there is a unique factorization result for polynomials that is analogous to the fundamental theorem of arithmetic for integers.[1]

Theorem 508 *Let F be a field and let $p(x) \in F[x]$ be a polynomial of degree at least 1. Then there exist distinct irreducible monic polynomials $q_1(x), q_2(x), \ldots, q_t(x)$ in $F[x]$, a scalar $c \in F$, and positive integers $k_1, k_2, \ldots, k_t$ such that*

$$p(x) = cq_1(x)^{k_1} q_2(x)^{k_2} \cdots q_t(x)^{k_t}.$$

Moreover, this factorization is unique up to the order of the irreducible polynomials.

[1] The fundamental theorem of arithmetic states that every positive integer n can be written as a product $p_1^{k_1} p_2^{k_2} \cdots p_t^{k_t}$, where $p_1, p_2, \ldots, p_t$ are distinct primes and $k_1, k_2, \ldots, k_t$ are positive integers. Moreover, this factorization is unique up to the order of the factors.

D

Summary of analysis in **R**

In this appendix, we list the main definitions and theorems of single-variable analysis. This material is developed in introductory texts on real analysis, such as [12], [27], or (at a more demanding level) [40].

D.0.1 Convergence

Definition 509 *A* sequence *of real numbers is an infinite (ordered) list of elements of* $\mathbf{R}$*:* $x_1, x_2, x_3, \ldots$. *We usually denote a sequence as* $\{x_i\}$ *or* $\{x_i\}_{i=1}^{\infty}$. *More precisely, a sequence is a function* $x : \mathbf{Z}_+ \to \mathbf{R}$*, where we denote values of the function as* x_i *instead of* $x(i)$*,* $i \in \mathbf{Z}_+$*. Here* $\mathbf{Z}$ *is the set of integers and* $\mathbf{Z}_+$ *is the set of positive integers. (Sometimes we index a sequence using the nonnegative integers:* $\{x_i\}_{i=0}^{\infty}$*. This does not change any part of the theory.)*

Definition 510 *Let* $\{x_k\}$ *be a sequence of real numbers, and let* $L \in \mathbf{R}$*. We say that* $\{x_k\}$ converges *to* L *if, for any* $\epsilon > 0$*, there exists a positive integer* N *such that*

$$n \geq N \;\Rightarrow\; |x_n - L| < \epsilon.$$

If $\{x_k\}$ *converges to* L*, we write*

$$L = \lim_{k \to \infty} x_k$$

or $x_k \to L$ *as* $k \to \infty$*. If there is no real number* L *such that* $x_k \to L$ *as* $k \to \infty$*, we say that* $\{x_k\}$ diverges.

Definition 511 *Let* $\{x_k\}$ *be a sequence of real numbers. If for each* $R \in \mathbf{R}$*, there exists a positive integer* N *such that*

$$k \geq N \;\Rightarrow\; x_k > R,$$

then we say that $\{x_k\}$ *converges to* ∞*, and we write* $x_k \to \infty$ *as* $k \to \infty$*.*

Similarly, if for each $R \in \mathbf{R}$*, there exists a positive integer* N *such that*

$$k \geq N \;\Rightarrow\; x_k < R,$$

then we say that $\{x_k\}$ *converges to* $-\infty$ *and write* $x_k \to -\infty$ *as* $k \to \infty$*.*

Definition 512 *Let* S *be a subset of* $\mathbf{R}$*. A real number* c *is an* accumulation point *of* S *if and only if for each* $\epsilon > 0$*, the interval* $(c - \epsilon, c + \epsilon)$ *contains infinitely many elements of* S*.*

Definition 513 *Let S be a subset of* $\mathbf{R}$, *and let* $f : S \to \mathbf{R}$ *be a function. Furthermore, let c be an accumulation point of S. We say that $f(x)$ converges to L as x converges to c if, for any $\epsilon > 0$, there exists $\delta > 0$ such that*

$$x \in S,\ |x - c| < \delta,\ x \neq c \ \Rightarrow\ |f(x) - L| < \epsilon.$$

If $f(x)$ converges to L as x converges to c, we write

$$L = \lim_{x \to c} f(x)$$

or $f(x) \to L$ as $x \to c$. If there is no real number L such that $f(x) \to L$ as $x \to c$, then we say that $f(x)$ diverges *as $x \to c$.*

D.0.2 Completeness of R

Definition 514 *Let S be a subset of* $\mathbf{R}$. *We say that S is* bounded above *if and only if there exists a real number R such that $x \leq R$ for all $x \in S$. We say that R is an* upper bound *for S. Similarly, S is* bounded below *if and only if there exists a real number r such that $r \leq x$ for all $x \in S$, in which case m is a* lower bound *for S. Finally, S is* bounded *if and only if it is both bounded above and bounded below. We can equivalently define S to be bounded if and only if there exists $R > 0$ such that $|x| \leq R$ for all $x \in S$.*

Axiom 515 *If S is a nonempty subset of* $\mathbf{R}$ *that is bounded above, then S has* least upper bound *M, that is, an upper bound M satisfying*

$$x \leq M' \text{ for all } x \in S \ \Rightarrow M \leq M'.$$

The least upper bound M is also called the supremum *of S and is denoted* $\sup(S)$. *It should be obvious that the least upper bound of S is unique (if M_1 and M_2 are both least upper bounds of S, then $M_1 \leq M_2$ and $M_2 \leq M_1$).*

Lemma 516 *Let S be a nonempty subset of* $\mathbf{R}$ *that is bounded above, and let* $M = \sup(S)$. *For all $\epsilon > 0$, there exists $s \in S$ such that $s > M - \epsilon$.*

Theorem 517 *Let S be a nonempty subset of* $\mathbf{R}$ *that is bounded below. Then S has a greatest lower bound.*

The greatest lower bound of S is also called the *infimum* of S, and is denoted $\inf(S)$. If S is not bounded above, then $\sup(S) = \infty$; similarly, if S is not bounded below, then $\inf(S) = -\infty$. Finally, by convention, $\sup(\emptyset) = -\infty$ and $\inf(\emptyset) = \infty$.

Theorem 518 *Let $\{x_k\}$ be a sequence in* $\mathbf{R}$ *that is bounded above and* monotonically increasing, *that is, $x_{k+1} \geq x_k$ for all $k = 1, 2, 3, \ldots$. Then there exists $x \in \mathbf{R}$ such that $x_k \to x$ as $k \to \infty$.*

Similarly, if $\{x_k\}$ is bounded below and monotonically decreasing, then there exists $x \in \mathbf{R}$ such that $x_k \to x$ as $k \to \infty$.

Definition 519 *Let $\{x_k\}$ be a sequence of real numbers. We say that $\{x_k\}$ is a* Cauchy *sequence if given any $\epsilon > 0$, there exists a positive integer N such that*

$$m, n \geq N \;\Rightarrow\; |x_m - x_n| < \epsilon.$$

Theorem 520 *Any Cauchy sequence of real numbers converges.*

D.0.3 Open and closed sets

Definition 521 *Let S be a subset of* **R**. *We say that S is* open *if for each $x \in S$, there exists $\epsilon > 0$ such that $(x - \epsilon, x + \epsilon) \subset S$.*

Theorem 522 *Let S be an open subset of* **R**, *let $x \in S$, and suppose $\{x_k\}$ is a sequence of real numbers converging to x. Then there exists a positive integer N such that*

$$k \geq N \;\Rightarrow\; x_k \in S.$$

Informally, we say that $x_k \in S$ for k sufficiently large.

Theorem 523 *Let A be any set, and for each $\alpha \in A$, let S_α be an open subset of* **R**. *Then*

$$\bigcup_{\alpha \in A} S_\alpha$$

is an open subset of **R**.

Definition 524 *Let S be a subset of* **R**. *We say that S is* closed *if and only if $\mathbf{R} \setminus S$ is open.*

The basic example of a closed set in **R** is a closed interval $[a, b]$. We have $\mathbf{R} \setminus [a, b] = (-\infty, a) \cup (b, \infty)$, which is the union of open sets and hence open.

Theorem 525 *Let S be a subset of* **R**. *Then S is closed if and only if, for each convergent sequence $\{x_k\}$ such that $x_k \in S$ for all k, the limit*

$$\lim_{k \to \infty} x_k$$

also lies in S. In short, S is closed if and only if it contains all of its limit points.

Theorem 526 *Let n be a positive integer, and for each $i = 1, 2, \ldots, n$, let S_i be a closed subset of* **R**. *Then*

$$\bigcap_{i=1}^{n} S_i$$

is a closed subset of **R**.

Definition 527 *Let S be a subset of* $\mathbf{R}$. *The* interior *of S is the subset*

$$\mathrm{int}(S) = \{x \in S : (x-\epsilon, x+\epsilon) \subset S \text{ for some } \epsilon > 0\}.$$

The reader should notice that $\mathrm{int}(S) \subset S$.

Theorem 528 *Let S be a subset of* $\mathbf{R}$. *Then:*

1. $\mathrm{int}(S)$ *is an open set;*
2. $\mathrm{int}(S) = S$ *if and only if S is open;*
3. $\mathrm{int}(S)$ *is the largest open subset of S, in the sense that if $T \subset S$ and T is open, then* $T \subset \mathrm{int}(S)$.

Definition 529 *Let $\{x_k\}$ be a sequence of real numbers. Let $\{k_j\}$ be a strictly increasing sequence of positive integers: $i > j$ implies that $k_i > k_j$. We say that $\{x_{k_j}\}_{j=1}^{\infty}$ is a* subsequence *of $\{x_k\}$.*

Theorem 530 *Let $\{x_k\}$ be a sequence of real numbers that converges to m (which may be finite or infinite). If $\{x_{k_j}\}$ is a subsequence of $\{x_k\}$, then $\{x_{k_j}\}$ also converges to m.*

Theorem 531 (Bolzano-Weierstrass) *Let S be any nonempty subset that is both closed and bounded, and let $\{x_k\}$ be a sequence of real numbers contained in S. Then there exists a subsequence $\{x_{k_j}\}$ that converges to a point $x \in S$.*

D.0.4 Continuous functions

Definition 532 *Let S be a subset of* $\mathbf{R}$, *and let* $f : S \to \mathbf{R}$ *be a function. We say that f is* continuous at $c \in S$ *if for any $\epsilon > 0$, there exists $\delta > 0$ such that*

$$x \in S, |x-c| < \delta \;\Rightarrow\; |f(x) - f(c)| < \epsilon.$$

We say that f is continuous on S, *or simply* continuous, *if it is continuous at every $x \in S$.*

Theorem 533 *Let S be a subset of* $\mathbf{R}$, *let* $f : S \to \mathbf{R}$, *and suppose f is continuous at $c \in S$. If $\{x_k\} \subset S$ is a sequence of real numbers converging to c, then*

$$\lim_{k\to\infty} f(x_k) = f(c).$$

Theorem 534 (Weierstrass) *Let S be a nonempty, closed, and bounded subset of* $\mathbf{R}$, *and let* $f : S \to \mathbf{R}$ *be continuous. Then there exist $m_1, m_2 \in S$ such that*

$$f(m_1) \le f(x) \le f(m_2) \text{ for all } x \in S.$$

In short, we say that f attains its maximum and minimum values on S.

Bibliography

[1] Norman Biggs. *Algebraic Graph Theory.* Cambridge University Press, Cambridge, 2nd edition, 1993.

[2] A. Blokhuis and A. R. Calderbank. Quasi-symmetric designs and the Smith normal form. *Designs, Codes, and Cryptography,* 2(2):189–206, 1992.

[3] Susanne C. Brenner and L. Ridgway Scott. *The Mathematical Theory of Finite Element Methods.* Springer-Verlag, New York, 2nd edition, 2002.

[4] W. L. Briggs and V. E. Henson. *The DFT: An Owner's Manual for the Discrete Fourier Transform.* Society for Industrial and Applied Mathematics, Philadelphia, 1995.

[5] V. Chvátal. *Linear Programming.* W. H. Freeman, New York, 1983.

[6] J. W. Cooley and J. W. Tukey. An algorithm for the machine computation of complex Fourier series. *Mathematics of Computation,* 19:297–301, 1965.

[7] J. E. Dennis Jr. and Robert B. Schnabel. *Numerical Methods for Unconstrained Optimization and Nonlinear Equations.* SIAM, Philadelphia, 1996.

[8] Gerald B. Folland. *Real Anaysis: Modern Techniques and their Applications.* John Wiley & Sons, New York, 1984.

[9] Leslie V. Foster. Gaussian elimination with partial pivoting can fail in practice. *SIAM Journal on Matrix Analysis and Applications,* 15(4):1354–1362, 1994.

[10] Michael R. Garey and David S. Johnson. *Computers and Intractibility: A Guide to the Theory of NP-Completeness.* W. H. Freeman and Company, San Francisco, 1979.

[11] Robert Garfinkel and George L. Nemhauser. *Integer Programming.* John Wiley & Sons, New York, 1972.

[12] Edward D. Gaughan. *Introduction to Analysis.* Brooks/Cole, Pacific Grove, CA, 5th edition, 1998.

[13] Mark S. Gockenbach. *Partial Differential Equations: Analytical and Numerical Methods.* SIAM, Philadelphia, 2002.

[14] Mark S. Gockenbach. *Understanding and Implementing the Finite Element Method.* SIAM, Philadelphia, 2006.

[15] Gene H. Golub and Charles F. Van Loan. *Matrix Computations.* Johns Hopkins University Press, Baltimore, 3rd edition, 1996.

[16] Carolyn Gordon, David L. Webb, and Scott Wolpert. One cannot hear the shape of a drum. *Bulletin of the American Mathematical Society (New Series)*, 27:134–138, 1992.

[17] G. Hadley. *Linear Programming.* Addison-Wesley, Reading, MA, 1962.

[18] Per Christian Hansen. Analysis of discrete ill-posed problems by means of the L-curve. *SIAM Review*, 34(4):561–580, 1992.

[19] Nicholas J. Higham. *Accuracy and Stability of Numerical Algorithms.* SIAM, Philadelphia, 2nd edition, 2002.

[20] F. B. Hildebrand. *Introduction to Numerical Analysis.* McGraw-Hill, New York, 2nd edition, 1974.

[21] Kenneth Hoffman and Ray Kunze. *Linear Algebra.* Prentice-Hall, Englewood Cliffs, NJ, 2nd edition, 1971.

[22] Claes Johnson. *Numerical Solution of Partial Differential Equations by the Finite Element Method.* Cambridge University Press, Cambridge, 1987.

[23] M. Kac. Can one hear the shape of a drum? *American Mathematical Monthly*, 73:1–23, 1966.

[24] Jean-Pierre Kahane. Jacques Hadamard. *Mathematical Intelligencer*, 13(1):23–29, 1991.

[25] David W. Kammler. *A First Course in Fourier Analysis.* Prentice-Hall, Upper Saddle River, NJ, 2000.

[26] David Kincaid and Ward Cheney. *Numerical Analysis.* Brooks/Cole, Pacific Grove, CA, 2nd edition, 1996.

[27] Witold Kosmala. *Advanced Calculus: A Friendly Approach.* Prentice-Hall, Upper Saddle River, NJ, 1999.

[28] E. S. Lander. *Symmetric Designs: An Algebraic Approach.* Cambridge University Press, Cambridge, 1983.

[29] Jerrold E. Marsden and Anthony J. Tromba. *Vector Calculus.* W. H. Freeman, New York, 4th edition, 1996.

[30] Neal H. McCoy and Gerald J. Janusz. *Introduction to Modern Algebra.* Allyn and Bacon, Newton, MA, 4th edition, 1987.

[31] N. Meyers and J. Serrin. $H = W$. *Proceedings of the National Academy of Sciences*, 51:1055–1056, 1964.

[32] Cleve Moler and Charles Van Loan. Nineteen dubious ways to compute the exponential of a matrix, twenty-five years later. *SIAM Review*, 45(1):3–49, 2003.

[33] Jorge Nocedal and Stephen J. Wright. *Numerical Optimization.* Springer, New York, 1999.

[34] Michael L. Overton. *Numerical Computing with IEEE Floating Point Arithmetic: Including One Theorem, One Rule of Thumb, and One Hundred and One Exercises.* SIAM, Philadelphia, 2001.

[35] V. J. Rayward-Smith. On computing the Smith normal form of an integer matrix. *ACM Transactions on Mathematical Software*, 5(4):451–456, 1979.

[36] Béatrice Rivière. *Discontinuous Galerkin Methods for Solving Elliptic and Parabolic Equations: Theory and Implementation.* SIAM, Philadelphia, 2008.

[37] Theodore J. Rivlin. *The Chebyshev Polynomials.* Wiley, New York, 1974.

[38] Kenneth Rosen. *Elementary Number Theory and its Applications.* Addison Wesley, Reading, MA, 5th edition, 2005.

[39] H. L. Royden. *Real Analysis.* MacMillan, New York, 2nd edition, 1968.

[40] Walter Rudin. *Principles of Mathematical Analysis.* McGraw-Hill, New York, 1976.

[41] Adi Shamir. How to share a secret. *Communications of the ACM*, 22(11):612–613, 1979.

[42] R. A. Tapia. An introduction to the algorithms and theory of constrained optimization. Unpublished.

[43] Lloyd N. Trefethen and David Bau. *Numerical Linear Algebra.* Society for Industrial and Applied Mathematics, Philadelphia, 1997.

[44] Geoffrey Vining and Scott M. Kowalski. *Statistical Methods for Engineers.* Thomson Brooks/Cole, Belmont, CA, 2nd edition, 2006.

[45] Stephen J. Wright. A collection of problems for which Gaussian elimination with partial pivoting is unstable. *SIAM Journal on Scientific and Statistical Computing*, 14(1):231–238, 1993.

[46] Dennis G. Zill. *A First Course in Differential Equations.* Brooks/Cole, Belmont, CA, 8th edition, 2005.

Index

For Product Safety Concerns and Information please contact our EU representative GPSR@taylorandfrancis.com Taylor & Francis Verlag GmbH, Kaufingerstraße 24, 80331 München, Germany

Printed and bound by CPI Group (UK) Ltd, Croydon, CR0 4YY
08/06/2025
01897011-0008